Antibiotics and Antibiotic Resistance

A subject collection from *Cold Spring Harbor Perspectives in Medicine*

Antibiotics and Antibiotic Resistance

A subject collection from *Cold Spring Harbor Perspectives in Medicine*

EDITED BY

Lynn L. Silver
LL Silver Consulting, LLC

Karen Bush
Indiana University

COLD SPRING HARBOR LABORATORY PRESS
Cold Spring Harbor, New York • www.cshlpress.org

Antibiotics and Antibiotic Resistance

A subject collection from *Cold Spring Harbor Perspectives in Medicine*
Articles online at www.perspectivesinmedicine.org

Executive Editor	Richard Sever
Managing Editor	Maria Smit
Senior Project Manager	Barbara Acosta
Permissions Administrator	Carol Brown
Production Editor	Diane Schubach
Production Manager/Cover Designer	Denise Weiss
Publisher	John Inglis

Front cover artwork: (*Top left*) Scanning electron micrograph of a human neutrophil ingesting methicillin-resistant *Staphylococcus aureus* (MRSA). (Image is from the National Institute of Allergy and Infectious Diseases [NIAID]. U.S. Government work, public domain.) (*Bottom left*) Colorized scanning electron micrograph showing carbapenem-resistant *Klebsiella pneumoniae* interacting with a human neutrophil. (Image is from the NIAID. U.S. Government work, public domain.) (*Top right*) Scanning electron micrograph of *Mycobacterium tuberculosis* bacteria. (Image credit: From NIAID, courtesy of Flickr and Creative Commons Licensing.) (*Bottom right*) Color-enhanced scanning electron micrograph showing *Salmonella typhimurium* (red) invading cultured human cells. (Image credit: Rocky Mountain Laboratories, NIAID, NIH. This image is a work of the National Institutes of Health, part of the United States Department of Health and Human Services. As a work of the U.S. Federal Government, the image is in the public domain and is therefore free of known restrictions under copyright law.)

Library of Congress Cataloging-in-Publication Data

Names: Silver, Lynn L., editor. | Bush, Karen (Biologist) editor.
Title: Antibiotics and antibiotic resistance/edited by Lynn L. Silver, LL Silver Consulting, LLC, and Karen Bush, Indiana University.
Description: Cold Spring Harbor, New York : Cold Spring Harbor Laboratory Press, [2016] | "A subject collection from Cold Spring Harbor perspectives in medicine." | Includes bibliographical references and index.
Identifiers: LCCN 2016011671 | ISBN 9781621821199 (hardback)
Subjects: LCSH: Antibiotics. | Drug resistance in microorganisms. | BISAC: MEDICAL / Infectious Diseases. | SCIENCE / Life Sciences / Biology / Microbiology. | MEDICAL / Microbiology.
Classification: LCC RM267 .A525 2016 | DDC 615.7/922--dc23
LC record available at https://lccn.loc.gov/2016011671

All World Wide Web addresses are accurate to the best of our knowledge at the time of printing.

For a complete catalog of all Cold Spring Harbor Laboratory Press publications, visit our website at www.cshlpress.org.

Contents

Contents

DNA/RNA REPLICATION AND INTERMEDIATES

PROTEIN SYNTHESIS

30S Ribosome Inhibitors

50S Ribosome Inhibitors

Elongation Factor G Inhibitor

Preface

A NTIBIOTIC RESISTANCE IS A GLOBAL ISSUE THAT HAS BECOME A CRITICAL concern of many health organizations, infectious disease specialists, and basic scientists in laboratories. Antibiotics have provided an accepted and inexpensive way for humans to counteract infections that previously would have killed them. However, with the increased use of antibiotics comes increased resistance, not only in numbers of resistant bacterial strains, but also as a result of increased varieties of resistance mechanisms. As we have seen multiple times since the introduction of antibiotics, the emergence of a new resistance mechanism in a relatively isolated geographical area is rapidly followed by identification of that resistance in a distant region due to rapid and unrestrained global travel. The Centers for Disease Control (CDC) has coined the phrase "Resistance anywhere is resistance everywhere" to describe this phenomenon. This proliferation of resistance mechanisms active against our mainstay antibiotics, and, as with the finding of transmissible colistin resistance, against antibiotics of last resort, has occasioned the publication of this volume.

Here we have assembled contributions from some of the foremost experts in antibiotic resistance to describe some of the basic, as well as the newest, resistance mechanisms for well-known and well-used antibiotics. In the first two contributions, Strachan and Davies and Perry et al. remind us that natural products from the ancient environment provided us with both the first antibiotics and the first antibiotic-resistance mechanisms. Judicious use of these agents will be critical for them to retain their utility in the future. Silver follows with comments on the nature of appropriate targets for antibacterial drugs, based on existing drug classes and on lessons learned from the (generally poor) results of the search for new antibiotics.

Useful antibiotics have been grouped into sections according to their targets associated with inhibition of bacterial growth. Among those agents that interfere with cell wall synthesis are the β-lactams, the most widely used class of antibiotics. The most important of these drugs are described historically by Bush and Bradford, as is the expansion of their therapeutic utility by the addition of β-lactamase inhibitors to overcome hydrolysis by β-lactamases, the major β-lactam-resistance mechanism in Gram-negative bacteria. Fisher and Mobashery follow with a discussion of β-lactam resistance in the Gram-positive bacteria and *Mycobacterium tuberculosis*, with special emphasis on interference with peptidoglycan synthesis on a molecular level. Bonomo then examines the most critical β-lactamases in Gram-negative pathogenic bacteria. Glycopeptides are presented by Zeng et al., who describe their mechanism of action and the emergence of resistance over several decades as a result of both chromosomal and plasmid-encoded mechanisms. Silver discusses fosfomycin, an antibiotic inhibiting cell wall synthesis at its first committed step, which has been long in use for treatment of urinary tract infections with relatively low incidence of resistance. The recent promotion of much broader use of fosfomycin against MDR pathogens raises the possibility of acceleration of the spread of fosfomycin resistance.

Drugs that act by interfering with membrane synthesis or membrane integrity include daptomycin and the polymyxins. The lipopeptide daptomycin is frequently used for the treatment of infections caused by multidrug-resistant staphylococci and enterococci. Miller et al. discuss its novel mechanism of action and the molecular basis for daptomycin resistance in these organisms. The cyclic cationic polypeptides polymyxin B and colistin have become essential components of many antibacterial regimens for treatment of multidrug-resistant Gram-negative bacterial infections. Trimble et al. provide new insight into the mechanism of action of these agents based, in part, on recently described resistance mechanisms, including horizontal transfer of the *mcr-1* gene among species of Enterobacteriaceae in both food animals and humans. Two chapters involve investigational single-target inhibitors of enzymes in the lipopolysaccharide biosynthetic pathway. Investigational inhibitors of the enoyl-acyl carrier protein reductase (FabI) are described by Yao and Rock, who note the ability to utilize structure-guided design to obtain pathogen-specific inhibitors, but also note the negative impact of a single-base-pair mutation on the eventual clinical utility of single-target inhibitors. Erwin presents information concerning inhibitors of LpxC, the second enzyme in the Lipid A biosynthetic pathway, and its low selection of resistance in vitro. Although various groups have attempted to identify clinically useful candidates that inhibit LpxC, only one agent has advanced into a therapeutic clinical trial.

DNA or RNA replication has been a major target for some of the most frequently prescribed antibiotics in both the hospital and community setting. Hooper and Jacoby describe the mechanism of action of the fluoroquinolones that specifically target bacterial topoisomerases, enzymes involved directly in DNA synthesis. Because of their widespread use, resistance to fluoroquinolones has arisen in a variety of ways, including both chromosomally mediated and plasmid-encoded mechanisms, thus limiting their therapeutic utility, especially against Gram-negative pathogens. In contrast, the rifamycins target RNA polymerase and are used primarily for treatment of Gram-positive and mycobacterial infections. As described by Rothstein, rifamycin clinical utility has been compromised by resistance issues, primarily involving mutations in the bacterial RNA polymerase.

Bacterial protein synthesis provides an attractive target for antibacterial drug discovery, especially now that structure-based design is able to be guided by the recent availability of x-ray crystallographic data for the bacterial ribosome. Arenz and Wilson describe how advances in structural and biochemical analyses of the ribosome can lead to major advances in the development of new protein synthesis inhibitors. Among the antibacterial agents that target the 30S ribosome are the aminoglycosides and the tetracyclines, two of the oldest and most widely used antibiotic classes. Krause et al. provide an overview of the history of the aminoglycosides and the emergence of both enzymatic and nonenzymatic resistance mechanisms. They then describe the development of new aminoglycosides that are microbiologically active against multidrug-resistant Gram-negative bacteria, with potentially safer dosing regimens compared to the older agents. Grossman describes the tetracyclines, agents with broad-spectrum activity and utility as both oral and systemic antibiotics. Although the older tetracyclines are subject to both intrinsic and acquired resistance mechanisms, newer analogs have been developed to circumvent many of these mechanisms. The 50S ribosome is the target of another set of agents, with macrolides representing the largest class of antibiotics in this set. Fyfe et al. focus on the various resistance mechanisms known for the class, with molecular and structural interpretations provided for phenotypic observations of resistance. The mechanism of action and mechanisms of resistance for the lincosamides, streptogramins, phenicols, and pleuromutilins are described by Schwarz et al. Because these four diverse classes of 50S ribosomal inhibitors are used in both human and veterinary medicine, the opportunity is increased for multiple resistance mechanisms that can be transferred horizontally. However, Paukner and Riedl present information about a novel pleuromutilin with low rates for selection of resistance in vitro that is in late-stage clinical development.

Other antibacterial agents described in this volume include the steroidal antibiotic fusidic acid, an inhibitor of elongation factor G that has been used extensively throughout much of the world as an oral agent to treat infections caused by methicillin-resistant Staphylococcus aureus. Fernandes describes its unique mechanism of action and the multiple resistance mechanisms that have arisen during its use for over 50 years. The next set of agents includes the antifolates, agents that inhibit bacterial folic acid synthesis. Among these are the sulfa drugs that were used therapeutically as early as World War II and the dihydrofolate reductase inhibitor trimethoprim. Estrada et al. provide descriptions of the mechanisms of action and resistance that accompany these drugs, and introduce preclinical and investigational agents that were selected in an attempt to overcome common resistance mechanisms. In the final chapter, Leeds tackles the challenge of developing agents that display activity against a single organism (e.g., *Clostridium difficile*). Mechanisms of action are discussed for the recently approved *C. difficile*–specific RNA polymerase inhibitor fidaxomicin and the exploratory EF-Tu inhibitor LFF571, together with associated resistance mechanisms for each agent.

Antibiotic resistance is a fact of life, and presents a challenge for any antibacterial agent that is used to inhibit the growth of bacteria. In this volume, the major classes of antibiotics, together with their modes of action and common mechanisms of resistance, are described. Old drugs, as well as investigational agents, have all selected for resistant bacteria in natural environments. Resistance may occur either as a result of a single nucleotide substitution in a chromosomal gene, or as the result of the transfer of mobile resistance determinants across species. Common resistance mechanisms such as efflux pumps or decreases in drug uptake due to altered outer membrane proteins result in cross-class resistance, whereas resistance mechanisms involving target alteration or drug inactivation may be drug-specific. In the contributions outlined above, resistance can be described for every molecule that has the capability of preventing bacterial growth. As a result, new antibacterial agents will continue to be needed. Antibiotic discovery and development efforts must proceed without interruption. Resistance will not be overcome long term, but efforts must be made to try to contain the resistant bacteria that already exist through good stewardship of existing drugs, appropriate dosing, and continued basic research into the discovery of new molecules. And, finally, we hope that this volume will provide insight into antibiotic-resistance mechanisms in order to aid ongoing discovery efforts—since these must take into account the propensity for and rapidity with which resistance to new agents can arise.

LYNN L. SILVER
LL Silver Consulting
Springfield, New Jersey

KAREN BUSH
Indiana University, Bloomington
Bloomington, Indiana

The Whys and Wherefores of Antibiotic Resistance

Cameron R. Strachan[1] and Julian Davies[2]

[1]MetaMixis Biologics, Vancouver, BC V6T 1Z3, Canada

[2]Department of Microbiology and Immunology, University of British Columbia, Vancouver, BC V6T 1Z3, Canada

Correspondence: jed@interchange.ubc.ca

The development and rapid dissemination of antibiotic-resistant bacterial pathogens has tarnished the dream of a world without infectious diseases. However, our understanding of these processes, paired with sequence information from terrestrial bacterial populations, indicates that there is no shortage of novel natural products that could be developed into new medicines. Regardless, their therapeutic success in the clinic will depend on the introduction of mandatory controls and use restrictions.

One solution to control the threat of antimicrobial resistance is scientific discovery.

—Dame Sally Davies

The history of man has been punctuated by many plagues and pestilences during existence on this planet (see Table 1 for a partial list). In certain instances, upward of 50% of the population of a city or a country may have perished while others survived, albeit seriously weakened. Plagues have shaped history at both local and worldwide levels from an economic point of view by crippling the workforce, and also from a military viewpoint, infections on one side in a conflict could lead to victory over the enemy. In recent times, the most devastating microbial infection was the Spanish flu pandemic in 1918–1920 that killed some 5% of the world's population. However, in the past 50 years, the world has seen the rapid evolution of a new plague—that of worldwide antibiotic-resistant (AR) microbes. Although not a disease

in itself, AR results from the failure to effectively prevent and treat many diseases, leading to widespread untreatable microbial infections and greatly increased morbidity and mortality: a plague of resistance genes (Davies and Davies 2010). AR transforms treatable infectious diseases into untreatable ones. Regrettably, we were warned and aware of the cause and consequences of AR development and dissemination in the 1960s, but nonetheless we let it happen!

Bacterial pathogens readily develop resistance on treatment with antibiotics; they are often caused by the results of mutation of specific target genes but primarily by the inheritance of plasmids carrying resistance gene clusters (Wright 2011). Multidrug-resistant pathogens have been clinically relevant since the 1950s and can be considered the leading cause of mortality worldwide since the introduction of antibiotics (Davies and Smith 1978). Currently, estimates indicate that upward of 10 million

Table 1. History of plagues

Death toll (estimate)	Location	Date	Comment	Disease
ca. 40% of population	Europe	541–542	Plague of Justinian, attributable to the name of the Byzantine emperor in power at the time	Bubonic plague
30% to 70% of population	Europe	1346–1350	"Black Death" or second plague pandemic, first return of the plague to Europe after the Justinianic plague of the 6th century	Plague
100,000	England	1665–1666	Great plague of London	Plague
76,000	Austria	1679	Great plague of Vienna	Plague
>50,000	Russia	1770–1772	Russian plague of 1770–1772	Plague
>>100,000	Asia, Europe	1816–1826	First cholera pandemic	Cholera
>>100,000	Asia, Europe, North America	1829–1851	Second cholera pandemic	Cholera
20,000+	Canada	1847–1848	Typhus epidemic of 1847	Epidemic typhus
1,000,000	Russia	1852–1860	Third cholera pandemic	Cholera
1,000,000	Worldwide	1889–1890	1889–1890 flu pandemic	Influenza
75,000,000	Worldwide	1918–1920	1918 flu pandemic	Influenza
2,000,000	Worldwide	1957–1958	Asian flu	Influenza
1,000,000	Worldwide	1968–1969	Hong Kong flu	Influenza
775	Asia	2002–2003	SARS	SARS coronavirus
14,286	Worldwide	2009	2009 flu pandemic	Influenza
Incalculable	Worldwide	1950-	Antibiotic resistance	Infectious diseases

deaths occur every year attributable to AR, and this number is increasing. The financial costs associated with treating these intractable infections are in the many billions of dollars, and the emotional cost has been enormous (Davies et al. 2014). Indeed, many infected patients in the population may well be treated successfully, but antibiotic overuse and misuse propagates resistance development and further mortality from worsening, untreatable conditions. We have been using antibiotics in ever-increasing amounts for the last half century, and it is estimated that the AR plague may have claimed >500 million lives worldwide.

The AR plague has broad sequelae: Domesticated animals such as pets and farm animals also die of infection with AR pathogens. This contributes to the global AR gene pool (Bush et al. 2011). Furthermore, we have little or no notion as to the effects of antibiotics and resistance on the global microbial ecosystem that provides endless services to humans (Colwell 1997). The worldwide economic consequences

of AR are difficult to evaluate; meanwhile, antibiotics have not eliminated a single microbial disease.

A BRIEF HISTORICAL PERSPECTIVE

In all probability, microbial resistance to toxic molecules occurred long before the modern era of antibiotics. Indeed, agents such as arsenicals and mercurials were used for centuries in treating disease even while the role of microbes in causing these afflictions was not recognized. Often, treatment failure was more likely caused by the toxicity of the potion rather than the disease. It is not coincidental that many antibiotic resistance plasmids isolated in the 1950s carried genes for resistance to salts of arsenic and mercury (Silver and Misra 1988). There is also a real possibility that genetically determined resistance was spread during the intense pollution in the late 18th and early 19th centuries, when the environmental microbiomes of the industrial world were exposed to huge amounts of toxic organic chemicals. Not everyone lived in Downton Abbey! Respiratory diseases were very common in industrial areas like Manchester, Pittsburgh, and in other countries throughout the world. Currently, many cities in Asia suffer the same problem. What immediate and lasting impacts might this pollution have on human microbiomes?

SOME ANTIMICROBIAL HISTORY

The true antibiotic era began with the discovery of penicillin (a fungal product) by Fleming in the late 1920s, but serious studies did not commence until the 1930s (Abraham and Chain 1988). By the 1930s, most of the common infectious diseases were identified as microbial in origin and specific treatments could be applied. The first significant treatment discovery was that of the sulfonamides, a class of synthetic antimicrobial agents introduced into clinical practice in 1935. Sulfonamides were inexpensive, easily produced in large amounts, and were widely used in different forms and, together with penicillin, played an incredibly important role for the Allied forces in the Second World War preventing many deaths. Winston Churchill was treated successfully for bacterial pneumonia with sulfapyridine; this was obviously of great importance! Sulfonamide resistance was undoubtedly encountered in certain cases but there are few descriptions in the early literature; the biochemical and genetic mechanism(s) involved were not studied at the time. The situation has changed and mechanisms of action and resistance to the sulfonamide drugs have been well characterized (Sköld 2000). For example, antibiotic resistance integrons encoding sulfonamide resistance were unrecognized until the isolation of multiply drug-resistant strains in postwar Japan (Davies 1995). Sulfonamides are still in use today although less frequently.

Trimethoprim is another synthetic agent that was developed in the early 1960s and, like the sulfa drugs, has enjoyed a long life. It is often used in combination with a sulfonamide for the treatment of urinary tract infections, and dual resistance was first identified in 1969. Within the context of this short article, it is worth noting that resistance to sulfa drugs and trimethoprim can occur by mutations in the target pathways or by inheritance of plasmids carrying altered, drug-insensitive variants of the target enzymes (Sköld 2001). Very few examples are required to see that AR mechanisms are extremely varied and widely distributed.

It is often forgotten that *Bacillus* strains were studied early on and found to produce some of the first bioactive molecules, of which a number have proven to be useful therapeutics. These included polymyxin (the universal topical treatment) and colistin, currently one of the few drugs available for certain multidrug-resistant *Pseudomonas* infections. These compounds have been used for more than 80 years, and plasmid-borne colistin resistance has recently been observed to be widespread in China (Liu et al. 2016) (next stop, Europe and the United States). Sadly, there appears to have been little effort devoted to the discovery and development of bioactive compounds from *Bacillus* sp. and related genera in recent times. For some reason, pharmaceutical companies do not favor peptide drugs. These and other "forgotten" bacterial genera should become more fashionable now that their genome sequences can be scanned for biosynthetic pathways in the search for novel classes of bioactive compounds. Another topic of interest concerns the evolution of antibiotic resistance in the case of *Mycobacterium tuberculosis* (Mtb). The sole source of resistance to anti-TB drugs is by mutation of the target genes and no transferable antibiotic resistance plasmids have been found to encode resistance in Mtb: This pathogen seems to take care of antimicrobials quite nicely without participating in sex (Musser 1995).

THE CHEMICAL DEVELOPMENT OF ANTIBIOTICS

Since the discovery of antibiotic resistance genes that modify (inactivate) antibiotics, there has been considerable effort to develop, by synthetic chemical methods, compounds that prevent, inhibit, or otherwise avoid antibiotic resistance (Davies 2014). With each new resistance mechanism, we begin a search for blocking agents to restore antibiotic activity and there have been some remarkable successes. This often consists of the removal or modification of functional sites of antibiotic "core" structures to avoid modification or inactivation by resistance enzymes. A good example of this approach is the chemical modification of aminoglycosides to prevent phosphorylation or acetylation of sugar hydroxyl groups or amino groups (Davies 2006). In the case of amikacin, a structural element from a less effective aminoglycoside was added synthetically to kanamycin and prevented specific enzymatic modifications. The discovery of compounds such as amikacin were landmarks, but unfortunately not all resistance modifications can be avoided in this way and "Achilles heels" remain on the modified drugs (Courvalin and Davies 1977). Another example is the chemical modification of β-lactam antibiotics to prevent hydrolytic cleavage of the β-lactam ring, which has been successful in the case of the modern β-lactam antimicrobials. But again, modified β-lactamase enzymes (>1000) have evolved in concert with the development of new semi-synthetic β-lactams, and the conflict continues (Bush and Jacoby 2010). The case of the quinolones is especially interesting. When the fluoroquinolones were introduced, it was claimed by some that no resistance modification other than mutation of this new class of completely synthetic antimicrobials would be possible. Little did they know—microbes responded to this challenge successfully by adapting another enzyme for the job, and transferable resistance to the fluoroquinolones is now common in Gram-negative pathogens (Strahilevitz et al. 2009). Resistance is inevitable.

RESISTOMES

One of the most remarkable environmental findings of recent years is that of antibiotic resistomes: conglomerations of putative resistance genes isolated from a variety of environmental sources (Forsberg et al. 2012). This descriptive term was coined by Wright and his colleagues and confirmed and extended by the characterization of resistomes from many different environments (D'Costa et al. 2006). They include soils, ancient caves, and, not surprisingly, gut microbiomes—(putative) antibiotic resistance genes are everywhere. The question is, do these reservoirs play any role in the determination of clinically significant antibiotic resistance? The putative AR genes from resistomes have been shown to be active by gene-expression studies, but this does not establish their natural function. Indeed, the presence of AR genes in bacteria has been shown to influence many other phenotypes. Is it possible that these antibiotic resistance genes have different properties in the wild? Is there any evidence for association with plasmids? To date, the presence of natural resistomes appears to have no causal relationship with the use of antibiotics.

GENE TRANSFER

Antibiotic resistance might have been less of a problem were it not for the fact that most AR genes are genetically mobile. Resistance transfer factors (R-plasmids) appeared on the environmental/clinical scene in the 1950s in Japan, the United States, and Europe: Where did they originate? The combined interaction of mutation and gene transfer must have been taking place (with respect to resistance development) since eternity. There seems to be endless diversity with rampant gene exchange occurring within related groups of microbes and plasmids that are but one of the transfer mechanisms (Polz et al. 2013). Bacterial gene transfer has become prominent because of the use of antibiotics, and equally powerful selections were likely operating during the period of the industrial revolution (toxic chemicals and poisons). There is still much to be learned about gene transfer mechanisms and origins of AR genes. What is clear is that microbes are able to recombine promiscuously and readily access local gene pools. It is interesting to note that some recent studies of plasmids with strictly environmental roles show that they can be rapidly assembled and disseminated (Xue et al. 2015). Even with the most modern sequence data and analyses of resistance islands, it is not possible to trace the AR genes back to their origins (Ashton et al. 2015). What are natural functions and origins of AR genes? If the resistomes have their supposed function, does one find antibiotic-producing organisms and plasmids in plenitude in resistome environments? There are resistomes in the human gut, but do their resistance phenotypes all relate to antibiotics being ingested? If a person has never been treated with antibiotics such as tetracycline, streptomycin, or chloramphenicol, etc., where did the resistance genes come from?

THE BIG MISTAKE

The current situation, wherein transferable multidrug resistance exists universally, was an inevitable consequence of the negligent treatment practices used (and to some extent continued) when antimicrobials were first introduced in the 1940s. Their use as therapeutics was generally successful, but there were many senseless practices. For example, using antibiotic fermentation residues (and crude antibiotics) as feed supplements in the beef, chicken, and fish industries have proven to be a contributor to the AR plague. Efforts have been made to control these misuses, but they have only been partly successful in a few countries, and the consequences of the worldwide commercial (not-health-related) use of antibiotics are now irreversible. Antibiotic production residues should never have been used in agriculture and as food supplements for animals and fish. In retrospect, it seems that everything possible was done to ensure that antimicrobial use was encouraged in as many nonhuman health practices as possible. These actions guaranteed that effective therapeutic applications would be severely limited. Alexander Fleming predicted this outcome but it is unlikely that even he realized what would happen on a global scale.

WHAT SHOULD BE DONE

1. Strict (legal) control of antibiotic use must be exercised (from compound discovery to the commercial release by the manufacturer) and all remaining residues must be eliminated. Proper hospital practices must be enforced to prevent overuse and disposal of antimicrobials in their active forms. Could antibiotics be destroyed before disposal? Constant monitoring by modern methods is needed within and without. But, how do you punish hospital personnel for noncompliance?

2. Nonhuman use of antibiotics: Be they for prevention, therapy, or for growth promotion, should all be universally banned except for specific agents that are structurally and completely unrelated in mode-of-action to compounds used for humans. Infractions must be disciplined appropriately. Under no circumstances should antibiotic use be permitted other than for human therapy. These measures should have already been adopted internationally.

3. Use of any agent that leads to the development of antibiotic cross-resistant strains, such as triclosan, should be banned, no matter what the proposed use. More research to identify chemicals that cross-select for resistance to antibiotics is essential. The same is true for agents where multiple resistance genes are becoming frequently cotransferred.

4. Should antibiotic use in genetic engineering been permitted? Significant quantities of antibiotics (although very small compared with human and animal use) and AR genes have been used for gene cloning by academia and industry since the early 1970s. The impact of this "misuse" on the advancement of biological science has been enormous and has led, and will lead, to future advances in medicine and biotechnology. The hullabaloo over genetic engineering practices is misplaced, but the extent to which it depends on the use of antibiotic and resistance selection should be considered.

In conclusion, is there reason for optimism for the future of antibiotic therapy? Of course there is, but the only solution is to generate collections of truly novel antibiotics with a narrow spectrum of action that can be combined with synthetic inhibitors of AR function. There is no shortage of potential therapeutic agents in nature; there are many new antibiotics to be discovered, and current methodology comes nowhere near exhausting searches of natural environments. Creative screening approaches that rely on natural properties such as signaling will lead to a renewable supply of novel compounds. The same can be anticipated with bioinformatic-heterologous expression approaches (Donia and Fischbach 2015). However, the compounds will have short useful lives unless there is strict control of their use. AR is an evolutionary response by microbes that has had drastic consequences for the human race. We need to study the origins of AR and elucidate their "natural" functions. A key component will be to understand how AR diversity is generated as a result of rapid gene transfer and turnover. Finally, it must be recognized that the AR plague was entirely man-made and could/should have been prevented and/or contained by stricter control of the use of antibiotics. Without appropriate compliance of regulations, it is unlikely that the spread of AR will ever be prevented.

ACKNOWLEDGMENTS

We thank Vivian Miao and the reviewers for helpful comments.

REFERENCES

Abraham EP, Chain E. 1988. An enzyme from bacteria able to destroy penicillin. 1940. *Rev Infect Dis* **10:** 677–678.

Ashton PM, Satheesh N, Dallman T, Rubino S, Rabsch W, Mwaigwisya S, Wain J, O'Grady J. 2015. MinION nanopore sequencing identifies the position and structure of a bacterial antibiotic resistance island. *Nat Biotechnol* **33:** 296–300.

Bush K, Jacoby GA. 2010. Updated functional classification of β-lactamases. *Antimicrob Agents Chemother* **54:** 969–976.

Bush K, Courvalin P, Dantas G, Davies J, Eisenstein B, Huovinen P, Jacoby GA, Kishony R, Kreiswirth BN, Kutter E, et al. 2011. Tackling antibiotic resistance. *Nat Rev Microbiol* **9:** 894–896.

Colwell RR. 1997. Microbial diversity: The importance of exploration and conservation. *J Ind Microbiol Biotechnol* **18:** 302–307.

Courvalin P, Davies J. 1977. Plasmid mediated aminoglycoside phosphotransferase of broad substrate range that phosphorylates amikacin. *Antimicrob Agents Chemother* **11:** 619–624.

D'Costa VM, McGrann KM, Hughes DW, Wright GD. 2006. Sampling the antibiotic resistome. *Science* **311:** 374–377.

Davies J. 1995. Vicious circles: Looking back on resistance plasmids. *Genetics* **139:** 1465–1468.

Davies JE. 2006. Aminoglycosides: Ancient and modern. *J Antibiot (Tokyo)* **59:** 529–532.

Davies J. 2014. The origin and evolution of antibiotics. In *Antimicrobials*, pp. 3–10. Springer, Berlin.

Davies J, Davies D. 2010. Origins and evolution of antibiotic resistance. *Microbiol Mol Biol Rev* **74:** 417–433.

Davies J, Smith DI. 1978. Plasmid-determined resistance to antimicrobial agents. *Annu Rev Microbiol* **3:** 469–518.

Davies S, Grant J, Catchpole M. 2014. *The drugs don't work: A global threat.* Viking, New York.

Donia MS, Fischbach MA. 2015. HUMAN MICROBIOTA. Small molecules from the human microbiota. *Science* **349:** 1254766.

Forsberg KJ, Reyes A, Wang B, Selleck EM, Sommer MO, Dantas G. 2012. The shared antibiotic resistome of soil bacteria and human pathogens. *Science* **337:** 1107–1111.

Liu YY, Wang Y, Walsh TR, Yi LX, Zhang R, Spencer J, Doi Y, Tian G, Dong B, Huang X, et al. 2016. Emergence of plasmid-mediated colistin resistance mechanism MCR-1 in animals and human beings in China: A microbiological and molecular biology study. *Lancet Infect Dis* **16:** 161–168.

Musser JM. 1995. Antimicrobial agent resistance in mycobacteria: Molecular genetic insights. *Clin Microbiol Rev* **8:** 496–514.

Polz MF, Alm EJ, Hanage WP. 2013. Horizontal gene transfer and the evolution of bacterial and archaeal population structure. *Trends Genet* **29:** 170–175.

Silver S, Misra TK. 1988. Plasmid-mediated heavy metal resistances. *Annu Rev Microbiol* **42:** 717–743.

Sköld O. 2000. Sulfonamide resistance: Mechanisms and trends. *Drug Resist Updat* **3:** 155–160.

Sköld O. 2001. Resistance to trimethoprim and sulfonamides. *Vet Res* **32:** 261–273.

Strahilevitz J, Jacoby GA, Hooper DC, Robicsek A. 2009. Plasmid-mediated quinolone resistance: A multifaceted threat. *Clin Microbiol Rev* **22:** 664–689.

Wright GD. 2011. Molecular mechanisms of antibiotic resistance. *Chem Commun* **47:** 4055–4061.

Xue H, Cordero OX, Camas FM, Trimble W, Meyer F, Guglielmini J, Rocha EP, Polz MF. 2015. Eco-evolutionary dynamics of episomes among ecologically cohesive bacterial populations. *MBio* **6:** 1–10.

Cite this article as *Cold Spring Harb Perspect Med* doi: 10.1101/cshperspect.a025171

The Prehistory of Antibiotic Resistance

Julie Perry, Nicholas Waglechner, and Gerard Wright

M.G. DeGroote Institute for Infectious Disease Research, Department of Biochemistry and Biomedical Sciences, DeGroote School of Medicine, McMaster University, Hamilton, Ontario L8S 4K1, Canada

Correspondence: wrightge@mcmaster.ca

Antibiotic resistance is a global problem that is reaching crisis levels. The global collection of resistance genes in clinical and environmental samples is the antibiotic "resistome," and is subject to the selective pressure of human activity. The origin of many modern resistance genes in pathogens is likely environmental bacteria, including antibiotic producing organisms that have existed for millennia. Recent work has uncovered resistance in ancient permafrost, isolated caves, and in human specimens preserved for hundreds of years. Together with bioinformatic analyses on modern-day sequences, these studies predict an ancient origin of resistance that long precedes the use of antibiotics in the clinic. Understanding the history of antibiotic resistance is important in predicting its future evolution.

The emergence of antibiotic-resistant human pathogens at a rate that exceeds new drug discovery threatens to end an age of unparalleled achievement in modern medicine. Beyond the management of infectious disease, advances in organ transplantation, orthopedic surgery, renal dialysis, and cancer chemotherapy all depend on our ability to control infection. The Center for Disease Control and Prevention (CDC) have declared 15 antibiotic-resistant human pathogens as "urgent" or "serious" threats to human health, and estimate that two million people are sickened every year in the United States alone by antibiotic-resistant infections (see www.cdc.gov/drugresistance/biggest_threats.html). Antibiotics have only been in clinical use for ∼80 years, yet we face the imminent possibility that they will be lost as therapeutics as a result of resistance. How did we get here so quickly? Were pathogens always antibiotic-resistant? If not, how did resistance develop? This review will explore the ancient nature of antibiotic resistance in the context of the concept of the antibiotic "resistome."

THE ANTIBIOTIC "RESISTOME"

The origins of clinical antibiotic resistance can be found in the environment. Most antibiotics in medical or agricultural use are derived from or produced by a group of soil-dwelling bacteria called the Actinomycetes (the most notable genus for antibiotic production being *Streptomyces*). These organisms are prolific producers of specialized metabolites (so-called "natural products"), including the antibiotics streptomycin, tetracycline, chloramphenicol, erythromycin, and vancomycin. Of course, these organisms must also be resistant to the antibiotics they produce, or they would succumb to their own toxic metabolites. In fact, antibiotic producers may have been the original sources for many of the antibiotic-resistance genes circulating in the clinic today (Benveniste and Davies 1973; Humeniuk et al. 2002). However, most *Streptomyces* are resistant to an average of seven to eight antibiotics, including newly developed and clinically important therapeutics (D'Costa et al. 2006). How did these environmental organisms become multidrug-resistant? Bacteria are unique in that they acquire genes from the parent microorganism during division (vertical gene transfer), but can also acquire genes from the community at large (horizontal gene transfer, first shown for aminoglycoside resistance [Benveniste and Davies 1973], and reviewed in Ochman et al. 2000; van Elsas and Bailey 2002; Aminov and Mackie 2007; Aminov 2011; Skippington and Ragan 2011; Stokes and Gillings 2011; Wiedenbeck and Cohan 2011). It is now clear that the environment is a vast source of new and emerging resistance genes (Wright 2010) accessible by members of the microbial community via horizontal gene transfer. The global collective of all resistance genes in the clinic, in the microbiome of humans and animals, and in environmental bacteria is the antibiotic "resistome." The diversity and extent of the resistome across both environmental and human microbiomes suggests a long evolutionary history, which has been explored in the recent studies discussed below.

Links to the Environment and Gene Mobilization

The permafrost found in the Canadian High North offers a genomic glimpse into the past, as its permanently frozen nature has preserved DNA for tens to hundreds of thousands of years. In a defining study, D'Costa et al. (2011) showed the presence of gene-encoding resistance to β-lactam, tetracycline, and glycopeptide antibiotics in metagenome samples of 30,000-year-old permafrost, which also contained DNA belonging to woolly mammoths and other animal and plant species unique to the Pleistocene. Detailed analysis of the vancomycin resistance gene cluster in this metagenome revealed conservation of

gene sequence and synteny with modern resistance clusters in the clinic as well as protein function and structure. Other studies have shown the presence of resistance genes in Siberian permafrost (Petrova et al. 2009) and showed working resistance genes from at least 5000 years ago using functional metagenomics (Perron et al. 2015). Isolated instances of resistance genes in human commensals and pathogens have also been documented. The gut microbiome of a pre-Columbian Andean mummy from Cuzco, Peru (^{14}C dating of 980–1170 AD) was recently found to harbor genes with homology to β-lactam, fosfomycin, chloramphenicol, aminoglycoside, macrolide, sulfa, quinolones, tetracycline, and vancomycin resistance genes (Santiago-Rodriguez et al. 2015). Antibiotic resistance elements were detected in the oral microbiome of four adult human skeletons from a medieval monastery (ca. 950–200 CE). Native resistance to aminoglycosides, β-lactams, bacitracin, bacteriocins, and macrolides and a near-complete plasmid-encoded conjugative transposon carrying efflux pump genes with high homology to CTn5 of *Clostridium difficile* were found in these samples, implying that the human microbiome has long served as a reservoir of resistance genes accessible to pathogens even in the absence of the strong selective pressure of modern antibiotic availability (Warinner et al. 2014). The first bacterial isolate deposited in the United Kingdom's National Collection of Type Cultures (NCTC) was a strain of the dysentery-causing bacterium *Shigella flexneri* that killed a young soldier in World War I. Despite being isolated in 1915 (long before the discovery and use of antibiotics), this human pathogen was found to carry resistance genes for penicillin and erythromycin when it was revived and sequenced in 2014 (Mather et al. 2014).

These instances of resistance, spanning roughly 30,000 years, establish that the resistome is indeed ancient. We know, however, that resistance in the environment has increased significantly because of the introduction of antibiotics in agriculture and in medicine (Knapp et al. 2010), and that sensitivity has been steadily replaced with resistance in the clinic (Chait et al. 2012). There is no question that the production and use of antibiotics in modern times has influenced the dissemination of resistance, and it is now a matter of staying one step ahead of the microbes. Although research has traditionally focused on the present-day clinical resistome, we have just begun to probe the vastness of the environmental and ancient resistomes. Understanding how resistance develops over time and moves among microorganisms will be of immense value in predicting the future course of resistance development. The remainder of this article will explore the ancient resistome in greater detail.

HOW OLD IS RESISTANCE PREDICTED TO BE?

Assigning an evolutionary "age" to a gene is no small feat, and requires the comparison of genes from different species that have evolved from a common ancestral gene ("orthologs") (Tatusov et al. 1997). Divergence times can be determined from the appearance of major transitions in the fossil record, and used to calibrate the expected level of identity among sets of conserved orthologs across major groups in the tree of life (Landan et al. 1990; Weigel et al. 1988). The result uses the Grishin equation $q = \ln(1+2D)/2D$ (Grishin 1995), an expression that relates the average fraction of unchanged residues q to a distance D, which can be converted to a calibrated time of the last common ancestor (LCA) of the compared sequences. This is a nonlinear relationship between q and D, and a linear relationship D to time ($0.088/t$ in millions of years) (Feng et al. 1997). These types of analyses require full genome sequence data from multiple organisms, and sophisticated informatics tools capable of sifting through large amounts of data. First attempts at dating resistance were made using traditional Sanger DNA sequencing on a limited number of biosynthetic genes (discussed below) (Landan et al. 1990). Although effective, these approaches offer a limited scope of analysis, because some prior knowledge of sequence is required to retrieve orthologs from multiple organisms. Advances in next-generation sequencing technologies and bioinformatics on the other hand are mak-

ing unprecedented amounts of information available and generate data without prior sequence knowledge. As a result, analyses based on next-generation sequencing data are less subject to the sample bias inherent to Sanger sequencing-based methods. Gene dating based on both methods is discussed in more detail in the following sections.

PREDICTIONS OF AGE BASED ON BIOSYNTHETIC CLUSTERS: TRADITIONAL SEQUENCING

For producers of "natural products" with antibiotic activity, resistance must be as old as biosynthesis. Dating biosynthetic gene clusters can, therefore, give us a reasonable estimate of the age of resistance. In one of the first studies attempting to date antibiotic biosynthesis, cloning and traditional sequencing methods were used to date the transfer of isopenicillin-N synthase (IPNS, one of the enzymes involved in the synthesis of the antibiotic penicillin) between a bacterium and three fungi (Weigel et al. 1988). The analysis included IPNS sequences from *Penicillium chrysogenum*, *Aspergillus nidulans*, and *Cephalosporium acremonium*, and the Gram-positive bacterium *Streptomyces lipmanii*, and was performed assuming a constant rate of 10^{-9} nucleotide changes per site per year, to arrive at an estimated 370 million years, since gene transfer from bacteria to fungi. With the later discovery of a new IPNS in *Flavobacterium*, a Gram-negative bacterial genus, the hypothesis was revised to require a single horizontal transfer from bacteria to fungi before the Gram-positive/-negative split (Landan et al. 1990). This date estimation depends on the accuracy of the predicted rate of change of these sequences, which is different for prokaryotes and eukaryotes and likely has not remained constant because the sequences diverged (Lynch 2010).

Although Landan et al. (1990) focused their analysis on a single biosynthetic enzyme, Richard Baltz (2005) asked more broadly whether biosynthesis to specific antibiotics could be dated. Baltz (2005) reasoned that older antibiotics would be encountered more frequently than younger antibiotics in a drug discovery program, because of the spread of biosynthetic genes vertically and horizontally in a population given evolutionary time. Baltz (2005) calculated that the average similarity among nine orthologs involved in streptomycin biosynthesis was 72.6%, resulting in an estimate of 610 ± 71 million years since their LCA. Erythromycin, another commonly encountered antibiotic, was calculated to be 880 ± 134 million years old, using 18 sequences with an average of 62.9% identity. Similarly, six orthologous sequences involved in the production of the glycopeptide natural products balhimycin and chloroeremomycin shared on average 88% identity, resulting in an estimate of 240 ± 12 million years since their LCA. Baltz noted that the age of glycopeptide antibiotics may be less than his estimate, as the producing organism is rare and likely encountered less frequently in the soil than members of *Streptomyces*. Last, Baltz analyzed the biosynthetic genes of two organisms producing daptomycin, a newly discovered lipopeptide antibiotic. He calculated an LCA of 30 ± 0.2 million years based on an identity of 98.7% between two producing organisms. The 10-fold difference in the predicted age of the rarely encountered daptomycin biosynthetic cluster versus the more frequently encountered antibiotics streptomycin and erythromycin supports the hypothesis that older antibiotics are frequently "rediscovered," because their biosynthetic clusters have had evolutionary time to disperse more widely in the metagenome (Baltz 2005). However, this analysis is based on detection of production under laboratory conditions. As more actinomycete genome sequences become available, we are increasingly recognizing that these strains harbor "silent" biosynthetic clusters and that laboratory production estimates of the frequency of production likely significantly underestimates the actual genetic diversity of biosynthesis.

The above-mentioned studies examine antibiotic biosynthesis, which is only an approximation of the age of resistance. Can resistance genes themselves been dated? In a series of papers, Barlow, Hall, and Salipante examined the origin and evolution of β-lactamases, the major resistance determinants of β-lactam antibiotics (Barlow and Hall 2002a,b, 2003a,b; Hall et al.

2003, 2004; Hall and Barlow 2004). The investigators proposed that the origin of the AmpC β-lactamase was chromosomal, because the phylogeny of chromosomal AmpC sequences matches the species phylogeny of the hosts. AmpC has presumably undergone multiple mobilizations to plasmids and to other species in the course of evolution (Barlow and Hall 2002a). A separate phylogenetic analysis of the OXA β-lactamase family also argues for multiple mobilizations to plasmids for this gene ($\sim 42 \pm 9$ and 116 ± 25 million years ago), and a transfer of an OXA gene between *Streptococcus* and *Bacillus* between 320 and 520 million years ago (Barlow and Hall 2002a). The relative dates of three different classes of serine β-lactamases were also determined by Hall and Barlow (2003) using a structural alignment based on pairwise root mean square (RMS) distances. Structural distance was used rather than a sequence alignment because of the lack of detectable homology to these sequences. To root this phylogeny, a DD-peptidase structure from *Streptomyces* was used along with the EstB esterase from a species of *Burkholderia*. According to this analysis, the class C β-lactamases were earliest to diverge, followed by class D. Both were outgroups compared with the class A β-lactamases, but because the RMS distances are not calibrated to a clock there is no way to provide dates for these splits (Hall and Barlow 2003). Rounding out the analysis of the β-lactamases, the subgroups of the class B metallo-β-lactamases were proposed to have two independent origins (Hall et al. 2004). Their tree of the B1+B2 subclass metallo-β-lactamases suggest that these sequences arose around one billion years ago, around the time the gammaproteobacteria diverged from the β-proteobacteria. The B3 subclass was proposed to date to 2.2 billion years ago, before the split of Gram-positive and Gram-negative Eubacteria.

The Impact of Technology on Our Ability to Predict the History of Antibiotics

Estimates of age based on a few sequences (like daptomycin, above) are less accurate than analyses generated with more sequencing data. This is an ideal time to revisit these estimates, as the genome sequencing of the class *Actinomycetes* has expanded dramatically with the advent of next-generation sequencing (1055 genome sequences listed at the National Center for Biotechnology Information [NCBI], along with further 7020 genome-sequencing projects as of December 2015, see ncbi.nlm.nih.gov/Taxonomy/Browser/wwwtax.cgi). However, the development of new methods that make use of the information contained in whole genomes in general, instead of just handfuls of identifiable orthologs, is of utmost importance. To that end, novel techniques like single-cell sequencing and metagenomic analyses (the analysis of all DNA isolated directly from an environment without prior culture) are improving access to branches on the tree of life that have been previously undersampled (Jansson and Tas 2014). A recent series of analyses by the Dantas laboratory using culture-independent techniques have yielded an enormous wealth of resistance gene sequence data that will help in the phylogenetics and dating of resistance (Forsberg et al. 2012, 2014; Moore et al. 2013; Pehrsson et al. 2013; Gibson et al. 2015). These studies have used "functional metagenomics," in which the entire microbial metagenome is cloned into an expression vector, transformed into an expression host (*Escherichia coli*) and selected on an antibiotic of interest. This technique captures novel resistance genes that would have otherwise been missed by traditional culture- or polymerase chain reaction (PCR)-based methods. Forsberg et al. (2012) first showed the power of this technique to link genes in the environment and the clinic, and offer evidence of horizontal transfer of resistance between these two environments. Increasing the number and diversity of sequences we can retrieve from an environment will lead to a better understanding of the evolutionary trajectory of current (and future) resistance elements.

Advances in sequencing go hand in hand with the software required for the detection and analysis of biosynthetic gene clusters (Blin et al. 2013; Cimermancic et al. 2014; Doroghazi et al. 2014) enabling finer analysis of this genome data in the identification of new clusters.

New proposals for the annotation and classification of biosynthetic gene clusters will provide the foundation to examine and reconstruct their evolutionary histories with unprecedented detail (Medema et al. 2015). As more gene sequences become available, phylogenies will improve accordingly; dating resistance determinants is often an accidental by-product of phylogenetic studies aimed at classifying newly discovered resistance determinants. The next section discusses the methodology of phylogenetics in greater detail.

The Phylogenetics of Ancient Resistance

It is possible to tie some of the evolutionary history of antibiotic resistance determinants to major evolutionary events inferred by other phylogeneticists, such as divergence of major lineages or the emergence of large groups of organisms (Battistuzzi and Hedges 2009). However, evolution is rarely linear, and gene gain, duplication, loss, and horizontal gene transfer in bacteria confound dating efforts with even the most sophisticated methods. Modern phylogenomic approaches have three goals: (1) inferring more accurate gene trees give a known species tree, (2) inferring a more accurate species tree given gene trees, and (3) joint estimation of both a species and gene tree (Boussau et al. 2013; Chan and Ragan 2013). The evolution of biosynthetic gene clusters should not be expected to occur in the same manner as independent sequences, and the method used to infer the phylogeny of whole clusters should be sensitive to the gain, loss, duplication, and innovation of new components. An excellent review by Aminov and Mackie (2007) provides further detail on the phylogenies of the *tet* genes (conferring tetracycline-resistance), the *erm* family of methyltransferases (conferring resistance to erythromycin), and the *vanHAX* cluster (encoding glycopeptide resistance genes).

EXPERIMENTAL EVIDENCE FOR OLD RESISTANCE

In silico analyses of orthologous gene sequences have predictive value, but is there experimental evidence that resistance pre-dates the clinical use of antibiotics? Besides the studies mentioned in the introduction to this review (Mather et al. 2014; Warinner et al. 2014), few instances of resistance have been found in the limited number of sequenced of ancient microorganisms from human samples. The genome sequence of *Vibrio cholera* from 19th century Philadelphia did not reveal any candidate resistance genes apart from efflux (Devault et al. 2014), nor were any specific resistance genes found in several strains of the plague-causing bacillus *Yersinia pestis* isolated from the plague of Justinian (541–543 AD) (Wagner et al. 2014). It is not altogether surprising that resistance genes would be found sparingly in the human microbiome before the clinical use of antibiotics, because there would have been no selective pressure to maintain them. In contrast, environmental microorganisms have coevolved with antibiotics produced in Actinobacteria, and are more likely to harbor examples of resistance genes from long ago. An unparalleled source of ancient DNA is the permanently frozen soil known as "permafrost," found under an estimated 25% of the earth's surface (Jansson and Tas 2014). Permafrost is defined as soil that has remained frozen for at least two consecutive years, but some Arctic and Antarctic permafrost has been frozen for 1–3 million years (Wagner et al. 2014). The DNA from permafrost can be isolated and queried experimentally for antibiotic resistance genes (among other things). The seminal work of D'Costa et al. (2011) on Beringian permafrost is complemented by studies showing that functional resistance genes can be retrieved from 5000-year-old DNA (Perron et al. 2015), and that resistance had mobilized to plasmids and transposons in ancient times (Mindlin et al. 2005; Petrova et al. 2011, 2014). Modern day microorganisms found in a cave that has been isolated from the surface for four million years have also been shown to harbor functional antibiotic resistance genes (Bhullar et al. 2012). A phylogenetic tree of macrolide phosphostransferases was generated using a sequence found in the genome of a cave organism (identified as *Brachybacterium paraconglomeratum*), and compared with a phylogeny of

macrolide phosphotransferases from a terrestrial species of *Brachybacterium* (*B. faecium* DSM 4810) and environmental *Bacillus cereus* (Wang et al. 2015). Analysis of 10 kb upstream of and downstream from the *mph* revealed that MPHs from *Brachybacterium* strains from both cave and terrestrial origin cluster together as a separate group among known MPHs (Bhullar et al. 2012). The results of these studies provide direct experimental evidence that antibiotic resistance is ancient, and provide a glimpse into the evolutionary history of a natural environmental phenomenon.

CONCLUSIONS

Using both in silico methods and direct experimental analysis, we are beginning to unravel the complex natural history of antibiotic resistance. The importance of studying ancient resistance is twofold. First, cataloging the vast environmental reservoir of resistance genes provides advance notice as to what forms of resistance have the potential to emerge in the clinic under selective pressure. Second, understanding evolution has predictive value for resistance genes that are currently emerging as clinically important. Winston Churchill said "the farther backward you can look, the farther forward you are likely to see." This is especially true for antibiotic resistance where the past predicts and informs the future.

ACKNOWLEDGMENTS

This work in G.W.'s laboratory is funded by the Canadian Institutes of Health Research (CIHR) and by a Canada Research Chair in Biochemistry (to G.W.). N.W. is the recipient of a CIHR Fellowship.

REFERENCES

Aminov RI. 2011. Horizontal gene exchange in environmental microbiota. *Front Microbiol* **2:** 158.

Aminov RI, Mackie RI. 2007. Evolution and ecology of antibiotic resistance genes. *FEMS Microbiol Lett* **271:** 147–161.

Baltz RH. 2005. Antibiotic discovery from actinomycetes: Will a renaissance follow the decline and fall? *SIM News* **55:** 186–196.

Barlow M, Hall BG. 2002a. Phylogenetic analysis shows that the OXA β-lactamase genes have been on plasmids for millions of years. *J Mol Evol* **55**: 314–321.

Barlow M, Hall BG. 2002b. Origin and evolution of the AmpC β-lactamases of *Citrobacter freundii*. *Antimicrob Agents Chemother* **46**: 1190–1198.

Barlow M, Hall BG. 2003a. Experimental prediction of the natural evolution of antibiotic resistance. *Genetics* **163**: 1237–1241.

Barlow M, Hall BG. 2003b. Experimental prediction of the evolution of cefepime resistance from the CMY-2 AmpC β-lactamase. *Genetics* **164**: 23–29.

Battistuzzi FU, Hedges SB. 2009. A major clade of prokaryotes with ancient adaptations to life on land. *Mol Biol Evol* **26**: 335–343.

Benveniste R, Davies J. 1973. Aminoglycoside antibiotic-inactivating enzymes in actinomycetes similar to those present in clinical isolates of antibiotic-resistant bacteria. *Proc Natl Acad Sci* **70**: 2276–2280.

Bhullar K, Waglechner N, Pawlowski A, Koteva K, Banks ED, Johnston MD, Barton HA, Wright GD. 2012. Antibiotic resistance is prevalent in an isolated cave microbiome. *PLoS ONE* **7**: e34953.

Blin K, Medema MH, Kazempour D, Fischbach MA, Breitling R, Takano E, Weber T. 2013. antiSMASH 2.0—A versatile platform for genome mining of secondary metabolite producers. *Nucleic Acids Res* **41**: W204–W212.

Boussau B, Szollosi GJ, Duret L, Gouy M, Tannier E, Daubin V. 2013. Genome-scale coestimation of species and gene trees. *Genome Res* **23**: 323–330.

Chait R, Vetsigian K, Kishony R. 2012. What counters antibiotic resistance in nature? *Nat Chem Biol* **8**: 2–5.

Chan CX, Ragan MA. 2013. Next-generation phylogenomics. *Biol Direct* **8**: 3.

Cimermancic P, Medema MH, Claesen J, Kurita K, Wieland Brown LC, Mavrommatis K, Pati A, Godfrey PA, Koehrsen M, Clardy J, et al. 2014. Insights into secondary metabolism from a global analysis of prokaryotic biosynthetic gene clusters. *Cell* **158**: 412–421.

D'Costa VM, McGrann KM, Hughes DW, Wright GD. 2006. Sampling the antibiotic resistome. *Science* **311**: 374–377.

D'Costa VM, King CE, Kalan L, Morar M, Sung WW, Schwarz C, Froese D, Zazula G, Calmels F, Debruyne R, et al. 2011. Antibiotic resistance is ancient. *Nature* **477**: 457–461.

Devault AM, Golding GB, Waglechner N, Enk JM, Kuch M, Tien JH, Shi M, Fisman DN, Dhody AN, Forrest S, et al. 2014. Second-pandemic strain of *Vibrio cholerae* from the Philadelphia cholera outbreak of 1849. *N Engl J Med* **370**: 334–340.

Doroghazi JR, Albright JC, Goering AW, Ju KS, Haines RR, Tchalukov KA, Labeda DP, Kelleher NL, Metcalf WW. 2014. A roadmap for natural product discovery based on large-scale genomics and metabolomics. *Nat Chem Biol* **10**: 963–968.

Feng DF, Cho G, Doolittle RF. 1997. Determining divergence times with a protein clock: Update and reevaluation. *Proc Natl Acad Sci* **94**: 13028–13033.

Forsberg KJ, Reyes A, Wang B, Selleck EM, Sommer MO, Dantas G. 2012. The shared antibiotic resistome of soil bacteria and human pathogens. *Science* **337**: 1107–1111.

Forsberg KJ, Patel S, Gibson MK, Lauber CL, Knight R, Fierer N, Dantas G. 2014. Bacterial phylogeny structures soil resistomes across habitats. *Nature* **509**: 612–616.

Gibson MK, Forsberg KJ, Dantas G. 2015. Improved annotation of antibiotic resistance determinants reveals microbial resistomes cluster by ecology. *ISME J* **9**: 207–216.

Grishin NV. 1995. Estimation of the number of amino acid substitutions per site when the substitution rate varies among sites. *J Mol Evol* **41**: 675–679.

Hall BG, Barlow M. 2003. Structure-based phylogenies of the serine β-lactamases. *J Mol Evol* **57**: 255–260.

Hall BG, Barlow M. 2004. Evolution of the serine β-lactamases: Past, present and future. *Drug Resist Updat* **7**: 111–123.

Hall BG, Salipante SJ, Barlow M. 2003. The metallo-β-lactamases fall into two distinct phylogenetic groups. *J Mol Evol* **57**: 249–254.

Hall BG, Salipante SJ, Barlow M. 2004. Independent origins of subgroup Bl+B2 and subgroup B3 metallo-β-lactamases. *J Mol Evol* **59**: 133–141.

Humeniuk C, Arlet G, Gautier V, Grimont P, Labia R, Philippon A. 2002. β-lactamases of *Kluyvera ascorbata*, probable progenitors of some plasmid-encoded CTX-M types. *Antimicrob Agents Chemother* **46**: 3045–3049.

Jansson JK, Tas N. 2014. The microbial ecology of permafrost. *Nat Rev Microbiol* **12**: 414–425.

Knapp CW, Dolfing J, Ehlert PA, Graham DW. 2010. Evidence of increasing antibiotic resistance gene abundances in archived soils since 1940. *Environ Sci Technol* **44**: 580–587.

Landan G, Cohen G, Aharonowitz Y, Shuali Y, Graur D, Shiffman D. 1990. Evolution of isopenicillin N synthase genes may have involved horizontal gene transfer. *Mol Biol Evol* **7**: 399–406.

Lynch M. 2010. Evolution of the mutation rate. *Trends Genet* **26**: 345–352.

Mather AE, Baker KS, McGregor H, Coupland P, Mather PL, Deheer-Graham A, Parkhill J, Bracegirdle P, Russell JE, Thomson NR. 2014. Bacillary dysentery from World War 1 and NCTC1, the first bacterial isolate in the National Collection. *Lancet* **384**: 1720.

Medema MH, Kottmann R, Yilmaz P, Cummings M, Biggins JB, Blin K, de Bruijn I, Chooi YH, Claesen J, Coates RC, et al. 2015. Minimum information about a biosynthetic gene cluster. *Nat Chem Biol* **11**: 625–631.

Mindlin S, Minakhin L, Petrova M, Kholodii G, Minakhina S, Gorlenko Z, Nikiforov V. 2005. Present-day mercury resistance transposons are common in bacteria preserved in permafrost grounds since the upper pleistocene. *Res Microbiol* **156**: 994–1004.

Moore AM, Patel S, Forsberg KJ, Wang B, Bentley G, Razia Y, Qin X, Tarr PI, Dantas G. 2013. Pediatric fecal microbiota harbor diverse and novel antibiotic resistance genes. *PLoS ONE* **8**: e78822.

Ochman H, Lawrence JG, Groisman EA. 2000. Lateral gene transfer and the nature of bacterial innovation. *Nature* **405**: 299–304.

Pehrsson EC, Forsberg KJ, Gibson MK, Ahmadi S, Dantas G. 2013. Novel resistance functions uncovered using functional metagenomic investigations of resistance reservoirs. *Front Microbiol* **4**: 1–11.

Perron GG, Whyte L, Turnbaugh PJ, Goordial J, Hanage WP, Dantas G, Desai MM. 2015. Functional characterization of bacteria isolated from ancient arctic soil exposes diverse resistance mechanisms to modern antibiotics. *PLoS ONE* **10:** e0069533.

Petrova M, Gorlenko Z, Mindlin S. 2009. Molecular structure and translocation of a multiple antibiotic resistance region of a *Psychrobacter psychrophilus* permafrost strain. *FEMS Microbiol Lett* **296:** 190–197.

Petrova M, Gorlenko Z, Mindlin S. 2011. Tn5045, a novel integron-containing antibiotic and chromate resistance transposon isolated from a permafrost bacterium. *Res Microbiol* **162:** 337–345.

Petrova M, Kurakov A, Shcherbatova N, Mindlin S. 2014. Genetic structure and biological properties of the first ancient multiresistance plasmid pKLH80 isolated from a permafrost bacterium. *Microbiology* **160:** 2253–2263.

Santiago-Rodriguez TM, Fornaciari G, Luciani S, Dowd SE, Toranzos GA, Marota I, Cano RJ. 2015. Gut microbiome of an 11th century AD Pre-Columbian Andean mummy. *PLoS ONE* **10:** e0138135.

Skippington E, Ragan MA. 2011. Lateral genetic transfer and the construction of genetic exchange communities. *FEMS Microbiol Rev* **35:** 707–735.

Stokes HW, Gillings MR. 2011. Gene flow, mobile genetic elements and the recruitment of antibiotic resistance genes into Gram-negative pathogens. *FEMS Microbiol Rev* **35:** 790–819.

Tatusov RL, Koonin EV, Lipman DJ. 1997. A genomic perspective on protein families. *Science* **278:** 631–637.

van Elsas JD, Bailey MJ. 2002. The ecology of transfer of mobile genetic elements. *FEMS Microbiol Ecol* **42:** 187–197.

Wagner DM, Klunk J, Harbeck M, Devault A, Waglechner N, Sahl JW, Enk J, Birdsell DN, Kuch M, Lumibao C, et al. 2014. *Yersinia pestis* and the plague of Justinian 541-543 AD: A genomic analysis. *Lancet Infect Dis* **14:** 319–326.

Wang C, Sui Z, Leclercq SO, Zhang G, Zhao M, Chen W, Feng J. 2015. Functional characterization and phylogenetic analysis of acquired and intrinsic macrolide phosphotransferases in the *Bacillus cereus* group. *Environ Microbiol* **17:** 1560–1573.

Warinner C, Rodrigues JF, Vyas R, Trachsel C, Shved N, Grossmann J, Radini A, Hancock Y, Tito RY, Fiddyment S, et al. 2014. Pathogens and host immunity in the ancient human oral cavity. *Nat Genet* **46:** 336–344.

Weigel BJ, Burgett SG, Chen VJ, Skatrud PL, Frolik CA, Queener SW, Ingolia TD. 1988. Cloning and expression in *Escherichia coli* of isopenicillin N synthetase genes from *Streptomyces lipmanii* and *Aspergillus nidulans*. *J Bacteriol* **170:** 3817–3826.

Wiedenbeck J, Cohan FM. 2011. Origins of bacterial diversity through horizontal genetic transfer and adaptation to new ecological niches. *FEMS Microbiol Rev* **35:** 957–976.

Wright GD. 2010. Antibiotic resistance in the environment: A link to the clinic? *Curr Opin Microbiol* **13:** 589–594.

Appropriate Targets for Antibacterial Drugs

Lynn L. Silver

LL Silver Consulting, Springfield, New Jersey 07081

Correspondence: silverly@comcast.net

Successful small-molecule antibacterial agents must meet a variety of criteria. Foremost is the need for selectivity and safety: It is easy to kill bacteria with chemicals, but difficult to do it without harming the patient. Other requirements are possession of a useful antibacterial spectrum, no cross-resistance with existing therapeutics, low propensity for rapid resistance selection, and pharmacological properties that allow effective systemic dosing. Choosing molecular targets for new antibiotics does seem a good basis for achieving these criteria, but this could be misleading. Although the presence of the target is necessary to insure the desired spectrum, it is not sufficient, as the permeability and efflux properties of various species, especially Gram-negatives, are critical determinants of antibacterial activity. Further, although essentiality (at least in vitro), lack of close human homologs, lack of target-based cross-resistance, and presence in important pathogens can be predicted based on the target, the choice of a single enzyme as a target may increase the likelihood of rapid resistance selection. In fact, it is likely that the low output of antibacterial target-based discovery is because of difficulty of endowing lead enzyme inhibitors with whole-cell activity and to the propensity for such inhibitors (if they can gain entry) to select rapidly for resistance. These potential problems must be reckoned with for success of novel target-based discovery.

The specter of increasing antibiotic resistance has raised recognition that the world is reliant on antibiotics for the maintenance and improvement of world health. Whether multidrug-resistant (MDR) pathogens become prevalent and lead to a postantibiotic era before we are able to put the brakes on the resistance phenomenon is uncertain. But, it is clear that there are movements underway to address the looming problem of antibiotic resistance by various means, including increased surveillance to map the course of resistance spread, development of rapid diagnostics to insure early selection of suitable therapeutics, improved regulatory pathways, incentives for discovery of new agents, and studies of alternate therapeutic routes.

However, discovery of new agents is not a given, even if incentivized. The rate of discovery of developable novel antibacterial agents has been decreasing. Although it is true that large pharmaceutical companies have been leaving the area, their exit was not only for financial and regulatory reasons, but it followed many years of focused pursuit of new antibiotics with little success. The highly productive years of screening natural products for discovery of new antibiotics via mostly empirical ("kill-the-

Cite this article as *Cold Spring Harb Perspect Med* doi: 10.1101/cshperspect.a030239

bug") screens had ended in the early 1980s and the output of novel useful antibacterial classes fell dramatically. More rational methods of drug discovery, using whole-cell-directed phenotypic screens were instituted (Singh et al. 2011; Mills and Dougherty 2012; Silver 2012), but this too was generally unproductive, turning up leads but few drugs. Fosfomycin, thienamycin, and the early β-lactamase inhibitors, found through such screens, were exceptions (Gadebusch et al. 1992; Silver 2012). Thus, when antibiotic resistance started its dramatic increase in the 1980s and 1990s, starting with methicillin-resistant *Staphylococcus aureus* (MRSA), the industry looked for new ways to attack the problem. One direction that proved productive was re-evaluation and development of previously discovered antibiotics targeting Gram-positives, for example, daptomycin, and modification of existing classes to cover the resistant species, as with the β-lactams active against MRSA (Livermore 2006). By the mid-1990s, however, genomics and bioinformatics, along with high-throughput technologies for chemical synthesis and compound screening, transformed the discovery process to the search for inhibitors of novel targets.

THE ADVENT OF GENOMICS

It is often remarked that the molecular targets of existing antibacterial agents are limited. And this is true; there are estimated to be about 40 targets of marketed agents (Lange et al. 2007). In the mid-1990s, under threat of the increase in antibiotic resistance and aided (or goaded) by the advent of genomics, this apparently narrow set of exploited targets led many in industry and academe to search for new targets, with the idea that identification of hitherto unexploited targets would lead to discovery of new agents that were not cross-resistant with existing classes of drugs. This proposed lack of cross-resistance was based on the assumption that cross-resistance is a function of the molecular target (which is not always true). Thousands of scholarly reviews (e.g., Chan et al. 2002; Isaacson 2002; Lerner and Beutel 2002; McDevitt et al. 2002; Mills 2006; Monaghan and Barrett

2006), described the new paradigm of antibiotic discovery based on identification and prioritization of new targets and the proliferation of high-throughput techniques to find and optimize inhibitors of these targets.

In fact, there are many but not an unlimited number of essential gene products in bacteria, 160 to 170 shared by a broad spectrum of bacteria (Forsyth et al. 2002; Payne et al. 2007) and 400 or so in *Salmonella typhimurium* and *Escherichia coli* (Schmid et al. 1989; Black and Hare 2000). Interestingly, few, if any, of these were discovered through genomics. Rather, they were known from microbial genetics studies starting in the 1960s, based on conditional lethal mutants that defined most of the essential functions of bacteria. Because the earliest methods for antibiotic discovery used empirical screens, all of the essential functions of bacteria should have been subject to discovery. It is not necessary to know the target of a new antibacterial a priori and very few of the antibacterial drugs in use were discovered by target-directed screening. Still, target-based discovery was a rational approach and it gained widespread adherence.

Genomics does provide a great deal of useful information on the distribution of essential genes among species and forms the basis for greatly improved understanding of mechanisms of antibiotic action and resistance, bacterial physiology, and metabolic networks. But the genomic approach of providing a fertile field of unexploited targets for discovery of inhibitors that could be turned into drugs has been unproductive. The reasons for this have been the subject of general reviews (Overbye and Barrett 2005; Gwynn et al. 2010; Livermore 2011; Silver 2011; Chopra 2012) and has been documented in reports of the extensive novel target-directed screening programs of GlaxoSmithKline (Philadelphia, PA) (Payne et al. 2007) and AstraZeneca (Wilmington, DE) (Tommasi et al. 2015). The main barriers have been (1) the high probability of rapid resistance arising to inhibitors of the single enzymes chosen via genomics studies (Silver 2007; Brotz-Oesterhelt and Brunner 2008), and (2) the inappropriateness of current chemical libraries as sources of antibacterials, especially because of the difficulty of compound entry into bacterial cells (Silver 2011; Brown et al. 2014; Tommasi et al. 2015).

Once it was recognized that the narrow focus on novel "unexploited" targets, although heavily pursued, was not productive, the area slowly turned to appreciation that the tried-and-true targets are indeed worthy, somehow privileged, and should be pursued with novel chemical matter (Projan 2002; Lange et al. 2007).

THE NATURE OF "GOOD" ANTIBACTERIAL TARGETS

The molecular targets of the successful antibacterials are relatively few but they are almost uniformly involved in pathways of macromolecular synthesis; indeed, these are the essential functions of bacteria that cannot be satisfied by feeding of intermediates. Notably, only a few targets of the main classes of antibacterials used in sys-

temic monotherapy are essential enzymes. As listed in Table 1, most of these systemic mono-therapeutic agents have nonprotein targets. Most target ribosomal RNA, intermediates in cell-wall synthesis, or membranes. Only the β-lactams and fluoroquinolones target enzymes and these, notably, each target at least two enzymes. On the other hand, there are many registered antibacterials, listed in Table 2, that do target single essential enzymes. In column 3 of Tables 1 and 2, the relative frequency and function of single-step endogenous chromosomal mutations to high-level resistance (more than sixteenfold minimum inhibitory concentration [MIC]) seen in vitro are shown, whereas in column 4 of both tables, the major forms of clinically important resistance are noted. It can be seen that, for the monotherapeutic agents of Table 1, the occurrence of single-step mutations giving significant resistance in vitro is rare to low, whereas for the single-targeted agents in Table 2, single-step in vitro resistance is stan-

Table 1. Antibacterial classes used in systemic monotherapy

Drug class	Target	High-level[a] single-step resistance in vitro	Major clinical resistance
β-Lactams	Multiple PBPs	Endogenous β-lactamase, porin loss, efflux	HGT β-lactamases, endogenous β-lactamases, porin loss, efflux
Glycopeptides	Lipid II	Rare	HGT vancomycin-resistance cassettes; stepwise cell-wall changes
Macrolides	50S RNA	Rare	HGT MLS$_B$ ribosome methylation, efflux
Oxazolidinones	50S RNA	Rare	HGT *cfr* ribosome methylation
Chloramphenicol	50S RNA	Rare	HGT modification
Pleuromutilins	50S RNA	Rare	HGT *cfr* target methylation
Streptogramins	50S RNA	Rare	HGT MLS$_B$ ribosome methylation, efflux, modification
Lincomycins	50S RNA	Rare	HGT MLS$_B$
Tetracyclines	30S RNA	Low (some efflux)	HGT efflux, ribosome protection
Aminoglycosides	30S RNA	Low (ribosomal proteins)	HGT modification
Fluoroquinolones	Gyrase, Topo IV	Rare	Stepwise two or more than two target mutations, efflux
Daptomycin	Membranes	Rare	Stepwise membrane changes
Polymyxins	Membranes	Low (mutations modifying LPS)	Stepwise LPS changes, HGT modification
Metronidazole	DNA (?)	Low (loss of reductase, entry)	HGT *nim* (alternate reductase) genes

PBPs, Penicillin binding proteins; HGT, horizontal gene transfer; LPS, lipopolysaccharide.
[a]Minimum inhibitory concentrations (MICs) raised more than sixteenfold.

Table 2. Antibacterial classes targeting single enzymes

Drug class	Target	High-level[a] single-step resistance in vitro	Major clinical resistance
Rifampicin	RNA polymerase	Changes in *rpoB* at many sites	Changes in *rpoB* at fewer sites
Trimethoprim	Dihydrofolate reductase	Altered DHFR (folA)	Altered DHFR, HGT altered DHFR
Sulfonamides	Dihydropterate synthase	Altered DHPS (folP)	Altered *folP*, HGT altered DHPS
Fidaxomicin	RNA polymerase	Changes in *rpoC, rpoB*	Low, *rpoC* or *rpoB*
Fosfomycin	MurA	Loss of permeases	HGT-modifying enzymes
Fusidic acid	Elongation factor G	Mutations in *fusA* (EfG)	Mutations in *fusA*; HGT target protection FusB,C,D
Fab I inhibitors	Enoyl (acyl carrier protein) reductase	Mutations in *fabI*	Mutations in *fabI*
LpxC inhibitors	Lipid A deacetylase	Mutations in *fabZ*, efflux, *lpxC*	Mutations in *fabZ*, efflux

DHFR, Dihydrofolate reductase; DHPS, dihydropteroate synthase; HGT, horizontal gene transfer.
[a]Minimum inhibitory concentrations (MICs) raised more than sixteenfold.

dardly seen. Clinical resistance to the agents in Table 1 is generally caused by horizontal gene transfer (HGT) with contribution of the in vitro types of resistance seen in only a few cases. These observations and other supporting evidence led to the multitarget hypothesis (Silver and Bostian 1993; Silver 2007; Brotz-Oesterhelt and Brunner 2008) that successful systemic monotherapeutic agents are those that have low levels of endogenous single-step resistance, and this is due to their targeting of the products of multiple genes or of pathways.

For the single-targeted agents, the types of resistance seen in vitro are generally the same as those seen in the clinic, with few exceptions. In fact, few of the agents in Table 2 are used in systemic monotherapy. Rifampicin is almost always used in combination, as are trimethoprim and sulfamethoxazole, presumably limiting the selection of resistance. Fidaxomicin is used in nonsystemic treatment of *Clostridium difficile*; although resistance due to target mutations can occur, dosing is extremely high and presumably explains why resistance is rarely seen. Fusidic acid has been used for treatment of *S. aureus* skin infections, generally in combination with rifampicin to retard resistance development. However, it is now being developed for mono-

therapeutic use with a high loading dose to decrease resistance development (Fernandes 2016).

Of the single-enzyme inhibitors, only fosfomycin is currently used as monotherapy and that mainly for urinary tract infections (UTIs). Interestingly, the prevalent in vitro mutations to fosfomycin resistance, caused by loss of permeases necessary for active transport across the cytoplasmic membrane, are only rarely seen in vivo (in UTI). It seems that these transport mutants have reduced growth rates in media and urine in the presence of fosfomycin and are not maintained in the bladder (Nilsson et al. 2003). Recently, it has been proposed to use fosfomycin for more systemic indications against MDR pathogens, and it remains to be seen whether the low resistance frequencies, caused by endogenous mutations, extend to other sites of infection (Karageorgopoulos et al. 2012).

Inhibitors of FabI and LpxC are in various stages of development, and progress in these two areas is reviewed in the literature by Yao and Rock (2016) and Erwin (2016), respectively. In each case, the propensity of inhibitors to select for resistance is carefully reviewed. According to Yao and Rock, for FabI inhibitors, which have a narrow spectrum (*Staphylococci* only), it may be possible to design extremely potent enzyme inhibitors that are able to be dosed at levels that overcome target-based MIC increases. For single-target inhibitors with broader spectra, the need to inhibit homologs across many species may limit the potency attainable. The potential for clinically important resistance will have to be critically monitored.

Not listed in Table 2 is the inhibitor of leucyl-tRNA synthetase, GSK2251052, which failed in a phase II trial for complicated UTIs due to high-level resistance seen after 1 day of treatment (O'Dwyer et al. 2014). The mutants were highly fit and were shown to arise at a high rate in vitro. This has proved a cautionary tale for the antibacterial discovery community and has fortunately increased attention to the possibility of rapid resistance development.

TARGET LOCATION

As noted above, one of the reasons for failure of target-based antibacterial discovery, especially that done by screening for inhibition of enzyme activity, is the inability to endow such enzyme inhibitors with the ability to enter and be retained in bacterial cells. Entry into Gram-positives is generally attainable, as their permeability barrier is the cytoplasmic membrane; thus, neutral, nonpolar compounds are preferred. It is Gram-negative entry that is highly problematic as there are no simple rules for the physicochemical properties that can guide the entry of a molecule through the outer membrane, avoiding efflux and penetrating the cytoplasmic membrane. Thus, cytoplasmic targets are generally a poor bet for Gram-negative agents. However, it is often possible to afford entry into the periplasm. Thus, extracellular or periplasmic targets are most attractive for Gram-negatives. Still, the resistance–nodulation–division (RND) efflux pumps exert their power in the periplasm and must be avoided. It may be that the β-lactam antibiotics owe much of their success, aside from their multiple targets, to periplasmic location and covalent interaction with those targets that would allow escape from the sweep of efflux pumps.

NONESSENTIAL TARGETS

It is often posited that virulence functions of pathogens would make good targets, even though they are nonessential to the survival of the bacteria, because inhibitors of those functions would exert little selective pressure for their loss. There is little proof of this and some evidence to the contrary. A recent review (Ruer et al. 2015) discusses in detail the probability of resistance development against antivirulence agents, noting that resistance has been shown to occur in some cases of quorum-sensing inhibitors (Maeda et al. 2012) but that there may be some virulence functions nonessential for survival that might have low resistance potential. Mutations will occur regardless of the presence of inhibitors and it is their selection that is responsible for maintenance in the population. If inhibitors of virulence genes have no effect on growth or adherence, then it is conceivable that selective pressure could be negligible. But more studies are required (Garcia-Contreras et al. 2015). As with all target/inhibitor pairs, rapid resistance development, its maintenance, and spread must be tested and monitored.

CONCLUSIONS

The standard list of target criteria should emphasize low-resistance potential and accessibility to inhibitors. The classically useful antibiotics (in Table 1) define targets that have shown their value and validity. For use as monotherapeutic systemic agents, it would seem that, with possible exceptions, a significant frequency of single-step spontaneous resistance is not acceptable because those mutations would be present in a sufficiently large infectious load and, if fit, could compete with susceptible siblings. The agents in Table 1 have very low single-step resistance rates, presumably caused by their so-called multitargeting. Notably, few of these classical targets are proteinaceous enzymes. New agents should be sought, using novel chemical matter that attack these validated targets. Inhibitors of single enzymes, which otherwise meet target criteria, may theoretically avoid resistance selection by use in combinations, by being safe enough for dosing at high enough levels to overcome resistance (above the mutant-prevention concentration), or by selecting mutants that have a high-fitness cost that prevents retention of resistance after removal of the selective pressure.

REFERENCES

*Reference is also in this collection.

Black T, Hare R. 2000. Will genomics revolutionize antimicrobial drug discovery? *Curr Opin Microbiol* **3**: 522–527.

Brotz-Oesterhelt H, Brunner NA. 2008. How many modes of action should an antibiotic have? *Curr Opin Pharmacol* **8**: 564–573.

Brown DG, May-Dracka TL, Gagnon MM, Tommasi R. 2014. Trends and exceptions of physical properties on antibacterial activity for Gram-positive and Gram-negative pathogens. *J Med Chem* **57**: 10144–10161.

Chan PF, Macarron R, Payne DJ, Zalacain M, Holmes DJ. 2002. Novel antibacterials: A genomics approach to drug discovery. *Curr Drug Targets Infect Disord* **2**: 291–308.

Chopra I. 2012. The 2012 Garrod Lecture: Discovery of antibacterial drugs in the 21st century. *J Antimicrob Chemother* **68**: 496–505.

* Erwin AL. 2016. Antibacterial drug discovery targeting the lipopolysaccharide biosynthetic enzyme LpxC. *Cold Spring Harb Perspect Med* doi: 10.1101/cshperspect.a025304.

* Fernandes P. 2016. Fusidic acid: A bacterial elongation factor inhibitor for the oral treatment of acute and chronic staphylococcal infections. *Cold Spring Harb Perspect Med* **6**: a025437.

Forsyth RA, Haselbeck RJ, Ohlsen KL, Yamamoto RT, Xu H, Trawick JD, Wall D, Wang L, Brown-Driver V, Froelich JM, et al. 2002. A genome-wide strategy for the identification of essential genes in *Staphylococcus aureus*. *Mol Microbiol* **43**: 1387–1400.

Gadebusch HH, Stapley EO, Zimmerman SB. 1992. The discovery of cell wall active antibacterial antibiotics. *Crit Rev Biotechnol* **12**: 225–243.

Garcia-Contreras R, Maeda T, Wood TK. 2015. Can resistance against quorum-sensing interference be selected? *ISME J* **10**: 4–10.

Gwynn MN, Portnoy A, Rittenhouse SF, Payne DJ. 2010. Challenges of antibacterial discovery revisited. *Ann NY Acad Sci* **1213**: 5–19.

Isaacson RE. 2002. Genomics and the prospects for the discovery of new targets for antibacterial and antifungal agents. *Curr Pharm Des* **8**: 1091–1098.

Karageorgopoulos DE, Wang R, Yu XH, Falagas ME. 2012. Fosfomycin: Evaluation of the published evidence on the emergence of antimicrobial resistance in Gram-negative pathogens. *J Antimicrob Chemother* **67**: 255–268.

Lange RP, Locher HH, Wyss PC, Then RL. 2007. The targets of currently used antibacterial agents: Lessons for drug discovery. *Curr Pharm Des* **13**: 3140–3154.

Lerner CG, Beutel BA. 2002. Antibacterial drug discovery in the post-genomics era. *Curr Drug Targets Infect Disord* **2**: 109–119.

Livermore DM. 2006. Can β-lactams be re-engineered to beat MRSA? *Clin Microbiol Infect* **12**: 11–16.

Livermore DM. 2011. Discovery research: The scientific challenge of finding new antibiotics. *J Antimicrob Chemother* **66**: 1941–1944.

Maeda T, Garcia-Contreras R, Pu M, Sheng L, Garcia LR, Tomas M, Wood TK. 2012. Quorum quenching quandary: Resistance to antivirulence compounds. *ISME J* **6**: 493–501.

McDevitt D, Payne DJ, Holmes DJ, Rosenberg M. 2002. Novel targets for the future development of antibacterial agents. *J Appl Microbiol* **92**: 28S–34S.

Mills SD. 2006. When will the genomics investment pay off for antibacterial discovery? *Biochem Pharmacol* **71**: 1096–1102.

Mills SD, Dougherty TJ. 2012. Cell-based screening in antibacterial discovery. In *Antibiotic discovery and development* (ed. Dougherty JT, Pucci JM), pp. 901–929. Springer, Boston.

Monaghan RL, Barrett JF. 2006. Antibacterial drug discovery—Then, now and the genomics future. *Biochem Pharmacol* **71**: 901–909.

Nilsson AI, Berg OG, Aspevall O, Kahlmeter G, Andersson DI. 2003. Biological costs and mechanisms of fosfomycin resistance in *Escherichia coli*. *Antimicrob Agents Chemother* **47**: 2850–2858.

O'Dwyer K, Spivak A, Ingraham K, Min S, Holmes DJ, Jakielaszek C, Rittenhouse S, Kwan A, Livi GP, Sathe G, et al. 2014. Bacterial resistance to leucyl-tRNA synthetase inhibitor GSK2251052 develops during treatment of complicated urinary tract infections. *Antimicrob Agents Chemother* **59**: 289–298.

Overbye KM, Barrett JF. 2005. Antibiotics: Where did we go wrong? *Drug Discov Today* **10**: 45–52.

Payne DJ, Gwynn MN, Holmes DJ, Pompliano DL. 2007. Drugs for bad bugs: Confronting the challenges of antibacterial discovery. *Nat Rev Drug Discov* **6**: 29–40.

Projan SJ. 2002. New (and not so new) antibacterial targets—From where and when will the novel drugs come? *Curr Opin Pharmacol* **2**: 513–522.

Ruer S, Pinotsis N, Steadman D, Waksman G, Remaut H. 2015. Virulence-targeted antibacterials: Concept, promise, and susceptibility to resistance mechanisms. *Chem Biol Drug Des* **86**: 379–399.

Schmid MB, Kapur N, Isaacson DR, Lindroos P, Sharpe C. 1989. Genetic analysis of temperature-sensitive lethal mutants of *Salmonella typhimurium*. *Genetics* **123**: 625–633.

Silver LL. 2007. Multi-targeting by monotherapeutic antibacterials. *Nat Rev Drug Discov* **6**: 41–55.

Silver LL. 2011. Challenges of antibacterial discovery. *Clin Microbiol Rev* **24**: 71–109.

Silver LL. 2012. Rational approaches to antibacterial discovery: Pre-genomic directed and phenotypic screening. In *Antibiotic discovery and development* (ed. Dougherty TJ, Pucci MJ), pp. 33–75. Springer, Boston.

Silver LL, Bostian KA. 1993. Discovery and development of new antibiotics: The problem of antibiotic resistance. *Antimicrob Agents Chemother* **37:** 377–383.

Singh SB, Young K, Miesel L. 2011. Screening strategies for discovery of antibacterial natural products. *Expert Rev Anti Infect Ther* **9:** 589–613.

Tommasi R, Brown DG, Walkup GK, Manchester JI, Miller AA. 2015. ESKAPEing the labyrinth of antibacterial discovery. *Nat Rev Drug Discov* **14:** 529–542

* Yao J, Rock CO. 2016. Resistance mechanisms and the future of bacterial enoyl-acyl carrier protein reductase (FabI) antibiotics. *Cold Spring Harb Perspect Med* **6:** a027045.

β-Lactams and β-Lactamase Inhibitors: An Overview

Karen Bush[1] and Patricia A. Bradford[2]

[1]Molecular and Cellular Biochemistry, Indiana University, Bloomington, Indiana 47405

[2]AstraZeneca Pharmaceuticals, Waltham, Massachusetts 02451

Correspondence: karbush@indiana.edu

β-Lactams are the most widely used class of antibiotics. Since the discovery of benzylpenicillin in the 1920s, thousands of new penicillin derivatives and related β-lactam classes of cephalosporins, cephamycins, monobactams, and carbapenems have been discovered. Each new class of β-lactam has been developed either to increase the spectrum of activity to include additional bacterial species or to address specific resistance mechanisms that have arisen in the targeted bacterial population. Resistance to β-lactams is primarily because of bacterially produced β-lactamase enzymes that hydrolyze the β-lactam ring, thereby inactivating the drug. The newest effort to circumvent resistance is the development of novel broad-spectrum β-lactamase inhibitors that work against many problematic β-lactamases, including cephalosporinases and serine-based carbapenemases, which severely limit therapeutic options. This work provides a comprehensive overview of β-lactam antibiotics that are currently in use, as well as a look ahead to several new compounds that are in the development pipeline.

When Alexander Fleming was searching for an antistaphylococcal bacteriophage in his laboratory in the 1920s, he deliberately left plates out on the bench to capture airborne agents that might also serve to kill staphylococci (Fleming 1929). His success was greater than he must have hoped for. His initial publication on benzylpenicillin described a substance that was unstable in aqueous solution but that might serve as an antiseptic or as a selective agent for isolation of Gram-negative bacteria that were present in mixed cultures of staphylococci and streptococci. As the potential utility of penicillin G as a parenteral therapeutic agent became more obvious, Fleming, Abraham, Florey, and a consortium of scientists from England and the United States were able to optimize the isolation and identification of benzylpenicillin to assist in the treatment of Allied soldiers in World War II (Macfarlane 1979). These activities set the stage for the launch of the most successful class of antibiotics in history.

β-Lactam antibiotics are currently the most used class of antibacterial agents in the infectious disease armamentarium. As shown in Figure 1, β-lactams account for 65% of all prescriptions for injectable antibiotics in the United States. Of the β-lactams, cephalosporins comprise nearly half of the prescriptions (Table 1). The β-lactams are well tolerated, efficacious,

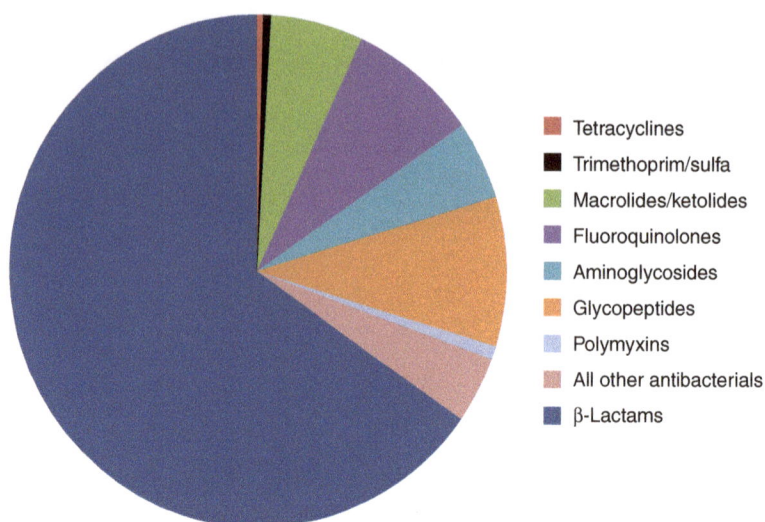

Figure 1. Proportion of prescriptions in the United States for injectable antibiotics by class for years 2004–2014. The percentage of standard units for each injectable antibiotic prescribed in the United States from 2004 to 2014 is shown as follows: β-lactams, 65.24%; glycopeptides, 9%; fluoroquinolones, 8%; macrolides/ketolides, 6%; aminoglycosides, 5%; polymyxins, 1%; trimethoprim/sulfamethoxazole, 0.5%; tetracyclines (excluding tigecycline), 0.4%; all other antibiotics (including daptomycin, linezolid, and tigecycline), 4.21%. (Data from the IMS MDART Quarterly Database on file at AstraZeneca.)

and widely prescribed. Their major toxicity is related to an allergic response in a small percentage of patients who react to related side chain determinants; notably, these reactions are most common with penicillins and cephalosporins with minimal reactivity caused by monobactams (Saxon et al. 1984; Moss et al. 1991). The bactericidal mechanism of killing

Table 1. Usage of parenteral β-lactams by class from 2004–2104 in the United States

Class of β-lactam	Percentage of prescriptions[a]
Narrow spectrum penicillins	3.12
Broad spectrum penicillins[b]	36.54
Cephalosporins	47.49
Monobactams	1.66
Carbapenems	11.20

[a]The percentage for each injectable antibiotic class prescribed in the United States from 2004 to 2014. (Data from the IMS MDART Quarterly Database on file at AstraZeneca.)

[b]Broad-spectrum penicillins include the β-lactam/β-lactam-inhibitor combinations piperacillin-tazobactam, ticarcillin-clavulanate, and ampicillin-sulbactam.

by β-lactams is perceived to be a major advantage in the treatment of serious infections. When these agents were threatened by the rapid emergence of β-lactamases, β-lactamase-stable agents were developed, as well as potent β-lactamase inhibitors (BLIs). In this introductory description of the β-lactams, the most commonly available β-lactams and BLIs will be presented, with a brief summary of their general characteristics. Occasional agents have been included for their historical or scientific importance. Note that resistance mechanisms will be discussed in detail in other articles in this collection.

MECHANISM OF ACTION

β-Lactam antibiotics are bactericidal agents that interrupt bacterial cell-wall formation as a result of covalent binding to essential penicillin-binding proteins (PBPs), enzymes that are involved in the terminal steps of peptidoglycan cross-linking in both Gram-negative and Gram-positive bacteria. Every bacterial species has its own distinctive set of PBPs that can range from three to eight enzymes per species (Georgopa-padakou and Liu 1980). The inhibition of bacterial peptidoglycan transpeptidation by penicillin was described mechanistically in a classical paper by Tipper and Strominger (1965), who noted a structural similarity of penicillin G to the terminal D-Ala-D-Ala dipeptide of the nascent peptidoglycan in the dividing bacterial cell. This mechanism is now known to involve binding of penicillin, or another β-lactam, to an active site serine found in all functional PBPs (Georgopapadakou et al. 1977). The resulting inactive acyl enzyme may then slowly hydrolyze the antibiotic to form a microbiologically inactive entity (Frère and Joris 1985). In addition to these functionalities, recent work has shown the binding of selected β-lactams, such as ceftaroline, to an allosteric site in PBP2a from *Staphylococcus aureus*, resulting in an increased sensitization of the organism to the antibiotic (Otero et al. 2013; Gonzales et al. 2015).

PBPs may be divided into classes according to molecular mass (Goffin and Ghuysen 1998; Massova and Mobashery 1998), with low-molecular-mass PBPs serving mainly as monofunctional D-Ala-D-Ala carboxypeptidases. High-molecular-mass PBPs have been divided into two subclasses, one of which (class A) includes bifunctional enzymes with both a transpeptidase and a transglycosylase domain, and the second of which (class B) encompasses D-Ala-D-Ala-dependent transpeptidases. At least one PBP is deemed to be essential in each species, with a unique specificity for β-lactam binding that varies among each species and each β-lactam class (Curtis et al. 1979; Georgopapadakou and Liu 1980). In Gram-negative bacteria, essential PBPs include the high-molecular-weight PBPs 1a and 1b that are involved in cell lysis, PBP2, the inhibition of which results in a cessation of cell division and the formation of spherical cells, and PBP3 for which inhibition arrests cell division, resulting in filamentation. Cell death may occur as a result of inhibiting one or more of these PBPs (Spratt 1977, 1983). The roles of PBPs in Gram-positive bacteria and *Mycobacterium tuberculosis* are discussed in detail in Fisher and Mobashery (2016).

PENICILLINS

Penicillin G (benzylpenicillin) was the first β-lactam to be used clinically, most frequently to treat streptococcal infections for which it had high potency (Rammelkamp and Keefer 1943; Hirsh and Dowling 1946). Another naturally occurring penicillin, penicillin V (phenoxymethylpenicillin), in an oral formulation is still used therapeutically and prophylactically for mild to moderate infections caused by susceptible *Streptococcus* spp., including use in pediatric patients (Pottegard et al. 2015). However, the selection of penicillin-resistant penicillinase-producing staphylococci in patients treated with penicillin G led to decreased use of this agent, and prompted the search for more penicillins with greater stability to the staphylococcal β-lactamases (Kirby 1944, 1945; Medeiros 1984). A list of historically important and clinically useful penicillins is provided in Table 2. Among the penicillinase-stable penicillins of clinical significance are methicillin, oxacillin, cloxacillin, and nafcillin, with the latter suggested as the β-lactam of choice for skin infections, catheter infections, and bacteremia caused by methicillin-susceptible *S. aureus* (Bamberger and Boyd 2005). All were used primarily for staphylococcal infections until the emergence of methicillin-resistant *S. aureus* (MRSA) in 1979–1980 (Hemmer et al. 1979; Saroglou et al. 1980).

Penicillins with improved activity against Gram-negative pathogens included the orally bioavailable ampicillin and amoxicillin, both of which were introduced in the 1970s. These agents were initially used for the treatment of infections caused by Enterobacteriaceae and did not effectively inhibit the growth of *Pseudomonas aeruginosa*, which became more of a concern during the late 1970s. Carbenicillin was the first antipseudomonal penicillin to be introduced, but lacked stability to β-lactamase hydrolysis and was less potent than piperacillin or ticarcillin, later antipseudomonal penicillins. These latter drugs were considered to be potent broad-spectrum penicillins that included penicillin-susceptible staphylococci, enteric bacteria, anaerobes, and *P. aeruginosa* in their

Table 2. Penicillins of current and historical utility

Name	R$_1$	R$_2$	Route of administration	Approval date[b,c]	Status
Benzylpenicillin (penicillin G)	—H	—H$_2$C– (benzyl)	IM or IV	1946	Approved worldwide
Phenoxymethylpenicillin (penicillin V)	—H	—O–CH$_2$– (phenoxymethyl)	Oral	1968	Approved worldwide
Methicillin	—H	(2,6-dimethoxyphenyl)	IV	1960	No longer available; of historical interest
Oxacillin	—H	(methylphenylisoxazolyl)	Oral, IV	1962	Widely available, but not in the United Kingdom
Cloxacillin	—H	(chlorophenylmethylisoxazolyl)	Oral, IV	1974	Widely available, but not in the United Kingdom
Ampicillin	—H	(aminophenylmethyl, NH$_2$)	Oral, IV	1963	Widely available

Cite this article as *Cold Spring Harb Perspect Med* doi: 10.1101/cshperspect.a025247

Name		R	Route	Date	Availability
Nafcillin		—H	IV	1970	Limited availability
Amoxicillin		—H	Oral, IV	1972	Widely available
Carbenicillin		—H	Oral	1972	Discontinued
Ticarcillin		—H	IV	1976	Limited availability
Piperacillin		—H	IV	1981	Widely available, primarily in combination with tazobactam
Temocillin		—OCH$_3$	IV	1985 in Europe (Harvengt 1985)	Limited availability (Europe)
Mecillinam			IV	1978	Limited availability

IM, Intramuscular; IV, intravenous.

[a]FDA approval unless otherwise noted.

[b]Dates were updated from Medeiros (1997) (www.accessdata.fda.gov/scripts/cder/drugsatfda; www.drugs.com).

spectrum of activity. They were used extensively to treat serious nosocomial infections, especially when combined with a β-lactamase inhibitor (see below).

Two parenteral penicillins with unusual chemical structures, mecillinam and temocillin (Table 2), were introduced to treat infections caused by enteric bacteria before the global emergence of extended-spectrum β-lactamases (ESBLs) in the late 1980s. Mecillinam (also known as amdinocillin), with a 6-β-amidino side chain, is a narrow-spectrum β-lactam that binds exclusively to PBP2 in enteric bacteria (Curtis et al. 1979). Because of this specificity, it shows synergy in vitro in combination with other β-lactams that bind to PBPs 1a/1b and/or PBP3 in Gram-negative bacteria (Hanberger et al. 1991), thus decreasing the possibility that a point mutation in a single PBP would lead to resistance (Hickman et al. 2014). Temocillin, the 6-α-methoxypenicillin analog of ticarcillin, had greater stability than ticarcillin to hydrolysis by serine β-lactamases, but lost antibacterial activity against Gram-positive bacteria, anaerobic Gram-negative pathogens, and some enteric bacteria that included the important pathogens *Enterobacter* spp. and *Serratia marcescens* (Martinez-Beltran et al. 1985). Mecillinam and temocillin are currently enjoying a resurgence in interest owing to their stability to many ESBLs (Livermore et al. 2006; Rodriguez-Villalobos et al. 2006), often resulting in greater than 90% susceptibility when tested against many contemporary ESBL-producing Enterobacteriaceae (Giske 2015; Zykov et al. 2016).

Because increasing numbers of β-lactamases have compromised the use of penicillins as single agents (Bush 2013), there is currently limited therapeutic use of the penicillins as monotherapy. Ampicillin, amoxicillin, piperacillin, and ticarcillin have continued to be useful, primarily as a result of their combination with an appropriate β-lactamase inhibitor (see below). However, even ampicillin, amoxicillin, penicillin G, and penicillin V are still active as monotherapy against Group A streptococci, and *Treponema pallidum*, two of the few bacterial species that do not produce β-lactamases (Schaar et al. 2014).

CEPHALOSPORINS

During the 1950s, the discovery of the naturally occurring penicillinase-stable cephalosporin C opened a new pathway to the development of hundreds of novel cephalosporins (Newton and Abraham 1956; Abraham 1987) to treat infections caused by the major penicillinase-producing pathogen of medical interest at that time, *S. aureus*. Dozens of cephalosporins were introduced into clinical practice (Abraham 1987), either as parenteral or oral agents. The molecules exhibited antibacterial activity with MICs often ≤4 μg/mL against not only staphylococci, but also *Streptococcus pneumoniae* and non-β-lactamase-producing enteric bacteria. The parenteral agents were generally eightfold more potent than the oral agents that were used in some cases to replace oral penicillins in penicillin-allergic patients. The early cephalosporins, for example, those in the cephalosporin I subclass (Bryskier et al. 1994) introduced before 1980, were labile to hydrolysis by many β-lactamases that emerged following their introduction into clinical practice, so that only a few of the early molecules remain in use (see Table 3), primarily to treat mild to moderate skin infections caused by methicillin-susceptible *S. aureus* (MSSA) (Giordano et al. 2006). Cefazolin with high biliary concentrations is still used for surgical prophylaxis and for treatment of abdominal infections (Sudo et al. 2014) and is effective as empiric therapy in 80% of Japanese children with their first upper urinary tract infection (Abe et al. 2016).

When the TEM-1 penicillinase began to appear on transmissible plasmids in *Neisseria gonorrhoeae* (Ashford et al. 1976) and *Haemophilus influenzae* (Gunn et al. 1974; Khan et al. 1974), it was quickly recognized that the penicillins and cephalosporins in medical use were becoming ineffective, not only in treating those TEM-1-producing organisms, but also for the enteric bacteria and *P. aeruginosa* that could all acquire this enzyme. Another surge of synthetic activity in the pharmaceutical industry provided both oral and parenteral cephalosporins with stability to this common enzyme. These agents tended to have decreased potency against the

Table 3. Cephalosporins of current clinical utility or of historical interest

Name	Subclass[a]	R_1	R_2	R_3	Route of administration	Approval date[b,c]	Status
Cephalexin	Cephalosporin I		–H	–CH$_3$	Oral	1971	Limited availability
Cefaclor	Cephalosporin I	–Cl	–H		Oral	1979	Widely available
Cefixime	Cephalosporin V		–H		Oral	1989	Widely available
Cefpodoxime	Cephalosporin IV		–H		Oral	1992	Widely available
Ceftibutin	Cephalosporin III	–H	–H		Oral	1995	Widely available

Continued

Table 3. *Continued*

Name	Subclass[a]	R₁	R₂	R₃	Route of administration	Approval date[b,c]	Status
Cefdinir	Cephalosporin V		–H		Oral	1997	Widely available
Cefazolin	Cephalosporin I		–H		IV	1973	Widely available
Cefuroxime	Cephalosporin II		–H		Oral,[d] IV	1983	Widely available
Cefotaxime	Cephalosporin III		–H		IV	1981	Widely available
Cefoperazone	Cephalosporin III		–H		IV	1982	Widely available
Ceftriaxone	Cephalosporin III		–H		IV	1984	Widely available

Name	Class				Route	Year	Availability
Ceftazidime	Cephalosporin III		–H		IV	1985	Widely available
Cefepime	Cephalosporin IV		–H		IV	1996	Widely available
Ceftaroline (fosamil)	Anti-MRSA cephalosporin		–H		IV	2010	Widely available
Ceftobiprole	Anti-MRSA cephalosporin		–H		IV	2013 (Europe)	Limited availability
Ceftolozane	Antipseudomonal cephalosporin–cephalosporin VI		–H		IV	2014	Limited availability

Continued

Table 3. *Continued*

Name	Subclass[a]	R$_1$	R$_2$	R$_3$	Route of administration	Approval date[b,c]	Status
S-649266	Siderophore cephalosporin–cephalosporin V	*(structure)*	–H	*(structure)*	IV	Not approved	Phase 2
Cefoxitin	Cephamycin	*(structure)*	–OCH$_3$	*(structure)*	IV	1978	Widely available
Moxalactam	Oxacephem	*(structure)*			IV	1982[e]	Limited availability

IM, Intramuscular; IV, intravenous.

[a]Subclasses assigned according to CLSI (2016), Bryskier et al. (1994), or Bryskier and Belfiglio (1999).

[b]FDA approved unless otherwise noted.

[c]Dates were updated from Medeiros (1997) (www.accessdata.fda.gov/scripts/cder/drugsatfda; www.drugs.com; www.price-rx.com/lists/lantibiotics.shtml).

[d]Oral when dosed as cefuroxime axetil.

[e]Anonymous (1982).

staphylococci, but gained antibacterial activity against Gram-negative pathogens. Cefuroxime, dosed parenterally or orally as the axetil ester, was the only member of the cephalosporin II class (Bryskier et al. 1994) with both oral and systemic dosage forms, but its stability to β-lactamase hydrolysis was diminished compared to later oral cephalosporins (Jacoby and Carreras 1990). As seen with cefuroxime, acceptable oral bioavailability of cefpodoxime required esterification through addition of a proxetil group to attain sufficient absorption for efficacy (Bryskier and Belfiglio 1999). Of the oral agents approved after 1983 in Table 3, cefdinir was generally more stable to hydrolysis, not only to the original TEM enzyme, but also to the AmpC cephalosporinases that are produced at a basal level in many enteric bacteria and *P. aeruginosa* (Payne and Amyes 1993; Labia and Morand 1994).

Among the parenteral agents introduced in the 1980s were the cephamycin cefoxitin, and cephalosporins in the cephalosporin III and cephalosporin IV subclasses (Bryskier et al. 1994), which continue to serve as important antibiotics for the treatment of serious infections caused by Gram-negative pathogens. The novel oxacephem moxalactam, or latamoxef, which had similar antimicrobial activity to the cephalosporin III/IV subclasses, has exquisite stability to hydrolysis by β-lactamases (Sato et al. 2015), but was not a highly successful antibiotic owing, in part, to a relatively high frequency of bleeding in patients treated with this drug (Brown et al. 1986). The cephamycin cefoxitin is notable for its characteristic 7-methoxy side chain that confers stability to the TEM-type β-lactamases, including ESBLs. It has useful antibacterial activity against MSSA and enteric bacteria that do not produce high levels of AmpC cephalosporinases (Jacoby and Han 1996). Cefotaxime, cefoperazone, ceftriaxone, and ceftazidime, designated as subclass cephalosporin III, and cefepime in the cephalosporin IV subclass, are also known as expanded-spectrum cephalosporins with increased hydrolytic stability to the common penicillinases, SHV-1 and TEM-1 β-lactamase (Martinez-Martinez et al. 1996). These agents have diminished activity against staphylococci and enterococci compared to earlier cephalosporins, but have more potent activity against Gram-negative organisms. Cefepime tends to have lower MICs against enteric bacteria than the other expanded-spectrum cephalosporins, attributed to greater penetration through the OmpF outer-membrane porin protein (Nikaido et al. 1990; Bellido et al. 1991). Cefotaxime and ceftriaxone are often used to treat susceptible streptococcal infections; all can be used to treat serious infections caused by enteric bacteria if the organisms test susceptible. Notably, ceftazidime and cefepime have maintained their observed activity against *P. aeruginosa*, with recent susceptibility rates exceeding 80% (Sader et al. 2015). A liability of the expanded-spectrum cephalosporins, however, began to emerge only a few years after the introduction of cefotaxime, when the ESBLs were identified with the ability to hydrolyze all of the β-lactams, with the exception of the carbapenems. These enzymes, in addition to both serine and metallo-carbapenemases, have severely compromised the activity of almost all penicillins and cephalosporins, necessitating the development of combination therapy with other β-lactams, β-lactamase inhibitors, or antibiotics from other classes.

Ceftolozane, recently approved in combination with tazobactam for the treatment of complicated urinary tract infections and complicated intraabdominal infections, shows potent antipseudomonal activity, and includes activity against enteric bacteria that produce some ESBLs (Zhanel et al. 2014), particularly CTX-M-producing isolates (Estabrook et al. 2014). Another recent addition to the cephalosporin family is the siderophore-substituted cephalosporin S-649266 with a catechol in the 3-position, thus allowing the molecule to enter the cells via an iron transport mechanism (Kohira et al. 2015). In addition to increased penetrability, the cephalosporin is stable to hydrolysis by many carbapenemases, resulting in activity against many β-lactam-resistant enteric bacteria (Kohira et al. 2015).

In the mid-1990s, reports began to emerge describing cephalosporins with MICs <4 μg/mL against MRSA (Hanaki et al. 1995) as a result of targeted binding to PBP2a. PBP2a is an acquired low-affinity PBP responsible for the observed lack of antibacterial activity of most β-lactams in MRSA isolates. Ceftobiprole (Hanaki et al. 1995; Hebeisen et al. 2001) and ceftaroline (Moisan et al. 2010), two cephalosporins with IC_{50} values <1 μg/mL for binding to the staphylococcal PBP2a, have been developed for clinical use (Table 3). Ceftaroline is approximately twofold to fourfold more potent than ceftobiprole in inhibiting staphylococcal and streptococcal growth (Karlowsky et al. 2011), but ceftobiprole is up to fourfold more potent against *Enterococcus faecalis* (Karlowsky et al. 2011). Ceftobiprole generally has at least fourfold to eightfold lower MICs than ceftaroline against enteric bacteria, *P. aeruginosa*, and *Acinetobacter* spp. (Pillar et al. 2008; Karlowsky et al. 2011). Neither cephalosporin is stable to hydrolysis by ESBLs or carbapenemases (Pillar et al. 2008; Castanheira et al. 2012), although the combination of ceftaroline with the β-lactamase inhibitor avibactam overcomes many of these issues (Mushtaq et al. 2010; Flamm et al. 2014) (see below). Both drugs are highly insoluble and have been derivatized as prodrugs for therapeutic use, as ceftaroline fosamil (Talbot et al. 2007) and ceftobiprole medocaril (Hebeisen et al. 2001), respectively.

CARBAPENEMS

Thienamycin was identified in the mid-1970s as a potent broad-spectrum antibiotic with the typical four-membered β-lactam structure fused to a novel five-membered ring in which carbon rather than sulfur was present at the 1-position (Kahan et al. 1979). Because of its chemical instability, this carbapenem was never developed as a therapeutic agent, but was stabilized by adding the *N*-formimidoyl group to the 2-position, resulting in imipenem (Table 4). Imipenem has been widely used for infections caused by Gram-positive, Gram-negative, nonfermentative, and anaerobic bacteria based on its sustained high activity against these organisms, particularly among non-carbapenemase-producing enteric bacteria (Bradley et al. 1999; Kiratisin et al. 2012). Carbapenems, in

general, bind strongly to PBP2 in Gram-negative bacteria, but may also bind to PBP1a, 1b, and 3, thus providing supplemental killing mechanisms that may serve to lessen the emergence of resistance (Sumita and Fukasawa 1995; Yang et al. 1995). Carbapenems are notable for their stability to most β-lactamases (Bonfiglio et al. 2002), with the exception of the emerging carbapenemases found primarily in Gram-negative bacteria (Bush 2013). Because of the lability of imipenem to hydrolysis by the human renal dehydropeptidase (DHP) causing inactivation of the drug (Kropp et al. 1982), it is dosed in combination with cilastatin, a DHP inhibitor that also acts as a nephroprotectant (Kahan et al. 1983).

Based on the potent broad-spectrum activity of the early carbapenems, other related agents, including meropenem, ertapenem, and doripenem, have been developed for global use, with generally the same group of organisms included in their activity spectrum (Baughman 2009). All these carbapenems are more stable chemically than imipenem, thus allowing for a longer shelf life for the formulated drug and the potential for prolonged infusion times (Cielecka-Piontek et al. 2008; Prescott et al. 2011). Like imipenem, they are stable to most β-lactamases, other than the carbapenemases (Bush 2013). Following the introduction of imipenem, later carbapenems contained a 1β-methyl group that conferred stability to the human DHP, thus negating the necessity for coadministration of an inhibitor such as cilastatin (Zhanel et al. 2007). In terms of antibacterial activity, meropenem is generally twofold to fourfold more potent that imipenem against enteric bacteria (Jorgensen et al. 1991), is similar in potency against *P. aeruginosa*, but may have twofold to eightfold less antibacterial activity against Gram-positive bacteria (Neu et al. 1989). In addition, meropenem and doripenem retain greater activity against isolates of *P. aeruginosa* lacking the outer membrane porin protein OprD than imipenem (Riera et al. 2011). Meropenem is the only carbapenem approved for use in meningitis because of its excellent penetration into the meninges (Dagan et al. 1994). Doripenem, a carbapenem with somewhat

Table 4. Carbapenems of current clinical utility

Name	R_1	R_2	Approval date[a,b]	Status
Imipenem	H		1985	Widely available
Meropenem	CH_3		1996	Widely available
Ertapenem	CH_3		2001	Widely available
Doripenem	CH_3		2007	Widely available
Biapenem	CH_3		2001 (Japan)	Available in Japan
Tebipenem[c]	CH_3		2009 (Japan)	Available in Japan

[a]FDA approved unless otherwise noted.

[b]Dates were updated from Medeiros (1997) (www.accessdata.fda.gov/scripts/cder/drugsatfda; www.drugs.com; adisin sight.springer.com/drugs/800010812).

[c]Formulated as the pivoxil ester.

higher chemical stability than imipenem or meropenem (Prescott et al. 2011), follows the antibacterial profile of meropenem, but is slightly more potent against Gram-negative organisms (Nordmann et al. 2011). Ertapenem, recognized for its long elimination half-life in humans because of its high protein binding (95%) (Majumdar et al. 2002), may be effectively administered once daily (Kattan et al. 2008) in contrast to the other carbapenems that are dosed most commonly two or three times a day. Although its antibacterial spectrum is similar to the other carbapenems against Enterobacteriaceae, ertapenem differs from imipenem, meropenem, and doripenem in that it has no useful activity against *P. aeruginosa* (Kohler et al. 1999). Two carbapenems approved for use only in Japan include biapenem, with an antimicrobial spectrum similar to meropenem and doripenem (Neu et al. 1992; Papp-Wallace et al.

2011), and tebipenem, which lacks appreciable antipseudomonal activity (Fujimoto et al. 2013) (Table 4). Tebipenem is notable for its dosing as the pivoxil ester, rendering it orally bioavailable for use in pediatric respiratory infections (Kato et al. 2010). Like the other carbapenems, they are stable to hydrolysis by most serine β-lactamases, but can be hydrolyzed by both serine and metallo-carbapenemases. Biapenem has been reported to have better hydrolytic stability to metallo-β-lactamases (MBLs) compared to imipenem or meropenem (Neu et al. 1992; Inoue et al. 1995; Yang et al. 1995) with at least fourfold lower MICs than imipenem when tested against organisms producing IMP, VIM, or NDM MBLs (Livermore and Mushtaq 2013).

MONOCYCLIC β-LACTAMS

Aztreonam, a monocyclic β-lactam with an N1-sulfonic acid substituent, originated as a derivative from a novel antibiotic isolated from the New Jersey Pine Barrens (Cimarusti and Sykes 1983) (Table 5), and is the only monobactam to gain regulatory approval for therapeutic use. It has targeted activity against aerobic enteric bacteria and *P. aeruginosa*, with MICs against *S. aureus*, *S. pneumoniae*, and *E. faecalis* $\geq$50 μg/mL (Sykes et al. 1982). It binds tightly to PBP3 in Gram-negative rods, with weaker binding to PBP1a, leading to filamentation followed by cell lysis (Sykes et al. 1982). At the time that it was introduced into clinical practice, aztreonam was stable to hydrolysis by all of the common β-lactamases (Sykes et al. 1982); the emergence of ESBLs and the serine carbapenemases has since rendered it less effective against multidrug-resistant β-lactamase-producing organisms (Wang et al. 2014). However, the monobactam nucleus is not a good substrate for hydrolysis by MBLs, thus leading to a unique opportunity for this monobactam to be used in combination therapy with a serine β-lactamase inhibitor to treat infections caused by multi-β-lactamase-producing bacteria (see below) (Wang et al. 2014).

BAL30072 is a novel monosulfactam with an N1-*O*-sulfate group, an activity-enhancing 3-dihydropyridone siderophore substituent, and a 4-gem-dimethyl substitution on the azetidinone ring (Page et al. 2010) (Table 5). Its spectrum of activity is similar to aztreonam, but supplemented with activity against addi-

Table 5. Monocyclic β-lactams

Name	Subclass	R_1	R_2	Approval date	Status
Aztreonam X = α-methyl	Monobactam	—SO$_3$H		1986[a]	Widely available
BAL30072 X = gem-dimethyl	Monosulfactam	—OSO$_3$H		Not approved	Phase 1

[a]U.S. approval date provided in Medeiros (1997).

tional nonfermentative bacteria. As a result of the increased penetration of BAL30072 via iron uptake mechanisms, it is more potent against some Gram-negative bacteria than other β-lactams, with activity against *Acinetobacter* spp. and *Burkholderia* spp. eightfold to >256-fold better than imipenem (Page et al. 2010). It is susceptible to hydrolysis by ESBLs and many carbapenemases, and has shown synergistic activity in combination with β-lactamase inhibitors (Mushtaq et al. 2013) or meropenem (Hofer et al. 2013; Hornsey et al. 2013). Like aztreonam, it is stable to hydrolysis by MBLs; additionally, it was hydrolyzed 3000-fold less efficiently by the KPC-2 serine carbapenemase compared to aztreonam (Page et al. 2010).

β-LACTAMASE INHIBITORS

Attempts to identify inhibitors of common β-lactamases began in the mid-1970s, triggered by the appearance of the transferable TEM-1 penicillinase in *Neisseria gonorrhoeae* (Ashford et al. 1976) and *Haemophilus influenzae* (Gunn et al. 1974; Khan et al. 1974). As the result of natural product screening, clavulanic acid with a novel clavam structure (Table 6) was identified as a broad spectrum inhibitor of the staphylococcal penicillinases and most of the recognized plasmid-encoded penicillinases found in enteric bacteria (Reading and Cole 1977; Cole 1982), including the highly prevalent TEM and SHV enzymes (Simpson et al. 1980). The TEM β-lactamase was shown to be inactivated by this suicide inhibitor that initially acylates the active site serine with transient inhibition that includes hydrolysis of the inhibitor before complete enzyme inactivation (Charnas et al. 1978; Charnas and Knowles 1981). The spectrum of the inhibitor is now recognized to include most class A β-lactamases, including ESBLs (Steward et al. 2001) and, to a lesser extent, serine carbapenemases (Nordmann and Poirel 2002; Yigit et al. 2003). Clavulanic acid acts synergistically with penicillins and cephalosporins against β-lactamase-producing enteric bacteria to inhibit sensitive β-lactamases, thus allowing the companion β-lactam to kill the bacteria. It has been combined with ticarcillin as a parenteral combination for nosocomial infections that include *P. aeruginosa* as a causative pathogen (Neu 1990), and with amoxicillin as an orally bioavailable formulation for therapeutic use especially in pediatric populations (Klein 2003). It is also used in phenotypic testing to determine the presence of ESBLs in *Escherichia coli* and *Klebsiella pneumoniae* (Steward et al. 2001).

Following the discovery of clavulanic acid, medicinal chemists synthesized a number of penicillanic acid sulfones (Table 6) with β-lactamase inhibitory activity (English et al. 1978; Fisher et al. 1981; Aronoff et al. 1984). Of these, sulbactam (English et al. 1978) and tazobactam (Aronoff et al. 1984) were successfully commercialized. Both had a similar spectrum of activity as clavulanic acid. Against class A β-lactamases, sulbactam had less inhibitory activity than clavulanic acid or tazobactam based on IC_{50} values, but both sulfones were better inhibitors of class C cephalosporinase β-lactamases (Bush et al. 1993). Each followed the same general inhibitory/inactivation-mechanism as for clavulanic acid (Easton and Knowles 1984; Bush et al. 1993). The number of hydrolytic events before inactivation was at least 25-fold higher for sulbactam than for clavulanic acid or tazobactam for the TEM-2 β-lactamase (Bush et al. 1993; Easton and Knowles 1984). In contrast to clavulanic acid, the sulfone inhibitors do not function as inducers of chromosomally mediated AmpC β-lactamase (Weber and Sanders 1990).

Sulbactam has been combined with ampicillin for general global use (Neu 1990) and with cefoperazone to provide additional synergistic activity against nonfermentative and anaerobic bacteria, primarily in Japan (Eliopoulos et al. 1989). Tazobactam has been combined with piperacillin and, more recently, with cefoperazone and ceftolozane for nosocomial infections, including those caused by *P. aeruginosa* (Lister 2000). In general, none of the inhibitors has useful antibacterial activity as monotherapy, although there are several notable exceptions. Clavulanic acid alone has been reported to have an MIC as low as 1 μg/mL against *N. gonorrhoeae* (Wise et al. 1978); sulbactam has modest activity against wild-type *Acinetobacter* spp. and *Burkholderia cepacia*, with MIC_{90} values

Table 6. β-lactamase inhibitors of current interest

Name	Structure	Subclass	Partner β-lactam	Approval date[a]	Status
Clavulanic acid[b]		Clavam	Amoxicillin	1984	Widely available
Sulbactam[c]		Penicillanic acid sulfone	Ampicillin	1986	Widely available
Tazobactam		Penicillanic acid sulfone	Piperacillin Ceftolozane	1993 2014	Widely available Available in the United States and Europe
Avibactam[d]		DBO[e]	Ceftazidime[d]	2015	Widely available
Relebactam		DBO	Imipenem	Not approved	Phase 3 in the United States
RG6080		DBO	Not selected	Not approved	Phase 1
RPX7009		Boronic acid	Meropenem	Not approved	Phase 3 in the United States

[a]Dates provided in Medeiros (1997) or www.accessdata.fda.gov/scripts/cder/drugsatfda.
[b]Also initially combined with ticarcilllin to provide parenteral activity against *Pseudomonas aeruginosa*.
[c]Also combined with cefoperazone outside the United States.
[d]Also in development with ceftaroline or aztreonam.
[e]Diazabicyclooctane.

≤8 and 10 µg/mL, respectively (Jacoby and Sutton 1989; Fass et al. 1990), but does not retain activity against isolates with multiple resistance mechanisms (Dong et al. 2014). None of these inhibitors is effective in inhibiting the hydrolytic activity of MBLs (Bush 2015), and their modest activity against serine carbapenemases does not translate into clinical susceptibility (Yigit et al. 2003; Woodford et al. 2004) owing, at least in part, to the presence of multiple β-lactamases in the producing organisms (Moland et al. 2007). Even the potent inhibitory activity against individual ESBLs that is observed with clavulanic acid and tazobactam is not sufficient to protect their accompanying penicillins in the presence of multiple β-lactamases (Jones-Dias et al. 2014).

Following a hiatus of approximately two decades, a unique class of non-β-lactam β-lactamase inhibitors emerged, based on a novel bridged diazabicyclooctane (DBO) structure (Table 6) (Coleman 2011). The first of these inhibitors, avibactam, has a broader spectrum of activity than clavulanic acid and the sulfone inhibitors. Not only are class A penicillinases, ESBLs, and serine carbapenemases potently inhibited, but class C cephalosporinases and some class D oxacillinases are also effectively inhibited (Ehmann et al. 2012, 2013). Unlike the previous inactivators described above, avibactam is a tight-binding, covalent, reversible inhibitor for most enzymes, with the KPC-2 enzyme, a notable exception for which slow avibactam hydrolysis was observed (Ehmann et al. 2012). In addition, avibactam does not induce AmpC β-lactamases at clinically relevant concentrations (Coleman 2011). Avibactam has been approved for therapeutic use in combination with ceftazidime, and is under development for ceftaroline–avibactam or aztreonam–avibactam combinations (Flamm et al. 2014; Biedenbach et al. 2015; Li et al. 2015). Other DBOs under development include RG6080 and relebactam (MK 7655), in combination with imipenem. The spectrum of relebactam shows a similar spectrum of activity to avibactam; however, it provides less potentiation against important class D β-lactamases such as OXA-48 (Livermore et al. 2013). RG6080 (formerly OP0565) is a DBO that has an inhibitory spectrum similar to the other DBOs but has the additional benefit of exhibiting some intrinsic antibacterial activity against enteric bacteria (Livermore et al. 2015).

The boronic acid inhibitor RPX7009 (Table 6) represents another novel class of synthetic non-β-lactam β-lactamase inhibitors (Hecker et al. 2015), although boronic acids have been known for many years to be effective inhibitors of serine β-lactamases (Kiener and Waley 1978). Despite the inhibitory activity of RPX7009 against many groups of serine β-lactamases

(Hecker et al. 2015), it is being developed in combination with meropenem to target pathogens producing serine carbapenemases (Lapuebla et al. 2015).

β-LACTAM RESISTANCE: CONCLUDING REMARKS

Resistance to the β-lactams continues to increase, especially in Gram-negative organisms (Vasoo et al. 2015), because of the widespread therapeutic dependence on these efficacious and safe antibiotics (see Fig. 1). Major resistance mechanisms will be expanded on in other articles in this collection. PBP acquisition or mutation is the major β-lactam-resistance mechanism in Gram-positive bacteria (see Fisher and Mobashery 2016). The most prevalent and most damaging resistance mechanisms among Gram-negative pathogens are represented by the β-lactamases (Babic et al. 2006; Livermore 2012), both chromosomally encoded enzymes that may be produced at high levels and transferable enzymes that travel on mobile elements among species (Bush 2013). When these targeted mechanisms are combined with decreased uptake or increased efflux of the β-lactam, high-level resistance becomes a major clinical problem (see Bonomo 2016). Perhaps the most encouraging prospect in counteracting resistance is the emergence of new classes of β-lactamase inhibitors that will provide protection for some of the most valuable antibiotics in clinical practice, at least for the present time.

REFERENCES

*Reference is also in this collection.

Abe Y, Wakabayashi H, Ogawa Y, Machida A, Endo M, Tamai T, Sakurai S, Hibino S, Mikawa T, Watanabe Y, et al. 2016. Validation of cefazolin as initial antibiotic for first upper urinary tract infection in children. *Global Ped Health* **3:** 1–7.

Abraham EP. 1987. Cephalosporins 1945–1986. *Drugs* **34:** 1–14.

Anonymous. 1982. FDA approves new antibiotic for surgical infections, meningitis. *Hospitals* **56:** 55–57.

Aronoff SC, Jacobs MR, Johenning S, Yamabe S. 1984. Comparative activities of the β-lactamase inhibitors YTR 830, sodium clavulanate, and sulbactam combined with amoxicillin or ampicillin. *Antimicrob Agents Chemother* **26:** 580–582.

Ashford WA, Golash RG, Hemming VG. 1976. Penicillinase-producing *Neisseria gonorrhoeae*. *Lancet* **308:** 657–658.

Babic M, Hujer AM, Bonomo RA. 2006. What's new in antibiotic resistance? Focus on β-lactamases. *Drug Resist Updat* **9:** 142–156.

Bamberger DM, Boyd SE. 2005. Management of *Staphylococcus aureus* infections. *Am Fam Physician* **72:** 2474–2481.

Baughman RP. 2009. The use of carbapenems in the treatment of serious infections. *J Intens Care Med* **24:** 230–241.

Bellido F, Pechère J-C, Hancock RW. 1991. Novel method for measurement of outer membrane permeability to new β-lactams in intact *Enterobacter cloacae* cells. *Antimicrob Agents Chemother* **35:** 68–72.

Biedenbach DJ, Kazmierczak K, Bouchillon SK, Sahm DF, Bradford PA. 2015. In vitro activity of aztreonam–avibactam against a global collection of Gram-negative pathogens from 2012 and 2013. *Antimicrob Agents Chemother* **59:** 4239–4248.

Bonfiglio G, Russo G, Nicoletti G. 2002. Recent developments in carbapenems. *Exp Opin Investig Drugs* **11:** 529–544.

* Bonomo RA. 2016. β-Lactamases: A focus on current challenges. *Cold Spring Harb Perspect Med* doi: 10.1101/cshperspect.a025239.

Bradley JS, Garau J, Lode H, Rolston KV, Wilson SE, Quinn JP. 1999. Carbapenems in clinical practice: A guide to their use in serious infection. *Int J Antimicrob Agents* **11:** 93–100.

Brown RB, Klar J, Lemeshow S, Teres D, Pastides H, Sands M. 1986. Enhanced bleeding with cefoxitin or moxalactam. Statistical analysis within a defined population of 1493 patients. *Archiv Intern Med* **146:** 2159–2164.

Bryskier AB, Belfiglio SR. 1999. Cephalosporins: Oral. In *Antimicrobial therapy and vaccines* (ed. Yu VL, et al.), pp. 703–747. Lippincott Williams & Wilkins, Philadelphia.

Bryskier A, Aszodi J, Chantot JF. 1994. Parenteral cephalosporin classification. *Exp Opin Investig Drugs* **3:** 145–171.

Bush K. 2013. Proliferation and significance of clinically relevant β-lactamases. *Ann NY Acad Sci* **1277:** 84–90.

Bush K. 2015. A resurgence of β-lactamase inhibitor combinations effective against multidrug-resistant Gram-negative pathogens. *Int J Antimicrob Agents* **46:** 483–493.

Bush K, Macalintal C, Rasmussen BA, Lee VJ, Yang Y. 1993. Kinetic interactions of tazobactam with β-lactamases from all major structural classes. *Antimicrob Agents Chemother* **37:** 851–858.

Castanheira M, Sader HS, Farrell DJ, Mendes RE, Jones RN. 2012. Activity of ceftaroline–avibactam tested against Gram-negative organism populations, including strains expressing one or more β-lactamases and methicillin-resistant *Staphylococcus aureus* carrying various staphylococcal cassette chromosome *mec* types. *Antimicrob Agents Chemother* **56:** 4779–4785.

Charnas RL, Knowles JR. 1981. Inactivation of RTEM β-lactamase from *Escherichia coli* by clavulanic acid and 9-deoxyclavulanic acid. *Biochemistry* **20:** 3214–3219.

Charnas RL, Fisher J, Knowles JR. 1978. Chemical studies on the inactivation of *Escherichia coli* RTEM β-lactamase by clavulanic acid. *Biochemistry* **17:** 2185–2189.

Cielecka-Piontek J, Zajac M, Jelinska A. 2008. A comparison of the stability of ertapenem and meropenem in pharmaceutical preparations in solid state. *J Pharm Biomed Anal* **46:** 52–57.

Cimarusti CM, Sykes RB. 1983. Monobactams—Novel antibiotics. *Chem Brit* **19:** 302–303.

CLSI. 2016. Performance standards for antimicrobial susceptibility testing. In *CLSI supplement M100S*, 26th ed. Clinical and Laboratory Standards Institute, Wayne, PA.

Cole M. 1982. Biochemistry and action of clavulanic acid. *Scott Med J* **27:** S10–S16.

Coleman K. 2011. Diazabicyclooctanes (DBOs): A potent new class of non-β-lactam β-lactamase inhibitors. *Curr Opin Microbiol* **14:** 550–555.

Curtis N, Orr D, Ross GW, Boulton MG. 1979. Affinities of penicillins and cephalosporins for the penicillin-binding proteins of *Escherichia coli* K-12 and their antibacterial activity. *Antimicrob Agents Chemother* **16:** 533–539.

Dagan R, Velghe L, Rodda JL, Klugman KP. 1994. Penetration of meropenem into the cerebrospinal fluid of patients with inflamed meninges. *J Antimicrob Chemother* **34:** 175–179.

Dong X, Chen F, Zhang Y, Liu H, Liu Y, Ma L. 2014. In vitro activities of rifampin, colistin, sulbactam and tigecycline tested alone and in combination against extensively drug-resistant *Acinetobacter baumannii*. *J Antibiotics* **67:** 677–680.

Easton CJ, Knowles JR. 1984. Correlation of the effect of β-lactamase inhibitors on the β-lactamase in growing cultures of Gram-negative bacteria with their effect on the isolated β-lactamase. *Antimicrob Agents Chemother* **26:** 358–363.

Ehmann DE, Jahic H, Ross PL, Gu RF, Hu J, Kern G, Walkup GK, Fisher SL. 2012. Avibactam is a covalent, reversible, non-β-lactam β-lactamase inhibitor. *Proc Natl Acad Sci* **109:** 11663–11668.

Ehmann DE, Jahić H, Ross PL, Gu RF, Hu J, Durand-Réville TF, Lahiri S, Thresher J, Livchak S, Gao N, et al. 2013. Kinetics of avibactam inhibition against class A, C, and D β-lactamases. *J Biol Chem* **288:** 27960–27971.

Eliopoulos GM, Klimm K, Ferraro MJ, Moellering RC Jr. 1989. In vitro activity of cefoperazone–sulbactam combinations against cefoperazone-resistant clinical bacterial isolates. *Eur J Clinl Microbiol Infect Dis* **8:** 624–626.

English AR, Retsema JA, Girard AE, Lynch JE, Barth WE. 1978. CP-45.899, a β-lactamase inhibitor that extends the antibacterial spectrum of β-lactams: Initial bacteriological characterization. *Antimicrob Agents Chemother* **14:** 414–419.

Estabrook M, Bussell B, Clugston SL, Bush K. 2014. In vitro activity of ceftolozane-tazobactam as determined by broth dilution and agar diffusion assays against recent U.S. *Escherichia coli* isolates from 2010 to 2011 carrying CTX-M-type extended-spectrum β-lactamases. *J Clin Microbiol* **52:** 4049–4052.

Fass RJ, Gregory WW, D'Amato RF, Matsen JM, Wright DN, Young LS. 1990. In vitro activities of cefoperazone and sulbactam singly and in combination against cefoperazone-resistant members of the family Enterobacteriaceae and nonfermenters. *Antimicrob Agents Chemother* **34:** 2256–2259.

* Fisher JF, Mobashery S. 2016. β-Lactam-resistance mechanisms: Gram-positive bacteria and *Mycobacterium tuberculosis*. *Cold Spring Harb Perspect Med* doi: 10.1101/cshperspect.a025221.

Fisher J, Charnas RL, Bradley SM, Knowles JR. 1981. Inactivation of the RTEM β-lactamase from *Escherichia coli*. Interaction of penam sulfones with enzyme. *Biochemistry* **20:** 2726–2731.

Flamm RK, Farrell DJ, Sader HS, Jones RN. 2014. Antimicrobial activity of ceftaroline combined with avibactam tested against bacterial organisms isolated from acute bacterial skin and skin structure infections in United States medical centers (2010–2012). *Diagn Microbiol Infect Dis* **78:** 449–456.

Fleming A. 1929. On the antibacterial action of cultures of a penicillium, with special reference to their use in the isolation of *B. influenzae*. *Br J Exp Pathol* **10:** 226–236.

Frère JM, Joris B. 1985. Penicillin-sensitive enzymes in peptidoglycan biosynthesis. *CRC Crit Rev Microbiol* **11:** 299–396.

Fujimoto K, Takemoto K, Hatano K, Nakai T, Terashita S, Matsumoto M, Eriguchi Y, Eguchi K, Shimizudani T, Sato K, et al. 2013. Novel carbapenem antibiotics for parenteral and oral applications: In vitro and in vivo activities of 2-aryl carbapenems and their pharmacokinetics in laboratory animals. *Antimicrob Agents Chemother* **57:** 697–707.

Georgopapadakou NH, Liu FY. 1980. Penicillin-binding proteins in bacteria. *Antimicrob Agents Chemother* **18:** 148–157.

Georgopapadakou N, Hammarstrom S, Strominger JL. 1977. Isolation of the penicillin-binding peptide from D-alanine carboxypeptidase of *Bacillus subtilis*. *Proc Natl Acad Sci* **74:** 1009–1012.

Giordano PA, Elston D, Akinlade BK, Weber K, Notario GF, Busman TA, Cifaldi M, Nilius AM. 2006. Cefdinir vs. cephalexin for mild to moderate uncomplicated skin and skin structure infections in adolescents and adults. *Curr Med Res Opin* **22:** 2419–2428.

Giske CG. 2015. Contemporary resistance trends and mechanisms for the old antibiotics colistin, temocillin, fosfomycin, mecillinam and nitrofurantoin. *Clin Microbiol Infect* **21:** 899–905.

Goffin C, Ghuysen JM. 1998. Multimodular penicillin-binding proteins: An enigmatic family of orthologs and paralogs. *Microbiol Mol Biol Rev* **62:** 1079–1093.

Gonzales PR, Pesesky MW, Bouley R, Ballard A, Biddy BA, Suckow MA, Wolter WR, Schroeder VA, Burnham CA, Mobashery S, et al. 2015. Synergistic, collaterally sensitive β-lactam combinations suppress resistance in MRSA. *Nature Chem Biol* **11:** 855–861.

Gunn BA, Woodall JB, Jones JF, Thornsberry C. 1974. Letter: Ampicillin-resistant *Haemophilus influenzae*. *Lancet* **2:** 845.

Hanaki H, Akagi H, Masaru Y, Otani T, Hyodo A, Hiramatsu K. 1995. TOC-39, a novel parenteral broad-spectrum cephalosporin with excellent activity against methicillin-resistant *Staphylococcus aureus*. *Antimicrob Agents Chemother* **39:** 1120–1126.

Hanberger H, Nilsson LE, Svensson E, Maller R. 1991. Synergic post-antibiotic effect of mecillinam, in combination with other β-lactam antibiotics in relation to morphology and initial killing. *J Antimicrob Chemother* **28:** 523–532.

Harvengt C. 1985. Drugs recently released in Belgium. Temocillin. *Acta Clin Belg* **40:** 398–399.

Hebeisen P, Heinze-Krauss I, Angehrn P, Hohl P, Page MG, Then RL. 2001. In vitro and in vivo properties of Ro 63–9141, a novel broad-spectrum cephalosporin with activity against methicillin-resistant staphylococci. *Antimicrob Agents Chemother* **45:** 825–836.

Hecker SJ, Reddy KR, Totrov M, Hirst GC, Lomovskaya O, Griffith DC, King P, Tsivkovski R, Sun D, Sabet M, et al. 2015. Discovery of a cyclic boronic acid β-lactamase inhibitor (RPX7009) with utility vs class A serine carbapenemases. *J Med Chem* **58:** 3682–3692.

Hemmer RJ, Vaudaux P, Waldvogel FA. 1979. Methicillin potentiates the effect of gentamicin on methicillin-resistant *Staphylococcus aureus*. *Antimicrob Agents Chemother* **15:** 34–41.

Hickman RA, Hughes D, Cars T, Malmberg C, Cars O. 2014. Cell-wall-inhibiting antibiotic combinations with activity against multidrug-resistant *Klebsiella pneumoniae* and *Escherichia coli*. *Clin Microbiol Infect* **20:** O267–O273.

Hirsh HL, Dowling HF. 1946. The treatment of *Streptococcus viridans* endocarditis with penicillin. *Southern Med J* **39:** 55–60.

Hofer B, Dantier C, Gebhardt K, Desarbre E, Schmitt-Hoffmann A, Page MG. 2013. Combined effects of the siderophore monosulfactam BAL30072 and carbapenems on multidrug-resistant Gram-negative bacilli. *J Antimicrob Chemother* **68:** 1120–1129.

Hornsey M, Phee L, Stubbings W, Wareham DW. 2013. In vitro activity of the novel monosulfactam BAL30072 alone and in combination with meropenem versus a diverse collection of important Gram-negative pathogens. *Int J Antimicrob Agents* **42:** 343–346.

Inoue K, Hamana Y, Mitsuhashi S. 1995. In vitro antibacterial activity and β-lactamase stability of a new carbapenem, BO-2727. *Antimicrob Agents Chemother* **39:** 2331–2336.

Jacoby GA, Carreras I. 1990. Activities of β-lactam antibiotics against *Escherichia coli* strains producing extended-spectrum β-lactamases. *Antimicrob Agents Chemother* **34:** 858–862.

Jacoby GA, Han P. 1996. Detection of extended-spectrum β-lactamases in clinical isolates of *Klebsiella pneumoniae* and *Escherichia coli*. *J Clin Microbiol* **34:** 908–911.

Jacoby GA, Sutton L. 1989. *Pseudomonas cepacia* susceptibility to sulbactam. *Antimicrob Agents Chemother* **33:** 583–584.

Jones-Dias D, Manageiro V, Ferreira E, Louro D, Antibiotic Resistance Surveillance Program in Portugal (ARSIP), Canica M. 2014. Diversity of extended-spectrum and plasmid-mediated AmpC β-lactamases in Enterobacteriaceae isolates from Portuguese health care facilities. *J Microbiol* **52:** 496–503.

Jorgensen JH, Maher LA, Howell AW. 1991. Activity of meropenem against antibiotic-resistant or infrequently encountered Gram-negative bacilli. *Antimicrob Agents Chemother* **35:** 2410–2414.

Kahan JS, Kahan FM, Goegelman R, Currie SA, Jackson M, Stapley EO, Miller TW, Miller AK, Hendlin D, Mochales S, et al. 1979. Thienamycin, a new β-lactam antibiotic. I: Discovery, taxonomy, isolation and physical properties. *J Antibiotics* **32:** 1–12.

Kahan FM, Kropp H, Sundelof JG, Birnbaum J. 1983. Thienamycin: Development of imipenem–cilastatin. *J Antimicrob Chemother* **12:** 1–35.

Karlowsky JA, Adam HJ, Decorby MR, Lagace-Wiens PR, Hoban DJ, Zhanel GG. 2011. In vitro activity of ceftaroline against Gram-positive and Gram-negative pathogens isolated from patients in Canadian hospitals in 2009. *Antimicrob Agents Chemother* **55:** 2837–2846.

Kato K, Shirasaka Y, Kuraoka E, Kikuchi A, Iguchi M, Suzuki H, Shibasaki S, Kurosawa T, Tamai I. 2010. Intestinal absorption mechanism of tebipenem pivoxil, a novel oral carbapenem: Involvement of human OATP family in apical membrane transport. *Mol Pharm* **7:** 1747–1756.

Kattan JN, Villegas MV, Quinn JP. 2008. New developments in carbapenems. *Clin Microbiol Infect* **14:** 1102–1111.

Khan W, Ross S, Rodriguez W, Controni G, Saz AK. 1974. *Haemophilus influenzae* type B resistant to ampicillin: A report of two cases. *JAMA* **229:** 298–301.

Kiener PA, Waley SG. 1978. Reversible inhibitors of penicillinases. *Biochemical J* **169:** 197–204.

Kiratisin P, Chongthaleong A, Tan TY, Lagamayo E, Roberts S, Garcia J, Davies T. 2012. Comparative in vitro activity of carbapenems against major Gram-negative pathogens: Results of Asia-Pacific surveillance from the COMPACT II study. *Int J Antimicrob Agents* **39:** 311–316.

Kirby WM. 1944. Extraction of a highly potent penicillin inactivator from penicillin resistant staphylococci. *Science* **99:** 452–453.

Kirby WM. 1945. Bacteriostatic and lytic actions of penicillin on sensitive and resistant staphylococci. *J Clin Invest* **24:** 165–169.

Klein JO. 2003. Amoxicillin/clavulanate for infections in infants and children: Past, present and future. *Pediatr Infect Dis J* **22:** S139–S148.

Kohira N, West J, Ito A, Ito-Horiyama T, Nakamura R, Sato T, Rittenhouse S, Tsuji M, Yamano Y. 2015. In vitro antimicrobial activity of siderophore cephalosporin S-649266 against Enterobacteriaceae clinical isolates including carbapenem-resistant strains. *Antimicrob Agents Chemother* **60:** 729–734.

Kohler J, Dorso KL, Young K, Hammond GG, Rosen H, Kropp H, Silver LL. 1999. In vitro activities of the potent, broad-spectrum carbapenem MK-0826 (L-749, 345) against broad-spectrum β-lactamase and extended-spectrum β-lactamase-producing *Klebsiella pneumoniae* and *Escherichia coli* clinical isolates. *Antimicrob Agents Chemother* **43:** 1170–1176.

Kropp H, Sundelof JG, Hadju R, Kahan FM. 1982. Metabolism of thienamycin and related carbapenem antibiotics by the renal dipeptidase, dehydropeptidase-I. *Antimicrob Agents Chemother* **22:** 62–70.

Labia R, Morand A. 1994. Interaction of cefdinir with β-lactamases. *Drugs Exp Clin Res* **20:** 43–48.

Lapuebla A, Abdallah M, Olafisoye O, Cortes C, Urban C, Quale J, Landman D. 2015. Activity of meropenem combined with RPX7009, a novel β-lactamase inhibitor, against Gram-negative clinical isolates in New York City. *Antimicrob Agents Chemother* **59:** 4856–4860.

Li H, Estabrook M, Jacoby GA, Nichols WW, Testa RT, Bush K. 2015. In vitro susceptibility of characterized β-lactamase-producing strains tested with avibactam combinations. *Antimicrob Agents Chemother* **59:** 1789–1793.

Lister PD. 2000. β-lactamase inhibitor combinations with extended-spectrum penicillins: Factors influencing antibacterial activity against Enterobacteriaceae and *Pseudomonas aeruginosa*. *Pharmacotherapy* **20:** 213S–218S; discussion 224S–228S.

Livermore DM. 2012. Current epidemiology and growing resistance of Gram-negative pathogens. *Korean J Intern Med* **27:** 128–142.

Livermore DM, Mushtaq S. 2013. Activity of biapenem (RPX2003) combined with the boronate β-lactamase inhibitor RPX7009 against carbapenem-resistant Enterobacteriaceae. *J Antimicrob Chemother* **68:** 1825–1831.

Livermore DM, Hope R, Fagan EJ, Warner M, Woodford N, Potz N. 2006. Activity of temocillin against prevalent ESBL- and AmpC-producing Enterobacteriaceae from south-east England. *J Antimicrob Chemother* **57:** 1012–1014.

Livermore DM, Warner M, Mushtaq S. 2013. Activity of MK-7655 combined with imipenem against Enterobacteriaceae and *Pseudomonas aeruginosa*. *J Antimicrob Chemother* **68:** 2286–2290.

Livermore DM, Mushtaq S, Warner M, Woodford N. 2015. Activity of OP0595/β-lactam combinations against Gram-negative bacteria with extended-spectrum, AmpC and carbapenem-hydrolysing β-lactamases. *J Antimicrob Chemother* **70:** 3032–3041.

Macfarlane G. 1979. *Howard Florey, the making of a great scientist*. Oxford University Press, London.

Majumdar AK, Musson DG, Birk KL, Kitchen CJ, Holland S, McCrea J, Mistry G, Hesney M, Xi L, Li SX, et al. 2002. Pharmacokinetics of ertapenem in healthy young volunteers. *Antimicrob Agents Chemother* **46:** 3506–3511.

Martinez-Beltran J, Loza E, Gomez-Alferez A, Romero-Vivas J, Bouza E. 1985. Temocillin. In vitro activity compared with other antibiotics. *Drugs* **29:** 91–97.

Martinez-Martinez L, Hernandez-Alles S, Alberti S, Tomas JM, Benedi VJ, Jacoby GA. 1996. In vivo selection of porin-deficient mutants of *Klebsiella pneumoniae* with increased resistance to cefoxitin and expanded-spectrum-cephalosporins. *Antimicrob Agents Chemother* **40:** 342–348.

Massova I, Mobashery S. 1998. Kinship and diversification of bacterial penicillin-binding proteins and β-lactamases. *Antimicrob Agents Chemother* **42:** 1–17.

Medeiros AA. 1984. β-lactamases. *Br Med Bull* **40:** 18–27.

Medeiros AA. 1997. Evolution and dissemination of β-lactamases accelerated by generations of β-lactam antibiotics. *Clin Infect Dis* **24:** S19–S45.

Moisan H, Pruneau M, Malouin F. 2010. Binding of ceftaroline to penicillin-binding proteins of *Staphylococcus aureus* and *Streptococcus pneumoniae*. *J Antimicrob Chemother* **65:** 713–716.

Moland ES, Hong SG, Thomson KS, Larone DH, Hanson ND. 2007. A *Klebsiella pneumoniae* isolate producing at least eight different β-lactamases including an AmpC and KPC β-lactamase. *Antimicrob Agents Chemother* **51:** 800–801.

Moss RB, McClelland E, Williams RR, Hilman BC, Rubio T, Adkinson NF. 1991. Evaluation of the immunologic cross-reactivity of aztreonam in patients with cystic fibrosis who are allergic to penicillin and/or cephalosporin antibiotics. *Rev Infect Dis* **13:** S598–S607.

Mushtaq S, Warner M, Williams G, Critchley I, Livermore DM. 2010. Activity of chequerboard combinations of ceftaroline and NXL104 versus β-lactamase-producing Enterobacteriaceae. *J Antimicrob Chemother* **65:** 1428–1432.

Mushtaq S, Woodford N, Hope R, Adkin R, Livermore DM. 2013. Activity of BAL30072 alone or combined with β-lactamase inhibitors or with meropenem against carbapenem-resistant Enterobacteriaceae and non-fermenters. *J Antimicrob Chemother* **68:** 1601–1608.

Neu HC. 1990. β-lactamases, β-lactamase inhibitors, and skin and skin-structure infections. *J Am Acad Dermatol* **22:** 896–904.

Neu HC, Novelli A, Chin NX. 1989. In vitro activity and β-lactamase stability of a new carbapenem, SM-7338. *Antimicrob Agents Chemother* **33:** 1009–1018.

Neu HC, Gu JW, Fang W, Chin NX. 1992. In vitro activity and β-lactamase stability of LJC 10,627. *Antimicrob Agents Chemother* **36:** 1418–1423.

Newton GGF, Abraham EP. 1956. Isolation of cephalosporin C, a penicillin-like antibiotic containing D-α amino-adipic acid. *Biochem J* **62:** 651–658.

Nikaido H, Liu W, Rosenberg EY. 1990. Outer membrane permeability and β-lactamase stability of dipolar ionic cephalosporins containing methoxyimino substituents. *Antimicrob Agents Chemother* **34:** 337–342.

Nordmann P, Poirel L. 2002. Emerging carbapenemases in Gram-negative aerobes. *Clin Microbiol Infect* **8:** 321–331.

Nordmann P, Picazo J, Mutters R, Korten V, Quintana A, Laeuffer J, Seak J, Flamm R, Morrissey I. 2011. Comparative activity of carbapenem testing: The COMPACT study. *J Antimicrob Chemother* **66:** 1070–1078.

Otero LH, Rojas-Altuve A, Llarrull LI, Carrasco-Lopez C, Kumarasiri M, Lastochkin E, Fishovitz J, Dawley M, Hesek D, Lee M, et al. 2013. How allosteric control of *Staphylococcus aureus* penicillin binding protein 2a enables methicillin resistance and physiological function. *Proc Natl Acad Sci* **110:** 16808–16813.

Page MGP, Dantier C, Desarbre E. 2010. In vitro properties of BAL30072, a novel siderophore sulfactam with activity against multiresistant Gram-negative bacilli. *Antimicrob Agents Chemother* **54:** 2291–2302.

Papp-Wallace KM, Endimiani A, Taracila MA, Bonomo RA. 2011. Carbapenems: Past, present, and future. *Antimicrob Agents Chemother* **55:** 4943–4960.

Payne DJ, Amyes SG. 1993. Stability of cefdinir (C1-983, FK482) to extended-spectrum plasmid-mediated β-lactamases. *J Med Microbiol* **38:** 114–117.

Pillar CM, Aranza MK, Shah D, Sahm DF. 2008. In vitro activity profile of ceftobiprole, an anti-MRSA cephalosporin, against recent Gram-positive and Gram-negative isolates of European origin. *J Antimicrob Chemother* **61:** 595–602.

Pottegard A, Broe A, Aabenhus R, Bjerrum L, Hallas J, Damkier P. 2015. Use of antibiotics in children: A Danish nationwide drug utilization study. *Pediatr Infect Dis J* **34:** e16–e22.

Prescott WA Jr, Gentile AE, Nagel JL, Pettit RS. 2011. Continuous-infusion antipseudomonal β-lactam therapy in patients with cystic fibrosis. *P&T* **36:** 723–763.

Rammelkamp CH, Keefer CS. 1943. Penicillin: Its antibacterial effect in whole blood and serum for the hemolytic *Streptococcus* and *Staphylococcus aureus*. *J Clin Investig* **22:** 649–657.

Reading C, Cole M. 1977. Clavulanic acid: A β-lactamase inhibitor from *Streptomyces clavuligerus*. *Antimicrob Agents Chemother* **11:** 852–857.

Riera E, Cabot G, Mulet X, Garcıa-Castillo M, del Campo R, Juan C, Canton R, Oliver A. 2011. *Pseudomonas aeruginosa* carbapenem resistance mechanisms in Spain: Impact on the activity of imipenem, meropenem and doripenem. *J Antimicrob Chemother* **66:** 2022–2027.

Rodriguez-Villalobos H, Malaviolle V, Frankard J, de Mendonca R, Nonhoff C, Struelens MJ. 2006. In vitro activity of temocillin against extended spectrum β-lactamase-producing *Escherichia coli*. *J Antimicrob Chemother* **57:** 771–774.

Sader HS, Castanheira M, Mendes RE, Flamm RK, Farrell DJ, Jones RN. 2015. Ceftazidime-avibactam activity against multidrug-resistant *Pseudomonas aeruginosa* isolated in U.S. medical centers in 2012 and 2013. *Antimicrob Agents Chemother* **59:** 3656–3659.

Saroglou G, Cromer M, Bisno AL. 1980. Methicillin-resistant *Staphylococcus aureus*: Interstate spread of nosocomial infections with emergence of gentamicin–methicillin resistant strains. *Infect Control* **1:** 81–89.

Sato T, Hara T, Horiyama T, Kanazawa S, Yamaguchi T, Maki H. 2015. Mechanism of resistance and antibacterial susceptibility in extended-spectrum β-lactamase phenotype *Klebsiella pneumoniae* and *Klebsiella oxytoca* isolated between 2000 and 2010 in Japan. *J Med Microbiol* **64:** 538–543.

Saxon A, Hassner A, Swabb EA, Wheeler B, Adkinson NF Jr. 1984. Lack of cross-reactivity between aztreonam, a monobactam antibiotic, and penicillin in penicillin-allergic subjects. *J Infect Dis* **149:** 16–22.

Schaar V, Uddback I, Nordstrom T, Riesbeck K. 2014. Group A streptococci are protected from amoxicillin-mediated killing by vesicles containing β-lactamase derived from *Haemophilus influenzae*. *J Antimicrob Chemother* **69:** 117–120.

Simpson IN, Harper PB, O'Callaghan CH. 1980. Principal β-lactamases responsible for resistance to β-lactam antibiotics in urinary tract infections. *Antimicrob Agents Chemother* **17:** 929–936.

Spratt BG. 1977. Properties of the penicillin-binding proteins of *Escherichia coli* K12. *Eur J Biochem* **72:** 341–352.

Spratt BG. 1983. Penicillin-binding proteins and the future of β-lactam antibiotics. *J Gen Microbiol* **129:** 1247–1260.

Steward CD, Rasheed JK, Hubert SK, Biddle JW, Raney PM, Anderson GJ, Williams PP, Brittain KL, Oliver A, McGowan JE Jr, et al. 2001. Characterization of clinical isolates of *Klebsiella pneumoniae* from 19 laboratories using the National Committee for Clinical Laboratory Standards extended-spectrum β-lactamase detection methods. *J Clin Microbiol* **39:** 2864–2872.

Sudo T, Murakami Y, Uemura K, Hashimoto Y, Kondo N, Nakagawa N, Ohge H, Sueda T. 2014. Perioperative antibiotics covering bile contamination prevent abdominal infectious complications after pancreatoduodenectomy in patients with preoperative biliary drainage. *World J Surg* **38:** 2952–2959.

Sumita Y, Fukasawa M. 1995. Potent activity of meropenem against *Escherichia coli* arising from its simultaneous binding to penicillin-binding proteins 2 and 3. *J Antimicrob Chemother* **36:** 53–64.

Sykes RB, Bonner DP, Bush K, Georgopapadakou NH. 1982. Azthreonam (SQ 26,776), a synthetic monobactam specifically active against aerobic Gram-negative bacteria. *Antimicrob Agents Chemother* **21:** 85–92.

Talbot GH, Thye D, Das A, Ge Y. 2007. Phase 2 study of ceftaroline versus standard therapy in treatment of complicated skin and skin structure infections. *Antimicrob Agents Chemother* **51:** 3612–3616.

Tipper DJ, Strominger JL. 1965. Mechanism of action of penicillins: A proposal based on their structural similarity to acyl-D-alanyl-D-alanine. *Proc Natl Acad Sci* **54:** 1133–1141.

Vasoo S, Barreto JN, Tosh PK. 2015. Emerging issues in Gram-negative bacterial resistance: An update for the practicing clinician. *Mayo Clin Proc* **90:** 395–403.

Wang X, Zhang F, Zhao C, Wang Z, Nichols WW, Testa R, Li H, Chen H, He W, Wang Q, et al. 2014. In vitro activities of ceftazidime–avibactam and aztreonam–avibactam against 372 Gram-negative bacilli collected in 2011 and 2012 from 11 teaching hospitals in China. *Antimicrob Agents Chemother* **58:** 1774–1778.

Weber DA, Sanders CC. 1990. Diverse potential of β-lactamase inhibitors to induce class I enzymes. *Antimicrob Agents Chemother* **34:** 156–158.

Wise R, Andrews J, Bedford K. 1978. In vitro study of clavulanic acid in combination with penicillin, amoxycillin, and carbenicillin. *Antimicrob Agents Chemother* **13:** 389–393.

Woodford N, Tierno PM Jr, Young K, Tysall L, Palepou MF, Ward E, Painter RE, Suber DF, Shungu D, Silver LL, et al. 2004. Outbreak of *Klebsiella pneumoniae* producing a new carbapenem-hydrolyzing class A β-lactamase, KPC-3, in a New York Medical Center. *Antimicrob Agents Chemother* **48:** 4793–4799.

Yang Y, Bhachech N, Bush K. 1995. Biochemical comparison of imipenem, meropenem and biapenem: Permeability, binding to penicillin-binding proteins, and stability to hydrolysis by β-lactamases. *J Antimicrob Chemother* **35:** 75–84.

Yigit H, Queenan AM, Rasheed JK, Biddle JW, Domenech-Sanchez A, Alberti S, Bush K, Tenover FC. 2003. Carbapenem-resistant strain of *Klebsiella oxytoca* harboring carbapenem-hydrolyzing β-lactamase KPC-2. *Antimicrob Chemother* **47:** 3881–3889.

Zhanel GG, Wiebe R, Dilay L, Thomson K, Rubinstein E, Hoban DJ, Noreddin AM, Karlowsky JA. 2007. Comparative review of the carbapenems. *Drugs* **67:** 1027–1052.

Zhanel GG, Chung P, Adam H, Zelenitsky S, Denisuik A, Schweizer F, Lagace-Wiens PR, Rubinstein E, Gin AS, Walkty A, et al. 2014. Ceftolozane/tazobactam: A novel cephalosporin/β-lactamase inhibitor combination with activity against multidrug-resistant Gram-negative bacilli. *Drugs* **74:** 31–51.

Zykov IN, Sundsfjord A, Smabrekke L, Samuelsen O. 2016. The antimicrobial activity of mecillinam, nitrofurantoin, temocillin and fosfomycin and comparative analysis of resistance patterns in a nationwide collection of ESBL-producing *Escherichia coli* in Norway 2010–2011. *Infect Dis (Lond)* **48:** 99–107.

β-Lactam-Resistance Mechanisms: Gram-Positive Bacteria and *Mycobacterium tuberculosis*

Jed F. Fisher and Shahriar Mobashery

Department of Chemistry and Biochemistry, University of Notre Dame, Notre Dame, Indiana 46556-5670

Correspondence: jfisher1@nd.edu; mobashery@nd.edu

The value of the β-lactam antibiotics for the control of bacterial infection has eroded with time. Three Gram-positive human pathogens that were once routinely susceptible to β-lactam chemotherapy—*Streptococcus pneumoniae*, *Enterococcus faecium*, and *Staphylococcus aureus*—now are not. Although a fourth bacterium, the acid-fast (but not Gram-positive-staining) *Mycobacterium tuberculosis*, has intrinsic resistance to earlier β-lactams, the emergence of strains of this bacterium resistant to virtually all other antibiotics has compelled the evaluation of newer β-lactam combinations as possible contributors to the multidrug chemotherapy required to control tubercular infection. The emerging molecular-level understanding of these resistance mechanisms used by these four bacteria provides the conceptual framework for bringing forward new β-lactams, and new β-lactam strategies, for the future control of their infections.

Bacteria exemplify extraordinary diversity of size and shape (Young 2006, 2007). For the eubacteria, both the shape and integrity of the cell are intimately related to the chemical bonding pattern of the peptidoglycan polymer of their cell walls. Given this direct correlation, it is hardly surprising that many antibiotics target the enzymes that assemble the peptidoglycan (Schneider and Sahl 2010; Silhavy et al. 2010; Bugg et al. 2011; Silver 2013). Accordingly, bacteria have evolved a myriad of mechanisms to subvert these antibiotics.

Clinical bacterial resistance to antibiotics is often the acquisition of a primary resistance mechanism, abetted in important ways by secondary mechanisms. For the clinically important β-lactam antibiotics (these include the penicillin, cephalosporin, carbapenem, and

monobactam subfamilies), the primary resistance mechanisms used by Gram-positive bacteria are different from those used by the Gram-negative bacteria. The primary mechanism for the Gram-negative bacteria is expression of enzyme(s) that hydrolytically destroy the β-lactam, whereas for the Gram-positive bacteria the primary mechanism is target modification. This latter mechanism whereby structural changes to the specific enzyme targets of the β-lactam antibiotics render these enzymes less reactive to the β-lactam is identical to the mechanism used by the bacterial producers of the β-lactams (Ogawara 2015). The target of the β-lactams is a family of enzymes still known today by the name given to these enzymes—penicillin-binding proteins (PBPs)—dating from the discovery that these enzymes are inactivated, by covalent modification, by the β-lactams. The PBPs are the primary catalysts for the synthesis and remodeling of the peptidoglycan cell wall of bacteria. All bacteria have a family of PBPs, with some PBPs essential and others not essential. Recognition of a β-lactam structure by essential PBPs, leading to a mechanism-based loss of its enzymatic activity through the β-lactam inactivation, invariably culminates in cell lysis (Tomasz 1979). Although the evolutionary basis for the selection of target modification as the primary resistance mechanism for the Gram-positive bacteria is uncertain, the absence of an exterior membrane (and, thus, the absence of a control mechanism for exposure to antibiotics) in the Gram-positive bacteria surely contributes to this mechanism.

Although the peptidoglycan of all eubacteria has an identical core structure—a repeating disaccharide for the glycan strand with a peptide stem on the alternate saccharides of the glycan—each eubacteria tailors its peptidoglycan structure to accommodate the structural requirements for its shape, for the mechanisms for the reproduction inter alia of that shape, and for antibiotic resistance (Vollmer et al. 2008; Turner et al. 2014). The peptidoglycan of Gram-positive bacteria is a multilayer exoskeleton above their single membrane, whereas the peptidoglycan of the Gram-negative bacteria is a thinner (one- or two-layered) structure located in the periplasmic space between the two membranes. Two events define the structure of the peptidoglycan polymer: glycan strand elongation and cross-linking of the glycan strands by interconnection of the peptide stems on the alternate glycans. This latter event uses the terminal −D−Ala−D-Ala structure of the stem as an acyl donor from one strand to an amine acceptor of an adjacent strand. The β-lactam antibiotics have nearly synonymous three-dimensional disposition to −D−Ala−D-Ala, but whereas the acyl-enzyme derived from the −D−Ala−D-Ala structure is a catalytic species en route to a reaction, the acyl-enzyme derived from the β-lactam is not. The transpeptidase enzymes of peptidoglycan biosynthesis are inactivated by the β-lactam. Figure 1 summarizes the molecular events of peptidoglycan

synthesis, with focus on the transpeptidation event, in the context of the peptidoglycan structure found in *Streptococcus pneumoniae*. The compelling rationale for understanding these mechanisms is the timeless value of the β-lactams as therapeutic agents and the progressive emergence of resistance in both Gram-negative and -positive bacteria that were once routinely contained by β-lactam chemotherapy but are now clinically resistant.

THE ENZYME TARGETS OF THE β-LACTAM ANTIBIOTICS

The Enzymology of Peptidoglycan Assembly

The events of peptidoglycan synthesis involve a multiprotein and multienzyme assembly, whose composition and performance are distinctive to each bacterial species. The catalytic core of this assembly contains the enzyme catalysts of transglycosylation and transpeptidation. As β-lactams inactivate the enzyme catalysts of transpeptidation, these enzymes are collectively referred to as PBPs (Waxman and Strominger 1983; Sauvage et al. 2008; Spratt 2012). The use of "penicillin" in this terminology is historical: The penicillins were the first family of the β-lactam antibiotics, whereas the β-lactam antibiotics now used clinically include also the cephalosporins, carbapenems, and monobactams. All bacteria have a family of PBP enzymes classified by catalytic function and mass. The low-molecular-mass PBPs (also called class C PBPs) are primarily -D-Ala−Ala carboxypeptidases and control the population of stems competent for cross-linking. The high-molecular-mass PBPs divide into two subclasses. The first (class A) is bifunctional, having (in separate domains) the two catalytic activities (transglycosylase and transpeptidase) required for peptidoglycan assembly. Class B PBPs are monofunctional catalysts of D-Ala−D-Ala-dependent transpeptidation. Although all PBPs show signature active-site sequence motifs organized by a characteristic tertiary structure, in other respects (such as remaining sequence and thus nuance of domain function) they are distinctive, both with respect to each

Cite this article as *Cold Spring Harb Perspect Med* doi: 10.1101/cshperspect.a025221

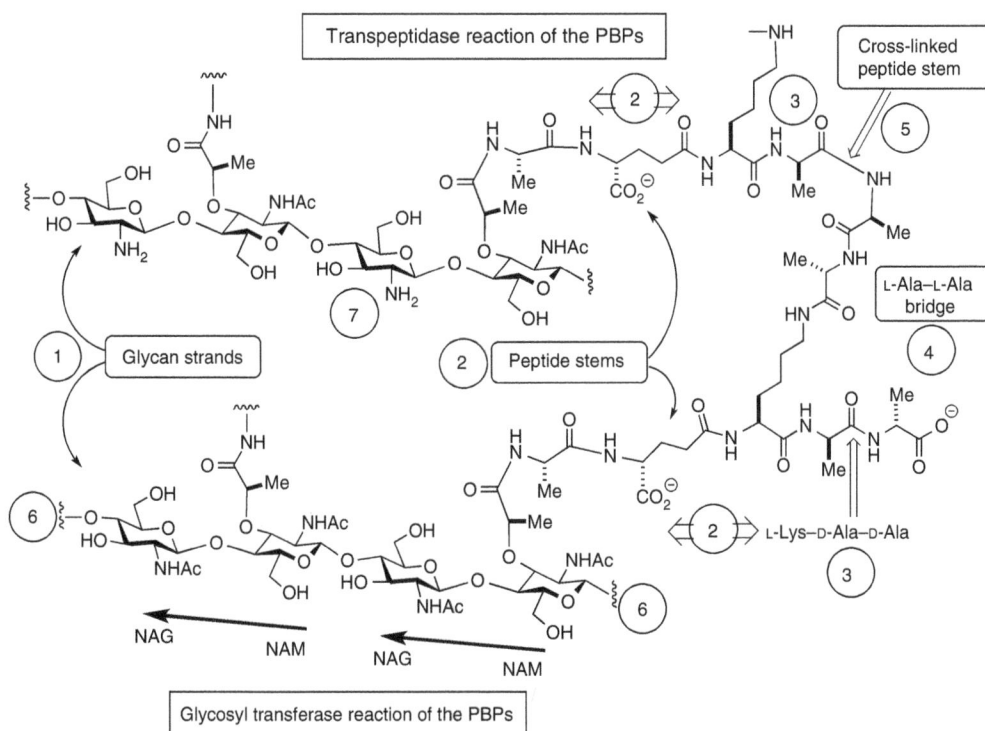

Figure 1. Schematic for the D,D-cross-link of the peptidoglycan of *Streptococcus pneumoniae*. This schematic indicates the structural connectivity for the normal cross-linking of the peptidoglycan and does not represent three-dimensional structure. The peptidoglycan is assembled by sequential actions of the penicillin-binding proteins (PBPs). The presumed first action occurs at the glycosyl transferase active site wherein a glycan strand (*1*) is assembled by polymerization of lipid II as the substrate. The second action is cross-linking of the peptide stems (*2*) of adjacent glycan strands through transpeptidase catalysis by the PBPs. Completion of both steps gives the polymeric structure of the peptidoglycan cell wall. Transpeptidase catalysis is the step blocked by the β-lactam antibiotics. This schematic shows two glycan strands and the completed cross-link between them. Key additional features of the peptidoglycan are emphasized. The sequence of the terminal amino acids of the stem is −L-Lys−D-Ala−D-Ala (in which the *3* identifies the penultimate D-Ala). In *S. pneumoniae*, an L-Ala-L-Ala dipeptide bridge (*4*) is added to the side-chain amine of this lysine. An L-Ser-L-Ala bridge is also encountered in this bacterium. An L-Ala-L-Ala bridge is also used by *Enterococcus faecium*, whereas *Staphylococcus aureus* uses a Gly$_5$ pentapeptide bridge. In all of these bacteria, the primary cross-linking event uses the central D-Ala of the stem as an acyl donor to a serine in the transpeptidase active site of the PBP, releasing the terminal D-Ala. This acyl-enzyme then transfers the acyl moiety to the terminal amino acid of the bridge, completing the cross-link (*5*). In this schematic, the upper glycan strand is the acyl-donor strand and the lower strand is the acyl-acceptor strand. The mechanisms controlling glycan strand length and the termination of glycan elongation (*6*) are poorly understood. In *S. pneumoniae*, a substantial portion of the N-acetyl groups of the glycan strand are hydrolytically removed (*7*) as a defensive measure against untoward peptidoglycan degradation, such as by lysozyme. An alternative pattern of cross-linking is used by *E. faecium* as a β-lactam-resistance method. PBPs, Penicillin-binding proteins; NAM, N-acetylmuramic acid; NAG, N-acetyl-2-amino-2-deoxyglucosamine.

other and with respect to the PBPs of other (even closely related) bacterial species. A critical corollary is that individual β-lactam antibiotics show distinctive patterns for PBP inactivation at the (often) subsaturating conditions encoun-tered in chemotherapy. This pattern determines the selection of the β-lactam to control infection and the development of β-lactam resistance. We examine the complexity of this nexus—involving β-lactam structure, PBP structure and function, and β-lactam resistance—for the four Gram-positive pathogens *S. pneumoniae*, *Enterococcus faecium*, *Staphylococcus aureus*, and *Mycobacterium tuberculosis*.

THE MOLECULAR MECHANSIMS OF β-LACTAM RESISTANCE BY *S. pneumoniae*

S. pneumoniae is a human commensal of the nasopharyngeal microbiota (Hakenbeck et al. 2012; Henriques-Normark and Tuomanen 2013; Fischetti and Ryan 2015). Notwithstanding the benefit of pneumococcal vaccines (which are themselves forces for clinical strain and resistance-determinant selections) (Angoulvant et al. 2015; van Tonder et al. 2015), invasive infection of *S. pneumoniae* among humans with compromised immune systems remains a significant cause of morbidity and mortality. In the 40 years since the first appearance of β-lactam resistance in *S. pneumoniae*, this bacterium has responded quickly to clinical interventions by recombination with other streptococci (Sauerbier et al. 2012; Jensen et al. 2015) to secure resistance not just to the β-lactams but to other clinically important antibiotics (Reinert 2009; Croucher et al. 2011). *S. pneumoniae* achieves β-lactam resistance through extensive and complementary "mosaic" mutation of three key PBP target enzymes with minimal fitness cost (Albarracin Orio et al. 2011) while retaining the genetic determinants that make this bacterium so successful for persistent human colonization (Nobbs et al. 2015). The challenge for successful chemotherapy against this bacterium is the invention of structure, whether antibiotic or vaccine, that subverts this balance. We address this challenge as it relates to β-lactam control of infection by *S. pneumoniae*, through consideration of the role of its PBPs in peptidoglycan biosynthesis and evaluation of the effect of the mutations to critical PBPs.

The PBPs of *S. pneumoniae*

S. pneumoniae is a fitting introduction to the topic of β-lactam resistance among the Gram-positive pathogens as this bacterium (compared with the others to be discussed) arguably presents our best (although far from complete)

understanding of Gram-positive peptidoglycan biosynthesis (Massidda et al. 2013). Three phases of peptidoglycan synthesis complete the ovococcal peptidoglycan of this bacterium. One phase is sidewall growth. The remaining two phases are septal growth and final septation. Septal growth occurs across the middle plane of the ovococcus and culminates with the synthesis of a distinctive circular growth—a surface annulus—of peptidoglycan growth immediately preceding septation (Wheeler et al. 2011). The regions of old and new peptidoglycan are distinctly demarcated, and *S. pneumoniae* grows without peptidoglycan turnover so as to presumably avoid detection of its peptidoglycan by the innate immune system (Boersma et al. 2015). *S. pneumoniae* apportions six PBPs to the task of constructing its peptidoglycan. As the different phases of peptidoglycan growth are interdependent (Philippe et al. 2015), functional complementarity (and redundancy) among these PBPs follows. Three (PBP1a, PBP1b, PBP2a) of the six PBPs are bifunctional class A PBPs, having both transglycosylase and transpeptidase activities. Two are class B transpeptidases (PBP2b and PBP2x), whereas the sixth PBP (PBP3) is a class C D,D-carboxypeptidase. The success of β-lactam chemotherapy in controlling *S. pneumoniae* infection establishes that certain of these PBPs, and certain events of peptidoglycan growth catalyzed by these PBPs, are essential to the bacterium. Unmasking this identity is central to the understanding of its β-lactam resistance.

The complexity of experimental design required for this task cannot be overstated. We emphasized previously the substantial sequence variability among the PBPs, apart from their signature catalytic motifs, within their conserved tertiary structure. Likewise, the β-lactam antibiotics also encompass structural diversity. Accordingly, the identification of β-lactam structure useful against a particular bacterium rests on the empirical discernment of the match between β-lactam structure and the structural space of the active site of the "essential" PBPs. It is thus not surprising that there is great breadth of β-lactam match, and mismatch, to the *S. pneumoniae* PBP ensemble (Kocaoglu

Cite this article as *Cold Spring Harb Perspect Med* doi: 10.1101/cshperspect.a025221

et al. 2015). Genetic analyses establish the monofunctional transpeptidases PBP2b (as a member of the elongosome complex used for peptidoglycan elongation) and PBP2x (as a member of the divisome complex synthesizing the peptidoglycan of the septum) as the two essential PBPs of *S. pneumoniae* (Berg et al. 2013; Philippe et al. 2014). Each of the bifunctional PBPs (PBP1a, PBP1b, PBP2) can be deleted individually. Paired PBP1a/PBP2 deletions as well as the triple deletion are lethal, whereas other pairwise deletions show morphological defects. Deletion of the low-molecular-mass PBP3 maintains viability but also at the cost of morphological defects (Hakenbeck et al. 2012). As PBP2b and PBP2x are monofunctional transpeptidases, completion of peptidoglycan synthesis requires their pairing with a transglycosylase. As a result of the functional redundancy among PBP1a, PBP1b, and PBP2, the identities for such pairings (and roles) are uncertain. Indeed, although the elongosome and divisome complexes colocalize at mid-cell at the start of the growth and division events of peptidoglycan synthesis, the PBPs of these complexes subsequently follow individual patterns of spatial localization (Land et al. 2013; Tsui et al. 2014). Resistance to β-lactams by *S. pneumoniae* is the consequence of mutations to either PBP2b or PBP2x. High-level resistance to the penicillins requires abetting mutations in PBP1a (Zapun et al. 2008).

The β-Lactam-Resistant PBPs of *S. pneumoniae*

Notwithstanding the preservation of the lysine-serine signature motifs in both the PBP2b and PBP2x D,D-transpeptidases, PBP2x is structurally distinct from PBP2b. PBP2b is a "typical" four-domain class A/B PBP: a short amino-terminal cytoplasmic tail; a transmembrane anchor; and protruding into the periplasm, a protein–protein interaction "pedestal" domain followed by the catalytic domain (Contreras-Martel et al. 2009). The cytoplasmic tail and transmembrane domains are also functional, presumably for protein–protein recognition (Berg et al. 2014). In addition to equivalents

of these four domains, PBP2x has two additional carboxy-terminal domains—termed "PBP and serine-threonine kinase-associated" (PASTA) domains—that follow the catalytic domain. The PASTA domains are suggested to respond to peptidoglycan structure under phosphorylation control (Maestro et al. 2011; Morlot et al. 2013). The importance of the PASTA domains for modulation of the catalytic domain of PBP2x (Dias et al. 2009; Maurer et al. 2012) and for its septal localization (Peters et al. 2014) has been shown.

The key questions with respect to β-lactam resistance are the mutation(s) to the essential PBPs, the mechanistic relevance of the mutations, and the additional contributors to *S. pneumoniae* β-lactam resistance. A recent genetic analysis addressed the last question. Genome-wide association of single nucleotide polymorphisms identified, in addition to the genes for PBP1a, PBP2b, and PBP2x, contributions from the *mraW* and *mraY* genes within the peptidoglycan synthesis pathway, as well as genes in the cell-division pathway (*ftsL*, *gpsB*), genes encoding chaperones (*clpL*, *clpX*), and a gene in the recombination pathway (*recU*) (Chewapreecha et al. 2014). The gene associations with penicillins were not coincident for the cephalosporins. The polymorphisms were distributed both in vaccine-targeted and non-vaccine-targeted lineages, suggested to explain why vaccination has failed to reduce β-lactam resistance (Hakenbeck 2014). A key determinant with respect to PBP mutation is the β-lactam used for resistance selection. For example, cefotaxime selectively inactivates the PBP2x (it is unreactive to PBP2b) of susceptible *S. pneumoniae* (Kocaoglu et al. 2015). A single-point mutation in the catalytic domain of PBP2x achieves clinical resistance to cefotaxime—but not to penicillins—as a result of a fourfold increase in the minimal-inhibitory concentration (MIC) of cefotaxime (Coffey et al. 1995). The more common experience is exemplified by the comparative structures of the PBP2b isolated from a penicillin-susceptible (minimum inhibitory concentration [MIC] of 0.01 μg/mL) and a PBP2b isolated from a highly penicillin-resistant (MIC of 6 μg/mL) clinical strain of *S. pneumoniae*. This latter enzyme had 58 mutations (Fig. 2A), presenting a mosaic pattern of alteration across both the pedestal and catalytic domains (Contreras-Martel et al. 2009). The differentiation between mutations that are incidental to resistance and those that contribute to resistance cannot be made solely from structural analysis (Hakenbeck et al. 2012). Although genetic analyses of the mosaic changes underscore notable "point" mutations

to PBP2b and PBP2x, as exemplified recently (Ip et al. 2015), the realization of β-lactam resistance has a greater dimension than that seen as amino acid change alone.

The intertwining of protein structure, catalytic function, and resistance pathways is exemplified by the "piperacillin paradox" observed for *S. pneumoniae* (Philippe et al. 2015). Piperacillin (a penicillin) is extensively used in the clinic against *S. pneumoniae*, and, although

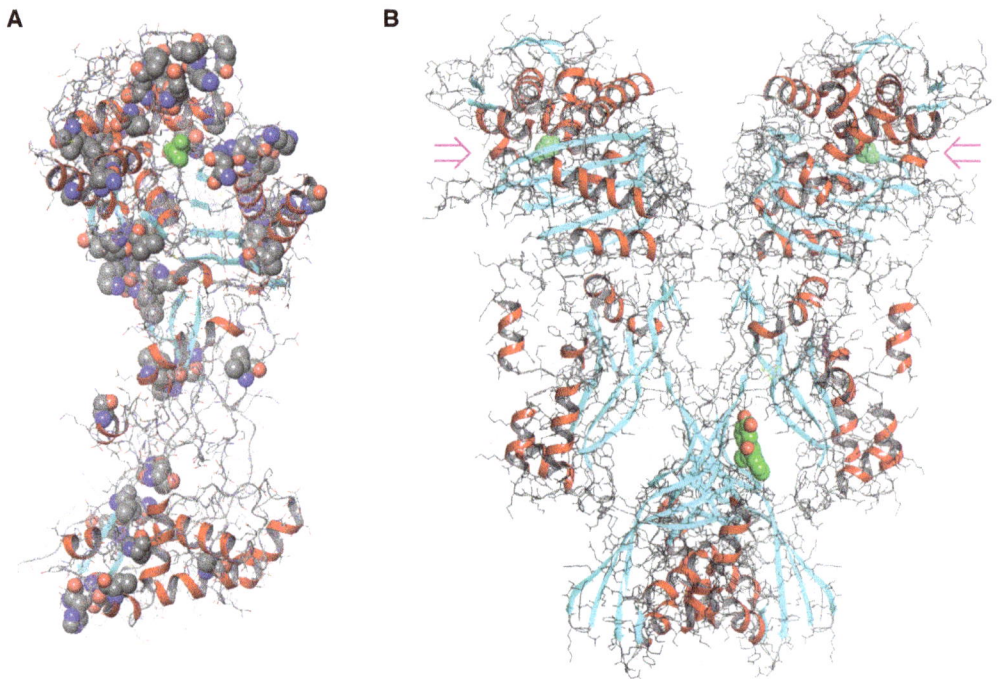

Figure 2. (*A*) The mosaic pattern of resistance mutations within the periplasm-located domains of *S. pneumococcus* PBP2b (PDB Code 2WAE; Contreras-Martel et al. 2009). The enzyme crystallizes as a monomer. The 58 residues that undergo mutation during the transformation of this enzyme from a β-lactam-susceptible to a β-lactam-resistant state are shown in space-filling depictions. The catalytic serine used in the acyl-transfer reactions of transpeptidation is depicted in green in a space-filling representation. Although mutations in the catalytic domain predominate, several distal mutations are implicated as critical (as evidenced by the frequency of their appearance). In general, the differentiation between mutations that are incidental and those mutations that contribute directly to resistance by favorably altering the loop "breathing motions" required to enable access to this serine is exceptionally challenging. (*B*) The structure of the periplasm-located domains of *S. aureus* PBP2a with a bound quinazolinone (non-β-lactam) allosteric effector (PDB Code 4CJN; Bouley et al. 2015). The enzyme crystallizes as a dimer. The allosteric effector has intrinsic antibacterial activity and is depicted in green in a space-filling depiction (*lower right* of the structure). A single molecule of the allosteric effector is bound in a groove between the allosteric domain and a so-called pedestal domain. The orientation of this PBP2a dimer is approximately 90° relative to that of the PBP2b of *A*. The two catalytic serines are shown in green in a space-filling depiction. The mauve-colored arrows direct attention to their location. These serines are ∼60 Å distant from the allosteric site. The structural change that occurs in the allosteric transformation that controls access to these active-site serines is understood (Otero et al. 2013).

it has greater potency against PBP2x (half minimal inhibitory concentration (IC_{50}) of 0.02 µg/mL in susceptible *S. pneumoniae*), it selects preferentially for resistance mutations in PBP2b (IC_{50} of 0.18 µg/mL) (Kocaoglu et al. 2015). The proposed explanation for this paradox is that partial loss of PBP2b function in elongation is sufficient to arrest growth, whereas the more complete loss of PBP2x function in septation does not. Acquisition of a low-affinity PBP2b, thus, suffices to maintain cell multiplication (Philippe et al. 2015).

The molecular basis for the low affinity in the PBP targets has been extensively explored by crystallography. Correlation of the observed mutations to the locations within these structures shows preference for mutation at or near the active site. The cumulative effect of these mutations is interpreted as a conformation adjustment within the cleft that opens so as to favor substrate over the β-lactam inactivator. This concept is discussed with respect to the particular structures of *S. pneumoniae* PBP1a (Contreras-Martel et al. 2006; Job et al. 2008), PBP1b (Macheboeuf et al. 2005), PBP2b (Contreras-Martel et al. 2009), and PBP2x (Gordon et al. 2000; Dessen et al. 2001; Chesnel et al. 2003; Carapito et al. 2006; Maurer et al. 2008; Yamada et al. 2008). In selected cases, the molecular basis of the mutation is understood. Cephalosporins have poor PBP2b affinity but good PBP2x affinity and, thus, select PBP2x mutations. In response to cefotaxime challenge, under both laboratory and clinical conditions, mutation within the active site of PBP2x of the noncatalytic threonine-550 residue to alanine confers cefotaxime resistance (T550A, 20-fold decrease in acylation). The basis for cefotaxime resistance in these PBP2x mutants is interpreted as a result of the loss of a key hydrogen bond that is used for recognition of the cephalosporins (Gordon et al. 2000). The T550A PBP2x is, however, more susceptible to inactivation by penicillins (Mouz et al. 1999). Computational evaluations are now used in which such simple correlations are not possible, as with clinical mosaic mutations (Ge et al. 2012; Ramalingam et al. 2013). An aspect missing from these analyses is the effect of mutations on the transpeptidase reaction itself (as in vitro PBP assay of this reaction is not possible) and on the integration of the PBP into the elongosome and divisome complexes (Zerfass et al. 2009).

Antibiotic resistance often corresponds to multiple adaptation mechanisms and, here as well, *S. pneumoniae* shows its capability for β-lactam resistance. Its *murMN* operon encodes transferases that add an L-Ala-L-Ala (or L-Ser-L-Ala) cross-bridge extension to the L-Lys residue of the peptidoglycan stem (Filipe et al. 2002). These extensions contribute to β-lactam resistance (Hakenbeck et al. 2012), possibly by improving the efficacy of PBP2b-dependent transpeptidation (Berg et al. 2013). MurM/MurN-dependent cross-bridge extension is also important for proper control of pneumolysin release from the peptidoglycan to support virulence (Greene et al. 2015). An additional enzyme of peptidoglycan biosynthesis, MurE (the catalyst of L-Lys addition to the stem in the course of lipid II biosynthesis), also enhances—for unknown reasons—the β-lactam resistance of *S. pneumoniae*, as it also does for *S. aureus* (Todorova et al. 2015). *S. pneumoniae* further exemplifies the increasing recognition that antibiotic discovery in the future will require evaluations beyond that of the interaction of the antibiotic with its target. Although we recognize that all targets (and especially the PBPs) function within a confluence of regulated metabolic pathways, we equally well recognize that we understand neither the key components of these pathways nor how these components interrelate. For example, the roles for Ser/Thr kinase control of not just PBP2x, but of the peptidoglycan biosynthetic pathway (Dias et al. 2009) and the pathways for cell growth and division (Falk and Weisblum 2013; Fleurie et al. 2014), remain essentially unknown. An example of the value provided by a whole bacterium perspective on antibiotic selection is the study of the growth response of antibiotic-susceptible *S. pneumoniae* following exposure to antibiotics that were either bacteriostatic, bactericidal, or bactericidal as a result of lysis (i.e., β-lactams) (Sorg and Veening 2015). Exposure of *S. pneumoniae* to these antibiotics at the "F10" concentration of the antibiotic (in which

F10 is the antibiotic concentration that achieves a 10-fold suppression of growth over a 10-h period and is a concentration typically somewhat greater than the MIC) confirmed the advantage of a bactericidal over a bacteriostatic mechanism. However, remarkable differences were seen among different antibiotics. Among the three β-lactams compared (ampicillin, MIC of 0.018 μg/mL; penicillin G, MIC of 0.015 μg/mL; and cephalexin, MIC of 0.22 μg/mL), ampicillin was the most efficacious. The basis for its superiority was interpreted in terms of its possession of a narrow concentration range for efficacy (corresponding to a narrow mutant selection window) coinciding with suppression of heterogeneous phenotype selection. In addition, the β-lactam data further suggested a relationship between heteroresistance and cell morphology (Sorg and Veening 2015). These observations, although fully consistent with the uniqueness of β-lactam structure coinciding with a unique profile for inhibition among the PBP family (Kocaoglu et al. 2015), affirm the growing recognition of the limitation of the MIC value (alone) as the criterion to define antibiotic efficacy.

THE MOLECULAR MECHANISMS OF β-LACTAM RESISTANCE BY *E. faecium*

Before the introduction of the antibiotics, the enterococci were innocuous commensal bacteria of the gut and were infrequently associated with infection (Arias and Murray 2012; Hendrickx et al. 2013; Werner et al. 2013; Kristich et al. 2014). Coincident with the introduction of the antibiotics, rapid genetic diversification culminated in the emergence 30 years ago of the enterococci as insidious, multidrug-resistant nosocomial pathogens. Whereas the key pathogen at the time of this emergence was *Enterococcus faecalis*, infections by *E. faecium* and *E. faecalis* now are equally prevalent. A key observation made during this transition was that the enterococci had intrinsic resistance to the cephalosporin class of β-lactam antibiotics. The transition of these bacteria from high- to low-penicillin susceptibility—to the point today in which nosocomial infection by the en-

terococci in the United States is presumed to be both β-lactam- and vancomycin-resistant—has been addressed from the vantages of genomics, structural biology, and enzymology (Palmer et al. 2010; Hendrickx et al. 2013; Lebreton et al. 2013; Werner et al. 2013). The origin of the β-lactam resistance of *E. faecium* (the more pathogenic of the two, in large part as a result of its greater β-lactam resistance) exemplifies the multifactorial resistance mechanisms now used by resistant bacteria.

Despite the identity of *S. pneumoniae*, the enterococci, and *S. aureus* as Gram-positive cocci, the structural similarities of their peptidoglycans, and the in vitro ability to functionally interchange the peptidoglycan biosynthetic pathways of these cocci (Arbeloa et al. 2004a), each chooses a different PBP mechanism to attain β-lactam resistance. The ability of the enterococci to assimilate pathways in support of antibiotic resistance and virulence was exemplified sharply by the appearance of vancomycin resistance as a result of the acquisition of the self-resistance mechanism used by vancomycin-producing bacteria. Vancomycin resistance results from the remodeling of peptidoglycan synthesis so as to replace the vancomycin-binding D-Ala-D-Ala stem terminus normally used for transpeptidation (Cattoir and Leclercq 2013). β-Lactam resistance by *E. faecium* involves a low-affinity PBP for catalysis of transpeptidation. Mutation of the essential (and intrinsically cephalosporin-unreactive) PBP5 of *E. faecium* gives the requisite low-affinity enzyme PBP5fm (Zorzi et al. 1996; Arbeloa et al. 2004b; Lebreton et al. 2013; Pietta et al. 2014). PBP5fm pairs with one of several transglycosylases to complete peptidoglycan synthesis (Rice et al. 2009). The basis for the low affinity, visualized from the vantage of the location of the mutations in the PBP5fm structure, is suggested as restricted motion in the loop that controls by an opening motion access to the active site (Sauvage et al. 2002). This mechanism parallels the observation for the key PBPs of *S. pneumoniae* and also parallels the resistance mechanism of PBP2a of *S. aureus*.

The contribution of PBP5fm to the β-lactam resistance of *E. faecium* is central but not exclusive. All bacteria revamp their metabolism in response to cell-wall stress. For the enterococci, these additional changes include mitigation of reactive species to attenuate the bactericidal effect of antibiotics in general (Ladjouzi et al. 2013; Djorić and Kristich 2015), consistent with the emerging hypotheses for the bactericidal effect of antibiotics (Lobritz et al. 2015). A genome-wide evaluation of ampicillin resistance in *E. faecium* confirmed the central role of PBP5fm and identified a supporting contribution arising from an alternative mechanism for peptidoglycan cross-linking (Zhang et al. 2012). Here, the cooperative activity of a PBP carboxypeptidase (to remove the terminal D-Ala from the peptide stem) and an L,D-peptidoglycan transpeptidase, enabling use of the L-Lys-D-Ala stem segment as the acyl-donor for peptidoglycan cross-linking and, thus, the bypass of the β-lactam-sensitive use of D-Ala-D-Ala as the acyl donor

(Fig. 3). An identical enzyme function contributes to β-lactam resistance in *M. tuberculosis*. An L,D-transpeptidase is encoded in yet other Gram-positive bacterium (such as *Clostridium difficile*) but is not present in *S. pneumoniae* and *E. faecalis*. The operation of this pathway is under both two-component and Ser/Thr kinase control (Sacco et al. 2014). The structure, mechanism, and versatility (with respect to peptidoglycan structure) of this L,D-transpeptidase are characterized in Mainardi et al. (2005), Biarrotte-Sorin et al. (2006), Cremniter et al. (2006), and Magnet et al. (2007). Notwithstanding substantive catalytic differences compared with the PBPs (a catalytic cysteine, rather than a serine), this L,D-transpeptidase is inactivated by carbapenems via acylation of the catalytic cysteine. Penicillins do not inactivate and cephalosporins weakly inactivate (Dubée et al. 2012; Lecoq et al. 2013; Triboulet et al. 2013, 2015).

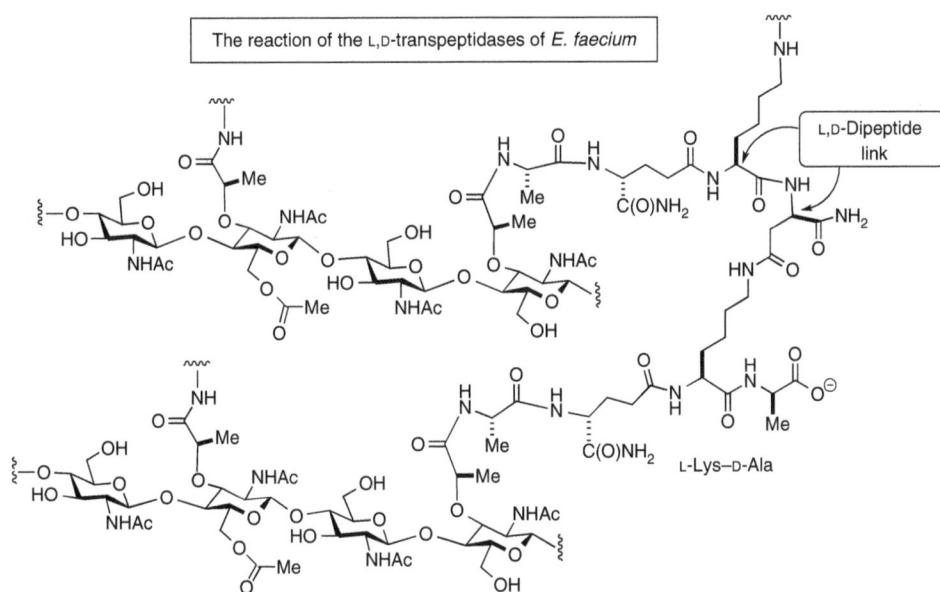

Figure 3. Schematic for the L,D-cross-link of the peptidoglycan of *E. faecium*. The peptidoglycan found in susceptible strains of *E. faecium* has D,D-cross-linking, wherein a −D-Ala−D-Ala-derived acyl moiety is transferred to the α-amine of the *iso*-D-Asx (depicted as *iso*-Asn) residue of the stem bridge. Following in vitro selection for ampicillin resistance, *E. faecium* that is fully resistant to both β-lactams and vancomycin is obtained by peptidoglycan synthesis using a non-PBP-dependent L,D-cross-link. Here, trimming of terminal D-Ala of the cross-link the stem permits the use of the −L-Lys−D-Ala moiety of the stem for acyl transfer to the α-amine of the *iso*-D-Asx to achieve the cross-link. The six-position of the MurNAc saccharide is acetylated, as occurs frequently in the enterococci as a lysozyme-resistance adaptation (Pfeffer et al. 2006).

The minimal fitness cost of these resistance pathways (Foucault et al. 2010; Starikova et al. 2013; Gilmore et al. 2015) and the ability of the enterococci to further assimilate resistance to new therapeutic agents (such as daptomycin) account for the modern challenge of enterococci chemotherapy (Kristich et al. 2014). Later cephalosporins against *E. faecalis* and β-lactam-containing antibiotic combinations against *E. faecium* exemplify the direction for future β-lactam therapy of enterococcal infection (Henry et al. 2013; Hindler et al. 2015; Smith et al. 2015a,b; Werth et al. 2015).

THE MOLECULAR MECHANISMS OF β-LACTAM RESISTANCE BY *S. aureus*

S. aureus is a Gram-positive coccus and human commensal (Missiakas and Schneewind 2015). Invasive infection by β-lactam-resistant *S. aureus* following surgery and increasingly within the community to the soft tissue remains as difficult a challenge for chemotherapeutic control today as in prior decades (de Lencastre et al. 2007; Stryjewski and Corey 2014; Knox et al. 2015). New β-lactams, again acting to compromise the integrity of the peptidoglycan of the cell wall, remain a critical means of surmounting the resistance mechanisms used by *S. aureus* (Holmes and Howden 2014; Peyrani and Ramirez 2015). These mechanisms correspond to a complexity of regulatory mechanisms and center on an acquired PBP known today as PBP2a. This acquired PBP2a completes the synthesis of the peptidoglycan of *S. aureus* following incapacitation of its other PBPs by a β-lactam. The cell wall of *S. aureus* is thicker than many other Gram-positive bacteria, and, indeed, increased cell-wall thickness is a resistance mechanism used by *S. aureus* against other cell-wall targeting antibiotics, notably vancomycin (Cazares-Dominguez et al. 2015). Its peptidoglycan is exceptionally cross-linked (as much as 80% of the available stem peptide). Growth of the peptidoglycan of *S. aureus* involves synthesis of a cross-wall septum to create a pair of nascent, hemisphere-shaped bacteria. Rapid remodeling of the peptidoglycan provides the spherical shape of the mature bacterium. Microscopic

analysis of this transformation suggests a concentric growth pattern, followed by loss of the concentric features coinciding with the peptidoglycan remodeling (Turner et al. 2010, 2014; Bailey et al. 2014). At the molecular level, a distinctive feature of the peptidoglycan of β-lactam-resistant *S. aureus* is the presence of a pentaglycine cross-bridge extension to lysine of the peptidoglycan stem. Addition of this pentaglycine extension involves catalysis by the enzymes of the Fem ("factors enhancing methicillin" resistance) pathway (Dare and Ibba 2012). A comparison of the mature cell wall of *S. aureus* and the cell wall made by *S. aureus* having disruptions in the Fem pathway shows distinct differences in the polymeric structure (Kim et al. 2015; Singh et al. 2015). Nonetheless, the difference at the molecular level for the correlation of the pentaglycine extension to methicillin resistance is not known.

There is no uncertainty for the core mechanism of β-lactam resistance by *S. aureus*. This bacterium acquired two resistance mechanisms, with each acquisition occurring early in the 50-year history of this resistance (Chambers 2005; Moellering 2012; Peacock and Paterson 2015; Tong et al. 2015). Following the clinical introduction of the earliest penicillins, penicillin-resistant *S. aureus* appeared as a result of expression of a penicillin-specific β-lactamase. This β-lactamase was, and remains today, a "penicillinase" of modest catalytic ability by comparison to the pan-β-lactam capability of the β-lactamases now endemic among the Gram-negative bacteria. This penicillinase nonetheless provided resistance to *S. aureus* against these early penicillin structures. In response, medicinal chemists discovered that penicillins substituted with sterically large aryl groups, exemplified by methicillin, were poor substrates of the penicillinase. The therapeutic value of methicillin was not long-lived. *S. aureus*, in short order, acquired methicillin resistance by acquisition from environmental cocci of a new PBP (Zhou et al. 2008; Antignac and Tomasz 2009; Tsubakishita et al. 2010). This acquired transpeptidase (PBP2a) integrates into the enzyme assembly for peptidoglycan synthesis as a β-lactam-insensitive catalyst. PBP2a

Cite this article as *Cold Spring Harb Perspect Med* doi: 10.1101/cshperspect.a025221

distinguishes between the peptidoglycan as a substrate and against the β-lactam as an inactivator. As this ability extends to all structural classes of β-lactams (not just methicillin), this second resistance mechanism has persevered as a powerful resistance mechanism against all but the newest guises of β-lactam structure. The empirically derived structures of the anti-MRSA (methicillin-resistant *S. aureus*) cephalosporins, exemplified by ceftobiprole and ceftaroline (Fernandez et al. 2014; Stryjewski and Corey 2014), are successful antibiotics against *S. aureus* (and other Gram-positive bacteria) as a direct result of their ability to evade this structural discrimination by PBP2a. However, although the framework for our discussion is PBP2a, we also provide a perspective on the role of the auxiliary mechanisms.

The central question is the uniqueness of PBP2a. Methicillin-sensitive *S. aureus* encodes eight enzymes for peptidoglycan synthesis. MRSA has nine (now including PBP2a). The core eight enzymes are PBP1 (a monofunctional transpeptidase, active in cell division and separation), PBP2 (a bifunctional transglycosylase and transpeptidase), PBP3 (a transpeptidase), PBP4 (a low-molecular-mass transpeptidase), two monofunctional transglycosylases, and two "auxiliary" transpeptidases (FmtA and FmtB). Genetic deletion of these activities, most recently using a MRSA strain, shows that only two—PBP1 and PBP2—are required for the normal growth and normal shape of *S. aureus*, albeit with loss of β-lactam resistance, increased lysozyme susceptibility, and decreased virulence (Reed et al. 2015). Assembly of the MRSA peptidoglycan in the presence of the β-lactam challenge requires cooperative PBP2-dependent transglycosylation and PBP2a-dependent transpeptidation (Pinho et al. 2001). The presumption that the peptidoglycan benefits from cross-linking by a second transpeptidase is supported by evidence implicating both PBP4 and FmtA. PBP4 is involved in peripheral wall peptidoglycan remodeling (Leski and Tomasz 2005; Loskill et al. 2014; Qiao et al. 2014; Gautam et al. 2015). PBP4 is essential for β-lactam resistance in the community strains of MRSA (Memmi et al. 2008). A β-lactam (cefoxitin) with high PBP4 affinity is synergistic with oxacillin (Memmi et al. 2008). Although the catalytic role of FmtA remains to be fully clarified, its presumed function is transpeptidation under conditions of cell-wall stress (Qamar and Golemi-Kotra 2012).

Peptidoglycan biosynthesis is highly regulated at numerous levels, including by kinase phosphorylation, stress-response pathways, and the transmembrane delivery of lipid II as the substrate for its PBPs (Lages et al. 2013). Notwithstanding the catalytic competence of PBP2a in MRSA, PBP2a expression is induced following the irreversible acylation of the exposed cell-surface domain of the transmembrane sensor protein MecR by a β-lactam antibiotic. PBP2a is made only when circumstances demand its presence. The MecR protein is structurally and functionally homologous to a BlaR sensor protein, which itself controls expression of the *S. aureus* penicillinase. Cross talk between the two pathways adds additional complexity to the poorly understood and multi-event cascades ultimately inducing PBP2a (and/or penicillinase) expression (Oliveira and de Lencastre 2011; Amoroso et al. 2012; Llarrull and Mobashery 2012; Peacock and Paterson 2015; Staude et al. 2015). A molecular-level understanding of these pathways is anticipated to identify new targets for antibiotic synergy with β-lactams. An important advance with respect to PBP2a is the discovery that its active site is under allosteric control with respect to two loop motions that open the active site in response to its peptidoglycan substrate (Otero et al. 2013; Fishovitz et al. 2014). β-Lactams that appropriately mimic peptidoglycan structure, such as ceftaroline, have good MRSA activity by their ability to effect this allosteric opening. The appreciable distance between the newly discovered allosteric site and transpeptidase active site is evident from the crystal structure of the non-β-lactam bound to PBP2a (Fig. 2B). As a consequence of their allosteric-induced opening of the active site, the concentrations achieved by these β-lactams in vivo coincide with the concentration required for PBP2a inactivation (Fishovitz et al. 2014, 2015). Subversion of the allosteric mechanism also has been achieved in vitro using a threefold β-lactam combination (Gonzales et al. 2015). Non-β-lactam allosteric effectors capable of β-lactam synergy have also been identified (Bouley et al. 2015).

Genomic technologies identify additional loci that synergize with β-lactams (Roemer et al. 2012), including within the peptidoglycan biosynthesis pathway (Mann et al. 2013), the FtsZ-dependent organization of the peptidoglycan biosynthetic machinery (Tan et al. 2012), and the coordination of the synthesis of the wall teichoic acids with that of the peptidoglycan (Atilano et al. 2010; Pasquina et al. 2013; Wang et al. 2013; Sewell and Brown 2014; Winstel et al. 2014). The targets identified within the wall teichoic acid pathway may have special significance as the synergistic pairing of intervention against both biosynthetic pathways may extend to other Gram-positive bacteria (Hendrickx et al. 2013). The translation of successful in vitro combination therapy into successful clinical therapy is never straightforward (Bush 2015). Nonetheless, intervention at coupled binding sites (such as by simultaneous occupancy of the allosteric site and active site of PBP2a by two β-lactams) and at intersecting pathways (such as synergy between simultaneous disruption of peptidoglycan and teichoic acid biosynthesis by two separate inhibitors) is emerging as a credible (if not yet viable, apart from β-lactam–β-lactamase pairs) strategy to control resistant bacterial pathogens.

THE MOLECULAR MECHANISMS OF β-LACTAM RESISTANCE BY *M. tuberculosis*

The non-Gram-positive staining mycobacteria possess a cell envelope structure that is fundamentally different from and structurally more complex compared with the cell envelope of either the Gram-positive or -negative bacterium (Jackson et al. 2013; Alderwick et al. 2015; Nataraj et al. 2015). A consequence of the nuanced layers of this cell envelope is impermeability to antibiotic structure. Indeed, the challenge of chemotherapeutic control of mycobacterial infection is legendary (Chakraborty and Rhee 2015). Important additional factors contributing toward the β-lactam insensitivity of

M. tuberculosis (in addition to impermeability) are the expression of efflux transporters, the versatility of this bacterium with respect to the synthesis of alternate peptidoglycan structures, and expression of a robust—sufficiently so, as to represent a possible means of detection of *M. tuberculosis* infection (Cheng et al. 2014)— β-lactamase. For these reasons, the β-lactams have never been among the many antibiotics used to control *M. tuberculosis* infection. Nonetheless, the combination of the emergence of highly resistant *M. tuberculosis* strains (Seung et al. 2015) with recent studies showing some promise for β-lactams against *M. tuberculosis* (Hugonnet et al. 2009; Hazra et al. 2014; Wivagg et al. 2014) has justified reconsideration of the β-lactams.

Although *M. tuberculosis* has the expected PBP family for peptidoglycan biosynthesis (Prigozhin et al. 2014), as a matter of routine, it uses both PBP-catalyzed (and, thus, β-lactam-sensitive) D,D-transpeptidation and the β-lactam-insensitive L,D-transpeptidation for this task (Goffin and Ghuysen 2002). The intrinsic β-lactam unreactivity by some of these PBPs (Bansal et al. 2015), a greater dependence on L,D-transpeptidation in the presence of β-lactams (Gupta et al. 2010; Kumar et al. 2012; Schoonmaker et al. 2014), and reactivity as BlaC substrates explain the historic therapeutic failure of the penicillin β-lactams. There is, however, promise for carbapenem combinations. As discussed previously for *E. faecium*, the carbapenems not only inactivate these L,D-transpeptidases (Lavollay et al. 2008; Triboulet et al. 2011; Cordillot et al. 2013; Schoonmaker et al. 2014; Brammer Basta et al. 2015; Wivagg et al. 2016) but have excellent activity against several of the essential PBPs of *M. tuberculosis* (Chambers et al. 2005; Kumar et al. 2012) and are poor substrates for its BlaC β-lactamase (Tremblay et al. 2010; Hazra et al. 2014; Horita et al. 2014). The sensitivity of BlaC-type enzymes to inactivation by clavulanate (Tremblay et al. 2008) and the diazabicyclooctanone class (Xu et al. 2012; Dubée et al. 2015) indicates promise for β-lactam pairing. What remains to be seen is whether clinical use would give a BlaC mutation that diminishes the efficacy of clavulanate (Veziris et al. 2011; Feiler et al. 2013; Kurz et al. 2013; Egesborg et al. 2015; Soroka et al. 2015), and whether a diazabicyclooctanone derivative will be identified having adequate clinical reactivity. Among the exploratory pairings reported are meropenem-clavulanate, faropenem-clavulanate, amoxacillin-clavulanate, ceftaroline-clavulanate, ceftaroline-avibactam, and meropenem–sulbactam (Hugonnet et al. 2009; Gonzalo and Drobniewski 2013; Solapure et al. 2013; Dhar et al. 2015; Dubée et al. 2015; Zhang et al. 2015). Although some data for these pairings are promising, it is certain that such pairs will require incorporation into a multidrug regimen. The composition of such regimen may correspond to proven, emerging, or new drugs as may be identified against new targets (some of unknown identity) as recognized by genomic synthetic lethality screening (Lun et al. 2014).

CONCLUSION

The power of the β-lactams as antibiotics has been diminished, but surely not lost, by the breadth of resistance mechanisms now encountered in both the Gram-positive and -negative bacteria. The most recent generations of cephalosporin and carbapenem structures (in some cases, now paired with β-lactamase inhibitors), in particular, show promise against many of the most resistance-capable bacteria now encountered. This outcome argues that the structural space of the β-lactams needed to subvert target-based (and other) resistance mechanisms, as has been discussed here for the Gram-positive bacteria, is not exhausted. We now understand that the target modification of the Gram-positive PBPs does not stand alone but is supported by underlying pathways that may secure synergy with β-lactams if cocompromised. Even a bacterium historically regarded as impervious to the β-lactams, M. tuberculosis, may be made vulnerable. Yet, even with the molecular dissection of the resistant targets and resistance pathways, the future exploration of the structural space around the β-lactams will still demand investment in empirical structure-activity development at a time when commercial interest in empirical antibiotic discovery has waned. This

obligation is no less true for the complementary targets of these bacteria. We arguably are entering an era in which the most intractable bacteria will be treated with multiantibiotic regimens, as always has been the case for M. tuberculosis. It is small comfort to offer assurance that when the impasse with respect to investing in antibiotic discovery and development is surmounted, the targets and strategies to secure the future place of yet-undiscovered β-lactam antibiotics will be in place. The substance of this review attests to this assurance, although we are compelled to omit the identity of such structures.

ACKNOWLEDGMENTS

The authors are supported by grants from the National Institutes of Health (AI104987 and GM61629).

REFERENCES

Albarracin Orio AG, Pinas GE, Cortes PR, Cian MB, Echenique J. 2011. Compensatory evolution of *pbp* mutations restores the fitness cost imposed by β-lactam resistance in *S. pneumoniae*. PLoS Pathog **7**: e1002000.

Alderwick LJ, Harrison J, Lloyd GS, Birch HL. 2015. The mycobacterial cell wall—Peptidoglycan and arabinogalactan. *Cold Spring Harb Perspec Med* **5**: a021113.

Amoroso A, Boudet J, Berzigotti S, Duval S, Teller N, Mengin-Lecreulx D, Luxen A, Simorre JP, Joris B. 2012. A peptidoglycan fragment triggers β-lactam resistance in *B. licheniformis*. PLoS Pathog **8**: e1002571.

Angoulvant F, Cohen R, Doit C, Elbez A, Werner A, Béchet S, Bonacorsi S, Varon E, Levy C. 2015. Trends in antibiotic resistance of *S. pneumoniae* and *H. influenzae* isolated from nasopharyngeal flora in children with acute otitis media in France before and after 13 valent pneumococcal conjugate vaccine introduction. *BMC Infect Dis* **15**: 236.

Antignac A, Tomasz A. 2009. Reconstruction of the phenotypes of methicillin-resistant *S. aureus* by replacement of the staphylococcal cassette chromosome *mec* with a plasmid-borne copy of *S. sciuri pbpD* gene. *Antimicrob Agents Chemother* **53**: 435–441.

Arbeloa A, Hugonnet JE, Sentilhes AC, Josseaume N, Dubost L, Monsempes C, Blanot D, Brouard JP, Arthur M. 2004a. Synthesis of mosaic peptidoglycan cross-bridges by hybrid peptidoglycan assembly pathways in Gram-positive bacteria. *J Biol Chem* **279**: 41546–41556.

Arbeloa A, Segal H, Hugonnet JE, Josseaume N, Dubost L, Brouard JP, Gutmann L, Mengin-Lecreulx D, Arthur M. 2004b. Role of class A PBPs in PBP5-mediated β-lactam resistance in *E. faecalis*. *J Bacteriol* **186**: 1221–1228.

Arias CA, Murray BE. 2012. The rise of the *Enterococcus*: Beyond vancomycin resistance. *Nat Rev Microbiol* **10**: 266–278.

Atilano ML, Pereira PM, Yates J, Reed P, Veiga H, Pinho MG, Filipe SR. 2010. Teichoic acids are temporal and spatial regulators of peptidoglycan cross-linking in *S. aureus*. *Proc Natl Acad Sci* **107:** 18991–18996.

Bailey RG, Turner RD, Mullin N, Clarke N, Foster SJ, Hobbs JK. 2014. The interplay between cell wall mechanical properties and the cell cycle in *S. aureus*. *Biophys J* **107:** 2538–2545.

Bansal A, Kar D, Murugan RA, Mallick S, Dutta M, Pandey SD, Chowdhury C, Ghosh AS. 2015. A putative low-molecular mass (LMM) PBP of *M. smegmatis* exhibits prominent physiological characters of DD-carboxypeptidase and β-lactamase. *Microbiology* **161:** 1081–1091.

Berg KH, Stamsås GA, Straume D, Håvarstein LS. 2013. Effects of low PBP2b levels on cell morphology and peptidoglycan composition in *S. pneumoniae* R6. *J Bacteriol* **195:** 4342–4354.

Berg KH, Straume D, Håvarstein LS. 2014. The function of the transmembrane and cytoplasmic domains of pneumococcal PBP2x and PBP2b extends beyond that of simple anchoring devices. *Microbiology* **160:** 1585–1598.

Biarrotte-Sorin S, Hugonnet JE, Delfosse V, Mainardi JL, Gutmann L, Arthur M, Mayer C. 2006. Crystal structure of a novel β-lactam-insensitive peptidoglycan transpeptidase. *J Mol Biol* **359:** 533–538.

Boersma MJ, Kuru E, Rittichier JT, VanNieuwenhze MS, Brun YV, Winkler ME. 2015. Minimal peptidoglycan (PG) turnover in wild-type and pg hydrolase and cell division mutants of *S. pneumoniae* D39 growing planktonically and in host-relevant biofilms. *J Bacteriol* **197:** 3472–3485.

Bouley R, Kumarasiri M, Peng Z, Otero LH, Song W, Suckow MA, Schroeder VA, Wolter WR, Lastochkin E, Antunes NT, et al. 2015. Discovery of antibiotic (*E*)-3-(3-carboxyphenyl)-2-(4-cyanostyryl)quinazolin-4(3H)-one. *J Am Chem Soc* **137:** 1738–1741.

Brammer Basta LA, Ghosh A, Pan Y, Jakoncic J, Lloyd EP, Townsend CA, Lamichhane G, Bianchet M. 2015. A loss of a functionally and structurally distinct LD-transpeptidase, LdtMt5, compromises cell wall integrity in *M. tuberculosis*. *J Biol Chem* **290:** 25670–25685.

Bugg TD, Braddick D, Dowson CG, Roper DI. 2011. Bacterial cell wall assembly: Still an attractive antibacterial target. *Trends Biotechnol* **29:** 167–173.

Bush K. 2015. Antibiotics: Synergistic MRSA combinations. *Nat Chem Biol* **11:** 832–833.

Carapito R, Chesnel L, Vernet T, Zapun A. 2006. Pneumococcal β-lactam resistance due to a conformational change in PBP2x. *J Biol Chem* **281:** 1771–1777.

Cattoir V, Leclercq R. 2013. Twenty-five years of shared life with vancomycin-resistant enterococci: Is it time to divorce? *J Antimicrob Chemother* **68:** 731–742.

Cazares-Dominguez V, Cruz-Cordova A, Ochoa SA, Escalona G, Arellano-Galindo J, Rodriguez-Leviz A, Hernandez-Castro R, Lopez-Villegas EO, Xicohtencatl-Cortes J. 2015. Vancomycin-tolerant, methicillin-resistant *S. aureus* reveals the effects of vancomycin on cell wall thickening. *PLoS ONE* **10:** e0118791.

Chakraborty S, Rhee KY. 2015. Tuberculosis drug development: History and evolution of the mechanism-based paradigm. *Cold Spring Harb Perspect Med* **5:** a021147.

Chambers HF. 2005. Community-associated MRSA—Resistance and virulence converge. *N Engl J Med* **352:** 1485–1487.

Chambers HF, Turner J, Schecter G, Kawamura M, Hopewell PC. 2005. Imipenem for treatment of tuberculosis in mice and humans. *Antimicrob Agents Chemother* **49:** 2816–2821.

Cheng Y, Xie H, Sule P, Hassounah H, Graviss EA, Kong Y, Cirillo JD, Rao J. 2014. Fluorogenic probes with substitutions at the 2 and 7 positions of cephalosporin are highly BlaC-specific for rapid *M. tuberculosis* detection. *Angew Chem, Int Ed* **53:** 9360–9364.

Chesnel L, Pernot L, Lemaire D, Champelovier D, Croizé J, Dideberg O, Vernet T, Zapun A. 2003. The structural modifications induced by the M339F substitution in PBP2x from *St. pneumoniae* further decreases the susceptibility to β-lactams of resistant strains. *J Biol Chem* **278:** 44448–44456.

Chewapreecha C, Marttinen P, Croucher NJ, Salter SJ, Harris SR, Mather AE, Hanage WP, Goldblatt D, Nosten FH, Turner C, et al. 2014. Comprehensive identification of single nucleotide polymorphisms associated with β-lactam resistance within pneumococcal mosaic genes. *PLoS Genet* **10:** e1004547.

Coffey TJ, Daniels M, McDougal LK, Dowson CG, Tenover FC, Spratt BG. 1995. Genetic analysis of clinical isolates of *S. pneumoniae* with high-level resistance to expanded-spectrum cephalosporins. *Antimicrob Agents Chemother* **39:** 1306–1313.

Contreras-Martel C, Job V, Di Guilmi AM, Vernet T, Dideberg O, Dessen A. 2006. Crystal structure of PBP1a reveals a mutational hotspot implicated in β-lactam resistance in *S. pneumoniae*. *J Mol Biol* **355:** 684–696.

Contreras-Martel C, Dahout-Gonzalez C, Martins Ados S, Kotnik M, Dessen A. 2009. PBP active site flexibility as the key mechanism for β-lactam resistance in pneumococci. *J Mol Biol* **387:** 899–909.

Cordillot M, Dubee V, Triboulet S, Dubost L, Marie A, Hugonnet JE, Arthur M, Mainardi JL. 2013. In vitro cross-linking of *M. tuberculosis* peptidoglycan by LD-transpeptidases and inactivation of these enzymes by carbapenems. *Antimicrob Agents Chemother* **57:** 5940–5945.

Cremniter J, Mainardi JL, Josseaume N, Quincampoix JC, Dubost L, Hugonnet JE, Marie A, Gutmann L, Rice LB, Arthur M. 2006. Novel mechanism of resistance to glycopeptide antibiotics in *E. faecium*. *J Biol Chem* **281:** 32254–32262.

Croucher NJ, Harris SR, Fraser C, Quail MA, Burton J, van der Linden M, McGee L, von Gottberg A, Song JH, Ko KS, et al. 2011. Rapid pneumococcal evolution in response to clinical interventions. *Science* **331:** 430–434.

Dare K, Ibba M. 2012. Roles of tRNA in cell wall biosynthesis. *Wiley Interdisc Rev RNA* **3:** 247–264.

de Lencastre H, Oliveira D, Tomasz A. 2007. Antibiotic resistant *S. aureus*: A paradigm of adaptive power. *Curr Opin Microbiol* **10:** 428–435.

Dessen A, Mouz N, Gordon E, Hopkins J, Dideberg O. 2001. Crystal structure of PBP2x from a highly penicillin-resistant *S. pneumoniae* clinical isolate: A mosaic framework containing 83 mutations. *J Biol Chem* **276:** 45106–45112.

Dhar N, Dubée V, Ballell L, Cuinet G, Hugonnet JE, Signorino-Gelo F, Barros D, Arthur M, McKinney JD. 2015.

Rapid cytolysis of *M. tuberculosis* by faropenem, an orally bioavailable β-lactam antibiotic. *Antimicrob Agents Chemother* **59:** 1308–1319.

Dias R, Felix D, Canica M, Trombe MC. 2009. The highly conserved serine threonine kinase StkP of *S. pneumoniae* contributes to penicillin susceptibility independently from genes encoding PBPs. *BMC Microbiol* **9:** 121.

Djorić D, Kristich CJ. 2015. Oxidative stress enhances cephalosporin resistance of *E. faecalis* through activation of a two-component signaling system. *Antimicrob Agents Chemother* **59:** 159–169.

Dubée V, Arthur M, Fief H, Triboulet S, Mainardi JL, Gutmann L, Sollogoub M, Rice LB, Etheve-Quelquejeu M, Hugonnet JE. 2012. Kinetic analysis of *E. faecium* L,D-transpeptidase inactivation by carbapenems. *Antimicrob Agents Chemother* **56:** 3409–3412.

Dubée V, Bernut A, Cortes M, Lesne T, Dorchene D, Lefebvre AL, Hugonnet JE, Gutmann L, Mainardi JL, Herrmann JL, et al. 2015. β-Lactamase inhibition by avibactam in *M. abscessus*. *J Antimicrob Chemother* **70:** 1051–1058.

Egesborg P, Carlettini H, Volpato JP, Doucet N. 2015. Combinatorial active-site variants confer sustained clavulanate resistance in BlaC β-lactamase from *M. tuberculosis*. *Protein Sci* **24:** 534–544.

Falk SP, Weisblum B. 2013. Phosphorylation of the *S. pneumoniae* cell wall biosynthesis enzyme MurC by a eukaryotic-like ser/thr kinase. *FEMS Microbiol Lett* **340:** 19–23.

Feiler C, Fisher AC, Boock JT, Marrichi MJ, Wright L, Schmidpeter PA, Blankenfeldt W, Pavelka M, Delisa MP. 2013. Directed evolution of *M. tuberculosis* β-lactamase reveals gatekeeper residue that regulates antibiotic resistance and catalytic efficiency. *PLoS ONE* **8:** e73123.

Fernandez R, Paz LI, Rosato RR, Rosato AE. 2014. Ceftaroline is active against heteroresistant methicillin-resistant *Staphylococcus aureus* clinical strains despite associated mutational mechanisms and intermediate levels of resistance. *Antimicrob Agents Chemother* **58:** 5736–5746.

Filipe SR, Severina E, Tomasz A. 2002. The *murMN* operon: A functional link between antibiotic resistance and antibiotic tolerance in *S. pneumoniae*. *Proc Natl Acad Sci* **99:** 1550–1555.

Fischetti VA, Ryan P. 2015. *Streptococcus*. In *Practical handbook of microbiology*, 3rd ed. (ed. Goldman E, Green LH), pp. 411–427. CRC, Boca Raton, FL.

Fishovitz J, Rojas-Altuve A, Otero LH, Dawley M, Carrasco-Lopez C, Chang M, Hermoso JA, Mobashery S. 2014. Disruption of allosteric response as an unprecedented mechanism of resistance to antibiotics. *J Am Chem Soc* **136:** 9814–9817.

Fishovitz J, Taghizadeh N, Fisher JF, Chang M, Mobashery S. 2015. The Tipper-Strominger hypothesis and triggering of allostery in PBP2a of methicillin-resistant *S. aureus* (MRSA). *J Am Chem Soc* **137:** 6500–6505.

Fleurie A, Manuse S, Zhao C, Campo N, Cluzel C, Lavergne JP, Freton C, Combet C, Guiral S, Soufi B, et al. 2014. Interplay of the serine/threonine-kinase StkP and the paralogs DivIVA and GpsB in pneumococcal cell elongation and division. *PLoS Genet* **10:** e1004275.

Foucault ML, Depardieu F, Courvalin P, Grillot-Courvalin C. 2010. Inducible expression eliminates the fitness cost of vancomycin resistance in *Enterococci*. *Proc Natl Acad Sci* **107:** 16964–16969.

Gautam S, Kim T, Spiegel DA. 2015. Chemical probes reveal an extraseptal mode of cross-linking in *S. aureus*. *J Am Chem Soc* **137:** 7441–7447.

Ge Y, Wu J, Xia Y, Yang M, Xiao J, Yu J. 2012. Molecular dynamics simulation of the complex PBP-2x with drug cefuroxime to explore the drug resistance mechanism of *S. suis* R61. *PLoS ONE* **7:** e35941.

Gilmore MS, Rauch M, Ramsey MM, Himes PR, Varahan S, Manson JM, Lebreton F, Hancock LE. 2015. Pheromone killing of multidrug-resistant *E. faecalis* V583 by native commensal strains. *Proc Natl Acad Sci* **112:** 7273–7278.

Goffin C, Ghuysen JM. 2002. Biochemistry and comparative genomics of SxxK superfamily acyltransferases offer a clue to the mycobacterial paradox: Presence of penicillin-susceptible target proteins versus lack of efficiency of penicillin as therapeutic agent. *Microbiol Mol Biol Rev* **66:** 702–738.

Gonzales PR, Pesesky MW, Bouley R, Ballard A, Biddy BA, Suckow MA, Wolter WR, Schroeder VA, Burnham CD, Mobashery S, et al. 2015. Synergistic, collaterally sensitive β-lactam combinations suppress resistance in MRSA. *Nat Chem Biol* **11:** 855–861.

Gonzalo X, Drobniewski F. 2013. Is there a place for β-lactams in the treatment of multidrug-resistant/extensively drug-resistant tuberculosis? Synergy between meropenem and amoxicillin/clavulanate. *J Antimicrob Chemother* **68:** 366–369.

Gordon E, Mouz N, Duée E, Dideberg O. 2000. The crystal structure of the penicillin-binding protein 2x from *S. pneumoniae* and its acyl-enzyme form: Implication in drug resistance. *J Mol Biol* **299:** 477–485.

Greene NG, Narciso AR, Filipe SR, Camilli A. 2015. Peptidoglycan branched stem peptides contribute to *S. pneumoniae* virulence by inhibiting pneumolysin release. *PLoS Pathog* **11:** e1004996.

Gupta R, Lavollay M, Mainardi JL, Arthur M, Bishai WR, Lamichhane G. 2010. The *M. tuberculosis* protein LdtMt2 is a nonclassical transpeptidase required for virulence and resistance to amoxicillin. *Nat Med* **16:** 466–469.

Hakenbeck R. 2014. Discovery of β-lactam-resistant variants in diverse pneumococcal populations. *Genome Med* **6:** 72.

Hakenbeck R, Bruckner R, Denapaite D, Maurer P. 2012. Molecular mechanisms of β-lactam resistance in *S. pneumoniae*. *Future Microbiol* **7:** 395–410.

Hazra S, Xu H, Blanchard JS. 2014. Tebipenem, a new carbapenem antibiotic, is a slow substrate that inhibits the β-lactamase from *M. tuberculosis*. *Biochemistry* **53:** 3671–3678.

Hendrickx APA, van Schaik W, Willems RJL. 2013. The cell wall architecture of *E. faecium*: From resistance to pathogenesis. *Future Microbiol* **8:** 993–1010.

Henriques-Normark B, Tuomanen EI. 2013. The pneumococcus: Epidemiology, microbiology, and pathogenesis. *Cold Spring Harb Perspect Med* **3:** a010215.

Henry X, Verlaine O, Amoroso A, Coyette J, Frere JM, Joris B. 2013. Activity of ceftaroline against *E. faecium* PBP5. *Antimicrob Agents Chemother* **57:** 6358–6360.

Hindler JA, Wong-Beringer A, Charlton CL, Miller SA, Kelesidis T, Carvalho M, Sakoulas G, Nonejuie P, Pogliano J, Nizet V, et al. 2015. In vitro activity of daptomycin in combination with β-lactams, gentamicin, rifampin, and tigecycline against daptomycin-nonsusceptible enterococci. *Antimicrob Agents Chemother* **59**: 4279–4288.

Holmes NE, Howden BP. 2014. What's new in the treatment of serious MRSA infection? *Curr Opin Infect Dis* **27**: 471–478.

Horita Y, Maeda S, Kazumi Y, Doi N. 2014. In vitro susceptibility of *M. tuberculosis* isolates to an oral carbapenem alone or in combination with β-lactamase inhibitors. *Antimicrob Agents Chemother* **58**: 7010–7014.

Hugonnet JE, Tremblay LW, Boshoff HI, Barry CE, Blanchard JS. 2009. Meropenem-clavulanate is effective against extensively drug-resistant *M. tuberculosis*. *Science* **323**: 1215–1218.

Ip M, Ang I, Liyanapathirana V, Ma H, Lai R. 2015. Genetic analyses of PBP determinants in multidrug-resistant *S. pneumoniae* serogroup 19 CC320/271 clone with high-level resistance to third-generation cephalosporins. *Antimicrob Agents Chemother* **59**: 4040–4045.

Jackson M, McNeil MR, Brennan PJ. 2013. Progress in targeting cell envelope biogenesis in *Mycobacterium tuberculosis*. *Future Microbiol* **8**: 855–875.

Jensen A, Valdórsson O, Frimodt-Møller N, Hollingshead S, Kilian M. 2015. Commensal streptococci serve as a reservoir for β-lactam resistance genes in *S. pneumoniae*. *Antimicrob Agents Chemother* **59**: 3529–3540.

Job V, Carapito R, Vernet T, Dessen A, Zapun A. 2008. Common alterations in PBP1a from resistant *S. pneumoniae* decrease its reactivity toward β-lactams: Structural insights. *J Biol Chem* **283**: 4886–4894.

Kim SJ, Chang J, Singh M. 2015. Peptidoglycan architecture of Gram-positive bacteria by solid-state NMR. *Biochim Biophys Acta* **1848**: 350–362.

Knox J, Uhlemann AC, Lowy FD. 2015. *S. aureus* infections: Transmission within households and the community. *Trends Microbiol* **23**: 437–444.

Kocaoglu O, Tsui HCT, Winkler ME, Carlson EE. 2015. Profiling of β-lactam selectivity for penicillin-binding proteins in *S. pneumoniae* D39. *Antimicrob Agents Chemother* **59**: 3548–3555.

Kristich CJ, Rice LB, Arias CA. 2014. Enterococcal infection—Treatment and antibiotic resistance. In *Enterococci: From commensals to leading causes of drug resistant infection* (ed. Gilmore MS, Clewell DB, Ike Y, Shankar N), pp. 1–46. Massachusetts Eye and Ear Infirmary, Boston.

Kumar P, Arora K, Lloyd JR, Lee IY, Nair V, Fischer E, Boshoff HI, Barry CE. 2012. Meropenem inhibits D,D-carboxypeptidase activity in *M. tuberculosis*. *Mol Microbiol* **86**: 367–381.

Kurz SG, Wolff KA, Hazra S, Bethel CR, Hujer AM, Smith KM, Xu Y, Tremblay LW, Blanchard JS, Nguyen L, et al. 2013. Can inhibitor resistant substitutions in the *M. tuberculosis* β-lactamase BlaC lead to clavulanate resistance? A biochemical rationale for the use of β-lactam β-lactamase inhibitor combinations. *Antimicrob Agents Chemother* **57**: 6085–6096.

Ladjouzi R, Bizzini A, Lebreton F, Sauvageot N, Rincé A, Benachour A, Hartke A. 2013. Analysis of the tolerance of pathogenic *Enterococci* and *S. aureus* to cell wall active antibiotics. *J Antimicrob Chemother* **68**: 2083–2091.

Lages MC, Beilharz K, Morales Angeles D, Veening J-W, Scheffers DJ. 2013. The localization of key *Bacillus subtilis* penicillin binding proteins during cell growth is determined by substrate availability. *Environ Microbiol* **15**: 3272–3281.

Land AD, Tsui HC, Kocaoglu O, Vella SA, Shaw SL, Keen SK, Sham LT, Carlson EE, Winkler ME. 2013. Requirement of essential Pbp2x and GpsB for septal ring closure in *S. pneumoniae* D39. *Mol Microbiol* **90**: 939–955.

Lavollay M, Arthur M, Fourgeaud M, Dubost L, Marie A, Veziris N, Blanot D, Gutmann L, Mainardi JL. 2008. The peptidoglycan of stationary-phase *M. tuberculosis* predominantly contains cross-links generated by L,D-transpeptidation. *J Bacteriol* **190**: 4360–4366.

Lebreton F, van Schaik W, McGuire AM, Godfrey P, Griggs A, Mazumdar V, Corander J, Cheng L, Saif S, Young S, et al. 2013. Emergence of epidemic multidrug-resistant *E. faecium* from animal and commensal strains. *MBio* **4**: e00534-13.

Lecoq L, Dubée V, Triboulet S, Bougault C, Hugonnet JE, Arthur M, Simorre JP. 2013. Structure of *E. faecium* L,D-transpeptidase acylated by ertapenem provides insight into the inactivation mechanism. *ACS Chem Biol* **8**: 1140–1146.

Leski TA, Tomasz A. 2005. Role of PBP2 in the antibiotic susceptibility and cell wall cross-linking of *S. aureus*: Evidence for the cooperative functioning of PBP2, PBP4, and PBP2A. *J Bacteriol* **187**: 1815–1824.

Llarrull LI, Mobashery S. 2012. Dissection of events in the resistance to β-lactam antibiotics mediated by the protein BlaR1 from *S. aureus*. *Biochemistry* **51**: 4642–4649.

Lobritz MA, Belenky P, Porter CBM, Gutierrez A, Yang JH, Schwarz EG, Dwyer DJ, Khalil AS, Collins JJ. 2015. Antibiotic efficacy is linked to bacterial cellular respiration. *Proc Natl Acad Sci* **112**: 8173–8180.

Loskill P, Pereira PM, Jung P, Bischoff M, Herrmann M, Pinho MG, Jacobs K. 2014. Reduction of the peptidoglycan crosslinking causes a decrease in stiffness of the *S. aureus* cell envelope. *Biophys J* **107**: 1082–1089.

Lun S, Miranda D, Kubler A, Guo H, Maiga MC, Winglee K, Pelly S, Bishai WR. 2014. Synthetic lethality reveals mechanisms of *M. tuberculosis* resistance to β-lactams. *mBio* **5**: e01767.14.

Macheboeuf P, Di Guilmi AM, Job V, Vernet T, Dideberg O, Dessen A. 2005. Active site restructuring regulates ligand recognition in class A PBPs. *Proc Natl Acad Sci* **102**: 577–582.

Maestro B, Novakova L, Hesek D, Lee M, Leyva E, Mobashery S, Sanz JM, Branny P. 2011. Recognition of peptidoglycan and β-lactam antibiotics by the extracellular domain of the Ser/Thr protein kinase StkP from *S. pneumoniae*. *FEBS Lett* **585**: 357–363.

Magnet S, Arbeloa A, Mainardi JL, Hugonnet JE, Fourgeaud M, Dubost L, Marie A, Delfosse V, Mayer C, Rice LB, et al. 2007. Specificity of L,D-transpeptidases from Gram-positive bacteria producing different peptidoglycan chemotypes. *J Biol Chem* **282**: 13151–13159.

Mainardi JL, Fourgeaud M, Hugonnet JE, Dubost L, Brouard JP, Ouazzani J, Rice LB, Gutmann L, Arthur M. 2005. A novel peptidoglycan cross-linking enzyme for a β-lactam-resistant transpeptidation pathway. *J Biol Chem* **280**: 38146–38152.

Mann PA, Muller A, Xiao L, Pereira PM, Yang C, Lee SH, Wang H, Trzeciak J, Schneeweis J, Dos Santos MM, et al. 2013. Murgocil is a highly bioactive staphylococcal-specific inhibitor of the peptidoglycan glycosyltransferase enzyme MurG. *ACS Chem Biol* **8:** 2442–2451.

Massidda O, Novakova L, Vollmer W. 2013. From models to pathogens: How much have we learned about *S. pneumoniae* cell division? *Environ Microbiol* **15:** 3133–3157.

Maurer P, Koch B, Zerfass I, Krauss J, van der Linden M, Frère JM, Contreras-Martel C, Hakenbeck R. 2008. PBP2x of *S. pneumoniae*: Three new mutational pathways for remodelling an essential enzyme into a resistance determinant. *J Mol Biol* **376:** 1403–1416.

Maurer P, Todorova K, Sauerbier J, Hakenbeck R. 2012. Mutations in *S.* PBP2x: Importance of the C-terminal PBP and serine/threonine kinase-associated domains for β-lactam binding. *Microb Drug Resist* **18:** 314–321.

Memmi G, Filipe SR, Pinho MG, Fu Z, Cheung A. 2008. *S. aureus* PBP4 is essential for β-lactam resistance in community-acquired methicillin-resistant strains. *Antimicrob Agents Chemother* **52:** 3955–3966.

Missiakas D, Schneewind O. 2015. *S. aureus* and related staphylococci. In *Practical handbook of microbiology*, 3rd ed. (ed. Goldman E, Green LH), pp. 383–409. CRC, Boca Raton, FL.

Moellering RC Jr. 2012. MRSA: The first half century. *J Antimicrob Chemother* **67:** 4–11.

Morlot C, Bayle L, Jacq M, Fleurie A, Tourcier G, Galisson F, Vernet T, Grangeasse C, Di Guilmi AM. 2013. Interaction of PBP2x and Ser/Thr protein kinase StkP, two key players in *S. pneumoniae* R6 morphogenesis. *Mol Microbiol* **90:** 88–102.

Mouz N, Di Guilmi AM, Gordon E, Hakenbeck R, Dideberg O, Vernet T. 1999. Mutations in the active site of penicillin-binding protein PBP2x from *S. pneumoniae*. Role in the specificity for β-lactam antibiotics. *J Biol Chem* **274:** 19175–19180.

Nataraj V, Varela C, Javid A, Singh A, Besra GS, Bhatt A. 2015. Mycolic acids: Deciphering and targeting the Achilles' heel of the tubercle bacillus. *Mol Microbiol* **98:** 7–16.

Nobbs AH, Jenkinson HF, Everett DB. 2015. Generic determinants of *Streptococcus* colonization and infection. *Infect Genet Evol* **33:** 361–370.

Ogawara H. 2015. PBPs in *Actinobacteria*. *J Antibiot* **68:** 223–245.

Oliveira DC, de Lencastre H. 2011. Methicillin-resistance in *S. aureus* is not affected by the overexpression in *trans* of the *mecA* gene repressor: A surprising observation. *PLoS ONE* **6:** e23287.

Otero LH, Rojas-Altuve A, Llarrull LI, Carrasco-López C, Kumarasiri M, Lastochkin E, Fishovitz J, Dawley M, Hesek D, Lee M, et al. 2013. How allosteric control of *S. aureus* PBP2a enables methicillin resistance and physiological function. *Proc Natl Acad Sci* **110:** 16808–16813.

Palmer KL, Kos VN, Gilmore MS. 2010. Horizontal gene transfer and the genomics of enterococcal antibiotic resistance. *Curr Opin Microbiol* **13:** 632–639.

Pasquina LW, Santa Maria JP, Walker S. 2013. Teichoic acid biosynthesis as an antibiotic target. *Curr Opin Microbiol* **16:** 531–537.

Peacock SJ, Paterson GK. 2015. Mechanisms of methicillin resistance in *S. aureus*. *Annu Rev Biochem* **84:** 577–601.

Peters K, Schweizer I, Beilharz K, Stahlmann C, Veening JW, Hakenbeck R, Denapaite D. 2014. *S pneumoniae* PBP2x mid-cell localization requires the C-terminal PASTA domains and is essential for cell shape maintenance. *Mol Microbiol* **92:** 733–755.

Peyrani P, Ramirez J. 2015. What is the best therapeutic approach to methicillin-resistant *Staphylococcus aureus* pneumonia. *Curr Opin Infect Dis* **28:** 164–170.

Pfeffer JM, Strating H, Weadge JT, Clarke AJ. 2006. Peptidoglycan *O*-acetylation and autolysin profile of *E. faecalis* in the viable but nonculturable state. *J Bacteriol* **188:** 902–908.

Philippe J, Vernet T, Zapun A. 2014. The elongation of ovococci. *Microb Drug Resist* **20:** 215–221.

Philippe J, Gallet B, Morlot C, Denapaite D, Hakenbeck R, Chen Y, Vernet T, Zapun A. 2015. Mechanism of β-lactam action in *S. pneumoniae*: The piperacillin paradox. *Antimicrob Agents Chemother* **59:** 609–621.

Pietta E, Montealegre MC, Roh JH, Cocconcelli PS, Murray BE. 2014. *E faecium* PBP5-S/R, the missing link between PBP5-S and PBP5-R. *Antimicrob Agents Chemother* **58:** 6978–6981.

Pinho MG, Filipe SR, de Lencastre H, Tomasz A. 2001. Complementation of the essential peptidoglycan transpeptidase function of PBP2 by the drug resistance protein PBP2A in *S. aureus*. *J Bacteriol* **183:** 6525–6531.

Prigozhin DM, Krieger IV, Huizar JP, Mavrici D, Waldo GS, Hung L, Sacchettini JC, Terwilliger TC, Alber T. 2014. Subfamily-specific variations in the structures of two PBPs from *M. tuberculosis*. *PLoS ONE* **9:** e116249.

Qamar A, Golemi-Kotra D. 2012. Dual roles of FmtA in *S. aureus* cell wall biosynthesis and autolysis. *Antimicrob Agents Chemother* **56:** 3797–3805.

Qiao Y, Lebar MD, Schirner K, Schaefer K, Tsukamoto H, Kahne D, Walker S. 2014. Detection of lipid-linked peptidoglycan precursors by exploiting an unexpected transpeptidase reaction. *J Am Chem Soc* **136:** 14678–14681.

Ramalingam J, Vennila J, Subbiah P. 2013. Computational studies on the resistance of PBP2B of wild-type and mutant strains of *S. pneumoniae* against β-lactam antibiotics. *Chem Biol Drug Des* **82:** 275–289.

Reed P, Atilano ML, Alves R, Hoiczyk E, Sher X, Reichmann NT, Pereira PM, Roemer T, Filipe SR, Pereira-Leal JB, et al. 2015. *S. aureus* survives with a minimal peptidoglycan synthesis machine but sacrifices virulence and antibiotic resistance. *PLoS Pathog* **11:** e1004891.

Reinert RR. 2009. The antimicrobial resistance profile of *S. pneumoniae*. *Clin Microbiol Infect* **15:** 7–11.

Rice LB, Carias LL, Rudin S, Hutton R, Marshall S, Hassan M, Josseaume N, Dubost L, Marie A, Arthur M. 2009. Role of class A PBPs in the expression of β-lactam resistance in *E. faecium*. *J Bacteriol* **191:** 3649–3656.

Roemer T, Davies J, Giaever G, Nislow C. 2012. Bugs, drugs and chemical genomics. *Nat Chem Biol* **8:** 46–56.

Sacco E, Cortes M, Josseaume N, Rice LB, Mainardi JL, Arthur M. 2014. Serine/threonine protein phosphatase-mediated control of the peptidoglycan cross-linking L,D-transpeptidase pathway in *E. faecium*. *mBio* **5:** e01446.14.

Sauerbier J, Maurer P, Rieger M, Hakenbeck R. 2012. *S pneumoniae* R6 interspecies transformation: Genetic analysis of penicillin resistance determinants and genome-wide recombination events. *Mol Microbiol* **86:** 692–706.

Sauvage E, Kerff F, Fonze E, Herman R, Schoot B, Marquette JP, Taburet Y, Prevost D, Dumas J, Leonard G, et al. 2002. The 2.4-Å crystal structure of the PBP-binding protein PBP5fm from *E. faecium* in complex with benzylpenicillin. *Cell Mol Life Sci* **59:** 1223–1232.

Sauvage E, Kerff F, Terrak M, Ayala JA, Charlier P. 2008. The PBPs: Structure and role in peptidoglycan biosynthesis. *FEMS Microbiol Rev* **32:** 234–258.

Schneider T, Sahl HG. 2010. An oldie but a goodie—Cell wall biosynthesis as antibiotic target pathway. *Int J Med Microbiol* **300:** 161–169.

Schoonmaker MK, Bishai WR, Lamichhane G. 2014. Nonclassical transpeptidases of *M. tuberculosis* alter cell size, morphology, the cytosolic matrix, protein localization, virulence, and resistance to β-lactams. *J Bacteriol* **196:** 1394–1402.

Seung KJ, Keshavjee S, Rich ML. 2015. Multidrug-resistant tuberculosis and extensively drug-resistant tuberculosis. *Cold Spring Harb Perspect Med* **5:** a017863.

Sewell EWC, Brown ED. 2014. Taking aim at wall teichoic acid synthesis: New biology and new leads for antibiotics. *J Antibiot (Tokyo)* **67:** 43–51.

Silhavy TJ, Kahne D, Walker S. 2010. The bacterial cell envelope. *Cold Spring Harb Perspect Biol* **2:** a000414.

Silver LL. 2013. Viable screening targets related to the bacterial cell wall. *Ann NY Acad Sci* **1277:** 29–53.

Singh M, Kim SJ, Sharif S, Preobrazhenskaya M, Schaefer J. 2015. REDOR constraints on the peptidoglycan lattice architecture of *S. aureus* and its FemA mutant. *Biochim Biophys Acta* **1848:** 363–368.

Smith JR, Barber KE, Raut A, Aboutaleb M, Sakoulas G, Rybak MJ. 2015a. β-Lactam combinations with daptomycin provide synergy against vancomycin-resistant *E. faecalis* and *E. faecium. J Antimicrob Chemother* **70:** 1738–1743.

Smith JR, Barber KE, Raut A, Rybak MJ. 2015b. β-Lactams enhance daptomycin activity against vancomycin-resistant *E. faecalis* and *E. faecium* in in vitro pharmacokinetic/pharmacodynamic models. *Antimicrob Agents Chemother* **59:** 2842–2848.

Solapure S, Dinesh N, Shandil R, Ramachandran V, Sharma S, Bhattacharjee D, Ganguly S, Reddy J, Ahuja V, Panduga V, et al. 2013. In vitro and in vivo efficacy of β-lactams against replicating and slowly growing/nonreplicating *M. tuberculosis. Antimicrob Agents Chemother* **57:** 2506–2510.

Sorg RA, Veening JW. 2015. Microscale insights into pneumococcal antibiotic mutant selection windows. *Nat Commun* **6:** 8773.

Soroka D, Li de la Sierra-Gallay I, Dubée V, Triboulet S, van Tilbeurgh H, Compain F, Ballell L, Barros D, Mainardi JL, Hugonnet JE, et al. 2015. Hydrolysis of clavulanate by *Mycobacterium tuberculosis* β-lactamase BlaC harboring a canonical SDN motif. *Antimicrob Agents Chemother* **59:** 5714–5720.

Spratt BG. 2012. The 2011 Garrod lecture: From penicillin-binding proteins to molecular epidemiology. *J Antimicrob Chem* **67:** 1578–1588.

Starikova I, Al-Haroni M, Werner G, Roberts AP, Sørum V, Nielsen KM, Johnsen PJ. 2013. Fitness costs of various mobile genetic elements in *E. faecium* and *E. faecalis. J Antimicrob Chem* **68:** 2755–2765.

Staude MW, Frederick TE, Natarajan SV, Wilson BD, Tanner CE, Ruggiero ST, Mobashery S, Peng JW. 2015. Investigation of signal transduction routes within the sensor/transducer protein BlaR1 of *S. aureus. Biochemistry* **54:** 1600–1610.

Stryjewski ME, Corey GR. 2014. Methicillin-resistant *S. aureus*: An evolving pathogen. *Clin Infect Dis* **58:** S10–S19.

Tan CM, Therien AG, Lu J, Lee SH, Caron A, Gill CJ, Lebeau-Jacob C, Benton-Perdomo L, Monteiro JM, Pereira PM, et al. 2012. Restoring methicillin-resistant *S. aureus* susceptibility to β-lactam antibiotics. *Sci Transl Med* **4:** 126ra35.

Todorova K, Maurer P, Rieger M, Becker T, Bui NK, Gray J, Vollmer W, Hakenbeck R. 2015. Transfer of penicillin resistance from *S. oralis* to *S. pneumoniae* identifies *murE* as resistance determinant. *Mol Microbiol* **97:** 866–880.

Tomasz A. 1979. The mechanism of the irreversible antimicrobial effects of penicillins: How the β-lactams kill and lyse bacteria. *Annu Rev Microbiol* **33:** 113–137.

Tong SYC, Davis JS, Eichenberger E, Holland TL, Fowler VG Jr. 2015. *S. aureus* infections: Epidemiology, pathophysiology, clinical manifestations, and management. *Clin Microbiol Rev* **28:** 603–661.

Tremblay LW, Hugonnet JE, Blanchard JS. 2008. Structure of the covalent adduct formed between *M. tuberculosis* β-lactamase and clavulanate. *Biochemistry* **47:** 5312–5316.

Tremblay LW, Fan F, Blanchard JS. 2010. Biochemical and structural characterization of *M. tuberculosis* β-lactamase with the carbapenems ertapenem and doripenem. *Biochemistry* **49:** 3766–3773.

Triboulet S, Arthur M, Mainardi JL, Veckerlé C, Dubée V, Nguekam-Moumi A, Gutmann L, Rice LB, Hugonnet JE. 2011. Inactivation kinetics of a new target of β-lactam antibiotics. *J Biol Chem* **286:** 22777–22784.

Triboulet S, Dubee V, Lecoq L, Bougault C, Mainardi JL, Rice LB, Etheve-Quelquejeu M, Gutmann L, Marie A, Dubost L, et al. 2013. Kinetic features of L,D-transpeptidase inactivation critical for β-lactam antibacterial activity. *PLoS ONE* **8:** e67831.

Triboulet S, Bougault CM, Laguri C, Hugonnet JE, Arthur M, Simorre J-P. 2015. Acyl acceptor recognition by *Enterococcus faecium* L,D-transpeptidase Ldt$_{fm}$. *Mol Microbiol* **98:** 90–100.

Tsubakishita S, Kuwahara-Arai K, Sasaki T, Hiramatsu K. 2010. Origin and molecular evolution of the determinant of methicillin resistance in *Staphylococci. Antimicrob Agents Chemother* **54:** 4352–4359.

Tsui HCT, Boersma MJ, Vella SA, Kocaoglu O, Kuru E, Peceny JK, Carlson EE, VanNieuwenhze MS, Brun YV, Shaw SL, et al. 2014. Pbp2x localizes separately from Pbp2b and other peptidoglycan synthesis proteins during later stages of cell division of *S. pneumoniae* D39. *Mol Microbiol* **94:** 21–40.

Cite this article as *Cold Spring Harb Perspect Med* doi: 10.1101/cshperspect.a025221

Turner RD, Ratcliffe EC, Wheeler R, Golestanian R, Hobbs JK, Foster SJ. 2010. Peptidoglycan architecture can specify division planes in *S. aureus*. *Nat Commun* **1:** 1025.

Turner RD, Vollmer W, Foster SJ. 2014. Different walls for rods and balls: The diversity of peptidoglycan. *Mol Microbiol* **91:** 862–874.

van Tonder AJ, Bray JE, Roalfe L, White R, Zancolli M, Quirk SJ, Haraldsson G, Jolley KA, Maiden MCJ, Bentley SD, et al. 2015. Genomics reveals the worldwide distribution of multidrug-resistant serotype 6E pneumococci. *J Clin Microbiol* **53:** 2271–2285.

Veziris N, Truffot C, Mainardi JL, Jarlier V. 2011. Activity of carbapenems combined with clavulanate against murine tuberculosis. *Antimicrob Agents Chemother* **55:** 2597–2600.

Vollmer W, Blanot D, de Pedro MA. 2008. Peptidoglycan structure and architecture. *FEMS Microbiol Rev* **32:** 149–167.

Wang H, Gill CJ, Lee SH, Mann P, Zuck P, Meredith TC, Murgolo N, She X, Kales S, Liang L, et al. 2013. Discovery of wall teichoic acid inhibitors as potential anti-MRSA β-lactam combination agents. *Chem Biol* **20:** 272–284.

Waxman DJ, Strominger JL. 1983. PBPs and the mechanism of action of β-lactam antibiotics. *Annu Rev Biochem* **52:** 825–869.

Werner G, Coque TM, Franz CMA, Grohmann E, Hegstad K, Jensen L, van Schaik W, Weaver K. 2013. Antibiotic resistant *Enterococci*—Tales of a drug resistance gene trafficker. *Int J Med Microbiol* **303:** 360–379.

Werth BJ, Barber KE, Tran KNT, Nonejuie P, Sakoulas G, Pogliano J, Rybak MJ. 2015. Ceftobiprole and ampicillin increase daptomycin susceptibility of daptomycin-susceptible and -resistant VRE. *J Antimicrob Chemother* **70:** 489–493.

Wheeler R, Mesnage S, Boneca IG, Hobbs JK, Foster SJ. 2011. Super-resolution microscopy reveals cell wall dynamics and peptidoglycan architecture in ovococcal bacteria. *Mol Microbiol* **82:** 1096–1109.

Winstel V, Xia G, Peschel A. 2014. Pathways and roles of wall teichoic acid glycosylation in *S. aureus*. *Int J Med Microbiol* **304:** 215–221.

Wivagg CN, Bhattacharyya RP, Hung DT. 2014. Mechanisms of β-lactam killing and resistance in the context of *M. tuberculosis*. *J Antibiot* **67:** 645–654.

Wivagg CN, Wellington S, Gomez JE, Hung DT. 2016. Loss of a class A penicillin-binding protein alters β-lactam susceptibilities in *M. tuberculosis*. *ACS Infect Dis* **2:** 104–110.

Xu H, Hazra S, Blanchard JS. 2012. NXL104 irreversibly inhibits the β-lactamase from *M. tuberculosis*. *Biochemistry* **51:** 4551–4557.

Yamada M, Watanabe T, Baba N, Takeuchi Y, Ohsawa F, Gomi S. 2008. Crystal structures of biapenem and tebipenem complexed with PBPs 2X and 1A from *S. pneumoniae*. *Antimicrob Agents Chemother* **52:** 2053–2060.

Young KD. 2006. The selective value of bacterial shape. *Microbiol Mol Biol Rev* **70:** 660–703.

Young KD. 2007. Bacterial morphology: Why have different shapes? *Curr Opin Microbiol* **10:** 596–600.

Zapun A, Contreras-Martel C, Vernet T. 2008. PPBs and β-lactam resistance. *FEMS Microbiol Rev* **32:** 361–385.

Zerfass I, Hakenbeck R, Denapaite D. 2009. An important site in PBP2x of penicillin-resistant clinical isolates of *S. pneumoniae*: Mutational analysis of Thr338. *Antimicrob Agents Chemother* **53:** 1107–1115.

Zhang X, Paganelli FL, Bierschenk D, Kuipers A, Bonten MJM, Willems RJL, van Schaik W. 2012. Genome-wide identification of ampicillin resistance determinants in *E. faecium*. *PLoS Genet* **8:** e1002804.

Zhang D, Wang Y, Lu J, Pang Y. 2015. In vitro activity of β-lactams in combination with β-lactamase inhibitors against multidrug-resistant *Mycobacterium tuberculosis* isolates. *Antimicrob Agents Chemother* **60:** 393–399.

Zhou Y, Antignac A, Wu SW, Tomasz A. 2008. PBPs and cell wall composition in β-lactam-sensitive and -resistant strains of *S. sciuri*. *J Bacteriol* **190:** 508–514.

Zorzi W, Zhou XY, Dardenne O, Lamotte J, Raze D, Pierre J, Gutmann L, Coyette J. 1996. Structure of the low-affinity PBP5fm in wild-type and highly penicillin-resistant strains of *E. faecium*. *J Bacteriol* **178:** 4948–4957.

β-Lactamases: A Focus on Current Challenges

Robert A. Bonomo[1,2]

[1]Department of Medicine, Case Western Reserve University School of Medicine, Louis Stokes Cleveland Department of Veterans Affairs Medical Center, Cleveland, Ohio 44120

[2]Departments of Pharmacology, Molecular Biology and Microbiology, Biochemistry, and Proteomics and Bioinformatics, Case Western Reserve University School of Medicine, Cleveland, Ohio 44120

Correspondence: robert.bonomo@va.gov

β-Lactamases, the enzymes that hydrolyze β-lactam antibiotics, remain the greatest threat to the usage of these agents. In this review, the mechanism of hydrolysis is discussed for both those β-lactamases that use serine at the active site and those that require divalent zinc ions for hydrolysis. The β-lactamases now include >2100 unique, naturally occurring amino acid sequences. Some of the clinically most important of these are the class A penicillinases, the extended-spectrum β-lactamases (ESBLs), the AmpC cephalosporinases, and the carbapenem-hydrolyzing enzymes in both the serine and metalloenzyme groups. Because of the versatility of these enzymes to evolve as new β-lactams are used therapeutically, new approaches to antimicrobial therapy may be required.

Despite the tremendous advancements in biomedicine, the production of β-lactam hydrolyzing enzymes, β-lactamases, by Gram-negative and -positive bacteria still remains one of the most significant threats to human health (Hauck et al. 2016). With the introduction of every new class of antibiotics, bacteria have continued to evolve resistance, as they are amazingly capable of responding to environmental pressure via selection of existing mutations and acquisition of new genes. The most significant threat has been faced by β-lactam antibiotics. The rapid evolution of β-lactamases, especially carbapenem hydrolyzing enzymes, makes each new drug obsolete in a very short period of time (Bush 2010a,b, 2014; Drawz and Bonomo 2010).

MECHANISM OF β-LACTAM ACTION

To properly appreciate the mechanisms by which β-lactamases have changed the status of β-lactams in our therapeutic armamentarium, it is important to briefly review how β-lactams kill bacteria. β-Lactam antibiotics show their bactericidal effects by inhibiting enzymes involved in cell-wall synthesis, that is, penicillin-binding proteins (PBPs). The integrity of the bacterial cell wall is essential to maintaining cell shape in a hypertonic and hostile environment such as serum, urine, lung mucus, or gastrointestinal tract. Osmotic stability is preserved by a rigid cell wall comprised of alternating N-acetylmuramic acid (NAM) and N-acetylglucosamine (NAG) units. These glycosidic units are linked by a transglycosidases. A pentapeptide is attached to each NAM unit; the PBPs act as transpeptidases to catalyze the cross-linking of two D-alanine-D-alanine NAM pentapeptides. This cross-linking of adjacent glycan strands confers the rigidity of the cell wall. In the 1960s, Strominger realized that the β-lactam ring is sterically similar to the D-alanine-D-alanine of the NAM pentapeptide (Drawz and Bonomo 2010; Fisher and Mobashery 2014; Fishovitz et al. 2015). As a result, PBPs "mistakenly" use the β-lactam as a substrate "building block" during cell-wall synthesis. This "error" results in acylation of the PBP, which renders the enzyme unable to further carry out transpeptidation reactions. As cell-wall synthesis slows, constitutive peptidoglycan autolysis continues as a result of amidases, bacterial autolytic enzymes. The breakdown of the murein sacculus, the peptidoglycan net that surrounds the bacterium, leads to cell-wall compromise and increased permeability. In this way, the β-lactam-mediated inhibition of transpeptidation causes cell lysis.

MECHANISMS OF RESISTANCE TO β-LACTAMS

There are four primary mechanisms by which bacteria can overcome β-lactam antibiotics (Drawz and Bonomo 2010; Papp-Wallace et al. 2011). First, changes in the active site of PBPs can lower the affinity for β-lactam antibiotics and subsequently increase resistance to these agents, such as in PBP2x of *Streptococcus pneumoniae*. In a similar manner, penicillin resistance in *Streptococcus sanguis*, *Streptococcus oralis*, and *Streptococcus mitis* developed from horizontal transfer of a PBP2b gene from *S. pneumoniae*.

Methicillin resistance in *Staphylococcus* spp. is another example of an altered PBP. Although the cause for this resistance is heterogeneous, it is often conferred by acquisition of the *mec* el-

ement, the *mec*A gene, which encodes PBP2a (also denoted PBP2′). This low-affinity transpeptidase can assemble new cell wall in the presence of high concentration of penicillins (i.e., methicillin) and cephalosporins.

Second, to access PBPs on the surface of the inner membrane, β-lactams must either diffuse through or directly traverse porin channels in the outer membrane of Gram-negative bacterial cell walls. Resistance to β-lactams can occur when these porin proteins are modified such that they are not produced in a fully active form. Some Gram-negative bacteria show resistance to carbapenems based on loss and or reduction of these outer membrane proteins, such as the loss of OprD, which is associated with resistance to imipenem and reduced susceptibility to meropenem in *Pseudomonas aeruginosa* (Papp-Wallace et al. 2011).

Third, multicomponent drug efflux pump systems (*mex*), as part of either an acquired or intrinsic resistance repertoire, are capable of exporting a wide-range of substrates from the periplasm of Gram-negative bacteria to the surrounding environment (Papp-Wallace et al. 2011). These pumps are an important determinant of multidrug resistance in many Gram-negative pathogens, particularly notable in *P. aeruginosa* and *Acinetobacter* spp. Other pumps are found in the enteric bacteria but will not be discussed in detail, as this is beyond the scope of this review. As an example of the role played by efflux pumps, increased production of the MexA–MexB system, in combination with the low intrinsic permeability of *P. aeruginosa*, can contribute to decreased susceptibility to penicillins, cephalosporins, carbapenems, as well as quinolones, tetracycline, and chloramphenicol.

Last, β-lactamases hydrolyze β-lactams. This is the most common and important mechanism of resistance in Gram-negative bacteria and will be the focus of this review. The description of β-lactamases conferring resistance to penicillins and cephalosporins has been extensively detailed (Drawz and Bonomo 2010; Papp-Wallace et al. 2011). This work will build on those reviews and highlight the role of carbapenemases as clinically important β-lactamases.

Emphasis will be placed on select class D oxacillinases as much still needs to be learned about these less popular β-lactamases.

β-LACTAMASE HISTORY

The first β-lactamase was identified in *Bacillus* (*Escherichia*) *coli* before the clinical use of penicillin (Abraham and Chain 1988). Within a decade, a significant clinical problem emerged when *S. aureus* was observed to be resistant to penicillin owing to the production of the staphylococcal penicillinase, PC1. As more β-lactams were developed by pharmaceutical companies and introduced into the clinic, resistance to each agent was observed; the growing number of β-lactam antibiotics increased the selective pressure on bacteria, promoting the survival of organisms with effective β-lactamases (Massova and Mobashery 1998). As a result, β-lactamases were discovered in a multitude of Gram-negative bacteria including the enteric bacteria such as *Klebsiellae* spp., *Enterobacter* spp., and nonfermenters such as *P. aeruginosa*.

Presently, >2100 naturally occurring β-lactamases are now identified and each possesses a unique amino acid sequence and characteristic hydrolysis profile (K Bush, pers. comm.). It is speculated that the rapid replication rate and high mutation frequency permit bacteria to adapt to novel β-lactams by evolution of these β-lactamases. Because we also enjoy the benefits of rapid whole genome sequencing methods, novel variants of a particular family of β-lactamases are constantly being discovered, for example, *Acinetobacter*-derived cephalosporinases, ADCs, and species-specific families of OXAs.

NAMING OF β-LACTAMASES

Jacoby (2006) has provided a very nice summary of this interesting practice of naming the β-lactamases. β-Lactamases have been named on the basis of molecular characteristics or functional properties. Earlier, β-lactamases were initially designated by the name of the bacteria or plasmid that produced them (e.g., PC1 or P99). Since these original descriptions, β-lactamases have been named after substrates

that are hydrolyzed (FOX), discovery location (OHIO), patient's names (TEM), or the names of the discoverers (HMS). Notably, a few β-lactamases have been given more than one name (e.g., ARI-1 and OXA-23; YOU-1 and TEM-26; YOU-2 and TEM-12; PIT-2 and SHV-1). The interested reader should refer to Jacoby (2006).

CLASSIFICATION

Two major classification schemes exist for categorizing β-lactamase enzymes (the Ambler and Bush–Jacoby systems). Ambler classes A through D use amino acid sequence homology to categorize β-lactamases (Bush and Jacoby 2010). The Bush–Jacoby system groups 1 through 4 are based on substrate hydrolysis profiles (penicillin, cephalosporin, extended-spectrum cephalosporin, carbapenem) and inhibitor profile (inhibition by β-lactamase inhibitors clavulanate and tazobactam) (shown in Table 1).

A "family portrait" reveals the structural similarity of class A, C, and D serine β-lactamases (Fig. 1). Class B β-lactamases ("a class apart") are metallo-β-lactamases (MBLs). These proteins possess either a single or pair of Zn^{2+} ions coordinated in their active sites. More details will be discussed below.

CLASS A HYDROLYTIC MECHANISM

β-Lactamases are bacterial hydrolases that bind and acylate β-lactam antibiotics, much like PBPs, and then use strategically positioned water molecules to hydrolyze and inactivate the antibiotic before it can reach its target (Drawz and Bonomo 2010). In this manner, the β-lactamase is regenerated and can inactivate additional antibiotic molecules. This reaction may be schematically represented by the following equation:

$$E + S \underset{k_{-1}}{\overset{k_1}{\rightleftharpoons}} E:S \overset{k_2}{\longrightarrow} E - S \overset{k_3}{\longrightarrow} E - P. \quad (1)$$

In this scheme, E is a β-lactamase, S is a β-lactam substrate, E:S is the Henri–Michaelis

Table 1. Comparison of nomenclature systems

Bush–Jacoby	Ambler	Defining substrates	Inhibited by EDTA	Inhibited by clavulanaic acid or tazobactam	Representatives
1	Class C	Cephalosporinases	(−)	No	P99
		Cephamycinases			FOX-4
2	Class A		(−)	Yes	
2a		Penicillins		Yes	PC1
2b				Yes	TEM-1, SHV-1
2be		Cephalosporins		Yes	TEM-10, SHV-2
2br				No	TEM-30
2ber				No	TEM-50
2ce				Yes	RTG-4
2d	Class D	Penicillins	(−)	Variable	OXA-1
2de		Cephalosporins			OXA-11
2df		Carbapenems			OXA-23
2e				Yes	CepA
2f		Carbapenems		Variable	KPC-2
3	Class B	Carbapenems	(+)	No	
3a	B1				NDM-1, VIM-2, IMP-1
3b	B2				CphA
3a	B3				L1

Based on data from Bush and Jacoby (2010).
EDTA, ethylenediaminetetraacetate.

complex, E–S is the acyl enzyme, and P is the product devoid of antibacterial activity. The rate constants for each step are represented by: k_1, k_{-1}, k_2, and k_3; k_1 and k_{-1} are association and dissociation rate constants for the preacylation complex, respectively; k_2 is acylation rate constant; and k_3 is deacylation rate constant.

Serine β-lactamases, for example, TEM-1, SHV-1, P99, and KPC-2, actually use a multistep process to inactivate β-lactams. First, after penicillin or cephalosporin binding, nucleophilic attack by the active site serine on the carbonyl group of the β-lactam antibiotic results in a high-energy acylation intermediate. Next, this intermediate "transitions" into a lower energy covalent acyl enzyme. Following this, a catalytic water molecule attacks the covalent complex and leads to a high-energy deacylation intermediate, with subsequent hydrolysis of the bond between the β-lactam carbonyl and the serine oxygen. Last, deacylation regenerates the active enzyme and renders the β-lactam inactive.

Both acylation and deacylation require the activation of the nucleophilic serine and hydro-lytic water, respectively. In class A enzymes, the ultra-high-resolution (0.85 Å) structure of TEM-1 in complex with an acylation transition state analogue revealed the protonated state of Glu166, supporting the hypothesis that Glu166 acts as the activating base for both acylation and deacylation.

CLASS C, B, AND D HYDROLYTIC MECHANISM

The mechanistic symmetry seen in the class A discussion above is mirrored in class C enzymes, in which Tyr150 likely behaves as the general base for both acylation and deacylation, increasing the nucleophilicity of Ser64 and the catalytic water, respectively (Chen et al. 2009). The hydrolysis reaction catalyzed by class C β-lactamases consists of two steps: acylation and deacylation. In the acylation half of the reaction, Ser64 attacks the β-lactam ring carbon and forms a covalent acyl-enzyme complex. Current structural and conformational evidence suggests that during acylation the catalytic nucleo-

Class A TEM-1 β-lactamase

Class B IMP-1 β-lactamase

Class C *E. coli* AmpC β-lactamase

Class D OXA-1 β-lactamase

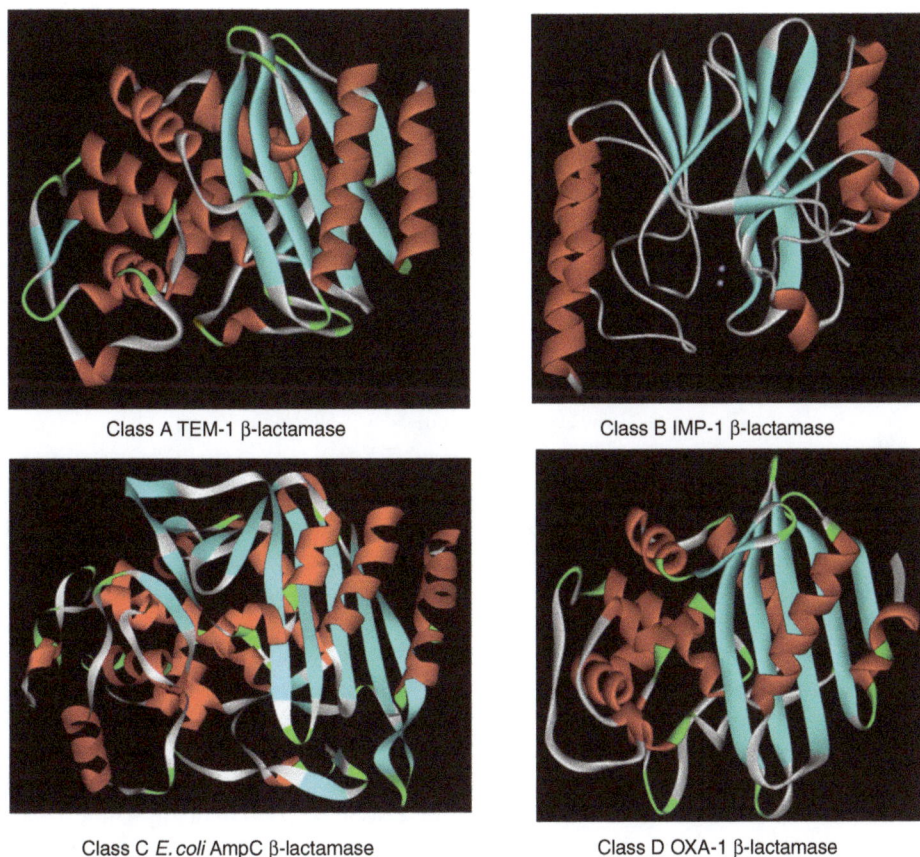

Figure 1. The structural similarity of class A, C, and D serine β-lactamases.

phile, Ser64, is deprotonated and a proton is also transferred to the leaving group, the β-lactam ring nitrogen. In deacylation, it is believed that a general base activates the structurally conserved deacylating water, whereas a general acid may be needed to reprotonate Ser64. In the deacylation step, the catalytic water reacts with the covalent linkage between the enzyme and the substrate, leading to the release of the hydrolyzed product. Both acylation and deacylation reactions proceed through a high-energy tetrahedral transition state.

Despite this similarity, debate exists about the role of Lys67 (Chen et al. 2009). It is generally agreed that many mechanisms may contribute to hydrolysis in class C β-lactamases, depending on the enzyme and the substrate, explaining why different variants and substrates seem to support different pathways and mechanisms, that

is, substrate assisted or conjugate base. For the wild-type enzyme itself, the conjugate base mechanism may be well favored. Substrate assisted catalysis may also occur; in this mechanism, the proton from the catalytic water is transferred to the substrate ring nitrogen, whereas Tyr150 stabilizes the water molecule. It is the opinion of this writer that the two different hypotheses are not mutually exclusive.

Class B includes Zn^{2+}-dependent enzymes that follow a different hydrolytic mechanism. These MBLs use the OH group from a water molecule that is coordinated by Zn^{2+} to hydrolyze the scissile amide bond of a β-lactam (Fig. 2). The importance of a high-energy anionic intermediate is currently favored as being essential to the reaction coordinate.

Class D enzymes hydrolyze β-lactams using a slightly different scheme by featuring a carba-

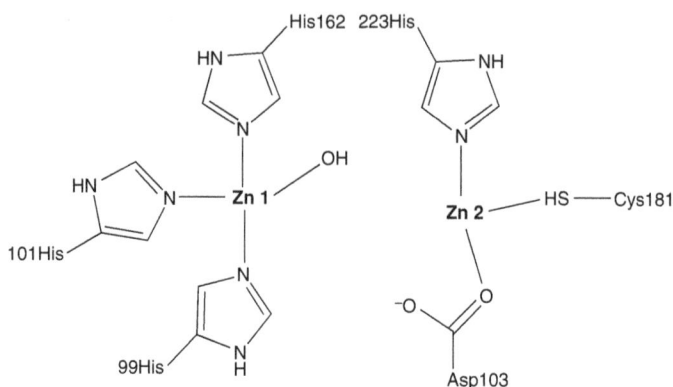

Figure 2. The active site of a metallo-β-lactamase.

mylated lysine. OXA β-lactamases use this carbamylated lysine to activate the nucleophilic serine used for β-lactam hydrolysis (like Glu166 in class A β-lactamases). The deacylating water molecule approaches the acyl-enzyme species, anchored at the nucleophilic serine using the carbamylated lysine as a chemical anchor. Carbapenem hydrolyzing class D β-lactamases have evolved the ability to hydrolyze imipenem, an important carbapenem in clinical use, by subtle structural changes in the active site. These changes may contribute to tighter binding of imipenem to the active site and removal of steric hindrances from the path of the deacylating water molecule as was shown in class A (Verma et al. 2011).

CLINICALLY IMPORTANT β-LACTAMASES

Class A—Serine Penicillinases TEM, SHV, and CTX-M and the Carbapenemases KPC, etc.

Class A β-lactamases are often plasmid-encoded, but can also be located on the bacterial chromosome. For example, bla_{SHV-1} is a chromosomal gene in *Klebsiella pneumoniae*, but may also be found on plasmids; *pen*A from *Burkholderia pseudomallei* is chromosomally encoded (Papp-Wallace et al. 2015). In general, class A enzymes are usually susceptible to inactivation by the clinically available β-lactamase inhibitors: clavulanate, sulbactam, tazobactam, and avibactam. TEM, SHV, and CTX-M β-lacta-

mases are mostly found in *E. coli* and *Klebsiellae* spp. Many class A β-lactamases have substrate profiles that include expanded-spectrum cephalosporins, and these extended-spectrum β-lactamases (ESBLs) have been discussed extensively (Paterson and Bonomo 2005; Perez et al. 2007). The widespread distribution of CTX-M β-lactamases, especially CTX-M-14 and CTX-M-15, in *E. coli* is responsible for the large part of the global advanced generation cephalosporin resistance seen in many clinical isolates. Other class A enzymes are encoded on integrons, for example, GES-1 from *K. pneumoniae* and VEB-1 in *P. aeruginosa* and *Acinetobacter baumannii* (Poirel et al. 2012).

Few Ambler class A β-lactamases show carbapenem-hydrolyzing activity. The major class A carbapenemases include KPC, GES, Nmc-A/IMI, and SME β-lactamases, with SME carbapenemases identified, to date, only in *Serratia marcescens*. With the notable exception of KPCs and GES, the clinical distribution of the types of carbapenemases is relatively limited (Bush 2010a).

Currently, most carbapenem resistance among Enterobacteriaceae in the United States is attributed to plasmid-mediated expression of a KPC-type (*K. pneumoniae* carbapenemase) carbapenemase. KPC-producing Enterobacteriaceae are considered endemic in many places, for example, in Greece, along with other carbapenemases, specifically VIM-type metallo-β-lactamases. KPC β-lactamases efficiently hydrolyze carbapenems as well as penicillins, cephalosporins, and aztreonam and are not overcome in vitro by clinically available β-lactamase inhibitors; in fact, clavulanic acid, sulbactam, and tazobactam are hydrolyzed. Avibactam inhibits KPC enzymes, but is hydrolyzed slowly (Nguyen et al. 2016).

Carbapenem resistance secondary to KPC production was first described in a *K. pneumoniae* isolate recovered in North Carolina in 1996 (Yigit et al. 2003). The bla_{KPC} gene has been mapped to a highly conserved Tn3-based transposon, Tn4401, and different isoforms of Tn4401 are described. Plasmids carrying bla_{KPC} are of various sizes and many carry additional genes conferring resistance to fluoroquinolones and aminoglycosides, thus limiting the antibiotics available to treat infections with KPC-producing pathogens; bla_{KPC} has rarely been mapped to a chromosomal location (Chen et al. 2009).

A predominant strain of *K. pneumoniae* appears responsible for outbreaks and the international spread of KPC-producing *K. pneumoniae*. A specific MLST sequence type (ST), ST258, has spread in the United States. Now there is widespread acknowledgment that two clades of ST258 exist (Deleo et al. 2014). A second sequence type, ST14, was also common in institutions in the midwestern region of the United States. These findings implied that certain strains of *K. pneumoniae* may be more apt to obtain and retain the bla_{KPC} gene.

KPC-production can confer variable levels of carbapenem resistance with reported minimum inhibitory concentrations (MICs) ranging from ≤1 μg/mL (susceptible) to ≥16 μg/mL. Analysis of isolates displaying high-level carbapenem resistance showed that increased phenotypic resistance may be caused by increased bla_{KPC} gene copy number or the loss of an outer membrane porin, Omp K35 and/or Omp K36. The highest level of imipenem resistance was seen with isolates lacking both porins and with augmented KPC enzyme production (Chen et al. 2009).

Nmc-A (non-metallo-carbapenemase-A) is a chromosomal carbapenemase originally isolated from *Enterobacter cloacae* in France. Currently, reports of this particular β-lactamase are still rare. IMI-1 was initially recovered from the chromosome of an *E. cloacae* isolate in the southwestern United States. A variant of IMI-1, IMI-2, has been identified on plasmids isolated from environmental strains of *Enterobacter absuriae* in U.S. rivers.

SME-1 (*S. marcescens* enzyme) was originally identified in an isolate of *S. marcescens* from a patient in London in 1982. SME-2 and SME-3 were subsequently isolated in the United States, Canada, and Switzerland. Chromosomally encoded SME-type carbapenemases continue to be isolated at a low frequency in North America (Naas et al. 2016). Although infrequent, it is currently recommended to screen for SME enzymes in carbapenem-resistant *S. marcescens* isolates (Bush et al. 2013).

The GES-type (Guiana extended-spectrum) β-lactamases are acquired β-lactamases recovered from *P. aeruginosa*, Enterobacteriaceae, and *A. baumannii*. The genes encoding these β-lactamase have often, but not exclusively, been identified within class 1 integrons residing on transferrable plasmids. GES-1 has a similar hydrolysis profile to other ESBLs, although they essentially spare monobactams. Several GES β-lactamases are described with six (i.e., GES-2, GES-4, GES-5, GES-6, GES-11, and GES-14), showing detectable carbapenemase activity in the setting of amino acid substitutions at their active sites (specifically at residue 104 and 170). These GES-type carbapenemases have been described in Europe, South Africa, Asia, and the Middle East.

Class B Metallo-β-Lactamases

Class B β-lactamases are bacterial enzymes that degrade β-lactam antibiotics with the help of a metal cofactor (divalent zinc in the natural form). Class B enzymes are Zn^{2+}-dependent β-lactamases that follow a different hydrolytic mechanism than the serine β-lactamases of classes A, C, and D. Although catalyzing the same overall reaction as serine-β-lactamases, that is, breaking the amide bond, class B MBLs are structurally and mechanistically unrelated to the serine enzymes, and their common function apparently represents an example of convergent evolution of different protein lineages within the bacterial domain (Mojica et al. 2015).

Organisms producing MBLs usually show resistance to penicillins, cephalosporins, carbapenems, and the clinically available β-lactamase inhibitors. MBL *bla* genes are located on the chromosome, plasmid, and integrons. Until a few years ago, the clinically important class B enzymes included those found in the nosocomial pathogen *Stenotrophomonas maltophilia*. The rapid emergence of NDM MBLs in the Enterobacteriaceae has changed this (Walsh et al. 2011).

Because of the dependence on Zn^{2+}, catalysis is inhibited in the presence of metal-chelating agents like EDTA. MBLs are not inhibited by the presence of commercially available β-lactamase inhibitors; however, hydrolytic stability of monobactams (i.e., aztreonam) leading to susceptibility of the producing organism appears to be preserved in the absence of concomitant expression of other resistance mechanisms (e.g., ESBL production). The more geographically widespread MBLs include IMP, VIM, and NDM (Mojica et al. 2015).

NDM-1 (New Delhi MBL) was first identified in 2008. NDM-1 was first discovered in Sweden in a patient of Indian descent previously hospitalized in India (Yong et al. 2009; Kumarasamy et al. 2010). The patient was colonized with a *K. pneumoniae* strain and an *E. coli*–carrying bla_{NDM-1} on transferrable plasmids. In the United Kingdom, an increase in the number of clinical isolates of carbapenem-resistant Enterobacteriaceae was also seen in both 2008 and 2009. A U.K. reference laboratory reported that at least 17 of 29 patients found to be harboring NDM-1 expressing Enterobacteriaceae had a history of recent travel to the Indian subcontinent with the majority having been hospitalized in those countries.

NDM-1 shares the most homology (32.4%) with VIM-1 and VIM-2 (Yong et al. 2009). It is a 28-kDa monomeric protein that shows tight binding to both penicillins and cephalosporins. Binding to carbapenems does not appear to be as avid as other MBLs, but catalytic efficiencies appear to be similar. Using ampicillin as a substrate allowed for detailed characterization of the interactions between NDM's active site and β-lactams as well as improved evaluation of MBLs unique mechanism of β-lactam hydrolysis. More recent crystal structures of NDM-1 reveal the molecular details of how carbapenem antibiotics are recognized by di-zinc-containing metallo-β-lactamases (King et al. 2012).

Because of its rapid international dissemination and its ability to be expressed by numerous Gram-negative pathogens, NDM is poised to become the most commonly isolated and distributed carbapenemase worldwide. Initial reports frequently showed an epidemiologic link to the Indian subcontinent where these MBLs are endemic. Indeed, retrospective analyses of stored isolates suggest that NDM-1 may have been circulating in the subcontinent as early as 2006. Despite initial controversy, the Balkans may be another area of endemicity for NDM-1. Sporadic recovery of NDM-1 in the Middle East suggests that this region may be an additional reservoir.

Like KPCs, the conveniences of international travel and medical tourism have quickly propelled this relatively novel MBL into a formidable public health threat. Gram-negative bacilli harboring bla_{NDM} have been identified worldwide (Nordmann et al. 2011).

European reports suggest that horizontal transfer of bla_{NDM-1} exists within hospitals outside of endemic areas. Of overwhelming concern are the reported cases without specific contact with the health care system locally or in endemic areas, suggesting autochthonous acquisition.

Surveillance of public water supplies in India indicates that exposure to NDM-1 may be environmental. Walsh et al. (2011) analyzed samples of public tap water and seepage water from sites around New Delhi. The results were disheartening in that bla_{NDM-1} was detected by polymerase chain reaction (PCR) in 4% of drinking water samples and 30% of seepage samples. In this survey, carriage of bla_{NDM-1} was noted in 11 species of bacteria not previously described, including virulent ones like *Shigella boydii* and *Vibrio cholerae*.

The rapid spread of NDM-1 highlights the fluidity and rapidity of gene transfer among bacterial species. Although bla_{NDM-1} was initially and repeatedly mapped to plasmids isolated from carbapenem-resistant *E. coli* and *K. pneumoniae*, reports of both plasmid and chromosomal expression of bla_{NDM-1} has been noted in other species of Enterobacteriaceae as well as *Acinetobacter* spp. and *P. aeruginosa* (Jones et al. 2014). Recently, bacteremia with an NDM-1 expressing *V. cholerae* has been described in a patient previously hospitalized in India colonized with a variety of Enterobacteriaceae previously known to be capable of carrying plasmids with bla_{NDM-1} (Darley et al. 2012).

In contrast to KPCs, the presence of a dominant clone among bla_{NDM-1}-carrying isolates remains elusive. NDM-1 expression in *E. coli* has been noted among sequence types previously associated with the successful dissemination of other β-lactamases, including ST101 and ST131. Mushtaq et al. analyzed a relatively large group of bla_{NDM-1} expressing *E. coli* from the United Kingdom, Pakistan, and India to potentially identify a predominant strain responsible for the rapid and successful spread of NDM-1 (Mushtaq et al. 2011). The most frequent sequence type identified was ST101. Another study examining a collection of carbapenem-resistant Enterobacteriaceae from India shows the diversity of strains capable of harboring bla_{NDM-1}. Carriage of bla_{NDM-1} was confirmed in ten different sequence types of *K. pneumoniae* and five sequence types of *E. coli*. This multiplicity was confirmed in a study looking at a collection of bla_{NDM-1} expressing Enterobacteriaceae from around the world (Poirel et al. 2011). Of most concern is that NDM-1 has been identified in *E. coli* ST131, the strain of *E. coli* credited with the global propagation of CTX-M-15 ESBLs. Similar to KPCs, NDM-1 expression portends variable levels of carbapenem resistance, and there is often concomitant carriage of a myriad of resistance determinants including other β-lactamases and carbapenemases as well as genes associated with resistance to fluoroquinolones and aminoglycosides.

To date, NDM-1 remains the most common NDM variant isolated. It is currently believed that bla_{NDM-1} is a chimeric gene that may have evolved from *A. baumannii*. Contributing to this theory is the presence of complete or variations of the insertion sequence, IS*Aba125*, upstream of the bla_{NDM-1} gene in both Enterobacteriaceae and *A. baumannii*. This insertion sequence has primarily been found in *A. baumannii*.

A recent evaluation of the genetic construct associated with bla_{NDM-1} has led to the discovery of a new bleomycin resistance protein, BRP_{MBL}. Evaluation of 23 isolates of $bla_{NDM-1/2}$ harboring Enterobacteriaceae and *A. baumannii* noted that the overwhelming majority of them possessed a novel bleomycin resistance gene, ble_{MBL}. Coexpression of bla_{NDM-1} and ble_{MBL} appears to be mediated by a common promoter (P_{NDM-1}), which includes portions of IS*Aba125*. It is postulated that BRP_{MBL} expression may contribute some sort of selective advantage allowing NDM-1 to persist in the environment.

A contemporary evaluation of recently recovered NDM-1 producing *A. baumannii* isolates from Europe shows that bla_{NDM-1} and bla_{NDM-2} genes are situated on the same chromosomally located transposon, Tn*125*. Dissemination of bla_{NDM} in *A. baumannii* seems be caused by different strains carrying Tn*125* or derivatives of Tn*125*, rather than plasmid-mediated or clonal (Poirel et al. 2012).

Before the description of NDM-1, frequently detected MBLs include IMP-type (imipenem-resistant) and VIM-type (Verona integron-encoded MBL) with VIM-2 being the most prevalent. These MBLs are embedded within a variety of genetic structures, most commonly integrons. When these integrons are associated with transposons or plasmids, they can readily be transferred among species (Mojica et al. 2015).

A more commonly recovered MBL is the VIM-type enzyme. VIM-1 was first described in Italy in 1997 in *P. aeruginosa*. VIM-2 was next discovered in southern France in *P. aeruginosa* cultured from a neutropenic patient in 1996. Although originally thought to be limited to nonfermenting Gram-negative bacilli, VIM-type MBLs are being increasingly identified in Enterobacteriaceae as well. Many variants of VIM have been described with VIM-2 being the most common MBL recovered worldwide (Mojica et al. 2015).

Other more geographically restricted MBLs include (1) SPM-1, Sao Paulo MBL, which has been associated with hospital outbreaks in Brazil; (2) GIM-1, German imipenemase, isolated in carbapenem-resistant *P. aeruginosa* isolates in Germany; (3) SIM-1, Seoul imipenemase, isolated from *A. baumannii* isolates in Korea; (4) KHM-1, Kyorin Health Science MBL, isolated from a *Citrobacter freundii* isolate in Japan (Sekiguchi et al. 2008); (5) AIM-1, Australian imipenemase, isolated from *P. aeruginosa* in Australia; (6) DIM-1, Dutch imipenemase, isolated from a clinical *Pseudomonas stutzeri* isolate in the Netherlands; (7) SMB-1, *S. marcescens* MBL, in *S. marcescens* in Japan; (8) TMB-1, Tripoli MBL, in *Achromobacter xylosoxidans* in Libya; and (9) FIM-1, Florence imipenemase, from a clinical isolate of *P. aeruginosa* in Italy. With the notable exception of SPM-1, which has been introduced into European hospitals by a Brazilian pediatric patient, these MBLs have remained confined to their countries/cities of origin.

Class C Cephalosporinases

Class C enzymes include the AmpC β-lactamases, which are usually encoded by *bla* genes located on the bacterial chromosome, although plasmid-borne AmpC enzymes have become more prevalent. Organisms expressing the AmpC β-lactamase are typically resistant to penicillins, β-lactamase inhibitors (clavulanate and tazobactam), and most cephalosporins including cefoxitin, cefotetan, ceftriaxone, and cefotaxime. AmpC enzymes poorly hydrolyze cefepime, an expanded-spectrum cephalosporin, and are readily inactivated by carbapenems. Notably, AmpC cephalosporinases are very susceptible to inactivation by avibactam. These enzymes are absent in *K. pneumoniae*, *Klebsiella oxytoca*, *Proteus mirabilis*, and *Salmonella* spp. as well as other bacteria (Perez et al. 2016).

Benzylpenicillin, ampicillin, amoxicillin, and cephalosporins such as cefazolin and cephalothin are very good inducers and good substrates for AmpC β-lactamase. Cefoxitin and imipenem are also strong inducers but are much more stable to hydrolysis. Cefotaxime, ceftriaxone, ceftazidime, cefepime, cefuroxime, piperacillin, and aztreonam are weak inducers and weak substrates but can be hydrolyzed if enough enzyme is expressed. Consequently, MICs of weakly inducing oxyimino-β-lactams are dramatically increased with AmpC hyperproduction. Conversely, MICs of agents that are strong inducers show little change with regulatory mutations because the level of induced *ampC* expression is already high.

Some β-lactamase inhibitors are also inducers, especially clavulanate, which has little inhibitory effect on AmpC β-lactamase activity, but can paradoxically appear to increase AmpC-mediated resistance in an inducible organism. The inducing effect of clavulanate is especially important for *P. aeruginosa*, in which clinically achieved concentrations of clavulanate by inducing AmpC expression have been shown to antagonize the antibacterial activity of ticarcillin (Papp-Wallace et al. 2014).

Production of AmpC enzymes in clinically important Gram-negative bacteria is normally at a low level ("repressed"), but can be "derepressed" by induction with certain β-lactams, particularly cefoxitin and imipenem. Sulbactam, but not tazobactam, is also a good inducer of AmpC β-lactamases. The genetic underpinnings of this regulation have been the subject of intense investigation, but are not the subject of this review. Members of the Enterobacteriaceae family, such as *Citrobacter*, *Salmonella*, and *Shigella*, are clinically relevant producers of AmpC enzymes that resist inhibition by clavulanate and sulbactam (Bauvois and Wouters 2007).

Class D Serine Oxacillinases

Class D β-lactamases were initially categorized as "oxacillinases" because of their ability to hydrolyze oxacillin at a rate of at least 50% that of benzylpenicillin, in contrast to the relatively slow hydrolysis of oxacillin by classes A and C. Different enzymes in this diverse class can also confer resistance to penicillins, cephalosporins, extended-spectrum cephalosporins (OXA-type ESBLs), and carbapenems (OXA-type carbapenemases) (Leonard et al. 2013).

Oxacillinases comprise a heterogeneous group of class D β-lactamases, which are able to hydrolyze amino- and carboxypenicillins. The majority of class D β-lactamases are not inhibited by commercially available β-lactamase inhibitors but are inhibited in vitro by NaCl. OXA enzymes are insensitive to inhibition by clavulanate, sulbactam, and tazobactam, with some exceptions; for example, OXA-2 and OXA-32 are inhibited by tazobactam, but not sulbactam and clavulanate; OXA-53 is inhibited by clavulanate. Interestingly, sodium chloride at concentrations >50–75 mM inhibits some carbapenem-hydrolyzing oxacillinases (e.g., OXA-25 and OXA-26). Site-directed mutagenesis studies suggest that susceptibility to inhibition by sodium chloride is related to the presence of a Tyr residue at position 144, which may facilitate sodium chloride binding better than the Phe residue found in resistant oxacillinases. Examples of OXA enzymes include those rapidly emerging in *A. baumannii* (e.g., OXA-23) and *P. aeruginosa* (e.g., OXA-50).

Carbapenem-Hydrolyzing Class D β-Lactamases

More than 490 types of oxacillinases are reported with a minority showing low levels of carbapenem-hydrolyzing activity (lahey.org/Studies/other.asp#table1). This select group of enzymes is also referred to as the carbapenem-hydrolyzing class D β-lactamases (CHDLs). CHDLs have been identified most frequently in *Acinetobacter* spp.; however, there has been increasing isolation among Enterobacteriaceae, specifically OXA-48 (Patel and Bonomo 2013).

With the exception of OXA-163, CHDLs efficiently inactivate penicillins, early cephalosporins, and β-lactam/β-lactamase inhibitor combinations, but spare expanded-spectrum cephalosporins. Carbapenem hydrolysis efficiency is lower than that of other carbapenemases, including the MBLs, and often additional resistance mechanisms are expressed in organisms showing higher levels of phenotypic carbapenem resistance. These include expression of other carbapenemases, alterations in outer membrane proteins (e.g., CarO, OmpK36), in-

creased transcription mediated by *IS* elements functioning as promoters, increased gene copy number, and amplified drug efflux. Many subgroups of CHDLs have been described. We will focus on those found in *A. baumannii* and Enterobacteriaceae: OXA-23 and -27; OXA-24/40, -25, and -26; OXA-48 variants; OXA-51, -66, and -69; OXA-58 and OXA-143.

CHDLs can be intrinsic or acquired (Patel and Bonomo 2013). *A. baumannii* does have naturally occurring but variably expressed chromosomal CHDLs, OXA-51, OXA-66, and OXA-69. For the most part, in isolation the phenotypic carbapenem resistance associated with these oxacillinases is low. However, levels of carbapenem resistance appear to be increased in the presence of specific insertion sequences promoting gene expression. Additional resistance to extended-spectrum cephalosporins can be seen in the setting of coexpression of ESBLs and/or other carbapenemases (Leonard et al. 2013).

The first reported acquired oxacillinase with appreciable carbapenem-hydrolyzing activity was OXA-23. OXA-23, or ARI-1, was identified from an *A. baumannii* isolate in Scotland in 1993, although the isolate was first recovered in 1985 (Paton et al. 1993). Subsequently, OXA-23 expression has been reported worldwide and both plasmid and chromosomal carriage of *bla*$_{OXA-23}$ are described. The OXA-23 group includes OXA-27, found in a single *A. baumannii* isolate from Singapore. With the exception of an isolate of *P. mirabilis* identified in France in 2002, this group of β-lactamases has been exclusively recovered from *Acinetobacter* species. Increased expression of OXA-23 has been associated with the presence of upstream insertion sequences (e.g., IS*Aba1* and IS*Aba4*) acting as strong promoters.

Another group of CHDLs includes OXA-24/40, OXA-25, and OXA-26. OXA-25 and OXA-26 are point mutation derivatives of OXA-24/40. Although primarily linked with clonal outbreaks in Spain and Portugal, OXA-24/40 β-lactamases has been isolated in other European countries and the United States. OXA-40 was in fact the first CHDL documented in the United States. OXA-58 has also only been detected in *Acinetobacter* spp. Initially identified in France, OXA-58 has been associated with institutional outbreaks and has been recovered from clinical isolates of *A. baumannii* worldwide (Leonard et al. 2013).

As civilian and military personnel began returning from Afghanistan and the Middle East, practitioners noted increasing recovery of *A. baumannii* from skin and soft tissue infections. Drug resistance was associated with expression of both OXA-23 and OXA-58. Many isolates carrying the *bla*$_{OXA-58}$ gene concurrently carry insertion sequences (e.g., IS*aba1*, IS*Aba2, or* IS*Aba3*) associated with increased carbapenemase production and, thus, higher levels of carbapenem resistance. In one report, increased gene copy number was also associated with a higher level of enzyme production and increased phenotypic carbapenem resistance (Hujer et al. 2006).

Spread of OXA-type carbapenemases among *A. baumannii* appears to be clonal, and in-depth reviews of the molecular epidemiology and successful dissemination of these clones has been published. Two MLST schemes with three loci in common exist for *A. baumannii*—the PubMLST scheme and the Pasteur scheme. Both schemes assign different sequence types into clonal complexes. Sequence types and clonal complexes (CCs) from both schemes can be further categorized into the international (European) clones I, II, and III. It should be noted, however, that the molecular taxonomy of *A. baumannii* continues to evolve. OXA-23 producing *A. baumannii* predominantly belong to international clones I and II with a notable proportion being part of CC92 (PubMLST). Similarly, *A. baumannii* isolates associated with epidemic spread of OXA-24/40 in Portugal and Spain appear to be incorporated in international clone II and ST56 (PubMLST). OXA-58 expressing *A. baumannii* have been associated with international clones I, II, and III and a variety of unrelated sequence types (Hujer et al. 2006).

OXA-48 was originally identified in a carbapenem-resistant isolate of *K. pneumoniae* in Turkey. Early reports suggested that this enzyme was geographically restricted to Turkey. In the past few years, however, the enzyme has been recovered from variety of Enterobacteriaceae

and has successfully circulated outside of Turkey with reports of isolation in the Middle East, North Africa, Europe, and, most recently, the United States. The Middle East and North Africa may be secondary reservoirs for these CHDLs. Indeed, the introduction of OXA-48 expressing Enterobacteriaceae in some countries has been from patients from the Middle East or Northern Africa. In the United States, the first clinical cases were associated with ST199 and ST43.

At least six OXA-48 variants (e.g., OXA-48, OXA-162, OXA-163, OXA-181, OXA-204, and OXA-232) have been identified. OXA-48 is by far the most globally dispersed and its epidemiology has been recently reviewed. Unlike KPCs and NDM-1, which have been associated with a variety of plasmids, a single 62-kb self-conjugative IncL/M-type plasmid has contributed to a large proportion of the distribution of *bla*$_{OXA-48}$ in Europe. Sequencing of this plasmid (pOXA-48a) notes that *bla*$_{OXA-48}$ had been integrated through the acquisition of a Tn*1999* composite transposon. *bla*$_{OXA-48}$ appears to be associated with a specific insertion sequence, IS*1999*. A variant of Tn*1999*, Tn*1999.2*, has been identified among isolates from Turkey and Europe. Tn*1999.2* harbors an IS*1R* element within the IS*1999*. OXA-48 appears to have the highest affinity for imipenem of the CHDLs specifically those harboring *bla*$_{OXA-48}$ within a Tn*1999.2* composite transposon. Three isoforms of the Tn*1999* transposon have been described.

Although much of the spread of OXA-48 is attributed to a specific plasmid, outbreak evaluations show that a variety of strains have contributed to dissemination of this emerging carbapenemase in *K. pneumoniae*. The same *K. pneumoniae* sequence type, ST395, harboring *bla*$_{OXA-48}$ was identified in Morocco, France, and the Netherlands. ST353 was associated with an outbreak of OXA-48 producing *K. pneumoniae* in London (Woodford et al. 2011) and ST221 with an outbreak of OXA-48 in Ireland. OXA-48 production in *K. pneumoniae*, like KPC expressing *K. pneumoniae*, has also been associated with ST14 (Poirel et al. 2004a) and a recent outbreak in Greece was associated with ST11. *bla*$_{OXA-48}$ is remarkably similar to *bla*$_{OXA-54}$, a β-lactamase gene intrinsic to *Shewanella onei-*

densis (Poirel et al. 2004b). *Shewanella* spp. are relatively ubiquitous waterborne Gram-negative bacilli and are proving to be a potential environmental reservoir for OXA-48 like carbapenemases as well as other resistance determinants.

OXA-163, a single amino acid variant of OXA-48, was identified in isolates of *K. pneumoniae* and *E. cloacae* from Argentina and is unique in that it has activity against expanded-spectrum cephalosporins (Poirel et al. 2011). OXA-163 also has been identified in Egypt, which has a relatively prevalence of OXA-48, in patients without epidemiologic links to Argentina (Abdelaziz et al. 2012).

OXA-181 was initially identified among carbapenem-resistant Enterobacteriaceae collected from India (Castanheira et al. 2011). OXA-181 differs from OXA-48 by four amino acids; however, it appears to be nestled in an entirely different genetic platform. The *bla*$_{OXA-181}$ gene has been mapped to a different group of plasmids, the ColE family, and has been associated with an alternative insertion sequence, IS*Ecp1*. The latter insertion sequence has been associated with the acquisition of other β-lactamases including CTX-M-like ESBLs. Like, OXA-48, it appears that OXA-181 may have evolved from a waterborne environmental species *Shewanella xiamenensis*. OXA-204 differs from OXA-48 by two amino acid substitution. It was recently identified in a clinical *K. pneumoniae* isolate from Tunisia. Its genetic construct appears to be similar to that of OXA-181. OXA-232 was recently identified among *K. pneumoniae* isolates in France. OXA-143 is a novel plasmid-borne carbapenem-hydrolyzing oxacillinase recovered from clinical *A. baumannii* isolates in Brazil. Information regarding its significance and prevalence continues to evolve.

CONCLUSIONS

This review briefly recapitulates our understanding of the history of this complex family of enzymes, reviews the mechanisms by which these hydrolases inactivate β-lactams, and highlights the current challenges that threaten our β-lactam armamentarium, especially the carbapenemases KPCs, NDMs, and OXAs. In 80 years, we have learned that the constant evolution of substrate specificity meets each new β-lactam introduced. The "long view" predicts these enzymes will continue to evolve in novel forms. Perhaps, we will eventually understand why this occurs and how to prevent this. The future of research in this arena needs to be open and mindful to new approaches. We have still not defined all the correlates of activity and resistance or have answered why novel structural variants emerge. The current challenges will have to be faced with new technologies.

ACKNOWLEDGMENTS

The author thanks Dr. Karen Bush for assistance with this chapter and for her suggestions.

REFERENCES

Abdelaziz MO, Bonura C, Aleo A, El-Domany RA, Fasciana T, Mammina C. 2012. OXA-163-producing *Klebsiella pneumoniae* in Cairo, Egypt, in 2009 and 2010. *J Clin Microbiol* **50:** 2489–2491.

Abraham EP, Chain E. 1988. An enzyme from bacteria able to destroy penicillin. *Rev Infect Dis* **10:** 677–678.

Bauvois C, Wouters J. 2007. Crystal structures of class C β-lactamases: Mechanistic implications and perspectives in drug design. In *Enzyme-mediated resistance to antibiotics: Mechanisms, dissemination, and prospects for inhibition* (ed. Bonomo RA, Tolmasky ME), pp. 145–161. ASM, Washington DC.

Bush K. 2010a. Alarming β-lactamase-mediated resistance in multidrug-resistant Enterobacteriaceae. *Curr Opin Microbiol* **13:** 558–564.

Bush K. 2010b. Bench-to-bedside review: The role of β-lactamases in antibiotic-resistant Gram-negative infections. *Crit Care* **14:** 224.

Bush K. 2014. Introduction to Antimicrobial therapeutics reviews: Infectious diseases of current and emerging concern. *Ann NY Acad Sci* **1323:** v–vi.

Bush K, Jacoby GA. 2010. Updated functional classification of β-lactamases. *Antimicrob Agents Chemother* **54:** 969–976.

Bush K, Pannell M, Lock JL, Queenan AM, Jorgensen JH, Lee RM, Lewis JS, Jarrett D. 2013. Detection systems for carbapenemase gene identification should include the SME serine carbapenemase. *Int J Antimicrob Agents* **41:** 1–4.

Castanheira M, Deshpande LM, Mathai D, Bell JM, Jones RN, Mendes RE. 2011. Early dissemination of NDM-1- and OXA-181-producing Enterobacteriaceae in Indian hospitals: Report from the SENTRY Antimicrobial Surveillance Program, 2006–2007. *Antimicrob Agents Chemother* **55:** 1274–1278.

Chen Y, McReynolds A, Shoichet BK. 2009. Re-examining the role of Lys67 in class C β-lactamase catalysis. *Protein Sci* **18:** 662–669.

Darley E, Weeks J, Jones L, Daniels V, Wootton M, MacGowan A, Walsh T. 2012. NDM-1 polymicrobial infections including *Vibrio cholerae. Lancet* **380:** 1358.

Deleo FR, Chen L, Porcella SF, Martens CA, Kobayashi SD, Porter AR, Chavda KD, Jacobs MR, Mathema B, Olsen RJ, et al. 2014. Molecular dissection of the evolution of carbapenem-resistant multilocus sequence type 258 *Klebsiella pneumoniae. Proc Natl Acad Sci* **111:** 4988–4993.

Drawz SM, Bonomo RA. 2010. Three decades of β-lactamase inhibitors. *Clin Microbiol Rev* **23:** 160–201.

Fisher JF, Mobashery S. 2014. The sentinel role of peptidoglycan recycling in the β-lactam resistance of the Gram-negative Enterobacteriaceae and *Pseudomonas aeruginosa. Bioorg Chem* **56:** 41–48.

Fishovitz J, Taghizadeh N, Fisher JF, Chang M, Mobashery S. 2015. The Tipper–Strominger hypothesis and triggering of allostery in penicillin-binding protein 2a of methicillin-resistant *Staphylococcus aureus* (MRSA). *J Am Chem Soc* **137:** 6500–6505.

Hauck C, Cober E, Richter SS, Perez F, Salata RA, Kalayjian RC, Watkins RR, Scalera NM, Doi Y, Kaye KS, et al. 2016. Spectrum of excess mortality due to carbapenem-resistant *Klebsiella pneumoniae* infections. *Clin Microbiol Infect* doi: 10.1016/j.cmi.2016.01.023.

Hujer KM, Hujer AM, Hulten EA, Bajaksouzian S, Adams JM, Donskey CJ, Ecker DJ, Massire C, Eshoo MW, Sampath R, et al. 2006. Analysis of antibiotic resistance genes in multidrug-resistant *Acinetobacter* sp. isolates from military and civilian patients treated at the Walter Reed Army Medical Center. *Antimicrob Agents Chemother* **50:** 4114–4123.

Jacoby GA. 2006. β-Lactamase nomenclature. *Antimicrob Agents Chemother* **50:** 1123–1129.

Jones LS, Toleman MA, Weeks JL, Howe RA, Walsh TR, Kumarasamy KK. 2014. Plasmid carriage of bla NDM-1 in clinical *Acinetobacter baumannii* isolates from India. *Antimicrob Agents Chemother* **58:** 4211–4213.

King DT, Worrall LJ, Gruninger R, Strynadka NC. 2012. New Delhi metallo-β-lactamase: Structural insights into β-lactam recognition and inhibition. *J Am Chem Soc* **134:** 11362–11365.

Kumarasamy KK, Toleman MA, Walsh TR, Bagaria J, Butt F, Balakrishnan R, Chaudhary U, Doumith M, Giske CG, Irfan S, et al. 2010. Emergence of a new antibiotic resistance mechanism in India, Pakistan, and the UK: A molecular, biological, and epidemiological study. *Lancet Infect Dis* **10:** 597–602.

Leonard DA, Bonomo RA, Powers RA. 2013. Class D β-lactamases: A reappraisal after five decades. *Acc Chem Res* **46:** 2407–2415.

Massova I, Mobashery S. 1998. Kinship and diversification of bacterial penicillin-binding proteins and β-lactamases. *Antimicrob Agents Chemother* **42:** 1–17.

Mojica MF, Bonomo RA, Fast W. 2015. B1-metallo-β-lactamases: Where do we stand? *Curr Drug Targets* **17:** 1029–1050

Mushtaq S, Irfan S, Sarma JB, Doumith M, Pike R, Pitout J, Livermore DM, Woodford N. 2011. Phylogenetic diversity of *Escherichia coli* strains producing NDM-type carbapenemases. *J Antimicrob Chemother* **66:** 2002–2005.

Naas T, Dortet L, Iorga BI. 2016. Structural and functional aspects of class A carbapenemases. *Curr Drug Targets* **17:** 1006–1028.

Nguyen NQ, Krishnan NP, Rojas LJ, Prati F, Caselli E, Romagnoli C, Bonomo RA, van den Akker F. 2016. Crystal structures of KPC-2 and SHV-1 β-lactamases in complex with the boronic acid transition state analog S02030. *Antimicrob Agents Chemother* **60:** 1760–1766.

Nordmann P, Poirel L, Walsh TR, Livermore DM. 2011. The emerging NDM carbapenemases. *Trends Microbiol* **19:** 588–595.

Papp-Wallace KM, Endimiani A, Taracila MA, Bonomo RA. 2011. Carbapenems: Past, present, and future. *Antimicrob Agents Chemother* **55:** 4943–4960.

Papp-Wallace KM, Winkler ML, Gatta JA, Taracila MA, Chilakala S, Xu Y, Johnson JK, Bonomo RA. 2014. Reclaiming the efficacy of β-lactam-β-lactamase inhibitor combinations: Avibactam restores the susceptibility of CMY-2-producing *Escherichia coli* to ceftazidime. *Antimicrob Agents Chemother* **58:** 4290–4297.

Papp-Wallace KM, Becka SA, Taracila MA, Winkler ML, Gatta JA, Rholl DA, Schweizer HP, Bonomo RA. 2015. Exposing a β-lactamase "twist": The mechanistic basis for the high level of ceftazidime resistance in the C69F variant of the *Burkholderia pseudomallei* PenI β-lactamase. *Antimicrob Agents Chemother* **60:** 777–788.

Patel G, Bonomo RA. 2013. "Stormy waters ahead": Global emergence of carbapenemases. *Front Microbiol* **4:** 48.

Paterson DL, Bonomo RA. 2005. Extended-spectrum β-lactamases: A clinical update. *Clin Microbiol Rev* **18:** 657–686.

Paton R, Miles RS, Hood J, Amyes SG. 1993. ARI 1: β-lactamase-mediated imipenem resistance in *Acinetobacter baumannii. Int J Antimicrob Agents* **2:** 81–87.

Perez F, Endimiani A, Hujer KM, Bonomo RA. 2007. The continuing challenge of ESBLs. *Curr Opin Pharmacol* **7:** 459–469.

Perez F, El Chakhtoura NG, Papp-Wallace KM, Wilson BM, Bonomo RA. 2016. Treatment options for infections caused by carbapenem-resistant Enterobacteriaceae: Can we apply "precision medicine" to antimicrobial chemotherapy? *Expert Opin Pharmacother* **17:** 761–781.

Poirel L, Heritier C, Tolun V, Nordmann P. 2004a. Emergence of oxacillinase-mediated resistance to imipenem in *Klebsiella pneumoniae. Antimicrob Agents Chemother* **48:** 15–22.

Poirel L, Heritier C, Nordmann P. 2004b. Chromosome-encoded ambler class D β-lactamase of *Shewanella oneidensis* as a progenitor of carbapenem-hydrolyzing oxacillinase. *Antimicrob Agents Chemother* **48:** 348–351.

Poirel L, Castanheira M, Carrer A, Rodriguez CP, Jones RN, Smayevsky J, Nordmann P. 2011. OXA-163, an OXA-48-related class D β-lactamase with extended activity toward expanded-spectrum cephalosporins. *Antimicrob Agents Chemother* **55:** 2546–2551.

Cite this article as *Cold Spring Harb Perspect Med* doi: 10.1101/cshperspect.a025239

Poirel L, Bonnin RA, Nordmann P. 2012. Genetic support and diversity of acquired extended-spectrum β-lactamases in Gram-negative rods. *Infect Genet Evol* **12:** 883–893.

Sekiguchi JI, Morita K, Kitao T, Watanabe N, Okazaki M, Miyoshi-Akiyama T, Kanamori M, Kirikae T. 2008. KHM-1, a novel plasmid-mediated metallo-β-lactamase from a *Citrobacter freundii* clinical isolate. *Antimicrob Agents Chemother* **52:** 4194–4197.

Verma V, Testero SA, Amini K, Wei W, Liu J, Balachandran N, Monoharan T, Stynes S, Kotra LP, Golemi-Kotra D. 2011. Hydrolytic mechanism of OXA-58 enzyme, a carbapenem-hydrolyzing class D β-lactamase from *Acinetobacter baumannii*. *J Biol Chem* **286:** 37292–37303.

Walsh TR, Weeks J, Livermore DM, Toleman MA. 2011. Dissemination of NDM-1 positive bacteria in the New Delhi environment and its implications for human health: An environmental point prevalence study. *Lancet Infect Dis* **11:** 355–362.

Woodford N, Turton JF, Livermore DM. 2011. Multiresistant Gram-negative bacteria: The role of high-risk clones in the dissemination of antibiotic resistance. *FEMS Microbiol Rev* **35:** 736–755.

Yigit H, Queenan AM, Rasheed JK, Biddle JW, Domenech-Sanchez A, Alberti S, Bush K, Tenover FC. 2003. Carbapenem-resistant strain of *Klebsiella oxytoca* harboring carbapenem-hydrolyzing β-lactamase KPC-2. *Antimicrob Agents Chemother* **47:** 3881–3889.

Yong D, Toleman MA, Giske CG, Cho HS, Sundman K, Lee K, Walsh TR. 2009. Characterization of a new metallo-β-lactamase gene, bla(NDM-1), and a novel erythromycin esterase gene carried on a unique genetic structure in *Klebsiella pneumoniae* sequence type 14 from India. *Antimicrob Agents Chemother* **53:** 5046–5054.

Approved Glycopeptide Antibacterial Drugs: Mechanism of Action and Resistance

Daina Zeng,[1] Dmitri Debabov,[2] Theresa L. Hartsell,[3] Raul J. Cano,[4,5] Stacy Adams,[6] Jessica A. Schuyler,[4] Ronald McMillan,[4] and John L. Pace[4,7,8]

[1]Agile Sciences, Raleigh, North Carolina 27606

[2]Emery Pharma Services, Emeryville, California 94608

[3]Department of Anesthesiology/Critical Care Medicine, The Johns Hopkins School of Medicine and Nursing, Baltimore, Maryland 21287

[4]ATCC Center for Translational Microbiology, Union, New Jersey 07083

[5]Biological Sciences Department, California Polytechnic State University, San Luis Obispo, California 93407

[6]Center for Skin Biology, GlaxoSmithKline, Durham, North Carolina 27703

[7]STEM Program, Kean University, Union, New Jersey 07083

[8]Biomanufacturing Research Institute and Technology Enterprise, North Carolina Central University, Durham, North Carolina 27707

Correspondence: daina.zeng@gmail.com

The glycopeptide antimicrobials are a group of natural product and semisynthetic glycosylated peptides that show antibacterial activity against Gram-positive organisms through inhibition of cell-wall synthesis. This is achieved primarily through binding to the D-alanyl-D-alanine terminus of the lipid II bacterial cell-wall precursor, preventing crosslinking of the peptidoglycan layer. Vancomycin is the foundational member of the class, showing both clinical longevity and a still preferential role in the therapy of methicillin-resistant *Staphylococcus aureus* and of susceptible *Enterococcus* spp. Newer lipoglycopeptide derivatives (telavancin, dalbavancin, and oritavancin) were designed in a targeted fashion to increase antibacterial activity, in some cases through secondary mechanisms of action. Resistance to the glycopeptides emerged in delayed fashion and occurs via a spectrum of chromosome- and plasmid-associated elements that lead to structural alteration of the bacterial cell-wall precursor substrates.

The glycopeptide antibiotics are a group of glycosylated cyclic or polycyclic nonribosomal peptides that inhibit Gram-positive bacterial cell-wall synthesis. Unlike other antimicrobial classes, these agents act as substrate binders (of cell-wall precursors) as opposed to active-site enzyme inhibitors (Barna and Williams 1984; Nagarajan 1994). Significant members of this class include vancomycin and the lipoglycopeptide teicoplanin, along with the semisynthetic lipoglycopeptide derivatives telavancin, dalbavancin, and oritavancin (Fig. 1). All

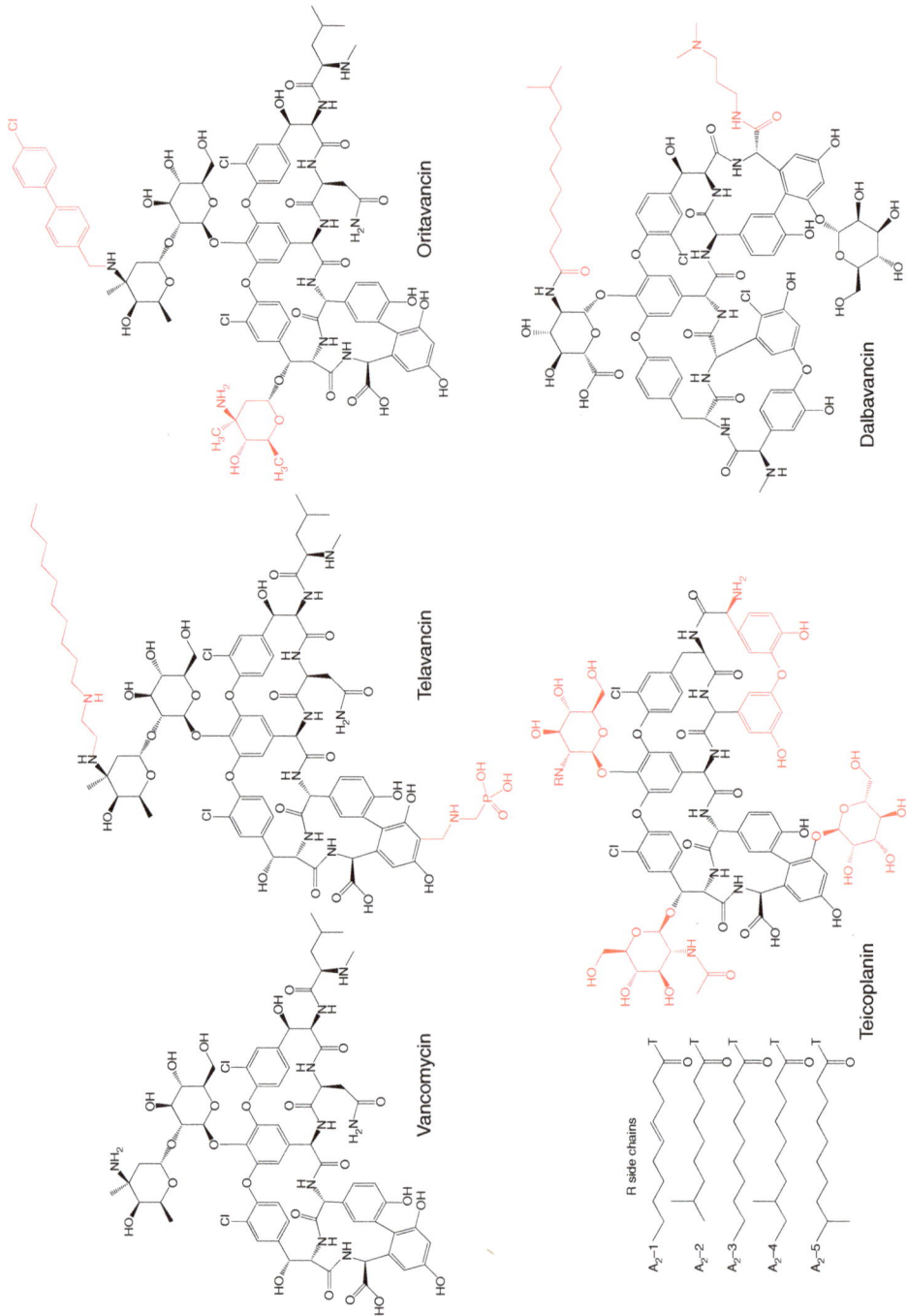

Figure 1. Glycopeptide and lipoglycopeptide antibacterial structures. Key structural features are highlighted in red.

act to prevent cross-linking of the bacterial cell-wall peptidoglycan layer by binding to the stochiometrically limited D-alanyl-D-alanine terminus of the lipid II monomer. Although the lipid II substrate is broadly expressed in most bacterial species, the antibacterial spectrum of this class is limited to Gram-positive pathogens. The physiochemical properties of the glycopeptide structure preclude transit through the outer membrane in Gram-negative species, thus blocking access to the lipid II target.

STRUCTURE AND ORIGIN OF GLYCOPEPTIDE AGENTS

Vancomycin (from the root word "vanquish") is the oldest glycopeptide antibiotic still in clinical use and has been named to the World Health Organization model list of essential medicines (see who.int/medicines/publications/essentialmedicines/EML2015_8-May-15.pdf). A natural product, vancomycin was isolated in 1953 from a strain of *Amycolatopsis orientalis* found in a soil sample (Nagarajan 1994; Kahne et al. 2005; Levine 2006) and first used clinically in 1955 with Food and Drug Administration (FDA) approval to treat penicillin-resistant staphylococci (McCormick et al. 1956). The glycopeptide class name derives from vancomycin's sugar-decorated heptapeptide structure, the hallmark of which is a central, highly cross-linked, multiring core that includes the polar uncharged amino acid asparagine and an amino-terminal *N*-methylleucine (Fig. 1) (Harris and Harris 1982). Other amino acid positions are variable across species, differentiating many of the other natural product glycopeptides that have not been pursued as drug candidates (Nagarajan 1994). Also common across these agents is the terminal vancosamine moiety of the disaccharide substituent.

The related natural product, teicoplanin, was isolated ∼30 years later from *Actinoplanes teichomyceticus* (Nagarajan 1994) and possesses a hydrophobic substituent that markedly differentiates its properties from vancomycin. Because of this difference, teicoplanin has been termed a lipoglycopeptide. Subsequent advances in medicinal and organic chemistry have yielded novel semisynthetic lipoglycopeptide derivatives of teicoplanin, vancomycin, and related compounds that show improved activity against methicillin-resistant *Staphylococcus aureus* (MRSA) and are bactericidal for enterococci. Specifically, *N*-alkyl modifications of the amino sugars and alteration of the terminal carboxyl by amide derivatization (Fig. 1) improve antimicrobial activity of these agents without directly reducing lipid II binding (Allen et al. 1997; Leadbetter et al. 2004; Wenzler and Rodvold 2015).

Such clinically relevant lipoglycopeptide derivatives include the second-generation agents dalbavancin (BI-387) and oritavancin (LY-333328), and the third-generation agent telavancin (TD-6424) (Judice and Pace 2003; Pace et al. 2003; Zhanel et al. 2010; Karlowsky et al. 2015; Leuthener et al. 2015). Dalbavancin is derived structurally from teicoplanin; the addition of an extended, lipophilic side-chain results in enhanced potency and an extended half-life, whereas an amidated carboxyl side group results in enhanced antistaphylococcal activity (Fig. 1) (Smith et al. 2015). Oritavancin is a synthetic derivative of the naturally occurring glycopeptide chloroeremomycin (A82846B) (Nagarajan 1994; Linsdell et al. 1996). The addition of a 4′-chlorobiphenylmethyl substituent to the disaccharide sugar, along with the additional monosaccharide moiety (4-epivancosamine) attached to the amino acid residue on ring 6, confers significantly enhanced activity against vancomycin-resistant enterococci, and vancomycin-intermediate and -resistant staphylococci (Mitra et al. 2015). Telavancin was derived from *N*-decylaminoethylvancomycin by regioselective reductive alkylation of the vancosamine nitrogen. For telavancin, the addition of the negatively charged, auxiliary hydrophilic group methylaminophosphonic acid to the resorcinol position results in a substantial increase in urinary clearance and decrease in kidney and liver distribution when compared with the parent derivative, but the newer agent also shows diminished bactericidal activity versus the original precursor (Leadbetter et al. 2004; JL Pace, unpubl.).

ANTIBACTERIAL SPECTRUM

Vancomycin possesses both in vitro and in vivo antibacterial activity against the clinically significant pathogens *Staphylococcus* spp., *Streptococcus* spp., susceptible *Enterococcus* spp., and many lesser species (Table 1). It is also bactericidal against *S. aureus* (including MRSA isolates) and the streptococci. This bactericidal activity led to the increased prevalence of vancomycin use as the incidence of multidrug-resistant MRSA burgeoned clinically, particularly before the approval of newer effective agents such as linezolid and the next-generation lipoglycopeptide agents. Because the physiochemical nature of the glyco- and lipoglycopeptide agents results in limited to no oral bioavailability, vancomycin has been restricted to parenteral administration, with the exception of oral (intraluminal) therapy against *Clostridium difficile* colitis.

Teicoplanin, which has been considered for development in this country but is currently only approved outside of the United States for clinical use, also possesses a Gram-positive antibacterial spectrum. However, in addition to the staphylococci, streptococci, other less signif-

icant species, and vancomycin-susceptible enterococci, teicoplanin shows antibacterial activity against VanB (vancomycin-resistant) enterococci. The spectrum of the semisynthetic derivative dalbavancin is similar (Zhanel et al. 2010). In contrast, oritavancin and telavancin also show markedly improved activity against VanA enterococci, albeit of questionable clinical significance (Zhanel et al. 2010; Arias et al. 2012).

MECHANISMS OF ACTION

Early investigations into glycopeptide activity showed vancomycin-mediated inhibition of amino acid incorporation into bacterial peptidoglycan, with intracellular accumulation of metabolic precursors including the lipid carrier-linked N-acetyl-muramic pentapeptide (MurNAc-pentapeptide; lipid II) (Anderson et al. 1965; Chatterjee and Perkins 1966; Sinha and Neuhaus 1968; Perkins 1969). This suggested interference with the transglycosylation step of peptidoglycan assembly. Distribution studies with iodinated vancomycin showed that binding was surface-localized in the bacterial cell

Table 1. Glycopeptide and lipoglycopeptide antibacterial spectrum

| Species | MIC$_{50}$/MIC$_{90}$ (μg/mL), ECOFF, or breakpoint | | | | |
	Vancomycin	Teicoplanin	Telavancin	Oritavancin	Dalbavancin
Staphylococcus aureus	1/1[a]	0.5/1[b]	0.06/0.06[c]	0.03/0.06[d]	0.06/0.06[e]
S. aureus (MRSA)	1/2[a]	1/2[b]	0.06/0.06[c]	0.03/0.06[d]	0.06/0.06[e]
S. aureus (VISA)	4–8[f]	2[g]	0.12/0.25[c]	–/1[h]	0.5/–[e]
Streptococcus pneumoniae	0.25/0.5[a]	0.12/0.12[b]	0.008/0.015[c]	0.002/0.004[a]	0.015/0.03[a]
Streptococcus pyogenes	0.5/1[a]	0.5[g]	0.03/0.03[c]	0.03/0.25[d]	≤0.03/≤0.03[e]
Streptococcus agalactiae	0.25/0.5[a]	0.12/0.12[b]	0.03/0.06[c]	0.03/0.12[d]	≤0.03/0.12[e]
Streptococcus anginosus		0.12/0.12[b]		≤0.008/0.015[d]	≤0.03/≤0.03[e]
Enterococcus faecium VSE	0.5/1[a]	0.25/1[b]	≤0.015/0.03[c]	≤0.008/0.015[d]	0.06/0.12[e]
E. faecium VRE	512/512[a]	64/>64[i]	1/>1[c]	0.008/0.06[d]	>4/>4[e]

VRE, Vancomycin-resistant *Enterococcus*; VSE, vancomycin-susceptible *Enterococcus*.
[a]Data from Zhanel et al. 2010.
[b]Data from Gales et al. 2005.
[c]Data from Karlowsky et al. 2015.
[d]Data from Mitra et al. 2015.
[e]Data from Smith et al. 2015.
[f]Data from CLSI 2016.
[g]Data from EUCAST 2015.
[h]Data from Das et al. 2013.
[i]Data from Shonekan et al. 1997.

(Bordet and Perkins 1970; Perkins and Nieto 1970; Nieto and Perkins 1971). Further, it was observed that vancomycin complexed with lipid II, and that peptidoglycan synthesis in *Gaffkyia homari* (which uses a distinct, tetrapeptide-linked precursor) was not inhibited (Hammes and Neuhaus 1974). Subsequently, it was concluded that vancomycin binds to the D-alanyl-D-alanine terminating pentapeptide moiety of lipid II, preventing both transpeptidation through substrate sequestration and transglycosylation via associated steric hindrance. The binding of vancomycin to the D-alanyl-D-alanine terminating pentapeptide is mediated by hydrogen binding (Table 2) (Cristofaro et al. 1995; Cooper et al. 1997, 2000; Cooper and Williams 1999; Walsh and Howe 2002).

In addition to the very distinct mechanism of substrate sequestration mediating its antibacterial properties, vancomycin's activity is influenced by noncovalent self-association of the antibiotic molecules (Gerhard et al. 1993; Mackay et al. 1994b; Beauregard et al. 1995; Westwell et al. 1996). Using a variety of antibiotic analogues and substrate proxies as models, the dimerization of vancomycin was determined to be promoted in part by interaction with the lipid II target (Mackay et al 1994a; Sharman et al. 1997; Loll et al. 1998, 2009). This antibiotic self-association is believed to be entropically beneficial and actually results in orders of magnitude greater affinity, as binding to a second lipid II moiety then requires lower energy because of colocalization (Williams et al. 1993, 1998; Rao et al. 1999a,b; Rekharsky et al. 2006). Cocrystalization studies with vancomycin and substrate target confirm a conformation of what

are termed back-to-back dimers (vancosamine–vancosamine [V–V]); in addition, there can be association of a third vancomycin molecule in a side–side configuration (Williamson and Williams 1981; Schaefer et al 1996; Loll et al 1997; Nitani et al. 2009). Studies with synthetic, covalently linked dimers further support these conclusions, with the V–V constructs showing the broadest in vitro antibacterial activity, often superior to that of vancomycin itself (Sundram et al. 1996; Rao et al. 1998; Stack et al. 1998; Staroske and Williams 1998; Arimoto et al. 1999; Nicolau et al. 2001a,b; Griffin et al. 2003; Shiozawa et al. 2003; Li and Xu 2005).

Vancomycin also antagonizes peptidoglycan remodeling, or "autolysis" (Sieradzki and Tomasz 1997; Peschel et al. 2000). The importance of this antagonism may explain the lack of bactericidal activity against otherwise susceptible enterococci, as these species show a low autolytic capacity. Almost certainly, the balance of effects on peptidoglycan synthesis (anabolism) and remodeling (autolysis, catabolism) is critical to antibacterial success for cell-wall inhibitors like the glycopeptides. This mechanism also likely mediates the concentration-independent bactericidal nature of vancomycin against the staphylococci; in fact, with some of the synthetic, covalently linked, vancomycin dimer analogues, treatment of bacteria with higher antibiotic concentrations leads to greater numbers of surviving cells (JL Pace, unpubl.).

Teicoplanin, like vancomycin, mediates its antibacterial activity by binding to the D-alanyl-D-alanine moiety and sequestration of the lipid II substrate, resulting in inhibition of bacterial peptidoglycan synthesis (Barna et al.

Table 2. Glyccopeptide and lipoglycopeptide mechanism(s) of action

| | Mechanism | | | |
| | Lipid II-binding | | | |
Antibiotic	Dimerization	Membrane anchoring	Transpeptidation (enzymatic)	Membrane (permeabilization–depolarization)
Vancomycin	Yes	No	No	No
Teicoplanin	No	Yes	No	No
Dalbavancin	No	Yes	No	No
Oritavancin	Yes	Yes	Yes	Yes
Telavancin	Yes	Yes	Unknown	Yes

1985; Nagarajan 1994; Beauregard et al. 1995). Importantly, distinct structural differences impact the self-associated dimer formation that is important for vancomycin target affinity but absent with teicoplanin. In contrast, a second evolutionarily beneficial property has been imparted to teicoplanin, which minimizes the lack of dimerization potential (Barna et al. 1985; Beauregard et al. 1995). The hydrophobic moiety of teicoplanin interacts with the lipid bilayer of the bacterial membrane resulting in localization (anchoring) in proximity to the lipid II substrate, thus also providing a potentially energetically favorable interaction that mediates antibacterial action (Barna et al. 1985; Beauregard et al. 1995; Economou et al. 2013). Likewise, dalbavancin mechanism of action is believed to be similar to that of teicoplanin.

Oritavancin, however, has additional mechanisms of action. The hydrophobic 4′-chlorobiphenylmethyl side chain on the disaccharide sugar of oritavancin facilitates direct interaction with the bacterial cell membrane, providing additional stability during oritavancin's interaction with lipid II (Allen and Nicas 2003). Oritavancin can also bind to D-alanyl-D-lactate, and this affinity is enhanced by its ability to form noncovalent dimers before attachment to the bacterial peptidoglycan cell wall. This dimerization is possible because of interactions between the disaccharides attached to residue 4, the chlorine on ring 2, and the 4-epi-vancosamine on ring 6, also found with the underivatized natural product chloroeremomycin (A28246B), which already shows superior dimerization to vancomycin (Linsdell et al. 1996; Allen et al. 2002). In addition to the above, oritavancin has been reported to inhibit transpeptidation. This action is attributed to the 4′-chlorobiphenylmethyl side chain, which allows binding of the drug to the pentaglycyl bridge, a secondary site in peptidoglycan, and likely contributes to oritavancin's activity against vancomycin-intermediate S. aureus (VISA) and vancomycin-resistant Enterococcus isolates (Leimkuhler et al. 2005; Mitra et al. 2015). The 4′-chlorobiphenylmethyl side chain of the molecule is also credited for Gram-positive bacterial cell death via membrane depolarization and increased membrane permeability. This mechanism, which leads to disruption of membrane ultrastructure, has been shown in vitro to effect both antibacterial activity against stationary-phase inocula of staphylococci and rapid killing of susceptible Gram-positive pathogens. Interestingly, oritavancin additionally shows activity against S. aureus biofilms in vitro (Belley et al. 2009). Finally, oritavancin may also have some effect on inhibition of RNA synthesis but a direct effect is questionable (Allen and Nicas 2003).

Telavancin differs from vancomycin by having a dual mechanism of action, thought to be responsible for its rapid, dose-dependent bactericidal activity (Higgins et al. 2005). Like vancomycin and the other members of this class, telavancin has a glycopeptide core that binds with high affinity to the D-alanyl-D-alanine terminus of cell-wall precursors through a network of hydrogen bonds and hydrophobic packing interactions (Karlowsky et al. 2015). Thus, telavancin inhibits synthesis of peptidoglycan by binding to late-stage precursors. This binding prevents both polymerization of the lipid II precursor and subsequent cross-linking (Lunde et al. 2009). In contrast to vancomycin, telavancin does not inhibit autolysis (JL Pace, unpubl.). However, like oritavancin, telavancin does bind to bacterial cell membranes and cause a rapid, concentration-dependent depolarization of the plasma membrane, increased permeability, and leakage of cellular adenosine triphosphate and potassium (Higgins et al. 2005). The timing of these membrane changes correlates with the rapid, concentration-dependent bacterial killing seen with this agent, suggesting that telavancin's bactericidal activity results from the dissipation of cell membrane potential and an increase in membrane permeability (Higgins et al. 2005). Telavancin is also 10 times more active than vancomycin in inhibiting both the transglycosylation and the synthesis of peptidoglycan because of its ability to bind to bacterial cell membranes (Debabov et al. 2002; Higgins et al. 2005; Lunde et al. 2009). An alternative mechanism has been proposed for lipoglycopeptides suggesting that the liposaccharide elements of lipoglycopeptides interact directly with and inhibit transglycosylase enzymes that mediate the polymerization of precursors into immature, un-cross-linked peptidoglycan (Ge et al. 1999; Goldman et al. 2000; Kerns et al. 2002; Chen et al. 2003); however, in the case of telavancin, this mechanism is not supported by observations that inhibition of transglycosylase is completely antagonized by the cell-wall-mimicking tripeptide dKAA (Higgins et al. 2005). Finally, it is hypothesized that telavancin inhibits bacterial lipid synthesis (Debabov et al. 2002).

MECHANISMS OF RESISTANCE

Gram-negative bacilli are intrinsically resistant to vancomycin and lipoglycopeptides, because the presence of the outer membrane prevents these large, complex molecules from entering the cell and binding to their target sites. In Gram-positive bacteria, the onset of vancomycin resistance was long delayed compared with that for other nonglycopeptide antibiotics. The first vancomycin-resistant clinically isolated strain was reported in 1987, more than 30 years after the introduction of vancomycin (Leclercq et al. 1988). Today more than 30% of intensive care unit (ICU) *Enterococcus faecalis* infections are vancomycin-resistant *Enterococcus* (VRE) and, following the emergence of vancomycin resistance in *Enterococcus*, rapid resistance developed in *S. aureus* (VRSA) as a result of horizontal gene transfer from resistant enterococci (Nebreda et al. 2007; Werner et al. 2008; Zhu et al. 2010; Fernandes et al. 2015). The first fully VRSA strain (vancomycin, minimum inhibitory concentration [MIC] > 256 µg/mL; teicoplanin, MIC = 128 µg/mL) was reported in 2002 (Howden et al. 2010) with an increasing number of cases each year in the United States (Walters et al. 2015).

Because vancomycin forms complexes with peptidoglycan precursors, its activity is determined by its binding affinity to precursor substrates. Indeed, the main mechanism of resistance to vancomycin is owing to the presence of enzymes that produce lower-affinity binding precursors in which the carboxy-terminal D-alanine residue is replaced by either D-lactate, resulting in a 1000-fold decrease in binding affin-

ity caused by the loss of a hydrogen bond interaction (Pootoolal et al. 2002), or D-serine, resulting in a sixfold loss of binding affinity likely caused by steric hindrance. There are six different molecular resistance elements that confer resistance to vancomycin by altering the structure of the peptidoglycan precursors (Table 3) (Cetinkaya et al. 2000; Courvalin 2006).

Van-A type resistance is the most frequently encountered mechanism of resistance in enterococci and is the only one detected in *S. aureus* to date (Perichon and Courvalin 2009). Transposon Tn*1546* and closely related elements encoding nine polypeptides responsible for vancomycin resistance mediate this by changing the carboxyl terminus peptidoglycan precursor from D-alanine to D-lactate. These proteins include those needed for the synthesis of modified peptidoglycan precursors (VanH and VanA), hydrolysis of normal precursors (VanX and VanY), regulation of expression of resistance (VanR and VanS), transposition (products of ORF1 and ORF2), and an unknown function (VanZ). VanH is a dehydrogenase that reduces pyruvate to D-Lac, and VanA is a ligase allowing the formation of a D-Ala-D-Lac depsipeptide that replaces the D-Ala-D-Ala dipeptide. Furthermore, indigenous concentrations of D-Ala-D-Ala are reduced by VanX and VanY. VanX is a D,D-dipeptidase that hydrolyzes the D-Ala-D-Ala formed by the host chromosomal D-Ala-D-Ala ligase, and VanY is a D,D-carboxypeptidase that hydrolyzes the terminal D-Ala of pentapeptide precursors that are produced if elimination of the D-Ala-D-Ala is not complete (Arthur et al. 1996). Inducible expression of this resistance is regulated by the VanR/VanS two-component system. VanS is the membrane-associated sensor that is activated by the presence of glycopeptides and transfers a phosphoryl group to VanR, which is the response regulator that activates the cotranscription of the *vanH*, *vanA*, *vanX*, and *vanY* genes by binding to the P_{RES} promoter (Arthur et al. 1992), and of the *vanR* and *vanS* genes by binding to the P_{REG} promoter (Depardieu et al. 2007). This transposition confers high-level resistance to vancomycin and teicoplanin as well as telavancin, although telavancin

Table 3. Glycopeptide and lipoglycopeptide antibacterial resistance determinants (Courvalin 2006)

Resistance type	Resistance element	Expression	Location	Modified target	MIC (μg/mL)			
					Vancomycin	Teicoplanin	Oritavancin	Telavancin
Van-A	Tn1546	Inducible	Plasmid	D-alanine D-lactate	64–1000	16–512	8–16[a]	32[a]
Van-B	Tn1547 or Tn1549	Inducible	Plasmid	D-alanine D-lactate	4–1000	0.5–1	8–16[a]	
Van-C		Constitutive Inducible	Chromosome	D-alanine D-serine	2–32	0.5–1	8–16[a]	
Van-D		Constitutive	Chromosome	D-alanine D-lactate	64–128	4–64		
Van-E		Inducible	Chromosome	D-alanine D-serine	8–32	0.5		
Van-G		Inducible	Chromosome	D-alanine D-serine	16	0.5		

[a]Selected in resistance passaging studies.

Cite this article as Cold Spring Harb Perspect Med doi: 10.1101/cshperspect.a026989

MICs remain 32- and 128-fold lower compared with those of vancomycin and teicoplanin (Karlowsky et al. 2015). The VanA-type resistance is transmittable from *Enterococcus* species to *S. aureus* and is the main resistance mechanism for VRSA (Perichon and Courvalin 2009).

Like VanA, VanB-type resistance is caused by the presence of a tranposition (Tn*1547* or Tn*1549*), resulting in the altered synthesis of the depsipeptide D-Ala-D-Lac to replace the dipeptide D-Ala-D-Ala and leading to decreased lipid II binding by the antimicrobials. Also, the organization and functionality of the vanB cluster are similar to that of vanA, encoding a dehydrogenase, ligase, and a dipeptidase, all of which have a high level of sequence identity (67%–76%) to the corresponding proteins of the *vanA* operon. However, the regulatory proteins in VanB type resistance only distantly relate to VanRS (34% and 24% identity) and are only induced by vancomycin but not teicoplanin; hence, strains with this phenotype have high-level resistance only to vancomycin (Courvalin 2006).

Likewise, VanD-type resistance also results in the production of peptidoglycan precursors ending in D-Ala-D-Lac. The organization of the *vanD* operon is similar to that of *vanA* and *vanB*; however, *vanD* is located exclusively on the bacterial chromosome and resistance is not transferable by conjugation to other enterococci (Depardieu et al. 2004). In addition, VanD-type strains do not have D,D-dipeptidase activity, which is not needed because various point mutations also result in null D-Ala-D-Ala ligase activity (Arthur et al. 1998; Depardieu et al. 2003b). Interestingly, because of this defect in indigenous D-Ala-D-Ala production, VanD-type strains require constitutive expression of these resistance elements to survive and achieve this via mutations in the $VanS_D$ sensor or $VanR_D$ regulator (Depardieu et al. 2003b, 2004).

The VanC-type resistance is found in *Enterococcus gallinarum*, *Enterococcus casseliflavus*, and *Enterococcus flavescens* and results in production of peptidoglycan precursors terminating in D-Ala-D-Ser (Reynolds and Courvalin 2005). This cluster contains three main genes that are needed to confer resistance: VanT is a membrane-bound serine racemase that produces D-Ser; VanC is a ligase that catalyzes synthesis of D-Ala-D-Ser; and $VanXY_C$ possesses both D,D-dipeptidase and D,D-carboxypeptidase activities and hydrolyzes precursors ending in D-Ala-D-Ala. The VanC phenotype is expressed constitutively or inducibly on the chromosome and confers resistance to low levels of vancomycin but not to other glycopeptides (Reynolds and Courvalin 2005).

VanE-type resistance is organized similarly to VanC and is found in intrinsically resistant *E. faecalis*. Unlike VanC, the VanE phenotype is only inducibly expressed (Courvalin 2006).

The VanG-type resistance consists of genes from various *van* operons: VanG is a D-Ala-D-Ser ligase (similar to VanC-type strains); $VanXY_G$ is a putative bifunctional D,D-peptidase and D,D-carboxypeptidase; $VanT_G$ is a serine racemase (similar to VanE-type strains); and $VanR_G$ and $VanS_G$ make up the two component regulatory response proteins (highest similarity to the VanD-type strains). Like VanC and VanE phenotypes, the VanG-type strains produce D-Ala-D-Ser and only have intermediate-level resistance to vancomycin (Depardieu et al. 2003a).

VISA strains, unlike those described above, do not carry imported foreign genetic elements; rather, resistance occurs owing to mutations that appear in the invading pathogen during treatment. Although the MIC value of these strains are modest (MIC = $2–16$ µg/mL), vancomycin treatment of VISA infections often ends in treatment failure (Linares 2001; Fridkin et al. 2003). Most commonly, these strains have emerged in patients with MRSA infections undergoing prolonged vancomycin therapy (Linares 2001; Chen et al. 2003; Howden et al. 2010). Although VISA isolates have differing point mutations, they do share common abnormal properties including increased cell-wall material, an aberrant separation of daughter cells at the end of cell division, and altered (usually decreased) rates of autolysis (Hiramatsu 2001; Walsh and Howe 2002; Sieradzki and Tomasz 2003; Pfeltz and Wilkinson 2004; Gardete et al. 2012; Howden et al. 2014; Karlowsky et al. 2015). The most frequent genetic changes in VISA strains are in *walkR* and *vraSR* (two-component sensory regulatory system controlling the transcription of genes in cell-wall synthesis), and *rpoB* (RNA polymerase) genes (Howden et al. 2008, 2011; Kato et al. 2010; Watanabe et al. 2011; Hafer et al. 2012). It is hypothesized that these point mutations decrease the cross-linking of peptidoglycan strands, which results in an increase in the number of free D-Ala-D-Ala residues in the cell wall that bind and sequester vancomycin (Sieradzki et al. 1999; Hiramatsu 2001; Sieradzki and Tomasz 2003). These D-Ala-D-Ala residues can be maintained in the mature peptidoglycan because of low carboxypeptidase activity, and vancomycin-resistant laboratory mutants have shown an increased capacity to bind vancomycin in their cell walls (Sieradzki and Tomasz 1997). Not only does this result in "decoy" targets, but binding of the large glycopeptide molecules would further hinder the progress of the antibiotic through the mesh-like cell wall to its lethal target, the lipid-linked peptidoglycan precursor, thus creating a "drug capture" (Perichon and Courvalin 2009) and "clogging" phenomenon (Sieradzki et al. 1999).

Reported in vitro mutational frequency of resistance against dalbavancin, oritavancin, and telavancin has been remarkably, although not unexpectedly, limited (Arthur et al. 1999; Krause et al. 2005; Sahm et al. 2006; Goldstein et al. 2007; Laohavaleeson et al. 2007; Kosowska-Shick et al. 2009; Song et al. 2013). Resistance to vancomycin required ~30 years in the clinic to become prevalent and required concomitant acquisition of several exogenous genes. With the semisynthetic lipoglycopeptide antimicrobials showing additional mechanisms of action, resistance frequency is expected to be lower.

Like teicoplanin, telavancin retains activity against strains that express vanB, but not vanA as it induces expression of the latter (Zhanel et al. 2010). In a study by Kosowska-Shick and colleagues that serially passaged 10 strains of MRSA to evaluate the potential for emergence of resistance, spontaneous mutants resistant to telavancin were detected in only 1 of the 10 MRSA strains tested. The MIC for that strain increased from 0.25 to 2 μg/mL after 43 days of serial passaging; no further increase in the

MIC of this mutant occurred when serial passaging was continued to a maximum of 50 days (Kosowska-Shick et al. 2009). In another study, six species of Gram-positive bacteria, including *S. aureus*, vancomycin-resistant enterococci, and multidrug-resistant *Streptococcus pneumoniae*, were exposed to sub-MIC concentrations of telavancin for 10 days to attempt to isolate resistant mutants. However, this failed to produce any organisms that were resistant to telavancin; mutation frequencies were $<1 \times 10^{-9}$ for all isolates tested (Sahm et al. 2006). Laohavaleeson et al. (2007) also reported that telavancin showed a low frequency of spontaneous resistance among staphylococci and enterococci, with low spontaneous resistance rates for *S. aureus* strains and both VanA *E. faecalis* and *Enterococcus faecium* ($<1 \times 10^{-9}$ and $<1 \times 10^{-7}$, respectively). Similarly, Krause et al. (2005) reported telavancin to have a low potential for resistant-mutant selection among *S. aureus* (MSSA, MRSA) and vancomycin-susceptible enterococci because it failed to select mutants with MICs >4 times those of parental strains following 20 passages. Two mutants with telavancin MICs 4 times that of the parental strain (32 μg/mL) were selected from vancomycin-resistant enterococci strains (VanA) used; mutant phenotypes were stable on extended subculture, and no changes were observed in growth properties of isolates with reduced susceptibility to telavancin. Song et al. (2013) sought to determine mechanisms of decreased telavancin susceptibility in a laboratory-derived mutant of *S. aureus*, TlvDSMED1952, and showed extensive changes in the mutant transcriptome compared with the susceptible parent strain, MED1951. Upregulated genes in TlvDSMED1952 included cofactor biosynthesis genes, cell-wall-related genes, fatty acid biosynthesis genes, and stress genes, whereas down-regulated genes included lysine operon biosynthesis genes and *lrgB* (induced by telavancin in susceptible strains), *agr* and *kdpDE* genes, various cell surface protein genes, phenol-soluble modulin genes, several protease genes, and genes involved in anaerobic metabolism. The investigators reported that TlvDSMED1952 showed various features similar to VISA and strains with decreased dapto-mycin susceptibility, but also showed its own unique features. Consistent with these findings was that in vitro resistance was more frequently selected in VanA *E. faecalis,* and that an increased IC_{50} for $[^{14}C]$-acetate incorporation into both cellular phospholipids and saponifiable fatty acids appeared to correlate with the increased MICs, whereas peptidoglycan synthesis inhibition was unchanged (Debabov et al. 2002; Higgins et al. 2005; JL Pace, unpubl.). Heteroresistance has not been detected in isolates of *S. aureus,* nor has resistance been detected in isolates of *S. aureus* or *S. pneumoniae* collected either through clinical trials or through global surveillance (Kosowska-Shick et al. 2009; Krause et al. 2012). MIC_{90} values of 0.25–0.5 µg/mL (*S. aureus* isolates including MRSA) and 0.03 µg/mL (*S. pneumoniae*) have been reported (Theravance 2014).

Similar to teicoplanin and telavancin, dalbavancin resistance is observed for strains that express the vanA operon, but not vanB, as this agent only induces expression of the former. Prospective worldwide surveillance from 2002 to 2012 of the in vitro potency of dalbavancin against >150,000 Gram-positive isolates found no evidence of glycopeptide-resistant or glycopeptide intermediate *S. aureus* by CLSI criteria (MIC_{90} was 0.06 µg/mL for *S. aureus*). Isolates from phase-III clinical trials showed similar susceptibility profiles to those detailed in the surveillance studies (Jones et al. 2013a,b). The potential for staphylococci to develop resistance to dalbavancin was studied by Goldstein et al. (2007) by means of direct selection and by serial passage at subtherapeutic concentrations. In direct selection experiments with the six strains tested (one MSSA strain, three MRSA strains, one VISA strain, and one methicillin-resistant *S. epidermidis* strain), no colonies with increased MICs were obtained on dalbavancin-containing plates (frequency $<10^{10}$). The same six strains were passaged in the presence of sub-inhibitory concentrations of dalbavancin. After 20 passages, four strains (including the VISA strain) had a dalbavancin MIC that was within one doubling dilution of the MIC obtained before the 20 passages. For two MRSA strains,

increases of two or three dilutions were seen (to 0.25 and 0.5 µg/mL). When these isolates were grown on drug-free medium for 3 consecutive days and retested, the MICs were equivalent or within one doubling dilution of the starting MICs.

Oritavancin resistance among clinical isolates has not been detected so far (Allen and Nicas 2003). However, a moderate level of resistance (MIC ≤ 16 µg/mL) to the drug has been observed in the laboratory among Enterococcus isolates showing the VanA and VanB phenotypes and can occur by various mechanisms (Arthur et al. 1999). Total replacement of the peptidoglycan precursors terminating in D-alanine by isolates capable of producing peptidoglycan precursors terminating in D-lactate can confer resistance to oritavancin. This can be achieved in vitro by either increasing resistance gene expression or reducing production of D-Ala-D-Ala. Resistance to the drug may also occur with expression of the *vanZ* gene. Additionally, mutations in the *vanSB* sensor gene of the *vanB* cluster confer cross-resistance to teicoplanin and oritavancin (Arthur et al. 1999), making it is quite likely that emergence of resistance to oritavancin may occur with widespread clinical use.

CONCLUDING REMARKS

Over the past 60 years, vancomycin has come to play a critical, lifesaving role in a physician's armamentarium against serious Gram-positive bacterial infections. This antibiotic's remarkable longevity of use and delayed onset of resistance speaks to the unique mode of action of the glycopeptide antibiotic class. Since 2011, three semisynthetic lipoglycopeptide antibacterials have gained FDA approval in the United States. The development of these new agents was guided by the extensive body of earlier research on vancomycin—its mechanism of action, mechanisms of resistance, and therapeutic applications. These targeted efforts have imparted new, beneficial properties to the glycopeptide class. One can only hope that, through stewardship and appropriate practice, the successful clinical use of vancomycin and the newer glycopeptide agents will continue far into the future, to the benefit of all.

ACKNOWLEDGMENTS

Dr. Dmitri Debabov worked on this review before his employment at Allergan, and the opinions stated within do not represent those of Allergan.

REFERENCES

Allen NE, Nicas TI. 2003. Mechanism of action of oritavancin and related glycopeptide antibiotics. *FEMS Microbiol Rev* **26:** 511–532.

Allen NE, LeTourneau DL, Hobbs N. 1997. Role of hydrophobic side chains as determinants of antibiotic activity of semi-synthetic glycopeptide antibiotics. *J Antibiot* **50:** 677–684.

Allen NE, LeTourneau DL, Hobbs N, Thompson RC. 2002. Hexapeptide derivatives of glycopeptide antibiotics: Tools for mechanism of action studies. *Antimicrob Agents Chemother* **46:** 2344–2348.

Anderson JS, Matsuhashi M, Haskin MA, Strominger JL. 1965. Lipid-phosphoacetylmuramyl-pentapeptide and lipid-phosphodisaccharide-pentapeptide: Presumed membrane transport intermediates in cell wall synthesis. *Proc Natl Acad Sci* **53:** 881–889.

Arias CA, Mendes RE, Stilwell MG, Jones RN, Murray BE. 2012. Unmet needs and prospects for oritavancin in the management and vancomycin-resistant enterococcal infections. *Clin Infect Dis* **54:** S233–S238.

Arimoto HK, Nishimura K, Hayakawa I, Kinumi T, Uemura D. 1999. Multi-valent polymer of vancomycin: Enhanced activity against VRE. *Chem Commun* 1361–1362.

Arthur M, Molinas C, Courvalin P. 1992. The VanS-VanR two-component regulatory system controls synthesis of depsipeptide peptidoglycan precursors in *Enterococcus faecium* BM4147. *J Bacteriol* **174:** 2582–2591.

Arthur M, Reynolds P, Courvalin P. 1996. Glycopeptide resistance in enterococci. *Trends Microbiol* **4:** 401–407.

Arthur M, Depardieu F, Cabanié L, Reynolds P, Courvalin P. 1998. Requirement of the VanY and VanX D,D-peptidases for glycopeptide resistance in enterococci. *Mol Microbiol* **30:** 819–830.

Arthur M, Depardieu F, Reynolds P, Courvalin P. 1999. Moderate-level resistance to glycopeptide LY333328 mediated by genes of the *vanA* and *vanB* clusters in enterococci. *Antimicrob Agents Chemother* **43:** 1875–1880.

Barna JCJ, Williams DH. 1984. The structure and mode of action of glycopeptide antibiotics of the vancomycin group. *Annu Rev Microbiol* **38:** 339–357.

Barna JCJ, Williams DH, Williamson MI. 1985. Structural features that affect the binding of teicoplanin, ristocetin A, and their derivatives to bacterial cell wall model *N*-acetyl-D-alaynyl-D-alanine. *J Chem Soc Chem Commun* **5:** 254–256.

Beauregard DA, Williams DH, Gwynn MN, Knowles DJC. 1995. Dimerization and membrane anchors in extracellular targeting of vancomycin group antibiotics. *Antimicrob Agents Chemother* **39:** 781–785.

Belley AE, Neesham-Grenon E, McKay G, Arhin FF, Harris R, Beveridge T, Parr TR, Moeck G. 2009. Oritavancin kills stationary-phase and biofilm *Staphylococcus aureus* cells in vitro. *Antimicrob Agents Chemother* **53:** 918–925.

Bordet C, Perkins HR. 1970. Iodinated vancomycin and mucopeptide biosynthesis by cell-free preparations from *Micrococcus lysodeikticus*. *Biochem J* **119:** 877–883.

Cetinkaya YP, Falk P, Mayhall CG. 2000. Vancomycin-resistant enterococci. *Clin Microbiol Rev* **13:** 686–707.

Chatterjee AN, Perkins HR. 1966. Compounds formed between nucleotides related to the biosynthesis of bacterial cell wall and vancomycin. *Biochem Biophys Res Commun* **24:** 489–494.

Chen LD, Walker D, Sun B, Hu Y, Walker S, Kahne D. 2003. Vancomycin analogues active against vanA-resistant strains inhibit bacterial transglycosylase without binding substrate. *Proc Natl Acad Sci* **100:** 5658–5663.

CLSI. 2016. *Performance Standards for Antimicrobial Susceptibility Testing*. CLSI supplement M100S26, 26th ed. Clinical and Laboratory Standards Institute, Wayne, PA.

Cooper MA, Williams DH. 1999. Binding of glycopeptide antibiotics to a model of a vancomycin-resistant bacterium. *Chem Biol* **6:** 891–899.

Cooper MA, Williams DH, Cho YR. 1997. Surface plasmon resonance analysis of glycopeptide antibiotic activity at a model membrane surface. *Chem Commun* 1625–1626.

Cooper MA, Fiorini MT, Abell C, Williams DH. 2000. Binding of vancomycin group antibiotics to D-alanine and D-lactate presenting self-assembled monolayers. *Bioorg Med Chem* **8:** 2609–2616.

Courvalin P. 2006. Vancomycin resistance in Gram-positive cocci. *Clin Infect Dis* **42:** S25–S34.

Cristofaro MF, Beauregard DA, Yan H, Osborn NJ, Williams DH. 1995. Cooperativity between non-polar and ionic forces in the binding of bacterial cell wall analogues by vancomycin in aqueous solution. *J Antibiotics* **48:** 805–810.

Das B, Sarkar C, Schachter J. 2013. Oritavancin—A new semisynthetic lipoglycopeptide agent to tackle the challenge of resistant Gram-positive pathogens. *Pak J Pharm Sci* **26:** 1045–1055.

Debabov DPJ, Kaniga K, Nodwell M, Farrington L, Campbell B, Karr D, Leadbetter M, Linsell M, Wu T, Krause K, et al. 2002. A novel bactericidal antibiotic inhibits bacterial lipid synthesis. *Interscience Conference on Antimicrobial Agents and Chemotherapy.* San Diego, CA, September 27–30.

Depardieu F, Bonora MG, Reynolds PE, Courvalin P. 2003a. The vanG glycopeptide resistance operon from *Enterococcus faecalis* revisited. *Mol Microbiol* **50:** 931–948.

Depardieu F, Reynolds PE, Courvalin P. 2003b. VanD-type vancomycin-resistant *Enterococcus faecium* 10/96A. *Antimicrob Agents Chemother* **47:** 7–18.

Depardieu FM, Kolbert M, Pruul H, Bell J, Couvalin P. 2004. VanD-type vancomycin-resistant *Enterococcus faecium* and *Enterococcus faecalis*. *Antimicrob Agents Chemother* **48:** 3892–3904.

Depardieu F, Podglajen I, Leclercq R, Collatz E, Couvalin P. 2007. Modes and modulations of antibiotic resistance gene expression. *Clin Microbiol Rev* **20:** 79–114.

Economou NJ, Zentner IJ, Lazo E, Jakoncic J, Stojanoff V, Weeks SD, Grasty KC, Cocklin S, Loll PJ. 2013. Structure of the complex between teicoplanin and a bacterial cell-wall peptide: Use of a carrier-protein approach. *Acta Crystallogr D Biol Crystallogr* **69:** 520–533.

European Committee on Antimicrobial Susceptibility Testing. 2015. *Breakpoint tables for interpretation of MIC and zone diameters*, version 4.0. ESCMID, Basel, Switzerland.

Fernandes MS, Fujimoto S, de Souza LP, Kabuki DY, da Silva MJ, Kuave AY. 2015. Dissemination of *Enterococcus faecalis* and *Enterococcus faecium* in a ricotta processing plant and evaluation of pathogenic and antibiotic resistance profiles. *J Food Sci* **80:** M765–775.

Fridkin SK, Hageman J, McDougal LK, Mohammed J, Jarvis WR, Peri TM, Tenover FC; Vancomycin-Intermediate *Staphylococcus aureus* Epidemiology Study Group. 2003. Epidemiological and microbiological characterization of infections caused by *Staphylococcus aureus* with reduced susceptibility to vancomycin, United States, 1997–2001. *Clin Infect Dis* **36:** 429–439.

Gales AC, Jones RN, Andrade SS, Pereira AS, Sader HS. 2005. In vitro activity of tigecycline, a new glycylcycline, tested against 1,326 clinical bacterial strains isolated from Latin America. *Braz J Infect Dis* **9:** 348–356.

Gardete SC, Kim C, Hartmann BM, Mwangi M, Roux CM, Dunman PM, Chambers HF, Tomasz A. 2012. Genetic pathway in acquisition and loss of vancomycin resistance in a methicillin resistant *Staphylococcus aureus* (MRSA) strain of clonal type USA300. *PLoS Pathog* **8:** e1002505.

Ge M, Chen Z, Onishi HR, Kohler J, Silver LL, Kerns R, Fukuzawa S, Thompson C, Kahne D. 1999. Vancomycin derivatives that inhibit peptidoglycan biosynthesis without binding D-Ala-D-Ala. *Science* **284:** 507–511.

Gerhard U, Mackay JP, Maplestone RA, Williams DH. 1993. The role of the sugar and chlorine substituents in the dimerization of vancomycin antibiotics. *J Am Chem Soc* **115:** 232–237.

Goldman RC, Baizman ER, Longley CB, Branstrom AA. 2000. Chlorobipehnyl-desleucyl-vancomycin inhibits the transglycosylation process required for peptidoglycan synthesis in bacteria in the absence of didpeptide binding. *FEMS Microbiol Lett* **183:** 209–214.

Goldstein BP, Draghi FC, Sheehan DJ, Hogan P, Sahm DF. 2007. Bactericidal activity and resistance development profiling of dalbavancin. *Antimicrob Agents Chemother* **51:** 1150–1154.

Griffin JH, Linsell MS, Nodwell MB, Chen Q, Pace JL, Quast KI, Krause KM, Farrington L, Wu TX, Higgins DL, et al. 2003. Multivalent drug design: Synthesis and in vitro analysis of an array of vancomycin dimers. *J Am Chem Soc* **125:** 6517–6531.

Hafer C, Lin Y, Kornblum J, Lowy FD, Uhlemann AC. 2012. Contribution of selected gene mutations to resistance in clinical isolates of vancomycin-intermediate *Staphylococcus aureus*. *Antimicrob Agents Chemother* **56:** 5845–5851.

Hammes WP, Neuhaus FC. 1974. On the mechanism of action of vancomycin: Inhibition of peptidoglycan synthesis in *Gaffkya homari*. *Antimicrob Agents Chemother* **6:** 722–728.

Harris CM, Harris TM. 1982. Structure of the glycopeptides antibiotic vancomycin, Evidence for an asparagine residue in the peptide. *J Am Chem Soc* **104:** 4293–4295

Higgins DL, Chang R, Debabov DV, Leung J, Wu T, Krause KM, Sandvik E, Hubbard JM, Kaniga K, Schmidt DE, et al. 2005. Telavancin, a multifunctional lipoglycopeptide, disrupts both cell wall synthesis and cell membrane integrity in methicillin-resistant *Staphylococcus aureus*. *Antimicrob Agents Chemother* **49:** 1127–1134.

Hiramatsu K. 2001. Vancomycin-resistant *Staphylococcus aureus*: A new model of antibiotic resistance. *Lancet Infect Dis* **1:** 147–155.

Howden BP, Stinear TP, Allen DL, Johnson PD, Ward PB, Davies JK. 2008. Genomic analysis reveals a point mutation in the two-component sensor gene *graS* that leads to intermediate vancomycin resistance in clinical *Staphylococcus aureus*. *Antimicrob Agents Chemother* **52:** 3755–3762.

Howden BP, Davies JK, Johnson PD, Stinear TP, Grayson ML. 2010. Reduced vancomycin susceptibility in *Staphylococcus aureus*, including vancomycin-intermediate and heterogeneous vancomycin-intermediate strains: Resistance mechanisms, laboratory detection, and clinical implications. *Clin Microbio Rev* **23:** 99–139.

Howden BP, McEvoy CR, Allen DL, Chua K, Gao W, Harrison PF, Bell J, Coombs G, Bennett-Wood V, Porter JL, et al. 2011. Evolution of multidrug resistance during *Staphylococcus aureus* infection involves mutation of the essential two component regulator WalKR. *PLoS Pathog* **7:** e1002359.

Howden BP, Peleg AY, Stinear TP. 2014. The evolution of vancomycin intermediate *Staphylococcus aureus* (VISA) and heterogenous-VISA. *Infect Genet Evol* **21:** 575–582.

Jones RN, Sader HS, Flamm RK. 2013a. Update of dalbavancin spectrum and potency in the USA: Report from the SENTRY Antimicrobial Surveillance Program (2011). *Diagn Microbiol Infect Dis* **75:** 304–307.

Jones RN, Sader HS, Mendes RE, Flamm RK. 2013b. Update on antimicrobial susceptibility trends among *Streptococcus pneumoniae* in the United States: Report of ceftaroline activity from the SENTRY antimicrobial surveillance program (1998–2011). *Diagn Microbiol Infect Dis* **75:** 107–109.

Judice JK, Pace JL. 2003. Semi-synthetic glycopeptide antibacterials. *Biorg Med Chem Lett* **13:** 4165–4168.

Kahne D, Leimkuhler C, Lu W, Walsh C. 2005. Glycopeptide and lipoglycopeptide antibiotics. *Chem Rev* **105:** 425–448.

Karlowsky JA, Nichol K, Zhanel GG. 2015. Telavancin: Mechanisms of action, in vitro activity, and mechanisms of resistance. *Clin Infect Dis* **61:** S58–S68.

Kato Y, Suzuki T, Ida T, Maebashi K. 2010. Genetic changes associated with glycopeptide resistance in *Staphylococcus aureus*: Predominance of amino acid substitutions in YvqF/VraSR. *J Antimicrob Chemother* **65:** 37–45.

Kerns R, Dong SD, Fukuzawa S, Carbeck J, Kohler J, Silver L, Kahne D. 2002. The role of hydrophobic substituents in the biological activity of glycopeptide antibiotics. *J Am Chem Soc* **122:** 12608–12609.

Kosowska-Shick K, Clark C, Pankush GA, McGhee P, Dewasse B, Beachel L, Appelbaum PC. 2009. Activity of telavancin against staphylococci and enterococci determined by MIC and resistance selection studies. *Antimicrob Agents Chemother* **53**: 4217–4224.

Krause KM, Benton BM, Higgins DL, Kaniga K, Renelli M, Humphrey PK. 2005. Telavancin possesses low potential for resistant mutant selection in serial passage studies of *Staphylococcus aureus* and enterococci. *15th European Congress of Clinical Microbiology and Infectious Diseases*, Abstract P1577. Copenhagen, Denmark, April 2–5.

Krause KM, Blais J, Lewis SR, Lunde CS, Barrier SL, Friedland HD, Kitt MM, Benton BM. 2012. In vitro activity of telavancin and occurrence of vancomycin heteroresistance in isolates from patients enrolled in phase 3 clinical trials of hospital-acquired pneumonia. *Diagn Microbiol Infect Dis* **74**: 429–431

Laohavaleeson S, Kuti JL, Nicolau DP. 2007. Telavancin: A novel lipoglycopeptide for serious Gram-positive infections. *Expert Opin Investig Drugs* **16**: 347–57.

Leadbetter MR, Adams SM, Bazzini B, Fatheree PR, Karr DE, Krause KM, Lam BM, Linsell MS, Nodwell MB, Pace JL, et al. 2004. Hydrophobic vancomycin derivatives with improved ADME properties: Discovery of telavancin (TD-6424). *J Antibiot (Tokyo)* **57**: 326–36.

Leclercq R, Derlot E, Duval J, Courvalin P. 1988. Plasmid-mediated resistance to vancomycin and teicoplanin in *Enterococcus faecium*. *N Engl J Med* **319**: 157–161.

Leimkuhler C, Chen L, Barrett D, Panzone G, Sun B, Falcone B, Oberthur M, Donadion S, Walker S, Kahne D. 2005. Differential inhibition of *Staphylococcus aureus* PBP2 by glycopeptide antibiotics. *J Am Chem Soc* **127**: 3250–3251.

Leuthener KD, Yuen A, Mao Y, Rahbar A. 2015. Dalbavancin (BI-387) for the treatment of complicated skin and skin structure infection. *Expert Rev Anti Infect Ther* **13**: 149–159.

Levine DP. 2006. Vancomycin: A history. *Clin Infect Dis* **42**: S5–S12.

Li L, Xu B. 2005. Multivalent vancomycins and related antibiotics against infectious diseases. *Curr Pharmaceut Design* **11**: 3111–3124.

Linares J. 2001. The VISA/GISA problem: Therapeutic implications. *Clin Microbiol Infect* **7**: 8–15.

Linsdell H, Toiron C, Bruix M, Rivas G, Menendez M. 1996. Dimerization of A82846B, vancomycin, and ristocetin: Influence on antibiotic complexation with cell wall model peptides. *J Antibiot* **49**: 181–193.

Loll PJ, Bevivino AE, Korty BD, Axelsen PH. 1997. Simultaneous recognition of a carboxylate-containing ligand and an intramolecular surrogate ligand in the crystal structure of an asymmetric vancomycin dimer. *J Am Chem Soc* **119**: 1516–1522.

Loll PJ, Miller R, Weeks CM, Axelsen PH. 1998. A ligand-mediated dimerization mode for vancomycin. *Chemist Biol* **5**: 293–298.

Loll PJ, Derhovanessian A, Shapovaloz MV, Kaplan J, Yang L, Axelsen PH. 2009. Vancomycin forms ligand-mediated supramolecular complexes. *J Mol Biol* **385**: 200–211.

Lunde CS, Hartouni SR, Janc JW, Mammen M, Humphrey PP, Benton BM. 2009. Telavancin disrupts the functional integrity of the bacterial membrane through targeted interaction with the cell wall precursor lipid II. *Antimicrob Agents Chemother* **53**: 3375–3383.

Mackay JP, Gerhard U, Beauregard DA, Maplestone RA, Williams DH. 1994a. Dissection of the contributions toward dimerization of glycopeptide antibiotics. *J Am Chem Soc* **116**: 4573–4580.

Mackay JP, Gerhard U, Beauregard DA, Westwell MS, Searle MS, Williams DH. 1994b. Glycopeptide antibiotic activity and the possible role of dimerization: A model for biological signaling. *J Am Chem Soc* **116**: 4581–4590.

McCormick MH, McGuire JM, Pittenger GE, Pittenger RC, Stark WM. 1956. Vancomycin, a new antibiotic. I: Chemical and biologic properties. *Antibiot Annu* **3**: 606–611.

Mitra S, Saeed U, Havlichek DH, Stein GE. 2015. Profile of oritavancin and its potential in the treatment of acute bacterial skin structure infections. *Infect Drug Resist* **8**: 189–197.

Mouton JW, Jansz AR. 2001. The DUEL study: A multicenter in vitro evaluation of linezolid compared with other antibiotics in the Netherlands. *Clin Microbiol Infect* **7**: 486–491.

Nagarajan R, 1994. Glycopeptide antibiotics. *Drugs and the pharmaceutical sciences*, Vol. 63. Marcel Dekker, New York.

Nebreda T, Oteo J, Aldea C, Garcia-Estebanez C, Gastelu-Iturri J, Bautista V, Garcia-Cobos S, Campos J. 2007. Hospital dissemination of a clonal complex 17 vanB2-containing *Enterococcus faecium*. *J Antimicrob Chemother* **59**: 806–807.

Nicolau KC, Cho SY, Hughes R, Winssinger N, Smethurst C, Labischinski H, Endermann R. 2001a. Solid- and solution-phase synthesis of vancomycin and vancomycin analogues with activity against vancomycin-resistant bacteria. *Chemistry* **7**: 3798–3823.

Nicolau KC, Hughes R, Cho SY, Winssinger N, Labischinski H, Endermann R. 2001b. Synthesis and biological evaluation of vancomycin dimers with potent activity against vancomycin-resistant bacteria: Target-accelerated combinatorial synthesis. *Chemistry* **7**: 3824–3843.

Nieto M, Perkins HR. 1971. Physicochemical properties of vancomycin and iodovancomycin and their complexes with diacetyl-L-lysyl-D-alanyl-D-alanine. *Biochem J* **123**: 773–787.

Nitani Y, Kikuchi T, Kakoi K, Hanamaki S, Fujisawa I, Aoki K. 2009. Crystal structures of the complexes between vancomycin and cell-wall precursor analogs. *J Mol Biol* **385**: 1422–1432.

Pace JL, Krause K, Johnston D, Debabov D, Wu T, Farrington L, Lane C, Higgins D, Christensen D, Judice K, et al. 2003. In vitro activity of TD-6424 against *Staphylococcus aureus*. *Antimicrob Agents Chemother* **47**: 3602–3604.

Perichon B, Courvalin P. 2009. VanA-type vancomycin-resistant *Staphylococcus aureus*. *Antimicrob Agents Chemother* **53**: 4580–4587.

Perkins HR. 1969. Specificity of combination between mucopeptide precursors and vancomycin or ristocetin. *Biochem J* **111**: 195–205.

Perkins HR, Nieto M. 1970. The preparation of iodinated vancomycin and its distribution in bacteria treated with the antibiotic. *Biochem J* **116**: 83–92.

Peschel A, Vuong C, Otto M, Gotz F. 2000. The D-alanine residues of *Staphylococcus aureus* teichoic acids alter the susceptibility to vancomycin and the activity of autolytic enzymes. *Antimicrob Agents Chemother* **44:** 2845–2847.

Pfeltz RF, Wilkinson BJ. 2004. The escalating challenge of vancomycin resistance in *Staphylococcus aureus*. *Curr Drug Targets Infect Disord* **4:** 273–294.

Pootoolal J, Neu J, Wright GD. 2002. Glycopeptide antibiotic resistance. *Annu Rev Pharmacol Toxicol* **42:** 381–408.

Rao J, Lahiri J, Isaacs L, Weis RW, Whitesides GM. 1998. A trivalent system from vancomycin-D-Ala-D-Ala with higher affinity than biotin-avidin. *Science* **280:** 708–711.

Rao J, Yan L, Lahiri J, Whitesides GM, Weis RM, Warren HS. 1999a. Binding of dimeric derivative of vancomycin to L-Lys-D-Ala-D-lactate in solution and at a surface. *Chem Biol* **6:** 353–359.

Rao J, Yan L, Xu B, Whitesides GM. 1999b. Using surface plasmon resonance to study the binding of vancomycin and its dimer to self-assembled monolayers presenting D-Ala-D-Ala. *J Am Chem Soc* **121:** 2629–2630.

Rekharsky M, Hesek D, Lee M, Meroueh SO, Inoue Y, Mobashery S. 2006. Thermodynamics of interactions of vancomycin and synthetic surrogates of bacterial cell wall. *J Am Chem Soc* **128:** 7736–7737.

Reynolds PE, Courvalin P. 2005. Vancomycin resistance in enterococci due to synthesis of precursors terminating in D-alanyl-D-serine. *Antimicrob Agents Chemother* **49:** 21–25.

Sahm DF, Benton BM, Cohen MA. 2006. Telavancin demonstrates low potential for in vitro selection of resistance among key target gram-positive species. *Interscience Conference on Antimicrobial Agents and Chemotherapy*. San Francisco, CA, September 27–30.

Samra Z, Ofer O, Shmuely H. 2005. Susceptibility of methicillin-resistant *Staphylococcus aureus* to vancomycin, teicoplanin, linezolid, pristinamycin and other antibiotics. *Israel Med Assoc J* **7:** 148–150.

Schaefer M, Scheider TR, Scheldrick GM. 1996. Crystal structure of vancomycin. *Structure* **4:** 1509–1515.

Sharman GJ, Try AC, Dancer RJ, Cho YR, Staroske T, Bardsley B, Maguire AJ, Cooper MA, O'Brie DP, Williams DH. 1997. The roles of dimerization and membrane anchoring in activity of glycopeptide antibiotics against vancomycin-resistant bacteria. *J Am Chem Soc* **119:** 12041–12047.

Shiozawa H, Zerela R, Bardsley B, Tuck KL, Williams DH. 2003. Noncovalent bond lengths and their cooperative shortening: Dimers of vancomycin group antibiotics in crystals and in solution. *Helv Chim Acta* **116:** 1359–1370.

Shonekan D, Handwerger S, Mildvan D. 1997. Comparative in-vitro activities of RP59500 (quinupristin/dalfopristin), CL 329,998, CL 331,002, trovafloxacin, clinafloxacin, teicoplanin and vancomycin against Gram-positive bacteria. *J Antimicrob Chemother* **39:** 405–409.

Sieradzki K, Tomasz A. 1997. Inhibition of cell wall turnover and autolysis by vancomycin in a highly vancomycin-resistant mutant of *Staphylococcus aureus*. *J Bacteriol* **179:** 2557–2566.

Sieradzki K, Tomasz A. 2003. Alterations of cell wall structure and metabolism accompany reduced susceptibility to vancomycin in an isogenic series of clinical isolates of *Staphylococcus aureus*. *J Bacteriol* **185:** 7103–7110.

Sieradzki K, Pinho MG, Tomasz A. 1999. Inactivated *pbp4* in highly glycopeptide-resistant laboratory mutants of *Staphylococcus aureus*. *J Biol Chem* **274:** 18942–18946.

Sinha RK, Neuhaus F. 1968. Reversal of the vancomycin inhibition of peptidoglycan synthesis by cell walls. *J Bacteriol* **96:** 374–382.

Smith JR, Roberts KD, Rybak MJ. 2015. Dalbavancin: A novel lipoglycopeptide antibiotic with extended activity against Gram-positive infections. *Infect Dis Ther* **4:** 245–258.

Song Y, Lunde CS, Benton BM, Wilkinson BJ. 2013. Studies on the mechanism of telavancin decreased susceptibility in a laboratory-derived mutant. *Microb Drug Resist* **19:** 247–255.

Stack DR, Letourneau DL, Mullen DL, Butler TF, Allen NE, Kline AD, Nicas TI, Thompson RC. 1998. Covalent glycopeptide dimers: Synthesis, physical characterization, and antibacterial activity. *Intersci Conf Antimicrob Agents Chemother* **37:** 146.

Staroske T, Williams DH. 1998. Synthesis of covalent head-to-tail dimers of vancomycin. *Tetrahedron Lett* **39:** 4917–4920.

Sundram UN, Griffin JH, Nicas TJ. 1996. Novel vancomycin dimers with activity against vancomycin-resistant enterococci. *J Am Chem Soc* **118:** 13107–13108.

Walsh TR, Howe RA. 2002. The prevalence and mechanisms of vancomycin resistance in *Staphylococcus aureus*. *Annu Rev Microbiol* **56:** 657–675.

Walters M, Lonsway D, Rasheed K, Albrecht V, McAllister S, Limbago B, Kallen A. 2015. *Investigation and control of vancomycin-resistant* Staphylococcus aureus: *A guide for health departments and infection control personnel*. Centers for Disease Control and Prevention, Atlanta, GA.

Watanabe Y, Cui L, Katayama Y, Kozue K, Hiramatsu K. 2011. Impact of rpoB mutations on reduced vancomycin susceptibility in *Staphylococcus aureus*. *J Clin Microbiol* **49:** 2680–2684.

Wenzler E, Rodvold KA. 2015. Telavancin: The long and winding road from discovery to food and drug administration approvals and future directions. *Clin Infect Dis* **61:** S38–S47.

Werner G, Klare I, Fleige C, Witte W. 2008. Increasing rates of vancomycin resistance among *Enterococcus faecium* isolated from German hospitals between 2004 and 2006 are due to wide clonal dissemination of vancomycin-resistant enterococci and horizontal spread of *vanA* clusters. *Int J Med Microbiol* **298:** 515–527.

Westwell MS, Gerhard U, Williams DH. 1995. Two conformers of the glycopeptide antibiotic teicoplanin with distinct ligand binding sites. *J Antibiot* **48:** 1292–1298.

Westwell MS, Bardsley B, Dancer RJ, Try AC, Williams DH. 1996. Cooperativity in ligand binding expressed at a model cell membrane by the vancomycin group antibiotics *Chem Commun* **5:** 589–590.

Williams DH. 1996. The glycopeptide story—How to kill the deadly "superbugs." *Nat Prod Rep* **13:** 469.

Williams DH, Williamson MP, Butcher DW, Hammond SJ. 1983. Detailed binding sites of the antibiotics vancomycin and ristocetin A: Determination of intermolecular distances in antibiotic/substrate complexes by use of the time-dependent NOE. *J Am Chem Soc* **105:** 1332–1339.

Williams DH, Searle MS, Mackay JP, Gerhard U, Maplestone RA. 1993. Toward an estimation of binding constants in aqueous solution: Studies of associations of vancomycin group antibiotics. *Proc Natl Acad Sci* **90:** 1172–1178.

Williams DH, Maguire AJ, Tsuzuki W, Westfall MS. 1998. An analysis of the origins of a cooperative binding energy of dimerization. *Science* **280:** 711–714.

Williamson MP, Williams DH. 1981. Structure revision of the antibiotic vancomycin. Use of nuclear Overhauser effect difference spectroscopy. *J Am Chem Soc* **103:** 6580–6585.

Zhanel GG, Calic D, Schweizer F, Zelenitsky S, Adam H, Lagace-Wiens PR, Rubinstein E, Gin AS, Hoban DJ, Karlowsky JA. 2010. New lipoglycopeptides: A comparative review of dalbavancin, oritavancin and telavancin. *Drugs* **70:** 859–886.

Zhu W, Murray PR, Huskins WC, Jernigan JA, McDonald LC, Clark NC, Anderson KF, McDougal LK, Hageman JC, Olsen-Rasmussen M, et al. 2010. Dissemination of an *Enterococcus* Inc18-like *vanA* plasmid associated with vancomycin-resistant *Staphylococcus aureus*. *Antimicrob Agents Chemother* **54:** 4314–4320.

Fosfomycin: Mechanism and Resistance

Lynn L. Silver

LL Silver Consulting, LLC, Springfield, New Jersey 07081

Correspondence: silverly@comcast.net

Fosfomycin, a natural product antibiotic, has been in use for >20 years in Spain, Germany, France, Japan, Brazil, and South Africa for urinary tract infections (UTIs) and other indications and was registered in the United States for the oral treatment of uncomplicated UTIs because of *Enterococcus faecalis* and *Escherichia coli* in 1996. It has a broad spectrum, is bactericidal, has very low toxicity, and acts as a time-dependent inhibitor of the MurA enzyme, which catalyzes the first committed step of peptidoglycan synthesis. Whereas resistance to fosfomycin arises rapidly in vitro through loss of active transport mechanisms, resistance is rarely seen during therapy of UTIs, seemingly because of the low fitness of the resistant organisms. Recently, interest has grown in the use of fosfomycin against multidrug-resistant (MDR) pathogens in other indications, prompting the advent of development in the United States of a parenteral formulation for use, initially, in complicated UTIs. Whereas resistance has not been problematic in the uncomplicated UTI setting, it remains to be seen whether resistance remains at bay with expansion to other indications.

Fosfomycin, originally called phosphonomycin, is a broad spectrum antibiotic first found in fermentation broths of *Streptomyces fradiae* (ATCC 21096) in Spain through a collaborative effort of Merck and the Compañía Española de Penicilina y Antibióticos (CEPA) (Hendlin et al. 1969). Fosfomycin was initially developed in Europe by CEPA and has been in use since the early 1970s, initially as an IV preparation of the disodium salt and later as an oral formulation of fosfomycin trometamol. Its primary use in Spain, Germany, France, Japan, Brazil, and South Africa has been as an oral treatment for urinary tract infections (UTIs) but it has also been used more broadly in other indications (Falagas et al. 2008, 2009, 2010a). Fosfomycin was approved for use in the United States (as Monurol, fosfomycin tromethamine [same as trometamol]) in 1996 for treatment by single-dose oral therapy of uncomplicated UTIs (acute cystitis) in women caused by *Escherichia coli* and *Enterococcus faecalis*. With the problem of increasing resistance to other antibiotics, parenteral use of fosfomycin has been studied in therapy of a variety of infections because it is active against many multidrug-resistant (MDR) pathogens (Falagas et al. 2009) and is now under development in the United States for parenteral treatment of complicated UTIs (Zavante 2016).

DISCOVERY AND SPECTRUM

Fosfomycin (Fig. 1A) is a phosphonic acid antibiotic discovered in Spain in a fermentation

A

B

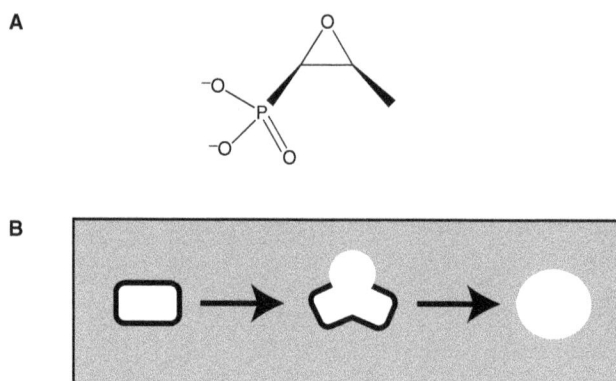

Figure 1. Fosfomycin and the screen in which it was discovered. (*A*) Structure of fosfomycin. (*B*) Schematic of SPHERO assay. Gram-negative rods grown in osmotically protective medium, when treated with inhibitors of peptidoglycan synthesis, will show morphological changes leading to the production of round cells called spheroplasts, which are highly differentiable microscopically from normal rods and debris. The assay was run in plastic trays with wells for growth of cells and the wells viewed, after incubation, with a stereomicroscope.

broth of *S. fradiae* by means of a Merck screen for inhibition of peptidoglycan synthesis, the SPHERO assay (Gadebusch et al. 1992). In this morphological assay, schematized in Figure 1B, Gram-negative bacilli are grown in osmotically protective medium and treated for several doublings with test samples. Inhibitors of steps in the synthesis of peptidoglycan will lead to the production of microscopically recognizable refractile spheroplasts. Fosfomycin is also produced by several other *Streptomyces*, including *Streptomyces viridochromogenes* (ATCC21240) and *Streptomyces wedmorensis* (ATCC 21239), as well as *Pseudomonas syringae* (Shoji et al. 1986), *Pseudomonas viridiflava*, and *Pseudomonas fluorescens* (Katayama et al. 1990).

Early tests indicated that fosfomycin was efficacious via IV dosing against intraperitoneal murine infections by specific isolates of *E. coli*, *Klebsiella pneumoniae*, *Pseudomonas aeruginosa*, *Proteus vulgaris*, *Salmonella schottmuelleri*, *Staphylococcus aureus*, and *Streptococcus pyogenes*—although ED$_{50}$s (median effective dose) against *K. pneumoniae*, *P. aeruginosa*, and *S. pyogenes* were high (although attainable and safe) (Hendlin et al. 1969). More recent reports note its broad spectrum, including activity against many important pathogens such as *S. aureus* (including methicillin-resistant *S. aureus* [MRSA]), *Staphylococcus epidermidis*, *Strepto-*

coccus pneumoniae, *E. faecalis*, *E. coli*, *Proteus* species, *K. pneumoniae*, *Enterobacter* species, *Serratia marcescens*, and *Salmonella typhi*. Whereas *P. aeruginosa* shows variable susceptibility, fosfomycin has shown anti-pseudomonal efficacy especially in combinations with cefepime, aztreonam, and meropenem (Falagas et al. 2008). *Acinetobacter*, *Vibrio fischeri*, *Chlamydia trachomatis*, and *Bacteroides* species are resistant to fosfomycin (Falagas et al. 2008; Karageorgopoulos et al. 2012). Fosfomycin is not cross-resistant with other antibiotics because of its unique structure and mechanism of action.

MECHANISM OF ACTION

Fosfomycin is an inhibitor of the MurA enzyme, UDP-*N*-acetylglucosamine-enolpyruvyltransferase, that catalyzes the first committed step in peptidoglycan synthesis, the reaction of UDP-*N*-acetylglucosamine (UDP-GlcNAc) with phosphoenolpyruvate (PEP) to form UDP-GlcNAc-enoylpyruvate plus inorganic phosphate (shown in Fig. 2). The finding that competition of fosfomycin inhibition by PEP implied that fosfomycin was acting as a PEP analog (Kahan et al. 1974). Fosfomycin covalently binds, in a time-dependent reaction involving nucleophilic attack by the cysteine 115 residue of *E. coli* MurA on the epoxide of fosfomycin (Marquardt et al.

Cite this article as *Cold Spring Harb Perspect Med* doi: 10.1101/cshperspect.a025262

1994), and C115 is considered responsible for the catalytic action on PEP in the synthetic reaction (Wanke and Amrhein 1993; Brown et al. 1994).

In vitro, the MurA reaction proceeds with UDP-GlcNAc binding to the so-called open form of MurA, leading to a structural change to a closed form to which PEP binds (Skarzynski et al. 1996) leading to a tetrahedral intermediate and product release (Eschenburg et al. 2005). Even though it acts as an analog of PEP, fosfomycin's interaction with MurA is highly selective (Kahan et al. 1974; Marquardt et al. 1994) and shows extremely low toxicity. Its LD_{50} (median lethal dose) in mice by intraperitoneal dosing of Na_2-fosfomycin was 4 g/kg, lethality likely being because of sodium content (Hendlin et al. 1969).

The existence of the MurA analog of *Mycobacterium tuberculosis* that is naturally resistant to fosfomycin and has an asparagine in the position comparable to the C115 led to studies revealing further details of MurA reactions. Replacement of the C115 of the *E. coli* enzyme by asparagine (to yield a C115D enzyme) leads to fosfomycin resistance (Kim et al. 1996) and, as expected, replacement of asparagine in the *M. tuberculosis* enzyme with cysteine endows it with fosfomycin sensitivity (De Smet et al. 1999). This indicates that C115 is not necessarily required for the catalytic activity of the enzyme, but this activity can be supplied by the asparagine acting as a general acid (Kim et al. 1996). Substitution of C115 of *Enterobacter cloacae* MurA by serine (C115S) yielded an enzyme with the reaction products, UDP-GlcNAc-enoylpyruvate and Pi, bound in the active site

(Eschenburg et al. 2005), leading the authors to conclude that C115 is necessary for turnover and release of the products. They reasoned that, in C115S, other residues could carry out the catalytic step and that, in the C115D enzyme, the flexible asparagine should be able to catalyze the release of the reaction products.

Recent work (Zhu et al. 2012) indicates that, in the cell, the activity of MurA is subject to regulation by the binding of UDP-NAc-muramic acid (UDP-MurNAc), the product of the MurB enzyme, to MurA. In the presence of bound UDP-MurNAc, PEP covalently attaches to C115, leading to the formation of a "locked" dormant tertiary complex (MurA:PEP-UDP-MurNAc), which is the predominant form of cellular MurA. This had apparently been missed previously because much crystallographic work with MurA had been done by diluting samples into phosphate buffer, which releases the ligands. Extensive crystallographic findings by these authors led them to a model for the regulatory and reaction schemes: when the UDP-MurNAc to UDP-GlcNAc ratio in the cell decreases, UDP-GlcNAc replaces UDP-MurNAc in the dormant complex in a rapidly reversible manner, leading to formation of the tetrahedral reaction intermediate that then yields the products, still bound in the active site. PEP enters, reacts with C115, and displaces Pi, stimulating the reversible exchange of UDP-GlcNAc with UDP-GlcNAc-enoylpyruvate, leading to its release to restart the cycle. The authors speculate that the existence of the locked dormant complex may explain why discovery of reversible, noncovalent inhibitors of MurA has been difficult. In this model, then, the covalent interac-

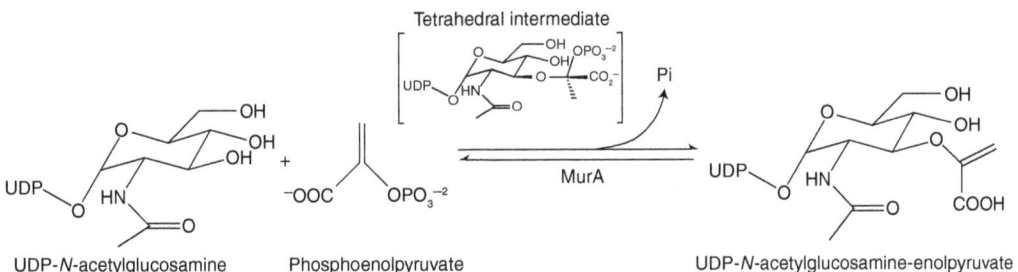

Figure 2. Overall reaction performed by the MurA enzyme.

tion of PEP with C115 is necessary for formation of the dormant complex and also for release of the reaction products (as had been proposed by Eschenburg et al. 2005) but is not necessary for catalysis by the enzyme that may be supplied by other residues (as had been proposed by Kim et al. 1996).

In low GC Gram-positives, there are two "*murA*" genes, *murA* and *murZ*, the products of which are structurally very similar (Du et al. 2000; Blake et al. 2009). In the case of the *S. pneumoniae* and *S. aureus* enzymes, each gene product is capable of sustaining peptidoglycan synthesis and both isozymes are sensitive to fosfomycin in vitro (Blake et al. 2009). Neither gene is essential but a double deletion is not viable. In *Bacillus subtilis* (Kobayashi et al. 2003; Kock et al. 2004) and *Bacillus anthracis* (Kedar et al. 2008), however, only the *murA* gene is essential and, evidently, the second enzyme cannot substitute.

FOSFOMYCIN UPTAKE

In *E. coli*, two carrier-dependent systems can actively transport fosfomycin. Initial work showed that fosfomycin could be transported by the system for uptake of α-glycerophosphate and that mutants in the *glpT* gene, encoding the α-glycerophosphate permease, were 30-fold more resistant than the isogenic parent strain. Whereas α-glycerophosphate will induce expression of GlpT, the basal levels of GlpT are sufficient for fosfomycin uptake (Kahan et al. 1974). The GlpT system is widespread, found at least in *P. aeruginosa*, *E. coli*, *Salmonella*, *Shigella flexneri*, *Klebsiella*, *Haemophilus influenzae*, *S. aureus*, *B. subtilis*, *E. faecalis*, and *Rickettsia prowazekii* (Kahan et al. 1974; Lemieux et al. 2004). It was also shown that the blood component, glucose-6-phosphate (G-6-P), could induce the hexose-phosphate uptake system, UhpT, to levels sufficient for fosfomycin transport (Zimmerman et al. 1969; Kahan et al. 1974). The UhpT system is limited to Enterobacteriaceae (with the exception of *Proteus* species) and *S. aureus* (Winkler 1973). Importantly, *Pseudomonas* does not have a UhpT system

and so solely uses *glpT* for uptake (Winkler 1973; Castañeda-García et al. 2009).

FOSFOMYCIN RESISTANCE

The mechanisms of resistance to fosfomycin have been recently reviewed (Karageorgopoulos et al. 2012; Castañeda-García et al. 2013; Nikolaidis et al. 2014). It should be noted that the breakpoints for fosfomycin susceptibility have been formalized for few species and are different for the Clinical and Laboratory Standards Institute (CLSI) and the European Committee on Antimicrobial Susceptibility Testing (EUCAST). EUCAST break points for Enterobacteriaceae (oral or IV) and *S. aureus* (IV only) are S ≤ 32 mg/L and R > 32 mg/L (in the presence of 25 mg/L G-6-P). Wild-type isolates of *Pseudomonas* species minimum inhibitory concentrations (MICs) ≤128 mg/L have been treated with combinations (European Committee on Antimicrobial Susceptibility Testing 2016). CLSI breakpoints for *E. coli* and *E. faecalis* are S ≤ 64 mg/L, I 128 mg/L, R ≥ 256 mg/L (also in the presence of G-6-P) (Performance Standards for Antimicrobial Susceptibility Testing 2012).

Mutational Resistance

Resistance to fosfomycin arises in *E. coli* at high frequencies in vitro, because of loss of transport systems required for uptake. As noted above, in *E. coli*, uptake of fosfomycin can be mediated by the GlpT permease system, whose main substrate is α-glycerophosphate, and also, when induced by G-6-P, by the UhpT system (Kahan et al. 1974). Both of these systems are positively regulated by cAMP, and cAMP levels can be lowered by mutations in the *ptsI* or *cyaA* genes (which will also affect catabolism of a variety of carbohydrates) (Alper and Ames 1978; Tsuruoka et al. 1978; Castañeda-García et al. 2009). For *E. coli* grown in the absence of G-6-P, in which only the GlpT permease is active, the frequency of resistance (ascertained by fluctuation test) is 10^{-7}, while in the presence of G-6-P the frequency is 10^{-8} (Nilsson et al. 2003). Most of the mutations seen in the Nilsson study were located in genes leading to a decrease in cAMP. Nilsson noted that despite this finding of high frequencies of resistance in vitro the rates of clinical resistance to fosfomycin in *E. coli* throughout Europe from 1999 to 2000 (Kahlmeter 2000) were uniformly low, in the range of 0.7% to 1.5%, with no differences between countries with a long history of fosfomycin use and those not using it. As opposed to their in vitro findings, 13 resistant clinical isolates tested by Nilsson were found to have mutations almost exclusively in *glpT* and/or *uhpA* or *uhpT*, with none found in the cAMP regulatory loci, perhaps indicating a selective disadvantage in vivo. Whereas mutants grew well in the absence of fosfomycin, it was found that growth rates of the three tested in vitro mutants and 12 of the 13 clinical isolates were severely reduced in growth rate in LB medium or urine in the presence of ≥8 mg/L of fosfomycin, which is significantly lower than the fosfomycin concentration in urine during treatment, normally greater than 128 mg/L. Mathematical modeling indicated that the growth rate retardation seen should be enough to prevent resistant strains from establishment in the bladder (Nilsson et al. 2003). Slow growth and the absence of the cAMP-related mutations in vivo could explain the discrepancy between high-frequency mutational resistance to fosfomycin in vitro and rates of resistance seen in the clinic.

Interestingly, very few instances have been reported of mutations in the fosfomycin target gene, *murA*. A mutation in *murA* of *E. coli* was isolated after mutagenesis (Wu and Venkateswaran 1974) and counterselection against transport mutants and two *murA* mutants of *E. coli* were reported among clinical isolates in a Japanese study (Takahata et al. 2010) It is likely that, in vitro, the frequency of transport mutants is so great relative to target mutants that they are not generally seen. The finding of *murA* mutants with clinical isolates of *E. coli* might reflect the low fitness of in vitro transport mutants.

Recent in vitro experiments (Couce et al. 2012) showed that overexpression of the *murA* gene by induction of a regulated promoter can lead to greatly increased MICs, to levels that would afford clinical resistance, while having relatively low effects on fitness (relative to mutations to fosfomycin resistance found in clinical isolates). However, this has not been noted yet in clinical isolates.

A study of a set of 441 Italian Gram-negative urinary isolates (Marchese et al. 2003) showed very high susceptibility to fosfomycin of *E. coli* isolates (99%), 87.5% susceptibility for *Proteus* species, whereas other species showed variable susceptibilities. Fosfomycin-resistant mutants (≥2000 mg/L) of sensitive *E. coli*, *K. pneumoniae*, and *P. aeruginosa* were obtained through stepwise selection on agar containing fosfomycin, and the growth rates of these resistant mutants were compared to their parental strains. In contrast to the mutants tested by Nilsson et al., these all showed significantly reduced growth rates in a variety of conditions (in the absence of fosfomycin) as well as reduced adhesion to uroepithelial cells and to urinary catheters.

A study of fosfomycin resistance in *Pseudomonas* (Rodríguez-Rojas et al. 2010b) showed that a *glpT* null mutant was equal in virulence to its *glp*⁺ parent in a mouse lung infection survival model, had equivalent biofilm-forming ability, and caused similar inflammation in histological studies. This and other data were interpreted by the authors to show that mutational resistance in *P. aeruginosa* may have no obvious fitness cost in vivo. It may well be that the site of infection contributes to the difference in apparent fitness seen between *P. aeruginosa* in the lung and *E. coli* in the bladder.

As there is interest in use of parenteral fosfomycin for treatment of indications in addition to uncomplicated UTI, recent hollow fiber pharmacokinetic/pharmacodynamic (PK/PD) studies were undertaken by the Ambrose group (VanScoy et al. 2015) to determine the PK/PD index for IV fosfomycin and to determine requirements for stasis, and 1- and 2-log reductions of colony-forming units. The results showed that, at least for the *E. coli* isolates studied, there was a large resistant subpopulation from the inception of infection. Thus, a new PK/PD index was instituted, %T > RIC—the percentage of the dosing interval in which the concentration was above the "resistance inhibitory concentration." For stasis, 1- and 2-log reductions, %T > RIC were 11.9%, 20.9%, and 32.8%, respectively. The authors note some caveats with this type of in vitro work: in vivo fitness of the mutants is unknown and longer treatment will be required to model clinical treatment duration.

In vitro pharmacodynamics of Na_2-fosfomycin against clinical 64 *P. aeruginosa* isolates was studied by an Australian group (Walsh et al. 2015). MICs from 1 to >512 were seen with 61% of isolates susceptible (MIC ≤ 64 mg/L), but all isolates tested had a resistant subpopulation as revealed by population analysis profiling. At low inocula ($\sim 10^6$ cfu/mL), there was moderate killing with regrowth by 24 h at most dosages. No killing was seen with high inocula ($\sim 10^8$ cfu/mL). This indicates that monotherapy of *P. aeruginosa* with fosfomycin may be problematic if this finding translates to in vivo conditions; furthermore, study of combination therapy is warranted.

Fosfomycin-Modifying Enzymes

Several fosfomycin-modifying enzymes have been found that inactivate the drug. The main enzymes described are three types of metalloenzymes, FosA, FosB, and FosX, and two kinases, FomA and FomB, as shown in Figure 3.

The metalloenzymes open the epoxide (oxirane ring) by the addition of various substrates as recently reviewed (Castañeda-García et al. 2013). FosA is a glutathione-S-transferase present on transposon TN2921 originally found on a plasmid in *S. marcescens* (Suárez and Mendoza 1991), using Mn^{2+} and K^+ as metal cofactors; other related glutathione transferase (FosA type) enzymes found to be plasmid-borne are

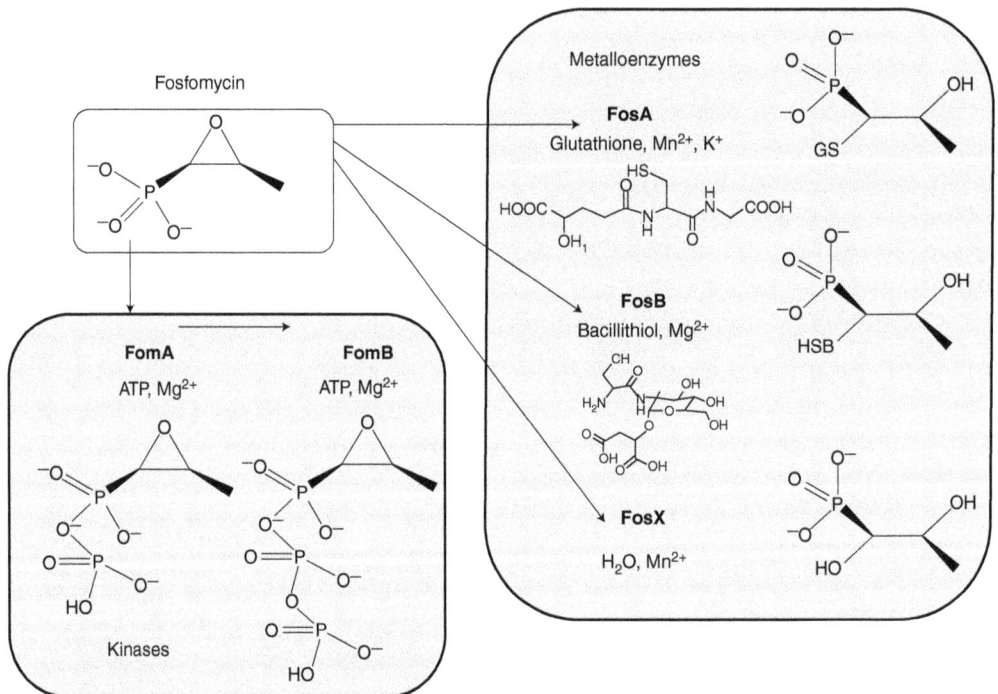

Figure 3. Fosfomycin-modifying enzymes leading to inactivation of fosfomycin and responsible for fosfomycin resistance. Enzyme names are in bold and are shown with their substrates, metal cofactors, and the end products of their modifying activity. (This figure is adapted from Castañeda-García et al. 2013 and is used under license from Creative Commons, creativecommons.org/licenses/by/3.0.)

Cite this article as *Cold Spring Harb Perspect Med* doi: 10.1101/cshperspect.a025262

FosA3, FosA4, FosA5, and FosC2. A related FosA enzyme is found encoded in the chromosome of *P. aeruginosa*.

Recent studies in China (Li et al. 2015) found that 7.8% of nonduplicate *E. coli* clinical isolates collected from 20 geographically dispersed hospitals from July 2009 to June 2010 were nonsusceptible to fosfomycin (MIC > 64 mg per liter). Of these, 80% carried the fosA3 gene; the fosA3 gene of 42% of those isolates was transferable, presumably via a conjugative plasmid. With growing interest in the possible use of fosfomycin to combat resistant Gram-negative infections, a study was done to ascertain the prevalence of fosfomycin resistance in 278 *K. pneumoniae* isolates carrying the *K. pneumoniae* carbapenemase (KPC) and in 80 extended-spectrum β-lactams (ESBLs) (non-KPC) *K. pneumoniae* from 12 hospitals in China (Jiang et al. 2015). Fosfomycin resistance was 60.8% in KPC producers, compared to 12.5% among ESBL producers. Ninety-four of the KPC isolates were found to carry *fosA3* genes and 92 of those appeared clonally related, with 71% belonging to a single clonal group. The distribution of fosfomycin resistance levels varied greatly among the different hospitals, but there was general correlation between levels of resistance and presence of *fos3* genes. Rather alarming was the finding that, in a representative isolate of the predominant clone, the *fosA3* and *bla*$_{KPC-2}$ genes were colocalized on a plasmid designated pFOS18. The *bla*$_{KPC-2}$ gene, located on a Tn3–Tn4401 structure, and *fosA3*, bracketed by IS26 sequences, are normally transmitted on plasmids. This data indicates that they can be linked on a single plasmid. The data from this recent study in China contrasts with earlier data from a U.S. study (Endimiani et al. 2010) in which fosfomycin susceptibility of 68 *K. pneumoniae* KPC-possessing strains was measured and 93% were found susceptible. It is likely that the Chinese study represents a clonal outbreak that has not (yet) spread to the United States. It will be necessary to monitor the spread of such clones.

FosA3 was also found on CTX-M plasmids of *E. coli* in Japan, flanked there as well by IS26 elements (Sato et al. 2013). An unexpected increase in fosfomycin resistance of *E. coli* CTX-M-15-carrying strains in Madrid was studied and found to be caused by two main clonal types (Oteo et al. 2009). In the larger clone, none of the normal mutational resistance alleles were found, whereas in the second clone of five strains, there was a small deletion in the uhpA gene and a single IS26 linked to *bla*$_{CTX-M-15}$. There was no mention of genes associated with the IS26. Is it possible that these clones could contain fosfomycin-modifying enzymes, perhaps Fos3 associated with CTX-M as in the Japanese isolates (Sato et al. 2013)?

The regulation of expression of the *fosA* gene that is resident in the genome of *P. aeruginosa* is not well studied, if at all. It would seem critically important to study the conditions under which *fosA* is expressed in *P. aeruginosa*, to ascertain what its effect is on basal MICs and whether sufficient expression to yield a resistant phenotype can be achieved by mutation or induction. One reference (De Groote et al. 2011) notes that fosfomycin resistance in *P. aeruginosa* can be caused by (engineered) overexpression of the resident FosA because of an inserted promoter. Whether turn-on of FosA in *P. aeruginosa* is a possible source of mutational resistance is not known. However, the main mutational resistance in *P. aeruginosa* is clearly mutations in the GlpT system (Castañeda-García et al. 2009), because no spontaneous fosfomycin-resistant colonies grew on plates containing α-glycerophosphate as a carbon source, and all 10 sequenced mutations in this study were missense mutations in *glpT*.

FosB type enzymes, first discovered in *S. epidermidis* (Zilhao and Courvalin 1990), are found in low GC Gram-positive bacteria such as *B. subtilis*, *B. anthracis*, *S. epidermidis*, and *S. aureus*, species that do not make glutathione but use bacillithiol (shown in Fig. 3) as a substitute thiol donor. These bacillithiol-*S*-transferases use Mg^{2+} as a cofactor. The FosB of *S. aureus* is chromosomally located and is responsible for the innate level of fosfomycin activity—as deletion of either the *fosB* gene or the bacillithiol synthetic machinery greatly increases the sensitivity to fosfomycin (Thompson et al. 2014).

FosX is related to the FosA and FosB enzymes, but it is an Mn^{2+}-dependent epoxide hydrolase, using water to break the ring. FosX enzymes are found in *Listeria monocytogenes*, *Clostridium botulinum*, and *Brucella melitensis* (Castañeda-García et al. 2013).

FomA and FomB are kinases from *S. wedmorensis*, a species that produces fosfomycin (Kuzuyama et al. 1996), that sequentially add phosphates to the phosphonate moiety of fosfomycin from ATP with Mg^{2+} as a cofactor. Presumably, these enzymes act to provide autoresistance to the producers. Another such kinase, originally called FosC and found in the fosfomycin producer *P. syringae*, is actually an ortholog of FomA (Kim et al. 2012).

CLINICAL CONSIDERATIONS ON RESISTANCE

As fosfomycin has low toxicity and allergenicity, a broad spectrum of activity, including MDR organisms, good pharmacokinetics, and is available in parenteral as well as oral formulations, its use in non-UTI indications for hard-to-treat infections has become attractive. But there is not a large body of data on results of controlled trials in other indications. The Falagas group has published a large number of meta-analyses of small studies of fosfomycin use in various settings. Initial data supported use of fosfomycin for treatment of UTIs caused by MDR Enterobacteriaceae, including those carrying ESBLs (Falagas et al. 2010a). Resistance data from studies after 2010 (Falagas et al. 2016) was mostly on pathogens from urine samples and showed that *E. coli* susceptibility ranged from 82% to 100% and *K. pneumoniae* from 15% to 100%. There was less data on other Enterobacteriaceae, but fosfomycin was active against 72% to 97.5%. Fosfomycin was 90.5% to 100% active against MDR Enterobacteriaceae. One study showed 80.6% susceptibility to fosfomycin of carbapenem-resistant *P. aeruginosa* strains. Carbapenem-resistant *Acinetobacter baumannii* was resistant. No major differences in resistance were seen between data published before 2010 and that published after for Gram-negatives and *S. aureus*.

The main question is whether resistance to fosfomycin will compromise its use in non-UTI indications. There is no adequate answer as yet. It is clear that the single-dose oral treatment of cystitis with fosfomycin trometamol/tromethamine is highly effective and equivalent to other therapies with low evolution of resistance seen (Falagas et al. 2010b). Most of these infections are caused by *E. coli*, and it may be that the high urinary levels of fosfomycin attainable and the lowered fitness of mutants with impaired fosfomycin uptake (as discussed above) contribute to the efficacy of fosfomycin in this setting.

Combinations

Checkerboard assays, kill curves, and in vitro models have been used to predict the efficacy and potential synergy of combinations of fosfomycin. Whereas monotherapy against *E. coli* and *E. faecalis* cystitis has been successful over the years, there is concern that treatment of other pathogens at other sites where local drug concentrations will likely be lower than in urine may lead to resistance selection. Thus, combination therapy is being evaluated.

Combinations with underused cell-wall inhibitors were tested against *E. coli* and *K. pneumoniae* (Hickman et al. 2013), including MDR strains, leading to the finding that a fosfomycin/aztreonam and a fosfomycin/aztreonam/mecillinam combination were effective in reducing the population in a variety of models, including a new in vitro kinetic model.

An examination of resistance selection by fosfomycin in combination with a number of drugs in both wild-type and a mutator strain (Rodríguez-Rojas et al. 2010a) gave the interesting result that, in wild-type, combinations with tobramycin, amikacin, meropenem, ceftazidime, ciprofloxacin, and colistin gave frequencies below the limit of detection ($<1 \times 10^{-10}$), but the combination of fosfomycin and imipenem had a frequency of 1.1×10^{-9}, higher than the product of the individual frequencies (3.5×10^{-13}). For the mutator strain, in which frequencies for the single drugs were ~100 fold higher than in wild-type, all combinations yielded frequencies below the limit of detection except for fosfomycin plus imipenem or ceftazidime, which were 1.1×10^{-7} and 1×10^{-8}, respectively. There was, apparently, some sort of antagonistic effect occurring.

A recent publication (Walsh et al. 2016) reported testing combinations of fosfomycin with several drugs in killing *P. aeruginosa*. Against fosfomycin-susceptible isolates, fosfomycin monotherapy led to efficient killing but rapid regrowth. Combination of fosfomycin with polymyxin B or tobramycin against these susceptible isolates and combination of ciprofloxacin plus fosfomycin against fosfomycin-resistant isolates led to increased killing (versus monotherapy), but did not prevent regrowth of resistant mutants. Certainly, repeat dosing should be tried to ascertain whether rebound could be prevented with further treatment, but this study is sobering.

CONCLUSIONS

Fosfomycin is an old antibiotic, but it has proven useful. In this age of increasing antibiotic resistance, fosfomycin is being reconsidered for use against MDR pathogens within its spectrum of action. But, if it is true that fosfomycin has proven effective for uncomplicated UTIs because of fitness cost of mutants and high urinary drug concentrations, then these are variables that must be addressed to move on to indications involving other body sites. The use of combinations may help to keep resistance at bay, but that will likely rely on matched pharmacokinetics. Furthermore, data from various tests of combinations against *P. aeruginosa* show the need to choose combinations carefully. Dosing of drug at "resistance-inhibitory" or "mutant-prevention" concentrations may be a more realistic approach. Whereas the frequencies of fosfomycin resistance seen over time have not changed very much, that is likely because of the fact that most of the present resistance seen is a result of mutations in the pathogen. The reports on plasmid-borne fosfomycin resistance forewarn of increased levels of resistance, likely through spread of problematic clones in which resistance to fosfomycin is linked to resistance

to other drugs, especially β-lactams. The Chinese report of FosA3 and KPC-2 on a single plasmid and its likely clonal spread (Jiang et al. 2015) are very worrying. Whereas drugs to treat the growing threat of MDR bacteria are sorely needed, and fosfomycin has some attractive properties that could favor its use for treatment of certain life-threatening infections, to retain its productive use as a single-dose oral therapy for cystitis, stewardship and monitoring the spread of plasmid-borne resistance will be needed.

REFERENCES

Alper MD, Ames BN. 1978. Transport of antibiotics and metabolite analogs by systems under cyclic AMP control: Positive selection of *Salmonella typhimurium cya* and *crp* mutants. *J Bacteriol* 133: 149–157.

Blake KL, O'Neill AJ, Mengin-Lecreulx D, Henderson PJ, Bostock JM, Dunsmore CJ, Simmons KJ, Fishwick CW, Leeds JA, Chopra I. 2009. The nature of *Staphylococcus aureus* MurA and MurZ and approaches for detection of peptidoglycan biosynthesis inhibitors. *Mol Microbiol* 72: 335–343.

Brown ED, Marquardt JL, Lee JP, Walsh CT, Anderson KS. 1994. Detection and characterization of a phospholactoyl-enzyme adduct in the reaction catalyzed by UDP-*N*-acetylglucosamine enolpyruvoyl transferase, MurZ. *Biochemistry* 33: 10638–10645.

Castañeda-García A, Rodríguez-Rojas A, Guelfo JR, Blázquez J. 2009. The glycerol-3-phosphate permease GlpT is the only fosfomycin transporter in *Pseudomonas aeruginosa*. *J Bacteriol* 191: 6968–6974.

Castañeda-García A, Blázquez J, Rodríguez-Rojas A. 2013. Molecular mechanisms and clinical impact of acquired and intrinsic fosfomycin resistance. *Antibiotics* 2: 217–236.

Couce A, Briales A, Rodríguez-Rojas A, Costas C, Pascual A, Blázquez J. 2012. Genomewide overexpression screen for fosfomycin resistance in *Escherichia coli*: MurA confers clinical resistance at low fitness cost. *Antimicrob Agents Chemother* 56: 2767–2769.

De Groote VN, Fauvart M, Kint CI, Verstraeten N, Jans A, Cornelis P, Michiels J. 2011. *Pseudomonas aeruginosa* fosfomycin resistance mechanisms affect non-inherited fluoroquinolone tolerance. *J Med Microbiol* 60: 329–336.

De Smet KAL, Kempsell KE, Gallagher A, Duncan K, Young DB. 1999. Alteration of a single amino acid residue reverses fosfomycin resistance of recombinant MurA from *Mycobacterium tuberculosis*. *Microbiology* 145: 3177–3184.

Du W, Brown JR, Sylvester DR, Huang J, Chalker AF, So CY, Holmes DJ, Payne DJ, Wallis NG. 2000. Two active forms of UDP-*N*-acetylglucosamine enolpyruvyl transferase in Gram-positive bacteria. *J Bacteriol* 182: 4146–4152.

Endimiani A, Patel G, Hujer KM, Swaminathan M, Perez F, Rice LB, Jacobs MR, Bonomo RA. 2010. In vitro activity of fosfomycin against bla_{KPC}-containing *Klebsiella pneumoniae* isolates, including those nonsusceptible to tigecycline and/or colistin. *Antimicrob Agents Chemother* **54**: 526–529.

Eschenburg S, Priestman M, Schonbrunn E. 2005. Evidence that the fosfomycin target Cys115 in UDP-*N*-acetylglucosamine enolpyruvyl transferase (MurA) is essential for product release. *J Biol Chem* **280**: 3757–3763.

European Committee on Antimicrobial Susceptibility Testing. 2016. Breakpoint tables for interpretation of MICs and zone diameters. Version 6. www.eucast.org/clinical_breakpoints.

Falagas ME, Giannopoulou KP, Kokolakis GN, Rafailidis PI. 2008. Fosfomycin: Use beyond urinary tract and gastrointestinal infections. *Clin Inf Dis* **46**: 1069–1077.

Falagas ME, Kastoris AC, Karageorgopoulos DE, Rafailidis PI. 2009. Fosfomycin for the treatment of infections caused by multidrug-resistant non-fermenting Gram-negative bacilli: A systematic review of microbiological, animal and clinical studies. *Int J Antimicrob Agents* **34**: 111–120.

Falagas ME, Kastoris AC, Kapaskelis AM, Karageorgopoulos DE. 2010a. Fosfomycin for the treatment of multidrug-resistant, including extended-spectrum β-lactamase producing, Enterobacteriaceae infections: A systematic review. *Lancet Inf Dis* **10**: 43–50.

Falagas ME, Vouloumanou EK, Togias AG, Karadima M, Kapaskelis AM, Rafailidis PI, Athanasiou S. 2010b. Fosfomycin versus other antibiotics for the treatment of cystitis: A meta-analysis of randomized controlled trials. *J Antimicrob Chemother* **65**: 1862–1877.

Falagas ME, Vouloumanou EK, Samonis G, Vardakas KZ. 2016. Fosfomycin. *Clin Microbiol Rev* **29**: 321–347.

Gadebusch HH, Stapley EO, Zimmerman SB. 1992. The discovery of cell wall active antibacterial antibiotics. *Crit Rev Biotechnol* **12**: 225–243.

Hendlin D, Stapley EO, Jackson M, Wallick H, Miller AK, Wolf FJ, Miller TW, Chaiet L, Kahan FM, Foltz EL, et al. 1969. Phosphonomycin, a new antibiotic produced by strains of *Streptomyces*. *Science* **166**: 122–123.

Hickman RA, Hughes D, Cars T, Malmberg C, Cars O. 2013. Cell-wall-inhibiting antibiotic combinations with activity against multidrug-resistant *Klebsiella pneumoniae* and *Escherichia coli*. *Clin Microbiol Inf* **20**: O267–O273.

Jiang Y, Shen P, Wei Z, Liu L, He F, Shi K, Wang Y, Wang H, Yu Y. 2015. Dissemination of a clone carrying a fosA3-harbouring plasmid mediates high fosfomycin resistance rate of KPC-producing *Klebsiella pneumoniae* in China. *Int J Antimicrob Agents* **45**: 66–70.

Kahan FM, Kahan JS, Cassidy PJ, Kropp H. 1974. The mechanism of action of fosfomycin (phosphonomycin). *Ann NY Acad Sci* **235**: 364–386.

Kahlmeter G. 2000. The ECO•SENS Project: A prospective, multinational, multicentre epidemiological survey of the prevalence and antimicrobial susceptibility of urinary tract pathogens—Interim report. *J Antimicrob Chemother* **46**: 15–22.

Karageorgopoulos DE, Wang R, Yu XH, Falagas ME. 2012. Fosfomycin: Evaluation of the published evidence on the emergence of antimicrobial resistance in Gram-negative pathogens. *J Antimicrob Chemother* **67**: 255–268.

Katayama N, Tsubotani S, Nozaki Y, Harada S, Ono H. 1990. Fosfadecin and fosfocytocin, new nucleotide antibiotics produced by bacteria. *J Antibiot (Tokyo)* **43**: 238–246.

Kedar GC, Brown-Driver V, Reyes DR, Hilgers MT, Stidham MA, Shaw KJ, Finn J, Haselbeck RJ. 2008. Comparison of the essential cellular functions of the two *murA* genes of *Bacillus anthracis*. *Antimicrob Agents Chemother* **52**: 2009–2013.

Kim DH, Lees WJ, Kempsell KE, Lane WS, Duncan K, Walsh CT. 1996. Characterization of a Cys115 to Asp substitution in the *Escherichia coli* cell wall biosynthetic enzyme UDP-GlcNAc enolpyruvyl transferase (MurA) that confers resistance to inactivation by the antibiotic fosfomycin. *Biochemistry* **35**: 4923–4928.

Kim SY, Ju K-S, Metcalf WW, Evans BS, Kuzuyama T, van der Donk WA. 2012. Different biosynthetic pathways to fosfomycin in *Pseudomonas syringae* and *Streptomyces* species. *Antimicrob Agents Chemother* **56**: 4175–4183.

Kobayashi K, Ehrlich SD, Albertini A, Amati G, Andersen KK, Arnaud M, Asai K, Ashikaga S, Aymerich S, Bessieres P. 2003. Essential *Bacillus subtilis* genes. *Proc Natl Acad Sci* **100**: 4678–4683.

Kock H, Gerth U, Hecker M. 2004. MurAA, catalysing the first committed step in peptidoglycan biosynthesis, is a target of Clp-dependent proteolysis in *Bacillus subtilis*. *Mol Microbiol* **51**: 1087–1102.

Kuzuyama T, Kobayashi S, O'Hara K, Hidaka T, Seto H. 1996. Fosfomycin monophosphate and fosfomycin diphosphate, two inactivated fosfomycin derivatives formed by gene products of *fomA* and *fomB* from a fosfomycin producing organism *Streptomyces wedmorensis*. *J Antibiot (Tokyo)* **49**: 502–504.

Lemieux MJ, Huang Y, Wang DN. 2004. Glycerol-3-phosphate transporter of *Escherichia coli*: Structure, function and regulation. *Res Microbiol* **155**: 623–629.

Li Y, Zheng B, Li Y, Zhu S, Xue F, Liu J. 2015. Antimicrobial susceptibility and molecular mechanisms of fosfomycin resistance in clinical *Escherichia coli* isolates in mainland China. *PLoS ONE* **10**: e0135269.

Marchese A, Gualco L, Debbia EA, Schito GC, Schito AM. 2003. In vitro activity of fosfomycin against Gram-negative urinary pathogens and the biological cost of fosfomycin resistance. *Int J Antimicrob Agents* **22**: 53–59.

Marquardt JL, Brown ED, Lane WS, Haley TM, Ichikawa Y, Wong CH, Walsh CT. 1994. Kinetics, stoichiometry, and Identification of the reactive thiolate in the inactivation of UDP-GlcNAc enolpyruvoyl transferase by the antibiotic fosfomycin. *Biochemistry* **33**: 10646–10651.

Nikolaidis I, Favini-Stabile S, Dessen A. 2014. Resistance to antibiotics targeted to the bacterial cell wall. *Protein Sci* **23**: 243–259.

Nilsson AI, Berg OG, Aspevall O, Kahlmeter G, Andersson DI. 2003. Biological costs and mechanisms of fosfomycin resistance in *Escherichia coli*. *Antimicrob Agents Chemother* **47**: 2850–2858.

Oteo J, Orden B, Bautista V, Cuevas O, Arroyo M, Martinez-Ruiz R, Perez-Vazquez M, Alcaraz M, Garcia-Cobos S, Campos J. 2009. CTX-M-15-producing urinary *Escherichia coli* O25b-ST131-phylogroup B2 has acquired resistance to fosfomycin. *J Antimicrob Chemother* **64**: 712–717.

Cite this article as *Cold Spring Harb Perspect Med* doi: 10.1101/cshperspect.a025262

Performance Standards for Antimicrobial Susceptibility Testing. 2012. *22nd informational supplement M100-S22.* Clinical and Laboratory Standards Institute, Wayne, PA.

Rodríguez-Rojas A, Couce A, Blázquez J. 2010a. Frequency of spontaneous resistance to fosfomycin combined with different antibiotics in *Pseudomonas aeruginosa.* *Antimicrob Agents Chemother* **54:** 4948–4949.

Rodríguez-Rojas A, Maciá MD, Couce A, Gómez C, Castañeda-García A, Oliver A, Blázquez J. 2010b. Assessing the emergence of resistance: The absence of biological cost in vivo may compromise fosfomycin treatments for *P. aeruginosa* infections. *PLoS ONE* **5:** e10193.

Sato N, Kawamura K, Nakane K, Wachino JI, Arakawa Y. 2013. First detection of fosfomycin resistance gene *fosA3* in CTX-M-producing *Escherichia coli* isolates from healthy individuals in Japan. *Microb Drug Res* **19:** 477–482.

Shoji J, Kato T, Hinoo H, Hattori T, Hirooka K, Matsumoto K, Tanimoto T, Kondo E. 1986. Production of fosfomycin (phosphonomycin) by *Pseudomonas syringae.* *J Antibiot (Tokyo)* **39:** 1011–1012.

Skarzynski T, Mistry A, Wonacott A, Hutchinson SE, Kelly VA, Duncan K. 1996. Structure of UDP-*N*-acetylglucosamine enolpyruvyl transferase, an enzyme essential for the synthesis of bacterial peptidoglycan, complexed with substrate UDP-*N*-acetylglucosamine and the drug fosfomycin. *Structure* **4:** 1465–1474.

Suárez JE, Mendoza MC. 1991. Plasmid-encoded fosfomycin resistance. *Antimicrob Agents Chemother* **35:** 791–795.

Takahata S, Ida T, Hiraishi T, Sakakibara S, Maebashi K, Terada S, Muratani T, Matsumoto T, Nakahama C, Tomono K. 2010. Molecular mechanisms of fosfomycin resistance in clinical isolates of *Escherichia coli.* *Int J Antimicrob Agents* **35:** 333–337.

Thompson MK, Keithly ME, Goodman MC, Hammer ND, Cook PD, Jagessar KL, Harp J, Skaar EP, Armstrong RN. 2014. Structure and function of the genomically encoded fosfomycin resistance enzyme, FosB, from *Staphylococcus aureus.* *Biochemistry* **53:** 755–765.

Tsuruoka T, Miyata A, Yamada Y. 1978. Two kinds of mutants defective in multiple carbohydrate utilization isolated from in vitro fosfomycin-resistant strains of *Escherichia coli* K-12. *J Antibiot (Tokyo)* **31:** 192–201.

VanScoy BD, McCauley J, Ellis-Grosse EJ, Okusanya OO, Bhavnani SM, Forrest A, Ambrose PG. 2015. Exploration of the pharmacokinetic–pharmacodynamic relationships for fosfomycin efficacy using an in vitro infection model. *Antimicrob Agents Chemother* **59:** 7170–7177.

Walsh CC, McIntosh MP, Peleg AY, Kirkpatrick CM, Bergen PJ. 2015. In vitro pharmacodynamics of fosfomycin against clinical isolates of *Pseudomonas aeruginosa.* *J Antimicrob Chemother* **70:** 3042–3050.

Walsh CC, Landersdorfer CB, McIntosh MP, Peleg AY, Hirsch EB, Kirkpatrick CM, Bergen PJ. 2016. Clinically relevant concentrations of fosfomycin combined with polymyxin B, tobramycin or ciprofloxacin enhance bacterial killing of *Pseudomonas aeruginosa,* but do not suppress the emergence of fosfomycin resistance. *J Antimicrob Chemother* doi: 10.1093/jac/dkw115.

Wanke C, Amrhein N. 1993. Evidence that the reaction of the UDP-*N*-acetylglucosamine 1-carboxyvinyltransferase proceeds through the *O*-phosphothioketal of pyruvic acid bound to Cys115 of the enzyme. *Eur J Biochem* **218:** 861–870.

Winkler HH. 1973. Distribution of an inducible hexose-phosphate transport system among various bacteria. *J Bacteriol* **116:** 1079–1081.

Wu HC, Venkateswaran PS. 1974. Fosfomycin-resistant mutant of *Escherichia coli.* *Ann NY Acad Sci* **235:** 587–592.

Zavante Therapeutics. 2016. Zavante initiates the ZEUS study for ZTI-01 for the treatment of complicated urinary tract infections, Zavante Therapeutics, Inc., www.zavante.com/news/zavante-initiates-the-zeus-study-for-zti-01-for-the-treatment-of-complicated-urinary-tract-infections.

Zhu JY, Yang Y, Han H, Betzi S, Olesen SH, Marsilio F, Schönbrunn E. 2012. Functional consequence of covalent reaction of phosphoenolpyruvate with UDP-N-acetylglucosamine 1-carboxyvinyltransferase (MurA). *J Biol Chem* **287:** 12657–12667.

Zilhao R, Courvalin P. 1990. Nucleotide sequence of the *fosB* gene conferring fosfomycin resistance in *Staphylococcus epidermidis.* *FEMS Microbiol Lett* **68:** 267–272.

Zimmerman S, Stapley E, Wallick H, Baldwin R. 1969. Phosphonomycin. IV: Susceptibility testing method and survey. *Antimicrob Agents Chemother (Bethesda)* **9:** 303–309.

Mechanism of Action and Resistance to Daptomycin in *Staphylococcus aureus* and Enterococci

William R. Miller,[1] Arnold S. Bayer,[2,3] and Cesar A. Arias[1,4,5,6]

[1]University of Texas Medical School at Houston, Department of Internal Medicine, Division of Infectious Diseases, Houston, Texas 77030

[2]Los Angeles Biomedical Research Institute at Harbor-UCLA Medical Center, Torrance, California 90502

[3]David Geffen School of Medicine at UCLA, Los Angeles, California 90095

[4]Department of Microbiology and Molecular Genetics, Houston, Texas 77030

[5]Molecular Genetics and Antimicrobial Resistance Unit, Universidad El Bosque, Bogota, Colombia

[6]International Center for Microbial Genomics, Universidad El Bosque, Bogota, Colombia

Correspondence: Cesar.Arias@uth.tmc.edu

Lipopeptides are natural product antibiotics that consist of a peptide core with a lipid tail with a diverse array of target organisms and mechanisms of action. Daptomycin (DAP) is an example of these compounds with specific activity against Gram-positive organisms. DAP has become increasingly important to combat infections caused by Gram-positive bacteria because of the presence of multidrug resistance in these organisms, particularly in methicillin-resistant *Staphylococcus aureus* (MRSA) and vancomycin-resistant enterococci (VRE). However, emergence of resistance to DAP during therapy is a well-described phenomenon that threatens the clinical use of this antibiotic, limiting further the therapeutic options against both MRSA and VRE. This work will review the historical aspects of the development of DAP, as well as the current knowledge on its mechanism of action and pathways to resistance in a clinically relevant context.

Lipopeptides refer to a diverse class of compounds that share the general structure of a peptide core attached to a lipid tail, and are produced by a variety of environmental microorganisms including soil bacteria and fungi. This class of compounds possesses a wide therapeutic potential as evidenced by drugs that are currently in clinical use as antimicrobials, including the polymixins (polymixin B and colistin), echinocandins (caspofungin, micafungin, and anidulafungin), and daptomycin. The first isolation of a lipopeptide antibiotic occurred in 1953 with the discovery of amphomycin (Heinemann et al. 1953). However, this compound (and related molecules) was not further developed as a result, in part, of complex chemical structure and, in some cases, concerns of toxicity. Of note, there has recently been a resurgence of interest in compounds similar to amphomycin, driven by increasing rates of antimicrobial resistance to more traditional therapeutic agents.

Daptomycin (DAP), a lipopeptide antibiotic with in vitro bactericidal activity against Gram-positive bacteria, received approval by the Food and Drug Administration (FDA) in 2003 for soft-tissue infections and in 2006 for *Staphylococcus aureus* bacteremia and right-sided endocarditis. DAP has become a front-line agent in the treatment of challenging infections caused by both methicillin-resistant *S. aureus* (MRSA) and vancomycin-resistant *Enterococcus faecium* (VRE) (Munita et al. 2015). Despite its increasing role in the treatment of serious infections by these organisms, details of the precise mechanism of action and the mechanisms by which bacteria develop resistance are incompletely understood. Here, we will provide a brief overview of the structure and synthesis of DAP, explore what is known about its mechanism of action, and discuss the genetic changes associated with DAP nonsusceptibility (hereafter referred to as daptomycin resistance [DAP-R]) in *S. aureus* and the enterococci.

HISTORY, STRUCTURE, AND SYNTHESIS OF DAPTOMYCIN

After the discovery of amphomycin, a variety of lipopeptides with antimicrobial properties were identified over the next decade, including crystallomycin (Lomakina and Brazhnikova 1959), aspertocin (Shay et al. 1960), glumamycin (Shibata et al. 1962), laspertomycin, and tsushimycin (Naganawa et al. 1968; Shoji et al. 1968). Further development of these compounds for study and use was limited by several factors, including (1) the heterogeneous mixture of related molecules isolated from the fermentation of source organisms, (2) the complex chemistry needed to manipulate isolated compounds, and (3) a lack of understanding of the genetics behind their production. By the late 1980s, several important breakthroughs would allow DAP to make the journey from drug discovery to the bedside.

DAP began its journey as a molecule identified as A21987C, a group of lipopeptides pro-

duced by an isolate of *Streptomyces roseosporus* collected from the soil of the slopes of Mount Ararat in Turkey (Eisenstein et al. 2010). It consists of a 13-amino-acid depsipeptide, which harbors a cyclic decapeptide core with three extra-cyclic amino acids attached to an amino-terminal fatty acid tail (Fig. 1A). A distinctive feature of the lipopeptides is the diverse nature of the peptide core. In the case of DAP, the core contains a variety of noncanonical amino acids (kynurenine, ornithine, and 3-methylglutamic acid) and L-enantiomers (D-alanine and D-serine) (Debono et al. 1987). Several of these residues, in particular, kynurenine (of which the carboxyl group is the site of cyclization) and 3-methylglutamic acid have been shown, via substitution, to be important in the antimicrobial activity of the molecule with altered peptides displaying an increase of up to five times the minimum inhibitory concentration (MIC) (Grünewald et al. 2004). Further, six acidic residues in the DAP peptide ring are conserved across other calcium-dependent antimicrobial lipopeptides, highlighting the importance of this inorganic ion in both the mechanism of action and resistance (see below) (Hojati et al. 2002). The fatty acid tail also plays an important role in the activity of the compound, particularly in regard to toxicity. A21987C was found by high-performance liquid chromatography (HPLC) to be a mix of three main constituents differing only in the lipid moiety (with chain lengths of 11, 12, and 13 carbons) at the amino terminus (Debono et al. 1988). It was noted that longer chain lengths correlated with increasing toxicity; however, batch fermentation and subsequent separation was a laborious and difficult task with inefficient yield. The discovery that a penicillin deacylase produced by *Actinoplanes utahensis* (Debono et al. 1988) could remove the lipid tail opened the door for further characterization of the molecule. Using this technique, a semisynthetic derivative of A21987C with an *n*-decanoyl tail, named daptomycin, was found to balance antimicrobial activity with toxicity in a mouse model. Large-scale biosynthesis of DAP was achieved by feeding a controlled amount of decanoic acid to cultures of *S. roseosporus* (Huber et al. 1988).

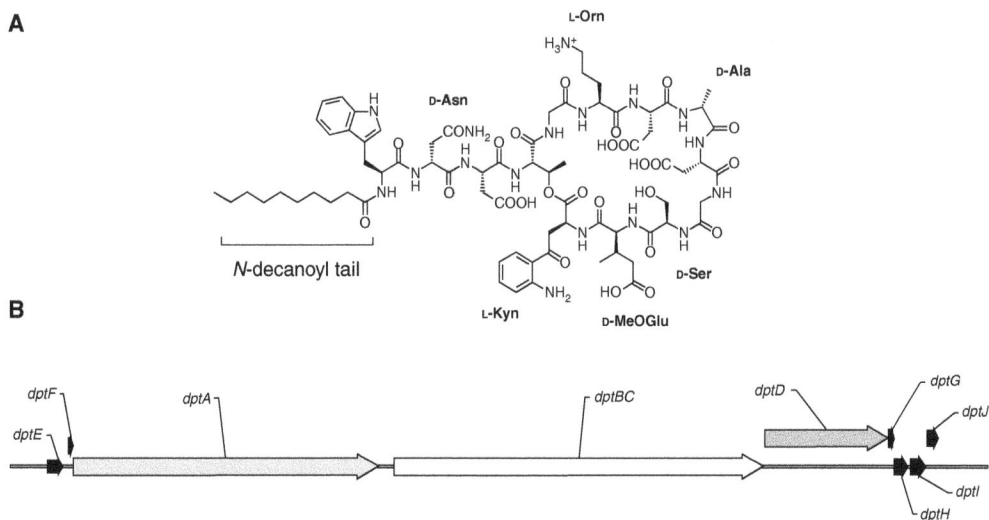

Figure 1. Structure of daptomycin and organization of the daptomycin biosynthesis gene cluster in *Streptomyces filamentosus*. (*A*) Chemical structure of daptomycin (DAP) with noncanonical amino acids and *N*-decanoyl fatty acid tail labeled. L-Kyn, L-Kynurenine; L-Orn, L-Ornithine; D-MeOGlu, D-3-methylglutamic acid. (*B*) Organization of the DAP biosynthesis gene cluster (see text for details). (Sequence information from NCBI database, accession number AY787762.1.)

Lipopeptides, similar to many other natural product antimicrobials, are produced by non-ribosomal peptide synthetases (NRPS). These large enzymatic complexes work in an assembly-line fashion to generate a specific peptide sequence. At their core is a series of three enzymatic activities (condensation, adenylation, and thiolation [CAT]) that perform a function analogous to ribosomal polypeptide synthesis, with amino acid specificity determined by the binding characteristics of each adenylation domain rather than an mRNA codon (Marahiel et al. 1997; Fischbach and Walsh 2006). The adenylation domain uses energy from adenosine triphosphate (ATP) to form an aminoacyladenosine monophosphate (AMP) intermediary from its cognate amino acid. Next, the AMP is displaced by the formation of a thioester bond coupling the amino acid to the thiolation domain carrier protein. The condensation domain then catalyzes the addition of the growing peptide chain to the amino acid monomer via an amide linkage, resulting in the passage of the nascent chain from one thiolation domain to the next module in the complex, wherein the process is repeated. Additional enzymes aug-

ment this core synthesis machinery, allowing modifications such as the incorporation of D-amino acids and allowing for the cyclic structure of DAP. In *S. roseosporus*, this machinery is organized into three multimodular subunits, DptA, DptBC, and DptD (Fig. 1B), which are responsible for the synthesis, modification, and cyclization of the 13 amino acid core (Baltz 2009). Two genes, *dptE* and *dptF*, located directly upstream of the primary peptide synthesis cluster, show similarity to acyl-CoA ligase and acyl carrier proteins, and are thus predicted to be involved in the addition of fatty acids to the amino-terminal end of DAP (Mchenney et al. 1998; Miao et al. 2005). Downstream are four accessory genes, one of which encodes a protein that shares identity with those known to metabolize tryptophan (a needed step for the synthesis of kynurenine). Another of the accessory genes is predicted to encode a glutamate methyltransferase presumably involved in the production of 3-methylglutarate (Miao et al. 2005).

The understanding of the genetic organization of the DAP NRPS machinery has opened the pathway to further drug modification and discovery. Although the *dpt* locus is transcribed as a single long mRNA, splitting the DptA, DptBC, and DptD submodules by deletion and subsequent reintroduction into different chromosomal locations (under control of a constitutive *erm* promoter) was not shown to adversely affect the production of DAP (Coëffet-Le Gal et al. 2006). Further, substitution of various CAT domains between lipopeptide synthesis clusters of different *Streptomyces* species has allowed the creation of altered peptide cores to screen for desired characteristics, such as increased activity in the presence of surfactant (Nguyen et al. 2006, 2010). Continued experimentation with novel arrangements of NRPS modules may offer further insights into DAP and may lead to discovery of novel compounds with improved activities.

MECHANISM OF ACTION

DAP shares structural similarities with a group of molecules produced by the mammalian innate immune system known as cationic antimicrobial peptides (CAMPs), specifically the human cathelicidin LL-37. These effectors of the innate immune response possess a wide spectrum of activity against bacteria, fungi, and some encapsulated viruses, and are thought to exert their effect by binding to and disrupting membrane integrity (Bals and Wilson 2003). The structural similarities between DAP and CAMPs have led investigators to postulate that they may share a common mechanism of membrane disruption, as DAP is known to bind the Gram-positive bacterial membrane and initiate a series of events that lead to cell death (Straus and Hancock 2006). Although the precise mechanism of action remains to be fully elucidated, there are at least two important interactions required for DAP to exert its bactericidal effect.

First is the interaction between DAP and calcium. Nuclear magnetic resonance (NMR) data of DAP in solution suggests that DAP complexes with calcium in a 1:1 molar ratio to form small (14–16 molecules) DAP micelles that may aid in antimicrobial delivery to the bacterial membrane (Scott et al. 2007). Changes in the NMR signal of the tryptophan at position 1 and the kynurenine at position 13 on the addition of calcium were thought to indicate a possible role for these residues in calcium binding or oligomerization of the molecule (Ho et al. 2008). Other divalent cations, such as magnesium, can induce micelle formation at higher concentrations (2.5:1 ratio), but result in decreased antimicrobial activity as evidenced by an increase in MICs by 64-fold (Ho et al. 2008).

The second important interaction takes place between DAP and the anionic phospholipid phosphatidylglycerol (PG). Once in proximity to the bacterial membrane, DAP undergoes a structural transition to insert into the cell membrane (Jung et al. 2004). This process appears to be dependent on the presence of PG in the target membrane (Muraih et al. 2011) and is facilitated by calcium ions, which decrease the DAP concentration needed for membrane insertion by ∼50-fold (Chen et al. 2014). Indeed, the presence of PG is an important mediator of DAP aggregation on model membranes. Using excimer fluorescence, excitation of DAP-perylene conjugants was seen in PG-containing membranes, but was absent from those made exclusively of phosphatidylcholine (Muraih et al. 2012). Thus, the first key steps of DAP membrane insertion and oligomerization rely on both calcium and PG as crucial mediators.

The dependence on specific phospholipids for the mechanism of action of DAP may also explain the antimicrobial spectrum of this drug, as it has potent activity against Gram-positive organisms, but none against Gram-negative organisms. This effect seems to be independent of the permeability barrier of the outer membrane (OM) of Gram-negative bacteria, as *Escherichia coli* protoplasts in which the OM was removed showed a fourfold reduction in MICs to vancomycin (a large glycopeptide antibiotic that would otherwise be excluded from the periplasmic space), but no change in DAP MIC (Randall et al. 2013). This observation has led some to suggest that Gram-negative bacteria are devoid of phospholipids that may interact with DAP. Indeed, the membrane lipid composition of *E. coli* includes 80% of phosphatidylethanolamine (PE) and only 15% PG, as compared with *S. aureus*, which lacks PE and contains 58% PG and 42% cardiolipin (CL)

(Epand et al. 2007). Thus, the differences of phospholipid content may explain the lack of activity of DAP against Gram-negative bacteria.

The series of events that occur after DAP gains access to the membrane are less clear. Early investigations into the mechanism of action of DAP observed that cell-wall synthesis was inhibited by a decreased intracellular pool of UDP-N-acetylmuramyl-pentapeptide in *Bacillus megaterium* (Mengin-Lecreulx et al. 1990), which the investigators attributed to inhibition of the enzymes involved in the formation of UDP-N-acetylglucosamine. This deficit was, however, subsequently found to be caused by impaired active transport of the amino acids required for murein synthesis, an effect associated with dissipation of the membrane electrochemical gradient (Allen et al. 1991). Analysis of major metabolic pathways and macromolecules showed DAP had little effect on the synthesis of DNA, RNA, or proteins. In contrast, cell envelope metabolism was consistently altered, with radiolabeled acetate incorporation into lipids decreased by 50% and lipoteichoic acid synthesis reduced by 93%. Moreover, both enterococci and *Bacillus* species displayed important morphologic changes (elongation) with a relative increase in sidewall synthesis (Canepari et al. 1990) on exposure to DAP. The finding that serial washes

with ethylenediaminetetraacetic acid (EDTA) was unable to remove DAP from bacterial membranes (Canepari et al. 1990) suggested that the cell membrane was the site of action, an observation that fit well with the amphipathic nature of the DAP molecule. Further, DAP is able to exert its bactericidal action against *S. aureus* in stationary phase, under conditions in which active metabolism is quenched and without requiring lysis of the target cell (Mascio et al. 2007; Cotroneo et al. 2008), consistent with disruption of the bacterial cell membrane, rather than inhibition of cell-wall or teichoic acid synthesis.

There are currently two proposed mechanisms of oligomeric DAP action (Fig. 2). One hypothesis, originating from the observed correlation of membrane depolarization and cell death in *S. aureus*, proposes that aggregates of DAP form an oligomeric pore like structure in the membrane, which results in ion leakage and subsequent dissipation of the membrane potential (Silverman et al. 2003). Experimental support for this hypothesis is derived from several studies. Initial stoichiometric calculations using Forester resonance energy transfer showed that ~7–8 DAP subunits associate in a PG-dependent manner for each oligomeric complex (Muraih and Palmer 2012; Zhang

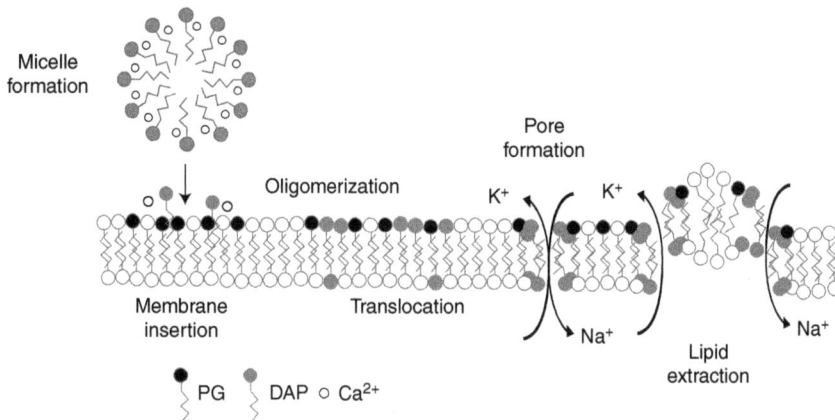

Figure 2. Proposed mechanisms for the action of daptomycin. In solution, daptomycin (DAP) complexes with calcium to form small micelles, and subsequent membrane insertion is dependent on both the presence of calcium and phosphatidylglycerol (PG). Once inserted, DAP oligomerizes and transitions to the inner membrane leaflet. These complexes then align on opposite sides of the membrane to form a pore channel permeable to small cations, or disrupt membrane integrity by extracting lipids and leading to transient ion leakage.

et al. 2014a). Further, the introduction of DAP into the outer leaflet induces a local membrane stress that increases levels of lipid flip-flop, an exchange of lipids between the inner and outer membrane leaflets (Jung et al. 2004), including the transition of DAP from the outer to inner leaflet. In the presence of PG, DAP associates into two oligomers of four units each opposite each other on the membrane, bending the membrane and establishing a pore like structure (Zhang et al. 2014a). Using model liposomes, exposure to DAP was found to make the membrane permeable to small cations such as sodium and potassium, and less so to anions or larger organic acids, suggesting that an influx of sodium ions abolished the membrane potential and served as the effector of DAP action (Zhang et al. 2014b). Interestingly, the presence of another phospholipid (PL), CL, in liposomes containing PG served to inhibit the translocation of DAP from the outer leaflet to the inner one, resulting in tetrameric complexes on the outer surface only (Zhang et al. 2014a). As we will discuss below, alterations of enzymes involved in PL metabolism are a common feature of resistance to DAP in some bacteria, consistent with the important role of PL metabolism in its mechanism of action.

A second hypothesis centers on a newly described phenomenon termed the lipid extracting effect. Using giant unilamellar vesicles (GUV), Chen et al. (2014) observed that DAP insertion into the membrane results in an initial expansion of vesicle surface area. As DAP concentrations continue to increase, there is a rapid aggregation of lipid on the membrane surface, while at the same time the overall surface area of the vesicle decreases, implying that the lipid clusters are extracted and "released" from the vesicle membrane. Interestingly, this phenomenon is dependent on both calcium and PG, and displays a threshold concentration of DAP required to initiate the membrane changes, which the investigators postulate may correlate with MIC values in bacterial isolates. Further, the extraction of lipids results in the formation of transient water pores, which could theoretically explain the ion leakage observed experimentally (Gurtovenko and Vattulainen 2007). This effect

may also explain the observations of Pogliano et al. (2012) in *Bacillus subtilis* showing that DAP binding to the membrane near the cell septum induced a patchy aggregate of lipid, altering cell morphology to a bent "L" shape and mislocalizing the essential cell division protein DivIVA. Indeed, abnormal septation and thickened cell walls are common features of DAP-R bacteria, and may be because of recognition of altered lipid membranes as signals for new peptidoglycan synthesis away from the septum.

It is important to note that the two hypotheses are not mutually exclusive because both pore formation and lipid extraction may be playing a role once DAP makes contact with the bacterial membrane and could explain the broad effects of the antibiotic in bacterial permeabilization, cell division, and metabolism.

DAPTOMYCIN RESISTANCE

DAP-R in *S. aureus* and the enterococci has been well documented and it is a serious concern for the treatment of serious infections caused by these organisms (Bayer et al. 2013; Miller et al. 2014). Given the clinical burden of disease that these organisms represent, an understanding of the mechanisms by which they subvert the DAP "attack" is likely to provide novel insights into the manner that bacteria protect their cell membrane and adapt to the antimicrobial challenge. Detailed analyses of both DAP-R laboratory and clinical isolates have revealed several common pathways associated with resistance, namely, alteration of regulatory systems responsible for the bacterial cell envelope stress response, as well as enzymes involved in phospholipid metabolism and membrane homeostasis. Despite the genetic similarities, the mechanisms by which these changes drive DAP-R seem quite varied and are adapted to the biology of each organism, a fascinating feature of bacterial evolution. Thus, we will discuss each relevant species separately.

DAP-R IN *Staphylococcus aureus*

S. aureus use several strategies to circumvent the DAP effect, the most common appears to in-

 Cite this article as *Cold Spring Harb Perspect Med* doi: 10.1101/cshperspect.a026997

volve the alteration of the cell-surface charge (Fig. 3A). Indeed, *S. aureus* seems to primarily respond to the DAP attack by producing a more positive overall cell-surface charge, presumably to prevent the positively charged DAP–calcium insertion by electrostatic repulsion. This phenotype is classically associated with mutations in *mprF* (multiple peptide resistance factor), which encodes a bifunctional enzyme that contains a carboxy-terminal cytoplasmic tail responsible for lysinylation of PG and an amino-terminal domain, which consists of eight transmembrane domains. The amino-terminal domain encodes a "flippase" activity, which is responsible for the translocation of lysyl-PG (LPG) from the inner to the outer membrane. A central domain of four transmembrane helices seems to assist with both lysinylation and flippase activities (Ernst et al. 2009).

In DAP-R *S. aureus*, a number of *mprF* mutations have been described that result in amino acid changes clustering in the central bifunctional region that overall confer a "gain-of-function" of the enzyme (Bayer et al. 2015). Thus, the net result is an increased synthesis and expression of positively charged LPG on the outer membrane. Strong evidence for the role of *mprF* in DAP-R are studies in which expression of *mprF* with DAP-R associated mutations (but not wild-type *mprF*) in *trans* could restore elevated DAP MICs to strains of *S. aureus* in which *mprF* had been deleted from the chromosome (Yang et al. 2013). Moreover, inhibition of MprF protein synthesis in DAP-R strains harboring gain-of-function mutations by antisense RNA (directed against *mprF* transcripts) was able to reverse DAP-R in vitro (Rubio et al. 2011).

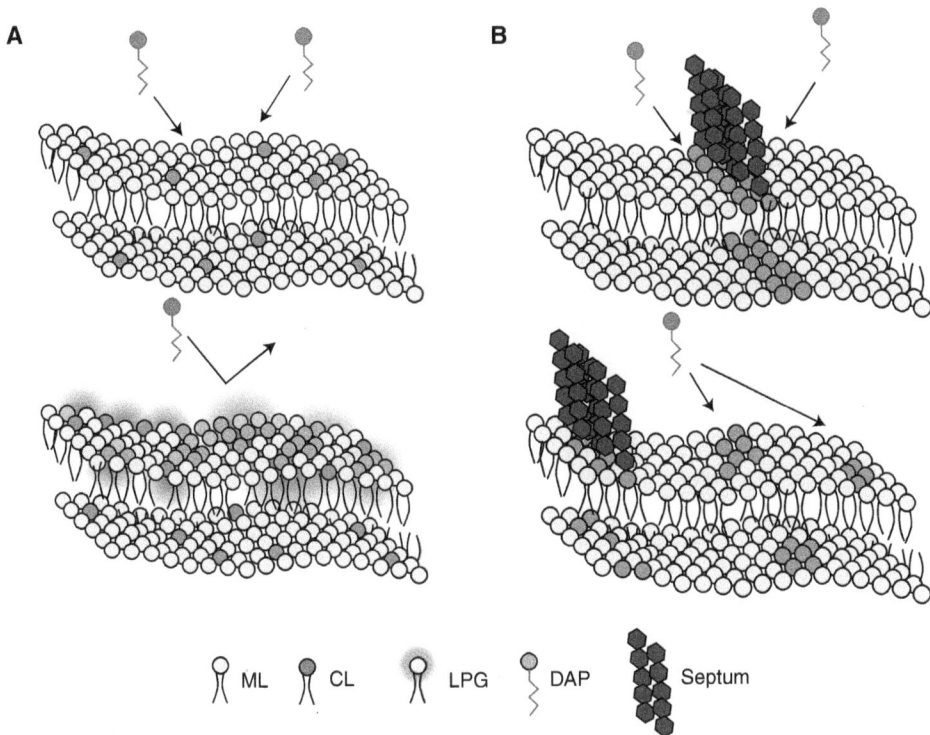

Figure 3. Strategies for resisting daptomycin membrane attack. (*A*) Repulsion: In *Staphylococcus aureus* and *Enterococcus faecium*, changes in cell-surface charge and membrane phospholipid content block daptomycin (DAP) membrane association and oligomerization. (*B*) Diversion: In *E. faecalis* sensitive to DAP cardiolipin (CL) clusters at the division septum. In resistant isolates, redistribution of CL microdomains "traps" DAP away from the septum. ML, membrane lipid; LPG, lysylphosphatidylglycerol.

An alternative pathway for DAP-R in *S. aureus* that results in an increase of cell-surface charge is the overexpression of the *dlt* operon (Yang et al. 2009; Cafiso et al. 2014). This operon produces the machinery responsible for attaching the positively charged amino acid alanine to cell-wall teichoic acid (WTA), leading to an increase in positive cell-surface charge in a manner similar to increased LPG synthesis. Upregulation of WTA synthesis (as observed by increased transcription of *tagA*) and the *dlt* operon were also associated with increased cell-wall mass, another common phenotype observed in DAP-R staphylococci (Bertsche et al. 2011, 2013). However, despite the strong association between mutations in *mprF* and increased expression of *dlt* with increases in net positive cell-surface charge, these changes do not seem to correlate with changes in DAP MICs in all strains (Mishra et al. 2014). Indeed, an in vitro study generated DAP-R isolates with alterations in both *mprF* and *dlt*, but the net positive charge of the DAP-R mutants was less than the parent strain (Mishra et al. 2009). Thus, additional characteristics must also play a role in mediating DAP-R in staphylococci.

A second mechanism associated with DAP-R in staphylococci is the alteration of membrane phospholipid composition, which is postulated to either decrease the amount of PG available at the membrane interface or to change the fluidity of the membrane, thus interfering with DAP binding and subsequent oligomerization. Interestingly, by analyzing the action of a membrane active antimicrobial peptide on GUVs, it was found that increases in LPG were not associated with decreased peptide binding (as might be expected in a charge repulsion mechanism) but rather with inhibition of intravesicular dye leakage after binding takes place, consistent with a membrane integrity protective effect (Kilelee et al. 2010). Further, as discussed above, other phospholipid species, such as CL, may also play a protective role in preventing DAP translocation once inserted in the membrane (Zhang et al. 2014a). The enzyme responsible for cardiolipin synthesis, cardiolipin synthase, joins two molecules of PG to make CL (Short and White 1972). Thus, it is tempting to speculate that mutations producing changes in enzyme function may play a role in DAP-R by altering the ratio of PG to CL in the cell membrane. Indeed, genomic analysis of 33 DAP-R strains indicated that, among others, mutations in *pgsA* (which encodes an enzyme involved in PG synthesis) and *cls2* (cardiolipin synthase) were associated with DAP-R (Peleg et al. 2012). Additionally, membrane fluidity (which is highly dependent on PL and fatty acid composition) may also be an important factor that influences the DAP-R phenotype in certain strains. (Jones et al. 2008; Mishra et al. 2011). Interestingly, membranes of DAP-R clinical isolates are more fluid, whereas laboratory isolates tend to have more rigid membranes (Mishra et al. 2009), suggesting that DAP requires an optimal membrane order for insertion and oligomerization and perturbations of this order to either side may be protective. Along these lines, changes (both increase and decrease) in the production of staphyloxanthin, the carotenoid responsible for the golden color of *S. aureus*, was associated with DAP-R and was postulated to be a result of its influence on membrane fluidity (Mishra and Bayer 2013).

Global regulatory changes in genes modulating cell envelope stress and maintenance in *S. aureus* have also been associated with development of DAP-R (Utaida et al. 2003; Rose et al. 2012). Interestingly, DAP challenge induces important changes in global gene expression. These genomic pathways are similar to those associated with resistance to other antibiotics such as vancomycin and seem to affect the expression of the cell-wall "stimulon." Two important two-component regulatory systems (TCS) have been involved in DAP-R, namely, VraSR and YycFG (Muthaiyan et al. 2008; Mehta et al. 2012). Of note, DAP was also found to induce a group of genes that was previously associated with exposure to carbonyl cyanide *m*-chlorophenylhydrazone, a proton ionophore, reflecting its ability to disrupt the membrane and induce ion leakage (Muthaiyan et al. 2008).

In general, TCS consist of a membrane-bound sensor histidine kinase (HK) responsible for detecting a particular stimulus or cellular perturbation, and a DNA-binding response regulatory (RR) that alters transcription of target genes (Dubrac et al. 2008). Mutations in these proteins can lead to altered expression of the system's regulon, profoundly affecting membrane homeostasis. The essential TCS YycFG (also known as WalKR) is involved in the control of peptidoglycan biosynthesis in *S. aureus*, mainly through the regulation of expression of two major autolysins, LytM and AltA (Dubrac and Msadek 2004). The genes encoding this system are clustered with two other genes, *yycHI*, that are "accessory" to the function of YycFG. Both YycH and YycI are amino-terminal transmembrane proteins with extracellular carboxy-terminal domains that in *B. subtilis* have been shown to repress the activity of the YycG HK (Szurmant et al. 2007). In nondividing cells, the entire YycFGHI complex remains in the peripheral cell wall, presumably in an inactive state. However, under growth conditions, YycG is recruited to the site of septal formation, whereas YycH and YycI remain in the periphery (Fukushima et al. 2011).

Using an inducible promoter to control YycFG expression, Dubrac et al. (2007) showed that low levels of YycFG expression were associated with decreased peptidoglycan turnover, increased cross-linking, and increased glycan chain length. Interestingly, low levels of YycFG were also associated with increased resistance to lysis by the detergent Triton X-100. By varying the temperature of model lipid membranes, it was shown that the activity of the HK YycG was impacted by membrane fluidity, with the system turned off under highly fluid conditions (Türck and Bierbaum 2012). The investigators suggested a mechanism by which YycG senses changes in membrane fluidity and responds by adjusting cell-wall cross-linking to compensate for stresses caused by osmotic pressure. Of note, in DAP-R isolates, several mutations in *yycFG* affecting multiple domains of both YycG HK and YycF RR (Friedman et al. 2006; Howden et al. 2011) have been described. Additionally, mutations in the accessory genes have also been noted. For example, a mutation resulting in a frameshift and truncation of ~10% of the accessory protein YycH (which in *B. subtilis* is associated

with regulating YycF signaling) was associated with DAP-R (Szurmant et al. 2005; Mwangi et al. 2007). Given that the DAP-R phenotype displays some similarities to the YycFG-deficient phenotype (e.g., thickened cell walls, increased membrane fluidity, and resistance to membrane disruption), it is tempting to speculate that the observed changes in YycFG impair the functioning of the operon, down-regulating cell-wall homeostasis to survive the DAP-mediated attack.

The VraSR TCS is orthologous to the LiaSR system of *B. subtilis* and enterococci (discussed below) and is conserved across the low G+C bacteria (Jordan et al. 2006). It is up-regulated by both vancomycin and DAP exposure, and is associated with cell-wall biosynthesis via transcription of *pbp2* (penicillin binding protein 2), *tagA* (WTA synthesis), *prsA* (a chaperone), and *murZ* (UDP-*N*-acetylglucosamine enolpyruvyl transferase), among others (Kuroda et al. 2003; Mwangi et al. 2007; Camargo et al. 2008). Structural studies have shown that on activation by phosphorylation, the VraR RR undergoes a conformational change allowing for dimerization and a subsequent increase in its binding affinity for target DNA (Leonard et al. 2013). Experimental evidence supports a role for this system in DAP-R, as deletion of the *vraSR* operon from a DAP-R strain of *S. aureus* resulted in a DAP-sensitive (DAP-S) phenotype, which could be reversed by supplying the genes in *trans* (Mehta et al. 2012). Additional mutations associated with the DAP-R phenotype include genes encoding the RNA polymerase subunits *rpoB* and *rpoC* (Friedman et al. 2006; Peleg et al. 2012). A mutation in *rpoB*, resulting in the amino acid change A621E, was associated with increased expression of the *dlt* operon and correlated with an increase in positive cell-surface charge, whereas RpoB mutations A621E and A477D were both linked to activation of cell-wall biosynthesis and increased cell-wall thickness (Cui et al. 2010; Bæk et al. 2015).

The pathway to DAP-R also results in significant cellular metabolic shifts. Analysis of six strain pairs of *S. aureus* under normal growth conditions and DAP exposure revealed that there is a decrease in activity of the tricarboxylic acid (TCA) cycle and, instead, carbon sources are redirected into the pentose phosphate pathway (Gaupp et al. 2015). This is corroborated by prior work that had shown levels of succinate dehydrogenase, an enzyme involved in the TCA cycle, were lower in a DAP-R strain when compared with its DAP-S counterpart (Fischer et al. 2011). Additionally, mutations noted upstream of acetyl-CoA synthetase in DAP-R isolates (Friedman et al. 2006) may affect the production of acetyl-CoA, which is involved in lipid synthesis and may also feed into the TCA cycle. Redirection of the flow of metabolites results in the formation of larger pools of amino sugar precursors, which can be used for peptidoglycan, teichoic acid, and nucleotide synthesis (Gaupp et al. 2015). Thus, a metabolic shift primes DAP-R isolates to build larger stores of cell envelope precursors allowing them to weather the storm of DAP-induced membrane stress.

DAP-R IN ENTEROCOCCI

The introduction of DAP provided clinicians with an agent that possessed in vitro bactericidal activity against enterococci, and it quickly became a front-line antibiotic for recalcitrant VRE infections, despite the lack of FDA approval for this indication. Even in early development of DAP, it was noted that longer acyl chain lengths (13–14 carbons) tended to improve activity against enterococci, but with the trade-off of increased toxicity (Debono et al. 1988). Thus, DAP (with its n-decanoyl fatty acyl side chain) is less potent against enterococci, a fact that is reflected in the clinical breakpoints, which are fourfold higher for enterococci compared with S. aureus (4 μg/mL vs. 1 μg/mL). Similar to what has been discussed in staphylococci, development of DAP-R in enterococci seems to affect two important groups of genes, namely, those controlling the cell membrane stress response and phospholipid metabolism. Despite the genetic similarities, the two clinically relevant species, Enterococcus faecalis and E. faecium, seem to display distinctive phenotypic differences in their response to DAP challenge and, thus, we will discuss them separately.

Daptomycin Resistance in *E. Faecalis*

The genetic bases of DAP-R in *E. faecalis* were mapped using whole-genome sequencing of both in vitro and clinical isolates that had developed resistance in the presence of the drug (Arias et al. 2011; Palmer et al. 2011). Using a strain pair from a patient with *E. faecalis* bacteremia who failed DAP therapy, Arias et al. (2011) mapped the genetic changes to genes encoding the LiaFSR system (a conserved TCS associated with DAP-R in *B. subtilis*) and two enzymes involved in phospholipid metabolism, cardiolipin synthase (Cls), and a glycerophosphoryl diester phosphodiesterase (GdpD). Phenotypic changes associated with the DAP-R phenotype included increased thickness of the cell-wall and abnormal septations. Additionally, the DAP-R derivative was found to have a decrease in the proportion of PG and increased rigidity of the cell membrane (Mishra et al. 2012). However, in contrast to both S. aureus and E. faecium, a distinct characteristic of DAP-R *E. faecalis* is a rearrangement of cell membrane PL microdomains. Indeed, DAP-S *E. faecalis* shows prominent concentration of anionic PLs (including CL) at the division septum and in polar areas. Development of DAP resistance markedly changes the architecture of these PL microdomains, moving them away from the division septum, the principle site of DAP action (Tran et al. 2014). This reorganization in *E. faecalis* seems to be crucial for full expression of the DAP-R phenotype. It is postulated that these PL aggregates may serve as "sink holes" for DAP, diverting the antibiotic away from the vital septal area of the membrane (the diversion hypothesis) (Fig. 3B). Indeed, compelling experimental data suggest that DAP-R *E. faecalis* strains do not "repel" DAP from the cell surface as shown previously by S. aureus (Tran et al. 2014).

Detailed studies on the molecular basis of the DAP-R phenotype in *E. faecalis* has identified the LiaFSR system as a major contributor to the adaptive response against DAP and antimicrobial peptide "attack." This system is conserved across the Firmicutes (VraSR is its ortholog in S. aureus, see above) and has been

Cite this article as *Cold Spring Harb Perspect Med* doi: 10.1101/cshperspect.a026997

well-characterized in the model organism *B. subtilis* (Jordan et al. 2006; Schrecke et al. 2013). The HK LiaS responds to as-yet-unidentified membrane stressors induced by DAP or other membrane active agents. LiaS phosphorylates its cognate RR LiaR, which contains a DNA-binding motif and alters expression of target genes. LiaF serves a regulatory role by inhibiting the activation of LiaR through interactions with LiaS in the absence of membrane stress. The *liaFSR* operon in *E. faecalis* consists of only three open reading frames. However, in *B. subtilis*, an additional three genes, *liaG*, *liaH*, and *liaI*, are targets of LiaR (Wolf et al. 2010) and mediate resistance to antimicrobial peptides via a response that appears to be similar to that described for the phage shock protein (PSP) response of Gram-negative organisms (Brissette et al. 1990; Yamaguchi et al. 2013).

Several lines of experimental evidence point to an activation of the LiaFSR system and its downstream effectors as mediators of DAP-R in *E. faecalis*. Mutations in the predicted inhibitor LiaF have been associated with increases in DAP MIC, presumably caused by increased activity of the system. A deletion of isoleucine at position 177 of LiaF, (identified in a clinical isolate of *E. faecalis*) was sufficient to increase the DAP MIC of a susceptible isolate from 1 to 4 μg/mL and resulted in redistribution of membrane phospholipid microdomains (Tran et al. 2014). Further, this same change was noted to abolish the bactericidal action of DAP in vitro (loss of a three $\log_{10}$ decrease in time–kill curve colony counts), despite the MIC being within the "susceptible" range (Munita et al. 2013). In an experimental evolution of a polymorphic population of *E. faecalis* maintained in continuous culture, changes in the LiaFSR system emerged as the first step in the pathway to DAP resistance (Miller et al. 2013). The most frequently observed mutations involved either insertion or deletion of the isoleucine at position 177 in LiaF (suggesting the importance of this residue for the inhibitory function of LiaF) and appeared after ~2 weeks as MICs rose into the 3–4 μg/mL range.

Because of the major role of LiaFSR in DAP and antimicrobial resistance, Davlieva et al.

(2015) sought to investigate the structural bases of DAP-R associated with mutations in LiaR (which have been commonly identified in clinical isolates of DAP-R enterococci). These studies showed that a substitution of asparagine for aspartate at position 191 of LiaR mimics phosphorylation and changes the oligomeric state of LiaR. Indeed, "wild-type" unphosphorylated LiaR seems to exist as a dimer. When the protein is phosphorylated or harbors mutations that mimic phosphorylation LiaR tetramerizes, increasing the binding affinity for its own and other promoters by 100-fold (Davlieva et al. 2015), resulting in constitutive activation of the LiaFSR system. Furthermore, deletion of the *liaR* gene results in a "hypersusceptible" phenotype (MICs of 0.047 μg/mL) that is independent of the genetic background into which it is introduced (Reyes et al. 2015). Thus, LiaFSR seems crucial in orchestrating the specific response to a variety of membrane active agents and overexpression of this system results in a membrane protective effect that results in DAP-R.

Once established, LiaFSR mutations allow the accrual of additional genetic changes resulting in the full resistance phenotype (Miller et al. 2013). Mutations in genes affecting membrane phospholipids, particularly *cls*, have been frequently associated with DAP-R. In *E. faecalis*, introduction of the altered *cls* alleles in *trans* bearing the R218Q substitution or the N77-Q79 deletion were able to confer resistance to the laboratory strain OG1RF (Palmer et al. 2011). Mutations in GdpD had no effect on DAP MICs in isolation, but when introduced along with LiaF mutations, they resulted in a fully resistant phenotype (Arias et al. 2011). Genes in the LiaR regulon bear similarities to the Psp system mediated by *liaI* and *liaH* in *B. subtilis*, although these genes (named *liaXYZ*) seem to be organized into an independent operon in the *E. faecalis* genome distant from *liaFSR* (Miller et al. 2013). Interestingly, point mutations in this group of three genes, specifically a frameshift disrupting the carboxy-terminal end of LiaX and a second frameshift mutation in LiaY have been associated with DAP-R in enterococci both in vitro and in clinical isolates (Palmer et al. 2011; Humphries et al.

2012). Additional mutations in *yybT*, a cyclic dinucleotide phosphodiesterase predicted to be involved in cell stress and signaling, and *gshF*, a glutathione synthase, have been described, although their contributions to DAP-R are not well understood (Miller et al. 2013).

Daptomycin Resistance in *E. Faecium*

Although a number of genetic determinants of DAP-R in *E. faecium* have been identified, the biochemical bases for their effect on the DAP resistance phenotype are not well understood. Unlike *E. faecalis*, *E. faecium* does not display a visible alteration or rearrangement of anionic phospholipids in the membrane, even in isolates with mutations in the LiaFSR system (Tran et al. 2015). Instead, it appears that the impact of mutations in *E. faecium* results in phenotypic changes that are more akin to those associated with DAP-R in *S. aureus* (Mishra et al. 2012). Indeed, the overall mechanism for DAP-R in *E. faecium* appear to involve repulsion of the antibiotic from the cell surface.

Analysis of the genomes of 19 clinical isolates of *E. faecium* with DAP MICs ranging from 3 to 48 µg/mL revealed that the majority of the strains harbored mutations in either LiaFSR or YycFG and that either pathway can lead to DAP-R (Diaz et al. 2014). In the LiaFSR system, the most common mutation was a W73C change in LiaR accompanied by a T120A substitution in LiaS, suggesting that these changes coevolve during the development of DAP-R (Munita et al. 2012). Four strains also harbored various mutations in LiaF, although these changes did not affect the isoleucine at position 177 as described in *E. faecalis*. The importance of LiaFSR changes in *E. faecium* was shown by deletion of the *liaR* gene from clinical strains harboring mutations in both the LiaFSR and YycFG pathway (Panesso et al. 2015). In both cases, strains developed a "hypersusceptible" phenotype with increased binding of fluorescently labeled DAP to the cell membrane in the absence of *liaR*. The presence of LiaRS substitution has also been associated with clinical failure of DAP and loss of bactericidal activity of the antibiotic (Munita et al. 2014). Changes in the YycFG pathway are

commonly localized to the YycG HK as well as both accessory proteins YycH and YycI (Diaz et al. 2014); however, the role of such mutations in the development of DAP-R remains to be established.

As in both *S. aureus* and *E. faecalis*, mutations in *cls*, the gene encoding cardiolipin synthase, are common in DAP-R *E. faecium*. They are often found with substitutions in LiaFSR or YycFG and, in this setting, they may contribute to the progression of an isolate from DAP-tolerant to DAP-resistant (Diaz et al. 2014). Exchange of the R218Q *cls* allele from a DAP-R strain into a susceptible one was not able to increase the MIC, further suggesting that this change alone is not sufficient for the development of a DAP-R phenotype in *E. faecium* (Tran et al. 2013). Biochemical characterization of Cls proteins from a susceptible and resistant strain pair of *E. feacium* showed that the R218Q and H215R substitutions mapped to the PLD1 phospholipase catalytic domain resulted in an increase in the V_{max} of the enzyme (Davlieva et al. 2013). This is consistent with an enzymatic gain-of-function and may allow for a more rapid depletion of the available PG by shunting this PL to the CL pool during times of membrane stress. Mutations in a *pspC*-like protein (the above mentioned LiaY), *cfa* (a cyclooxygenase that catalyzes the addition of a methyl group to unsaturated fatty acids), *dlt*, and *mprF*, among others, have been associated with DAP-R *E. faecium*. However, they appear to be rare in clinical isolates and their role in resistance is currently difficult to assess (Humphries et al. 2012; Tran et al. 2013; Diaz et al. 2014).

CONCLUDING REMARKS

Over the last decade, the increase of multidrug-resistant Gram-positive organisms has brought DAP into the spotlight as a therapeutic option for severe infections. DAP has potent bactericidal activity and a unique mechanism of action, which have made it a useful addition to the clinician's antibiotic repertoire. As its clinical use continues to increase reports of resistance are becoming more common. To preserve the use of this and other antimicrobial compounds,

Cite this article as *Cold Spring Harb Perspect Med* doi: 10.1101/cshperspect.a026997

a deeper understanding of the robust and redundant pathways that mediate the mechanism of resistance may shed light on the biology of bacterial membrane adaptation, including the response to the innate immune system. With continued efforts to unravel the complex networks that mediate DAP-R, additional insights into the coordination of cell envelope synthesis machinery are sure to provide new therapeutic targets to exploit against recalcitrant Gram-positive infections in the future.

REFERENCES

Allen NE, Alborn WE Jr, Hobbs JN Jr. 1991. Inhibition of membrane potential-dependent amino acid transport by daptomycin. *Antimicrob Agents Chemother* **35:** 2639–2642.

Arias CA, Panesso D, McGrath DM, Qin X, Mojica MF, Miller C, Diaz L, Tran TT, Rincon S, Barbu EM, et al. 2011. Genetic basis for in vivo daptomycin resistance in enterococci. *N Engl J Med* **365:** 892–900.

Bæk KT, Thøgersen L, Mogenssen RG, Mellergaard M, Thomsen LE, Petersen A, Skov S, Cameron DR, Peleg AY, Frees D. 2015. Stepwise decrease in daptomycin susceptibility in clinical *Staphylococcus aureus* isolates associated with an initial mutation in *rpoB* and a compensatory inactivation of the *clpX* gene. *Antimicrob Agents Chemother* **59:** 6983–6991.

Bals R, Wilson JM. 2003. Cathelicidins—A family of multifunctional antimicrobial peptides. *Cell Mol Life Sci* **60:** 711–720.

Baltz RH. 2009. Daptomycin: Mechanisms of action and resistance, and biosynthetic engineering. *Curr Opin Chem Biol* **13:** 144–151.

Bayer AS, Schneider T, Sahl HG. 2013. Mechanisms of daptomycin resistance in *Staphylococcus aureus*: Role of the cell membrane and cell wall. *Ann NY Acad Sci* **1277:** 139–158.

Bayer AS, Mishra NN, Chen L, Kreiswirth BN, Rubio A, Yang SJ. 2015. Frequency and distribution of single-nucleotide polymorphisms within *mprF* in methicillin-resistant *Staphylococcus aureus* clinical Isolates and their role in cross-resistance to daptomycin and host defense antimicrobial peptides. *Antimicrob Agents Chemother* **59:** 4930–4937.

Bertsche U, Weidenmaier C, Kuehner D, Yang SJ, Baur S, Wanner S, Francois P, Schrenzel J, Yeaman MR, Bayer AS. 2011. Correlation of daptomycin resistance in a clinical *Staphylococcus aureus* strain with increased cell wall teichoic acid production and D-alanylation. *Antimicrob Agents Chemother* **55:** 3922–3928.

Bertsche U, Yang SJ, Kuehner D, Wanner S, Mishra NN, Roth T, Nega M, Schneider A, Mayer C, Grau T, et al. 2013. Increased cell wall teichoic acid production and D-alanylation are common phenotypes among daptomycin-resistant methicillin-resistant *Staphylococcus aureus* (MRSA) clinical isolates. *PLoS ONE* **8:** e67398.

Brissette JL, Russel M, Weiner L, Model P. 1990. Phage shock protein, a stress protein of *Escherichia coli*. *Proc Natl Acad Sci* **87:** 862–866.

Cafiso V, Bertuccio T, Purrello S, Campanile F, Mammina C, Sartor A, Raglio A, Stefani S. 2014. *dltA* overexpression: A strain-independent keystone of daptomycin resistance in methicillin-resistant *Staphylococcus aureus*. *Int J Antimicrob Agents* **43:** 26–31.

Camargo IL, Neoh HM, Cui L, Hiramatsu K. 2008. Serial daptomycin selection generates daptomycin-nonsusceptible *Staphylococcus aureus* strains with a heterogeneous vancomycin-intermediate phenotype. *Antimicrob Agents Chemother* **52:** 4289–4299.

Canepari P, Boaretti M, Lleó MM, Satta G. 1990. Lipoteichoic acid as a new target for activity of antibiotics: Mode of action of daptomycin (LY146032). *Antimicrob Agents Chemother* **34:** 1220–1226.

Chen YF, Sun TL, Sun Y, Huang HW. 2014. Interaction of daptomycin with lipid bilayers: A lipid extracting effect. *Biochemistry* **53:** 5384–5392.

Coëffet-Le Gal MF, Thurston L, Rich P, Miao V, Baltz RH. 2006. Complementation of daptomycin *dptA* and *dptD* deletion mutations in *trans* and production of hybrid lipopeptide antibiotics. *Microbiology* **152:** 2993–3001.

Cotroneo N, Harris R, Perlmutter N, Beveridge T, Silverman JA. 2008. Daptomycin exerts bactericidal activity without lysis of *Staphylococcus aureus*. *Antimicrob Agents Chemother* **52:** 2223–2225.

Cui L, Isii T, Fukuda M, Ochiai T, Neoh HM, Camargo IL, Watanabe Y, Shoji M, Hishinuma T, Hiramatsu K. 2010. An RpoB mutation confers dual heteroresistance to daptomycin and vancomycin in *Staphylococcus aureus*. *Antimicrob Agents Chemother* **54:** 5222–5233.

Davlieva M, Zhang W, Arias CA, Shamoo Y. 2013. Biochemical characterization of cardiolipin synthase mutations associated with daptomycin resistance in enterococci. *Antimicrob Agents Chemother* **57:** 289–296.

Davlieva M, Shi Y, Leonard PG, Johnson TA, Zianni MR, Arias CA, Ladbury JE, Shamoo Y. 2015. A variable DNA recognition site organization establishes the LiaR-mediated cell envelope stress response of enterococci to daptomycin. *Nucleic Acids Res* **43:** 4758–4773.

Debono M, Barnhart M, Carrell CB, Hoffmann JA, Occolowitz JL, Abbott BJ, Fukuda DS, Hamill RL, Biemann K, Herlihy WC. 1987. A21978C, a complex of new acidic peptide antibiotics: Isolation, chemistry, and mass spectral structure elucidation. *J Antibiot (Tokyo)* **40:** 761–777.

Debono M, Abbott BJ, Molloy RM, Fukuda DS, Hunt AH, Daupert VM, Counter FT, Ott JL, Carrell CB, Howard LC, et al. 1988. Enzymatic and chemical modifications of lipopeptide antibiotic A21978C: The synthesis and evaluation of daptomycin (LY146032). *J Antibiot (Tokyo)* **41:** 1093–1105.

Diaz L, Tran TT, Munita JM, Miller WR, Rincon S, Carvajal LP, Wollam A, Reyes J, Panesso D, Rojas NL, et al. 2014. Whole-genome analyses of *Enterococcus faecium* isolates with diverse daptomycin MICs. *Antimicrob Agents Chemother* **58:** 4527–4534.

Dubrac S, Msadek T. 2004. Identification of genes controlled by the essential YycG/YycF two-component system of *Staphylococcus aureus*. *J Bacteriol* **186:** 1175–1181.

Dubrac S, Boneca IG, Poupel O, Msadek T. 2007. New insights into the WalK/WalR (YycG/YycF) essential signal transduction pathway reveal a major role in controlling cell wall metabolism and biofilm formation in *Staphylococcus aureus*. *J Bacteriol* **189:** 8257–8269.

Dubrac S, Bisicchia P, Devine KM, Msadek T. 2008. A matter of life and death: Cell wall homeostasis and the WalKR (YycGF) essential signal transduction pathway. *Mol Microbiol* **70:** 1307–1322.

Eisenstein BI, Oleson FB, Baltz RH. 2010. Daptomycin: From the mountain to the clinic, with essential help from Francis Tally, MD. *Clin Infect Dis* **50:** S10–S15.

Epand RF, Savage PB, Epand RM. 2007. Bacterial lipid composition and the antimicrobial efficacy of cationic steroid compounds (ceragenins). *Biochim Biophys Acta* **1768:** 2500–2509.

Ernst CM, Staubitz P, Mishra NN, Yang SJ, Hornig G, Kalbacher H, Bayer AS, Kraus D, Peschel A. 2009. The bacterial defensin resistance protein MprF consists of separable domains for lipid lysinylation and antimicrobial peptide repulsion. *PLoS Pathog* **5:** e1000660.

Fischbach MA, Walsh CT. 2006. Assembly-line enzymology for polyketide and nonribosomal peptide antibiotics: Logic, machinery, and mechanisms. *Chem Rev* **106:** 3468–3496.

Fischer A, Yang SJ, Bayer AS, Vaezzadeh AR, Herzig S, Stenz L, Girard M, Sakoulas G, Scherl A, Yeaman MR, et al. 2011. Daptomycin resistance mechanisms in clinically derived *Staphylococcus aureus* strains assessed by a combined transcriptomics and proteomics approach. *J Antimicrob Chemother* **66:** 1696–1711.

Friedman L, Alder JD, Silverman JA. 2006. Genetic changes that correlate with reduced susceptibility to daptomycin in *Staphylococcus aureus*. *Antimicrob Agents Chemother* **50:** 2137–2145.

Fukushima T, Furihata I, Emmins R, Daniel RA, Hoch JA, Szurmant H. 2011. A role for the essential YycG sensor histidine kinase in sensing cell division. *Mol Microbiol* **79:** 503–522.

Gaupp R, Lei S, Reed JM, Peisker H, Boyle-Vavra S, Bayer AS, Bischoff M, Herrmann M, Daum RS, Powers R, et al. 2015. *Staphylococcus aureus* metabolic adaptations during the transition from a daptomycin susceptibility phenotype to a daptomycin nonsusceptibility phenotype. *Antimicrob Agents Chemother* **59:** 4226–4238.

Grünewald J, Sieber SA, Mahlert C, Linne U, Marahiel MA. 2004. Synthesis and derivatization of daptomycin: A chemoenzymatic route to acidic lipopeptide antibiotics. *J Am Chem Soc* **126:** 17025–17031.

Gurtovenko AA, Vattulainen I. 2007. Ion leakage through transient water pores in protein-free lipid membranes driven by transmembrane ionic charge imbalance. *Biophys J* **92:** 1878–1890.

Heinemann B, Kaplan MA, Muir RD, Hooper IR. 1953. Amphomycin, a new antibiotic. *Antibiot Chemother (Northfield)* **3:** 1239–1242.

Ho SW, Jung D, Calhoun JR, Lear JD, Okon M, Scott WR, Hancock RE, Straus SK. 2008. Effect of divalent cations on the structure of the antibiotic daptomycin. *Eur Biophys J* **37:** 421–433.

Hojati Z, Milne C, Harvey B, Gordon L, Borg M, Flett F, Wilkinson B, Sidebottom PJ, Rudd BA, Hayes MA, et al. 2002. Structure, biosynthetic origin, and engineered biosynthesis of calcium-dependent antibiotics from *Streptomyces coelicolor*. *Chem Biol* **9:** 1175–1187.

Howden BP, McEvoy CR, Allen DL, Chua K, Gao W, Harrison PF, Bell J, Coombs G, Bennett-Wood V, Porter JL, et al. 2011. Evolution of multidrug resistance during *Staphylococcus aureus* infection involves mutation of the essential two component regulator WalKR. *PLoS Pathog* **7:** e1002359.

Huber FM, Pieper RL, Tietz AJ. 1988. The formation of daptomycin by supplying decanoic acid to *Streptomyces roseosporus* cultures producing the antibiotic complex A21978C. *J Biotechnol* **7:** 283–292.

Humphries RM, Kelesidis T, Tewhey R, Rose WE, Schork N, Nizet V, Sakoulas G. 2012. Genotypic and phenotypic evaluation of the evolution of high-level daptomycin nonsusceptibility in vancomycin-resistant *Enterococcus faecium*. *Antimicrob Agents Chemother* **56:** 6051–6053.

Jones T, Yeaman MR, Sakoulas G, Yang SJ, Proctor RA, Sahl HG, Schrenzel J, Xiong YQ, Bayer AS. 2008. Failures in clinical treatment of *Staphylococcus aureus* infection with daptomycin are associated with alterations in surface charge, membrane phospholipid asymmetry, and drug binding. *Antimicrob Agents Chemother* **52:** 269–278.

Jordan S, Junker A, Helmann JD, Mascher T. 2006. Regulation of LiaRS-dependent gene expression in *Bacillus subtilis*: Identification of inhibitor proteins, regulator binding sites, and target genes of a conserved cell envelope stress-sensing two-component system. *J Bacteriol* **188:** 5153–5166.

Jung D, Rozek A, Okon M, Hancock RE. 2004. Structural transitions as determinants of the action of the calcium-dependent antibiotic daptomycin. *Chem Biol* **11:** 949–957.

Kilelee E, Pokorny A, Yeaman MR, Bayer AS. 2010. Lysyl-phosphatidylglycerol attenuates membrane perturbation rather than surface association of the cationic antimicrobial peptide 6W-RP-1 in a model membrane system: Implications for daptomycin resistance. *Antimicrob Agents Chemother.* **54:** 4476–4479.

Kuroda M, Kuroda H, Oshima T, Takeuchi F, Mori H, Hiramatsu K. 2003. Two-component system VraSR positively modulates the regulation of cell-wall biosynthesis pathway in *Staphylococcus aureus*. *Mol Microbiol* **49:** 807–821.

Leonard PG, Golemi-Kotra D, Stock AM. 2013. Phosphorylation-dependent conformational changes and domain rearrangements in *Staphylococcus aureus* VraR activation. *Proc Natl Acad Sci* **110:** 8525–8530.

Lomakina NN, Brazhnikova MG. 1959. Chemical composition of crystallomycin. *Biokhimiia* **24:** 425–431.

Marahiel MA, Stachelhaus T, Mootz HD. 1997. Modular peptide synthetases involved in nonribosomal peptide synthesis. *Chem Rev* **97:** 2651–2674.

Mascio CT, Alder JD, Silverman JA. 2007. Bactericidal action of daptomycin against stationary-phase and nondividing *Staphylococcus aureus* cells. *Antimicrob Agents Chemother* **51:** 4255–4260.

Mchenney MA, Hosted TJ, Dehoff BS, Rosteck PR Jr, Baltz RH. 1998. Molecular cloning and physical mapping of the daptomycin gene cluster from *Streptomyces roseosporus*. *J Bacteriol* **180:** 143–151.

Mehta S, Cuirolo AX, Plata KB, Riosa S, Silverman JA, Rubio A, Rosato RR, Rosato AE. 2012. VraSR two-component regulatory system contributes to *mprF*-mediated decreased susceptibility to daptomycin in in vivo-selected clinical strains of methicillin-resistant *Staphylococcus aureus*. *Antimicrob Agents Chemother* 56: 92–102.

Mengin-Lecreulx D, Allen NE, Hobbs JN, van Heijenoort J. 1990. Inhibition of peptidoglycan biosynthesis in *Bacillus megaterium* by daptomycin. *FEMS Microbiol Lett* 57: 245–248.

Miao V, Coëffet-Legal MF, Brian P, Brost R, Penn J, Whiting A, Martin S, Ford R, Parr I, Bouchard M, et al. 2005. Daptomycin biosynthesis in *Streptomyces roseosporus*: Cloning and analysis of the gene cluster and revision of peptide stereochemistry. *Microbiology* 151: 1507–1523.

Miller C, Kong J, Tran TT, Arias CA, Saxer G, Shamoo Y. 2013. Adaptation of *Enterococcus faecalis* to daptomycin reveals an ordered progression to resistance. *Antimicrob Agents Chemother* 57: 5373–5383.

Miller WR, Munita JM, Arias CA. 2014. Mechanisms of antibiotic resistance in enterococci. *Expert Rev Anti Infect Ther* 12: 1221–1236.

Mishra NN, Bayer AS. 2013. Correlation of cell membrane lipid profiles with daptomycin resistance in methicillin-resistant *Staphylococcus aureus*. *Antimicrob Agents Chemother* 57: 1082–1085.

Mishra NN, Yang SJ, Sawa A, Rubio A, Nast CC, Yeaman MR, Bayer AS. 2009. Analysis of cell membrane characteristics of in vitro-selected daptomycin-resistant strains of methicillin-resistant *Staphylococcus aureus*. *Antimicrob Agents Chemother* 53: 2312–2318.

Mishra NN, McKinnell J, Yeaman MR, Rubio A, Nast CC, Chen L, Kreiswirth BN, Bayer AS. 2011. In vitro cross-resistance to daptomycin and host defense cationic antimicrobial peptides in clinical methicillin-resistant *Staphylococcus aureus* isolates. *Antimicrob Agents Chemother* 55: 4012–4018.

Mishra NN, Bayer AS, Tran TT, Shamoo Y, Mileykovskaya E, Dowhan W, Guan Z, Arias CA. 2012. Daptomycin resistance in enterococci is associated with distinct alterations of cell membrane phospholipid content. *PLoS ONE* 7: e43958.

Mishra NN, Bayer AS, Weidenmaier C, Grau T, Wanner S, Stefani S, Cafiso V, Bertuccio T, Yeaman MR, Nast CC, et al. 2014. Phenotypic and genotypic characterization of daptomycin-resistant methicillin-resistant *Staphylococcus aureus* strains: Relative roles of *mprF* and *dlt* operons. *PLoS ONE* 9: e107426.

Munita JM, Panesso D, Diaz L, Tran TT, Reyes J, Wanger A, Murray BE, Arias CA. 2012. Correlation between mutations in liaFSR of *Enterococcus faecium* and MIC of daptomycin: Revisiting daptomycin breakpoints. *Antimicrob Agents Chemother* 56: 4354–4359.

Munita JM, Tran TT, Diaz L, Panesso D, Reyes J, Murray BE, Arias CA. 2013. A *liaF* codon deletion abolishes daptomycin bactericidal activity against vancomycin-resistant *Enterococcus faecalis*. *Antimicrob Agents Chemother* 57: 2831–2833.

Munita JM, Mishra NN, Alvarez D, Tran TT, Diaz L, Panesso D, Reyes J, Murray BE, Adachi JA, Bayer AS, et al. 2014. Failure of high-dose daptomycin for bacteremia caused by daptomycin-susceptible *Enterococcus faecium* harboring LiaSR substitutions. *Clin Infect Dis* 59: 1277–1280.

Munita JM, Bayer AS, Arias CA. 2015. Evolving resistance among Gram-positive pathogens. *Clin Infect Dis.* 61: S48–S57.

Muraih JK, Palmer M. 2012. Estimation of the subunit stoichiometry of the membrane-associated daptomycin oligomer by FRET. *Biochim Biophys Acta* 1818: 1642–1647.

Muraih JK, Pearson A, Silverman J, Palmer M. 2011. Oligomerization of daptomycin on membranes. *Biochim Biophys Acta* 1808: 1154–1160.

Muraih JK, Harris J, Taylor SD, Palmer M. 2012. Characterization of daptomycin oligomerization with perylene excimer fluorescence: Stoichiometric binding of phosphatidylglycerol triggers oligomer formation. *Biochim Biophys Acta* 1818: 673–678.

Muthaiyan A, Silverman JA, Jayaswal RK, Wilkinson BJ. 2008. Transcriptional profiling reveals that daptomycin induces the *Staphylococcus aureus* cell wall stress stimulon and genes responsive to membrane depolarization. *Antimicrob Agents Chemother* 52: 980–990.

Mwangi MM, Wu SW, Zhou Y, Sieradzki K, de Lencastre H, Richardson P, Bruce D, Rubin E, Myers E, Siggia ED, et al. 2007. Tracking the in vivo evolution of multidrug resistance in *Staphylococcus aureus* by whole-genome sequencing. *Proc Natl Acad Sci* 104: 9451–9456.

Naganawa H, Hamada M, Maeda K, Okami Y, Takeushi T. 1968. Laspartomycin, a new anti-staphylococcal peptide. *J Antibiot (Tokyo)* 21: 55–62.

Nguyen KT, Ritz D, Gu JQ, Alexander D, Chu M, Miao V, Brian P, Baltz RH. 2006. Combinatorial biosynthesis of novel antibiotics related to daptomycin. *Proc Natl Acad Sci* 103: 17462–17467.

Nguyen KT, He X, Alexander DC, Li C, Gu JQ, Mascio C, Van Praagh A, Mortin L, Chu M, Silverman JA, et al. 2010. Genetically engineered lipopeptide antibiotics related to A54145 and daptomycin with improved properties. *Antimicrob Agents Chemother* 54: 1404–1413.

Palmer KL, Daniel A, Hardy C, Silverman J, Gilmore MS. 2011. Genetic basis for daptomycin resistance in enterococci. *Antimicrob Agents Chemother* 55: 3345–3356.

Panesso D, Reyes J, Gaston EP, Deal M, Londoño A, Nigo M, Munita JM, Miller WR, Shamoo Y, Tran TT, et al. 2015. Deletion of *liaR* reverses daptomycin resistance in *Enterococcus faecium* independent of the genetic background. *Antimicrob Agents Chemother* 59: 7327–7334.

Peleg AY, Miyakis S, Ward DV, Earl AM, Rubio A, Cameron DR, Pillai S, Moellering RC Jr, Eliopoulos GM. 2012. Whole genome characterization of the mechanisms of daptomycin resistance in clinical and laboratory derived isolates of *Staphylococcus aureus*. *PLoS ONE* 7: e28316.

Pogliano J, Pogliano N, Silverman JA. 2012. Daptomycin-mediated reorganization of membrane architecture causes mislocalization of essential cell division proteins. *J Bacteriol* 194: 4494–4504.

Randall CP, Mariner KR, Chopra I, O'Neill AJ. 2013. The target of daptomycin is absent from *Escherichia coli* and other Gram-negative pathogens. *Antimicrob Agents Chemother* 57: 637–639.

Reyes J, Panesso D, Tran TT, Mishra NN, Cruz MR, Munita JM, Singh KV, Yeaman MR, Murray BE, Shamoo Y, et al.

2015. A liaR deletion restores susceptibility to daptomycin and antimicrobial peptides in multidrug-resistant *Enterococcus faecalis*. *J Infect Dis* **211**: 1317–1325.

Rose WE, Fallon M, Moran JJ, Vanderloo JP. 2012. Vancomycin tolerance in methicillin-resistant *Staphylococcus aureus*: Influence of vancomycin, daptomycin, and telavancin on differential resistance gene expression. *Antimicrob Agents Chemother* **56**: 4422–4427.

Rubio A, Conrad M, Haselbeck RJ, Kedar GC, Brown-Driver V, Finn J, Silverman JA. 2011. Regulation of mprF by antisense RNA restores daptomycin susceptibility to daptomycin-resistant isolates of *Staphylococcus aureus*. *Antimicrob Agents Chemother* **55**: 364–367.

Schrecke K, Jordan S, Mascher T. 2013. Stoichiometry and perturbation studies of the LiaFSR system of *Bacillus subtilis*. *Mol Microbiol* **87**: 769–788.

Scott WR, Baek SB, Jung D, Hancock RE, Straus SK. 2007. NMR structural studies of the antibiotic lipopeptide daptomycin in DHPC micelles. *Biochim Biophys Acta* **1768**: 3116–3126.

Shay AJ, Adam J, Martin JH, Hausmann WK, Shu P, Bohonos N. 1960. Aspartocin. I: Production, isolation, and characteristics. *Antibiot Annu* **7**: 194–198.

Shibata M, Kanzaki T, Nakazawa K, Inoue M, Hitomi H, Mizuno K, Fujino M, Akira M. 1962. On glumamycin, a new antibiotic. *J Antibiot (Tokyo)* **15**: 1–6.

Shoji JI, Kozuki S, Okamoto S, Sakazaki R, Otsuka H. 1968. Studies on tsushimycin. I: Isolation and characterization of an acidic acylpeptide containing a new fatty acid. *J Antibiot (Tokyo)* **21**: 439–443.

Short SA, White DC. 1972. Biosynthesis of cardiolipin from phosphatidylglycerol in *Staphylococcus aureus*. *J Bacteriol* **109**: 820–826.

Silverman JA, Perlmutter NG, Shapiro HM. 2003. Correlation of daptomycin bactericidal activity and membrane depolarization in *Staphylococcus aureus*. *Antimicrob Agents Chemother* **47**: 2538–2544.

Straus SK, Hancock RE. 2006. Mode of action of the new antibiotic for Gram-positive pathogens daptomycin: Comparison with cationic antimicrobial peptides and lipopeptides. *Biochim Biophys Acta* **1758**: 1215–1223.

Szurmant H, Nelson K, Kim EJ, Perego M, Hoch JA. 2005. YycH regulates the activity of the essential YycFG two-component system in *Bacillus subtilis*. *J Bacteriol* **187**: 5419–5426.

Szurmant H, Mohan MA, Imus PM, Hoch JA. 2007 YycH and YycI interact to regulate the essential YycFG two-component system in *Bacillus subtilis*. *J Bacteriol* **189**: 3280–3289.

Tran TT, Panesso D, Gao H, Roh JH, Munita JM, Reyes J, Diaz L, Lobos EA, Shamoo Y, Mishra NN, et al. 2013. Whole-genome analysis of a daptomycin-susceptible *Enterococcus faecium* strain and its daptomycin-resistant variant arising during therapy. *Antimicrob Agents Chemother* **57**: 261–268.

Tran TT, Panesso D, Mishra NN, Mileykovskaya E, Guan Z, Munita JM, Reyes J, Diaz L, Weinstock GM, Murray BE, Shamoo Y, et al. 2014. Daptomycin-resistant *Enterococcus faecalis* diverts the antibiotic molecule from the division septum and remodels cell membrane phospholipids. *mBio* **4**: e00281–13.

Tran TT, Munita JM, Arias CA. 2015. Mechanisms of drug resistance: Daptomycin resistance. *Ann NY Acad Sci* **1354**: 32–53.

Türck M, Bierbaum G. 2012. Purification and activity testing of the full-length YycFGHI proteins of *Staphylococcus aureus*. *PLoS ONE* **7**: e30403.

Utaida S, Dunman PM, Macapagal D, Murphy E, Projan SJ, Singh VK, Jayaswal RK, Wilkinson BJ. 2003. Genome-wide transcriptional profiling of the response of *Staphylococcus aureus* to cell-wall-active antibiotics reveals a cell-wall-stress stimulon. *Microbiology* **149**: 2719–2732.

Wolf D, Kalamorz F, Wecke T, Juszczak A, Mäder U, Homuth G, Jordan S, Kirstein J, Hoppert M, Voigt B, et al. 2010. In-depth profiling of the LiaR response of *Bacillus subtilis*. *J Bacteriol* **192**: 4680–4693.

Yamaguchi S, Reid DA, Rothenberg E, Darwin AJ. 2013. Changes in Psp protein binding partners, localization and behaviour upon activation of the *Yersinia enterocolitica* phage shock protein response. *Mol Microbiol* **87**: 656–671.

Yang SJ, Kreiswirth BN, Sakoulas G, Yeaman MR, Xiong YQ, Sawa A, Bayer AS. 2009. Enhanced expression of dltABCD is associated with the development of daptomycin non-susceptibility in a clinical endocarditis isolate of *Staphylococcus aureus*. *J Infect Dis* **200**: 1916–1920.

Yang SJ, Mishra NN, Rubio A, Bayer AS. 2013. Causal role of single nucleotide polymorphisms within the mprF gene of *Staphylococcus aureus* in daptomycin resistance. *Antimicrob Agents Chemother* **57**: 5658–5664.

Zhang T, Muraih JK, Tishbi N, Herskowitz J, Victor RL, Silverman J, Uwumarenogie S, Taylor SD, Palmer M, Mintzer E. 2014a. Cardiolipin prevents membrane translocation and permeabilization by daptomycin. *J Biol Chem* **289**: 11584–11591.

Zhang T, Muraih JK, MacCormick B, Silverman J, Palmer M. 2014b. Daptomycin forms cation- and size-selective pores in model membranes. *Biochim Biophys Acta* **1838**: 2425–2430.

Polymyxin: Alternative Mechanisms of Action and Resistance

Michael J. Trimble,[1] Patrik Mlynárčik,[2] Milan Kolář,[2] and Robert E.W. Hancock[1]

[1]Department of Microbiology and Immunology, University of British Columbia, Vancouver, BC V6T 1Z4, Canada

[2]Department of Microbiology, Faculty of Medicine and Dentistry, Palacký University, 771 47 Olomouc, Czech Republic

Correspondence: bob@hancocklab.com

Antibiotic resistance among pathogenic bacteria is an ever-increasing issue worldwide. Unfortunately, very little has been achieved in the pharmaceutical industry to combat this problem. This has led researchers and the medical field to revisit past drugs that were deemed too toxic for clinical use. In particular, the cyclic cationic peptides polymyxin B and colistin, which are specific for Gram-negative bacteria, have been used as "last resort" antimicrobials. Before the 1980s, these drugs were known for their renal and neural toxicities; however, new clinical practices and possibly improved manufacturing have made them safer to use. Previously suggested to primarily attack the membranes of Gram-negative bacteria and to not easily select for resistant mutants, recent research exploring resistance and mechanisms of action has provided new perspectives. This review focuses primarily on the proposed alternative mechanisms of action, known resistance mechanisms, and how these support the alternative mechanisms of action.

Public news outlets, academic articles (primary literature and reviews), government documents, and countless other sources echo the alarm regarding multiple antibiotic resistance development in nosocomial pathogens (Fernández et al. 2011; CDC 2013; Taylor et al. 2014; Berendonk et al. 2015). Unfortunately, recent antibiotic research has encountered numerous obstacles, notably limited discovery of new antibiotic compounds. This relates to the poor economic incentives for pharmaceutical companies to invest in antibiotic development given pharmacoeconomic considerations (a combination of poor success rates, limited markets because of the large number of different antibiotics on the market, and the fact that antibiotics are generally used acutely and, thus, do not have "repeat customers") and limited funding for academic laboratories working in the field. Even though occasional breakthroughs are made, for example, the antibiotic teixobactin (Ling et al. 2015), it is noteworthy that while chemically novel, this compound addresses a known target and has yet to be examined clinically. The medical community has recognized the critical importance of combating bacterial resistance, and this has led to the revival of disfavored antibacterials such as the cyclic peptide polymyxin B and its relatives (Falagas and Kasiakou 2006; Landman et al. 2008; Falagas et al. 2010).

Polymyxin was first isolated in 1947 from the Gram-positive soil bacterium *Bacillus polymyxa*, which was reclassified as *Paenibacillus polymyxa* in 1993 (Storm et al. 1977; Ash et al. 1994). Fifteen different molecular variations are known and are produced by *P. polymyxa* subspecies, including polymyxin E (colistin) from ssp. *Colistinus* and polymyxin M (mattacin) from *Paenibacillus kobensis* (Storm et al. 1977; Martin et al. 2003; Choi et al. 2009b; Tambadou et al. 2015). The polymyxins are small lipopeptide molecules of ~1200 Da in mass and are characterized by a polycationic peptide ring with a short protruding peptide attached to a hydrophobic fatty acid tail (Fig. 1A) (Newton 1956; Evans 1999; Nation et al. 2014). The clinically used polymyxins, polymyxin B (a mixture

of composed of polymyxins B3, B6, and minor components B1, B1-I, B2) and colistin (containing two major components colistin A and B and ~30 minor components) (Orwa et al. 2001), differ primarily by a single D-phenylalanine replaced by a D-leucine within the peptide ring (Landman et al. 2008; Yu et al. 2015). The cationic ring makes these drugs soluble in aqueous environments, whereas the hydrophobic acyl chain facilitates insertion into bacterial membranes (Evans 1999; Nation et al. 2014).

All polymyxins are nonribosomally produced (i.e., produced by large enzymes called nonribosomal peptide synthetases) (Finking and Marahiel 2004), and although their structures have been known for years, the mechanisms for their in vivo synthesis are still being elucidated. The peptides are formed in steps by enzyme modules with specific domains that govern adenylation, thiolation (peptidyl carrier

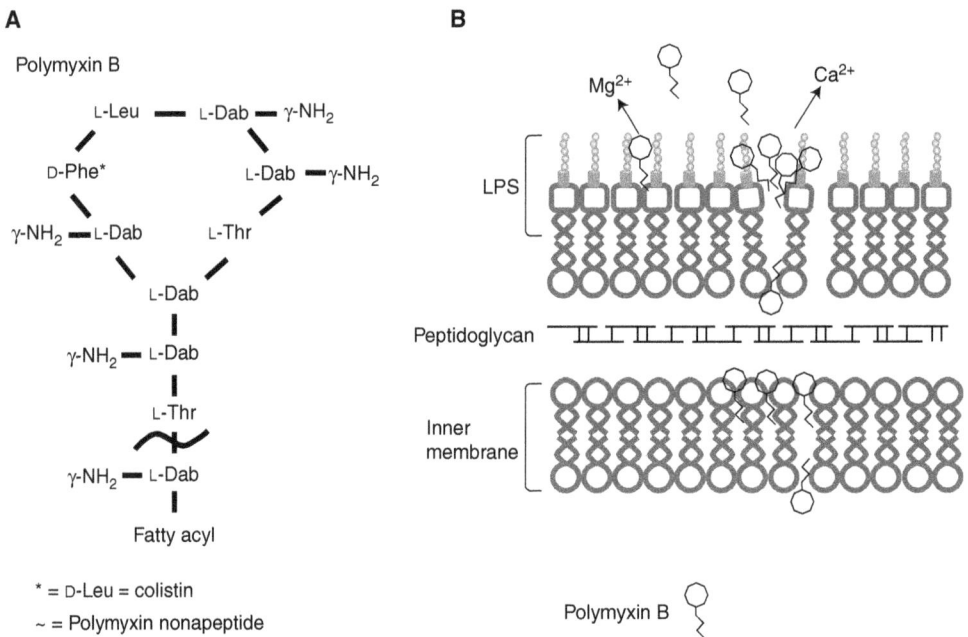

Figure 1. The structure of polymyxin B and likely mode of membrane interaction. (*A*) The general structure of the cyclic cationic peptide polymyxin B. Colistin (polymyxin E) replaces the phenylalanine (D-Phe*) in polymyxin B with a leucine. Attenuated polymyxin nonapeptide has the fatty acyl tail removed adjacent to the threonine outside the ring structure (~). (*B*) Polymyxin B interacts with the lipid A portion of the lipopolysaccharide (LPS) outer membrane. The peptides cross the outer membrane through a "self-promoted uptake" mechanism and then interact with the cytoplasmic membrane to inhibit cellular energization, and possibly cause inhibition of cell division and/or cytoplasmic membrane permeabilization and subsequent cell death.

Cite this article as *Cold Spring Harb Perspect Med* doi: 10.1101/cshperspect.a025288

protein [PCP]), and condensation (Finking and Marahiel 2004; Choi et al. 2009b). To date, the nonribosomal peptide synthetases have been identified for polymyxin M (Martin et al. 2003), polymyxin A (Choi et al. 2009b), variants of polymyxin B (Shaheen et al. 2011), and polymyxin P (Niu et al. 2013).

All of the polymyxin peptides have similar bactericidal activities with efficacy against Gram-negative organisms and a few Gram-positive species (see section ExPortal) (Newton 1956; Storm et al. 1977; Vega and Caparon 2012). Polymyxin and its relatives were commonly used in clinical treatment until the 1980s, when their nephrotoxicity caused a decline in usage (Falagas and Kasiakou 2006; Landman et al. 2008; Falagas et al. 2010). The nephrotoxicity is drug-related acute tubular necrosis leading to acute renal failure and stems from accumulation and persistence in the body, with preferences for kidney and brain tissues (Kunin and Bugg 1971; Evans 1999). Similar to the drug's interaction with the bacterial outer membrane, it permeabilizes eukaryotic membranes leading to swelling and lysis (Berg et al. 1996, 1998; Lewis and Lewis 2004; Falagas and Kasiakou 2006). However, recent studies have indicated that the incidence of nephrotoxicity is less common and severe compared with older studies (Berg et al. 1996, 1998; Lewis and Lewis 2004; Falagas and Kasiakou 2006) and, intriguingly, polymyxin B was found to be less nephrotoxic than the prodrug colistimethate in which the positive charges on colistin are neutralized (Phe et al. 2014). In addition, polymyxins can bind to neurons and block the release of acetylcholine, preventing neurotransmission and affecting muscular activity, although neurotoxic effects are usually mild and resolve on discontinuation of therapy (Falagas and Kasiakou 2006).

After usage was reduced because of toxicity concerns, polymyxins were reserved for treatment of persistent infections in cystic fibrosis patients, ophthalmic conjunctivitis infections, and in over-the-counter topical antibiotic ointments (Hancock and Chapple 1999; Hancock 2000; Falagas et al. 2010). The usage of polymyxins has seen a recent upsurge despite issues with toxicity because of increasing resistance to other drugs. Currently, they are the go-to drug for serious multidrug-resistant (MDR) Gram-negative bacterial infections, particularly those caused by MDR *Pseudomonas aeruginosa*, *Acinetobacter baumannii*, and *Klebsiella pneumoniae*, in which they have become the last-line treatment for infections that are resistant to other available antibiotics (Nation et al. 2015). Polymyxin B is administered directly as an antibiotic, whereas colistin is administered as the prodrug colistin methanesulfonate (in which diaminobenzoate residues are derivatized as methane sulphonates), which is hydrolyzed into colistin in vivo (Zavascki et al. 2007; Landman et al. 2008; Nation et al. 2014). As mentioned above, recent examinations into the reported renal and neural toxicities indicate that the severity and incidences of these effects remain high but not nearly as high as previously suggested (Falagas and Kasiakou 2006; Kwa et al. 2008; Landman et al. 2008). Thus, although renal and neural toxicity remain an adverse side effect, our current handling of the drugs has more positive outcomes (Nation et al. 2015). Correct formulations, dosage (lower than past usage), discontinuation on adverse symptoms, avoidance of coadministration with other potential nephro/neurotoxic dugs, and overall better critical care and, possibly, better manufacturing have led to lower incidences of nephro and neural toxicities (Falagas and Kasiakou 2006). Consensus recommendations for their clinical usage were recently published (Nation et al. 2015).

Disruption and/or permeabilization of Gram-negative bacterial cytoplasmic membrane has been suggested as the main mechanism of action by polymyxin B and colistin (Teuber 1974). However, doubt regarding the singular nature of this mechanism has been raised (Zhang et al. 2000). Because of resurgence of their usage, research into the mechanisms of action and resistance of polymyxin B and colistin has intensified. The purpose of this review is to highlight some of the alternative and less characterized mechanisms of action of polymyxin B and colistin as well as the bacterial resistance to these drugs.

MECHANISMS OF ACTION

Membrane Target and Evidence for Alternative Mechanisms of Action

Since their discovery many years ago, much has been learned regarding the mechanism of action of polymyxins. Newton (1956) originally suggested that the mechanism of polymyxin B involved its "ability to combine with and disorganize structures of the bacterial cell which are responsible for the maintenance of the osmotic equilibrium of the cell." It is generally accepted that the Gram-negative selectivity is mediated by initial interaction with the outer membrane of Gram-negative bacteria. Early studies showed that Gram-negative bacteria pretreated with polymyxin B were more susceptible to lysozyme treatment, indicating that polymyxin B disrupted the outer membrane, forming visible protrusions, thereby exposing the underlying peptidoglycan layer to lysozyme (Warren et al. 1957; Koike et al. 1969). Additionally, treatment of *Escherichia coli* with polymyxin B increases susceptibility to β-lactam antibiotics, which target the peptidoglycan synthesis machinery (Rosenthal and Storm 1977). It was subsequently proposed and shown that the cationic charges in the polypeptide portion of polymyxin electrostatically bind to the negatively charged lipopolysaccharide (LPS) that is the predominant (or only) surface lipid of the outer membrane in Gram-negative bacteria, assisted by the interaction of the lipid tail with the fatty acids of the lipid A moiety of the LPS molecule (Fig. 1B). When it binds to phosphate residues of lipid A, polymyxin displaces the membrane-stabilizing magnesium and calcium ions that cross-bridge adjacent lipid A molecules and stabilize the outer membrane (Moore et al. 1986; Evans 1999; Falagas and Kasiakou 2005; Landman et al. 2008; Fernández et al. 2013). By displacing these divalent cations with the bulkier polycationic polymyxin, the membrane becomes weakened and the permeability barrier is disrupted allowing for uptake of previously nonpermeating or weakly permeating molecules and leakage of periplasmic proteins (Hancock 1984). The displacement of divalent cations and permeabilization allows for self-promoted uptake of the

polymyxin molecule itself enabling it to penetrate the periplasm and approach the cytoplasmic membrane (Hancock and Bell 1988; Hancock 1997). Polymyxin B nonapeptide (PMBN), lacking the fatty acyl tail, had virtually no antibiotic activity but was still able to compromise the outer membrane (Vaara and Vaara 1983; Daugelavicius et al. 2000; Zhang et al. 2000; Lu et al. 2014), indicating that outer membrane permeabilization could occur because of charge:charge interactions alone, although larger concentrations of this attenuated polymyxin were required for permeabilization, suggesting that the hydrophobic acyl tail promotes outer membrane permeabilization. The lack of activity of PMBN, however, indicates that the outer membrane is a site of interaction, but it is not the killing target. In addition to binding to LPS, a polymyxin B photoprobe was recently shown to bind to unspecified outer membrane proteins (van der Meijden and Robinson 2015). However, for killing to occur, according to the membrane target model, the cytoplasmic membrane must be compromised, whereas other studies have suggested alternative membrane-associated or cytoplasmic targets (Zhang et al. 2000).

The membrane-active mechanism of action for polymyxins has been well studied, but it is unclear how polymyxin interacts with and disrupts the cytoplasmic membrane (Landman et al. 2008; Falagas et al. 2010; Yu et al. 2015). Early experiments showed that high concentrations of polymyxin caused the release of cytoplasmic material (Cerny and Teuber 1971; Schindler and Teuber 1975; Dixon and Chopra 1986; Landman et al. 2008). It was shown that polymyxin B caused lipid exchange between outer and inner membrane leaflet binding in experiments with simulated Gram-negative membrane (Clausell et al. 2007). Similarly, Berglund et al. (2015) used computer modeling to propose that polymyxin B initially aggregates at the LPS surface by burying their acyl tails into micelle/ pore-like structures. Conversely, polymyxin B did not form aggregates in the inner membrane model, suggesting that the membrane interactions are dependent on the membrane architecture of the inner and outer membranes in Gram-negative bacteria.

Cite this article as *Cold Spring Harb Perspect Med* doi: 10.1101/cshperspect.a025288

Despite this evidence and overall assumption for the membrane as the sole target for the polymyxins, there is evidence for alternative mechanisms of action. An early study investigating how different environmental ions affected polymyxin killing of *P. aeruginosa* (Klemperer et al. 1979) showed that lysis is not required for cell death. In *P. aeruginosa*, at concentrations greater than the minimum inhibitory concentration (MIC), polymyxin led to substantial cell death while only marginally increasing cytoplasmic membrane permeability, as assessed by reduction in the transmembrane potential gradient (Zhang et al. 2000). Similarly, the cytoplasmic membrane of polymyxin-treated *E. coli* was permeabilized, as determined by the measured release of ions, only at concentrations well above the minimal bactericidal range (Daugelavicius et al. 2000). This high cell death at concentrations showing little increase in cytoplasmic membrane permeability indicates that alternative or additional mechanisms of action are likely to contribute to the antibacterial activity of polymyxins. Studies with a modified, fluorescently labeled polymyxin B that retained activity showed that it coalesced at the outer membrane before penetration into the periplasm, cytoplasmic membrane, and, finally, the cytoplasm in *K. pneumoniae* (Deris et al. 2014), indicating a plethora of potential sites of action. Some of the alternative mechanisms of activity that have been proposed are described below.

Ribosome Binding

In early investigations of the mechanism of action of polymyxins and colistin, it was discovered that polymyxin could precipitate *E. coli* ribosomes (Nakajima and Kawamata 1966; Teuber 1967), similar to other polycationic antibiotics. Despite these seminal papers, there was no follow-up for ~50 years. Recently, McCoy et al. (2013) reestablished this link by examining the relationship of polymyxin to aminoglycosides, which are cationic saccharide bactericidal antibiotics that also bind to ribosomes. They posited that the shared cationic nature, nephrotoxicity, and self-promoted uptake (Hancock 1997) needed further exploration. Specifically,

they examined the ability of polymyxin to bind to bacterial 16S ribosomal RNA using a modified fluorescence resonance energy transfer assay. Polymyxin showed moderate binding to the 16S A-site of *E. coli* ribosomes, 10 times weaker than kanamycin yet 10 times stronger than the negative control, an inactive colistin derivative with sulfomethylated (charge-neutralized) amino groups. Intriguingly, an analog of polymyxin, with a biphenyl acyl tail and a D-octylglycine at the phenylalanine position, showed similar binding affinity to kanamycin. Also, PMBN showed two times better binding than intact polymyxin B. This is interesting because, in vivo, the nonapeptide apparently cannot pass the outer membrane (Vaara and Vaara 1983). Despite these abilities to bind to the ribosome, bacterial translation was not affected; however, in contrast, eukaryotic translation was negatively affected. Nevertheless, we deduce that ribosome binding is not an alternative or additional mechanism because there were apparently no obvious consequences of ribosome binding on prokaryotic translation nor evidence that polymyxin penetrates sufficiently into the cytoplasm to inhibit the large number of ribosomes in bacterial cells.

Bacterial Respiration

Strictly speaking, membrane permeabilization has many cellular consequences beyond a compromised cell structure or lysis. Indeed, disruption of cytoplasmic membrane integrity can impact cellular energetics, cell division, and cell wall biosynthesis. Bacterial respiration, for instance, requires an intact membrane to function properly (Storm et al. 1977). Consistent with this concept, sublethal concentrations of polymyxin B inhibited oxygen consumption in *A. baumannii* without penetrating the cytoplasmic membrane (Saugar et al. 2002).

Despite intrinsic resistance of Gram-positive bacteria to polymyxin, Tochikubo et al. (1986) examined nicotinamide adenine dinucleotide (NADH) oxidase activity following germination of *Bacillus subtilis* spores. They determined that polymyxin B inhibited NADH oxidase activity as well as NADH cytochrome *c* reductase activity, but not NADH dehydrogenase activity. It was proposed that the spore pericortex prevents direct interaction with the inner membrane, but somehow the activity of polymyxin still inhibited respiration (Tochikubo et al. 1986). Another group has shown similar effects on bacterial respiration machinery as polymyxin B inhibited the activity of the alternative NADH dehydrogenase and the malate: quinone oxidoreductase in *Mycobacterium smegmatis* (Mogi et al. 2009).

A screen for polymyxin B–resistant mutants in the susceptible Gram-positive *Streptococcus pyogenes*, identified the fatty acid biosynthesis master regulator, *fabT* (Dalebroux and Swanson 2012; Port et al. 2014). Fatty acid biosynthesis is known to activate the stringent response. Similarly, polymyxin treated cells show an increase in ppGpp, but not pppGpp (Cortay and Cozzone 1983b; de la Fuente-Núñez et al. 2014). This increase was independent of RelA, which is the normal producer of (p)ppGpp (Cortay and Cozzone 1983a). It was suspected that the increase in ppGpp was because of a lack of cellular ATP powering the enzymatic degradation by SpoT (Cortay and Cozzone 1983b). Indeed, inhibition of respiration by polymyxin leads to a decrease in the ATP pool of the cell (Storm et al. 1977). These reports provide compelling evidence that respiration inhibition is a target of polymyxin beyond membrane disruption. Similarly, other bactericidal cyclic cationic peptides are known to permeabilize the outer membrane and cause inhibition of respiration at their minimal effective doses (Skerlavaj et al. 1990; Wu and Hancock 1999; Spindler et al. 2011). Nevertheless, inhibition of respiration has not been shown to be a bactericidal target so it is likely that it contributes to, rather than causes, cell killing.

Cell Division

Bacterial cell division is an essential process that depends on cytoplasmic membrane–associated machinery and the membrane potential (Strahl and Hamoen 2010). For example, it is well established that many third-generation cephalosporin antibiotics act directly on penicillin binding protein 3 to inhibit correct cell division

leading the formation of long filamentous bodies in which individual cells have grown but not separated. A recent paper using atomic force microscopy (AFM) on colistin-treated *P. aeruginosa* provided evidence that colistin also affects cells division (Mortensen et al. 2009). A significant decrease in dividing cells was observed when cells were treated with sublethal concentrations of colistin in addition to a slight decrease in colony-forming units, but no significant killing (Mortensen et al. 2009). The investigators also observed an increase in cellular rigidity and speculated that colistin binding to peptidoglycan stiffens the bacterial cell wall and speculated that this prevented proper cell division (Mortensen et al. 2009). This conclusion, however, contrasts to another AFM study that showed reduced membrane integrity (Lu et al. 2014). Thus, an alternative possibility might be that polymyxin inserts into the cytoplasmic membrane and interferes with the cell division machinery. A cell division defect was also reported in *Mycobacterium aurum* in which sub-MIC doses of colistin reversibly arrested cell division without killing cells (David and Rastogi 1985). It was proposed that growth arrest arose because of membrane perturbation that disrupted genome attachment and inhibited DNA replication (David and Rastogi 1985), a process that has also been postulated to explain the bactericidal activity of the cationic aminoglycosides (Hancock 1981).

Other cationic amphipathic peptides have also been proposed to affect cell division. The cationic antimicrobial peptide indolicidin has been shown to induce filamentation in *E. coli* (Subbalakshmi and Sitaram 1998), although the investigators proposed that filamentation is a secondary effect because of inhibition of DNA synthesis by indolicidin. Conversely, it was shown that the mouse cathelicidin antimicrobial peptide (CRAMP) caused filamentation in *B. subtilis* by inhibiting FtsZ-mediated ring formation, which is a critical step in cell division (Handler et al. 2008).

At this stage, we cannot conclude that cell division is a direct target of the polymyxins. However, cell division is an extremely complex process, with many moving parts, and much of the cell division machinery lies within the inner membrane (Margolin 2005; Egan and Vollmer 2015; Gray et al. 2015). Therefore, it is important to further examine whether polymyxin may directly (e.g., through FtsZ) or indirectly (through antagonism of energetics) affect cell division as part of its overall mechanism of action.

ExPortal

The precise mechanism for Gram-positive bacteria resistance to polymyxin and colistin is largely unknown (Storm et al. 1977). It is believed that the large peptidoglycan layer prevents interaction with the cytoplasmic membrane, because *B. subtilis* and *Staphylococcus aureus* protoplasts are susceptible to polymyxin (LaPorte et al. 1977; Xiong et al. 2005). Interestingly, there are a few noted instances of Gram-positives showing polymyxin susceptibility. For example, *S. pyogenes* is susceptible to polymyxin B, likely because of the absence of the *mprF* gene, which encodes an enzyme that modifies the negative charge of phosphatidylglycerol on its capsular polysaccharide (Vega and Caparon 2012). It was observed that polymyxin B targeted the ExPortal structure of *S. pyogenes* (Vega and Caparon 2012). The ExPortal is a microdomain for the Sec secretion system (Rosch and Caparon 2005). Each cell has only one ExPortal found adjacent to the future division site (Rosch and Caparon 2005). Sublethal concentrations of polymyxin B disrupted the lipid structure around the ExPortal and impeded the secretion of the cysteine protease, SpeB, and the streptolysin O cytolysin (Rosch et al. 2007; Vega and Caparon 2012). Interestingly, despite the disruption in the lipids by sublethal doses, membrane integrity was not compromised (Vega and Caparon 2012), supporting the idea of an alternative mechanism of action different from direct membrane disruption. It was hypothesized that the defect in the secretion mechanism and/or another unknown internal target leads to the demise of the cells (Vega and Caparon 2012).

Reactive Oxygen Species

Recently, a highly controversial mode of bacterial cell death by bactericidal antibiotics was proposed. Some groups suggested that these antibiotics, including polymyxin B, colistin, and kanamycin kill through oxidative stress and reactive oxygen species (ROS) generation (Dwyer et al. 2007, 2009; Brochmann et al. 2014; Dong et al. 2015). However, these overall conclusions regarding this proposed common mechanism were disputed by several investigators, in particular, by showing that antibiotic killing was identical under aerobic and anaerobic conditions indicating independence of oxygen and hydroxyl radicals (Kindrachuk et al. 2011; Keren et al. 2013; Liu and Imlay 2013; Paulander et al. 2014). Similarly, ROS independent killing by colistin has been shown (Brochmann et al. 2014). Conversely, Dong et al. (2015) showed that polymyxin B induced transcription in *E. coli* of the oxidative stress response gene *soxS*, and caused an increase in ROS, although they did not imply that this explained the mechanism of killing. A variety of reports have discussed potential mechanisms of resistance mediated by oxidative enzymes but alternative explanations should be sought (Antonic et al. 2013; Heindorf et al. 2014; Pournaras et al. 2014). Despite all the studies and evidence for mechanisms of killing, clearly the exact mechanism of polymyxin action remains to be determined.

POLYMYXIN RESISTANCE

It has been observed that, compared with many other antibiotics, resistance to polymyxins is difficult to attain in the laboratory. Because polymyxins are being used as a last line of defense in treating MDR bacterial infections, it is of considerable interest to examine mechanisms of resistance that have arisen in the clinic or laboratory (Falagas et al. 2010; Fernández et al. 2011; Olaitan et al. 2014). Known resistance mechanisms include intrinsic, mutational, and adaptive, although there are no known horizontally acquired resistance mechanisms. The majority of known mechanisms address the first point of attack by polymyxin, namely, the outer membrane LPS. Overall resistance mechanisms include alterations to reduce the net negative charge or fluidity of LPS, increase in drug efflux, reduced porin pathway, capsule formation, and hypervesiculation. For brevity, this section will focus on the most studied mechanisms of resistance and ones related to alternative mechanisms of action.

LPS–Lipid A Modifications

In most Gram-negative bacteria, the two-component PhoP/Q and PmrA/B (response regulator/sensor kinase) regulatory systems mediate polymyxin resistance by governing the mechanisms that modify LPS (Fig. 2; Table 1). These systems regulate cationic antimicrobial resistance in response to low environmental Mg^{2+} and Ca^{2+} as well as cationic antimicrobial peptides (CAMPs) and other inducers, such as low pH, excess Fe^{3+}, excess Al^{3+}, and macrophage phagosomes (Brown and Melling 1969; McPhee et al. 2003; Nishino et al. 2006; Prost et al. 2007; Gunn 2008; Gooderham et al. 2009). In Salmonella enterica serovar Typhimurium, PhoP activates the expression of pmrD whose product was shown to activate the PmrA/B system (Kox et al. 2000; Kato et al. 2003), although PmrD is absent in P. aeruginosa and instead PhoP directly interacts with the pmrAB promoter while the two response regulators PhoP and PmrA bind to and maybe compete for binding to the same promoter site(s) on the LPS modification operon. Other two-component systems like ParR/S, CprR/S, and ColR/S in P. aeruginosa have been also implicated in the regulation of lipid A modification enzymes (McPhee et al. 2003; Moon and Gottesman 2009; Fernández et al. 2010; Muller et al. 2011; Fernández et al. 2012; Gutu et al. 2013). In addition, P. aeruginosa has a two-component sensor kinase cbrA that responds to carbon source/catabolite repression and adaptively affects polymyxin B resistance by altering the transcriptional expression of the oprH-phoPQ operon, the pmrAB operon, and the LPS modification arn operon (Yeung et al. 2011). The most likely physiological adaptive resistance mechanism is because of the direct re-

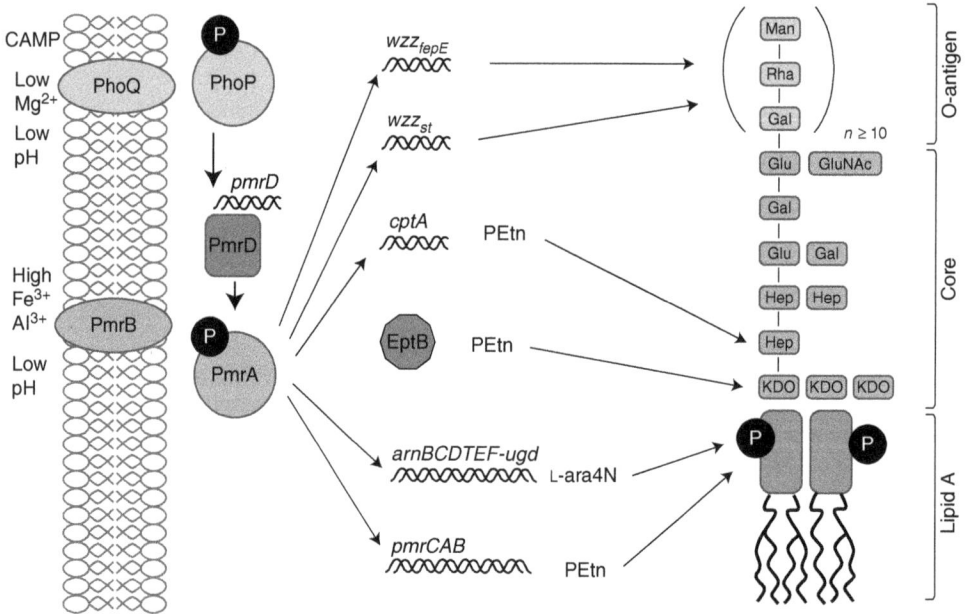

Figure 2. General mechanisms of resistance to polymyxin by lipopolysaccharide (LPS) modifications in Salmonella (mechanisms differ in Pseudomonas). The two-component system PhoQP, activates expression of pmrD. PmrD activates PmrA, which activates cptA and the pmr and arn operons. Along with EptB, CptA modifies the LPS core. The products pmr and arn substitute the phosphates the lipid A portion of the LPS for PEtn and L-Ara4N, respectively. Collectively, the net charge of the outer membrane changes, resulting in repulsion of polymyxin.

Cite this article as Cold Spring Harb Perspect Med doi: 10.1101/cshperspect.a025288

Table 1. Lipopolysaccharide (LPS) modifications leading to polymyxin resistance or supersusceptibility

Modification	Function	Genes[a]	Bacterial spp.	References
L-Ara4N or PEtn modification of lipid A	Two-component system (TCS) kinase/response regulator (K/RR)	phoP/phoQ	E. coli	Moon and Gottesman 2009
			S. Typhimurium	Gunn et al. 1998
			P. aeruginosa	Macfarlane et al. 2000
		pmrA/pmrB	P. aeruginosa	McPhee et al. 2003
			S. Typhimurium	Gunn et al. 1998
			A. baumannii	Arroyo et al. 2011
			K. pneumoniae	Choi and Ko 2014
			E. coli	Trent et al. 2001
		parR/parS	P. aeruginosa	Fernández et al. 2010
		cprR/cprS	P. aeruginosa	Fernández et al. 2012
		colR/colS	P. aeruginosa	Gutu et al. 2013
		Rcs system	S. Typhimurium	Mouslim and Groisman 2003
		cbrAB	P. aeruginosa	Yeung et al. 2011
	Proposed response regulator	rppA	P. mirabilis	Wang et al. 2008
	Activates pmrAB	pmrD	S. Typhimurium	Kato et al. 2003
	UDP-glucose dehydrogenase	pmrE (pagA or ugd)	P. mirabilis	Jiang et al. 2010a
			S. Typhimurium	Mouslim and Groisman 2003
		arnBCADTEF	P. aeruginosa	Gunn et al. 1998
		(pmrHFIJKLMpbgP)	E. coli	Muller et al. 2011
			S. Typhimurium	Yan et al. 2007
			P. mirabilis	Gunn et al. 2000
			Y. pseudotuberculosis	Lee et al. 2004
			K. pneumoniae	Jiang et al. 2010b
				Marceau et al. 2004
				Cheng et al. 2010
	L-Ara4N transferase	arnT	E. coli	Subashchandrabose et al. 2013
		(pmrK)	S. Typhimurium	Trent et al. 2001
		(pqaB)	S. typhi	Baker et al. 1999

Continued

Table 1. *Continued*

Modification	Function	Genes[a]	Bacterial spp.	References
	Mannosyltransferase	pbgE1	P. luminescens	Bennett and Clarke 2005
	UDP-glucose dehydrogenase	ugdBCAL2946	B. cenocepacia	Loutet et al. 2009
	Ugd phosphorylation	etk	E. coli	Lacour et al. 2008
	Oxidoreductase	DsbA3, DsbA2, and DsbA3	N. meningitidis	Piek et al. 2014
PEtn modification of LOS/LPS core	TCS (kinase/response regulator)	misS/misR	N. meningitidis	Tzeng et al. 2004
	Transferase	eptB	E. coli	Reynolds et al. 2005
			S. Typhimurium	Gibbons et al. 2008
	PEtn transferase	pmrC (eptA and lptA)	S. Typhimurium	Lee et al. 2004
			A. baumannii	Arroyo et al. 2011
			E. coli	Herrera et al. 2010
	Phosphotransferase	cptA	S. Typhimurium	Tamayo et al. 2005a
	sRNA	mgrR	E. coli	Moon and Gottesman 2009
	Transferase (also affects PEtn modification of flagella and glycans)	eptC	C. jejuni	Cullen et al. 2012, 2013; Scott et al. 2012
Deacylation of lipid A	Deacetylase	naxD	F. tularensis	Llewellyn et al. 2012
	Deacetylase	pagL	S. enterica	Kawasaki et al. 2007
	Deacetylase	yfbH (pmrI)	E. coli	Subashchandrabose et al. 2013
	sRNA	lpxR	S. Typhimurium	Reynolds et al. 2006
Repressor of phoPQ expression	Transmembrane regulator	mgrB (yobG)	K. pneumoniae	Lopez-Camacho et al. 2014
	MerR-like regulator	brlR	P. aeruginosa	Chambers and Sauer 2013
	sRNA	micA	E. coli	Coornaert et al. 2010
L-Lys modification of cell membrane phosphatidylglycerol	Transferase	mprF (lpiA)	S. aureus, E. faecalis, and R. tropici	Yang et al. 2012; Bao et al. 2012; Sohlenkamp et al. 2007
	Lysinylation	lysX	M. tuberculosis	Maloney et al. 2009

Function	Type	Gene	Organism	Reference
Phosphorylation of lipid A	Phosphotransferase	lpxT	S. Typhimurium	Herrera et al. 2010
			E. coli	Touze et al. 2008
		waaP (rfaP)	E. coli	Yethon et al. 2000
			S. Typhimurium	Tsai et al. 2012
			P. aeruginosa	Delucia et al. 2011
Dephosphorylation of lipid A	Phosphatase	lpxE and lpxF	H. pylori,	Tran et al. 2006
			F. novicida, and	Wang et al. 2006
			R. etli	Ingram et al. 2010
	Phosphatase	PG1587 and PG1773	P. gingivalis	Coats et al. 2009
	Transcriptional activator	slyA	D. dadantii	Haque et al. 2009
			S. Typhimurium	Shi et al. 2004b
		ugtL	S. Typhimurium	Shi et al. 2004a,b
Glycylation of lipid A	TCS (K/RR)	vprA/vprB	V. cholerae	Herrera et al. 2014
	TCS (K/RR)	carR/carS	V. cholerae	Bilecen et al. 2015
		almEFG	V. cholerae	Hankins et al. 2012
Addition of amide-linked acyl chains in lipid A	Dehydrogenase and transaminase	gmmA and gmmB	C. jejuni	van Mourik et al. 2010
Glucosamine modification		lgmABC	B. pertussis	Shah et al. 2014
		pgaABCD	E. coli	Amini et al. 2009
			A. baumannii	Choi et al. 2009a

[a]For brevity, not all genes are mentioned in the text. Gene contained within parentheses are alternative names or homologous genes.

sponse of the sensor kinases of two component regulators to the presence of peptides, both natural host defense (antimicrobial) peptides and polymyxins, which interact directly to activate these sensor kinases (Bader et al. 2005). In *Salmonella*, the peptide-specific sensor kinase is PhoQ, whereas in *Pseudomonas*, a pair of sensor kinases, ParS and CprS, respond to polymyxin and/or peptides (Fernández et al. 2010, 2012). Thus, adaptive resistance can be a result of both self-induced resistance during polymyxin therapy and the antimicrobial peptide component of host defense responses to infections.

The output targets of the two component systems are diverse, but commonly make modifications to the phosphate groups of lipid A in LPS to amine substituents, such as 4-amino-4-deoxy-L-arabinose (L-Ara4N), or phosphoethanolamine (PEtn). The addition of L-Ara4N or PEtn reduces the net negative charge of the bacterial surface and decreases the electrostatic interaction with polycationic peptides, such as the polymyxins, thus inhibiting self-promoted uptake. The *arnBCADTEF* operon, activated by ParR/S and PmrA/B in *P. aeruginosa* and *Salmonella*, etc. (Subashchandrabose et al. 2013) adds an L-Ara4N modification to lipid A (Aguirre et al. 2000; Gunn et al. 2000; Tamayo et al. 2005b; Raetz et al. 2007; Barrow and Kwon 2009; Gooderham et al. 2009; Fernández et al. 2010).

Conversely, in *Salmonella pmrC* (also named *eptA* and *lptA*), which is also activated by PmrA, is involved in PEtn modification of lipid A (Lee et al. 2004). Similarly, the *pmrC* gene encodes a phosphoethanolamine transferase and modifies LPS with PEtn in *A. baumannii* (Arroyo et al. 2011). The LPS core may also be modified with PEtn by the *eptB* (analogous to *pmrC*) and *cptA* gene products. EptB regulates PEtn addition to 3-deoxy-D-manno-oct-2-ulosonic acid (Kdo) in *E. coli* (Reynolds et al. 2005) and CptA in *Salmonella* modifies phosphorylated heptose-I residue of the LPS core (Tamayo et al. 2005a). Paradoxically, activation of PhoP/Q system activates the sRNA *mgrR*, which is a repressor of *eptB* (Moon and Gottesman 2009). Various oxidoreductases, such as DsbA1, DsbA2, and DsbA3, contribute to

PEtn modifications of lipid A in *Neisseria meningitidis*. Specifically, DsbA3 was shown to be important for the activity of LptA, a PEtn transferase (Piek et al. 2014).

Independent of the PhoP/Q and PmrA/B, the Rcs system (RcsC–YojN–RcsB) in *S. enterica* has been shown to regulate the expression of *ugd*, which encodes a UDP-glucose dehydrogenase, an enzyme important in the synthesis of L-Ara4N (Mouslim and Groisman 2003). Also the tyrosine-kinase *etk* in *E. coli* phosphorylates Ugd independently of the two-component systems (Lacour et al. 2008).

In addition to L-Ara4N or PEtN modifications, deacylation of lipid A occurs to decrease the fluidity of LPS and, thus, the ability of cationic peptides to partition into the outer membrane leading to self-promoted uptake (Olaitan et al. 2014). In *Salmonella*, the activation of PhoP/Q system activates *pagL*, the product of which deacylates lipid A (Kawasaki et al. 2007). Conversely, the *naxD* gene in *Francisella tularensis* has also been shown to be responsible for deacylation of lipid A (Llewellyn et al. 2012). Further, PmrA-activated genes include *lpxR*, which is associated with deacylation of lipid A and is repressed by the sRNA *micF* (Papenfort et al. 2006; Reynolds et al. 2006; Corcoran et al. 2012). Additional modifications to lipid A include phosphorylation, dephosphorylation, glycylation, and glucosylation. It has also been shown that *misR* mutants in *N. meningitidis* lose all of their lipooligosaccharide (LOS) core HepII PEtn decorations leading to an increase in polymyxin B sensitivity (Tzeng et al. 2004). Similar to LPS, LOS is essential for maintaining the integrity of the outer membrane and provides protection from antimicrobial peptides. The *waaP* (*rfaP*) gene product is also involved in LPS core modification by phosphorylation of lipid A and was required for polymyxin resistance in *S. enterica* and *P. aeruginosa* (Yethon et al. 2000; Delucia et al. 2011; Tsai et al. 2012).

Although it is more difficult to obtain polymyxin resistance than resistance to other antibiotics, the increased use of polymyxin has started to lead to clinical resistance. Mutations in *phoQ* genes have been identified in colistin-resistant clinical isolates of *P. aeruginosa* and

K. pneumoniae (Choi and Ko 2014; Olaitan et al. 2014). Further mutations in the *pmrA* or *pmrB* genes resulted in activation of *arnB* (*pmrH*) or other PmrA-activated loci that are required for polymyxin resistance (Sun et al. 2009). Mutations in *arnT* (also known as *pmrK* and *pqaB*) have been shown to affect resistance to polymyxin B in *E. coli* and *S. enterica* (Trent et al. 2001). Conversely, *S. enterica* and *Yersinia pseudotuberculosis* mutants in the *pmrF* gene, which is necessary for L-Ara4N substitution on bactoprenol phosphate, displayed an increased susceptibility to polymyxin B (Gunn et al. 2000; Marceau et al. 2004). The most exotic polymyxin resistance mutation is the complete loss of LPS by clinical isolates of *A. baumannii* (Moffatt et al. 2010, 2011), although many polymyxin resistant isolates in this bacterium have *pmrB* mutations influencing expression of the upstream *pmrC* gene (Arroyo et al. 2011). Another interesting clinical mutant was in the *parR* gene (one of the polymyxin receptors of *Pseudomonas*), which influenced both inducible and constitutive multidrug resistance to four different classes of antibiotics through the activation of three distinct mechanisms (efflux, porin loss, and LPS modification) (Muller et al. 2011).

The *tolA* gene is involved in LPS production and this gene has been shown to be necessary for polymyxin B resistance in *S. typhimurium* (Paterson et al. 2009). Further, LPS modification loci—wzz_{st} and wzz_{fep} activated by PmrA/P, *wbaP* (*rfbP*), *waaG* (*rfaG*), *waaI* (*rfaI*), *rfaH*, *waaJ* (*rfaJ*), *waaL* (*rfaL*), or *wzy* (*rfc*)—are involved in synthesis of the LPS rough core in *S. enterica* and appear to be required for polymyxin B resistance, likely because of effects on phosphorylation and/or acylation of LPS (Delgado et al. 2006; Holzer et al. 2009; Ilg et al. 2009; Kong et al. 2011; Pescaretti et al. 2011). Similar mutations in *waaL* and *rfbA* (*rffH*) contribute to resistance to colistin in *K. pneumoniae* (Sassera et al. 2014). Similarly, in *E. coli*, five lipid A–modifying genes, *ais* (*b2252*), *b2253* (*yfbE, orf1*), *b2254* (*yfbF, orf2*), *b2256* (*yfbH, orf4*), and *crcA* were associated with antimicrobial peptide resistance (Kruse et al. 2009). Overexpression of *K. pneumoniae* RamA, which regulates the *lpxC*, *lpxL-2*, and *lpxO* genes associated with lipid A biosynthesis, resulted in increased polymyxin B resistance because of LPS alterations (De Majumdar et al. 2015). An LPS synthesis gene encoding the glycosyltransferase, *lpsB*, has also been implicated in colistin resistance in *A. baumannii* (Hood et al. 2013). In addition, the *S. typhimurium* PhoP-activated gene, *mgtA*, is also required for resistance to the cationic antibiotic polymyxin B. MgtA controls modification of phosphate residues in the lipid A portion of the LPS (Park and Groisman 2014).

Stress Responses

Biofilm formation is thought to be a response to stress (de la Fuente-Núñez et al. 2014). The *P. aeruginosa* genes *psrA* and *cbrA*, and *V. cholerae* *carR*, regulate biofilm formation and polymyxin B resistance (Gooderham et al. 2008; Bilecen et al. 2015). Interestingly, *psrA* responds to cationic peptides, selected antibiotics (Gooderham et al. 2008) as well as an increase in fatty acids and is also an activator of the stationary phase σ factor, RpoS (Kang et al. 2009). As mentioned above, mutations in FabT, the fatty acid biosynthesis master regulator of *S. pyogenes*, led to an increase in fatty acids and polymyxin resistance (Port et al. 2014). The increase in fatty acids may activate RpoS through PsrA, which in turn activates the stringent response, biofilm formation, and resistance. Additionally, PhoP/Q has been shown to stabilize RpoS in *S. enterica* (Tu et al. 2006). Moreover, biofilm genes of *P. aeruginosa* have been correlated with increased resistance toward colistin because of the complex architecture of biofilms (Folkesson et al. 2008). Interestingly, it was shown that *P. aeruginosa* biofilm cells with high metabolic activity are more resistant to colistin through up-regulation of the *arn* operon (Pamp et al. 2008); in principle, this could be mediated through either PsrA or the carbon stress triggered regulator CbrA.

σ Factor RpoE in *Burkholderia cenocepacia* and other pathogens has been shown to contribute to polymyxin B resistance, likely through the regulation of genes controlling envelope and oxidative stress responses (Sikora et al. 2009; Loutet and Valvano 2011). Moreover, stationary phase σ factor RpoS regulon genes (*katE, osmY, wrbA, yciF*) are required for polymyxin resistance in *E. coli* and *S. enterica* (Oh et al. 1998, 2000; Bader et al. 2003). In *S. enterica*, an *rpoN* (nitrogen responsive σ factor σ[54]) mutant increased polymyxin resistance. It was suggested that *rpoN* induced resistance to polymyxin via regulation of PTS transporters (Barchiesi et al. 2009).

Polymyxin Efflux

Multidrug efflux pumps play an important role in antibiotic and polymyxin resistance in bacteria but have only been shown to influence polymyxin resistance in a small number of instances. In different organisms, the MtrC–MtrD–MtrE, RosAB, AcrAB–TolC, NorM, KpnEF, and VexAB efflux pumps have been described to confer tolerance toward polymyxin B (Bengoechea and Skurnik 2000; Fehlner-Gardiner and Valvano 2002; Tzeng et al. 2005; Bina et al. 2008; Padilla et al. 2010; Warner and Levy 2010; Tsai et al. 2012; Srinivasan et al. 2014). Further, *sap* has been implicated in polymyxin B resistance in *E. coli* because of its homology with ABC effluxers that function against antibiotics (Subashchandrabose et al. 2013). The *marRAB* operon (activator of efflux) was associated with polymyxin B resistance in *E. coli* through interactions with Rob and up-regulation of the AcrAB–TolC efflux pump (Warner and Levy 2010). Similarly, overexpression of RamA in *K. pneumoniae* resulted in increased polymyxin B resistance by mechanisms, including modulation of efflux pump genes such as *acrAB*, *oqxAB*, and *yrbB-F* (De Majumdar et al. 2015). Mutation of the outer membrane porins, PorB and OmpU, conferred resistance to polymyxin B in *N. meningitidis* and *V. cholerae*, suggesting a potential role in uptake, perhaps by influencing an active efflux process (Mathur and Waldor 2004; Tzeng et al. 2005).

Plasmid-Mediated Resistance

Until now, polymyxin and colistin resistance was thought to be strictly limited to chromo-

somal mutations. Very recently, Liu et al. (2016) noted an increase of colistin resistance among food-borne commensal *E. coli* in China. They identified the resistance gene and defined it as *mcr-1*. The gene product, MCR-1, was found to share sequence similarities to phosphoethanolamine transferases suggesting resistance by modifying the phosphate groups of lipid A in LPS. An expanded search identified the *mcr-1* gene in *E. coli* and *K. pneumoniae* from both animal and human isolates. The initial publication of this work online immediately drew the attention of research groups worldwide initiating searches for the *mcr-1* gene in public and local databases. By the time this initial report was published, other instances of *mcr-1* in Gram-negative bacteria had been found in North America, Europe, South America, and Africa (Arcilla et al. 2016; Hu et al. 2016; Olaitan et al. 2016; Tse and Yuen 2016; Webb et al. 2016). The potential for widespread uncontrolled transmission is undeniable, and every precaution should be made to prevent further dissemination of resistance.

CONCLUDING REMARKS

The cyclic cationic peptides polymyxin B and colistin are potent bactericidal agents against Gram-negative bacteria. Despite their traditional record of neural and renal toxicities, they have performed well as the last line of defense against persistent MDR infections. Although neural and renal toxicities still occur, physicians are more knowledgeable and better equipped to manage toxicity (Falagas and Kasiakou 2006), which has led to a resurgence in the interest and use of this class of antibiotics to combat serious infections.

As discussed in this review, the widely accepted mode of action of polymyxins involves targeting the membranes of Gram-negative bacteria, leading to permeabilized outer and inner membranes and resulting in lysis and cell death. However, there are substantial reasons to doubt this mechanism of action as outlined here. Conversely, groups have explored the ability of polymyxins to bind ribosomes, prevent cell division, and inhibit bacterial respiration. It is hoped that a more profound understanding of the mechanisms of action and resistance will improve our ability to design and develop more potent and less toxic derivatives of polymyxin and colistin. Indeed, some groups are exploring these novel polymyxins (Katsuma et al. 2009; Velkov et al. 2014) and leading the charge in combating bacterial resistance.

ACKNOWLEDGMENTS

We thank Dr. Evan Haney for critical reading of the manuscript. We also recognize funding support No. CZ.1.07/2.3.00/30.0004 to P.M. and funding from the Canadian Institutes for Health Research and Cystic Fibrosis Canada to R.E.W.H., who was also supported by a Canada Research Chair in Health and Genomics.

REFERENCES

Aguirre A, Lejona S, Vescovi EG, Soncini FC. 2000. Phosphorylated PmrA interacts with the promoter region of *ugd* in *Salmonella enterica* serovar Typhimurium. *J Bacteriol* **182:** 3874–3876.

Amini S, Goodarzi H, Tavazoie S. 2009. Genetic dissection of an exogenously induced biofilm in laboratory and clinical isolates of *E. coli*. *PLoS Pathog* **5:** e1000432.

Antonic V, Stojadinovic A, Zhang B, Izadjoo MJ, Alavi M. 2013. *Pseudomonas aeruginosa* induces pigment production and enhances virulence in a white phenotypic variant of *Staphylococcus aureus*. *Infect Drug Resist* **6:** 175–186.

Arcilla MS, van Hattem JM, Matamoros S, Melles DC, Penders J, de Jong MD, Schultsz C, consortium C. 2016. Dissemination of the *mcr-1* colistin resistance gene. *Lancet Infect Dis* **16:** 147–149.

Arroyo LA, Herrera CM, Fernandez L, Hankins JV, Trent MS, Hancock RE. 2011. The *pmrCAB* operon mediates polymyxin resistance in *Acinetobacter baumannii* ATCC 17978 and clinical isolates through phosphoethanolamine modification of lipid A. *Antimicrob Agents Chemother* **55:** 3743–3751.

Ash C, Priest FG, Collins MD. 1994. Molecular identification of rRNA group 3 bacilli (Ash, Farrow, Wallbanks and Collins) using a PCR probe test. *Antonie van Leeuwenhoek* **64:** 253–260.

Bader MW, Navarre WW, Shiau W, Nikaido H, Frye JG, McClelland M, Fang FC, Miller SI. 2003. Regulation of *Salmonella typhimurium* virulence gene expression by cationic antimicrobial peptides. *Mol Microbiol* **50:** 219–230.

Bader MW, Sanowar S, Daley ME, Schneider AR, Cho U, Xu W, Klevit RE, Le Moual H, Miller SI. 2005. Recognition of antimicrobial peptides by a bacterial sensor kinase. *Cell* **122:** 461–472.

Baker SJ, Gunn JS, Morona R. 1999. The *Salmonella typhi* melittin resistance gene *pqaB* affects intracellular growth in PMA-differentiated U937 cells, polymyxin B resistance and lipopolysaccharide. *Microbiology* **145:** 367–378.

Bao Y, Sakinc T, Laverde D, Wobser D, Benachour A, Theilacker C, Hartke A, Huebner J. 2012. Role of *mprF1* and *mprF2* in the pathogenicity of *Enterococcus faecalis*. *PLoS ONE* **7:** e38458.

Barchiesi J, Espariz M, Checa SK, Soncini FC. 2009. Down-regulation of RpoN-controlled genes protects *Salmonella* cells from killing by the cationic antimicrobial peptide polymyxin B. *FEMS Microbiol Lett* **291:** 73–79.

Barrow K, Kwon DH. 2009. Alterations in two-component regulatory systems of *phoPQ* and *pmrAB* are associated with polymyxin B resistance in clinical isolates of *Pseudomonas aeruginosa*. *Antimicrob Agents Chemother* **53:** 5150–5154.

Bengoechea JA, Skurnik M. 2000. Temperature-regulated efflux pump/potassium antiporter system mediates resistance to cationic antimicrobial peptides in *Yersinia*. *Mol Microbiol* **37:** 67–80.

Bennett HPJ, Clarke DJ. 2005. The *pbgPE* operon in *Photorhabdus luminescens* is required for pathogenicity and symbiosis. *J Bacteriol* **187:** 77–84.

Berendonk TU, Manaia CM, Merlin C, Fatta-Kassinos D, Cytryn E, Walsh F, Bürgmann H, Sørum H, Norström M, Pons MN, et al. 2015. Tackling antibiotic resistance: The environmental framework. *Nat Rev Microbiol* **13:** 310–317.

Berg JR, Spilker CM, Lewis SA. 1996. Effects of polymyxin B on mammalian urinary bladder. *J Membr Biol* **154:** 119–130.

Berg JR, Spilker CM, Lewis SA. 1998. Modulation of polymyxin B effects on mammalian urinary bladder. *Am J Physiol Renal Physiol* **275:** F204–F215.

Berglund NA, Piggot TJ, Jefferies D, Sessions RB, Bond PJ, Khalid S. 2015. Interaction of the antimicrobial peptide polymyxin B1 with both membranes of *E. coli*: A molecular dynamics study. *PLoS Comput Biol* **11:** e1004180.

Bilecen K, Fong JC, Cheng A, Jones CJ, Zamorano-Sanchez D, Yildiz FH. 2015. Polymyxin B resistance and biofilm formation in *Vibrio cholerae* are controlled by the response regulator CarR. *Infect Immun* **83:** 1199–1209.

Bina XR, Provenzano D, Nguyen N, Bina JE. 2008. *Vibrio cholerae* RND family efflux systems are required for antimicrobial resistance, optimal virulence factor production, and colonization of the infant mouse small intestine. *Infect Immun* **76:** 3595–3605.

Brochmann RP, Toft A, Ciofu O, Briales A, Kolpen M, Hempel C, Bjarnsholt T, Høiby N, Jensen PØ. 2014. Bactericidal effect of colistin on planktonic *Pseudomonas aeruginosa* is independent of hydroxyl radical formation. *Int J Antimicrob Agents* **43:** 140–147.

Brown MR, Melling J. 1969. Role of divalent cations in the action of polymyxin B and EDTA on *Pseudomonas aeruginosa*. *J Gen Microbiol* **59:** 263–274.

CDC. 2013. Antibiotic resistance threats in the United States, 2013. Centers for Disease Control and Prevention, Atlanta.

Cerny G, Teuber M. 1971. Differential release of periplasmic versus cytoplasmic enzymes from *Escherichia coli* B by polymyxin B. *Arch Mikrobiol* **78:** 166–179.

Chambers JR, Sauer K. 2013. The MerR-like regulator BrlR impairs *Pseudomonas aeruginosa* biofilm tolerance to colistin by repressing PhoPQ. *J Bacteriol* **195:** 4678–4688.

Cheng HY, Chen YF, Peng HL. 2010. Molecular characterization of the PhoPQ–PmrD–PmrAB mediated pathway regulating polymyxin B resistance in *Klebsiella pneumoniae* CG43. *J Biomed Sci* **17:** 60.

Choi MJ, Ko KS. 2014. Mutant prevention concentrations of colistin for *Acinetobacter baumannii*, *Pseudomonas aeruginosa* and *Klebsiella pneumoniae* clinical isolates. *J Antimicrob Chemother* **69:** 275–277.

Choi AHK, Slamti L, Avci FY, Pier GB, Maira-Litran T. 2009a. The *pgaABCD* locus of *Acinetobacter baumannii* encodes the production of poly-β-1-6-*N*-acetylglucosamine, which is critical for biofilm formation. *J Bacteriol* **191:** 5953–5963.

Choi SK, Park SY, Kim R, Kim SB, Lee CH, Kim JF, Park SH. 2009b. Identification of a polymyxin synthetase gene cluster of *Paenibacillus polymyxa* and heterologous expression of the gene in *Bacillus subtilis*. *J Bacteriol* **191:** 3350–3358.

Clausell A, Garcia-Subirats M, Pujol M, Busquets MA, Rabanal F, Cajal Y. 2007. Gram-negative outer and inner membrane models: Insertion of cyclic cationic lipopeptides. *J Phys Chem B* **111:** 551–563.

Coats SR, Jones JW, Do CT, Braham PH, Bainbridge BW, To TT, Goodlett DR, Ernst RK, Darveau RP. 2009. Human Toll-like receptor 4 responses to *P. gingivalis* are regulated by lipid A 1- and 4′-phosphatase activities. *Cell Microbiol* **11:** 1587–1599.

Coornaert A, Lu A, Mandin P, Springer M, Gottesman S, Guillier M. 2010. MicA sRNA links the PhoP regulon to cell envelope stress. *Mol Microbiol* **76:** 467–479.

Corcoran CP, Podkaminski D, Papenfort K, Urban JH, Hinton JC, Vogel J. 2012. Superfolder GFP reporters validate diverse new mRNA targets of the classic porin regulator, MicF RNA. *Mol Microbiol* **84:** 428–445.

Cortay JC, Cozzone AJ. 1983a. A study of bacterial response to polypeptide antibiotics. *FEBS Lett* **157:** 307–310.

Cortay JC, Cozzone AJ. 1983b. Accumulation of guanosine tetraphosphate induced by polymixin and gramicidin in *Escherichia coli*. *Biochim Biophys Acta* **755:** 467–473.

Cullen TW, Madsen JA, Ivanov PL, Brodbelt JS, Trent MS. 2012. Characterization of unique modification of flagellar rod protein FlgG by *Campylobacter jejuni* lipid A phosphoethanolamine transferase, linking bacterial locomotion and antimicrobial peptide resistance. *J Biol Chem* **287:** 3326–3336.

Cullen TW, O'Brien JP, Hendrixson DR, Giles DK, Hobb RI, Thompson SA, Brodbelt JS, Trent MS. 2013. EptC of *Campylobacter jejuni* mediates phenotypes involved in host interactions and virulence. *Infect Immun* **81:** 430–440.

Dalebroux ZD, Swanson MS. 2012. ppGpp: Magic beyond RNA polymerase. *Nat Rev Microbiol* **10:** 203–212.

Daugelavicius R, Bakiene E, Bamford DH. 2000. Stages of polymyxin B interaction with the *Escherichia coli* cell envelope. *Antimicrob Agents Chemother* **44:** 2969–2978.

David HL, Rastogi N. 1985. Antibacterial action of colistin (polymyxin E) against *Mycobacterium aurum*. *Antimicrob Agents Chemother* **27:** 701–707.

de la Fuente-Núñez C, Reffuveille F, Haney EF, Straus SK, Hancock REW. 2014. Broad-spectrum anti-biofilm peptide that targets a cellular stress response. *PLoS Path* **10:** e1004152.

Delgado MA, Mouslim C, Groisman EA. 2006. The PmrA/PmrB and RcsC/YojN/RcsB systems control expression of the *Salmonella* O-antigen chain length determinant. *Mol Microbiol* **60:** 39–50.

Delucia AM, Six DA, Caughlan RE, Gee P, Hunt I, Lam JS, Dean CR. 2011. Lipopolysaccharide (LPS) inner-core phosphates are required for complete LPS synthesis and transport to the outer membrane in *Pseudomonas aeruginosa* PAO1. *MBio* **2:** e00142–11.

De Majumdar S, Yu J, Fookes M, McAteer SP, Llobet E, Finn S, Spence S, Monahan A, Monaghan A, Kissenpfennig A, et al. 2015. Elucidation of the RamA regulon in *Klebsiella pneumoniae* reveals a role in LPS regulation. *PLoS Pathog* **11:** e1004627.

Deris ZZ, Swarbrick JD, Roberts KD, Azad MAK, Akter J, Horne AS, Nation RL, Rogers KL, Thompson PE, Velkov T, et al. 2014. Probing the penetration of antimicrobial polymyxin lipopeptides into Gram-negative bacteria. *Bioconjug Chem* **25:** 750–760.

Dixon RA, Chopra I. 1986. Polymyxin B and polymyxin B nonapeptide alter cytoplasmic membrane permeability in *Escherichia coli*. *J Antimicrob Chemother* **18:** 557–563.

Dong TG, Dong S, Catalano C, Moore R, Liang X, Mekalanos JJ. 2015. Generation of reactive oxygen species by lethal attacks from competing microbes. *Proc Natl Acad Sci* **112:** 2181–2186.

Dwyer DJ, Kohanski MA, Hayete B, Collins JJ. 2007. Gyrase inhibitors induce an oxidative damage cellular death pathway in *Escherichia coli*. *Mol Syst Biol* **3:** 91.

Dwyer DJ, Kohanski MA, Collins JJ. 2009. Role of reactive oxygen species in antibiotic action and resistance. *Curr Opin Microbiol* **12:** 482–489.

Egan AJF, Vollmer W. 2015. The stoichiometric divisome: A hypothesis. *Front Microbiol* **6:** 455.

Evans M. 1999. Polymyxin B sulfate and colistin: Old antibiotics for emerging multiresistant Gram-negative bacteria. *Ann Pharmacother* **33:** 960–967.

Falagas ME, Kasiakou SK. 2005. Colistin: The revival of polymyxins for the management of multidrug-resistant Gram-negative bacterial infections. *Clin Infect Dis* **40:** 1333–1341.

Falagas ME, Kasiakou SK. 2006. Toxicity of polymyxins: A systematic review of the evidence from old and recent studies. *Crit Care* **10:** R27.

Falagas ME, Rafailidis PI, Matthaiou DK. 2010. Resistance to polymyxins: Mechanisms, frequency and treatment options. *Drug Resist Updat* **13:** 132–138.

Fehlner-Gardiner CC, Valvano MA. 2002. Cloning and characterization of the *Burkholderia vietnamiensis norM* gene encoding a multi-drug efflux protein. *FEMS Microbiol Lett* **215:** 279–283.

Fernández L, Gooderham WJ, Bains M, McPhee JB, Wiegand I, Hancock REW. 2010. Adaptive resistance to the "last hope" antibiotics polymyxin B and colistin in *Pseu*-

domonas aeruginosa is mediated by the novel two-component regulatory system ParR–ParS. *Antimicrob Agents Chemother* **54:** 3372–3382.

Fernández L, Breidenstein EBM, Hancock REW. 2011. Creeping baselines and adaptive resistance to antibiotics. *Drug Resist Updat* **14:** 1–21.

Fernández L, Jenssen H, Bains M, Wiegand I, Gooderham WJ, Hancock REW. 2012. The two-component system CprRS senses cationic peptides and triggers adaptive resistance in *Pseudomonas aeruginosa* independently of ParRS. *Antimicrob Agents Chemother* **56:** 6212–6222.

Fernández L, Alvarez-Ortega C, Wiegand I, Kocíncová D, Lam JS, Martínez JL, Hancock REW. 2013. Characterization of the polymyxin B resistome of *Pseudomonas aeruginosa*. *Antimicrob Agents Chemother* **57:** 110–119.

Finking R, Marahiel MA. 2004. Biosynthesis of nonribosomal peptides1. *Annu Rev Microbiol* **58:** 453–488.

Folkesson A, Haagensen JA, Zampaloni C, Sternberg C, Molin S. 2008. Biofilm induced tolerance towards antimicrobial peptides. *PLoS ONE* **3:** e1891.

Gibbons HS, Reynolds CM, Guan Z, Raetz CR. 2008. An inner membrane dioxygenase that generates the 2-hydroxymyristate moiety of *Salmonella* lipid A. *Biochemistry* **47:** 2814–2825.

Gooderham WJ, Bains M, McPhee JB, Wiegand I, Hancock REW. 2008. Induction by cationic antimicrobial peptides and involvement in intrinsic polymyxin and antimicrobial peptide resistance, biofilm formation, and swarming motility of PsrA in *Pseudomonas aeruginosa*. *J Bacteriol* **190:** 5624–5634.

Gooderham WJ, Gellatly SL, Sanschagrin F, McPhee JB, Bains M, Cosseau C, Levesque RC, Hancock REW. 2009. The sensor kinase PhoQ mediates virulence in *Pseudomonas aeruginosa*. *Microbiology* **155:** 699–711.

Gray AN, Egan AJ, Van't Veer IL, Verheul J, Colavin A, Koumoutsi A, Biboy J, Altelaar AFM, Damen MJ, Huang KC, et al. 2015. Coordination of peptidoglycan synthesis and outer membrane constriction during *Escherichia coli* cell division. *eLife* **4:** e07118.

Gunn JS. 2008. The *Salmonella* PmrAB regulon: Lipopolysaccharide modifications, antimicrobial peptide resistance and more. *Trends Microbiol* **16:** 284–290.

Gunn JS, Lim KB, Krueger J, Kim K, Guo L, Hackett M, Miller SI. 1998. PmrA–PmrB-regulated genes necessary for 4-aminoarabinose lipid A modification and polymyxin resistance. *Mol Microbiol* **27:** 1171–1182.

Gunn JS, Ryan SS, Van Velkinburgh JC, Ernst RK, Miller SI. 2000. Genetic and functional analysis of a PmrA–PmrB-regulated locus necessary for lipopolysaccharide modification, antimicrobial peptide resistance, and oral virulence of *Salmonella enterica* serovar Typhimurium. *Infect Immun* **68:** 6139–6146.

Gutu AD, Sgambati N, Strasbourger P, Brannon MK, Jacobs MA, Haugen E, Kaul RK, Johansen HK, Høiby N, Moskowitz SM. 2013. Polymyxin resistance of *Pseudomonas aeruginosa phoQ* mutants is dependent on additional two-component regulatory systems. *Antimicrob Agents Chemother* **57:** 2204–2215.

Hancock RE. 1981. Aminoglycoside uptake and mode of action—With special reference to streptomycin and gen-

tamicin. II: Effects of aminoglycosides on cells. *J Antimicrob Chemother* **8:** 429–445.

Hancock RE. 1984. Alterations in outer membrane permeability. *Annu Rev Microbiol* **38:** 237–264.

Hancock RE. 1997. Peptide antibiotics. *Lancet* **349:** 418–422.

Hancock RE. 2000. Cationic antimicrobial peptides: Towards clinical applications. *Expert Opin Investig Drugs* **9:** 1723–1729.

Hancock REW, Bell A. 1988. Antibiotic uptake into Gram-negative bacteria. *Eur J Clin Microbiol Infect Dis* **7:** 713–720.

Hancock RE, Chapple DS. 1999. Peptide antibiotics. *Antimicrob Agents Chemother* **43:** 1317–1323.

Handler AA, Lim JE, Losick R. 2008. Peptide inhibitor of cytokinesis during sporulation in *Bacillus subtilis*. *Mol Microbiol* **68:** 588–599.

Hankins JV, Madsen JA, Giles DK, Brodbelt JS, Trent MS. 2012. Amino acid addition to *Vibrio cholerae* LPS establishes a link between surface remodeling in Gram-positive and Gram-negative bacteria. *Proc Natl Acad Sci* **109:** 8722–8727.

Haque MM, Kabir MS, Aini LQ, Hirata H, Tsuyumu S. 2009. SlyA, a MarR family transcriptional regulator, is essential for virulence in *Dickeya dadantii* 3937. *J Bacteriol* **191:** 5409–5418.

Heindorf M, Kadari M, Heider C, Skiebe E, Wilharm G. 2014. Impact of *Acinetobacter baumannii* superoxide dismutase on motility, virulence, oxidative stress resistance and susceptibility to antibiotics. *PLoS ONE* **9:** e101033.

Herrera CM, Hankins JV, Trent MS. 2010. Activation of PmrA inhibits LpxT-dependent phosphorylation of lipid A promoting resistance to antimicrobial peptides. *Mol Microbiol* **76:** 1444–1460.

Herrera CM, Crofts AA, Henderson JC, Pingali SC, Davies BW, Trent MS. 2014. The *Vibrio cholerae* VprA–VprB two-component system controls virulence through endotoxin modification. *MBio* **5:** e02283-14.

Holzer SU, Schlumberger MC, Jackel D, Hensel M. 2009. Effect of the O-antigen length of lipopolysaccharide on the functions of type III secretion systems in *Salmonella enterica*. *Infect Immun* **77:** 5458–5470.

Hood MI, Becker KW, Roux CM, Dunman PM, Skaar EP. 2013. Genetic determinants of intrinsic colistin tolerance in *Acinetobacter baumannii*. *Infect Immun* **81:** 542–551.

Hu Y, Liu F, Lin IY, Gao GF, Zhu B. 2016. Dissemination of the *mcr-1* colistin resistance gene. *Lancet Infect Dis* **16:** 146–147.

Ilg K, Endt K, Misselwitz B, Stecher B, Aebi M, Hardt WD. 2009. O-antigen-negative *Salmonella enterica* serovar Typhimurium is attenuated in intestinal colonization but elicits colitis in streptomycin-treated mice. *Infect Immun* **77:** 2568–2575.

Ingram BO, Sohlenkamp C, Geiger O, Raetz CR. 2010. Altered lipid A structures and polymyxin hypersensitivity of *Rhizobium etli* mutants lacking the LpxE and LpxF phosphatases. *Biochim Biophys Acta* **1801:** 593–604.

Jiang SS, Lin TY, Wang WB, Liu MC, Hsueh PR, Liaw SJ. 2010a. Characterization of UDP-glucose dehydrogenase and UDP-glucose pyrophosphorylase mutants of *Proteus mirabilis*: Defectiveness in polymyxin B resistance, swarming, and virulence. *Antimicrob Agents Chemother* **54:** 2000–2009.

Jiang SS, Liu MC, Teng LJ, Wang WB, Hsueh PR, Liaw SJ. 2010b. Proteus mirabilis pmrI, an RppA-regulated gene necessary for polymyxin B resistance, biofilm formation, and urothelial cell invasion. *Antimicrob Agents Chemother* **54:** 1564–1571.

Kang Y, Lunin VV, Skarina T, Savchenko A, Schurr MJ, Hoang TT. 2009. The long-chain fatty acid sensor, PsrA, modulates the expression of *rpoS* and the type III secretion *exsCEBA* operon in *Pseudomonas aeruginosa*. *Mol Microbiol* **73:** 120–136.

Kato A, Latifi T, Groisman EA. 2003. Closing the loop: The PmrA/PmrB two-component system negatively controls expression of its posttranscriptional activator PmrD. *Proc Natl Acad Sci* **100:** 4706–4711.

Katsuma N, Sato Y, Ohki K, Okimura K, Ohnishi K, Sakura N. 2009. Development of des-fatty acyl-polymyxin B decapeptide analogs with *Pseudomonas aeruginosa*-specific antimicrobial activity. *Chem Pharm Bull (Tokyo)* **57:** 332–336.

Kawasaki K, China K, Nishijima M. 2007. Release of the lipopolysaccharide deacylase PagL from latency compensates for a lack of lipopolysaccharide aminoarabinose modification-dependent resistance to the antimicrobial peptide polymyxin B in *Salmonella enterica*. *J Bacteriol* **189:** 4911–4919.

Keren I, Wu Y, Inocencio J, Mulcahy LR, Lewis K. 2013. Killing by bactericidal antibiotics does not depend on reactive oxygen species. *Science* **339:** 1213–1216.

Kindrachuk KN, Fernandez L, Bains M, Hancock RE. 2011. Involvement of an ATP-dependent protease, PA0779/AsrA, in inducing heat shock in response to tobramycin in *Pseudomonas aeruginosa*. *Antimicrob Agents Chemother* **55:** 1874–1882.

Klemperer RM, Gilbert P, Meier AM, Cozens RM, Brown MR. 1979. Influence of suspending media upon the susceptibility of *Pseudomonas aeruginosa* NCTC 6750 and its spheroplasts to polymyxin B. *Antimicrob Agents Chemother* **15:** 147–151.

Koike M, Iida K, Matsuo T. 1969. Electron microscopic studies on mode of action of polymyxin. *J Bacteriol* **97:** 448–452.

Kong Q, Yang J, Liu Q, Alamuri P, Roland KL, Curtiss R III. 2011. Effect of deletion of genes involved in lipopolysaccharide core and O-antigen synthesis on virulence and immunogenicity of *Salmonella enterica* serovar Typhimurium. *Infect Immun* **79:** 4227–4239.

Kox LF, Wösten MM, Groisman EA. 2000. A small protein that mediates the activation of a two-component system by another two-component system. *EMBO J* **19:** 1861–1872.

Kruse T, Christensen B, Raventos D, Nielsen AK, Nielsen JD, Vukmirovic N, Kristensen HH. 2009. Transcriptional profile of *Escherichia coli* in response to novispirin G10. *Int J Pept Res Ther* **15:** 17–24.

Kunin CM, Bugg A. 1971. Binding of polymyxin antibiotics to tissues: The major determinant of distribution and persistence in the body. *J Infect Dis* **124:** 394–400.

Kwa AL, Tam VH, Falagas ME. 2008. Polymyxins: A review of the current status including recent developments. *Ann Acad Med Singapore* **37:** 870–883.

Lacour S, Bechet E, Cozzone AJ, Mijakovic I, Grangeasse C. 2008. Tyrosine phosphorylation of the UDP-glucose dehydrogenase of *Escherichia coli* is at the crossroads of colanic acid synthesis and polymyxin resistance. *PLoS ONE* **3:** e3053.

Landman D, Georgescu C, Martin DA, Quale J. 2008. Polymyxins revisited. *Clin Microbiol Rev* **21:** 449–465.

LaPorte DC, Rosenthal KS, Storm DR. 1977. Inhibition of *Escherichia coli* growth and respiration by polymyxin B covalently attached to agarose beads. *Biochemistry* **16:** 1642–1648.

Lee H, Hsu FF, Turk J, Groisman EA. 2004. The PmrA-regulated *pmrC* gene mediates phosphoethanolamine modification of lipid A and polymyxin resistance in *Salmonella enterica*. *J Bacteriol* **186:** 4124–4133.

Lewis JR, Lewis SA. 2004. Colistin interactions with the mammalian urothelium. *Am J Physiol Cell Physiol* **286:** C913–C922.

Ling LL, Schneider T, Peoples AJ, Spoering AL, Engels I, Conlon BP, Mueller A, Schäberle TF, Hughes DE, Epstein S, et al. 2015. A new antibiotic kills pathogens without detectable resistance. *Nature* **517:** 455–459.

Liu Y, Imlay JA. 2013. Cell death from antibiotics without the involvement of reactive oxygen species. *Science* **339:** 1210–1213.

Liu YY, Wang Y, Walsh TR, Yi LX, Zhang R, Spencer J, Doi Y, Tian G, Dong B, Huang X, et al. 2016. Emergence of plasmid-mediated colistin resistance mechanism MCR-1 in animals and human beings in China: A microbiological and molecular biological study. *Lancet Infect Dis* **16:** 161–168.

Llewellyn AC, Zhao J, Song F, Parvathareddy J, Xu Q, Napier BA, Laroui H, Merlin D, Bina JE, Cotter PA, et al. 2012. NaxD is a deacetylase required for lipid A modification and *Francisella pathogenesis*. *Mol Microbiol* **86:** 611–627.

Lopez-Camacho E, Gomez-Gil R, Tobes R, Manrique M, Lorenzo M, Galvan B, Salvarelli E, Moatassim Y, Salanueva IJ, Pareja E, et al. 2014. Genomic analysis of the emergence and evolution of multidrug resistance during a *Klebsiella pneumoniae* outbreak including carbapenem and colistin resistance. *J Antimicrob Chemother* **69:** 632–636.

Loutet SA, Valvano MA. 2011. Extreme antimicrobial peptide and polymyxin B resistance in the genus *Burkholderia*. *Front Microbiol* **2:** 159.

Loutet SA, Bartholdson SJ, Govan JRW, Campopiano DJ, Valvano MA. 2009. Contributions of two UDP-glucose dehydrogenases to viability and polymyxin B resistance of *Burkholderia cenocepacia*. *Microbiology* **155:** 2029–2039.

Lu S, Walters G, Parg R, Dutcher JR. 2014. Nanomechanical response of bacterial cells to cationic antimicrobial peptides. *Soft Matter* **10:** 1806–1815.

Macfarlane EL, Kwasnicka A, Hancock RE. 2000. Role of *Pseudomonas aeruginosa* PhoP–PhoQ in resistance to antimicrobial cationic peptides and aminoglycosides. *Microbiology* **146:** 2543–2554.

Maloney E, Stankowska D, Zhang J, Fol M, Cheng QJ, Lun S, Bishai WR, Rajagopalan M, Chatterjee D, Madiraju MV. 2009. The two-domain LysX protein of *Mycobacterium tuberculosis* is required for production of lysinylated phosphatidylglycerol and resistance to cationic antimicrobial peptides. *PLoS Pathog* **5:** e1000534.

Marceau M, Sebbane F, Ewann F, Collyn F, Lindner B, Campos MA, Bengoechea JA, Simonet M. 2004. The *pmrF* polymyxin-resistance operon of *Yersinia pseudotuberculosis* is upregulated by the PhoP–PhoQ two-component system but not by PmrA–PmrB, and is not required for virulence. *Microbiology* **150:** 3947–3957.

Margolin W. 2005. FtsZ and the division of prokaryotic cells and organelles. *Nat Rev Mol Cell Biol* **6:** 862–871.

Martin NI, Hu H, Moake MM, Churey JJ, Whittal R, Worobo RW, Vederas JC. 2003. Isolation, structural characterization, and properties of mattacin (polymyxin M), a cyclic peptide antibiotic produced by *Paenibacillus kobensis* M. *J Biol Chem* **278:** 13124–13132.

Mathur J, Waldor MK. 2004. The *Vibrio cholerae* ToxR-regulated porin OmpU confers resistance to antimicrobial peptides. *Infect Immun* **72:** 3577–3583.

McCoy LS, Roberts KD, Nation RL, Thompson PE, Velkov T, Li J, Tor Y. 2013. Polymyxins and analogues bind to ribosomal RNA and interfere with eukaryotic translation in vitro. *ChemBioChem* **14:** 2083–2086.

McPhee JB, Lewenza S, Hancock REW. 2003. Cationic antimicrobial peptides activate a two-component regulatory system, PmrA–PmrB, that regulates resistance to polymyxin B and cationic antimicrobial peptides in *Pseudomonas aeruginosa. Mol Microbiol* **50:** 205–217.

Moffatt JH, Harper M, Harrison P, Hale JDF, Vinogradov E, Seemann T, Henry R, Crane B, St Michael F, Cox AD, et al. 2010. Colistin resistance in *Acinetobacter baumannii* is mediated by complete loss of lipopolysaccharide production. *Antimicrob Agents Chemother* **54:** 4971–4977.

Moffatt JH, Harper M, Adler B, Nation RL, Li J, Boyce JD. 2011. Insertion sequence ISAba11 is involved in colistin resistance and loss of lipopolysaccharide in *Acinetobacter baumannii. Antimicrob Agents Chemother* **55:** 3022–3024.

Mogi T, Murase Y, Mori M, Shiomi K, Omura S, Paranagama MP, Kita K. 2009. Polymyxin B identified as an inhibitor of alternative NADH dehydrogenase and malate: Quinone oxidoreductase from the Gram-positive bacterium *Mycobacterium smegmatis. J Biochem* **146:** 491–499.

Moon K, Gottesman S. 2009. A PhoQ/P-regulated small RNA regulates sensitivity of *Escherichia coli* to antimicrobial peptides. *Mol Microbiol* **74:** 1314–1330.

Moore RA, Bates NC, Hancock RE. 1986. Interaction of polycationic antibiotics with *Pseudomonas aeruginosa* lipopolysaccharide and lipid A studied by using dansylpolymyxin. *Antimicrob Agents Chemother* **29:** 496–500.

Mortensen NP, Fowlkes JD, Sullivan CJ, Allison DP, Larsen NB, Molin S, Doktycz MJ. 2009. Effects of colistin on surface ultrastructure and nanomechanics of *Pseudomonas aeruginosa* cells. *Langmuir* **25:** 3728–3733.

Mouslim C, Groisman EA. 2003. Control of the *Salmonella ugd* gene by three two-component regulatory systems. *Mol Microbiol* **47:** 335–344.

Muller C, Plésiat P, Jeannot K. 2011. A two-component regulatory system interconnects resistance to polymyxins, aminoglycosides, fluoroquinolones, and β-lactams in *Pseudomonas aeruginosa. Antimicrob Agents Chemother* **55:** 1211–1221.

Nakajima K, Kawamata J. 1966. Studies on the mechanism of action of colistin. IV: Activation of "latent" ribonuclease in *Escherichia coli* by colistin. *Biken J* **9:** 115–123.

Nation RL, Velkov T, Li J. 2014. Colistin and polymyxin B: Peas in a pod, or chalk and cheese? *Clin Infect Dis* **59:** 88–94.

Nation RL, Li J, Cars O, Couet W, Dudley MN, Kaye KS, Mouton JW, Paterson DL, Tam VH, Theuretzbacher U, et al. 2015. Framework for optimisation of the clinical use of colistin and polymyxin B: The Prato polymyxin consensus. *Lancet Infect Dis* **15:** 225–234.

Newton BA. 1956. The properties and mode of action of the polymyxins. *Bacteriol Rev* **20:** 14–27.

Nishino K, Hsu FF, Turk J, Cromie MJ, Wösten MMSM, Groisman EA. 2006. Identification of the lipopolysaccharide modifications controlled by the *Salmonella* PmrA/PmrB system mediating resistance to Fe(III) and Al(III). *Mol Microbiol* **61:** 645–654.

Niu B, Vater J, Rueckert C, Blom J, Lehmann M, Ru JJ, Chen XH, Wang Q, Borriss R. 2013. Polymyxin P is the active principle in suppressing phytopathogenic *Erwinia* spp. by the biocontrol rhizobacterium *Paenibacillus polymyxa* M-1. *BMC Microbiol* **13:** 137.

Oh JT, Van Dyk TK, Cajal Y, Dhurjati PS, Sasser M, Jain MK. 1998. Osmotic stress in viable *Escherichia coli* as the basis for the antibiotic response by polymyxin B. *Biochem Biophys Res Commun* **246:** 619–623.

Oh JT, Cajal Y, Skowronska EM, Belkin S, Chen J, Van Dyk TK, Sasser M, Jain MK. 2000. Cationic peptide antimicrobials induce selective transcription of *micF* and *osmY* in *Escherichia coli. Biochim Biophys Acta* **1463:** 43–54.

Olaitan AO, Morand S, Rolain JM. 2014. Mechanisms of polymyxin resistance: Acquired and intrinsic resistance in bacteria. *Front Microbiol* **5:** 643.

Olaitan AO, Chabou S, Okdah L, Morand S, Rolain JM. 2016. Dissemination of the *mcr-1* colistin resistance gene. *Lancet Infect Dis* **16:** 147.

Orwa JA, Govaerts C, Busson R, Roets E, Van Schepdael A, Hoogmartens J. 2001. Isolation and structural characterization of colistin components. *J Antibiot (Tokyo)* **54:** 595–599.

Padilla E, Llobet E, Domenech-Sanchez A, Martinez-Martinez L, Bengoechea JA, Alberti S. 2010. *Klebsiella pneumoniae* AcrAB efflux pump contributes to antimicrobial resistance and virulence. *Antimicrob Agents Chemother* **54:** 177–183.

Pamp SJ, Gjermansen M, Johansen HK, Tolker-Nielsen T. 2008. Tolerance to the antimicrobial peptide colistin in *Pseudomonas aeruginosa* biofilms is linked to metabolically active cells, and depends on the pmr and *mexAB–oprM* genes. *Mol Microbiol* **68:** 223–240.

Papenfort K, Pfeiffer V, Mika F, Lucchini S, Hinton JC, Vogel J. 2006. σ^E-dependent small RNAs of *Salmonella* respond to membrane stress by accelerating global *omp* mRNA decay. *Mol Microbiol* **62:** 1674–1688.

Park SY, Groisman EA. 2014. Signal-specific temporal response by the *Salmonella* PhoP/PhoQ regulatory system. *Mol Microbiol* **91:** 135–144.

Paterson GK, Northen H, Cone DB, Willers C, Peters SE, Maskell DJ. 2009. Deletion of *tolA* in *Salmonella typhimurium* generates an attenuated strain with vaccine potential. *Microbiology* **155:** 220–228.

Paulander W, Wang Y, Folkesson A, Charbon G, Lobner-Olesen A, Ingmer H. 2014. Bactericidal antibiotics increase hydroxyphenyl fluorescein signal by altering cell morphology. *PLoS ONE* **9**: e92231.

Pescaretti MdLM, Lopez FE, Morero RD, Delgado MA. 2011. The PmrA/PmrB regulatory system controls the expression of the *wzzfepE* gene involved in the O-antigen synthesis of *Salmonella enterica* serovar Typhimurium. *Microbiology* **157**: 2515–2521.

Phe K, Lee Y, McDaneld PM, Prasad N, Yin T, Figueroa DA, Musick WL, Cottreau JM, Hu M, Tam VH. 2014. In vitro assessment and multicenter cohort study of comparative nephrotoxicity rates associated with colistimethate versus polymyxin B therapy. *Antimicrob Agents Chemother* **58**: 2740–2746.

Piek S, Wang Z, Ganguly J, Lakey AM, Bartley SN, Mowlaboccus S, Anandan A, Stubbs KA, Scanlon MJ, Vrielink A, et al. 2014. The role of oxidoreductases in determining the function of the neisserial lipid a phosphoethanolamine transferase required for resistance to polymyxin. *PLoS ONE* **9**: e106513.

Port GC, Vega LA, Nylander AB, Caparon MG. 2014. *Streptococcus pyogenes* polymyxin B–resistant mutants display enhanced ExPortal integrity. *J Bacteriol* **196**: 2563–2577.

Pournaras S, Poulou A, Dafopoulou K, Chabane YN, Kristo I, Makris D, Hardouin J, Cosette P, Tsakris A, De E. 2014. Growth retardation, reduced invasiveness, and impaired colistin-mediated cell death associated with colistin resistance development in *Acinetobacter baumannii*. *Antimicrob Agents Chemother* **58**: 828–832.

Prost LR, Daley ME, Le Sage V, Bader MW, Le Moual H, Klevit RE, Miller SI. 2007. Activation of the bacterial sensor kinase PhoQ by acidic pH. *Mol Cell* **26**: 165–174.

Raetz CRH, Reynolds CM, Trent MS, Bishop RE. 2007. Lipid A modification systems in Gram-negative bacteria. *Annu Rev Biochem* **76**: 295–329.

Reynolds CM, Kalb SR, Cotter RJ, Raetz CRH. 2005. A phosphoethanolamine transferase specific for the outer 3-deoxy-D-*manno*-octulosonic acid residue of *Escherichia coli* lipopolysaccharide. Identification of the *eptB* gene and Ca^{2+} hypersensitivity of an *eptB* deletion mutant. *J Biol Chem* **280**: 21202–21211.

Reynolds CM, Ribeiro AA, McGrath SC, Cotter RJ, Raetz CRH, Trent MS. 2006. An outer membrane enzyme encoded by *Salmonella typhimurium lpxR* that removes the 3′-acyloxyacyl moiety of lipid A. *J Biol Chem* **281**: 21974–21987.

Rosch JW, Caparon MG. 2005. The ExPortal: An organelle dedicated to the biogenesis of secreted proteins in *Streptococcus pyogenes*. *Mol Microbiol* **58**: 959–968.

Rosch JW, Hsu FF, Caparon MG. 2007. Anionic lipids enriched at the ExPortal of *Streptococcus pyogenes*. *J Bacteriol* **189**: 801–806.

Rosenthal KS, Storm DR. 1977. Disruption of the *Escherichia coli* outer membrane permeability barrier by immobilized polymyxin B. *J Antibiot (Tokyo)* **30**: 1087–1092.

Sassera D, Comandatore F, Gaibani P, D'Auria G, Mariconti M, Landini MP, Sambri V, Marone P. 2014. Comparative genomics of closely related strains of *Klebsiella pneumoniae* reveals genes possibly involved in colistin resistance. *Ann Microbiol* **64**: 887–890.

Saugar JM, Alarcon T, Lopez-Hernandez S, Lopez-Brea M, Andreu D, Rivas L. 2002. Activities of polymyxin B and cecropin A-melittin peptide CA(1-8)M(1-18) against a

multiresistant strain of *Acinetobacter baumannii*. *Antimicrob Agents Chemother* **46**: 875–878.

Schindler PRG, Teuber M. 1975. Action of polymyxin B on bacterial membranes: Morphological changes in the cytoplasm and in the outer membrane of *Salmonella typhimurium* and *Escherichia coli* B. *Antimicrob Agents Chemother* **8**: 95–104.

Scott NE, Nothaft H, Edwards AVG, Labbate M, Djordjevic SP, Larsen MR, Szymanski CM, Cordwell SJ. 2012. Modification of the *Campylobacter jejuni* N-linked glycan by EptC protein-mediated addition of phosphoethanolamine. *J Biol Chem* **287**: 29384–29396.

Shah NR, Hancock RE, Fernandez RC. 2014. *Bordetella pertussis* lipid A glucosamine modification confers resistance to cationic antimicrobial peptides and increases resistance to outer membrane perturbation. *Antimicrob Agents Chemother* **58**: 4931–4934.

Shaheen M, Li J, Ross AC, Vederas JC, Jensen SE. 2011. *Paenibacillus polymyxa* PKB1 produces variants of polymyxin B–type antibiotics. *Chem Biol* **18**: 1640–1648.

Shi Y, Cromie MJ, Hsu FF, Turk J, Groisman EA. 2004a. PhoP-regulated *Salmonella* resistance to the antimicrobial peptides magainin 2 and polymyxin B. *Mol Microbiol* **53**: 229–241.

Shi Y, Latifi T, Cromie MJ, Groisman EA. 2004b. Transcriptional control of the antimicrobial peptide resistance *ugtL* gene by the *Salmonella* PhoP and SlyA regulatory proteins. *J Biol Chem* **279**: 38618–38625.

Sikora AE, Beyhan S, Bagdasarian M, Yildiz FH, Sandkvist M. 2009. Cell envelope perturbation induces oxidative stress and changes in iron homeostasis in *Vibrio cholerae*. *J Bacteriol* **191**: 5398–5408.

Skerlavaj B, Romeo D, Gennaro R. 1990. Rapid membrane permeabilization and inhibition of vital functions of Gram-negative bacteria by bactenecins. *Infect Immun* **58**: 3724–3730.

Sohlenkamp C, Galindo-Lagunas KA, Guan Z, Vinuesa P, Robinson S, Thomas-Oates J, Raetz CRH, Geiger O. 2007. The lipid lysyl-phosphatidylglycerol is present in membranes of *Rhizobium tropici* CIAT899 and confers increased resistance to polymyxin B under acidic growth conditions. *Mol Plant Microbe Interact* **20**: 1421–1430.

Spindler EC, Hale JDF, Giddings TH, Hancock REW, Gill RT. 2011. Deciphering the mode of action of the synthetic antimicrobial peptide Bac8c. *Antimicrob Agents Chemother* **55**: 1706–1716.

Srinivasan VB, Singh BB, Priyadarshi N, Chauhan NK, Rajamohan G. 2014. Role of novel multidrug efflux pump involved in drug resistance in *Klebsiella pneumoniae*. *PLoS ONE* **9**: e96288.

Storm DR, Rosenthal KS, Swanson PE. 1977. Polymyxin and related peptide antibiotics. *Annu Rev Biochem* **46**: 723–763.

Strahl H, Hamoen LW. 2010. Membrane potential is important for bacterial cell division. *Proc Natl Acad Sci* **107**: 12281–12286.

Subashchandrabose S, Smith SN, Spurbeck RR, Kole MM, Mobley HLT. 2013. Genome-wide detection of fitness genes in uropathogenic *Escherichia coli* during systemic infection. *PLoS Pathog* **9**: e1003788.

Subbalakshmi C, Sitaram N. 1998. Mechanism of antimicrobial action of indolicidin. *FEMS Microbiol Lett* **160:** 91–96.

Sun S, Negrea A, Rhen M, Andersson DI. 2009. Genetic analysis of colistin resistance in *Salmonella enterica* serovar Typhimurium. *Antimicrob Agents Chemother* **53:** 2298–2305.

Tamayo R, Choudhury B, Septer A, Merighi M, Carlson R, Gunn JS. 2005a. Identification of *cptA*, a PmrA-regulated locus required for phosphoethanolamine modification of the *Salmonella enterica* serovar Typhimurium lipopolysaccharide core. *J Bacteriol* **187:** 3391–3399.

Tamayo R, Prouty AM, Gunn JS. 2005b. Identification and functional analysis of *Salmonella enterica* serovar Typhimurium PmrA-regulated genes. *FEMS Immunol Med Microbiol* **43:** 249–258.

Tambadou F, Caradec T, Gagez AL, Bonnet A, Sopéna V, Bridiau N, Thiéry V, Didelot S, Barthélémy C, Chevrot R. 2015. Characterization of the colistin (polymyxin E1 and E2) biosynthetic gene cluster. *Arch Microbiol* **197:** 521–532.

Taylor PK, Yeung ATY, Hancock REW. 2014. Antibiotic resistance in *Pseudomonas aeruginosa* biofilms: Towards the development of novel anti-biofilm therapies. *J Biotechnol* **191:** 121–130.

Teuber M. 1967. Precipitation of ribosomes from *E. coli* B by polymyxin B. *Die Naturwissenschaften* **54:** 71.

Teuber M. 1974. Action of polymyxin B on bacterial membranes. III: Differential inhibition of cellular functions in *Salmonella typhimurium*. *Arch Microbiol* **100:** 131–144.

Tochikubo K, Yasuda Y, Kozuka S. 1986. Decreased particulate NADH oxidase activity in *Bacillus subtilis* spores after polymyxin B treatment. *Microbiology* **132:** 277–287.

Touze T, Tran AX, Hankins JV, Mengin-Lecreulx D, Trent MS. 2008. Periplasmic phosphorylation of lipid A is linked to the synthesis of undecaprenyl phosphate. *Mol Microbiol* **67:** 264–277.

Tran AX, Whittimore JD, Wyrick PB, McGrath SC, Cotter RJ, Trent MS. 2006. The lipid A 1-phosphatase of *Helicobacter pylori* is required for resistance to the antimicrobial peptide polymyxin. *J Bacteriol* **188:** 4531–4541.

Trent MS, Ribeiro AA, Lin S, Cotter RJ, Raetz CR. 2001. An inner membrane enzyme in *Salmonella* and *Escherichia coli* that transfers 4-amino-4-deoxy-L-arabinose to lipid A: Induction on polymyxin-resistant mutants and role of a novel lipid-linked donor. *J Biol Chem* **276:** 43122–43131.

Tsai MH, Wu SR, Lee HY, Chen CL, Lin TY, Huang YC, Chiu CH. 2012. Recognition of mechanisms involved in bile resistance important to halting antimicrobial resistance in nontyphoidal *Salmonella*. *Int J Antimicrob Agents* **40:** 151–157.

Tse H, Yuen KY. 2016. Dissemination of the *mcr-1* colistin resistance gene. *Lancet Infect Dis* **16:** 145–146.

Tu X, Latifi T, Bougdour A, Gottesman S, Groisman EA. 2006. The PhoP/PhoQ two-component system stabilizes the alternative σ factor RpoS in *Salmonella enterica*. *Proc Natl Acad Sci* **103:** 13503–13508.

Tzeng YL, Datta A, Ambrose K, Lo M, Davies JK, Carlson RW, Stephens DS, Kahler CM. 2004. The MisR/MisS two-component regulatory system influences inner core structure and immunotype of lipooligosaccharide in *Neisseria meningitidis*. *J Biol Chem* **279:** 35053–35062.

Tzeng YL, Ambrose KD, Zughaier S, Zhou X, Miller YK, Shafer WM, Stephens DS. 2005. Cationic antimicrobial peptide resistance in *Neisseria meningitidis*. *J Bacteriol* **187:** 5387–5396.

Vaara M, Vaara T. 1983. Polycations sensitize enteric bacteria to antibiotics. *Antimicrob Agents Chemother* **24:** 107–113.

van der Meijden B, Robinson JA. 2015. Synthesis of a polymyxin derivative for photolabeling studies in the Gram-negative bacterium *Escherichia coli*. *J Pept Sci* **21:** 231–235.

van Mourik A, Steeghs L, van Laar J, Meiring HD, Hamstra HJ, van Putten JPM, Wosten MMSM. 2010. Altered linkage of hydroxyacyl chains in lipid A of *Campylobacter jejuni* reduces TLR4 activation and antimicrobial resistance. *J Biol Chem* **285:** 15828–15836.

Vega LA, Caparon MG. 2012. Cationic antimicrobial peptides disrupt the *Streptococcus pyogenes* ExPortal. *Mol Microbiol* **85:** 1119–1132.

Velkov T, Roberts KD, Nation RL, Wang J, Thompson PE, Li J. 2014. Teaching "old" polymyxins new tricks: New-generation lipopeptides targeting Gram-negative "superbugs." *ACS Chem Biol* **9:** 1172–1177.

Wang X, McGrath SC, Cotter RJ, Raetz CR. 2006. Expression cloning and periplasmic orientation of the *Francisella novicida* lipid A 4'-phosphatase LpxF. *J Biol Chem* **281:** 9321–9330.

Wang WB, Chen IC, Jiang SS, Chen HR, Hsu CY, Hsueh PR, Hsu WB, Liaw SJ. 2008. Role of RppA in the regulation of polymyxin b susceptibility, swarming, and virulence factor expression in *Proteus mirabilis*. *Infect Immun* **76:** 2051–2062.

Warner DM, Levy SB. 2010. Different effects of transcriptional regulators MarA, SoxS and Rob on susceptibility of *Escherichia coli* to cationic antimicrobial peptides (CAMPs): Rob-dependent CAMP induction of the *marRAB* operon. *Microbiology* **156:** 570–578.

Warren GH, Gray J, Yurchenco JA. 1957. Effect of polymyxin on the lysis of *Neisseria catarrhalis* by lysozyme. *J Bacteriol* **74:** 788–793.

Webb HE, Granier SA, Marault M, Millemann Y, den Bakker HC, Nightingale KK, Bugarel M, Ison SA, Scott HM, Loneragan GH. 2016. Dissemination of the *mcr-1* colistin resistance gene. *Lancet Infect Dis* **16:** 144–145.

Wu M, Hancock REW. 1999. Interaction of the cyclic antimicrobial cationic peptide bactenecin with the outer and cytoplasmic membrane. *J Biol Chem* **274:** 29–35.

Xiong YQ, Mukhopadhyay K, Yeaman MR, Adler-Moore J, Bayer AS. 2005. Functional interrelationships between cell membrane and cell wall in antimicrobial peptide-mediated killing of *Staphylococcus aureus*. *Antimicrob Agents Chemother* **49:** 3114–3121.

Yan A, Guan Z, Raetz CR. 2007. An undecaprenyl phosphate-aminoarabinose flippase required for polymyxin resistance in *Escherichia coli*. *J Biol Chem* **282:** 36077–36089.

Yang S-J, Bayer AS, Mishra NN, Meehl M, Ledala N, Yeaman MR, Xiong YQ, Cheung AL. 2012. The *Staphylococcus aureus* two-component regulatory system, GraRS, senses and confers resistance to selected cationic antimicrobial peptides. *Infect Immun* **80:** 74–81.

Yethon JA, Gunn JS, Ernst RK, Miller SI, Laroche L, Malo D, Whitfield C. 2000. *Salmonella enterica* serovar Typhimurium *waaP* mutants show increased susceptibility to polymyxin and loss of virulence in vivo. *Infect Immun* **68:** 4485–4491.

Yeung ATY, Bains M, Hancock REW. 2011. The sensor kinase CbrA is a global regulator that modulates metabolism, virulence, and antibiotic resistance in *Pseudomonas aeruginosa*. *J Bacteriol* **193:** 918–931.

Yu Z, Qin W, Lin J, Fang S, Qiu J. 2015. Antibacterial mechanisms of polymyxin and bacterial resistance. *Biomed Res Int* **2015:** 679109.

Zavascki AP, Goldani LZ, Li J, Nation RL. 2007. Polymyxin B for the treatment of multidrug-resistant pathogens: A critical review. *J Antimicrob Chemother* **60:** 1206–1215.

Zhang L, Dhillon P, Yan H, Farmer S, Hancock REW. 2000. Interactions of bacterial cationic peptide antibiotics with outer and cytoplasmic membranes of *Pseudomonas aeruginosa*. *Antimicrob Agents Chemother* **44:** 3317–3321.

Resistance Mechanisms and the Future of Bacterial Enoyl-Acyl Carrier Protein Reductase (FabI) Antibiotics

Jiangwei Yao and Charles O. Rock

Department of Infectious Diseases, St. Jude Children's Research Hospital, Memphis, Tennessee 38105

Correspondence: charles.rock@stjude.org

Missense mutations leading to clinical antibiotic resistance are a liability of single-target inhibitors. The enoyl-acyl carrier protein reductase (FabI) inhibitors have one intracellular protein target and drug resistance is increased by the acquisition of single-base-pair mutations that alter drug binding. The spectrum of resistance mechanisms to FabI inhibitors suggests criteria that should be considered during the development of single-target antibiotics that would minimize the impact of missense mutations on their clinical usefulness. These criteria include high-affinity, fast on/off kinetics, few drug contacts with residue side chains, and no toxicity. These stringent criteria are achievable by structure-guided design, but this approach will only yield pathogen-specific drugs. Single-step acquisition of resistance may limit the clinical application of broad-spectrum, single-target antibiotics, but appropriately designed pathogen-specific antibiotics have the potential to overcome this liability.

The emergence of resistance to most clinically deployed antibiotics has stimulated considerable interest in developing new therapeutics. Bacterial fatty acid biosynthesis is an energy-intensive process that is essential for the formation of biological membranes (Zhang and Rock 2008). The importance of the pathway in bacterial physiology is highlighted by the existence of multiple natural products that target different points in fatty acid biosynthesis (Heath et al. 2001; Parsons and Rock 2011). These developments have led to a significant effort in academia and industry to develop antibiotics that target individual proteins in the fatty acid biosynthetic pathway (Campbell and Cronan 2001; Zhang et al. 2006). One concern about drugs that target

fatty acid synthesis is that fatty acids are abundant in the mammalian host, raising the concern that fatty acid synthesis inhibitors would be bypassed in vivo (Brinster et al. 2009). Although all bacteria studied are capable of incorporating extracellular fatty acids into their membranes, recent research shows that exogenous fatty acids cannot circumvent the inhibition of fatty acid synthesis in many major pathogens (Parsons et al. 2011; Yao and Rock 2015).

A greater concern is the fact that fatty acid synthesis inhibitors are designed to target individual steps in the pathway. Historically successful antibiotics used as monotherapy have multiple cellular targets (Silver 2011, 2007). Multitarget antibacterials are not subject to resistance arising from single missense mutations that can render the drug clinically useless in a single step. These considerations led to the "multitarget hypothesis," which posits that antibiotics with multiple cellular targets are superior to single-target drugs because of their ability to avoid single-step acquisition of resistance (Silver 2011, 2007). Missense mutations occur at frequencies approximating the error rate in DNA replication of about one in 10^9 cells, but environmental stresses can increase the mutation rate (Meyerovich et al. 2010). This aspect of bacterial physiology creates a reservoir of altered proteins in the bacterial population that leads to the emergence of resistant bacteria. If the resistance-causing missense mutation is a polymorphism that is normally observed in the environmental bacterial population, the evolution of resistance would be accelerated. Thus, a major liability of drugs with a single cellular target is the potential for the rapid evolution of clinical resistance.

This review focuses on a drug target in bacterial fatty acid synthesis that has received considerable attention. This target is the enoyl-acyl carrier protein (ACP) reductase (FabI) of bacterial fatty acid synthesis. FabI inhibitors are directed against a single cellular target and are subject to resistance arising from the acquisition of single-point mutations in the *fabI* gene. The extensive research on the mechanisms of acquired resistance to FabI-directed antibiotics coupled with insights from the clinical experience with the drugs provides a case study to evaluate the relevance of missense mutations to the utility of single-target drugs. Taken together, the experience with FabI inhibitors suggests criteria that should be considered in the development of single-target antibiotics to minimize or prevent the single-step evolution of clinical resistance.

ENOYL-ACP REDUCTASE (FabI) INHIBITORS

Each of the enzymatic steps in bacterial fatty acid synthesis (FASII) is essential so, in principle, each is a candidate for drug discovery. However, research has focused on those enzymes that catalyze key regulatory steps in the pathway (Parsons and Rock 2011). The enoyl-acyl carrier protein reductase (FabI) catalyzes the reduction of the *trans*-2-enoyl-ACP to acyl-ACP using NAD(P)H in the last step of the elongation cycle of fatty acid synthesis (Parsons and Rock 2013). Because the preceding step is an equilibrium enzyme, FabI activity is responsible for pulling each cycle of elongation to completion (Heath and Rock 1995, 1996). Thus, FabI is a rate-controlling step in fatty acid elongation, providing the solid rationale for targeting this enzyme.

FabI therapeutics provide an interesting case study in bacterial-resistance mechanisms against single-target inhibitors. There are three types of FabI inhibitors (Fig. 1). Isoniazid is the example of a bisubstrate inhibitor that is in clinical use for the treatment of *Mycobacterium tuberculosis*. Isoniazid is a prodrug with a complex mode of activation (Fig. 1A) and there is a complex array of isoniazid-resistance mechanisms. Triclosan and its closely related derivatives bind the NAD(P)–FabI product complex (Fig. 1B), whereas the AFN-1252 (Debio-1452) and CG400549 inhibitors bind to the NAD(P)H–FabI substrate complex (Fig. 1C). In both cases, resistance arises from missense mutations in the *fabI* gene, which leads to altered FabI proteins. Initially, it was thought that FabI inhibitors were broad-spectrum antibiotics because triclosan inhibits the growth of all bacteria. Subsequently, it was discovered that many Firmicutes do not have a FabI, but rather use a flavoprotein reductase called FabK to reduce enoyl-ACP (Heath and Rock 2000). The ability of triclosan to potently inhibit the growth of bacteria that depend on FabK is attributed to triclosan acting on other, yet to be identified, cellular target(s) (Heath and Rock 2000; Marrakchi et al. 2003). Also, some bacteria contain FabL or FabV enoyl-ACP reductases that are sufficiently different from FabI to render organisms expressing these enzymes refractory to therapeutics designed against FabI (Heath et al. 2000; Massengo-Tiasse and Cronan 2008; Zhu et al. 2010). Thus, FabI inhibitors target a select group of pathogens.

Bisubstrate FabI Inhibitors

Isoniazid is a frontline medication for the treatment of *M. tuberculosis* (Pinto and Menzies

A   **Bisubstrate FabI inhibitor**

B   **Inhibitors binding to the NAD(P)-enzyme complex**

C   **Inhibitors binding to the NAD(P)H-enzyme complex**

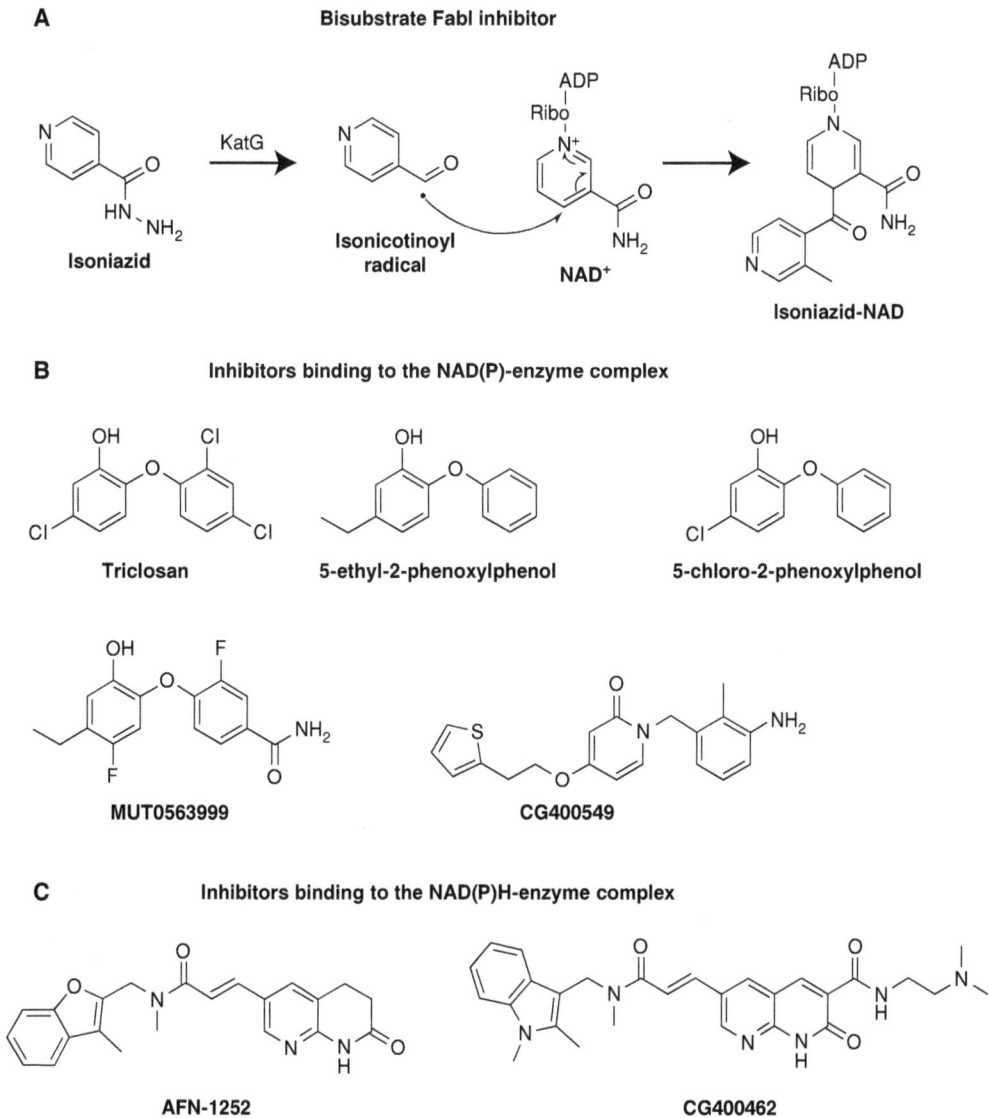

Figure 1. Structures of enoyl-ACP reductase inhibitors. (*A*) The mechanism of activating isoniazid, the prodrug, into the covalent isoniazid-NAD adduct, the slow-binding bisubstrate inhibitor of *M. tuberculosis* InhA (FabI). (*B*) Inhibitors that bind to the NAD(P)–FabI product complex. Triclosan is the prototypical inhibitor in this class. These compounds are slow-binding inhibitors. (*C*) Inhibitors that binds to the NAD(P)H–FabI substrate-binding complex of FabI. The scaffold for these two inhibitors was discovered through high-throughput screening against *Staphylococcus aureus* FabI. These compounds are high-affinity, fast on/off inhibitors.

2011). Daily regimens of isoniazid for 9 months is a standard treatment for latent *M. tuberculosis* infections, and isoniazid in combination with other drugs, such as rifampin, ethambutol, or pyrazinamide, is the standard treatment for active *M. tuberculosis* infections. Isoniazid was

first synthesized in 1912, and shown to be effective against *M. tuberculosis* in the 1950s (Bernstein et al. 1952). The mode of isoniazid inhibition has a controversial history, with disruption of cell permeability, inhibition of DNA synthesis, altered NAD metabolism, and inhibition of mycolic acid biosynthesis, all proposed as potential mechanisms of action (Vilchèze and Jacobs 2007). Vilchèze et al. (2006) showed that a point mutation in the *M. tuberculosis* enoyl-ACP reductase InhA confers resistance to isoniazid and rescued mycolic acid biosynthesis, definitively establishing the *M. tuberculosis* enoyl-ACP reductase InhA as the relevant target of isoniazid (Vilchèze et al. 2006).

M. tuberculosis synthesizes mycolic acids, a family of complex, long-chain fatty acids, via two carbon elongations similarly to how other bacteria synthesize fatty acids via the FASII system (Marrakchi et al. 2014). Mycolic acid synthesis is essential in *M. tuberculosis* (Marrakchi et al. 2014) just as fatty acid synthesis is essential in other bacteria (Rock and Jackowski 2002). The *M. tuberculosis inhA* encodes for the enoyl-ACP reductase involved in mycolic acid synthesis and is homologous to the FabI of FASII (Quemard et al. 1995). Isoniazid is a prodrug, complicating the spectrum of potential resistance mutations. The *M. tuberculosis* catalase-peroxidase enzyme KatG activates isoniazid into an isonicotinoyl radical that covalently interacts with NAD to form an isoniazid-NAD covalent adduct (Fig. 1A) (Vilchèze and Jacobs 2007). The isoniazid-NAD adduct is a slow-binding inhibitor of InhA (Rawat et al. 2003).

M. tuberculosis resistance to isoniazid is well studied with hundreds of publications that are reviewed in depth elsewhere (Hazbón et al. 2006; Seifert et al. 2015). Although the mechanism of a small percent of isoniazid-resistant, clinically isolated specimens are unknown, all known mechanisms of resistance are related to changing the rate of formation of the isoniazid-NAD covalent adduct, increasing the expression of InhA, or decreasing the affinity of InhA for the isoniazid-NAD adduct. The most frequent isoniazid resistance mechanism is mutation in the 315th amino acid in the *katG* gene found in 64.2% of resistance specimens (Seifert et al. 2015). Of these *katG* mutations, 95.3% cause a serine-to-threonine mutation. The rest of the mutations cause amino acid changes to asparagine (3.6%), isoleucine (0.5%), arginine (0.4%), and glycine (0.2%). The KatG(S315T) mutant enzyme does not suffer defects in the

catalase or peroxidase activity, but rather has decreased rate for the formation of the isoniazid-NAD adduct (Wengenack et al. 1998). The KatG(S315T) mutant appears to have a small growth defect, with a doubling time of 19.1 h versus 14.3 h for the Erdman laboratory strain and 15.8 h for the H37Rv laboratory strain (O'Sullivan et al. 2010). Epidemiological studies found that the KatG(S315T) mutant leads to secondary cases of tuberculosis as often as the isoniazid-susceptible strains, consistent with the fitness defect having a minor impact on virulence (van Soolingen et al. 2000). The other mutations to the KatG protein associated with isoniazid resistance occur at lower frequencies and the molecular mechanism(s) of resistance are not characterized.

The second-most-common isoniazid-resistance mechanism is mutations in the promoter region (-15) of the *inhA* gene. The change from a cytosine to a thymidine (19.2%) (Seifert et al. 2015) correlates with a 20-fold increase in *inhA* mRNA levels (Vilchèze et al. 2006). The *inhA*-15 mutants showed no statistically significant change in growth rate (O'Sullivan et al. 2010). Other mutations (-8, -17, and -47) in other locations within the *inhA* promoter have also been associated with resistance (Seifert et al. 2015) and are anticipated to also cause the up-regulation of *inhA* expression, although definitive experiments are needed. Mutations in the InhA protein itself are found in isoniazid-resistant *M. tuberculosis*, although the frequency observed in clinically resistant samples is low (Seifert et al. 2015). The most frequent (1.2%) and best-characterized mutation is InhA (S94A), and this mutation reduces the affinity of the isoniazid-NAD adduct to 323 $\pm$ 41 nM from 19 $\pm$ 10 nM for the wild-type (Vilchèze et al. 2006). Crystallographic studies found that the loss of the serine residue caused the movement of an ordered water molecule that disrupted the polar interactions with the isoniazid-NAD adduct (Vilchèze et al. 2006). Mutations in other residues of InhA, including residues Gly3, Ile21, Ala190, Ala194, and Ile258, have been found. Three of these interact with the isoniazid-NAD complex and the others do not (Fig. 2A). How mutations that do not directly contact the

Cite this article as *Cold Spring Harb Perspect Med* doi: 10.1101/cshperspect.a027045

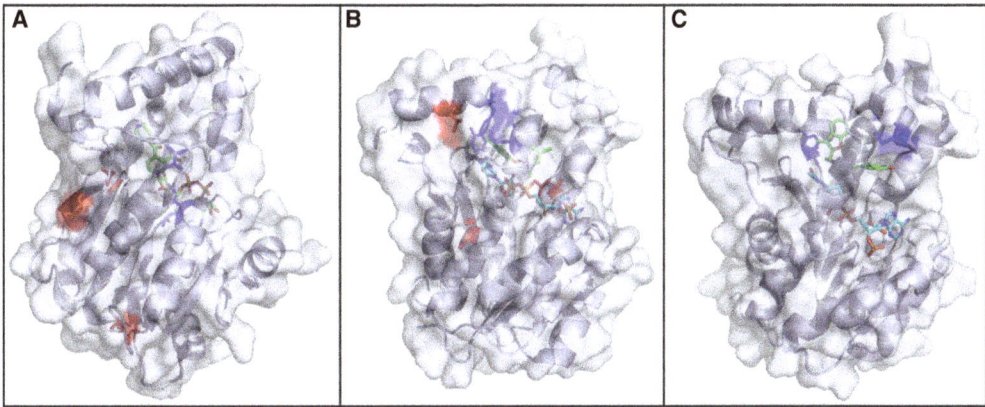

Figure 2. The locations of missense mutations in FabI conferring resistance against the three classes of FabI inhibitors are mapped onto the respective drug–protein complex structures. The respective protein structures are represented as a silver cartoon model overlaid with a transparent silver protein surface. Mutations in residues that directly interact with the inhibitor are shown in stick representation and colored blue. Mutations in residues that are not directly interacting with the inhibitor are also shown in stick representations and are colored red. The inhibitor carbons are green, nitrogens are blue, oxygens are red, and chlorines are green. The NAD(P) carbons are cyan, phosphorus is orange, nitrogens are blue, and oxygens are red. (A) The isoniazid–NAD–InhA ternary complex structure (*M. tuberculosis*) (PDB: 2NV6). (B) The triclosan–NADP–FabI ternary complex structure (*S. aureus*) (PDB: 4ALI). (C) The AFN–1252–NADPH–FabI ternary complex structure (*S. aureus*) (PDB: 4FS3).

bisubstrate drug impact resistance remain largely undefined (Seifert et al. 2015).

Mutations in other genetic locations are associated with isoniazid resistance and include mutations in the sigma factor SigI (Lee et al. 2012), mutations in activators or regulators of NADH/NAD metabolism (Vilchèze et al. 2005), and mutations within the *ahpC-oxyR* intergenic region (Seifert et al. 2015). The isoniazid resistance of the Δ*sigI* mutant is caused by the down-regulated expression of *katG* (Lee et al. 2012) and, thus, the mutation confers resistance by decreasing the rate of isoniazid-NAD formation. Although KatG catalase production is decreased in the Δ*sigI* mutant, this mutant is not attenuated in cell culture infections and is actually hypervirulent in mice (Lee et al. 2012). The prevalence of this mutation is not known because the association of SigI with isoniazid resistance was only recently discovered. Mutations that alter NADH/NAD metabolism cause resistance through changing the concentration of the substrates necessary for the formation of the active drug (Vilchèze et al. 2005), and also give rise to resistance to ethionamide, a drug

used against multidrug-resistant *M. tuberculosis*. However, this mutation was found at low frequencies among clinical samples (Hazbón et al. 2006). The mechanism by which mutations in the *ahpC-oxyR* intergenic region are associated with resistance in *M. tuberculosis* remains unclear (Sreevatsan et al. 1997). However, in *Escherichia coli*, deletion of *oxyR*, a regulator of hydrogen peroxide inducible genes, causes sensitivity to isoniazid (Rosner and Storz 1994).

How isoniazid is deployed against latent and active *M. tuberculosis* highlights the practical implications of drug resistance in antibiotic therapy. Because of the mode of action of isoniazid, a host of different mutations in different pathways triggers resistance, giving rise to an elevated rate of missense-resistance mutations against isoniazid as high as one in 10^8 (David 1970). Furthermore, the two most prevalent resistance mechanisms against isoniazid both require only a single-base-pair mutation and impose a minimal fitness cost to the mutant strain. Isoniazid is deployed as a monotherapy to treat latent *M. tuberculosis*, but the frequency of resistance mutations renders isoniazid ineffective as a monotherapy against active *M. tuberculosis* because each cavity contains more than 10^8 bacteria (Pinto and Menzies 2011).

Inhibitors Binding to the NAD(P)–FabI Product Complex

Triclosan is a broad-spectrum antibacterial and antifungal agent added to a variety of consumer products, including soaps, mouthwashes, and antiseptics (Russell 2004). Triclosan has two modes of action. First, triclosan is a nonspecific biocide that disrupts cell physiology by a poorly defined mechanism (Russell 2003). Second, triclosan is a slow-binding FabI inhibitor that binds to the FabI–NAD(P) complex of the enzyme formed after acyl-ACP product release (Heath et al. 1998, 1999; McMurry et al. 1998a). For bacteria that do not encode a FabI enoyl-ACP reductase, such as *Streptococcus pneumoniae*, triclosan inhibition is through the nonspecific biocide mechanism (Heath and Rock 2000). For bacteria that encode for a triclosan-sensitive FabI, such as *E. coli* or *S. aureus*, the minimal inhibitory concentration (MIC) for FabI inhibition occurs at ng/mL concentration, whereas the inhibitory concentration of triclosan mediated by the nonspecific biocide mechanism occurs at μg/mL concentrations (Heath et al. 1999; Heath and Rock 2000; Parsons et al. 2011; Yao et al. 2013).

The widespread inclusion of triclosan in consumer products along with the environmental persistence of triclosan has created an interesting environment for resistance development. Triclosan resistance is well studied in *E. coli* and *S. aureus*, and many missense mutants have been isolated (Fig. 2B). In *E. coli*, resistance to triclosan is mediated through the G93V missense mutation in the *fabI* gene and by accelerating efflux through the increased expression of *marA*, *soxS*, *acrAB*, and *acrF*. Triclosan is 64 times less potent against the *E. coli* FabI(G93V) mutant strain compared with the wild-type strain (Heath et al. 1999). Structurally, the Cγ of the Gly93 residue is 3.9 Å away from the chlorine on the phenyl ring of triclosan (Levy et al. 1999; Stewart et al. 1999). Mutating the glycine to valine is predicted to introduce a

side chain that causes a steric clash with the chlorine atom in triclosan, decreasing the binding affinity. The FabI(G93V) mutation imposes a minor fitness cost to growth (Curiao et al. 2015). Elevated expression of the *marA*, *soxS*, *acrAB*, and *acrF* genes leads to increased efflux, which is a second mechanism for triclosan resistance in *E. coli* (McMurry et al. 1998b; Curiao et al. 2015). Increased expression of these genes did not impose an observable fitness cost to the strains (Curiao et al. 2015). The overexpression of efflux through these genes also causes coresistance against other antibiotics, including ampicillin, tetracyclines, and fluoroquinolones (Curiao et al. 2015). No mutations in the *fabI* gene are observed in triclosan-resistant *Klebsiella pneumoniae*, and resistance was mediated through the elevated expression of the *marA*, *soxS*, *acrAB*, and *acrF* efflux pumps (Curiao et al. 2015).

Methicillin-resistant *S. aureus* was a major clinical concern in the last decade leading to the development of triclosan-derived FabI inhibitors targeting *S. aureus* (Fig. 1B). Two inhibitors, MUT056399 (Escaich et al. 2011) and CG400549 (Park et al. 2007a), have entered clinical trials, and active research is being conducted on other triclosan analogs. High-level resistance against triclosan and triclosan-derived inhibitors in *S. aureus* is mediated by mutations in the *fabI* gene. Missense mutations give rise to G23S, F204C, Y147H, A198G, L208F, and A95V FabI variants that confer resistance to triclosan in *S. aureus* (Fan et al. 2002; Brenwald and Fraise 2003). The F204L mutation was identified as resistant to CG400549 (Park et al., 2007a), and the A95V, F204S, and I193F mutations led to resistance against MUT056399 (Escaich et al. 2011). Strains expressing A95V, I193S, and F204S FabI variants are also resistant to 5-ethyl-2-phenoxyphenol and 5-chloro-2-phenoxyphenol compounds (Xu et al. 2008). Phe204, Ala198, and Tyr147 interact with the triclosan molecule, whereas direct interactions with Ala95, Ile193, Leu208, and Gly23 are absent. In particular, Gly23 is located on the surface of the protein far from the active site (Fig. 2B). Because triclosan is a slow-binding inhibitor requiring a conformational change in FabI for tight binding, the mutations that do not directly contact the drug are likely to be modifiers of the protein conformational transition. Triclosan resistance in *Staphylococcus epidermidis* is mediated by similar mutations to the *fabI* gene, as well as the elevated expression of *fabI* through mutations in the putative promoter region (Skovgaard et al. 2013). However, resistance arising from alterations in gene expression are usually only a few-fold, and single-target antibiotics should be designed with sufficient potency to make these relatively minor genetic mechanisms for resistance irrelevant in the clinic.

Inhibitors Binding to the NAD(P)H–FabI Substrate Complex

Two inhibitors derived from the same scaffold, CG400462 (Park et al. 2007b) and AFN-1252 (Fig. 1C) (Karlowsky et al. 2009; Kaplan et al. 2012, 2013a), bind to the NAD(P)H–FabI substrate complex (Fig. 2C). The best studied of this class of inhibitors is AFN-1252, which has moved through phase II clinical trials. AFN-1252 was discovered through target-based screening and structure-based optimization against the *S. aureus* FabI and has no detectable mammalian toxicity (Payne et al. 2002; Karlowsky et al. 2007, 2009; Kaplan et al. 2012). Using structure-based design to optimize the inhibition of *S. aureus*, FabI achieves low ng/mL MICs against this pathogen (Kaplan et al. 2012). AFN-1252 is a fast on/off inhibitor that binds to the FabI–NADPH complex of *S. aureus* FabI with low nanomolar affinity (Yao et al. 2013). AFN-1252 treatment causes a dose-dependent decrease in fatty acid synthesis and an accumulation of medium-chain acyl-ACP (Parsons et al. 2011; Yao et al. 2013). Exogenous fatty acids cannot save *S. aureus* from AFN-1252 inhibition because the inhibition of de novo fatty acid synthesis ties up all the free ACP as acyl-ACP (Parsons et al. 2011). Extracellular fatty acids can bypass AFN-1252 inhibition only when de novo fatty acid synthesis is completely disabled using an acetyl-CoA carboxylase knockout strain (Parsons et al. 2011). The inability of acetyl-CoA carboxylase mutants to initiate fatty acid synthesis means that there are

no substrates for FabI, and free ACP is available for the uptake and incorporation of exogenous fatty acids. This result illustrates that AFN-1252 does not have another cellular target (Parsons et al. 2011; Yao et al. 2013). However, *S. aureus* with inactivating mutations in acetyl-CoA carboxylase have significantly decreased growth rates and do not proliferate in mice demonstrating that inactivating acetyl-CoA carboxylase is not a biologically viable mechanism to bypass FabI inhibition in *S. aureus* (Parsons et al. 2013a).

Laboratory-directed evolution has defined the mechanisms for the development of *S. aureus* mutants resistant to AFN-1252 (Fig 2C). Three separate studies found only two missense mutations in the *fabI* gene that give rise to a greater than fourfold increase in resistance against AFN-1252 (Parsons et al. 2011; Kaplan et al. 2012; Yao et al. 2013). The same mutations confer resistance to CG400462 (Park et al. 2007b). The Tyr147 residue interacts with the 3-methylbenzofuran portion of AFN-1252 (Kaplan et al. 2012), and the Y147H mutation increased the MIC from 3.9 ng/mL to 500 ng/mL (Fig. 2C) (Yao et al. 2013). However, Tyr147 is predicted to bind the thioester carboxyl of the substrate to stabilize the transition state, and FabI(Y147H) has a greater than 10-fold decrease in catalytic activity (Yao et al. 2013). This catalytic defect is reflected in a significantly decreased growth rate of strains expressing the FabI(Y147H) mutant. Strains expressing FabI(Y147H) are cross-resistant to triclosan, with the MIC increased to 500 ng/mL from 62.5 ng/mL in the wild-type strain. Met99 interacts with the oxotetrahydronaphthyridine portion of AFN-1252. FabI(M99T) bound to AFN-1252 with a 17-fold decrease in affinity compared with the wild-type enzyme, and the mutant strain increased the MIC from 3.9 ng/mL to 250 ng/mL. This mutant enzyme has no apparent catalytic deficiency in vitro and the mutant strain does not have a growth defect. The genes encoding FASII components are up-regulated in the FabI(M99T)-expressing mutant strains, illustrating that there is a small fitness cost. Saturation mutagenesis was conducted at the Met99 residue, and the mutation to threonine gave the highest level of resistance. Interestingly, FabI(M99T)-expressing strains are ~16-fold more susceptible to triclosan, although the Met99 side chain does not interact with triclosan. Strains carrying either the FabI(M99T) or FabI(Y147H) mutations have an MIC that is significantly lower than the plasma concentrations of AFN-1252 that are achieved in published murine infection models (Kaplan et al. 2013b; Parsons et al. 2013b). The fitness of these two mutants in an animal model is unknown. Attempts to select for additional mutations yielding higher levels of resistance against AFN-1252 starting from these two single-point mutants were unsuccessful (Yao et al. 2013). The FabI enzyme containing both the M99T and Y147H mutations had lower catalytic activity than the Y147H mutant alone and was unable to support growth. Experiments are needed to determine whether strains harboring FabI(M99T) protein are susceptible to the doses of AFN-1252 used in the standard therapy for *S. aureus* infections. Drug dosing at AFN-1252 levels required to eliminate strains with the FabI(M99T) mutation would mitigate any impact of the mutation on the therapeutic usefulness of the antibiotic.

CONCLUSIONS

The multitarget hypothesis argues that antibiotics designed against single gene targets will fail because the bacterial load is high enough in human infections to select for missense mutants that will render the antibiotic ineffective (Silver and Bostian 1993; Silver 2007, 2011). This sound reasoning argues against developing single-target inhibitors as monotherapies, but the experience with resistance profiles of FabI inhibitors suggests key drug properties that should be optimized to diminish the rapid evolution of clinical resistance in single-target inhibitors. First, it is critical that the drug have very high potency (low MIC). This high potency must be coupled with a low toxicity profile that allows clinical dosing at concentrations that exceed the amounts required to eradicate any missense mutations that may arise. The AFN-1252 case illustrates that these stringent criteria for

single-target drugs can be achieved by structure-aided design. However, drugs with this high potency will necessarily be pathogen specific. Creating a broad-spectrum antibiotic requires compromising the potency against a single species to achieve broader clinical utility. This process inevitably results in lower drug potency against all species within the antibiotic spectrum. This compromised potency increases the potential for single-step resistance to shift the antibiotic dose needed to cure the resistant strains to concentrations that cannot be achieved and/or tolerated. Second, the drug should be designed to minimize the number of interactions with residue side chains while preserving high-affinity binding (Fig. 2C). Each side-chain interaction provides an opportunity for a missense mutation to arise that will compromise drug binding. One potential exception would be those side chains that are important for catalysis. If the drug target is a rate-controlling step in the pathway, then missense mutations in catalytic residues will introduce a fitness cost to the cells. These considerations mean that the experimental evaluation of drug resistance must be an early and integral part of the antibiotic development processes rather than something that is evaluated at the end.

The analysis of the resistance profiles of FabI inhibitors also reveals a liability associated with slow-binding inhibitors. These inhibitors have slow off rates caused by protein conformation-dependent transitions that lead to tight drug binding. The improvements in potency and target selectivity provided by slow-binding inhibitors have been clearly articulated (Lu and Tonge 2010; Walkup et al. 2015); however, if the slow-binding conformational change is critical to drug efficacy, resistance mutations will arise at residue positions that are key to initiating and/or maintaining the desired conformational change. The experience with triclosan and isoniazid, two slow-binding FabI inhibitors, reveals this vulnerability (Fig. 2A,B). Of the triclosan-resistant missense mutations discussed in this review, three of seven occur at residues with side chains that contact the drug, whereas the other four are found at other locations. Many isoniazid-resistant InhA mutations (three of six) are located in positions that do not directly contact the drug. These mutations are located in protein sequences that are predicted to participate in the conformational change responsible for its slow-binding characteristics. The spectrum of resistance mutations to isoniazid and triclosan is in sharp contrast to the number of mutations conferring AFN-1252 resistance. AFN-1252 is a fast, tight-binding inhibitor, and the only mutations that have been detected are in the two residues with side chains that contact the drug (Yao et al. 2013). Thus, advantages gained in optimizing the slow-binding characteristics of an antibiotic are offset by the higher number of missense mutations that can arise with the potential to limit the clinical usefulness of the drug.

Recent advances in human gut microbiome research suggest that taking a pathogen selective approach has merit. Broad-spectrum antibiotic treatment causes severe disturbances to the human microbiome and has been linked to both short-term health consequences such as reduced infection resistance, as well as long-term health consequences such as the development of allergic and metabolic diseases (Blaser and Falkow 2009; Croswell et al. 2009). In particular, broad-spectrum antibiotic treatment early in life is highly associated with the development of type 2 diabetes (Boursi et al. 2015), obesity (Cho et al. 2012), and celiac disease later in life (Mårild et al. 2013). One of the most difficult-to-treat infections, *Clostridium difficile* colitis, also arises once treatment with broad-spectrum antibiotics eliminates the gut microbiome (Buffie et al. 2012). In this respect, FabI is a fascinating target. Broad-spectrum antibiotics cause a several-thousand-fold reduction in the gut microbial load (Croswell et al. 2009). Many important species of the gut microbiome (Firmicutes) are predicted to be immune to FabI inhibitors because they encode for a different enoyl-ACP reductase isoform (FabK) (Yao and Rock 2015). A number of pathogens with significant medical impact do encode a FabI, including *Neisseria*, various bacteria from the class Gammaproteobacteria (*Enterobacteriaceae, Pseudomonas, Salmonella, Shigella, Acinetobacter*), *Campylobacter, S. aureus*, and *Mycobacterium* species. It will be important to experimentally test the prediction that FabI inhibitors will have minimal impact on the gut microbiome, but the deployment of potent, pathogen-selective antibiotics is clearly predicted to minimize collateral damage to the host microbiome.

ACKNOWLEDGMENTS

This work is supported by National Institutes of Health Grants GM034496 (C.O.R.), Cancer Center Support Grant CA21765, and the American Lebanese Syrian Associated Charities. A portion of our laboratory research is funded by Debiopharm Group.

REFERENCES

Bernstein J, Lott WA, Steinberg BA, Yale HL. 1952. Chemotherapy of experimental tuberculosis. V: Isonicotinic acid hydrazide (nydrazid) and related compounds. *Am Rev Tuberc* **65:** 357–364.

Blaser MJ, Falkow S. 2009. What are the consequences of the disappearing human microbiota? *Nat Rev Micro* **7:** 887–894.

Boursi B, Mamtani R, Haynes K, Yang YX. 2015. The effect of past antibiotic exposure on diabetes risk. *Eur J Endocrinol* **172:** 639–648.

Brenwald NP, Fraise AP. 2003. Triclosan resistance in methicillin-resistant *Staphylococcus aureus* (MRSA). *J Hosp Infect* **55:** 141–144.

Brinster S, Lamberet G, Staels B, Trieu-Cuot P, Gruss A, Poyart C. 2009. Type II fatty acid synthesis is not a suitable antibiotic target for Gram-positive pathogens. *Nature* **458:** 83–86.

Buffie CG, Jarchum I, Equinda M, Lipuma L, Gobourne A, Viale A, Ubeda C, Xavier J, Pamer EG. 2012. Profound alterations of intestinal microbiota following a single dose of clindamycin results in sustained susceptibility to *Clostridium difficile*-induced colitis. *Infect Immun* **80:** 62–73.

Campbell JW, Cronan JE Jr. 2001. Bacterial fatty acid biosynthesis: Targets for antibacterial drug discovery. *Annu Rev Microbiol* **55:** 305–332.

Cho I, Yamanishi S, Cox L, Methe BA, Zavadil J, Li K, Gao Z, Mahana D, Raju K, Teitler I, et al. 2012. Antibiotics in early life alter the murine colonic microbiome and adiposity. *Nature* **488:** 621–626.

Croswell A, Amir E, Teggatz P, Barman M, Salzman NH. 2009. Prolonged impact of antibiotics on intestinal microbial ecology and susceptibility to enteric *Salmonella* infection. *Infect Immun* **77:** 2741–2753.

Curiao T, Marchi E, Viti C, Oggioni MR, Baquero F, Martinez JL, Coque TM. 2015. Polymorphic variation in susceptibility and metabolism of triclosan-resistant mutants of *Escherichia coli* and *Klebsiella pneumoniae* clinical strains obtained after exposure to biocides and antibiotics. *Antimicrobiol Agents Chem* **59:** 3413–3423.

David HL. 1970. Probability distribution of drug-resistant mutants in unselected populations of *Mycobacterium tuberculosis*. *Appl Microbiol* **20**: 810–814.

Escaich S, Prouvensier L, Saccomani M, Durant L, Oxoby M, Gerusz V, Moreau F, Vongsouthi V, Maher K, Morrissey I, et al. 2011. The MUT056399 inhibitor of FabI is a new antistaphylococcal compound. *Antimicrob Agents Chemother* **55**: 4692–4697.

Fan F, Yan K, Wallis NG, Reed S, Moore TD, Rittenhouse SF, DeWolf JW Jr, Huang J, McDevitt D, Miller WH, et al. 2002. Defining and combating the mechanisms of triclosan resistance in clinical isolates of *Staphylococcus aureus*. *Antimicrob Agents Chemother* **46**: 3343–3347.

Hazbón MH, Brimacombe M, Bobadilla del Valle M, Cavatore M, Guerrero MI, Varma-Basil M, Billman-Jacobe H, Lavender C, Fyfe J, García-García L, et al. 2006. Population genetics study of Isoniazid resistance mutations and evolution of multidrug-resistant *Mycobacterium tuberculosis*. *Antimicrobiol Agents Chem* **50**: 2640–2649.

Heath RJ, Rock CO. 1995. Enoyl-acyl carrier protein reductase (*fabI*) plays a determinant role in completing cycles of fatty acid elongation in *Escherichia coli*. *J Biol Chem* **270**: 26538–26542.

Heath RJ, Rock CO. 1996. Regulation of fatty acid elongation and initiation by acyl–acyl carrier protein in *Escherichia coli*. *J Biol Chem* **271**: 1833–1836.

Heath RJ, Rock CO. 2000. A triclosan-resistant bacterial enzyme. *Nature* **406**: 145–146.

Heath RJ, Yu YT, Shapiro MA, Olson E, Rock CO. 1998. Broad spectrum antimicrobial biocides target the FabI component of fatty acid synthesis. *J Biol Chem* **273**: 30316–30321.

Heath RJ, Rubin JR, Holland DR, Zhang E, Snow ME, Rock CO. 1999. Mechanism of triclosan inhibition of bacterial fatty acid synthesis. *J Biol Chem* **274**: 11110–11114.

Heath RJ, Su N, Murphy CK, Rock CO. 2000. The enoyl-[acyl-carrier-protein] reductases FabI and FabL from *Bacillus subtilis*. *J Biol Chem* **275**: 40128–40133.

Heath RJ, White SW, Rock CO. 2001. Lipid biosynthesis as a target for antibacterial agents. *Prog Lipid Res* **40**: 467–497.

Kaplan N, Albert M, Awrey D, Bardouniotis E, Berman J, Clarke T, Dorsey M, Hafkin B, Ramnauth J, Romanov V, et al. 2012. Mode of action, in vitro activity, and in vivo efficacy of AFN-1252, a selective antistaphylococcal FabI inhibitor. *Antimicrob Agents Chemother* **56**: 5865–5874.

Kaplan N, Awrey D, Bardouniotis E, Berman J, Yethon J, Pauls HW, Hafkin B. 2013a. In vitro activity (MICs and rate of kill) of AFN-1252, a novel FabI inhibitor, in the presence of serum and in combination with other antibiotics. *J Chemother* **25**: 18–25.

Kaplan N, Garner C, Hafkin B. 2013b. AFN-1252 in vitro absorption studies and pharmacokinetics following microdosing in healthy subjects. *Eur J Pharm Sci* **50**: 440–446.

Karlowsky JA, Laing NM, Baudry T, Kaplan N, Vaughan D, Hoban DJ, Zhanel GG. 2007. In vitro activity of API-1252, a novel FabI inhibitor, against clinical isolates of *Staphylococcus aureus* and *Staphylococcus epidermidis*. *Antimicrob Agents Chemother* **51**: 1580–1581.

Karlowsky JA, Kaplan N, Hafkin B, Hoban DJ, Zhanel GG. 2009. AFN-1252, a FabI inhibitor, demonstrates a *Staphylococcus*-specific spectrum of activity. *Antimicrob Agents Chemother* **53**: 3544–3548.

Lee JH, Ammerman NC, Nolan S, Geiman DE, Lun S, Guo H, Bishai WR. 2012. Isoniazid resistance without a loss of fitness in *Mycobacterium tuberculosis*. *Nat Commun* **3**: 753.

Levy CW, Roujeinikova A, Sedelnikova S, Baker PJ, Stuitje AR, Slabas AR, Rice DW, Rafferty JB. 1999. Molecular basis of triclosan activity. *Nature* **398**: 383–384.

Lu H, Tonge PJ. 2010. Drug-target residence time: Critical information for lead optimization. *Curr Opin Chem Biol* **14**: 467–474.

Mårild K, Ye W, Lebwohl B, Green PH, Blaser MJ, Card T, Ludvigsson JF. 2013. Antibiotic exposure and the development of coeliac disease: A nationwide case-control study. *BMC Gastroenterol* **13**: 109.

Marrakchi H, DeWolf C Jr, Quinn C, West J, Polizzi BJ, So CY, Holmes DJ, Reed SL, Heath RJ, Payne DJ, et al. 2003. Characterization of *Streptococcus pneumoniae* enoyl-[acyl carrier protein] reductase (FabK). *Biochem J* **370**: 1055–1062.

Marrakchi H, Lanéelle MA, Daffé M. 2014. Mycolic acids: Structures, biosynthesis, and beyond. *Chem Biol* **21**: 67–85.

Massengo-Tiasse RP, Cronan JE. 2008. *Vibrio cholerae fabV* defines a new class of enoyl acyl-carrier-protein reductase. *J Biol Chem* **283**: 1308–1316.

McMurry LM, Oethinger M, Levy S. 1998a. Triclosan targets lipid synthesis. *Nature* **394**: 531–532.

McMurry LM, Oethinger M, Levy SB. 1998b. Overexpression of *marA*, *soxS*, or *acrAB* produces resistance to triclosan in laboratory and clinical strains of *Escherichia coli*. *FEMS Microbiol Lett* **166**: 305–309.

Meyerovich M, Mamou G, Ben-Yehuda S. 2010. Visualizing high error levels during gene expression in living bacterial cells. *Proc Natl Acad Sci* **107**: 11543–11548.

O'Sullivan DM, McHugh TD, Gillespie SH. 2010. Mapping the fitness of *Mycobacterium tuberculosis* strains: A complex picture. *J Med Microbiol* **59**: 1533–1535.

Park HS, Yoon YM, Jung SJ, Kim CM, Kim JM, Kwak JH. 2007a. Antistaphylococcal activities of CG400549, a new bacterial enoyl-acyl carrier protein reductase (FabI) inhibitor. *J Antimicrob Chemother* **60**: 568–574.

Park HS, Yoon YM, Jung SJ, Yun IN, Kim CM, Kim JM, Kwak JH. 2007b. CG400462, a new bacterial enoyl-acyl carrier protein reductase (FabI) inhibitor. *Int J Antimicrob Agents* **30**: 446–451.

Parsons JB, Rock CO. 2011. Is bacterial fatty acid synthesis a valid target for antibacterial drug discovery? *Curr Opin Microbiol* **14**: 544–549.

Parsons JB, Rock CO. 2013. Bacterial lipids: Metabolism and membrane homeostasis. *Prog Lipid Res* **52**: 249–276.

Parsons JB, Frank MW, Subramanian C, Saenkham P, Rock CO. 2011. Metabolic basis for the differential susceptibility of Gram-positive pathogens to fatty acid synthesis inhibitors. *Proc Natl Acad Sci* **108**: 15378–15383.

Parsons JB, Frank MW, Rosch JW, Rock CO. 2013a. *Staphylococcus aureus* fatty acid auxotrophs do not proliferate in mice. *Antimicrob Agents Chemother* **57**: 5729–5732.

Cite this article as *Cold Spring Harb Perspect Med* doi: 10.1101/cshperspect.a027045

Parsons JB, Kukula M, Jackson P, Pulse M, Simecka JW, Valtierra D, Weiss WJ, Kaplan N, Rock CO. 2013b. Perturbation of *Staphylococcus aureus* gene expression by the enoyl-acyl carrier protein reductase inhibitor AFN-1252. *Antimicrob Agents Chemother* **57:** 2182–2190.

Payne DJ, Miller WH, Berry V, Brosky J, Burgess WJ, Chen E, DeWolf JW Jr, Fosberry AP, Greenwood R, Head MS, et al. 2002. Discovery of a novel and potent class of FabI-directed antibacterial agents. *Antimicrob Agents Chemother* **46:** 3118–3124.

Pinto L, Menzies D. 2011. Treatment of drug-resistant tuberculosis. *Infect Drug Resist* **4:** 129–135.

Quemard A, Sacchettini JC, Dessen A, Vilchèze C, Bittman R, Jacobs WR, Blanchard JS. 1995. Enzymic characterization of the target for Isoniazid in *Mycobacterium tuberculosis*. *Biochemistry* **34:** 8235–8241.

Rawat R, Whitty A, Tonge PJ. 2003. The isoniazid-NAD adduct is a slow, tight-binding inhibitor of InhA, the *Mycobacterium tuberculosis* enoyl reductase: Adduct affinity and drug resistance. *Proc Natl Acad Sci* **100:** 13881–13886.

Rock CO, Jackowski S. 2002. Forty years of fatty acid biosynthesis. *Biochem Biophys Res Commun* **292:** 1155–1166.

Rosner JL, Storz G. 1994. Effects of peroxides on susceptibilities of *Escherichia coli* and *Mycobacterium smegmatis* to isoniazid. *Antimicrobiol Agents Chem* **38:** 1829–1833.

Russell AD. 2003. Similarities and differences in the responses of microorganisms to biocides. *J Antimicrob Chemother* **52:** 750–763.

Russell AD. 2004. Whither triclosan? *J Antimicrob Chemother* **53:** 693–695.

Seifert M, Catanzaro D, Catanzaro A, Rodwell TC. 2015. Genetic mutations associated with isoniazid resistance in *Mycobacterium tuberculosis*: A systematic review. *PLoS ONE* **10:** e0119628.

Silver LL. 2007. Multi-targeting by monotherapeutic antibacterials. *Nat Rev Drug Discov* **6:** 41–55.

Silver LL. 2011. Challenges of antibacterial discovery. *Clin Microbiol Rev* **24:** 71–109.

Silver LL, Bostian KA. 1993. Discovery and development of new antibiotics: The problem of antibiotic resistance. *Antimicrob Agents Chemother* **37:** 377–383.

Skovgaard S, Nielsen LN, Larsen MH, Skov RL, Ingmer H, Westh H. 2013. *Staphylococcus epidermidis* isolated in 1965 are more susceptible to Triclosan than current isolates. *PLoS ONE* **8:** e62197.

Sreevatsan S, Pan X, Zhang Y, Deretic V, Musser JM. 1997. Analysis of the *oxyR-ahpC* region in isoniazid-resistant and -susceptible *Mycobacterium tuberculosis* complex organisms recovered from diseased humans and animals in diverse localities. *Antimicrobiol Agents Chem* **41:** 600–606.

Stewart MJ, Parikh S, Xiao G, Tonge PJ, Kisker C. 1999. Structural basis and mechanism of enoyl reductase inhibition by triclosan. *J Mol Biol* **290:** 859–865.

van Soolingen D, de Haas PEW, van Doorn HR, Kuijper E, Rinder H, Borgdorff MW. 2000. Mutations at amino acid position 315 of the *katG* gene are associated with high-level resistance to Isoniazid, other drug resistance, and successful transmission of *Mycobacterium tuberculosis* in the Netherlands. *J Infect Dis* **182:** 1788–1790.

Vilchèze C, Jacobs WR Jr. 2007. The mechanism of Isoniazid killing: Clarity through the scope of genetics. *Annu Rev Microbiol* **61:** 35–50.

Vilchèze C, Weisbrod TR, Chen B, Kremer L, Hazbón MH, Wang F, Alland D, Sacchettini JC, Jacobs WR. 2005. Altered NADH/NAD$^+$ ratio mediates coresistance to isoniazid and ethionamide in mycobacteria. *Antimicrobiol Agents Chem* **49:** 708–720.

Vilchèze C, Wang F, Arai M, Hazbon MH, Colangeli R, Kremer L, Weisbrod TR, Alland D, Sacchettini JC, Jacobs WR. 2006. Transfer of a point mutation in *Mycobacterium tuberculosis inhA* resolves the target of isoniazid. *Nat Med* **12:** 1027–1029.

Walkup GK, You Z, Ross PL, Allen EK, Daryaee F, Hale MR, O'Donnell J, Ehmann DE, Schuck VJ, Buurman ET, et al. 2015. Translating slow-binding inhibition kinetics into cellular and in vivo effects. *Nat Chem Biol* **11:** 416–423.

Wengenack NL, Todorovic S, Yu L, Rusnak F. 1998. Evidence for differential binding of isoniazid by *Mycobacterium tuberculosis* KatG and the isoniazid-resistant mutant KatG(S315T). *Biochemistry* **37:** 15825–15834.

Xu H, Sullivan TJ, Sekiguchi J, Kirikae T, Ojima I, Stratton CF, Mao W, Rock FL, Alley MR, Johnson F, et al. 2008. Mechanism and inhibition of saFabI, the enoyl reductase from *Staphylococcus aureus*. *Biochemistry* **47:** 4228–4236.

Yao J, Rock CO. 2015. How bacterial pathogens eat host lipids: Implications for the development of fatty acid synthesis therapeutics. *J Biol Chem* **290:** 5940–5946.

Yao J, Maxwell JB, Rock CO. 2013. Resistance to AFN-1252 arises from missense mutations in *Staphylococcus aureus* enoyl-acyl carrier protein reductase (FabI). *J Biol Chem* **288:** 36261–36271.

Zhang YM, Rock CO. 2008. Membrane lipid homeostasis in bacteria. *Nat Rev Microbiol* **6:** 222–233.

Zhang YM, White SW, Rock CO. 2006. Inhibiting bacterial fatty acid synthesis. *J Biol Chem* **281:** 17541–17544.

Zhu L, Lin J, Ma J, Cronan JE, Wang H. 2010. Triclosan resistance of *Pseudomonas aeruginosa* PAO1 is due to FabV, a triclosan-resistant enoyl-acyl carrier protein reductase. *Antimicrobiol Agents Chem* **54:** 689–698.

Antibacterial Drug Discovery Targeting the Lipopolysaccharide Biosynthetic Enzyme LpxC

Alice L. Erwin

Erwin Consulting, Ruston, Washington 98407

Correspondence: alice.erwin@rcn.com

The enzyme LpxC (UDP-3-*O*-(*R*-3-hydroxymyristoyl)-*N*-acetylglucosamine deacetylase) is broadly conserved across Gram-negative bacteria and is essential for synthesis of lipid A, the membrane anchor of the lipopolysaccharides (LPSs), which are a major component of the outer membrane in nearly all Gram-negative bacteria. LpxC has been the focus of target-directed antibiotic discovery projects in numerous pharmaceutical and academic groups for more than 20 years. Despite intense effort, no LpxC inhibitor has been approved for therapeutic use, and only one has yet reached human studies. This article will summarize the history of LpxC as a drug target and the parallel history of research on LpxC biology. Both academic and industrial researchers have used LpxC inhibitors as tool compounds, leading to increased understanding of the differing mechanisms for regulation of LPS synthesis in *Escherichia coli* and *Pseudomonas aeruginosa*.

RAETZ AND THE LIPID A BIOSYNTHETIC PATHWAY

Our current understanding of the Gram-negative outer membrane is as being physically different from a cytoplasmic membrane and constituting a permeability barrier developed during the 1960s and 1970s (Leive 1974). Research on the chemical structure of lipopolysaccharide (LPS) and its synthesis was being conducted at the same time. A general outline of LPS synthesis and assembly was complete by the mid-1970s (Osborn et al. 1972). At that time, the chemical structure of lipid A was still uncertain, and none of the enzymes involved in its synthesis had been identified. The LPS transporter MsbA and the Lpt export sys-tem were not discovered until much later (Ruiz et al. 2009).

The biochemical pathway for synthesis of lipid A was determined by Christian H.R. Raetz almost singlehandedly (Raetz 1993). Each of the enzymes was discovered by research groups he directed, working successively at the University of Wisconsin (1976–1987), Merck Research Laboratories (Rahway, NJ) (1987–1993), and Duke University (1993–2011) (Dowhan 2011; Kresge et al. 2011). During the time Raetz was at Merck, his group identified the deacetylase now known as LpxC as the first committed step in lipid A synthesis (Fig. 1) (Anderson et al. 1993). Purification of the enzyme catalyzing this activity led to the recognition that it was

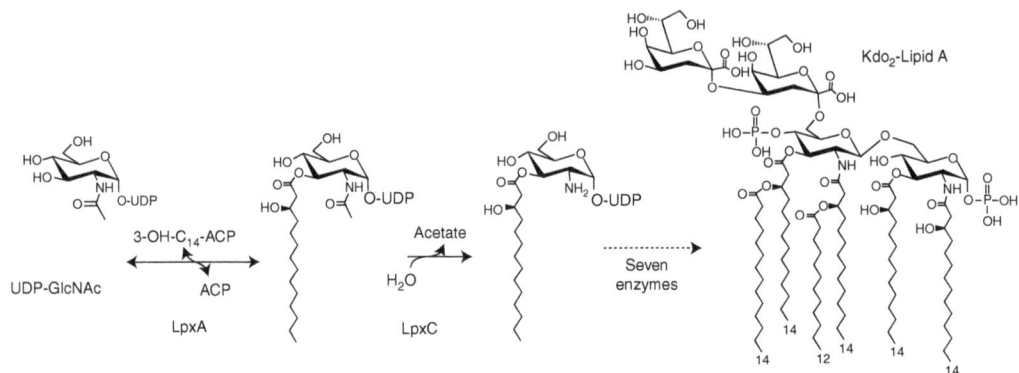

Figure 1. Lipid A biosynthesis. (Reprinted, with permission, from Mdluli et al. 2006.)

encoded by the genetic locus previously known as *envA* (Young et al. 1995).

EARLY LpxC INHIBITORS DISCOVERED IN BACTERIAL CELL SCREENS

The history of LpxC as a drug target began in the mid-1980s, before the discovery of the enzyme itself. The antibacterial discovery group at Merck Research Laboratories used a *galE* mutant of *Salmonella* to screen a library of chemical compounds, measuring LPS synthesis by monitoring incorporation of radiolabeled galactose into bacterial macromolecules. Among the compounds identified in the screen was the small oxazoline hydroxamic acid L-573,655 (Fig. 2), which had a minimum inhibitory concentration (MIC) for wild-type *Escherichia coli* of 200–400 µg/mL. This molecule was later found to be an inhibitor of LpxC, with an IC_{50} for the *E. coli* enzyme of 8.5 µM. Approximately 200 analogs were synthesized, increasing potency ~100-fold. Antibacterial activity improved in parallel. The most active compound, L-161,140, had an IC_{50} of 0.03 µM and an MIC for wild-type *E. coli* of 1–3 µg/mL. None of these compounds was active against *Pseudomonas aeruginosa*. This was a critical issue because at the time it was believed that the market for a Gram-negative antibiotic would be very limited unless it was active against *P. aeruginosa* as well as enteric bacteria. Recognition that broad Gram-negative activity might be very difficult to achieve led to

termination of this first generation of LpxC chemistry and subsequent publication of the work (Onishi et al. 1996; Chen et al. 1999).

British Biotech screened a library of metalloenzyme inhibitors for antibacterial activity using *E. coli* strain D22, an *envA1* mutant. Strains carrying the point mutation *envA1* (H19Y) have a defective envelope in which the outer membrane is unusually permeable to solvents and other hydrophobic compounds, conferring hypersensitivity to many antibiotics (Normark et al. 1969; Beall and Lutkenhaus 1987). Following the recognition that *envA* gene encodes LpxC, it had been shown that an *envA1* mutant has an 18-fold reduction in LpxC activity, compared with wild-type strains (Young et al. 1995).

The partial loss of LpxC function conferred by the *envA1* mutation would be expected to make the strain particularly sensitive to LpxC inhibitors, and the general hypersensitivity of the strain would increase the chances of identifying inhibitors of other targets. Two related compounds identified in this screen were found to be inhibitors of LpxC. Like the Merck series and, indeed, all potent LpxC inhibitors that have been described, both compounds are hydroxamic acid derivatives. The more active of the two, BB-78485, has an IC_{50} of 160 nM versus the purified *E. coli* LpxC enzyme and an MIC of 1 µg/mL for *E. coli*. It was active against a wide variety of other Gram-negative species (MIC 2–4 µg/mL), with the exception of *P. aeruginosa* (MIC >32 µg/mL for ATCC 27853; 4 µg/mL

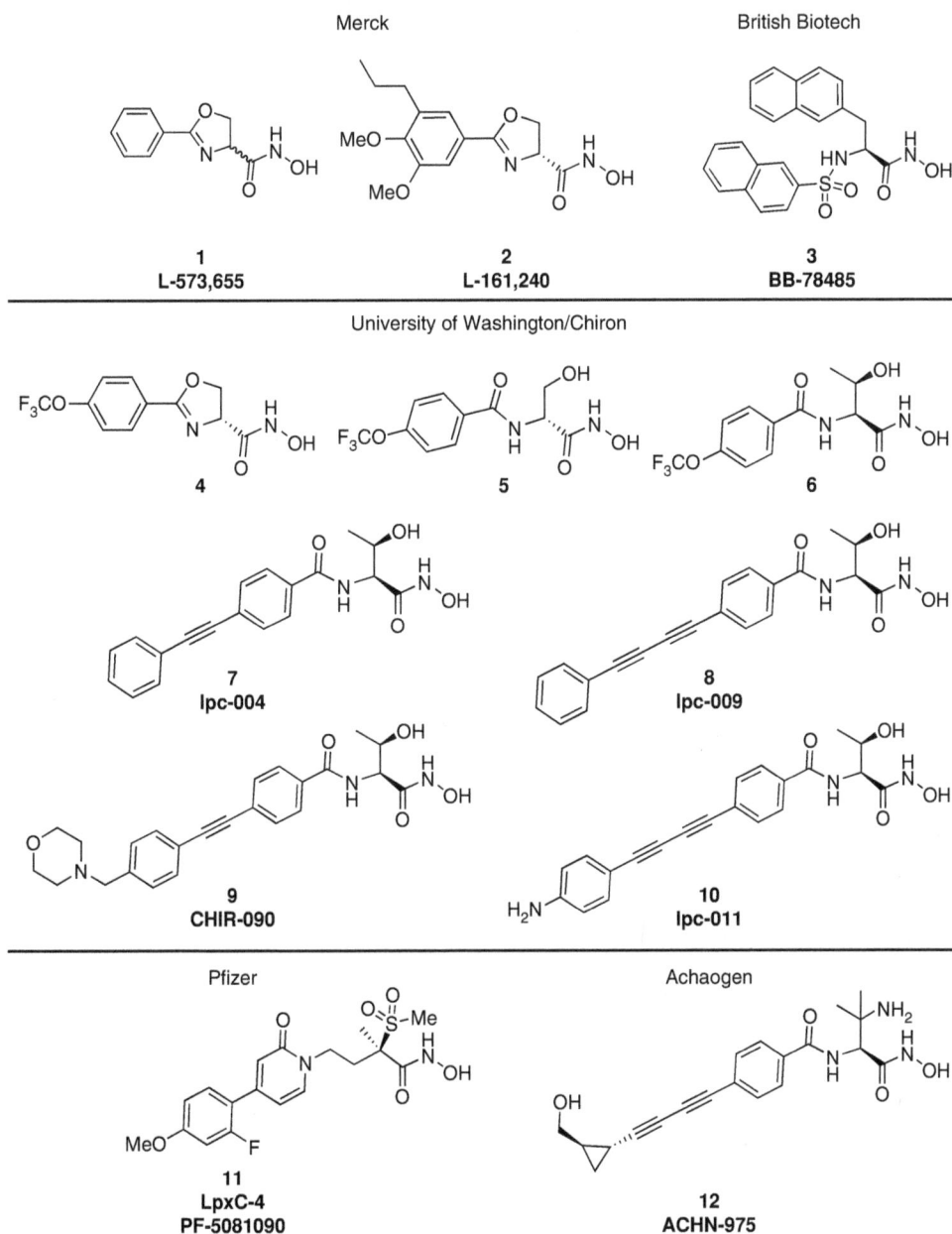

Figure 2. LpxC inhibitors. Merck: L-573,655 (compound 1) and L-161,240 (compound 2) (Onishi et al. 1996; Chen et al. 1999). British Biotech (Oxford): BB-78485 (compound 3) (Clements et al. 2002). University of Washington (UW)/Chiron (Emeryville, CA): compounds 4 and 5, previously designated 26 and 69 (Kline et al. 2002); compounds 6–10 (Andersen et al. 2011). In publications characterizing UW/Chiron compounds 7–10, they have been designated Lpc-004, CHIR-090, Lpc-009, and Lpc-011, respectively (McClerren et al. 2005; Lee et al. 2011; Liang et al. 2011). Pfizer (New York): LpxC-4 (PF-5081090) (compound 11) (Tomaras et al. 2014), previously compound 17-v (Montgomery et al. 2012), and PF1090 (Bulitta et al. 2011). Achaogen (South San Francisco): ACHN-975 (compound 12) (Kasar et al. 2012; Serio et al. 2013).

for a "leaky" strain, C53). As expected for inhibitors of LPS synthesis, the compounds had little or no Gram-positive activity (MIC for *Staphylococcus aureus* 32 or >32 µg/mL) (Clements et al. 2002).

FOCUS ON INHIBITION OF *P. aeruginosa* ENZYME LED TO DISCOVERY OF BROAD-SPECTRUM LpxC INHIBITORS

The first LpxC inhibitors able to inhibit the growth of *P. aeruginosa* were discovered by researchers from the University of Washington (UW) and Chiron, in a medicinal chemistry program funded by the Cystic Fibrosis Foundation (Andersen et al. 2011). Compounds were evaluated in an in vitro enzyme assay using LpxC from *P. aeruginosa*, rather than using the *E. coli* enzyme as in other early projects (Onishi et al. 1996; Raju et al. 2004). This strategy was based on the unexpected finding, discussed further below, that the reason L-161,240 does not inhibit growth of *P. aeruginosa* is that it is a poor inhibitor of the *P. aeruginosa* enzyme (Mdluli et al. 2006).

Approximately 1200 compounds were synthesized, of which the most active had MICs under 1 µg/mL for both *P. aeruginosa* and *E. coli*. Several compounds with MICs of 3 µg/mL or less were found to be efficacious in mouse models of systemic infection, with $ED_{50}s$ of 10 to 50 mg/kg for *P. aeruginosa* and 1.2 to 10 mg/kg for *E. coli*. Chiron terminated its antibacterial discovery program in early 2003, and data on the UW/Chiron LpxC compounds were presented at two conferences later that year (Anderson 2003; Erwin 2003).

EXPANSION OF PHARMACEUTICAL AND ACADEMIC LpxC RESEARCH

The reports of LpxC inhibitors active against *P. aeruginosa* led to initiation of LpxC programs at numerous companies. Most of these have not been described in the scientific literature, and public knowledge is available only through patent applications.

Between 2004 and 2013, patent applications claiming LpxC inhibitors were filed by numerous pharmaceutical companies, including Achaogen (South San Francisco, CA), Actelion Pharmaceuticals (Alschwil, Switzerland), AstraZeneca AB (Södertälje, Sweden), Novartis (Basel, Switzerland), Pfizer (New York), Schering Corporation (Kenilworth, NJ), Taisho Pharmaceuticals (Tokyo), and Vicuron Pharmaceuticals (New York) (Takashima et al. 2008; Benenato et al. 2010; Jain et al. 2011; Mansoor et al. 2011a; Kasar et al. 2012; Fu et al. 2014; Gauvin et al. 2015). Other companies have had varying levels of effort on LpxC programs that have not (yet) led to patent applications or to publications. The status of most of these programs is not known.

Medicinal chemistry was aided by LpxC biological research in both academic and industrial laboratories. In 2005, the Raetz laboratory and the University of Washington published the first report on the LpxC inhibitor they designated CHIR-090 (Fig. 2), identifying it as one of the most active UW/Chiron compounds (McClerren et al. 2005). This molecule, along with L-161,140 and BB-78485, was widely used as a tool compound in both academic and industrial laboratories for studies of LpxC enzymology, structural biology, and microbiology.

ADVANCING TOWARD CLINICAL CANDIDATES

From the limited biological data provided in patent applications, it does not appear that antibacterial activity per se is the major barrier to development of LpxC inhibitors as drugs. It is now routine to make compounds with in vitro antibacterial activity of MIC 1 µg/mL or less for both *E. coli* and *P. aeruginosa*. Data from both Achaogen and Pfizer (discussed below) show that it is possible to achieve good coverage (low MIC_{90}) of both these species and of additional species associated with nosocomial infections or with cystic fibrosis. The sparse information on activity for other bacterial species suggests that there is interest in developing LpxC inhibitors that could be used for gonorrhea and for infections with biothreat agents, such as *Francisella tularensis* (Zhou et al. 2015b). There is some variation in antibacterial spectrum from one chemical series to another. For example, the iso-xazole series from Pfizer includes compounds that are more active against *P. aeruginosa* than *E. coli* (Abramite et al. 2014).

Achieving antibacterial activity against *Acinetobacter baumannii* has been a challenge. *A. baumannii* is one of very few Gram-negative bacterial species for which LPS synthesis is not essential. Mutants in which *lpxA*, *lpxC*, or *lpxD* is inactivated have been found to be viable (Moffatt et al. 2010). This initially surprising observation has led some to conclude that even very potent inhibitors of *A. baumannii* LpxC would not inhibit bacterial growth. This does not appear to be the case. Although three compounds described by Pfizer described are far less active for this species (MIC$_{90}$ 32 μg/mL or higher) than for *E. coli* and *P. aeruginosa*, this is consistent with their poor in vitro potency against the *A. baumannii* enzyme (e.g., IC$_{50}$ of 183 nM for PF-5081090) (Tomaras et al. 2014). Some of the other compounds in the same chemical series have MICs for the *A. baumannii* reference strain of 4–16 μg/mL (Brown et al. 2014a). More recently, patent applications from Actelion described compounds that are as active against *A. baumannii* as against *P. aeruginosa*, many with MIC 1–8 μg/mL or lower (Gauvin et al. 2015; Hubschwerlen et al. 2015).

Apart from antibacterial activity, there is very little information on the limitations of LpxC inhibitors with regard to other characteristics required of drug candidates, such as pharmacokinetic or toxicology profile. All of the LpxC inhibitors that have been described are hydroxamic acid derivatives, raising concerns about the toxicity that could result if mammalian metalloenzymes are also inhibited. However, several series have been shown to be tolerated in mice well enough to allow evaluation of efficacy in experimental infection. The UW/Chiron compounds have limited solubility, high protein binding, and poor pharmacokinetics. A series of publications from Pfizer described the discovery of the pyridone series through a systematic effort to improve solubility and protein binding (Brown et al. 2012; McAllister et al. 2012; Montgomery et

al. 2012). This strategy led to the advanced compound PF-5081090.

ACHN-975 AND PF-5081090

The two most advanced compounds for which extensive microbiological characterization has been published are ACHN-975 from Achaogen (Kasar et al. 2012; Serio et al. 2013) and PF-5081090 from Pfizer (Brown et al. 2014a; Tomaras et al. 2014). Data for these compounds are summarized in Table 1, and the chemical structures are shown in Figure 2. Both have good in vitro activity against *P. aeruginosa* and *E. coli*, with MIC$_{90}$ of 1 μg/mL or less. Although both compounds are active against a variety of Gram-negative nosocomial pathogens (with the notable exception of *A. baumannii*, discussed above), they differ with respect to the cystic fibrosis pathogens. PF-5081090 has good activity against *Burkholderia cepacia* and *Stenotrophomonas maltophilia*, whereas ACHN-975 is essentially inactive against these species (Badal et al. 2013). Little is known about LpxC from either of these species, apart from a report that isolates of the *B. cepacia* complex vary widely in susceptibility to CHIR-090 (Bodewits et al. 2010). The reason for this is not known; it is possible that efflux pumps in these isolates differ in substrate specificity and that PF-5081090 is less subject to efflux by CF pathogens than is ACHN-975.

Both PF-5081090 and ACHN-975 were reported to be bactericidal and to be efficacious in mouse models of infection. For ACHN-975, cidality was described as concentration-dependent against *P. aeruginosa* and time-dependent against *E. coli* and *Klebsiella pneumoniae* (Serio et al. 2013). This was supported by the results of dose-fractionation studies in neutropenic thigh infection with each of these species (Reyes et al. 2013). PF-5081090 was reported to be efficacious in septicemia and neutropenic thigh infection models with both *P. aeruginosa* and *K. pneumoniae* and in a *P. aeruginosa* lung infection model in neutropenic mice (Tomaras et al. 2014).

No toxicology has been reported for PF-5081090. ACHN-975 was reported to induce bradycardia in mice (Bornheim et al. 2013). Phase I evaluation in humans was initiated in

Table 1. Comparison of advanced compounds PF-5081090 and ACHN-975 with CHIR-090

		CHIR-090	PF-5081090	ACHN-975
		9	11	12
Enzyme IC$_{50}$, nM				
P. aeruginosa		<2.1	1.1	0.05
K. pneumoniae		N.D.	0.069	N.D.
A. baumannii		N.D.	183	N.D.
Antibacterial activity, MIC or MIC$_{90}$ (range), μg/mL [number of isolates]				
P. aeruginosa	ATCC 27853	N.D.	N.D.	0.2
	PAO1	1	0.5	0.06
	MIC$_{90}$	4 [138]	1 [138]	0.5 (0.008–0.5) [100]
E. coli	ATCC 25922	N.D.	N.D.	0.125
	MIC$_{90}$	0.25 [79]	0.25 [79]	0.5 (0.03–2) [100]
K. pneumoniae	ATCC 43816	N.D.	1	0.5
	MIC$_{90}$	N.D.	1 [98]	2 (0.25–4) [113]
A. baumannii	MIC$_{90}$	>64 [31]	>64 [31]	>64 (4–64) [28]
B. cepacia	MIC$_{90}$	>64 [30]	0.5 [30]	16 (≤0.03–16) [26]
S. maltophilia	MIC$_{90}$	>64 [30]	2 [30]	>16 (16–16) [26]

Data for CHIR-090 and PF-5081090 (from Tomaras et al. 2014). Data for ACHN-975 compiled from 2013 ICAAC presentations (Badal et al. 2013; Serio et al. 2013); posters downloaded from www.achaogen.com/media-all on April 22, 2015. N.D., Not determined.

2012, but clinical studies were halted because of inflammation at the injection site (see clinical-trials.gov/ct2/show/NCT01597947; www.sec.gov/Archives/edgar/data/1301501/000119312514020548/d623715ds1.htm).

As of late in 2015, ACHN-975 is the only LpxC inhibitor known to have entered clinical trials. The patents covering ACHN-975 and PF-5081090 have been followed by additional patent applications from Achaogen and Pfizer, respectively (Abramite et al. 2014; Brown et al. 2014b; Linsell et al. 2014; Reilly et al. 2014; Patterson et al. 2015a,b). A clinical candidate may yet emerge from one of these companies or from one of the several other active groups of LpxC researchers. In 2013 and 2014, at least five laboratories filed new patent applications (Fu et al. 2014, 2015; Linsell et al. 2014; Chapoux et al. 2015; Cohen et al. 2015; Gauvin et al. 2015; Zhou et al. 2015a,b,c).

SELECTION FOR RESISTANT MUTANTS

Resistance to LpxC inhibitors has been studied by several groups, usually by plating bacteria onto agar containing drug at a concentration that is four times or eight times the MIC. Single-step mutations conferring resistance are rare but occur at measurable frequencies. For *E. coli*, separate studies of resistance to L-161,240, BB-78484, or CHIR-090 each reported a frequency of ~10^{-9} (Onishi et al. 1996; Rafanan et al. 2000; Clements et al. 2002; Zeng et al. 2013). The frequencies of resistance to PF-5081090 were reported to be <5.0×10^{-10} for *P. aeruginosa* and 9.6×10^{-8} for *K. pneumoniae* (Tomaras et al. 2014). A study from Novartis obtained *P. aeruginosa* mutants resistant to CHIR-090 by serial passage on drug; frequency of resistance was not reported (Caughlan et al. 2012). Resistance mechanisms were identified for mutants in most of these studies.

For both *E. coli* and *K. pneumoniae*, the most common finding was point mutation of *fabZ*, which encodes R-3-hydroxymyristoyl acyl carrier protein dehydrase (Clements et al. 2002; Zeng et al. 2013; Tomaras et al. 2014). As discussed below, it is thought that reducing the rate of phospholipid synthesis allows the cell to tolerate a reduction in LPS synthesis. For *P. aeruginosa*, *fabZ* mutants have not been observed, but several of the CHIR-090-resistant mutants isolated by Novartis were found to have mutations in a different fatty acid biosynthetic gene, *fabG* (Caughlan et al. 2012).

For *P. aeruginosa*, the most commonly seen mechanism of resistance was up-regulation of multidrug efflux pumps. Novartis found *P. aeruginosa* mutants resistant to CHIR-090 with mutations in *mexR* or *nfxB*, genes encoding repressors for the RND pumps MexAB-OprM and MexCD-OprJ, respectively. These mutants had reduced susceptibility to several antibiotics as well as to CHIR-090, as expected for isolates overexpressing multidrug efflux pumps (Caughlan et al. 2012). Pfizer reported evidence that several of the *P. aeruginosa* isolates selected for resistance to PF-5081090 are probably also overexpressors of RND pumps. Susceptibility to PF-5081090 was restored by treatment with the efflux pump inhibitor PAβN, and sequencing revealed mutations in efflux pump repressor genes (Tomaras et al. 2014). For enteric bacteria, efflux pump overexpression has not been reported as a mechanism of resistance to LpxC inhibitors.

Mutations that confer resistance by increasing LpxC activity have not been described in *E. coli*, but have been observed in *P. aeruginosa*. Both Pfizer and Novartis isolated resistant mutants with a C-to-A mutation 11 bp upstream of the *lpxC* start codon and showed that laboratory constructs with this mutation were resistant. Western blots showed an increase in the amount of LpxC protein (Caughlan et al. 2012). The Pfizer report noted that the mutated base is within a recently identified small RNA designated PA4406.1 and that this is the first description of a molecular mechanism for LpxC regulation in *P. aeruginosa* (Tomaras et al. 2014).

Notably, very few of the resistant mutants characterized have mutations within the coding region of the *lpxC* gene. *E. coli* isolates with *lpxC* mutations were found to be resistant to early LpxC inhibitors (Rafanan et al. 2000; Clements et al. 2002), but have not been reported for the more advanced molecules. For *P. aeruginosa*, Pfizer described a M62R substitution that confers resistance to their inhibitor LpxC-2 but not to their more advanced compound PF-5081090 (Tomaras et al. 2014). Novartis reported L18V

substitution in isolates of a hypermutator strain of *P. aeruginosa* selected for resistance to CHIR-090 (Caughlan et al. 2012).

CHEMISTRY AND STRUCTURAL BIOLOGY

Medicinal Chemistry

Nearly all LpxC inhibitors that have been described are related to the series that was discovered by the UW/Chiron program. Early compounds in this program were designed as loose analogs of the Merck series, with a heterocyclic linker connecting an aromatic moiety to the hydroxamate warhead (Kline et al. 2002). A key observation was that acyclic precursors of oxazaline compounds were more active than their cyclic products. For example, the oxazoline shown as compound 4 in Figure 2 had an in vitro IC_{50} of 5 μM for *P. aeruginosa* LpxC and did not inhibit growth of either *P. aeruginosa* or *E. coli*. The corresponding aroylserine (compound 5) had an in vitro IC_{50} of 1.5 μM. Replacing the D-Ser (αR stereochemistry) with L-Ser (αS) improved potency, producing the first compound active against an efflux-deficient strain of *P. aeruginosa* (MIC 12.5 μg/mL for mutant PAO200). Potency was increased further by addition of a methyl group to the β carbon.

The most active threonine stereoisomer was αS, βR (compound 6), with an MIC of 50 μg/mL for wild-type *P. aeruginosa*. Further exploration of L-Thr derivatives led to the molecule known as CHIR-090 as well as to dozens of other molecules with single-digit MICs for both *P. aeruginosa* and *E. coli* (Andersen et al. 2011). Antibacterial activity for *E. coli* emerged in this series along with *P. aeruginosa* activity, suggesting that for this scaffold, compounds had similar potency for both enzymes.

Although the UW/Chiron chemistry has not been published apart from the patent, many of the key features of the structure–activity relationship of this series are known from subsequent work published by other laboratories. Their mechanism of inhibition became apparent when the LpxC protein structure was solved.

The solution structure of LpxC from the thermophile *Aquifex aeolicus* in complex with a substrate-mimic inhibitor (TU-514) and the crystal structure of *A. aeolicus* LpxC were reported by two academic groups (Coggins et al. 2003; Whittington et al. 2003). The enzyme contains an unusual tunnel, open to solvent at both ends, into which the fatty acyl chain of the substrate is inserted. This observation suggested that the UW/Chiron inhibitors to the enzyme mimics that of the substrate, as indeed proved to be the case.

The solution structure of *A. aeolicus* LpxC in complex with CHIR-090 identified conserved amino acids that interact with the methyl and hydroxyl groups of the L-Thr, positioning the hydroxamic acid moiety in close proximity to the zinc residue in the catalytic site (Barb et al. 2007a). This is consistent with chemical data showing that L-Thr is more active than its stereoisomers (Liang et al. 2011; Hale et al. 2013).

A high-resolution crystal structure of *P. aeruginosa* LpxC with the BB-78485 inhibitor bound was first published by a group at Pfizer Global Research and Development (Mochalkin et al. 2008). With reported success in other groups following quickly, structure-based drug design is now widely used as part of LpxC medicinal chemistry programs, as described in publications from AstraZeneca, Merck, and Pfizer (Benenato et al. 2010; Mansoor et al. 2011b; Brown et al. 2012).

Nearly all reported LpxC inhibitors use hydroxamic acid as the chelating moiety. Replacing the hydroxamic acid with other warheads can reduce potency by a 100-fold or more. However, one patent application described nonhydroxamate LpxC inhibitors with submicromolar IC_{50} versus the *E. coli* enzyme (Cohen et al. 2015). A press release dated December 22, 2015 from Forge Therapeutics (www.forgetherapeutics.com/news-articles) referred to a nonhydroxamate LpxC inhibitor from the same researchers as efficacious in experimental bacterial infections.

Replacement of the L-Thr with other amino acids often leads to reduction in potency (Hale et al. 2013), although not always. ACHN-975 is one example of successful replacement. Alternatives to the amino acid scaffold have been identified by several groups (Mansoor et al. 2011b; Brown et al. 2012; McAllister et al. 2012; Murphy-Benenato et al. 2014). None of the most potent compounds reaches very far into the UDP pocket of the enzyme, although several studies have attempted to gain potency through interactions in this region (Barb et al. 2009; Hale et al. 2013; Liang et al. 2013).

Many of the most active UW/Chiron compounds have two aromatic rings separated by one or two triple bonds (compounds 7–10). *para* or *meta* substitution of the distal ring by a hydrophilic group is tolerated, improving solubility without loss of activity (compounds 9 and 10). Later structural analysis showed that these molecules occupy the full length of the hydrophobic tunnel in LpxC, with the morpholino group of CHIR-090 and the amino group of Lpc-011 protruding into solvent (Barb et al. 2007a; Liang et al. 2011). Most LpxC chemistry published by other groups has used similar hydrophobic tails. The Pfizer class exemplified by PF-5081090 is unusual in being extremely potent without such an extension.

Structural Basis of Species Specificity and Kinetics

In the first description of CHIR-090, the Raetz laboratory described it as a slow, tight binder of LpxC from *A. aeolicus* (McClerren et al. 2005). This report increased the level of interest in LpxC, as inhibitors with slow off-rates have the possibility of being more effective in vivo. This idea was explored more recently in a study modeling PK/PD parameters for a series of LpxC inhibitors varying in off-rate and post-antibiotic effect (Walkup et al. 2015).

Similar kinetics were reported for binding of CHIR-090 to LpxC from *E. coli* and *P. aeruginosa*. In contrast, binding to LpxC from *Rhizobium leguminosarum* differed in being not only much weaker but rapidly reversible. Two amino acid residues within the hydrophobic tunnel were found to be critical for susceptibility and time dependence (Barb et al. 2007a,b). Later structural studies showed that the hydrophobic

Cite this article as *Cold Spring Harb Perspect Med* doi: 10.1101/cshperspect.a025304

tunnel of the *R. leguminosarum* enzyme is narrower than that of *E. coli* and the other enzymes studied. Diacetylene compounds, such as lpc-009, are less bulky than CHIR-090 and are potent inhibitors of *R. leguminosarum* LpxC (Lee et al. 2011; Liang et al. 2011).

Most LpxC inhibitors described in the past decade are like the UW/Chiron compounds in being active versus both *E. coli* and *P. aeruginosa*. However, as noted above, this was not the case for the early inhibitors discovered by Merck and British Biotech. These compounds fail to inhibit growth of *P. aeruginosa* because they are poor inhibitors of the *P. aeruginosa* enzyme. Scientists at the PathoGenesis Corporation (later Chiron) constructed chimeric strains with controlled expression of *lpxC* genes derived from either *E. coli* or *P. aeruginosa*. Either *E. coli* or *P. aeruginosa* was susceptible to L-161,240 if the active *lpxC* gene was derived from *E. coli*, but resistant if the active *lpxC* gene was derived from *P. aeruginosa* (Mdluli et al. 2006).

This observation was confirmed and extended by researchers at Duke University, replacing the endogenous *lpxC* gene in *E. coli* with that of *P. aeruginosa* and showing that the MIC of several LpxC inhibitors was dependent on the source of the gene (Table 2). They published X-ray crystal structures of the *E. coli* and *P. aeruginosa* enzymes in complex with various inhibitors (Lee et al. 2014). The insert I region of the *E. coli* enzyme contains a flexible loop ($\beta a - \beta b$) that flips over to accommodate bulky compounds like L-161,240 and BB-78485. The $\beta a - \beta b$ loop in insert I of the *P. aeruginosa* en-

zyme is much more rigid. The bulky inhibitors can bind the *P. aeruginosa* enzyme, but with much lower affinity than for *E. coli* LpxC. Lpc-009, a slender molecule similar to CHIR-090, fits readily into either type of enzyme without distorting the $\beta a - \beta b$ loop. It now appears that the issue is not that the early compounds are narrow in spectrum but that the *E. coli* enzyme is unusually broad. Indeed, the report describing these findings was entitled "Structural basis of the promiscuous inhibitor susceptibility of *E. coli* LpxC" (Lee et al. 2014).

BIOLOGY OF LpxC INHIBITION

Bacterial Adaptation to Alteration of LPS Synthesis

As indicated above, one of the most common mechanisms of resistance to LpxC inhibitors is mutation in *fabZ*, which encodes *R*-3-hydroxymyristoyl acyl carrier protein dehydrase. In the first description of such mutants, Clements et al. (2002) noted that *R*-3-hydroxymyristic acid is a common precursor for both lipid A and phospholipids. They suggested that reduction of FabZ activity would divert *R*-3-hydroxymyristoyl-ACP away from phospholipid synthesis and allow the bacterial cell to maintain a normal level of LPS synthesis despite the presence of LpxC inhibitors (Clements et al. 2002).

More recent research suggests a slightly different interpretation. Zeng et al. (2013) reported that in *fabZ* mutants resistant to LpxC inhibitors, LpxC activity is reduced rather than

Table 2. Susceptibility of *E. coli* to LpxC inhibitors is dependent on the source of *lpxC* gene

Species and strain *lpxC* source	MIC (µg/mL)		
	E. coli W3110 *E. coli*	*E. coli* PA3110[a] *P. aeruginosa*	*P. aeruginosa* PAO1 *P. aeruginosa*
Inhibitor			
L-161,240	6.1	>100	>100
BB-78485	6.1	>100	>100
Lpc-009	0.05	0.7	0.7
CHIR-090	0.2	1.3	1.6
Lpc-011	0.03	0.32	0.32

Adapted, with permission, from Lee et al. (2014).
[a]LpxC replacement strain (*E. coli lpxC* gene replaced by *P. aeruginosa lpxC* gene).

increased. In contrast to *envA1* mutants, in which reduced LpxC activity confers antibiotic hypersensitivity, the *fabZ* mutants have normal susceptibility to polymyxin. These observations suggest that in *fabZ* mutants, the mechanism of resistance to LpxC inhibition is restoration of the balance between LPS synthesis and synthesis of phospholipids, reducing flux through both pathways while maintaining the permeability barrier of the outer membrane. Overexpression of *fabZ* was shown to increase the amount of LpxC protein in cells, supporting the idea that FabZ and LpxC activities are coregulated. In the same study, *fabZ* mutants subjected to a second round of selection were found to have mutations in *thrS* that conferred a further fourfold increase in MIC. The double mutants grew slowly, presumably as a result of an overall reduction in protein synthesis. The investigators describe this mechanism of resistance to LpxC inhibition as rebalancing cellular homeostasis (Zeng et al. 2013).

The study reported by Zeng et al. (2013) is consistent with previous evidence that in *E. coli*, LpxC activity is primarily controlled at the level of protein turnover. Incubation of *E. coli* with the early LpxC inhibitor L-573,655 resulted in an increase in bacterial LpxC content. This response was shown to be posttranscriptional and is likely to result from a transient reduction in FtsH activity (Sorensen et al. 1996). LpxC is subject to continuous degradation by the protease FtsH. During incubation at 42°C, the mutant *ftsH1* produces aberrant membrane structures within the periplasm and becomes nonviable. This phenotype can be suppressed by certain *fabZ* mutations (Ogura et al. 1999). Overexpression of plasmid-borne *lpxC* is toxic, particularly if the gene is mutated to produce a protein lacking the carboxy-terminal domain recognized by FtsH (Führer et al. 2006). Overproduction of LPS is, thus, extremely detrimental to the cell.

It currently appears that in *E. coli*, LPS synthesis is controlled primarily at the second step in lipid A synthesis through balancing FtsH, FabZ, and LpxC activities. A recently published quantitative model of the lipid A pathway suggests that tight control at this step thwarts the effect of LpxC inhibitors. LpxK is identified as a second rate-limiting step that is not subject to regulation and might be a better drug target (Emiola et al. 2014).

In *P. aeruginosa*, much less is known about regulation of LpxC activity or of LPS synthesis in general. The *P. aeruginosa* LpxC lacks a carboxy-terminal recognition tag and is not a substrate for FtsH (Langklotz et al. 2011). As noted above, characterization of mutants resistant to LpxC inhibitors led to identification of an sRNA that appears to control LpxC activity, although the mechanism is not yet fully understood (Tomaras et al. 2014). These mutants have elevated levels of LpxC protein. Mutants with this phenotype have not been identified in enteric organisms and indeed might not be viable.

Both *E. coli* and *P. aeruginosa* are able to tolerate moderate reductions in LPS synthesis and export. The *envA1* mutant discussed above is one such example. A second example is an *E. coli* mutant with a short in-frame deletion in a gene now known as *lptD*. Because of its defect in LPS export, this mutant (*imp-4213*) is like *envA1* in being hypersensitive to detergents and many antibiotics (Sampson et al. 1989; Ruiz et al. 2009). The *P. aeruginosa* mutant Z61 (ATCC 35151) has a similar phenotype (Zimmermann 1980). This strain is now known to have mutations in *lptE* and *oprM*; thus, it is defective in both outer membrane barrier and efflux (Shen et al. 2014).

Effect of LpxC Inhibition on Bacterial Defenses In Vivo

One of the potential advantages of LpxC as an antibiotic target is the possibility that sublethal concentrations of LpxC inhibitors might increase the susceptibility of bacteria to other antibiotics and perhaps also to host defenses. In in vitro "checkerboard" assays, LpxC inhibitors have been shown to be synergistic with rifampin, vancomycin, tetracycline, and other antibiotics for which the outer membrane barrier limits antibacterial activity. There have been some efforts to evaluate the extent to which this phenomenon can be exploited to improve therapy.

Co-administration of an LpxC inhibitor with another antibiotic might allow each drug to be dosed at a lower level to reduce toxicity or side effects, or allow dosing above the minimum effective level to minimize selection of resistant mutants. Pfizer reported synergy between polymyxin B nonapeptide and PF-5081090 in a mouse model of *P. aeruginosa* infection (Bulitta et al. 2011). Achaogen described synergy of both rifampin and vancomycin with LpxC inhibitors in mouse models of *P. aeruginosa* and *K. pneumoniae* (Patten and Armstrong 2012).

A limitation of the data obtained by in vitro selection and characterization of resistant mutants, as described above, is that they do not necessarily predict the risk for selection of resistant isolates in patients or the response of such isolates to therapy. Pfizer reported that, in mice infected with a *fabZ* mutant of *K. pneumoniae* and treated with PF-5081090, the AUC/MIC for stasis and 1-log kill were much lower than for mice infected with wild-type bacteria, suggesting that the *fabZ* mutants are more susceptible in vivo than their in vitro MICs would suggest (Tomaras et al. 2014). The *fabG* mutants of *P. aeruginosa* found by Novartis to be resistant to CHIR-090 had such a severe growth defect in vitro (Caughlan et al. 2012) that it seems unlikely they would survive in vivo.

To study selection of drug-resistant mutants during treatment, researchers at Achaogen infected mice with a higher inoculum than usual and treated with a suboptimal dose of an LpxC inhibitor. Tissue homogenates contained $>10^3$ resistant colonies per thigh. Co-administration of an additional drug (vancomycin) to improve bacterial clearance not only reduced the total number of colonies recovered but appeared to reduce the proportion of bacteria resistant to the LpxC inhibitor (Patten and Armstrong 2012). Although no characterization of the resistant bacteria was described, these observations suggest that bacterial variants with reduced susceptibility to LpxC inhibitors can emerge in vivo and that this phenomenon can be affected by dosing regimen.

The idea that sublethal concentrations of an LpxC inhibitor may sensitize bacteria to host defenses, such as complement or antimicrobial peptides, is attractive but has been little studied. Lin et al. (2012) reported that, for *A. baumannii*, activation of TLR4-mediated responses to endotoxin is a key aspect of virulence. They studied a Pfizer LpxC inhibitor (LpxC-1) that apparently inhibited LPS synthesis in *A. baumannii* (assessed by production of TLR4-activating material), although it failed to inhibit bacterial growth in vitro (MIC > 512 μg/mL). Treatment of *A. baumannii* with LpxC-1 was shown to reduce induction of cellular inflammatory responses in vitro and to increase the susceptibility of bacteria to phagocytosis. Treatment of *A. baumannii*-infected mice with LpxC-1 increased bacterial clearance and improved survival, with a substantial reduction in serum cytokines and tumor necrosis factor α (TNF-α) (Lin et al. 2012).

CONCLUDING REMARKS

LpxC programs have been much more successful than other target-directed antibiotic discovery efforts of the past several decades, although no compound has yet been tested in human infections. The collective efforts of academic and industrial LpxC researchers have added to our understanding of the regulation of LPS synthesis in *E. coli* and *P. aeruginosa*.

ACKNOWLEDGMENTS

I thank James Aggen for assistance with Figure 2. I thank the following for critical reading of the manuscript: James Aggen, Karen Bush, Cheryl Quinn, Alisa Serio, and Andrew Tomaras. I was involved in the University of Washington/Chiron LpxC program as an employee of Pathogenesis Corporation and Chiron Corporation, and am an inventor on the resulting patent (Andersen et al. 2011). I was later employed by Vertex Pharmaceuticals. I do not currently own stock in any of these companies. I have not received any financial support. The data presented here are from sources that are publicly available, although in some cases not peer-reviewed (e.g., conference presentations and patent applications).

REFERENCES

Abramite JA, Brown MF, Chen JM, Melnick MJ, Montgomery JI, Reilly U. 2014. Isoxazole derivatives useful as antibacterial agents. U.S. Patent 2014024690 (A1).

Andersen NH. 2003. Design of LpxC, UDP-(3-O-acyl)-N-acetylglucosamine deacetylase, inhibitors and their antibacterial activity. 43rd Interscience Conference on Antimicrobial Agents and Chemotherapy (ICAAC). Chicago, September 14–17.

Andersen NH, Bowman J, Erwin A, Harwood E, Kline T, Mdluli K, Ng S, Pfister KB, Shawar R, Wagman AS, et al. 2011. Antibacterial agents. U.S. Patent 2011172174 (A1).

Anderson MS, Bull HG, Galloway SM, Kelly TM, Mohan S, Radika K, Raetz CR. 1993. UDP-N-acetylglucosamine acyltransferase of Escherichia coli. The first step of endotoxin biosynthesis is thermodynamically unfavorable. J Biol Chem 268: 19858–19865.

Badal R, Hoban D, Hackel M, Bouchillon S, Serio A. 2013. In vitro activity of ACHN-975 against 1,050 non-fermentative Gram-negative bacilli. 53rd Interscience Conference on Antimicrobial Agents and Chemotherapy (ICAAC), Abstract E-614. Denver, September 10–13.

Barb AW, Jiang L, Raetz CRH, Zhou P. 2007a. Structure of the deacetylase LpxC bound to the antibiotic CHIR-090: Time-dependent inhibition and specificity in ligand binding. Proc Natl Acad Sci 104: 18433–18438.

Barb AW, McClerren AL, Snehelatha K, Reynolds CM, Zhou P, Raetz CRH. 2007b. Inhibition of lipid A biosynthesis as the primary mechanism of CHIR-090 antibiotic activity in Escherichia coli. Biochemistry 46: 3793–3802.

Barb AW, Leavy TM, Robins LI, Guan Z, Six DA, Zhou P, Hangauer MJ, Bertozzi CR, Raetz CRH. 2009. Uridine-based inhibitors as new leads for antibiotics targeting Escherichia coli LpxC. Biochemistry 48: 3068–3077.

Beall B, Lutkenhaus J. 1987. Sequence analysis, transcriptional organization, and insertional mutagenesis of the envA gene of Escherichia coli. J Bacteriol 169: 5408–5415.

Benenato K, Choy A, Hale M, Hill P, Marone V, Miller M. 2010. Hydroxamic acid derivatives as Gram-negative antibacterial agents. Patent WO2010100475 (A1).

Bodewits K, Raetz CR, Govan JR, Campopiano DJ. 2010. Antimicrobial activity of CHIR-090, an inhibitor of lipopolysaccharide biosynthesis, against the Burkholderia cepacia complex. Antimicrob Agents Chemother 54: 3531–3533.

Bornheim L, McKinnell J, Fuchs-Knotts T, Boggs J, Kostrub CF. 2013. Preclinical safety evaluation of the novel LpxC inhibitor ACHN-975 in rat and monkey. 53rd Interscience Conference on Antimicrobial Agents and Chemotherapy (ICAAC), Abstract F-1231. Denver, September 10–13.

Brown MF, Reilly U, Abramite JA, Arcari JT, Oliver R, Barham RA, Che Y, Chen JM, Collantes EM, Chung SW, et al. 2012. Potent inhibitors of LpxC for the treatment of Gram-negative infections. J Med Chem 55: 914–923.

Brown MF, Che Y, Marfat A, Melnick MJ, Montgomery JI, Reilly U. 2014a. N-link hydroxamic acid derivatives useful as antibacterial agents. U.S. Patent 2014343031 (A1).

Brown MF, Chen JM, Melnick MJ, Montgomery JI, Reilly U. 2014b. Imidazole, pyrazole, and triazole derivatives useful as antibacterial agents. U.S. Patent 2014038975 (A1).

Bulitta JB, Kuhn M, Finegan S, George D, Barham R, Haddish-Berhane N, Betts A, Brown M, Hanna D, Montgomery J, et al. 2011. Optimizing synergy between a new LpxC inhibitor (PF1090) and polymyxin B nonapeptide (PBN) by mechanism-based modeling to efficiently support drug development. 51st Interscience Conference on Antimicrobial Agents and Chemotherapy (ICAAC), Abstract A2-1170. Chicago, September 17–20.

Caughlan RE, Jones AK, DeLucia AM, Woods AL, Xie L, Ma B, Barnes SW, Walker JR, Sprague ER, Yang X, et al. 2012. Mechanisms decreasing in vitro susceptibility to the LpxC inhibitor CHIR-090 in the Gram-negative pathogen Pseudomonas aeruginosa. Antimicrob Agents Chemother 56: 17–27.

Chapoux G, Gauvin J-C, Panchaud P, Specklin J-L, Surivet J-P, Schmitt C. 2015. 1,2-dihydro-3H-pyrrolo[1,2-C]imidazol-3-one derivatives and their use as antibacterial agents. Patent WO2015132228 (A1).

Chen MH, Steiner MG, de Laszlo SE, Patchett AA, Anderson MS, Hyland SA, Onishi HR, Silver LL, Raetz CR. 1999. Carbohydroxamido-oxazolidines: Antibacterial agents that target lipid A biosynthesis. Bioorg Med Chem Lett 9: 313–318.

Clements JM, Coignard F, Johnson I, Chandler S, Palan S, Waller A, Wijkmans J, Hunter MG. 2002. Antibacterial activities and characterization of novel inhibitors of LpxC. Antimicrob Agents Chemother 46: 1793–1799.

Coggins BE, Li X, McClerren AL, Hindsgaul O, Raetz CRH, Zhou P. 2003. Structure of the LpxC deacetylase with a bound substrate-analog inhibitor. Nat Struct Biol 10: 645–651.

Cohen SM, Puerta DT, Perez C. 2015. Inhibitors of LpxC. Patent WO2015085238 (A1).

Dowhan W. 2011. The Raetz pathway for lipid A biosynthesis: Christian Rudolf Hubert Raetz, MD PhD, 1946–2011. J Lipid Res 52: 1857–1860.

Emiola A, George J, Andrews SS. 2014. A complete pathway model for lipid A biosynthesis in Escherichia coli. PLoS ONE 10: e0121216.

Erwin AL. 2003. LpxC inhibitors: New class of antibiotics for serious Gram-negative infections. In 8th International Antibacterial Drug Discovery & Development Summit, Princeton, NJ, March 24–25.

Fu J, Karur S, Madera AM, Pecchi S, Sweeney ZK, Tjandra M, Yifru A. 2014. Hydroxamic acid derivatives as Lpxc inhibitors for the treatment of bacterial infections. Patent WO2014160649 (A1).

Fu J, Lee P, Madera AM, Sweeney ZK. 2015. Oxazolidinone hydroxamic acid compounds for the treatment of bacterial infections. Patent WO2015066413 (A1).

Führer F, Langklotz S, Narberhaus F. 2006. The C-terminal end of LpxC is required for degradation by the FtsH protease. Mol Microbiol 59: 1025–1036.

Gauvin J-C, Mirre A, Ochala E, Surivet J-P. 2015. Antibacterial 2H-indazole derivatives. Patent WO2015036964 (A1).

Hale MR, Hill P, Lahiri S, Miller MD, Ross P, Alm R, Gao N, Kutschke A, Johnstone M, Prince B, et al. 2013. Exploring the UDP pocket of LpxC through amino acid analogs. Bioorg Med Chem Lett 23: 2362–2367.

Cite this article as Cold Spring Harb Perspect Med doi: 10.1101/cshperspect.a025304

Hubschwerlen C, Ochala E, Specklin J-L, Surivet J-P. 2015. Antibacterial 1H-indazole and 1H-indole derivatives. Patent WO2015091741 (A1).

Jain R, Gordeev M, Lewis J, Francavilla C. 2011. N-hydroxyamide derivatives possessing antibacterial activity. Patent EP2338878 (A2).

Kasar R, Linsell MS, Aggen JB, Lu QJ, Wang D, Church T, Moser HE, Patten PA. 2012. Hydroxamic acid derivatives and their use in the treatment of bacterial infections. Patent WO2012154204 (A1).

Kline T, Andersen NH, Harwood EA, Bowman J, Malanda A, Endsley S, Erwin AL, Doyle M, Fong S, Harris AL, et al. 2002. Potent, novel in vitro inhibitors of the *Pseudomonas aeruginosa* deacetylase LpxC. *J Med Chem* **45**: 3112–3129.

Kresge N, Simoni RD, Hill RL. 2011. The lipid A assembly pathway: The work of Christian Raetz. *J Biol Chem* **286**: e6–e8.

Langklotz S, Schakermann M, Narberhaus F. 2011. Control of lipopolysaccharide biosynthesis by FtsH-mediated proteolysis of LpxC is conserved in enterobacteria but not in all Gram-negative bacteria. *J Bacteriol* **193**: 1090–1097.

Lee C-J, Liang X, Chen X, Zeng D, Joo SH, Chung HS, Barb AW, Swanson SM, Nicholas RA, Li Y, et al. 2011. Species-specific and inhibitor-dependent conformations of LpxC: Implications for antibiotic design. *Chem Biol* **18**: 38–47.

Lee C-J, Liang X, Gopalaswamy R, Najeeb J, Ark ED, Toone EJ, Zhou P. 2014. Structural basis of the promiscuous inhibitor susceptibility of *Escherichia coli* LpxC. *ACS Chem Biol* **9**: 237–246.

Leive L. 1974. The barrier function of the gram-negative envelope. *Ann NY Acad Sci* **235**: 109–129.

Liang X, Lee C-J, Chen X, Chung HS, Zeng D, Raetz CRH, Li Y, Zhou P, Toone EJ. 2011. Syntheses, structures and antibiotic activities of LpxC inhibitors based on the diacetylene scaffold. *Bioorg Med Chem* **19**: 852–860.

Liang X, Lee C-J, Zhao J, Toone EJ, Zhou P. 2013. Synthesis, structure and antibiotic activity of aryl-substituted LpxC inhibitors. *J Med Chem* **56**: 6954–6966.

Lin L, Tan B, Pantapalangkoor P, Ho T, Baquir B, Tomaras A, Montgomery JI, Reilly U, Barbacci EG, Hujer K, et al. 2012. Inhibition of LpxC protects mice from resistant *Acinetobacter baumannii* by modulating inflammation and enhancing phagocytosis. *mBio* **3**: e00312-12.

Linsell MS, Aggen JB, Dozzo P, Hildebrandt DJ, Cohen F, Kasar RA, Kane TR, Gliedt MJ, McEnroe GA. 2014. Antibacterial agents. Patent WO2014165075 (A1).

Mansoor UF, Reddy PA, Siddiqui MA. 2011a. Urea derivatives as antibacterial agents. U.S. Patent 2011212080 (A1).

Mansoor UF, Vitharana D, Reddy PA, Daubaras DL, McNicholas P, Orth P, Black T, Siddiqui MA. 2011b. Design and synthesis of potent Gram-negative specific LpxC inhibitors. *Bioorg Med Chem Lett* **21**: 1155–1161.

McAllister LA, Montgomery JI, Abramite JA, Reilly U, Brown MF, Chen JM, Barham RA, Che Y, Chung SW, Menard CA, et al. 2012. Heterocyclic methylsulfone hydroxamic acid LpxC inhibitors as Gram-negative antibacterial agents. *Bioorg Med Chem Lett* **22**: 6832–6838.

McClerren AL, Endsley S, Bowman JL, Andersen NH, Guan Z, Rudolph J, Raetz CRH. 2005. A slow, tight-binding inhibitor of the zinc-dependent deacetylase LpxC of lipid A biosynthesis with antibiotic activity comparable to ciprofloxacin. *Biochemistry* **44**: 16574–16583.

Mdluli KE, Witte PR, Kline T, Barb AW, Erwin AL, Mansfield BE, McClerren AL, Pirrung MC, Tumey LN, Warrener P, et al. 2006. Molecular validation of LpxC as an antibacterial drug target in *Pseudomonas aeruginosa*. *Antimicrob Agents Chemother* **50**: 2178–2184.

Mochalkin I, Knafels JD, Lightle S. 2008. Crystal structure of LpxC from *Pseudomonas aeruginosa* complexed with the potent BB-78485 inhibitor. *Protein Sci* **17**: 450–457.

Moffatt JH, Harper M, Harrison P, Hale JDF, Vinogradov E, Seemann T, Henry R, Crane B, St Michael F, Cox AD, et al. 2010. Colistin resistance in *Acinetobacter baumannii* is mediated by complete loss of lipopolysaccharide production. *Antimicrob Agents Chemother* **54**: 4971–4977.

Montgomery JI, Brown MF, Reilly U, Price LM, Abramite JA, Arcari J, Barham R, Che Y, Chen JM, Chung SW, et al. 2012. Pyridone methylsulfone hydroxamate LpxC inhibitors for the treatment of serious Gram-negative infections. *J Med Chem* **55**: 1662–1670.

Murphy-Benenato KE, Olivier N, Choy A, Ross PL, Miller MD, Thresher J, Gao N, Hale MR. 2014. Synthesis, structure, and SAR of tetrahydropyran-based LpxC inhibitors. *ACS Med Chem Lett* **5**: 1213–1218.

Normark S, Boman HG, Matsson E. 1969. Mutant of *Escherichia coli* with anomalous cell division and ability to decrease episomally and chromosomally mediated resistance to ampicillin and several other antibiotics. *J Bacteriol* **97**: 1334–1342.

Ogura T, Inoue K, Tatsuta T, Suzaki T, Karata K, Young K, Su LH, Fierke CA, Jackman JE, Raetz CR, et al. 1999. Balanced biosynthesis of major membrane components through regulated degradation of the committed enzyme of lipid A biosynthesis by the AAA protease FtsH (HflB) in *Escherichia coli*. *Mol Microbiol* **31**: 833–844.

Onishi HR, Pelak BA, Gerckens LS, Silver LL, Kahan FM, Chen MH, Patchett AA, Galloway SM, Hyland SA, Anderson MS, et al. 1996. Antibacterial agents that inhibit lipid A biosynthesis. *Science* **274**: 980–982.

Osborn MJ, Gander JE, Parisi E. 1972. Mechanism of assembly of the outer membrane of *Salmonella typhimurium*. Site of synthesis of lipopolysaccharide. *J Biol Chem* **247**: 3973–3986.

Patten PA, Armstrong ES. 2012. Antibacterial compositions. U.S. Patent 2012283175 (A1).

Patterson B, Lu Q, Aggen J, Dozzo P, Kasar R, Linsell M, Kane T, Gliedt M, Hildebrandt D, McEnroe G, et al. 2015a. Antibacterial agents. Patent EP2847162 (A1), abstract of corresponding document WO2013170030 (A1).

Patterson BD, Lu Q, Aggen JB, Dozzo P, Kasar RA, Linsell MS, Kane TR, Gliedt MJ, Hildebrandt DJ, McEnroe Ga, et al. 2015b. Antibacterial agents. Patent EP2847168 (A1), abstract of corresponding document WO2013170165 (A1).

Raetz CR. 1993. Bacterial endotoxins: Extraordinary lipids that activate eucaryotic signal transduction. *J Bacteriol* **175**: 5745–5753.

Rafanan N, Lopez S, Hackbarth C, Maniar M, Margolis P, Wang W, Yuan Z, Jain R, Jacobs J, Trias J. 2000. Resistance in *E.coli* to LpxC inhibitor L-161,240 is due to mutations in the *lpxC* gene. *40th Interscience Conference on Antimicrobial Agents and Chemotherapy (ICAAC)*, Abstract 2026. Toronto, September 17–20.

Raju B, Odowd H, Gao H, Patel D, Trias J. 2004. *N*-hydroxyamide derivatives possessing antibacterial activity. WO2004007444 (A2).

Reilly U, Melnick MJ, Brown MF, Plummer MS, Montgomery JI, Che Y, Price LM. 2014. Fluoro-pyridinone derivatives useful as antibacterial agents. U.S. Patent 2014057919 (A1).

Reyes N, McKinnell J, Fuchs-Knotts T, Kostrub CF, Cirz R. 2013. ACHN-975 demonstrates potent efficacy against Gram-negative bacteria in the neutropenic mouse thigh infection model. *53rd Interscience Conference on Antimicrobial Agents and Chemotherapy (ICAAC)*, Abstract F-1229. Denver, September 10–13.

Ruiz N, Kahne D, Silhavy TJ. 2009. Transport of lipopolysaccharide across the cell envelope: The long road of discovery. *Nat Rev Microbiol* **7**: 677–683.

Sampson BA, Misra R, Benson SA. 1989. Identification and characterization of a new gene of *Escherichia coli* K-12 involved in outer membrane permeability. *Genetics* **122**: 491–501.

Serio AW, Kubo A, Lopez S, Gomez M, Corey VC, Andrews L, Schwartz MA, Kasar R, McEnroe G, Aggen J, et al. 2013. Structure, potency and bactericidal activity of ACHN-975, a first-in-class LpxC inhibitor. *53rd Interscience Conference on Antimicrobial Agents and Chemotherapy (ICAAC)*, Abstract F-1226. Denver, CO, September 10–13.

Shen X, Johnson NV, Jones AK, Barnes SW, Walker JR, Ranjitkar S, Woods AL, Six DA, Dean CR. 2014. Genetic characterization of the hypersusceptible *Pseudomonas aeruginosa* strain Z61; identification of a defect in LptE. *54th Interscience Conference on Antimicrobial Agents and Chemotherapy (ICAAC)*, Abstract C-105. Washington, DC, September 5–9.

Sorensen PG, Lutkenhaus J, Young K, Eveland SS, Anderson MS, Raetz CR. 1996. Regulation of UDP-3-*O*-[*R*-3-hydroxymyristoyl]-*N*-acetylglucosamine deacetylase in *Escherichia coli*. The second enzymatic step of lipid A biosynthesis. *J Biol Chem* **271**: 25898–25905.

Takashima H, Yoshinaga M, Ushiki Y, Tsuruta R, Urabe H, Tanikawa T, Tanabe K, Baba Y, Yokotani M, Kawaguchi Y, et al. 2008. Novel hydroxamic acid derivative. Patent WO2008105515 (A1).

Tomaras AP, McPherson CJ, Kuhn M, Carifa A, Mullins L, George D, Desbonnet C, Eidem TM, Montgomery JI, Brown MF, et al. 2014. LpxC inhibitors as new antibacterial agents and tools for studying regulation of lipid A biosynthesis in Gram-negative pathogens. *mBio* **5**: e01551–01514.

Walkup GK, You Z, Ross PL, Allen EKH, Daryaee F, Hale MR, O'Donnell J, Ehmann DE, Schuck VJA, Buurman ET, et al. 2015. Translating slow-binding inhibition kinetics into cellular and in vivo effects. *Nat Chem Biol* **11**: 416–423.

Whittington DA, Rusche KM, Shin H, Fierke CA, Christianson DW. 2003. Crystal structure of LpxC, a zinc-dependent deacetylase essential for endotoxin biosynthesis. *Proc Natl Acad Sci* **100**: 8146–8150.

Young K, Silver LL, Bramhill D, Cameron P, Eveland SS, Raetz CR, Hyland SA, Anderson MS. 1995. The *envA* permeability/cell division gene of *Escherichia coli* encodes the second enzyme of lipid A biosynthesis. UDP-3-*O*-(*R*-3-hydroxymyristoyl)-*N*-acetylglucosamine deacetylase. *J Biol Chem* **270**: 30384–30391.

Zeng D, Zhao J, Chung HS, Guan Z, Raetz CRH, Zhou P. 2013. Mutants resistant to LpxC inhibitors by rebalancing cellular homeostasis. *J Biol Chem* **288**: 5475–5486.

Zhou P, Toone E, Nicholas R. 2015a. 2-Piperidinyl substituted *N*,3-dihydroxybutanamides. Patent WO2015024016 (A2).

Zhou P, Toone E, Nicholas R. 2015b. Antibacterial compounds. Patent WO2015024021 (A2).

Zhou P, Toone E, Nicholas R. 2015c. Substituted hydroxamic acid compounds. Patent WO2015024010 (A2).

Zimmermann W. 1980. Penetration of β-lactam antibiotics into their target enzymes in *Pseudomonas aeruginosa*: Comparison of a highly sensitive mutant with its parent strain. *Antimicrob Agents Chemother* **18**: 94–100.

Topoisomerase Inhibitors: Fluoroquinolone Mechanisms of Action and Resistance

David C. Hooper[1] and George A. Jacoby[2]

[1]Division of Infectious Diseases, Massachusetts General Hospital, Boston, Massachusetts 02114
[2]Lahey Hospital and Medical Center, Burlington, Massachusetts 01805

Correspondence: dhooper@partners.org

Quinolone antimicrobials are widely used in clinical medicine and are the only current class of agents that directly inhibit bacterial DNA synthesis. Quinolones dually target DNA gyrase and topoisomerase IV binding to specific domains and conformations so as to block DNA strand passage catalysis and stabilize DNA–enzyme complexes that block the DNA replication apparatus and generate double breaks in DNA that underlie their bactericidal activity. Resistance has emerged with clinical use of these agents and is common in some bacterial pathogens. Mechanisms of resistance include mutational alterations in drug target affinity and efflux pump expression and acquisition of resistance-conferring genes. Resistance mutations in one or both of the two drug target enzymes are commonly in a localized domain of the GyrA and ParC subunits of gyrase and topoisomerase IV, respectively, and reduce drug binding to the enzyme–DNA complex. Other resistance mutations occur in regulatory genes that control the expression of native efflux pumps localized in the bacterial membrane(s). These pumps have broad substrate profiles that include other antimicrobials as well as quinolones. Mutations of both types can accumulate with selection pressure and produce highly resistant strains. Resistance genes acquired on plasmids confer low-level resistance that promotes the selection of mutational high-level resistance. Plasmid-encoded resistance is because of Qnr proteins that protect the target enzymes from quinolone action, a mutant aminoglycoside-modifying enzyme that also modifies certain quinolones, and mobile efflux pumps. Plasmids with these mechanisms often encode additional antimicrobial resistances and can transfer multidrug resistance that includes quinolones.

Quinolones have been used extensively for a wide range of clinical applications (Owens and Ambrose 2000; Kim and Hooper 2014). Nalidixic acid, a related naphthyridone structure and the first member of the class used clinically, was discovered by George Lesher as a byproduct of chloroquine synthesis in 1962. Its use was limited to treatment of urinary tract infections and resistance emerged quickly (Lesher et al. 1962). Medicinal chemists from a number of companies, however, subsequently modified the core quinolone and related chemical scaffolds, generating compounds with greater potency, broader spectra of activity, improved pharmacokinetics, and lower frequency of development of resistance (Domagala and Hagen 2003). An important addition of a fluorine substituent at position 6 added to potency and became the common feature of the fluoroquinolone class with the introductions of norfloxacin in 1986 and ciprofloxacin in 1987 that showed substantially greater potency against Gram-negative bacteria. Subsequently, other fluoroquinolones, such as levofloxacin, gatifloxacin, moxifloxacin, and gemifloxacin were developed with increased activity against Gram-positive bacteria. Because of their potency, spectrum of activity, oral bioavailability, and generally good safety profile, fluoroquinolones were used widely for a range of clinical indications worldwide. Although still clinically valuable, fluoroquinolone use has been compromised by the emergence of bacterial resistance because of mutation and acquisition of plasmid-encoded genes. In the sections that follow, we review the mechanisms of quinolone action and resistance, which are summarized in Table 1.

QUINOLONE MECHANISM OF ACTION

Quinolones target two essential bacterial type II topoisomerase enzymes, DNA gyrase and DNA topoisomerase IV (Hooper 1997). Both enzymes are heterotetramers with two subunits, gyrase being constituted as $GyrA_2GyrB_2$ and topoisomerase IV as $ParC_2ParE_2$. GyrA is homologous to ParC, and GyrB homologous to ParE (Wang 1996). In *Staphylococcus aureus*, topoisomerase IV subunits historically have also been referred to as GrlA and GrlB. Both enzymes act by catalyzing a DNA double-strand break, passing another DNA strand through the break, and resealing the break (Aldred et al. 2014). DNA strand passing occurs via domains that are localized in GyrA and ParC, whereas ATPase activity, which is required for enzyme catalysis, occurs via domains localized in GyrB and ParE.

Quinolones bind reversibly to the complexes of DNA with gyrase and topoisomerase IV at the interface between protein and DNA near the active site tyrosine (Tyr122 for GyrA, Tyr120 for ParC in *Escherichia coli* numbering), which is transiently covalently linked to DNA during DNA strand passage, with intercalation into the cleaved DNA (Laponogov et al. 2009,

2013; Bax et al. 2010; Wohlkonig et al. 2010). In GyrA, a noncatalytic Mg^{2+} ion coordinated with four water molecules appears as a bridge for hydrogen bonding between quinolone and Ser83 and Asp87 (Wohlkonig et al. 2010), the two most commonly mutated amino acids in quinolone-resistant mutants. Although quinolones can bind to mycobacterial gyrase in the absence of DNA (Kumar et al. 2014), in *E. coli*, the gyrase–DNA complex shows increases in the amount and specificity of quinolone binding relative to gyrase alone (Shen et al. 1989; Willmott and Maxwell 1993).

Quinolones inhibit enzyme function by blocking the resealing of the DNA double-strand break, but, in addition, this process stabilizes a catalytic intermediate covalent complex of enzyme and DNA that serves as a barrier to movement of the DNA replication fork (Wentzell and Maxwell 2000) or transcription complexes (Willmott et al. 1994) and can be converted to permanent double-strand DNA breaks (Drlica et al. 2008), thereby functioning as topoisomerase poisons (Kreuzer and Cozzarelli 1979). Quinolone interactions with DNA gyrase appear to result in more rapid inhibition of DNA replication than quinolone interactions with topoisomerase IV (Khodursky et al. 1995; Fournier et al. 2000), possibly relating to the overall proximity to the DNA replication complex of enzyme-binding sites on chromosomal DNA, with gyrase positioned ahead of the complex and topoisomerase IV behind it (Khodursky and Cozzarelli 1998). Quinolones can differ in their potency for the two enzymes, with a general pattern among quinolones in clinical use that there is greater activity against DNA gyrase in Gram-negative bacteria and greater activity against topoisomerase IV in Gram-positive bacteria; but exceptions occur, and some quinolones have similar potency against both enzymes (Blanche et al. 1996; Pan and Fisher 1997; Strahilevitz and Hooper 2005).

DNA strand breaks trigger bacterial SOS DNA repair responses (Piddock et al. 1990; Drlica et al. 2008) and, to the extent that repair is incomplete, generate quinolone bactericidal activity (Hiasa et al. 1996; Drlica and Zhao 1997; Drlica et al. 2008, 2009). Inhibition of

Table 1. Mechanisms of quinolone resistance

Mechanism	Gene alteration	Effector genes	Species	Other factors
Altered target with reduced drug binding	Mutation in chromosomal structural genes	*gyrA* *gyrB* *parC* (*grlA*) *parE* (*grlB*)	All Most except mycobacteria, *Campylobacter*, *Treponema*	Primary and secondary targets vary by drug and species with *gyrA* and/or *parC* most often mutated; mutations in both targets with additive resistance
Increased expression of efflux pumps	Mutation in chromosomal genes of transcriptional regulators of expression of chromosomally encoded pumps	MFS family: *norA*, *norB*, *norC*, *mdeA*, *lmrS*, *sdrM*, *qacB*(III) MATE family: *mepA*	*Staphylococcus aureus*	Regulators include *mgrA*, *norG*, *mepR*, *arlRS*
		MFS family: *bmr*, *bmr3*, *blt* MFS family: *lmrP* ABC family: *lmrA* MFS family: *pmrA* ABC family: *patAB* ABC family: *satAB* MFS family: *lde* MATE family: *fepA*	*Bacillus subtilis* *Lactococcus lactis* *Streptococcus pneumoniae* *Streptococcus suis* *Listeria monocytogenes*	Regulator *fepR*
		RND family: *acrAB-tolC* MFS family: *emrAB-tolC*, *mdfA*	*Escherichia coli*	Regulators include *marR*, *soxRS*, *rob*, *emrR*
		RND family: *acrAB-tolC* RND family: *oqxAB-tolC* RND family: *acrAB-tolC* RND family: *cmeABC* MATE family: *norM*	*Salmonella* spp. *Klebsiella pneumoniae* *Enterobacter aerogenes* *Campylobacter jejuni* *Vibrio parahaemolyticus*	Regulators include *oqxR*, *rarA*
		RND family: *mexAB-oprM*, *mexCD-oprJ*, *mexEF-oprN*, *mexXY-oprM*	*Pseudomonas aeruginosa*	Regulators include *mexR*, *nfxB*, *nfxC*, *mvaT*, *mexZ*
		RND family: *adeIJK*, *adeABC*, *adeFGH*	*Acinetobacter baumannii*	Regulators include *adeRS*, *adeL*
		RND family: *smeDEF* MATE family: *norA*	*Stenotrophomonas maltophilia* *Bacteroides fragilis*	

Continued

Table 1. *Continued*

Mechanism	Gene alteration	Effector genes	Species	Other factors
	Plasmid-acquired efflux pump gene	MATE family: *bexA* RND family: *oqxAB, qepA*	*Bacteroides thetaiotaomicron* Multiples species of Enterobacteriaceae	
Target protection	Chromosomal genes	*mfpA*	*Mycobacterium tuberculosis* and *Mycobacterium smegmatis*	Protection of gyrase and topoisomerase IV from quinolone action
	(among others)	*ahqnr* *efsqnr* *qnrVC*	*Aeromonas hydrophila* *Enterococcus faecalis* *Vibrio cholerae*	
	Plasmid-acquired genes	*qnrA, qnrB, qnrS, qnrC, qnrD, qnrVC*	Multiple species of Enterobacteriaceae, also *Acinetobacter, Aeromonas, Pseudomonas,* and *Vibrio* spp.	
Drug modification	Plasmid-acquired gene	*aac(6′)-Ib-cr*	Multiples species of Enterobacteriaceae	Acetylation of the secondary amine of the piperazinyl group of ciprofloxacin and norfloxacin

MFS, Major facilitator superfamily; MATE, multiple antibiotic and toxin extrusion; ABC, ATP-binding cassette; RND, resistance–nodulation–division.

protein synthesis does not interfere with quinolone-mediated inhibition of DNA replication but can reduce bacterial killing, with variation in the magnitude of reduction seen with different quinolones (Lewin et al. 1991; Howard et al. 1993). These phenomena suggest that events in addition to DNA replication inhibition that may affect DNA or other cellular damage may also contribute to quinolone bactericidal activity, but the molecular mechanisms are not yet understood.

QUINOLONE RESISTANCE BECAUSE OF MUTATIONAL ALTERATION IN TARGET ENZYMES

Single amino acid changes in either gyrase or topoisomerase IV that confer quinolone resistance have been most commonly localized to the amino-terminal domains of GyrA (residues 67 to 106 for *E. coli* numbering) or ParC (residues 63 to 102). These domains are near the active site tyrosines (Tyr122 for GyrA, Tyr120 for ParC) of both enzymes (Morais Cabral et al. 1997; Laponogov et al. 2009, 2013; Wohlkonig et al. 2010). These domains have been termed the "quinolone resistance-determining region" (QRDR) of GyrA and ParC (Yoshida et al. 1990). The most common site of mutation in GyrA of *E. coli* is at Ser83 followed by Asp87, which as noted above are key residues in the binding of quinolones to GyrA or ParC. There is conservation of a Ser and another acidic residue separated by four amino acids in both GyrA and ParC in other species, and mutation in these residues can be frequently selected with quinolones in laboratory mutants and are frequently found in resistant clinical isolates (Hooper 2003; Drlica et al. 2009; Aldred et al. 2014). Ser83Trp and Ser83Leu mutations of *E. coli* GyrA have been associated with reduced binding of the quinolone norfloxacin and enoxacin to gyrase–DNA complexes (Willmott and Maxwell 1993; Yoshida et al. 1993; Willmott et al. 1994). A Ser81Phe resistance mutation in ParC of *Bacillus anthracis* appears also to show decreased quinolone binding to the enzyme–DNA complex based on competition experiments with quinazolinediones, another gyr-

ase-targeting class of compounds (Aldred et al. 2012). Mutations in the Ser and nearby acidic residues differ in their effects on catalytic efficiency of gyrase and topoisomerase IV, with mutation of the Ser residue generally having little effect and mutation in the acidic residue resulting in a five- to 10-fold decrease in catalytic efficiency (Hiasa 2002; Aldred et al. 2014).

Resistance mutations in GyrB and ParE are substantially less common than those in GyrA and ParC in resistant clinical isolates. In *E. coli*, mutations at Asp426 and Lys447 of GyrB and Leu445 of ParE, as well as mutations at similar positions in other species, can cause resistance (Yoshida et al. 1991; Ito et al. 1994; Gensberg et al. 1995; Breines et al. 1997; Weigel et al. 2001). Binding of enoxacin to enzyme–DNA complexes constituted with resistant mutant GyrB is reduced (Yoshida et al. 1993). The crystal structures of some conformations of yeast topoisomerase II show proximity of the regions homologous to the QRDRs of GyrA and GyrB (Fass et al. 1999). In *E. coli* crystal structures, the basic substituents at position C7 of ciprofloxacin and moxifloxacin were shown to be facing the GyrB subunit, and they could be cross-linked to GyrB Cys466 (Mustaev et al. 2014). In addition, in another crystal structure Arg418 of *Acinetobacter baumannii* topoisomerase IV is in proximity to the moxifloxacin C7 basic substituent (Wohlkonig et al. 2010). Notably, resistance mutations in acidic residues in this domain of GyrB in *E. coli* (Asp426Asn) as well as in ParC, suggest that drug–enzyme contacts could be mediated by charge interactions (Wohlkonig et al. 2010). Thus, resistance mutations in the QRDRs of both GyrA/ParC and GyrB/ParE appear to act by reducing the affinity of quinolones for the enzyme–DNA complex. Although direct drug binding data are not yet available for mutant topoisomerase IV–DNA complexes, the similarity of structures between gyrase and topoisomerase IV and the conservation of key residues predict that resistance is similarly mediated by reduced affinity for both enzyme–DNA complexes.

The magnitude of resistance caused by a single-target mutation in one of the subunits of gyrase or topoisomerase IV varies by quinolone and bacterial species (Pan and Fisher 1997; Fournier et al. 2000). Because quinolone interaction with either target enzyme–DNA complex is sufficient to block cell growth and trigger cell death (Drlica and Zhao 1997), the level of susceptibility of a wild-type bacterium is determined by the more sensitive of the two target enzymes, as noted above, often gyrase in Gram-negative bacteria and topoisomerase IV in Gram-positive bacteria (Blanche et al. 1996; Pan and Fisher 1997). Thus, under quinolone selection pressure, resistance mutation in the more sensitive or primary target enzyme will generally occur first because mutation in the secondary less sensitive target enzyme alone does not have a resistance phenotype because of the dominance of the quinolone–primary target interaction (Trucksis et al. 1991; Ng et al. 1996; Breines et al. 1997). The magnitude of the increase in resistance from such a first-step mutation in the primary target is then determined by either the magnitude of the effect of the mutation on enzyme sensitivity or the intrinsic level of sensitivity of the secondary target enzyme. Thus, the level of sensitivity of the secondary target enzyme can set a cap on the magnitude of resistance conferred by mutation in the primary target enzyme. The dominance of sensitivity over resistance in the two target enzymes also implies that quinolones with similar potency against both gyrase and topoisomerase IV in an organism may require mutations in both enzymes before the mutant bacterium shows a substantial resistance phenotype (Pan and Fisher 1998, 1999; Strahilevitz and Hooper 2005). Fluoroquinolones currently in clinical use generally have differences in potency between the two target enzymes, and single target mutations produce eight- to 16-fold increases in resistance.

Accumulating mutations in both target enzymes have been shown to cause increasing quinolone resistance. In many species, high-level quinolone resistance is generally associated with mutations in both gyrase and topoisomerase IV (Schmitz et al. 1998). In several species, *Mycobacterium tuberculosis*, *Helicobacter pylori*, and *Treponema pallidum*, there is no topoisomerase IV, and gyrase provides the functions of

both enzymes and is the only quinolone target (Hooper 2003). Thus, selection of mutations with substantial resistance phenotypes is predicted to occur readily in these pathogens, a prediction consistent with the frequent occurrence of resistance with clinical use of quinolones without use of other active agents to treat infections with *M. tuberculosis* and *H. pylori* (Tsukamura et al. 1985; Mégraud 1998).

QUINOLONE RESISTANCE BECAUSE OF DECREASED DRUG ACCESS TO TARGET ENZYMES

Quinolones must cross the bacterial envelope to interact with their cytoplasmic gyrase and topoisomerase IV targets. Active quinolone efflux, reductions in influx, or both can decrease cytoplasmic quinolone concentrations and confer resistance. In Gram-positive bacteria, reduced diffusion across the cytoplasmic membrane has not been found to cause resistance, but active efflux transporters that include quinolones in their substrate profiles have been shown to cause low-level resistance. In contrast, in Gram-negative bacteria, reduced diffusion through outer membrane porin diffusion channels can contribute to resistance. Reduced influx often acts in concert with basal or increased expression of efflux transporters with both contributing additively to resistance (Lomovskaya et al. 1999; Li and Nikaido 2009). Quinolones themselves generally do not induce expression of efflux pumps. With the exception of plasmid-mediated quinolone resistance discussed later, acquired quinolone resistance by altered drug permeation occurs largely by mutations in genes encoding regulatory proteins that control the transcription of efflux pump or porin genes (Grkovic et al. 2002). Uncommonly, mutations in efflux pump structural genes have caused changes in pump substrate profiles that add quinolones (Blair et al. 2015). The levels of quinolone resistance because of regulatory mutation and pump overexpression are often limited to about four- to eightfold increases in inhibitory concentrations, likely because of counterbalancing regulatory factors and cellular toxicities of high levels of pump overexpression.

Altered Permeation in Gram-Positive Bacteria

In Gram-positive bacteria, the major facilitator superfamily (MFS) of transporters contains the largest number of efflux transporters that include quinolones in their substrate profiles. These efflux pumps are transporters energized by the proton gradient across the bacterial membrane and are generally antiporters with exchange of substrate and protons in opposite directions. The Nor MFS pumps of *S. aureus* have been most extensively studied (Li and Nikaido 2009; Schindler et al. 2015). NorA (Ubukata et al. 1989; Yu et al. 2002), NorB (Truong-Bolduc et al. 2005), and NorC (Truong-Bolduc et al. 2006) efflux pumps cause four- to eight-fold increases in resistance to quinolones when overexpressed. NorA confers resistance to hydrophilic quinolones, such as norfloxacin and ciprofloxacin, whereas NorB and NorC each confer resistance to both hydrophilic quinolones and hydrophobic quinolones, such as sparfloxacin and moxifloxacin (Yu et al. 2002; Truong-Bolduc et al. 2005, 2006); these pumps also have structurally unrelated substrates in addition to quinolones, in keeping with broad substrate profiles of many MFS transporters. Although there are natural quinolone-like compounds (Heeb et al. 2011), it is unlikely that synthetic antibacterial quinolones are themselves the natural pump substrates, which are as yet unknown for Gram-positive quinolone resistance pumps.

Regulation of expression of these transporters is complex and involves several transcriptional regulators. MgrA, the most studied, acts as a positive regulator of *norA* expression and a negative regulator of *norB* and *norC* expression (Ingavale et al. 2005; Truong-Bolduc et al. 2005). Posttranslational phosphorylation of MgrA by the PknB kinase results in the loss of the ability of MgrA dimers to bind the *norA* promoter and an increase in their binding to the *norB* promoter (Truong-Bolduc et al. 2008; Truong-Bolduc and Hooper 2010). Acidic conditions alter the proportions of phosphorylated and unphosphorylated MgrA, and oxidative and aeration conditions also affect dimerization and promoter binding (Chen et al. 2006; Truong-Bolduc et al. 2011a, 2012). Thus, relative levels of expression of NorA, NorB, and NorC are modified in response to a variety of environmental conditions. Notably, *norB* expression is selectively increased in an abscess environment in response to low-free iron conditions and contributes to fitness and bacterial survival in abscesses (Ding et al. 2008), a common form of *S. aureus* infection. The natural substrate of NorB, transport of which may contribute to improving fitness in an abscess environment, is not known. In addition, physiologic increased expression of NorB at the site of infection would suggest that susceptibility testing under clinical laboratory conditions may not fully reflect susceptibility at the site of infection.

NorG, a member of the GntR-like transcriptional regulators, can also modulate pump expression and levels of quinolone resistance; it is a direct activator of *norA* and *norB* expression but a direct repressor of *norC* expression (Truong-Bolduc and Hooper 2007; Truong-Bolduc et al. 2011b). ArlRS, a two-component regulatory system, has been shown to affect expression of *norA* as well (Fournier and Hooper 2000; Fournier et al. 2001). There are often hierarchies in regulatory networks, and other regulators can affect expression of MgrA and NorG. Such complex regulatory networks affecting pump expression imply the importance of modulation of pump functions in cellular physiology and may contribute to different bacterial responses to quinolones in different environments that affect pump expression.

Other MFS efflux transporters that can contribute to quinolone resistance in *S. aureus* include MdeA (norfloxacin and ciprofloxacin) (Huang et al. 2004), SdrM (norfloxacin) (Yamada et al. 2006), QacB(III) (norfloxacin and ciprofloxacin) (Nakaminami et al. 2010), and LmrS (gatifloxacin) (Floyd et al. 2010). MFS transporters in other Gram-positive bacteria have also been shown to include quinolones in their substrate profiles. These transporters include Bmr, Bmr3, and Blt of *B. subtilis* (Klyachko et al. 1997; Ohki and Murata 1997); PmrA of *Streptococcus pneumoniae* (Gill et al. 1999); LmrP of *Lactococcus lactis* (Bolhuis et al. 1995), and Lde of *Listeria monocytogenes* (Godreuil et al. 2003).

In addition to the MFS transporters, lesser numbers of efflux pumps of the multiple antibiotic and toxin extrusion (MATE) and ATP-binding cassette (ABC) families have been shown to confer quinolone resistance in Gram-positive bacteria. MATE family pumps, like those of the MFS, are secondary transporters energized by the membrane electrochemical gradient. MepA confers resistance to norfloxacin, ciprofloxacin, moxifloxacin, and sparfloxacin, as well as other antimicrobials and dyes (Kaatz et al. 2006). MepA is negatively regulated by MepR, and pentamidine, a MepA substrate, reduces MepR binding to the *mepA* promoter, thereby increasing *mepA* expression (Kumaraswami et al. 2009; Schindler et al. 2013). In *L. monocytogenes*, the FepA MATE family pump is overexpressed in quinolone-resistant strains and is regulated by the FepR transcriptional regulator, a member of the TetR family. Mutation in FepR causes FepA overexpression and resistance to norfloxacin and ciprofloxacin (Guerin et al. 2014).

Members of the ABC family of transporters are, in contrast to the other pump families discussed, energized by ATP hydrolysis. PatAB of *S. pneumoniae* (norfloxacin and ciprofloxacin) (Boncoeur et al. 2012), SatAB of *S. suis* (norfloxacin and ciprofloxacin) (Escudero et al. 2011), and LmrA of *L. lactis* (ciprofloxacin and ofloxacin) (Poelarends et al. 2000; Putman et al. 2000) all have been shown to confer resistance to some quinolones.

ALTERED PERMEATION IN GRAM-NEGATIVE BACTERIA

In Gram-negative bacteria, the majority of efflux pumps that can effect quinolone resistance are members of the resistance–nodulation–division (RND) superfamily (Li et al. 2015). The RND pumps are secondary antiporters composed of a pump protein localized in the cytoplasmic membrane, an outer membrane channel protein, and a membrane fusion protein that links the pump and the outer membrane protein (Du et al. 2014). Some outer membrane components may link to more than one pump–fusion protein pair, enabling

export of substrates across both inner and outer membranes (Li and Nikaido 2009). The best-studied systems have been in *E. coli* and *Pseudomonas aeruginosa*.

In *E. coli*, the AcrAB-TolC pump complex has been extensively studied. Crystal structures of the complex have revealed a trimer of AcrB pump monomers that rotate around a central axis perpendicular to the membrane, with each monomer as its rotation position changes assuming a different conformation mediating different steps in substrate binding and extrusion through the channel (Nikaido and Takatsuka 2009). Substrates enter the vestibule of AcrB from the periplasmic space between the inner and outer membranes or the outer leaflet of the inner membrane. Binding sites for ciprofloxacin and other substrates of diverse chemical types have been identified in the central cavity of the periplasmic domain of AcrB (Yu et al. 2003, 2005; Li and Nikaido 2004). Fluoroquinolones, which are zwitterionic, are presumed to cross the outer membrane through OmpF and OmpC porin diffusion channels, down-regulation or mutation of which may amplify resistance. Mutations in the MarR regulator can result in both an increase in *acrB* expression as well as a decrease in *ompF* expression, dually contributing to quinolone resistance (Alekshun and Levy 1999). Mutations in the *E. coli* SoxRS (Miller et al. 1994; Chou et al. 1998) and Rob (Jair et al. 1996) regulons can also effect resistance to fluoroquinolones in part related to reductions in OmpF and in a manner that is dependent on AcrAB-TolC. Expression of AcrAB-TolC also confers resistance to bile salts and is induced by bile salts, likely one of its natural substrates (Rosenberg et al. 2003). Thus, AcrAB supports the ability of *E. coli* to survive in its natural habitat, the lower gastrointestinal tract, and perhaps only incidentally affects quinolone susceptibility.

In *P. aeruginosa*, the OprF porin channel has permeability a 100-fold lower than that of OmpF in *E. coli* (Nikaido et al. 1991), contributing to its intrinsic resistance to quinolones and other antimicrobial agents relative to *E. coli* and other enteric bacteria. The MexAB-OprM efflux pump, a RND pump similar in structure to AcrAB-TolC and expressed in wild-type strains, acts in concert with the low permeability of OprF to augment the intrinsic level of resistance of *P. aeruginosa* to fluoroquinolones (Li et al. 2000b). Mutations in *mexA* and *oprM* cause increased uptake of norfloxacin and increased susceptibility to fluoroquinolones (Poole et al. 1996b). Increased expression of MexAB-OprM because of mutations in the MexR negative regulator causes increased resistance to ciprofloxacin and nalidixic acid, and *mexR* mutants can be selected with exposure to fluoroquinolones (Poole et al. 1993). *P. aeruginosa* also has three other efflux pump systems that include quinolones in their substrate profiles, MexCD-OprJ, MexEF-OprN, and MexXY-OprM (Masuda et al. 2000). These pumps vary in expression levels in wild-type strains (Li et al. 2000a), but resistant mutants overexpressing these pumps can be selected with fluoroquinolones and other antimicrobial substrates (Köhler et al. 1997b). Mutation in the NfxB repressor, which is encoded upstream of the *mexCD-oprJ* operon, causes increased expression of MexCD-OprJ and resistance to fluoroquinolones (Poole et al. 1996a). Mutation in *nfxC* results in overexpression of MexEF-OprN, but the exact regulatory mechanism is not yet known (Köhler et al. 1997a). Mutations in the global regulator MvaT, which affects quorum sensing and virulence, also causes increased expression of *mexEF-oprM* and resistance to norfloxacin (Westfall et al. 2006). Expression of both MexEF-OprN and MexCD-OprJ vary inversely with the level of expression of MexAB-OprM, but the mechanisms underlying this property have not yet been elucidated (Li et al. 2000a). Mutations in the MexZ repressor cause increased expression of MexXY-OprM and resistance to fluoroquinolones, aminoglycosides, and other pump substrates (Matsuo et al. 2004; Hay et al. 2013). Specific quinolones differ in the mutations they most commonly select (Köhler et al. 1997b). Most quinolones in clinical use have a fluorine at position 6 and a positively charged substituent at position 7 (e.g., norfloxacin, ciprofloxacin, levofloxacin, and moxifloxacin) and tend to select *nfxB*-type mutants. In contrast, quinolones lacking a positive charge at position 7 (e.g., nalidixic acid) often select *mexR* and *nfxC*-type mutants, differences presumably reflecting differences in the resistance profiles of the regulated pumps.

Additional RND pumps that cause quinolone resistance have been found in a broad range of Gram-negative bacteria. *Salmonella* spp. (Baucheron et al. 2002) and *Enterobacter aerogenes* (Pradel and Pagès 2002) have AcrAB homologs, and their increased expression has been associated with quinolone resistance. The CmeABC RND pump of *Campylobacter jejuni* contributes to the resistance of mutants selected with enrofloxacin, a veterinary quinolone similar to ciprofloxacin (Lin et al. 2002; Luo et al. 2003). In *Klebsiella pneumoniae*, the OqxAB-TolC pump is encoded on the chromosome (Kim et al. 2009b) and was originally identified in *E. coli* isolates from pigs as a cause of plasmid-mediated resistance to olaquindox, a growth promotant used in swine production; it also confers resistance to quinolones.

Among nonenteric bacteria, in *A. baumannii*, the AdeIJK RND pump (Fernando et al. 2014) is constitutively expressed and its broad resistance profile includes fluoroquinolones. In addition, overexpression of the AdeABC and AdeFGH RND pumps because of mutation in their respective regulators, AdeRS, a two-component sensor-regulator system, and AdeL, a LysR family transcriptional regulator, also confer a similarly broad resistance profile containing fluoroquinolones (Yoon et al. 2013, 2015). In *Stenotrophomonas maltophilia*, the SmeDEF pump (Alonso et al. 2000; Zhang et al. 2001) has been shown to contribute to quinolone resistance based on pump knockout mutants with increased susceptibility, resistant isolates with increased pump gene expression, and its ability to confer resistance when overexpressed in *E. coli*.

Non-RND efflux pumps are much less common in Gram-negative bacteria. A few cases MFS and MATE pumps associated with quinolone resistance have been identified. Among MFS pumps, in *E. coli*, EmrAB-TolC, an MFS pump that functions in tripartite structure like the RND pumps, is negatively regulated by EmrR and confers resistance to nalidixic acid but not fluoroquinolones (Lomovskaya et al. 1995). MdfA, originally termed CmlA, confers resistance to both chloramphenicol and fluoroquinolones (Yang et al. 2003). MATE pumps include the NorM pump, which can confer quinolone resistance in *Vibrio parahaemolyticus* (Morita et al. 2000), and in anaerobic Gram-negative bacteria the NorA pump of *Bacteroides fragilis* (Miyamae et al. 1998) and the BexA pump of *B. thetaiotaomicron* (Miyamae et al. 2001), which have been shown to efflux fluoroquinolones.

There are additional examples in both Gram-positive and Gram-negative bacteria in which there is evidence of efflux in quinolone-resistant isolates determined by either reduction in resistance with addition of a broad efflux pump inhibitor or reduced quinolone accumulation in resistant cells, but the contributing pump or its regulator have not been identified (Li and Nikaido 2009; Nikaido and Takatsuka 2009). Information on efflux mechanisms and resistance in >50 bacterial species has recently been extensively reviewed (Li and Nikaido 2009; Li et al. 2015) and is beyond the scope of this review. Thus, efflux-mediated resistance to quinolones and many other antimicrobials is widespread. The broad substrate profiles of these pumps link quinolone resistance to multidrug resistance and constitute mechanisms by which use of non-quinolone antimicrobials can also increase quinolone resistance. A similar linkage to multidrug resistance occurs with plasmid-mediated quinolone resistance, which is discussed in the next section.

PLASMID-MEDIATED QUINOLONE RESISTANCE

Plasmid-mediated quinolone resistance (PMQR) was reported in 1998, 31 years after nalidixic acid began to be used clinically and 12 years after modern fluoroquinolones were approved for use (Martínez-Martínez et al. 1998). Transferable nalidixic acid resistance had been sought unsuccessfully in the 1970s (Burman 1977), and plasmid-mediated resistance was thought unlikely to exist because quinolones are synthetic compounds, and adequate resistance can arise

by chromosomal mutations (Courvalin 1990). The first PMQR was discovered in a multiresistant urinary isolate of *K. pneumoniae* from Alabama that could transfer low-level ciprofloxacin resistance to a variety of Gram-negative bacteria. When the responsible gene, named *qnr* and later *qnrA*, was cloned and sequenced facilitating its identification by PCR (Tran and Jacoby 2002), *qnr* was soon found at low-frequency on plasmids in Gram-negative isolates around the world. One *qnrA* plasmid from Shanghai conferred an unusually high-level of resistance and further study disclosed that it carried an additional mechanism for PMQR, namely, modification of certain quinolones by a variant of the common aminoglycoside-modifying acetyltransferase AAC(6′)-Ib (Robicsek et al. 2006). A third mechanism for PMQR was the discovery of two plasmid-encoded quinolone efflux pumps: OqxAB and QepA (Sorensen et al. 2003, Périchon et al. 2007; Yamane et al. 2007). In the last decade, PMQR genes have been found in bacterial isolates worldwide. They reduce bacterial susceptibility to quinolones, usually not to the level of clinical non-susceptibility, but facilitate the selection of mutants with higher level quinolone resistance and promote treatment failure.

Qnr Structure and Function

Cloning and sequencing *qnrA* revealed that it encoded a 218-residue protein with a tandemly repeating unit of five amino acids that indicated membership in the many thousand-member pentapeptide repeat family of proteins (Tran and Jacoby 2002). Further searches led to the discovery of related genes for plasmid-mediated pentapeptide repeat proteins *qnrS* (Hata et al. 2005), *qnrB* (Jacoby et al. 2006), *qnrC* (Wang et al. 2009), *qnrD* (Cavaco et al. 2009), and *qnrVC* (Fonseca and Vicente 2013), as well as chromosomal *qnr* genes in bacteria from a variety of clinical and environmental sources (Rodríguez-Martínez et al. 2008a; Sánchez et al. 2008). These new *qnr* genes generally differed by 35% or more in sequence from *qnrA* and each other. Allelic varieties that differ by 10% or less have been described in almost all families: currently seven for QnrA, 78 for QnrB, one for QnrC, two for QnrD, nine for QnrS, and six for QnrVC (see lahey.org/qnrstudies) (Jacoby et al. 2008).

The first pentapeptide-repeat protein to have its structure determined by X-ray crystallography was MfpA, which is encoded on the chromosome of *M. smegmatus* and other mycobacteria (Hegde et al. 2005) and implicated in quinolone resistance (Montero et al. 2001). MfpA is a dimer, linked carboxyl terminus to carboxyl terminus, and folded into a right-handed quadrilateral β helix with size, shape, and charge mimicking the B-form of DNA and just the size to fit into the cationic G segment DNA-binding saddle of DNA gyrase and topoisomerase IV. In vitro, MfpA inhibits DNA supercoiling by gyrase and, although it fails to block gyrase inhibition by quinolone, it can still confer quinolone resistance to whole cells by competing with DNA to reduce the number of lethal double-strand breaks produced by quinolone (Hegde et al. 2005).

As shown first with purified QnrA1 (Tran and Jacoby 2002), and subsequently with plasmid-encoded QnrB1 (Jacoby et al. 2006) and QnrS1 (Tavio et al. 2014), and with chromosomally encoded AhQnr from *Aeromonas hydrophila* (Xiong et al. 2011) and EfsQnr from *E. faecalis* (Hegde et al. 2011), Qnr proteins do protect DNA gyrase from quinolone inhibition and only inhibit the enzyme at high concentration. Like MfpA, they form rod-like dimers but have additional structural features. The structure of QnrB1 is shown in Figure 1. The quadrilateral β-helix is stabilized by interactions between the middle, usually hydrophobic, amino acid (i) of the pentapeptide repeat and the first polar or hydrophobic residue (i-2), which point inward, whereas the remaining amino acids (i-1, i+1, i+2) are oriented outward, forming a generally anionic surface. Hydrogen bonding between backbone atoms of neighboring coils stabilizes the helix.

The monomers of QnrB1 and AhQnr have projecting loops of eight and 12 amino acids that are important for their activity. Deletion of the small A loop reduces quinolone protection, whereas deletion of the larger B loop or both loops destroys protective activity (Vetting et al. 2011; Xiong et al. 2011). Removal of even a single amino acid in the larger loop compromises protection. Other essential residues in QnrB are found in pentapeptide repeat positions i and i-2, in which alanine substitution for the native amino acid eliminates protection as does deletion of >10 amino acids at the amino terminus or as few as three amino acids from the dimerization module at the carboxyl terminus (Jacoby et al. 2013). EfsQnr lack loops, but EfsQnr differs from MfpA in having a 25-amino-acid flexible extension required for full protective activity (Hegde et al. 2011).

In vitro, more Qnr is required to protect DNA gyrase as the inhibiting concentration of quinolone is increased (Tran and Jacoby 2002). In a gel-displacement assay (Tran et al. 2005) or bacterial two-hybrid system (Kim et al. 2015), Qnr binds to both gyrase holoenzyme and its A and B subunits. Binding to GyrA is reduced by the same amino- and carboxy-terminal and loop B deletions in QnrB that destroy its protective activity, whereas subinhibitory concentrations of ciprofloxacin reduce binding to GyrA but not to GyrB, suggesting that Qnr protects gyrase by blocking access of quinolone to GyrA sites essential for its lethal action.

Many naturally occurring antibiotics and synthetic agents also target DNA gyrase. Qnr protects against compounds with a somewhat quinolone-like structure (Jacoby et al. 2015), for example, 2-pyridone (Flamm et al. 1995), quinazoline-2,4-dione (Huband et al. 2007), or spiropyrimidinetrione (Kern et al. 2011), so it is not strictly quinolone-specific. Qnr, however, does not block agents acting on the GyrB subunit, and it also does not block simocyclinone D8, which, like quinolones, binds to the amino terminus of GyrA and blocks DNA binding (Hearnshaw et al. 2014).

Qnr ORIGIN

Qnr homologs can be found encoded on the chromosome of many Gram-positive as well as Gram-negative bacteria, including species of *Bacillus, Enterococcus, Listeria,* and *Mycobacterium,* and anaerobes such as *Clostridium difficile* and *Clostridium perfringens* (Rodríguez-Martí-

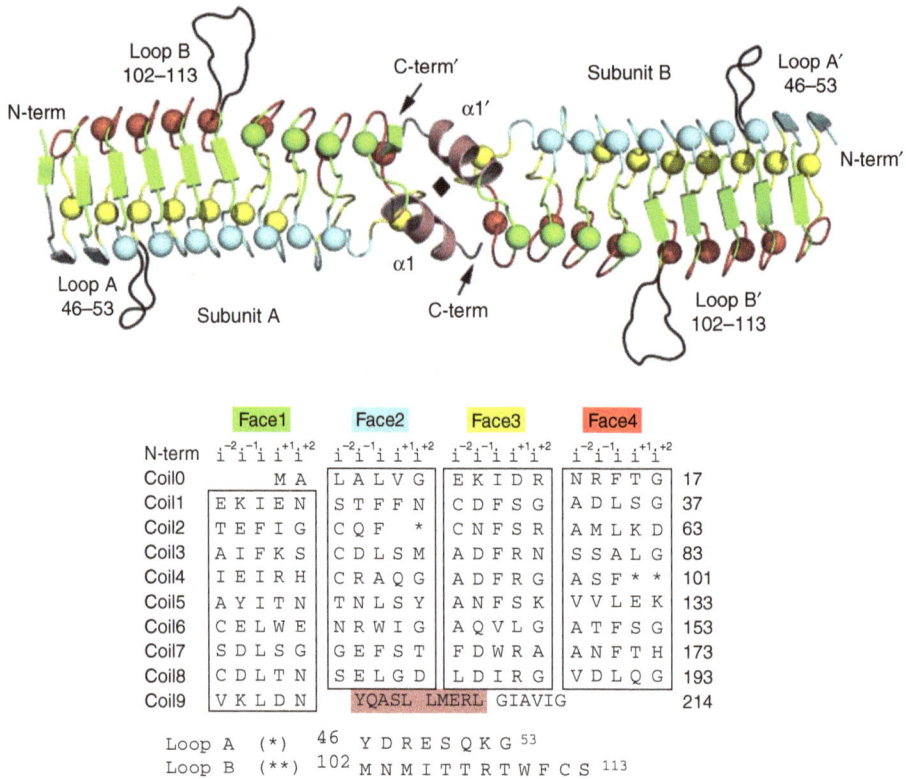

	Face1	Face2	Face3	Face4	
N-term	$i^{-2} i^{-1} i\, i^{+1} i^{+2}$	$i^{-2} i^{-1} i\, i^{+1} i^{+2}$	$i^{-2} i^{-1} i\, i^{+1} i^{+2}$	$i^{-2} i^{-1} i\, i^{+1} i^{+2}$	
Coil0	M A	L A L V G	E K I D R	N R F T G	17
Coil1	E K I E N	S T F F N	C D F S G	A D L S G	37
Coil2	T E F I G	C Q F *	C N F S R	A M L K D	63
Coil3	A I F K S	C D L S M	A D F R N	S S A L G	83
Coil4	I E I R H	C R A Q G	A D F R G	A S F * *	101
Coil5	A Y I T N	T N L S Y	A N F S K	V V L E K	133
Coil6	C E L W E	N R W I G	A Q V L G	A T F S G	153
Coil7	S D L S G	G E F S T	F D W R A	A N F T H	173
Coil8	C D L T N	S E L G D	L D I R G	V D L Q G	193
Coil9	V K L D N	YQASL LMERL GIAVIG			214

Loop A (*)   46 Y D R E S Q K G 53
Loop B (**) 102 M N M I T T R T W F C S 113

Figure 1. The rod-like structure of the QnrB1 dimer is shown (*above*) with the sequence of the monomer (*below*). The sequence is divided into four columns representing the four faces of the right-handed quadrilateral β-helix. Face names and color are shown at the *top* along with the naming convention for the five residues of the pentapeptide repeats. Loops A and B are indicated by one and two asterisks, respectively, with their sequences indicated *below* and the loops shown as black traces on the diagram. The carboxy-terminal α-helix is colored salmon. The molecular twofold symmetry is indicated with a black diamond. Type II turn containing faces are shown as spheres and type IV-containing faces as strands. N-term, Amino terminal; C-term, carboxy terminal. (From Jacoby et al. 2014; reproduced, with permission, from the authors.)

nez et al. 2008b; Sánchez et al. 2008; Boulund et al. 2012; Jacoby and Hooper 2013). Aquatic bacteria are especially well represented, including species of *Aeromonas*, *Photobacterium*, *Shewanella*, and *Vibrio* (Poirel et al. 2005a,b). QnrA1 is 98% identical to the chromosomally determined Qnr of *Shewanella algae* (Poirel et al. 2005a). QnrS1 is 97% identical to Qnr from *Vibrio parahemolyticus* S022 (GenBank accession number WP_029823919) or *Vibrio mytili* (GenBank WP_041155100), and QnrC is 97% identical to Qnr in *V. parahemolyticus* S145 (GenBank WP_025518018). QnrB homologs, on the other hand, are encoded on the chromosome of members of the *Citrobacter freundii*

complex (Jacoby et al. 2011; Ribeiro et al. 2015). The small, nonconjugative plasmids that carry *qnrD* are especially likely to be found in *Proteeae*, such as *Proteus mirabilis*, *P. vulgaris*, and *Providencia rettgeri* and may have originated there (Guillard et al. 2012, 2014; Zhang et al. 2013).

The worldwide distribution of *qnr* suggests an origin well before quinolones were discovered. Indeed, *qnrB* genes and pseudogenes have been discovered on the chromosome of *C. freundii* strains collected in the 1930s (Saga et al. 2013). What the native function of Qnr may have been is an as-yet unanswered question.

AAC(6′)-Ib-cr

AAC(6′)-Ib-cr is a bifunctional variant of a common acetyltransferase, providing resistance to such aminoglycosides as amikacin, kanamycin, and tobramycin, but also able to acetylate those fluoroquinolones with an amino nitrogen on the piperazinyl ring such as ciprofloxacin and norfloxacin (Robicsek et al. 2006). Compared with other AAC(6′)-Ib enzymes, the -cr variant has two unique amino acid substitutions—Trp102Arg and Asp179Tyr—both of which are required for quinolone acetylating activity. Models of enzyme action suggest that the Asp179Tyr replacement is particularly important in permitting π-stacking interactions with the quinolone ring to facilitate quinolone binding. The role of Trp102Arg is to position the Tyr face for optimal interaction (Vetting et al. 2008) or to hydrogen bond to keto or carboxyl groups of the quinolone to fix it in place (Maurice et al. 2008). The *aac(6′)-Ib-cr* gene is usually found in a cassette as part of an integron in a multiresistance plasmid, which may contain other PMQR genes. Several alleles have been described (Quiroga et al. 2015). The gene has been found worldwide in a variety of Enterobacteriaceae and even in *P. aeruginosa* (Ogbolu et al. 2011). Association with extended spectrum β-lactamase (ESBL) CTX-M-15 is particularly common (Oteo et al. 2009; Sabtcheva et al. 2009).

OqxAB AND QepA

OqxAB was first identified as a plasmid-mediated efflux pump conferring resistance to the olaquindox, a food additive enhancing growth in pigs (Sorensen et al. 2003) and later shown to confer resistance to ciprofloxacin and norfloxacin as well as other antimicrobials, including chloramphenicol, nitrofurantoin, and trimethoprim (Hansen et al. 2007; Ho et al. 2016). Genes for *oqxAB* are commonly found on the chromosome of *K. pneumoniae* and *Enterobacter* spp. but are expressed at a much higher level when captured on plasmids, often in association with the IS26 mobilizing element (Rodríguez-Martínez et al. 2013; Wong et al. 2015). In a study of fecal samples from animals and farmworkers in China, where olaquindox is used, 40% of *E. coli* isolated from animals and 30% from humans carried plasmid-mediated *oqxAB*, making it much more common than other types of PMQR in this setting (Zhao et al. 2010).

The QepA efflux pump was reported in 2007 by investigators in Japan and France on plasmids in clinical isolates of *E. coli* often associated with aminoglycoside resistance because of ribosomal methylase *rmtB* (Périchon et al. 2007; Yamane et al. 2007). It has subsequently been found worldwide (Habeeb et al. 2014; Zhao et al. 2015).

QUINOLONE RESISTANCE PLASMIDS

Genes for quinolone resistance have been found on plasmids varying in size and incompatibility specificity, indicating that the spread of multiple plasmids has been responsible for the dissemination of this resistance around the world and that plasmid acquisition of *qnr* and other quinolone resistance determinants has occurred independently multiple times. A mobile or transposable element is almost invariably associated with *qnr* genes, especially ISCR1 and IS26. *qnrD* and *qnrS2* are located within mobile insertion cassettes, elements with bracketing inverted repeats but lacking a transposase (Picão et al. 2008; Guillard et al. 2014), whereas *qnrVC* is so far the only *qnr* gene located in a cassette with a linked *attC* site (Fonseca et al. 2008; Belotti et al. 2015).

qnr genes are usually found in multiresistance plasmids linked to other resistance determinants. β-lactamase genes, including genes for ESBLs, AmpC enzymes, and carbapenemases, have been conspicuously common (Jacoby et al. 2014).

PMQR genes have been found in a variety of Enterobacteriaceae, especially *E. coli* and species of *Enterobacter*, *Klebsiella*, and *Salmonella* (Jacoby et al. 2014). They have rarely been found in nonfermenters but have occasionally been reported in *P. aeruginosa* and *A. baumannii*. *qnr* genes are also found in a variety of Gram-positive organisms but are chromosomal and not plasmid-mediated. The earliest known *qnr* outside of *Citrobacter* spp. dates from 1988 (Jacoby et al. 2009). Studies in the last decade suggest that the prevalence of PMQR is increasing (Kim et al. 2009a). In a recent study of >500 isolates of *E. coli* from 30 county hospitals in China, 37.3% carried at least one PMQR gene, including 19.7% with *aac(6′)-Ib-cr*, 14.4% with *qepA*, 3.8% with *oqxAB*, and 3.7% with *qnr* (Zhao et al. 2015).

RESISTANCE BECAUSE OF PMQR DETERMINANTS

Table 2 shows the effect on susceptibility of a common *E. coli* host of various PMQR genes. *qnr* alleles decrease ciprofloxacin or levofloxacin susceptibility 30-fold to a level similar to that of the common Ser83Leu GyrA mutation but have much less effect on nalidixic acid resistance. The other PMQR genes reduce susceptibility even less but with more specificity. Levofloxacin and nalidixic acid are unaffected by AAC(6′)-Ib-cr because they lack the C7 amino target for acetylation, and QepA affects ciprofloxacin susceptibility more than that of the other quinolones (Périchon et al. 2007). By themselves none reaches the CLSI breakpoint for loss of susceptibility. In combination, however, they may do so, and all facilitate selection of higher-level quinolone resistance (Martínez-Martínez et al. 1998; Robicsek et al. 2006; Rodríguez-Martínez et al. 2007).

From a PMQR-free *E. coli* strain selection on ciprofloxacin at ≥3 times minimum inhibitory concentration (MIC) commonly yields mutants with alterations in the QRDR of *gyrA*. Surprisingly, gyrase mutants are rarely selected from the same strain carrying *qnr* (Cesaro et al. 2008). Rather, mutants have increased expression of *acrAB*, *mdtEF*, or *ydhE* pumps, which efflux quinolones, or alterations in genes of lipopolysaccharide core biosynthesis, which may reduce quinolone entry via reduction in porin expression (Vinué et al. 2015).

Despite the modest effect of PMQR genes on susceptibility, their presence makes infections in animal models harder to treat (Rodríguez-Martínez et al. 2008a; Jakobsen et al. 2012), and there is some evidence for poorer outcomes from human infections with *qnr* containing pathogens (Chong et al. 2010; Liao et al. 2013).

SUMMARY

Resistance to quinolones has increased to substantial levels, despite these agents being synthetic and having two essential bacterial targets. The microbial resistance has been affected by an impressive diversity of mechanisms that have linked quinolone resistance to multidrug resistance, compounding the current public health and medical challenges of broadly resistant bacteria. Dual topoisomerase targets with differing sensitivities to many quinolones in clinical use create a pathway for additive mutational resistance to high levels. This core target-based resistance has been further supplemented and

Table 2. Minimum inhibitory concentrations (MICs) produced in *E. coli*

E. coli strain	MIC (μg/mL)		
	Ciprofloxacin	Levofloxacin	Nalidixic acid
J53	0.008	0.015	4
J53 *gyrA* S83L	0.25	0.5	≥512
J53 pMG252 (*qnrA1*)	0.25	0.5	16
J53 pMG299 (*qnrB1*)	0.25	0.5	16
J53 pMG306 (*qnrS1*)	0.25	0.38	16
J53 pMG320 (*aac(6′)-Ib-cr*)	0.06	0.015	4
J53 pAT851 (*qepA*)	0.064	0.032	4
CLSI susceptibility breakpoint	≤1.0	≤2.0	≤16

From Jacoby et al. 2014.
MIC, Minimum inhibitory concentration.

Cite this article as *Cold Spring Harb Perspect Med* doi: 10.1101/cshperspect.a025320

facilitated by more insidious low-level resistance. This low-level resistance has come from both overexpression of native multidrug efflux pumps as well as the unanticipated emergence of multiple mechanisms of plasmid-mediated quinolone resistance, which may not be readily detected in clinical laboratories but can facilitate selection of higher-level resistance and adds a plasmid linkage to multidrug resistance.

ACKNOWLEDGMENTS

This work is supported by Grants R01 AI057576 (to D.C.H and G.A.J.), R37 AI023988 (to D.C.H.), and P01 AI083214 (to D.C.H.) from the U.S. Public Health Service, National Institutes of Health.

REFERENCES

Aldred KJ, McPherson SA, Wang P, Kerns RJ, Graves DE, Turnbough CL Jr, Osheroff N. 2012. Drug interactions with *Bacillus anthracis* topoisomerase IV: Biochemical basis for quinolone action and resistance. *Biochemistry* **51**: 370–381.

Aldred KJ, Kerns RJ, Osheroff N. 2014. Mechanism of quinolone action and resistance. *Biochemistry* **53**: 1565–1574.

Alekshun MN, Levy SB. 1999. The *mar* regulon: Multiple resistance to antibiotics and other toxic chemicals. *Trends Microbiol* **7**: 410–413.

Alonso A, Martinez JL. 2000. Cloning and characterization of SmeDEF, a novel multidrug efflux pump from *Stenotrophomonas maltophilia*. *Antimicrob Agents Chemother* **44**: 3079–3086.

Baucheron S, Imberechts H, Chaslus-Dancla E, Cloeckaert A. 2002. The AcrB multidrug transporter plays a major role in high-level fluoroquinolone resistance in *Salmonella enterica* serovar typhimurium phage type DT204. *Microb Drug Resist* **8**: 281–289.

Bax BD, Chan PF, Eggleston DS, Fosberry A, Gentry DR, Gorrec F, Giordano I, Hann MM, Hennessy A, Hibbs M, et al. 2010. Type IIA topoisomerase inhibition by a new class of antibacterial agents. *Nature* **466**: 935–940.

Belotti PT, Thabet L, Laffargue A, Andre C, Coulange-Mayonnove L, Arpin C, Messadi A, M'Zali F, Quentin C, Dubois V. 2015. Description of an original integron encompassing *bla*VIM-2, *qnrVC1* and genes encoding bacterial group II intron proteins in *Pseudomonas aeruginosa*. *J Antimicrob Chemother* **70**: 2237–2240.

Blair JM, Bavro VN, Ricci V, Modi N, Cacciotto P, Kleinekathfer U, Ruggerone P, Vargiu AV, Baylay AJ, Smith HE, et al. 2015. AcrB drug-binding pocket substitution confers clinically relevant resistance and altered substrate specificity. *Proc Natl Acad Sci* **112**: 3511–3516.

Blanche F, Cameron B, Bernard FX, Maton L, Manse B, Ferrero L, Ratet N, Lecoq C, Goniot A, Bisch D, et al.

1996. Differential behaviors of *Staphylococcus aureus* and *Escherichia coli* type II DNA topoisomerases. *Antimicrob Agents Chemother* **40**: 2714–2720.

Bolhuis H, Poelarends G, Van Veen HW, Poolman B, Driessen AJM, Konings WN. 1995. The lactococcal *lmrP* gene encodes a proton motive force-dependent drug transporter. *J Biol Chem* **270**: 26092–26098.

Boncoeur E, Durmort C, Bernay B, Ebel C, Di Guilmi AM, Croize J, Vernet T, Jault JM. 2012. PatA and PatB form a functional heterodimeric ABC multidrug efflux transporter responsible for the resistance of *Streptococcus pneumoniae* to fluoroquinolones. *Biochemistry* **51**: 7755–7765.

Boulund F, Johnning A, Pereira MB, Larsson DG, Kristiansson E. 2012. A novel method to discover fluoroquinolone antibiotic resistance (qnr) genes in fragmented nucleotide sequences. *BMC Genomics* **13**: 695.

Breines DM, Ouabdesselam S, Ng EY, Tankovic J, Shah S, Soussy CJ, Hooper DC. 1997. Quinolone resistance locus *nfxD* of *Escherichia coli* is a mutant allele of *parE* gene encoding a subunit of topoisomerase IV. *Antimicrob Agents Chemother* **41**: 175–179.

Burman LG. 1977. Apparent absence of transferable resistance to nalidixic acid in pathogenic Gram-negative bacteria. *J Antimicrob Chemother* **3**: 509–516.

Cavaco LM, Hasman H, Xia S, Aarestrup FM. 2009. *qnrD*, a novel gene conferring transferable quinolone resistance in *Salmonella enterica* serovar Kentucky and Bovismorbificans strains of human origin. *Antimicrob Agents Chemother* **53**: 603–608.

Cesaro A, Bettoni RR, Lascols C, Merens A, Soussy CJ, Cambau E. 2008. Low selection of topoisomerase mutants from strains of *Escherichia coli* harbouring plasmid-borne qnr genes. *J Antimicrob Chemother* **61**: 1007–1015.

Chen PR, Bae T, Williams WA, Duguid EM, Rice PA, Schneewind O, He C. 2006. An oxidation-sensing mechanism is used by the global regulator MgrA in *Staphylococcus aureus*. *Nat Chem Biol* **2**: 591–595.

Chong YP, Choi SH, Kim ES, Song EH, Lee EJ, Park KH, Cho OH, Kim SH, Lee SO, Kim MN, et al. 2010. Bloodstream infections caused by qnr-positive Enterobacteriaceae: Clinical and microbiologic characteristics and outcomes. *Diagn Microbiol Infect Dis* **67**: 70–77.

Chou JH, Greenberg JT, Demple B. 1998. Postranscriptional repression of *Escherichia coli* OmpF protein in response to redox stress: Positive control of the *micF* antisense RNA by the *soxRS* locus. *J Bacteriol* **175**: 1026–1031.

Courvalin P. 1990. Plasmid-mediated 4-quinolone resistance: A real or apparent absence? *Antimicrob Agents Chemother* **34**: 681–684.

Ding Y, Onodera Y, Lee JC, Hooper DC. 2008. NorB, an efflux pump in *Staphylococcus aureus* MW2, contributes to bacterial fitness in abscesses. *J Bacteriol* **190**: 7123–7129.

Domagala JM, Hagen SE. 2003. Structure-activity relationships of the quinolone antibacterials in the new millennium: Some things change and some do not. In *Quinolone antimicrobial agents*, 3rd ed. (ed. Hooper DC, Rubinstein E), pp. 3–18. ASM, Washington, DC.

Drlica K, Zhao XL. 1997. DNA gyrase, topoisomerase IV, and the 4-quinolones. *Microbiol Rev* **61**: 377–392.

Drlica K, Malik M, Kerns RJ, Zhao X. 2008. Quinolone-mediated bacterial death. *Antimicrob Agents Chemother* **52:** 385–392.

Drlica K, Hiasa H, Kerns R, Malik M, Mustaev A, Zhao X. 2009. Quinolones: Action and resistance updated. *Curr Top Med Chem* **9:** 981–998.

Du D, Wang Z, James NR, Voss JE, Klimont E, Ohene-Agyei T, Venter H, Chiu W, Luisi BF. 2014. Structure of the AcrAB-TolC multidrug efflux pump. *Nature* **509:** 512–515.

Escudero JA, San MA, Gutierrez B, Hidalgo L, La Ragione RM, AbuOun M, Galimand M, Ferrandiz MJ, Dominguez L, de la Campa AG, et al. 2011. Fluoroquinolone efflux in *Streptococcus suis* is mediated by SatAB and not by SmrA. *Antimicrob Agents Chemother* **55:** 5850–5860.

Fass D, Bogden CE, Berger JM. 1999. Quaternary changes in topoisomerase II may direct orthogonal movement of two DNA strands. *Nature Struct Biol* **6:** 322–326.

Fernando DM, Xu W, Loewen PC, Zhanel GG, Kumar A. 2014. Triclosan can select for an AdeIJK-overexpressing mutant of *Acinetobacter baumannii* ATCC 17978 that displays reduced susceptibility to multiple antibiotics. *Antimicrob Agents Chemother* **58:** 6424–6431.

Flamm RK, Vojtko C, Chu DT, Li Q, Beyer J, Hensey D, Ramer N, Clement JJ, Tanaka SK. 1995. In vitro evaluation of ABT-719, a novel DNA gyrase inhibitor. *Antimicrob Agents Chemother* **39:** 964–970.

Floyd JL, Smith KP, Kumar SH, Floyd JT, Varela MF. 2010. LmrS is a multidrug efflux pump of the major facilitator superfamily from *Staphylococcus aureus*. *Antimicrob Agents Chemother* **54:** 5406–5412.

Fonseca EL, Vicente AC. 2013. Epidemiology of *qnrVC* alleles and emergence out of the Vibrionaceae family. *J Med Microbiol* **62:** 1628–1630.

Fonseca EL, Dos Santos Freitas F, Vieira VV, Vicente AC. 2008. New *qnr* gene cassettes associated with superintegron repeats in *Vibrio cholerae* O1. *Emerg Infect Dis* **14:** 1129–1131.

Fournier B, Hooper DC. 2000. A new two-component regulatory system involved in adhesion autolysis, and extracellular proteolytic activity of *Staphylococcus aureus*. *J Bacteriol* **182:** 3955–3964.

Fournier B, Zhao X, Lu T, Drlica K, Hooper DC. 2000. Selective targeting of topoisomerase IV and DNA gyrase in *Staphylococcus aureus*: Different patterns of quinolone-induced inhibition of DNA synthesis. *Antimicrob Agents Chemother* **44:** 2160–2165.

Fournier B, Klier A, Rapoport G. 2001. The two-component system ArlS-ArlR is a regulator of virulence gene expression in *Staphylococcus aureus*. *Mol Microbiol* **41:** 247–261.

Gensberg K, Jin YF, Piddock LJ. 1995. A novel *gyrB* mutation in a fluoroquinolone-resistant clinical isolate of *Salmonella typhimurium*. *FEMS Microbiol Lett* **132:** 57–60.

Gill MJ, Brenwald NP, Wise R. 1999. Identification of an efflux pump gene, *pmrA*, associated with fluoroquinolone resistance in *Streptococcus pneumoniae*. *Antimicrob Agents Chemother* **43:** 187–189.

Godreuil S, Galimand M, Gerbaud G, Jacquet C, Courvalin P. 2003. Efflux pump Lde is associated with fluoroquinolone resistance in *Listeria monocytogenes*. *Antimicrob Agents Chemother* **47:** 704–708.

Grkovic S, Brown MH, Skurray RA. 2002. Regulation of bacterial drug export systems. *Microbiol Mol Biol Rev* **66:** 671–701.

Guerin F, Galimand M, Tuambilangana F, Courvalin P, Cattoir V. 2014. Overexpression of the novel MATE fluoroquinolone efflux pump FepA in *Listeria monocytogenes* is driven by inactivation of its local repressor FepR. *PLoS ONE* **9:** e106340.

Guillard T, Cambau E, Neuwirth C, Nenninger T, Mbadi A, Brasme L, Vernet-Garnier V, Bajolet O, de Champs C. 2012. Description of a 2,683-base-pair plasmid containing *qnrD* in two *Providencia rettgeri* isolates. *Antimicrob Agents Chemother* **56:** 565–568.

Guillard T, Grillon A, de Champs C, Cartier C, Madoux J, Bercot B, Lebreil AL, Lozniewski A, Riahi J, Vernet-Garnier V, et al. 2014. Mobile insertion cassette elements found in small non-transmissible plasmids in *Proteeae* may explain *qnrD* mobilization. *PLoS ONE* **9:** e87801.

Habeeb MA, Haque A, Iversen A, Giske CG. 2014. Occurrence of virulence genes, 16S rRNA methylases, and plasmid-mediated quinolone resistance genes in CTX-M-producing *Escherichia coli* from Pakistan. *Eur J Clin Microbiol Infect Dis* **33:** 399–409.

Hansen LH, Jensen LB, Sorensen HI, Sorensen SJ. 2007. Substrate specificity of the OqxAB multidrug resistance pump in *Escherichia coli* and selected enteric bacteria. *J Antimicrob Chemother* **60:** 145–147.

Hata M, Suzuki M, Matsumoto M, Takahashi M, Sato K, Ibe S, Sakae K. 2005. Cloning of a novel gene for quinolone resistance from a transferable plasmid in *Shigella flexneri* 2b. *Antimicrob Agents Chemother* **49:** 801–803.

Hay T, Fraud S, Lau CH, Gilmour C, Poole K. 2013. Antibiotic inducibility of the *mexXY* multidrug efflux operon of *Pseudomonas aeruginosa*: Involvement of the MexZ anti-repressor ArmZ. *PLoS ONE* **8:** e56858.

Hearnshaw SJ, Edwards MJ, Stevenson CE, Lawson DM, Maxwell A. 2014. A new crystal structure of the bifunctional antibiotic simocyclinone D8 bound to DNA gyrase gives fresh insight into the mechanism of inhibition. *J Mol Biol* **426:** 2023–2033.

Heeb S, Fletcher MP, Chhabra SR, Diggle SP, Williams P, Camara M. 2011. Quinolones: From antibiotics to autoinducers. *FEMS Microbiol Rev* **35:** 247–274.

Hegde SS, Vetting MW, Roderick SL, Mitchenall LA, Maxwell A, Takiff HE, Blanchard JS. 2005. A fluoroquinolone resistance protein from *Mycobacterium tuberculosis* that mimics DNA. *Science* **308:** 1480–1483.

Hegde SS, Vetting MW, Mitchenall LA, Maxwell A, Blanchard JS. 2011. Structural and biochemical analysis of the pentapeptide repeat protein *Efs*Qnr, a potent DNA gyrase inhibitor. *Antimicrob Agents Chemother* **55:** 110–117.

Hiasa H. 2002. The Glu-84 of the ParC subunit plays critical roles in both topoisomerase IV-quinolone and topoisomerase IV-DNA interactions. *Biochemistry* **41:** 11779–11785.

Hiasa H, Yousef DO, Marians KJ. 1996. DNA strand cleavage is required for replication fork arrest by a frozen topoisomerase-quinolone-DNA ternary complex. *J Biol Chem* **271:** 26424–26429.

Ho PL, Ng KY, Lo WU, Law PY, Lai EL, Wang Y, Chow KH. 2016. Plasmid-mediated OqxAB is an important mechanism for nitrofurantoin resistance in *Escherichia coli*. *Antimicrob Agents Chemother* **60:** 537–543.

Hooper DC. 1997. Bacterial topoisomerases, anti-topoisomerases, and anti-topoisomerase resistance. *Clin Infect Dis* **27:** S54–S63.

Hooper DC. 2003. Mechanisms of quinolone resistance. In *Quinolone antimicrobial agents*, 3rd ed. (ed. Hooper DC, Rubinstein E), pp. 41–67. ASM, Washington, DC.

Howard BM, Pinney RJ, Smith JT. 1993. 4-Quinolone bactericidal mechanisms. *Arzneimittel-Forschung* **43:** 1125–1129.

Huang J, O'Toole PW, Shen W, Amrine-Madsen H, Jiang X, Lobo N, Palmer LM, Voelker L, Fan F, Gwynn MN, et al. 2004. Novel chromosomally encoded multidrug efflux transporter MdeA in *Staphylococcus aureus*. *Antimicrob Agents Chemother* **48:** 909–917.

Huband MD, Cohen MA, Zurack M, Hanna DL, Skerlos LA, Sulavik MC, Gibson GW, Gage JW, Ellsworth E, Stier MA, et al. 2007. In vitro and in vivo activities of PD 0305970 and PD 0326448, new bacterial gyrase/topoisomerase inhibitors with potent antibacterial activities versus multidrug-resistant gram-positive and fastidious organism groups. *Antimicrob Agents Chemother* **51:** 1191–1201.

Ingavale S, Van Wamel W, Luong TT, Lee CY, Cheung AL. 2005. Rat/MgrA, a regulator of autolysis, is a regulator of virulence genes in *Staphylococcus aureus*. *Infect Immun* **73:** 1423–1431.

Ito H, Yoshida H, Bogaki-Shonai M, Niga T, Hattori H, Nakamura S. 1994. Quinolone resistance mutations in the DNA gyrase *gyrA* and *gyrB* genes of *Staphylococcus aureus*. *Antimicrob Agents Chemother* **38:** 2014–2023.

Jacoby GA, Hooper DC. 2013. Phylogenetic analysis of chromosomally determined Qnr and related proteins. *Antimicrob Agents Chemother* **57:** 1930–1934.

Jacoby GA, Walsh KE, Mills DM, Walker VJ, Oh H, Robicsek A, Hooper DC. 2006. *qnrB*, another plasmid-mediated gene for quinolone resistance. *Antimicrob Agents Chemother* **50:** 1178–1182.

Jacoby G, Cattoir V, Hooper D, Martínez-Martínez L, Nordmann P, Pascual A, Poirel L, Wang M. 2008. *qnr* gene nomenclature. *Antimicrob Agents Chemother* **52:** 2297–2299.

Jacoby GA, Gacharna N, Black TA, Miller GH, Hooper DC. 2009. Temporal appearance of plasmid-mediated quinolone resistance genes. *Antimicrob Agents Chemother* **53:** 1665–1666.

Jacoby GA, Griffin CM, Hooper DC. 2011. *Citrobacter* spp. as a source of *qnrB* alleles. *Antimicrob Agents Chemother* **55:** 4979–4984.

Jacoby GA, Corcoran MA, Mills DM, Griffin CM, Hooper DC. 2013. Mutational analysis of quinolone resistance protein QnrB1. *Antimicrob Agents Chemother* **57:** 5733–5736.

Jacoby GA, Strahilevitz J, Hooper DC. 2014. Plasmid-mediated quinolone resistance. *Microbiol Spectr* doi: 10.1128/microbiolspec.PLAS-0006-2013.

Jacoby GA, Corcoran MA, Hooper DC. 2015. The protective effect of Qnr on agents other than quinolones that target

DNA gyrase. *Antimicrob Agent Chemother* **59:** 6689–6695.

Jair KW, Yu X, Skarstad K, Thöny B, Fujita N, Ishihama A, Wolf RE Jr, 1996. Transcriptional activation of promoters of the superoxide and multiple antibiotic resistance regulons by Rob, a binding protein of the *Escherichia coli* origin of chromosomal replication. *J Bacteriol* **178:** 2507–2513.

Jakobsen L, Cattoir V, Jensen KS, Hammerum AM, Nordmann P, Frimodt-Moller N. 2012. Impact of low-level fluoroquinolone resistance genes *qnrA1*, *qnrB19* and *qnrS1* on ciprofloxacin treatment of isogenic *Escherichia coli* strains in a murine urinary tract infection model. *J Antimicrob Chemother* **67:** 2438–2444.

Kaatz GW, DeMarco CE, Seo SM. 2006. MepR, a repressor of the *Staphylococcus aureus* MATE family multidrug efflux pump MepA, is a substrate-responsive regulatory protein. *Antimicrob Agents Chemother* **50:** 1276–1281.

Kern G, Basarab GS, Andrews B, Schuck V, Stone G, Kutschke A, Beaudoin M-E, San Martin M, Brassil P, Fan JH, et al. 2011. A DNA gyrase inhibitor with a novel mode of inhibition and in vivo efficacy. *51st Interscience Conference on Antimicrobial Agents and Chemotherapy (ICAAC)*, Abstract F1-1840. Chicago, September 17–20.

Khodursky AB, Cozzarelli NR. 1998. The mechanism of inhibition of topoisomerase IV by quinolone antibacterials. *J Biol Chem* **273:** 27668–27677.

Khodursky AB, Zechiedrich EL, Cozzarelli NR. 1995. Topoisomerase IV is a target of quinolones in *Escherichia coli*. *Proc Natl Acad Sci* **92:** 11801–11805.

Kim ES, Hooper DC. 2014. Clinical importance and epidemiology of quinolone resistance. *Infect Chemother* **46:** 226–238.

Kim HB, Park CH, Kim CJ, Kim EC, Jacoby GA, Hooper DC. 2009a. Prevalence of plasmid-mediated quinolone resistance determinants over a 9-year period. *Antimicrob Agents Chemother* **53:** 639–645.

Kim HB, Wang M, Park CH, Kim EC, Jacoby GA, Hooper DC. 2009b. *oqxAB* encoding a multidrug efflux pump in human clinical isolates of Enterobacteriaceae. *Antimicrob Agents Chemother* **53:** 3582–3584.

Kim ES, Chen C, Braun M, Kim HY, Okumura R, Wang Y, Jacoby GA, Hooper DC. 2015. Interactions between QnrB, QnrB mutants, and DNA gyrase. *Antimicrob Agents Chemother* **59:** 5413–5419.

Klyachko KA, Schuldiner S, Neyfakh AA. 1997. Mutations affecting substrate specificity of the *Bacillus subtilis* multidrug transporter Bmr. *J Bacteriol* **179:** 2189–2193.

Köhler T, Michea-Hamzehpour M, Henze U, Gotoh N, Curty L, Pechère JC. 1997a. Characterization of MexE–MexF–OprN, a positively regulated multidrug efflux system of *Pseudomonas aeruginosa*. *Mol Microbiol* **23:** 345–354.

Köhler T, Michea-Hamzehpour M, Plesiat P, Kahr AL, Pechère JC. 1997b. Differential selection of multidrug efflux systems by quinolones in *Pseudomonas aeruginosa*. *Antimicrob Agents Chemother* **41:** 2540–2543.

Kreuzer KN, Cozzarelli NR. 1979. *Escherichia coli* mutants thermosensitive for deoxyribonucleic acid gyrase subunit A: Effects on deoxyribonucleic acid replication, transcription, and bacteriophage growth. *J Bacteriol* **140:** 424–435.

Kumar R, Shankar MB, Nagaraja V. 2014. Molecular basis for the differential quinolone susceptibility of mycobacterial DNA gyrase. *Antimicrob Agents Chemother.*

Kumaraswami M, Schuman JT, Seo SM, Kaatz GW, Brennan RG. 2009. Structural and biochemical characterization of MepR, a multidrug binding transcription regulator of the *Staphylococcus aureus* multidrug efflux pump MepA. *Nucleic Acids Res* **37:** 1211–1224.

Laponogov I, Sohi MK, Veselkov DA, Pan XS, Sawhney R, Thompson AW, McAuley KE, Fisher LM, Sanderson MR. 2009. Structural insight into the quinolone-DNA cleavage complex of type IIA topoisomerases. *Nat Struct Mol Biol* **16:** 667–669.

Laponogov I, Veselkov DA, Crevel IM, Pan XS, Fisher LM, Sanderson MR. 2013. Structure of an "open" clamp type II topoisomerase-DNA complex provides a mechanism for DNA capture and transport. *Nucleic Acids Res* **41:** 9911–9923.

Lesher GY, Forelich ED, Gruet MD, Bailey JH, Brundage RP. 1962. 1,8-Naphthyridine derivatives. A new class of chemotherapeutic agents. *J Med Pharm Chem* **5:** 1063–1068.

Lewin CS, Howard BM, Smith JT. 1991. Protein- and RNA-synthesis independent bactericidal activity of ciprofloxacin that involves the A subunit of DNA gyrase. *J Med Microbiol* **34:** 19–22.

Li XZ, Nikaido H. 2004. Efflux-mediated drug resistance in bacteria. *Drugs* **64:** 159–204.

Li XZ, Nikaido H. 2009. Efflux-mediated drug resistance in bacteria: An update. *Drugs* **69:** 1555–1623.

Li XZ, Barré N, Poole K. 2000a. Influence of the MexA–MexB–OprM multidrug efflux system on expression of the MexC–MexD–OprJ and MexE–MexF–OprN multidrug efflux systems in *Pseudomonas aeruginosa*. *J Antimicrob Chemother* **46:** 885–893.

Li XZ, Zhang L, Poole K. 2000b. Interplay between the MexA–MexB–OprM multidrug efflux system and the outer membrane barrier in the multiple antibiotic resistance of *Pseudomonas aeruginosa*. *J Antimicrob Chemother* **45:** 433–436.

Li XZ, Plesiat P, Nikaido H. 2015. The challenge of efflux-mediated antibiotic resistance in Gram-negative bacteria. *Clin Microbiol Rev* **28:** 337–418.

Liao CH, Hsueh PR, Jacoby GA, Hooper DC. 2013. Risk factors and clinical characteristics of patients with *qnr*-positive *Klebsiella pneumoniae* bacteraemia. *J Antimicrob Chemother* **68:** 2907–2914.

Lin J, Michel LO, Zhang QJ. 2002. CmeABC functions as a multidrug efflux system in *Campylobacter jejuni*. *Antimicrob Agents Chemother* **46:** 2124–2131.

Lomovskaya O, Lewis K, Matin A. 1995. EmrR is a negative regulator of the *Escherichia coli* multidrug resistance pump EmrAB. *J Bacteriol* **177:** 2328–2334.

Lomovskaya O, Lee A, Hoshino K, Ishida H, Mistry A, Warren MS, Boyer E, Chamberland S, Lee VJ. 1999. Use of a genetic approach to evaluate the consequences of inhibition of efflux pumps in *Pseudomonas aeruginosa*. *Antimicrob Agents Chemother* **43:** 1340–1346.

Luo N, Sahin O, Lin J, Michel LO, Zhang QJ. 2003. In vivo selection of *Campylobacter* isolates with high levels of fluoroquinolone resistance associated with *gyrA* muta-

tions and the function of the CmeABC efflux pump. *Antimicrob Agents Chemother* **47:** 390–394.

Martínez-Martínez L, Pascual A, Jacoby GA. 1998. Quinolone resistance from a transferable plasmid. *Lancet* **351:** 797–799.

Masuda N, Sakagawa E, Ohya S, Gotoh N, Tsujimoto H, Nishino T. 2000. Substrate specificities of MexAB–OprM, MexCD–OprJ, and MexXY–OprM efflux pumps in *Pseudomonas aeruginosa*. *Antimicrob Agents Chemother* **44:** 3322–3327.

Matsuo Y, Eda S, Gotoh N, Yoshihara E, Nakae T. 2004. MexZ-mediated regulation of *mexXY* multidrug efflux pump expression in *Pseudomonas aeruginosa* by binding on the *mexZ*-*mexX* intergenic DNA. *FEMS Microbiol Lett* **238:** 23–28.

Maurice F, Broutin I, Podglajen I, Benas P, Collatz E, Dardel F. 2008. Enzyme structural plasticity and the emergence of broad-spectrum antibiotic resistance. *EMBO Rep* **9:** 344–349.

Mégraud F. 1998. Epidemiology and mechanism of antibiotic resistance in *Helicobacter pylori*. *Gastroenterology* **115:** 1278–1282.

Miller PF, Gambino L, Sulavik MC, Gracheck SJ. 1994. Genetic relationship between *soxRS* and *mar* loci in promoting multiple antibiotic resistance in *Escherichia coli*. *Antimicrob Agents Chemother* **38:** 1773–1779.

Miyamae S, Nikaido H, Tanaka Y, Yoshimura F. 1998. Active efflux of norfloxacin by *Bacteroides fragilis*. *Antimicrob Agents Chemother* **42:** 2119–2121.

Miyamae S, Ueda O, Yoshimura F, Hwang J, Tanaka Y, Nikaido H. 2001. A MATE family multidrug efflux transporter pumps out fluoroquinolones in *Bacteroides thetaiotaomicron*. *Antimicrob Agents Chemother* **45:** 3341–3346.

Montero C, Mateu G, Rodriguez R, Takiff H. 2001. Intrinsic resistance of *Mycobacterium smegmatis* to fluoroquinolones may be influenced by new pentapeptide protein MfpA. *Antimicrob Agents Chemother* **45:** 3387–3392.

Morais Cabral JH, Jackson AP, Smith CV, Shikotra N, Maxwell A, Liddington RC. 1997. Crystal structure of the breakage-reunion domain of DNA gyrase. *Nature* **388:** 903–906.

Morita Y, Kataoka A, Shiota S, Mizushima T, Tsuchiya T. 2000. NorM of *Vibrio parahaemolyticus* is an Na$^+$-driven multidrug efflux pump. *J Bacteriol* **182:** 6694–6697.

Mustaev A, Malik M, Zhao X, Kurepina N, Luan G, Oppegard LM, Hiasa H, Marks KR, Kerns RJ, Berger JM, et al. 2014. Fluoroquinolone-gyrase-DNA complexes: Two modes of drug binding. *J Biol Chem* **289:** 12300–12312.

Nakaminami H, Noguchi N, Sasatsu M. 2010. Fluoroquinolone efflux by the plasmid-mediated multidrug efflux pump QacB variant QacBIII in *Staphylococcus aureus*. *Antimicrob Agents Chemother* **54:** 4107–4111.

Ng EY, Trucksis M, Hooper DC. 1996. Quinolone resistance mutations in topoisomerase IV: Relationship of the *flqA* locus and genetic evidence that topoisomerase IV is the primary target and DNA gyrase the secondary target of fluoroquinolones in *Staphylococcus aureus*. *Antimicrob Agents Chemother* **40:** 1881–1888.

Nikaido H, Takatsuka Y. 2009. Mechanisms of RND multidrug efflux pumps. *Biochim Biophys Acta* **1794:** 769–781.

Nikaido H, Nikaido K, Harayama S. 1991. Identification and characterization of porins in *Pseudomonas aeruginosa*. *J Biol Chem* **266:** 770–779.

Ogbolu DO, Daini OA, Ogunledun A, Alli AO, Webber MA. 2011. High levels of multidrug resistance in clinical isolates of Gram-negative pathogens from Nigeria. *Int J Antimicrob Agents* **37:** 62–66.

Ohki R, Murata M. 1997. *bmr3*, a third multidrug transporter gene of *Bacillus subtilis*. *J Bacteriol* **179:** 1423–1427.

Oteo J, Cuevas O, López-Rodríguez I, Banderas-Florido A, Vindel A, Pérez-Vázquez M, Bautista V, Arroyo M, García-Caballero J, Marin-Casanova P, et al. 2009. Emergence of CTX-M-15-producing *Klebsiella pneumoniae* of multilocus sequence types 1, 11, 14, 17, 20, 35 and 36 as pathogens and colonizers in newborns and adults. *J Antimicrob Chemother* **64:** 524–528.

Owens RC Jr, Ambrose PG. 2000. Clinical use of the fluoroquinolones. *Med Clin N Am* **84:** 1447–1469.

Pan XS, Fisher LM. 1997. Targeting of DNA gyrase in *Streptococcus pneumoniae* by sparfloxacin: Selective targeting of gyrase or topoisomerase IV by quinolones. *Antimicrob Agents Chemother* **41:** 471–474.

Pan XS, Fisher LM. 1998. DNA gyrase and topoisomerase IV are dual targets of clinafloxacin action in *Streptococcus pneumoniae*. *Antimicrob Agents Chemother* **42:** 2810–2816.

Pan XS, Fisher LM. 1999. *Streptococcus pneumoniae* DNA gyrase and topoisomerase IV: Overexpression, purification, and differential inhibition by fluoroquinolones. *Antimicrob Agents Chemother* **43:** 1129–1136.

Périchon B, Courvalin P, Galimand M. 2007. Transferable resistance to aminoglycosides by methylation of G1405 in 16S rRNA and to hydrophilic fluoroquinolones by QepA-mediated efflux in *Escherichia coli*. *Antimicrob Agents Chemother* **51:** 2464–2469.

Picão RC, Poirel L, Demarta A, Silva CS, Corvaglia AR, Petrini O, Nordmann P. 2008. Plasmid-mediated quinolone resistance in *Aeromonas allosaccharophila* recovered from a Swiss lake. *J Antimicrob Chemother* **62:** 948–950.

Piddock LJ, Walters RN, Diver JM. 1990. Correlation of quinolone MIC and inhibition of DNA, RNA, and protein synthesis and induction of the SOS response in *Escherichia coli*. *Antimicrob Agents Chemother* **34:** 2331–2336.

Poelarends GJ, Mazurkiewicz P, Putman M, Cool RH, Veen HW, Konings WN. 2000. An ABC-type multidrug transporter of *Lactococcus lactis* possesses an exceptionally broad substrate specificity. *Drug Resist Updat* **3:** 330–334.

Poirel L, Liard A, Rodriguez-Martinez JM, Nordmann P. 2005a. Vibrionaceae as a possible source of Qnr-like quinolone resistance determinants. *J Antimicrob Chemother* **56:** 1118–1121.

Poirel L, Rodriguez-Martinez JM, Mammeri H, Liard A, Nordmann P. 2005b. Origin of plasmid-mediated quinolone resistance determinant QnrA. *Antimicrob Agents Chemother* **49:** 3523–3525.

Poole K, Krebes K, McNally C, Neshat S. 1993. Multiple antibiotic resistance in *Pseudomonas aeruginosa*: Evidence for involvement of an efflux operon. *J Bacteriol* **175:** 7363–7372.

Poole K, Gotoh N, Tsujimoto H, Zhao QX, Wada A, Yamasaki T, Neshat S, Yamagishi JI, Li XZ, Nishino T. 1996a. Overexpression of the *mexC–mexD–oprJ* efflux operon in *nfxB*- type multidrug-resistant strains of *Pseudomonas aeruginosa*. *Mol Microbiol* **21:** 713–724.

Poole K, Tetro K, Zhao QX, Neshat S, Heinrichs DE, Bianco N. 1996b. Expression of the multidrug resistance operon *mexA–mexB–oprM* in *Pseudomonas aeruginosa*: *mexR* encodes a regulator of operon expression. *Antimicrob Agents Chemother* **40:** 2021–2028.

Pradel E, Pagès JM. 2002. The AcrAB-TolC efflux pump contributes to multidrug resistance in the nosocomial pathogen *Enterobacter aerogenes*. *Antimicrob Agents Chemother* **46:** 2640–2643.

Putman M, Van Veen HW, Konings WN. 2000. Molecular properties of bacterial multidrug transporters. *Microbiol Mol Biol Rev* **64:** 672–693.

Quiroga MP, Orman B, Errecalde L, Kaufman S, Centron D. 2015. Characterization of Tn*6238* with a new allele of *aac(6′)-Ib-cr*. *Antimicrob Agents Chemother* **59:** 2893–2897.

Ribeiro TG, Novais A, Branquinho R, Machado E, Peixe L. 2015. Phylogeny and comparative genomics unveil independent diversification trajectories of *qnrB* and genetic platforms within particular *Citrobacter* species. *Antimicrob Agents Chemother* **59:** 5951–5958.

Robicsek A, Strahilevitz J, Jacoby GA, Macielag M, Abbanat D, Park CH, Bush K, Hooper DC. 2006. Fluoroquinolone-modifying enzyme: A new adaptation of a common aminoglycoside acetyltransferase. *Nat Med* **12:** 83–88.

Rodríguez-Martínez JM, Velasco C, García I, Cano ME, Martínez-Martínez L, Pascual A. 2007. Mutant prevention concentrations of fluoroquinolones for Enterobacteriaceae expressing the plasmid-carried quinolone resistance determinant *qnrA1*. *Antimicrob Agents Chemother* **51:** 2236–2239.

Rodríguez-Martínez JM, Pichardo C, García I, Pachón-Ibañez ME, Docobo-Pérez F, Pascual A, Pachón J, Martínez-Martínez L. 2008a. Activity of ciprofloxacin and levofloxacin in experimental pneumonia caused by *Klebsiella pneumoniae* deficient in porins, expressing active efflux and producing QnrA1. *Clin Microbiol Infect* **14:** 691–697.

Rodríguez-Martínez JM, Velasco C, Briales A, García I, Conejo MC, Pascual A. 2008b. Qnr-like pentapeptide repeat proteins in Gram-positive bacteria. *J Antimicrob Chemother* **61:** 1240–1243.

Rodríguez-Martínez JM, Díaz de Alba P, Briales A, Machuca J, Lossa M, Fernández-Cuenca F, Rodríguez Baño J, Martínez-Martínez L, Pascual A. 2013. Contribution of OqxAB efflux pumps to quinolone resistance in extended-spectrum-β-lactamase-producing *Klebsiella pneumoniae*. *J Antimicrob Chemother* **68:** 68–73.

Rosenberg EY, Bertenthal D, Nilles ML, Bertrand KP, Nikaido H. 2003. Bile salts and fatty acids induce the expression of *Escherichia coli* AcrAB multidrug efflux pump through their interaction with Rob regulatory protein. *Mol Microbiol* **48:** 1609–1619.

Sabtcheva S, Kaku M, Saga T, Ishii Y, Kantardjiev T. 2009. High prevalence of the *aac(6′)-Ib-cr* gene and its dissemination among Enterobacteriaceae isolates by CTX-M-15 plasmids in Bulgaria. *Antimicrob Agents Chemother* **53:** 335–336.

Saga T, Sabtcheva S, Mitsutake K, Ishii Y, Tateda K, Yama-guchi K, Kaku M. 2013. Characterization of *qnrB*-like genes in *Citrobacter* species of the American type culture collection. *Antimicrob Agents Chemother* **57**: 2863–2866.

Sánchez MB, Hernández A, Rodríguez-Martínez JM, Mar-tínez-Martínez L, Martínez JL. 2008. Predictive analysis of transmissible quinolone resistance indicates *Stenotro-phomonas maltophilia* as a potential source of a novel family of Qnr determinants. *BMC Microbiol* **8**: 148–161.

Schindler BD, Patel D, Seo SM, Kaatz GW. 2013. Mutagen-esis and modeling to predict structural and functional characteristics of the *Staphylococcus aureus* MepA multi-drug efflux pump. *J Bacteriol* **195**: 523–533.

Schindler BD, Frempong-Manso E, DeMarco CE, Kosmidis C, Matta V, Seo SM, Kaatz GW. 2015. Analyses of multi-drug efflux pump-like proteins encoded on the *Staph-ylococcus aureus* chromosome. *Antimicrob Agents Chemo-ther* **59**: 747–748.

Schmitz FJ, Jones ME, Hofmann B, Hansen B, Scheuring S, Lückefahr MFA, Verhoef J, Hadding U, Heinz HP, Köhrer K. 1998. Characterization of *grlA*, *grlB*, *gyrA*, and *gyrB* mutations in 116 unrelated isolates of *Staphylococcus au-reus* and effects of mutations on ciprofloxacin MIC. *Anti-microb Agents Chemother* **42**: 1249–1252.

Shen LL, Kohlbrenner WE, Weigl D, Baranowski J. 1989. Mechanism of quinolone inhibition of DNA gyrase. Ap-pearance of unique norfloxacin binding sites in enzyme-DNA complexes. *J Biol Chem* **264**: 2973–2978.

Sorensen AH, Hansen LH, Johannesen E, Sorensen SJ. 2003. Conjugative plasmid conferring resistance to olaquindox. *Antimicrob Agents Chemother* **47**: 798–799.

Strahilevitz J, Hooper DC. 2005. Dual targeting of topo-isomerase IV and gyrase to reduce mutant selection: Di-rect testing of the paradigm by using WCK-1734, a new fluoroquinolone, and ciprofloxacin. *Antimicrob Agents Chemother* **49**: 1949–1956.

Tavio MM, Jacoby GA, Hooper DC. 2014. QnrS1 structure–activity relationships. *J Antimicrob Chemother* **69**: 2102–2109.

Tran JH, Jacoby GA. 2002. Mechanism of plasmid-mediated quinolone resistance. *Proc Natl Acad Sci* **99**: 5638–5642.

Tran JH, Jacoby GA, Hooper DC. 2005. Interaction of the plasmid-encoded quinolone resistance protein Qnr with *Escherichia coli* DNA gyrase. *Antimicrob Agents Chemo-ther* **49**: 118–125.

Trucksis M, Wolfson JS, Hooper DC. 1991. A novel locus conferring fluoroquinolone resistance in *Staphylococcus aureus*. *J Bacteriol* **173**: 5854–5860.

Truong-Bolduc QC, Hooper DC. 2007. Transcriptional reg-ulators NorG and MgrA modulate resistance to both quinolones and β-lactams in *Staphylococcus aureus*. *J Bacteriol* **189**: 2996–3005.

Truong-Bolduc QC, Hooper DC. 2010. Phosphorylation of MgrA and its effect on expression of the NorA and NorB efflux pumps of *Staphylococcus aureus*. *J Bacteriol* **192**: 2525–2534.

Truong-Bolduc QC, Dunman PM, Strahilevitz J, Projan SJ, Hooper DC. 2005. MgrA is a multiple regulator of two new efflux pumps in *Staphylococcus aureus*. *J Bacteriol* **187**: 2395–2405.

Truong-Bolduc QC, Strahilevitz J, Hooper DC. 2006. NorC, a new efflux pump regulated by MgrA of *Staphylococcus aureus*. *Antimicrob Agents Chemother* **50**: 1104–1107.

Truong-Bolduc QC, Ding Y, Hooper DC. 2008. Posttransla-tional modification influences the effects of MgrA on *norA* expression in *Staphylococcus aureus*. *J Bacteriol* **190**: 7375–7381.

Truong-Bolduc QC, Bolduc GR, Okumura R, Celino B, Bevis J, Liao CH, Hooper DC. 2011a. Implication of the NorB efflux pump in the adaptation of *Staphylococcus aureus* to growth at acid pH and in resistance to moxi-floxacin. *Antimicrob Agents Chemother* **55**: 3214–3219.

Truong-Bolduc QC, Dunman PM, Eidem T, Hooper DC. 2011b. Transcriptional profiling analysis of the global regulator NorG, a GntR-like protein of *Staphylococcus aureus*. *J Bacteriol* **193**: 6207–6214.

Truong-Bolduc QC, Liao C-H, Villet R, Bolduc GR, Esta-brooks Z, Taguezem GF, Hooper DC. 2012. Reduced aer-ation affects the expression of the NorB efflux pump of *Staphylococcus aureus* by posttranslational modification of MgrA. *J Bacteriol* **194**: 1823–1834.

Tsukamura M, Nakamura E, Yoshii S, Amano H. 1985. Therapeutic effect of a new antibacterial substance oflox-acin (DL8280) on pulmonary tuberculosis. *Am Rev Resp Dis* **131**: 352–356.

Ubukata K, Itoh-Yamashita N, Konno M. 1989. Cloning and expression of the *norA* gene for fluoroquinolone resis-tance in *Staphylococcus aureus*. *Antimicrob Agents Che-mother* **33**: 1535–1539.

Vetting MW, Park CH, Hegde SS, Jacoby GA, Hooper DC, Blanchard JS. 2008. Mechanistic and structural analysis of aminoglycoside *N*-acetyltransferase AAC(6′)-Ib and its bifunctional fluoroquinolone-active AAC(6′)-Ib-cr variant. *Biochemistry* **47**: 9825–9835.

Vetting MW, Hegde SS, Wang M, Jacoby GA, Hooper DC, Blanchard JS. 2011. Structure of QnrB1, a plasmid-me-diated fluoroquinolone resistance factor. *J Biol Chem* **286**: 25265–25273.

Vinué L, Corcoran MA, Hooper DC, Jacoby GA. 2015. Mu-tations that enhance the ciprofloxacin resistance of *Es-cherichia coli* with *qnrA1*. *Antimicrob Agents Chemother* **60**: 1537–1545.

Wang JC. 1996. DNA topoisomerases. *Annu Rev Biochem* **65**: 635–692.

Wang M, Guo Q, Xu X, Wang X, Ye X, Wu S, Hooper DC, Wang M. 2009. New plasmid-mediated quinolone resis-tance gene, *qnrC*, found in a clinical isolate of *Proteus mirabilis*. *Antimicrob Agents Chemother* **53**: 1892–1897.

Weigel LM, Anderson GJ, Facklam RR, Tenover FC. 2001. Genetic analyses of mutations contributing to fluoro-quinolone resistance in clinical isolates of *Streptococcus pneumoniae*. *Antimicrob Agents Chemother* **45**: 3517–3523.

Wentzell LM, Maxwell A. 2000. The complex of DNA gyrase and quinolone drugs on DNA forms a barrier to the T7 DNA polymerase replication complex. *J Mol Biol* **304**: 779–791.

Westfall LW, Carty NL, Layland N, Kuan P, Colmer-Hamood JA, Hamood AN. 2006. *mvaT* mutation modifies the ex-pression of the *Pseudomonas aeruginosa* multidrug efflux operon *mexEF-oprN*. *FEMS Microbiol Lett* **255**: 247–254.

Willmott CJ, Maxwell A. 1993. A single point mutation in the DNA gyrase A protein greatly reduces binding of fluoroquinolones to the gyrase–DNA complex. *Antimicrob Agents Chemother* **37:** 126–127.

Willmott CJ, Critchlow SE, Eperon IC, Maxwell A. 1994. The complex of DNA gyrase and quinolone drugs with DNA forms a barrier to transcription by RNA polymerase. *J Mol Biol* **242:** 351–363.

Wohlkonig A, Chan PF, Fosberry AP, Homes P, Huang J, Kranz M, Leydon VR, Miles TJ, Pearson ND, Perera RL, et al. 2010. Structural basis of quinolone inhibition of type IIA topoisomerases and target-mediated resistance. *Nat Struct Mol Biol* **17:** 1152–1153.

Wong MH, Chan EW, Chen S. 2015. Evolution and dissemination of OqxAB-like efflux pumps, an emerging quinolone resistance determinant among members of Enterobacteriaceae. *Antimicrob Agents Chemother* **59:** 3290–3297.

Xiong X, Bromley EH, Oelschlaeger P, Woolfson DN, Spencer J. 2011. Structural insights into quinolone antibiotic resistance mediated by pentapeptide repeat proteins: Conserved surface loops direct the activity of a Qnr protein from a Gram-negative bacterium. *Nucleic Acids Res* **39:** 3917–3927.

Yamada Y, Hideka K, Shiota S, Kuroda T, Tsuchiya T. 2006. Gene cloning and characterization of SdrM, a chromosomally encoded multidrug efflux pump, from *Staphylococcus aureus*. *Biol Pharm Bull* **29:** 554–556.

Yamane K, Wachino J, Suzuki S, Kimura K, Shibata N, Kato H, Shibayama K, Konda T, Arakawa Y. 2007. New plasmid-mediated fluoroquinolone efflux pump, QepA, found in an *Escherichia coli* clinical isolate. *Antimicrob Agents Chemother* **51:** 3354–3360.

Yang S, Clayton SR, Zechiedrich EL. 2003. Relative contributions of the AcrAB, MdfA and NorE efflux pumps to quinolone resistance in *Escherichia coli*. *J Antimicrob Chemother* **51:** 545–556.

Yoon EJ, Courvalin P, Grillot-Courvalin C. 2013. RND-type efflux pumps in multidrug-resistant clinical isolates of *Acinetobacter baumannii*: Major role for AdeABC overexpression and AdeRS mutations. *Antimicrob Agents Chemother* **57:** 2989–2995.

Yoon EJ, Chabane YN, Goussard S, Snesrud E, Courvalin P, De E, Grillot-Courvalin C. 2015. Contribution of resistance-nodulation-cell division efflux systems to antibiotic resistance and biofilm formation in *Acinetobacter baumannii*. *MBio* **6:** e00309–e00315.

Yoshida H, Bogaki M, Nakamura M, Nakamura S. 1990. Quinolone resistance-determining region in the DNA gyrase *gyrA* gene of *Escherichia coli*. *Antimicrob Agents Chemother* **34:** 1271–1272.

Yoshida H, Bogaki M, Nakamura M, Yamanaka LM, Nakamura S. 1991. Quinolone resistance-determining region in the DNA gyrase *gyrB* gene of *Escherichia coli*. *Antimicrob Agents Chemother* **35:** 1647–1650.

Yoshida H, Nakamura M, Bogaki M, Ito H, Kojima T, Hattori H, Nakamura S. 1993. Mechanism of action of quinolones against *Escherichia coli* DNA gyrase. *Antimicrob Agents Chemother* **37:** 839–845.

Yu JL, Grinius L, Hooper DC. 2002. NorA functions as a multidrug efflux protein in both cytoplasmic membrane vesicles and reconstituted proteoliposomes. *J Bacteriol* **184:** 1370–1377.

Yu EW, Aires JR, Nikaido H. 2003. AcrB multidrug efflux pump of *Escherichia coli*: Composite substrate-binding cavity of exceptional flexibility generates its extremely wide substrate specificity. *J Bacteriol* **185:** 5657–5664.

Yu EW, Aires JR, McDermott G, Nikaido H. 2005. A periplasmic drug-binding site of the AcrB multidrug efflux pump: A crystallographic and site-directed mutagenesis study. *J Bacteriol* **187:** 6804–6815.

Zhang L, Li XZ, Poole K. 2001. SmeDEF multidrug efflux pump contributes to intrinsic multidrug resistance in *Stenotrophomonas maltophilia*. *Antimicrob Agents Chemother* **45:** 3497–3503.

Zhang S, Sun J, Liao XP, Hu QJ, Liu BT, Fang LX, Deng H, Ma J, Xiao X, Zhu HQ, et al. 2013. Prevalence and plasmid characterization of the *qnrD* determinant in Enterobacteriaceae isolated from animals, retail meat products, and humans. *Microb Drug Resist* **19:** 331–335.

Zhao J, Chen Z, Chen S, Deng Y, Liu Y, Tian W, Huang X, Wu C, Sun Y, Zeng Z, et al. 2010. Prevalence and dissemination of *oqxAB* in *Escherichia coli* Isolates from animals, farmworkers, and the environment. *Antimicrob Agents Chemother* **54:** 4219–4224.

Zhao L, Zhang J, Zheng B, Wei Z, Shen P, Li S, Li L, Xiao Y. 2015. Molecular epidemiology and genetic diversity of fluoroquinolone-resistant *Escherichia coli* isolates from patients with community-onset infections in 30 Chinese county hospitals. *J Clin Microbiol* **53:** 766–770.

Rifamycins, Alone and in Combination

David M. Rothstein

David Rothstein Consulting LLC, Lexington, Massachusetts 02421

Correspondence: dmrothstein@gmail.com

Rifamycins inhibit RNA polymerase of most bacterial genera. Rifampicin remains part of combination therapy for treating tuberculosis (TB), and for treating Gram-positive prosthetic joint and valve infections, in which biofilms are prominent. Rifabutin has use for AIDS patients in treating mycobacterial infections TB and *Mycobacterium avium* complex (MAC), having fewer drug–drug interactions that interfere with AIDS medications. Rifabutin is occasionally used in combination to eradicate *Helicobacter pylori* (peptic ulcer disease). Rifapentine has yet to fulfill its potential in reducing time of treatment for TB. Rifaximin is a monotherapeutic agent to treat gastrointestinal (GI) disorders, such as hepatic encephalopathy, irritable bowel syndrome, and travelers' diarrhea. Rifaximin is confined to the GI tract because it is not systemically absorbed on oral dosing, achieving high local concentrations, and showing anti-inflammatory properties in addition to its antibacterial activity. Resistance issues are unavoidable with all the rifamycins when the bioburden is high, because of mutations that modify RNA polymerase.

The four rifamycins approved for clinical use, rifampicin, rifabutin, rifapentine, and rifaximin (Fig. 1), are available as orally formulated agents derived from rifamycin SV, the natural product of *Amycolatopsis mediterranei* (alias *Streptomyces mediterranei*) (Tupin et al. 2010). The rifamycins are transcriptional inhibitors, and bind specifically to the β subunit of RNA polymerases from a broad range of bacteria while showing little or no activity against human RNA polymerases (Chen and Kaye 2009; Forrest and Tamura 2010).

Because rifamycins bind and inhibit most bacterial RNA polymerases, their spectra of activity are largely dictated by entry or exclusion from the bacterial cytoplasm. Rifampicin has potent activity against a variety of pathogens, including mycobacteria, Gram-positive cocci (notably staphylococci and streptococci), *Clostridium difficile*, and has activity against select Gram-negative pathogens *Neiserria meningitides*, *N. gonorrhoeae*, and *Hemophilus influenza*. The majority of Gram-negative pathogens are not susceptible to rifampicin (Chen and Kaye 2009; Forrest and Tamura 2010). The other rifamycins have similar spectra and potency (Table 1), although they differ substantially in pharmacokinetic properties (Table 2).

Rifamycins induce expression of human P450 cytochrome oxidases, notably CYP3A4, as well as the human P glycoprotein ABC transporter, which can cause drug–drug interactions that are a major complication in therapy (Burman et al. 2001). In a fortunate irony, the inducing properties of rifaximin, a drug well designed to treat gastrointestinal (GI) disorders, is

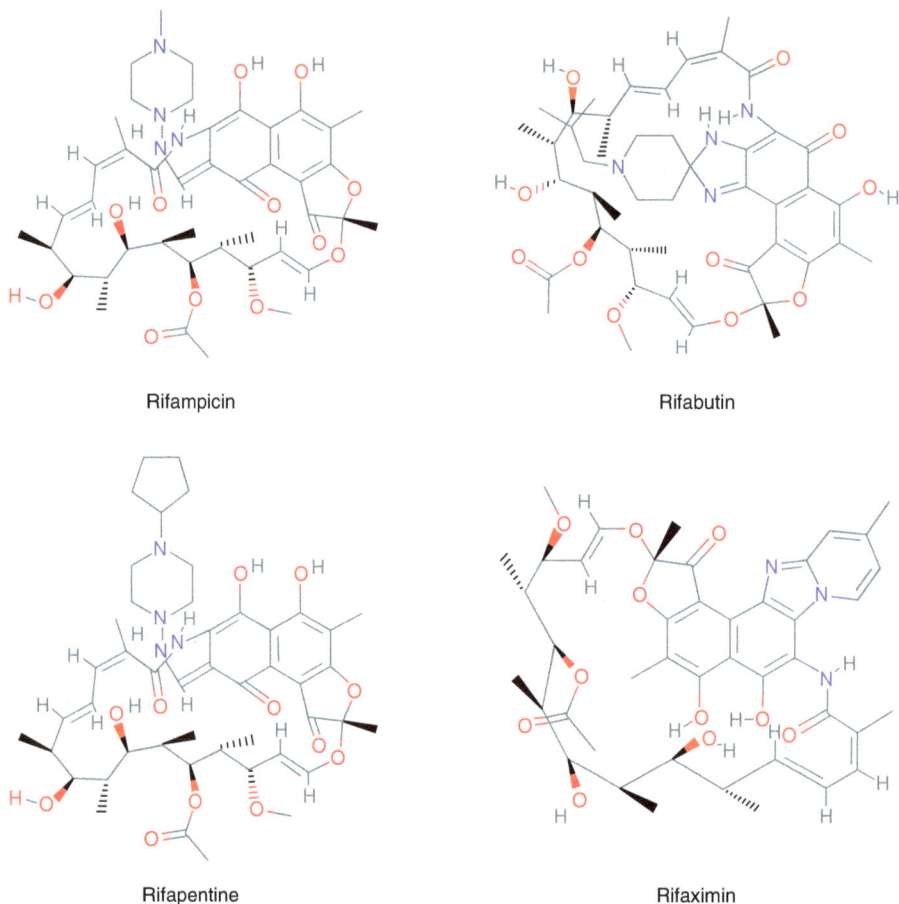

Figure 1. Chemical structures of approved rifamycins. Clockwise (from *top left*): rifampicin, rifabutin, rifaximin, and rifapentine.

probably an asset to therapy (see "rifaximin" section below.)

RIFAMYCIN-RESISTANCE POTENTIAL AND ITS INFLUENCE ON RIFAMYCIN DRUG ADMINISTRATION

The high frequency of rifamycin-resistance development among all susceptible bacteria is a pervasive concern. In particular, rifamycins are prone to "endogenous resistance development" (Silver 2011), resulting from mutations in *rpoB* encoding the β subunit of RNA polymerase, the target of rifamycin binding of all susceptible species. Mutations arise during DNA replication, unavoidable mistakes that encode RNA polymerase, which bind rifamycins less tightly. When the bioburden of susceptible bacteria exceeds 10^8, then mutant variants inevitably contain RNA polymerase that fails to bind rifamycins effectively.

Mutations conferring strong rifampicin resistance are cross-resistant to other approved rifamycins (Williams et al. 1998; Wichelhaus et al. 1999; Tupin et al. 2010; Goldstein 2014). No matter how susceptible the original bacteria are, the mutants will take over a population exposed to rifamycins, unless another antibacterial agent is also present to nullify their selective advantage. Hence, whenever the bioburden is assumed to be greater than 10^8, and the goal of therapy is elimination of a pathogen, rifamycins are routinely administered in combination. Monotherapy with approved rifamycins is a

Cite this article as *Cold Spring Harb Perspect Med* doi: 10.1101/cshperspect.a027011

Table 1. Minimum inhibitory concentrations (MICs in μg/mL) of rifamycins against bacterium causing the disorders indicated

	Rifampicin		Rifabutin		Rifapentine		Rifaximin	
	MIC$_{50}$	MIC$_{90}$	MIC$_{50}$	MIC$_{90}$	MIC$_{50}$	MIC$_{90}$	MIC$_{50}$	MIC$_{90}$
Mycobacteria; TB, MAC infections								
Mycobacterium tuberculosis	0.5	0.5	0.125	0.125	0.125			
M. avium complex	1	1.0	0.125		0.25			
Gram-negative (possible use in meningitis prophylaxis)								
Hemophilus influenzae	1.0	1.0						
Neisseria meningitides	0.03	0.5						
Peptic ulcer disease								
Helicobacter pylori		0.5		0.0078				
Gram-positive systemic infections								
Staphylococcus aureus	0.015	0.015	0.01		0.05	0.2	≤0.015	0.015
Streptococcus pyogenes	0.12	0.12	0.005		0.1	0.2	0.12	0.25
Propionebacterium acnes	0.007							
Enteric bacteria; GI disorders								
Enterococcus faecalis	2.0	8.0					4	16
Campylobacter	≥128	≥128					≥128	≥128
Escherichia coli	16	16	12.5				16	32
Salmonella	16	16					4	16
Shigella	16	32					8	32
Clostridium difficile colitis								
C. difficile	<0.002	4					0.008	256

Rifampicin MICs are adopted from Thornsberry et al. 1983; Klemens et al. 1994; Kunin 1996; Rastogi et al. 2000; Jiang et al. 2010; Hopkins et al. 2014. Rifabutin MICs are adopted from Klemens et al. 1994; Kunin 1996. Rifapentine MICs are adopted from Neu 1983; Klemens et al. 1994; Rastogi et al. 2000. Rifaximin MICs are adopted from Hoover et al. 1993; Rastogi et al. 2000; O'Connor et al. 2008; Jiang et al. 2010; Pistiki et al. 2014. TB, Tuberculosis; MAC, *Mycobacterium avium* complex; GI, gastrointestinal.

Table 2. Characteristics of the four approved rifamycins, which have similar potency and spectrum

Rifamycin	Pharmacokinetic (PK) distribution	Plasma half-life	3A4 inducer	Drug–drug interactions	Safety issues	Approved indications	Off-label use
Rifampicin	Balanced in plasma and tissues	2–5 h	Very strong	Increased metabolism of 3A4 substrates	≤5% GI discomfort (nausea, vomiting), ≤5% flu-like syndrome, <1% hepatotoxicity	1971 TB in combination 1971 prevent bacterial meningitis	1. Serious Gram-positive infections (e.g., PJI) 2. Prophylactic treatment of latent TB infection 3. Brucellosis
Rifabutin	Partitions into tissues ≫ plasma	32–67 h	Weak	Less metabolism of 3A4 substrates; note rifabutin levels are an issue[a]	Similar to rifampicin, but hepatotoxicity more an issue if DDI increases rifabutin half-life	1998 MAC infections	1. Treat TB in AIDS patients (avoid DDI) 2. *Helicobacter pylori* eradication in combination (ulcers)
Rifapentine	Partitions into tissue > plasma; more drug exposure than rifampicin	14–18 h	Strong	Increased metabolism of 3A4 substrates, but intermittent dosing attenuates DDI	Similar to rifampicin	1998 TB in combination (possibility to simplify and/or shorten therapy)	Prophylactic treatment of latent TB infection
Rifaximin	Confined to GI tract; <0.4% bioavailability; cirrhotic patients can have 20 times systemic exposure	NR[b]	Strong; note beneficial effects of 3A4 induction in GI tract	DDI mostly eliminated because negligible systemic absorption	Negligible systemic absorption eliminates systemic toxicities; possible GI discomfort	2004 traveler's diarrhea; 2008 HE recurrence; 2015 IBS(D)	1. IBD 2. *Clostridium difficile* colitis

Data from Burman et al. 2001; Munsiff et al. 2006; Chen and Kaye 2009; Mitnick et al. 2009; Gisbert and Calvet 2012; Rivkin and Gim 2011; Mitchison and Davies 2012.

3A4, P450 CYP3A4 enzyme; PJI, prosthetic joint infection; MAC, *Mycobacterium avium* complex; DDI, drug–drug interaction; GI, gastrointestinal; HE, hepatic encephalopathy; IBS(D), irritable bowel syndrome (diarrhea type); TB, tuberculosis; IBD, inflammatory bowel disease.

[a]Rifabutin levels can vary because rifabutin is metabolized by CYP3A4 enzyme leading to lower rifabutin levels, but some co-administered drugs can act as inhibitors of CYP3A4 enzyme, raising rifabutin levels.

[b]Not relevant; almost all rifaximin is passing through the GI tract.

consideration when the bioburden is assumed to be low, for example, as prophylactic therapy.

DNA sequence analysis of resistant isolates selected in vitro indicates that resistance was mediated by a single modification in one of 12 codons of the *rpoB* gene of *Staphylococcus aureus*, resulting in a single change in one of at least 25 nucleotide sites (Wichelhaus et al. 1999; Murphy et al. 2006; Goldstein 2014). Analysis of rifamycin-resistant clinical isolates of *S. aureus* revealed that resistant strains predominantly contained mutations in one or more of the 12 codons in the *rpoB* gene shown to mediate rifamycin resistance in vitro. Among the resistant clinical isolates, single mutational events in the *rpoB* gene have been detected within seven of these critical codons found by in vitro selection. When clinical isolates containing multiple mutations in the *rpoB* gene are included in the analysis, modifications in all but one of the 12 critical codons defined by in vitro selection experiments have been identified among clinical isolates (Aubry-Damon et al. 1998; Wichelhaus et al. 1999; O'Neill et al. 2006). One potential mutagenic hotspot was detected among clinical isolates at codon 481 mediating a histidine to asparagine change (Wichelhaus et al. 2002; O'Neill et al. 2006). Thus, in *S. aureus*, resistance is mediated in the laboratory and in the clinic primarily by mutations within the *rpoB* gene, mostly residing in cluster I of the rifampin-resistance-determining region (Wichelhaus et al. 2002).

Competitive fitness assays, in which a rifamycin-resistant strain and its isogenic rifamycin-sensitive parent are cogrown in the absence of rifampicin for multiple generations, reveal that the proportion of the resistant strain diminishes, indicating a fitness cost of these *rpoB* mutations (Wichelhaus et al. 2002; O'Neill et al. 2006). Strains carrying a mutation in codon 481 of the *S. aureus rpoB* gene may not show this fitness deficit (Wichelhaus et al. 2002), which may explain its prevalence among clinical isolates. However, others have observed diminished fitness of strains carrying this allele as well as other *rpoB* mutations (O'Neill et al. 2006).

These competitive fitness experiments predict that rifamycin-resistant strains would not compete well in the longer run, in an environment lacking rifamycins. However, laboratory experiments might exaggerate the long-term fitness deficits of rifamycin-resistant strains, because of the possible acquisition of compensatory mutations that would reduce the handicap of rifamycin-resistant alleles (O'Neill et al. 2006).

Rifamycin resistance in *Mycobacterium tuberculosis*, the pathogen responsible for TB, is also predominantly mediated by mutations in its *rpoB* gene. Rifamycin resistance selected in vitro results in modifications in homologous regions of the rifampin-resistance-determining region compared with *S. aureus* or *Escherichia coli* (Murphy et al. 2006; Goldstein 2014). Clinical isolates of resistant strains of *M. tuberculosis* similarly contain mutations in the key codons encoding the capacity to bind rifamycins (Goldstein 2014). Rifamycin-resistant strains of *M. tuberculosis* were also associated with a competitive fitness disadvantage (Billington et al. 1999; Mariam et al. 2004).

The most important consideration of endogenous resistance potential, in all therapeutic areas, can be summarized as follows. If the infecting population of rifamycin-susceptible bacteria exceeds 10^8, then there are a handful of rifamycin-resistant mutants in the infecting population. In this situation, selection clearly favors the resistant mutant(s), resulting in their rapid overgrowth despite any minor growth deficits of the rifamycin-resistant strains.

TUBERCULOSIS (TB)

Rifampicin was approved by the Food and Drug Administration (FDA) in 1971 for the treatment of TB. Combination therapy had already been established as the standard of care, with the discovery that streptomycin monotherapy, first tested in 1946, resulted in short-lived improvements, with the frequent emergence of streptomycin-resistant bacteria (Fox et al. 1999; Nuermberger et al. 2010; Field et al. 2012). Combination therapy of streptomycin together with *para*-aminosalicylic acid and, subsequently, with isoniazid in the early 1950s, made for a considerably more robust therapy, although 18 to 24 mo of treatment were required

(Fox et al. 1999; Mitchison and Davies 2012). The inclusion of rifampicin resulted in fewer relapses, and it was found that therapy could be shortened to 9 mo, and with subsequent improvements to the current 6-mo regimen.

Rifampicin was found to have unique bactericidal activity against persistent *M. tuberculosis* using in vitro and cell culture systems (Fox et al. 1999). Human studies of multiple combinations of drugs confirmed the importance of rifampicin (Jindani et al. 2003). Isoniazid was particularly active in the first week against metabolically active bacteria, whereas rifampicin contributed to bactericidal activity during the first week. Rifampicin continued to act in the coming months against the persistent, metabolically quiescent population of bacteria, often residing inside of host cells. More recently, pyrazinamide was found to contribute bactericidal activity against the persistent population (Mitchison and Davies 2012). These studies led to the combination therapy that is used to treat susceptible *M tuberculosis* infections, that is, 2 mo of isoniazid, rifampicin, pyrazinamide, and ethambutol, followed by 4 mo of continuation therapy consisting of daily dosing with isoniazid and rifampicin (Jindani et al. 2014).

Nausea is one common side effect of combination therapy, although generally it is not sufficient to stop therapy. Drug-induced hepatitis is rare, occurring in 2.5% of patients, and is generally attributed to the isoniazid and/or pyrazinamide components. Flu-like syndrome is experienced by a small minority of patients exposed to rifampicin and, rarely among this minority, can include thrombocytopenia, hemolytic anemia, and acute renal failure, resulting in discontinuation of therapy. The occurrence of these events is not dose related and may involve immune response to the drug (Burman et al. 2001; Mitnick et al. 2009).

With such a complex regimen and its side effects, adherence becomes an important issue and the main reason for treatment failure. Shortening therapy would be a major advance. One strategy is to raise the dose of rifampicin, which at 600 mg/d in the standard formulation, is at the low end of the curve to achieve optimal efficacy (Mitnick et al. 2009; Mitchison and Da-

vies 2012), with the additional knowledge that higher doses generally are tolerable (Boeree et al. 2015). Another possibility for shortening therapy is the incorporation of other drugs.

Rifapentine, approved by the FDA in 1998 for TB therapy, is a slightly more potent rifamycin derivative than is rifampicin, and has a five-fold increased exposure as a result of a longer half-life (Burman et al. 2001; Mitnick et al. 2009; Nuermberger et al. 2010). Preclinical studies suggested that if rifapentine replaced rifampicin, TB therapy might be shortened (Mitchison and Davies 2012). The results of a recent phase 3 clinical trial showed that the last 4 mo of the 6-mo therapy could be simplified with weekly dosing of rifapentine and moxifloxacin (a quinolone), but not shortened (Jindani et al. 2014).

Rifabutin is a potent antimycobacterial agent that partitions mostly into tissues as opposed to plasma (Burman et al. 2001; Mitnick et al. 2009). Rifabutin was approved in 1994 for the treatment of infections caused by *Mycobacterium avium* complex (MAC) (O'Brien and Vernon 1998; Mitnick et al. 2009). Rifabutin is a weaker inducer of CYP3A4, thereby diminishing drug–drug interactions, and is therefore often chosen for tuberculosis therapy for AIDS patients. In essence, patients are administered a watered-down version of TB chemotherapy in an attempt to minimize difficult drug–drug interactions; rifabutin can increase the elimination of essential antiretroviral medications, while the retroviral medications can increase the half-life of rifabutin, which can result in hepatotoxicity (Burman et al. 2006; Zhang et al. 2011; Regazzi et al. 2014).

Adherence issues leading to interrupted TB therapy can result in incomplete killing of persistent bacteria, requiring prolonged therapy. Interrupted therapy can also result in the selection of resistant *M. tuberculosis* mutants, for example, rifamycin-resistant mutants. Selection of resistant mutants leads to treatment failure, requiring a change from first-line agents to second-line agents.

Undertreatment can also select for mutants. For example, when AIDS patients were treated with intermittent rifabutin-based therapy, the exposure to rifamycin was diminished to avoid drug–drug interactions with AIDS medications. The majority of these AIDS patients were cured. For those patients who experienced relapses, however, usually a rifamycin-resistant mutant strain had been selected during therapy (Burman et al. 2006). There are alternative antibacterials to treat drug-resistant TB (Mitnick et al. 2009; Nuermberger et al. 2010; Field et al. 2012), but to lose the use of rifamycins, the most powerful killing agent, is a serious setback for an AIDS patient.

THE USE OF RIFAMPICIN AS A SINGLE AGENT FOR PROPHYLACTIC TREATMENT OF LATENT TB

A positive skin test for TB in the absence of clinical symptoms may indicate latent TB, a symptom-free quiescent infection. In the United States, most cases of TB are the result of reactivated infection (Horsburgh and Rubin 2011). The expectation is that 5% to 15% of latent carriers in the United States are expected to develop an active infection during their lifetimes (Getahun et al. 2015), an estimate that may be a moving target depending on immigration, the proportion of immune-compromised individuals, etc. Therefore, prophylactic treatment to prevent active outbreak is a consideration.

Because the bioburden is low, latent TB is often treated by shortened therapies of 3 to 4 mo. In fact, a variety of treatment options is used, including rifampicin and isoniazid for 3 mo or for 4 mo, rifapentine and isoniazid weekly for 3 mo, isoniazid alone for 3, 4, 6, or 9 mo, and rifampicin monotherapy for 3 or for 4 mo (Stagg et al. 2014; Getahun et al. 2015). Although these treatments are probably effective (i.e., result in a reduced conversion to active TB compared with untreated patients), it has been challenging to standardize treatment. It is probably difficult to determine the medical superiority of a particular treatment to prevent an outbreak of active TB, an infrequent event, as the end point of clinical trials. As mentioned previously, rifampicin is a well-tolerated drug, so there is a temptation to use rifampicin monotherapy, diminishing adherence issues

and costs. A word of caution, however, is that the failure of latent TB treatment may be rare and may not appear immediately, but the negative consequence of failure could be severe, such as a lifetime of battling an *M. tuberculosis* strain that has been selected for resistance to rifamycins. Although the bioburden for any one patient harboring a latent TB infection is low, the communal bioburden of a group of such patients can be predicted to include a rifampicin-resistant mutant, which could be selected in the unlucky individual undergoing rifampicin as a stand-alone therapy. If the severity of selecting a resistant pathogen (and not simply a failed therapy) has not been properly taken into account, the rifampicin stand-alone strategy perhaps should be abandoned.

The 3-mo isoniazid-rifampicin daily therapy is a more intuitively appealing choice to treat latent TB, because isoniazid would protect against development of a rifampicin-resistant active infection. However, this strategy will not be secure if the frequency of isoniazid resistance rises, as currently the susceptibility of the latently infecting bacteria is not determined before treatment.

RIFAMYCINS IN COMBINATION THERAPY BEYOND THE REALM OF MYCOBACTERIA; TREATING PEPTIC ULCER DISEASE

It is generally accepted that combination antibiotic therapy is necessary to eradicate *Helicobacter pylori* from stomach tissue, to cure peptic ulcer disease. Rifabutin, with its tissue penetrating ability, has been shown to be effective in combination with amoxicillin and other antibiotics in a series of small clinical trials (Gisbert and Calvet 2012). A practical strategy is to use rifabutin in combination as a backup if first-line therapy fails, knowing that rifamycin resistance is currently infrequently encountered in *H. pylori* strains. Amoxicillin resistance is also very rare. Gisbert and Calvet envision the role of rifabutin to be one reserved for the more recalcitrant cases of peptic ulcer disease, not to be overused, to prevent selection of rifampicin-resistant *M. tuberculosis* in the populace, for example, among latent carriers of TB, during the 10–12 d required for therapy.

RIFAMPICIN AND SYSTEMIC GRAM-POSITIVE INFECTIONS

The possibility of using rifampicin in combination therapy against Gram-positive infections was recognized early after its approval for treating TB. Rifamycins have potent activity against both staphylococci and streptococci, the predominant pathogens causing systemic infections, such as bacteremia, abscesses, endocarditis, and foreign body infections. If a partnering drug could prevent the emergence of rifampicin-resistant mutants, then the combination could provide great benefit (Sanders 1976). It is very important that the pharmacokinetic properties of the partnering drug are suitable, that is, that the second drug is always sufficiently present to prevent outgrowth of the tiny minority of rifampicin-resistant organisms, while allowing rifampicin, with its favorable potency and tissue penetration, to exert its activity (Achermann et al. 2013).

In vitro evaluation of rifampicin combinations did not provide clear-cut guidance in choosing a second drug. Isobolograms or time-kill curves often showed apparent synergy at sub-minimum inhibitory concentration (MIC) levels of two drugs, while showing interference or antagonism at higher concentrations, particularly when rifampicin was tested in combination with bactericidal drugs (Perlroth et al. 2008; Forrest and Tamura 2010). In general, these studies minimized the apparent benefit of pairing rifampicin with bactericidal agents, while exaggerating benefits of bacteriostatic agents, such as minocycline (Forrest and Tamura 2010).

Animal models did not provide compelling support for rifampicin adjunct therapy, except for foreign-body models described in the next section. Native valve endocarditis models and bacteremia models (mouse, rat, and rabbit) have provided inconsistent support for rifampicin adjunct therapy with linezolid, daptomycin, or fusidic acid. However, there was no evidence for greater efficacy of rifampicin as an adjunct with vancomycin, the mainstay drug for serious Gram-positive infections (Perlroth et al. 2008; Forrest and Tamura 2010).

Clinical publications have included case studies of one or a handful of patients undergoing combination therapy, including rifampicin, to treat native valve endocarditis and/or bacteremia, with mixed results (Forrest and Tamura 2010). In addition, several small prospective clinical combination trials were performed (Van der Auwera et al. 1985; Dworkin et al. 1989; Levine et al. 1991; Heldman et al. 1996; Schrenzel et al. 2004). The difficulty with all of these trials is they were underpowered to be convincing, given that the baseline benefit of monotherapy was substantial. How could a trial convincingly show benefit when each arm of the study contained fewer than 20 subjects, and when monotherapy was effective in most patients? However, one clinical trial showed approximate equivalence in medical outcome and a qualitative advantage of orally administered combination therapy (rifampicin with fleroxacin, a quinolone) to treat bacteremia, compared with IV-administered vancomycin, reducing hospitalizations by 11 d per patient (Schrenzel et al. 2004). The American Heart Association has not endorsed rifampicin combination therapy for treating native valve endocarditis or bacteremia. Perhaps a well-powered clinical study, with a carefully designed and optimized protocol, would clarify this issue.

RIFAMPICIN IN COMBINATION THERAPY TO TREAT PROSTHETIC JOINT INFECTIONS

Rifampicin combination therapy has more obvious advantages in treating prosthetic joint infections (PJIs), mostly resulting from staphylococcal infections following hip or knee replacements (Zimmerli et al. 2004). The foreign body provides a surface devoid of host innate defenses, and an opportunity for biofilms to develop. The bacteria secrete exopolysaccharides that can adhere to the device surface, and provide a barrier, protecting bacteria from phagocytes, antibodies, and antibiotics (Costerton et al. 1999; Jacqueline and Caillon 2014). MICs for most drugs are 100–1000 times higher in biofilms (Jacqueline and Caillon 2014). However, rifampicin retains more activity against biofilms of staphylococci than other antibiotics (Widmer et al. 1990; Amorena et al. 1999; Saginur et al. 2006; Perlroth et al. 2008; Gomes et al. 2012; Jacqueline and Caillon 2014). Rifampicin, in addition, shows activity against biofilms of *Propionibacterium acnes*, another prominent PJI pathogen (Furustrand Tafin et al. 2012).

Animal models of foreign body infections reinforce the idea that rifampicin has a unique antibacterial role in PJI. In the guinea pig cage model, staphylococci are injected into cages that have been surgically implanted subcutaneously in guinea pigs (Zimmerli et al. 1982). This model system can test for efficacy of antibiotics against both the planktonic cells, and biofilms on cage surfaces. The combinations of rifampicin with linezolid (Baldoni et al. 2009), with daptomycin (John et al. 2009), and with fosfomycin (Mihailescu et al. 2014), showed improved results compared with monotherapy, and potential for clinical efficacy. An experimental rabbit model of PJI also showed that rifampicin and daptomycin were an effective pair against methicillin-resistant *S. aureus* (MRSA) infection of a joint implant compared with daptomycin monotherapy, resulting in the suppression of resistance to daptomycin and to rifampicin. Similar results were observed for vancomycin paired with rifampicin (Saleh-Mghir et al. 2011). An experimental rat model, using a titanium wire implanted into the tibia to simulate a foreign body implant, suggested that rifampicin combinations with both linezolid and vancomycin reduced MRSA more effectively than monotherapy with linezolid or vancomycin (Vergidis et al. 2011). It is encouraging and consistent that rifampicin combinations shown efficacy in these three distinct animal models of PJI.

Approximately 1% of patients having a prosthetic implant experience PJI. Clinical evidence showed that some PJI infections could be successfully treated by using combination antibiotic therapy with rifampicin playing a central role. For the 11 patients in the first study, removal of the implant to treat the infection (the standard of care) was not practical (Widmer et al. 1992). Before antibiotic administration, patients had a surgical procedure to physically clean the area surrounding the implant, removing hopelessly infected and inflamed tissue (debridement) while leaving the prosthesis in place. Treatment of PJIs with a rifampicin/ciprofloxacin combination was successful in nine of 11 patients who showed no symptoms of infection after 2 yr of follow-up. These results were sufficiently encouraging to launch a prospective, blinded randomized trial in which patients were enrolled with early PJI infections, mostly of *S. aureus*, to determine whether adjunct therapy with rifampicin was beneficial (Zimmerli et al. 1998). All patients were treated initially with an IV course of the β-lactam flucloxacillin or vancomycin for 2 wk. The test group was also treated with rifampicin during the initial 2-wk treatment. For the long-term continuation phase, the control group was administered oral ciprofloxacin monotherapy, while the test group was administered both oral ciprofloxacin and rifampicin. All 12 patients able to complete the test regimen were cured of the infection, whereas only seven of 12 patients administered ciprofloxacin monotherapy were cured. More recently, a clinical study of 43 patients infected primarily with MRSA reinforced the concept that prosthetic retention also applies to MRSA infections (Peel et al. 2013).

The clinical evidence supporting PJI regimens includes numerous additional case studies, but still does not carry the statistical power of pivotal clinical trials for obtaining drug approval. In the absence of this well-documented approval process, with a defined FDA label prescribing drug administration, the medical profession has resorted to guidances representing the current thinking of experts in the field. In the Infectious Disease Society of America (IDSA) guidance (Osmon et al. 2013), there was consensus in the definition of PJI, methods of detection of infections, the use of antibacterial therapy, and when to attempt to salvage a prosthesis without surgical removal. There was recognition of alternative preferences for particular procedures and antibiotic strategies. However, on the idea of the centrality of rifampicin in treating these biofilm-prone infections, the committee reached consensus.

Prosthetic valve endocarditis, similar to PJI, involves infection of a prosthetic surface prone to biofilm infections. The latest guidelines from The European Society of Cardiology (ESC) has endorsed rifampicin combination therapy to treat endocarditis associated with prosthetic valves, but not native valve endocarditis (Habib et al. 2015).

RIFAXIMIN TREATMENT OF GI DISORDERS; GENERAL PROPERTIES OF RIFAXIMIN

Rifaximin is an oral drug confined to the GI tract. Its oral bioavailability is <0.4% (Darkoh et al. 2010). Stool samples contained 8000 μg/mL of rifaximin after 3 d of treatment at 800 mg/d (Jiang et al. 2000). It has been shown that bile salts increase rifaximin solubility, suggesting that there may be a gradient of rifaximin activity that diminishes as the drug proceeds through the GI tract as bile salts diminish in concentration (Darkoh et al. 2010).

The activity spectrum of rifaximin in standard MIC testing is similar to that of rifampicin: very potent activity against Gram-positive staphylococci and streptococci, *C. difficile*, and Neisseria, and modest activity against *H. influenza* (Table 1). Rifaximin has high MICs against most Gram-negative bacteria, such as Enterobacteriaceae, and these strains would normally be considered resistant, for example, MICs in the range of 16 μg/mL, 32 μg/mL, or higher (Rivkin and Gim 2011). However, rifaximin is a broad-spectrum agent in the GI tract, because its very high nominal concentration greatly exceeds the MICs of Gram-negative bacteria.

Rifaximin has few downside effects, given its GI localization. In clinical trials, it has been difficult to distinguish adverse events observed in the rifaximin test group and the placebo group even after 6 mo of dosing, and rifaximin is devoid of drug–drug interactions characteristic of rifampicin (Rivkin and Gim 2011). The use of rifaxamin has been explored in several GI indications, with growing formal FDA approvals: traveler's diarrhea (TD) in 2004, reduction in recurrence of hepatic encephalopathy (HE) in 2010, and irritable bowel syndrome (IBS) in May 2015.

RIFAXIMIN AS AN EFFECTOR OF THE PREGNANE RECEPTOR IN THE GI TRACT

Rifaximin, like rifampicin, is classified as a strong inducer of the CYP3A4 enzyme in hepatocyte cell culture testing. This induction is mediated by the binding of either rifamycin to the pregnane X receptor, a master regulator of genes involved in xenobiotic detoxification, bile biosynthesis, and other functions (Hirota 2015). However, because rifaximin is confined to the GI tract in vivo, there is no significant induction of CYP3A4 in the liver in humans, and consequently, rifaximin causes no drug–drug interactions characteristic of rifampicin (Rivkin and Gim 2011). However, when rifaximin binds to the pregnane receptor in the GI tract, it inhibits NF-κB, a transcription factor, preventing it from activating proinflammatory genes of its pathway (Hirota 2015).

An elegant set of preclinical studies using the mouse model of inflammatory bowel disease (IBD) suggests the importance of rifaximin as an effector of the pregnane receptor. Mice that were subjected to chemical insult were protected from the worst signs of colon damage by rifaximin administration. However, this protection only occurred in the isogenic strain of mouse humanized for the pregnane receptor gene; the rodent pregnane receptor fails to respond to rifamycins (Ma et al. 2007; Cheng et al. 2010). The most likely explanation for the different response of the isogenic mouse containing the humanized pregnane receptor is that rifaximin, acting as agonist of the pregnane receptor, was responsible for the protection from colon damage.

Additional support for the involvement of the pregnane receptor in IBD comes from human studies. The expression of pregnane receptor target genes was significantly reduced in patients having IBD (Langmann et al. 2004).

RIFAXIMIN AND THE TREATMENT OF IBD

IBD is actually composed of two multifactorial diseases having some common symptoms and genetic predispositions (Cho and Brant 2011). Ulcerative colitis (UC) causes inflammation and ulcers specifically in the lining of the colon, and Crohn's disease (CD) can afflict the colon or other regions of the GI tract, and all of its layers. Antibacterial treatments are a consideration because aspects of the inflammatory response may be directed by interactions with the microbiome. The pregnane receptor gene has specific polymorphisms that raise the risk of contracting UC, CD, or both (Dring et al. 2006). This specificity of genetic predisposition suggests that interaction of rifaximin with the pregnane receptor could have a non-antibacterial benefit (described in the previous section), in addition to its antibacterial activity in treating both diseases.

Eighty-three patients were enrolled in a double-blind randomized trial to determine whether rifaximin ameliorated CD symptoms. Progress was monitored with the CD activity index (CDAI) self-assessment. The group that received rifaximin for 12 wk experienced a 52% remission rate, whereas placebo remission rate was 33%. The difference was not statistically significant, although the subset of patients who entered the study with a high C-reactive protein (CRP) score (a biomarker for inflammation) did show a significant difference (Prantera et al. 2006).

In a second double-blind randomized trial of 402 patients, treated either with rifaximin as described above or placebo, a 62% remission was observed in the rifaximin group compared with 43% in the placebo group (Prantera et al. 2012). After an additional 12 wk, the remission rate in the rifaximin group was 45%, whereas the remission rate in the placebo group was 29%. The rifaximin treatment showed some lasting benefit compared with placebo. However, the diminished rate of the test group raises the possibility that additional rifaximin treatments might be necessary to sustain benefit. Rifaximin was generally well tolerated in both these studies, with no serious adverse events. Again, the subgroup that had the high CRP level at the initiation of the study (indicating a stronger baseline inflammatory response) experienced a more significant benefit, suggesting that the anti-inflammatory properties of rifaximin could be beneficial to this subgroup.

Several small clinical studies of UC patients refractory to steroid use were tested with rifaximin as an adjunct therapy. All patients were treated with mesalazine, a nonsteroidal anti-inflammatory agent that localizes in the colon. The addition of rifaximin resulted in reduction in stool frequency, diminished rectal bleeding, and in sigmoidoscopic score compared with the placebo group, objective criteria that were encouraging of future testing (Guslandi 2011).

Rifaximin treatment resulted in significant relief of IBD symptoms, however transient these changes might be. Rifaximin treatment was well tolerated with few serious side effects. Positive effects could be a consequence of rifaximin's antibacterial activity, and its anti-inflammatory activity as effector of the pregnane receptor, particularly among patients who may be prone to inflammation (i.e., patients having high CRP levels).

RIFAXIMIN AS A TREATMENT IN IRRITABLE BOWEL SYNDROME (IBS)

IBS is characterized by irregular bowel patterns, recurrent abdominal pain, bloating, and flatulence, but is devoid of the physical signs of inflammation and ulcers that are characteristic of IBD. Although IBS is not as serious or life threatening as IBD is, IBS afflicts up to 15% of the U.S. population and is a major source of morbidity (Iorio et al. 2015). Lactulose gas tests indicated that subjects having IBS were more prone to bacterial overgrowth in the small intestine (SIBO) (Saadi and McCallum 2013). Antibacterials are among a variety of medicines prescribed for IBS because of the possibility of correcting this imbalance in IBS patients (Kassinen et al. 2007).

Pivotal double-blind clinical trials led to approval of rifaximin for treating IBS(D) (diarrhea type), which was announced by the FDA in May 2015. A total of 1260 IBS subjects were enrolled in these two parallel randomized double-blind phase 3 studies. Four weeks after treatment in the first study, 40.8% of subjects administered rifaximin for 2 wk reported adequate relief from IBS symptoms (the primary endpoint) compared with 32.2% for the placebo group. In the second study, 40.7% of the test group reported relief from IBS symptoms compared with 31.7% in the placebo control. Similar observations were reported for specific symptoms such as bloating. All of these results had statistically significant differences among groups. Ten weeks after treatment, the test groups continued to report a higher level of relief than the placebo groups, although the percentage reporting adequate relief from all groups had diminished (Pimentel et al. 2011a).

An important issue for the often-chronic condition of IBS is the potential for retreatment. In fact, a retrospective study of patients who had received multiple treatments suggested that retreatment with rifaximin provided comparable benefit to some subjects (Pimentel et al. 2011b). It is difficult to rule out possible biases of self-selection in patients who decide to seek retreatment. For example, patients who particularly benefited from the first treatment might preferentially have enrolled in the trial, or alternatively, patients who responded best might not have relapsed, and then they would not have enrolled in the second trial. In any case, rifaximin seems to have provided at least temporary benefit to IBS subjects, with potential benefits of retreatment.

RIFAXIMIN TREATMENT OF HEPATIC ENCEPHALOPATHY (HE)

HE affects up to 80% of patients with cirrhotic liver disease, primarily caused by hepatitis C infection, excessive alcohol uptake, and fatty liver disease. Thirty to 45% of patients experience overt HE, which often requires hospitalization and is manifested by mental and personality changes, impaired cognition, and decreased hand–eye coordination (Scott 2014). Toxic levels of ammonia are thought to be the direct cause of HE. Although there is not a strict correlation of plasma ammonia levels and cognitive symptoms, successful therapies of nonabsorbable disaccharides and/or nonabsorbable antibiotics, correlated with reduction of ammonia in plasma (Sussman 2015). Lactulose, the approved disaccharide in the United States, passes into the colon where it may shift bac-

terial metabolism away from ammonia production. Lactulose fermentation in the colon may also create a more acid environment, resulting in ammonia conversion to less permeable NH_4^+ ions.

Rifaximin was approved in 2010 to prevent recurrence of overt HE. For the pivotal phase 3 trial, 299 patients who had experienced at least two episodes of overt HE in the previous 6 mo were enrolled in the trial, and either treated with rifaximin for up to 6 mo, or were assigned to the placebo-controlled arm. Lactulose treatment was not a criterion for enrollment but was continued for 91% of the patients taking this medication, in both arms of the trial. During these 6 mo of treatment, 22.1% of the rifaximin group had a breakthrough HE event compared with 45.9% of the placebo group, a statistically significant difference (Bass et al. 2010). The expectation from this trial is that rifaximin will prevent one patient from experiencing an HE breakthrough for every four patients treated for a 6-mo period.

The data on safety of the rifaximin and placebo arms were comparable (Bass et al. 2010). Small safety differences would be difficult to discern, because lactulose would be expected to contribute more than rifaximin to adverse events. Two cases, however, were terminated from the study because of *C. difficile* infections, only in the treatment arm. Although the 1% rate of *C. difficile* infection was not unusual for this population group, the result is one that should engender special attention in future investigations because *C. difficile*–associated colitis is life threatening.

The same investigators then performed an open-label clinical trial, in which all 392 patients were treated with rifaximin for 2 yr. Enrollment criteria required that all subjects had experienced an episode of HE within the previous year. The baseline characteristics of the subjects were similar to the previous trial, and the protocol was identical to the previous trial except for the longer treatment time. The subjects of this longer trial had a recurrence rate that was comparable to that of the rifaximin arm of the previous study. The rate of recurrence was considerably lower than the placebo arm of the previous study, suggesting that rifaximin treatment showed benefit for this prolonged treatment time (Mullen et al. 2014).

However, once again, 1% of the subjects (six patients) of this trial contracted *C. difficile*. As in the last trial, this 1% figure was approximately the rate anticipated for patients in a comparable condition (Scott 2014). Again, the appearance of *C. difficile* infections following rifaximin treatment is something to monitor closely in the future to determine whether there is any contribution of rifaximin in promoting *C. difficile*–associated colitis.

In comparing antibiotic treatment options for HE patients, rifaximin may be the most beneficial (Patidar and Bajaj 2013). Other antibiotics of low bioavailability, such as neomycin, have higher systemic absorption and more serious safety issues. A reasonable medical approach may be to treat initially with lactulose, and then to treat patients having a breakthrough occurrence of HE with rifaximin. However, it has been shown that generic rifaximin, which contains the amorphous molecule, is absorbed systemically up to 5 times the level when compared with the brand molecule in the rifaximin-α crystal form (Blandizzi et al. 2014), indicating that branded rifaximin is the better choice.

TRAVELER'S DIARRHEA (TD)

TD is caused by eating or drinking contaminated food or water, usually while visiting a developing country (see wwwnc.cdc.gov/travel/page/travelers-diarrhea). The most common agents causing TD are Enterobacteriaceae, such as *E. coli* (ETEC), which secrete toxins. Although usually a self-limiting disease, especially for the majority of noninvasive Enterobacteriaceae, occasionally TD initiates postinfectious IBS. TD was the first approval by the FDA for rifaximin in 2004 to treat uncomplicated cases of TD—no fever or blood in stools—with a 3-d treatment of rifaximin. The studies that led to approval consisted of three randomized and double-blind trials that enrolled subjects visiting Mexico, Guatemala, Jamaica, and Kenya. These trials showed a sig-

nificant reduction in duration of diarrhea compared with placebo, or an equivalent response compared with ciprofloxacin (Adachi and DuPont 2006). A subsequent clinical trial again showed significant benefit of rifaximin to treat patients for 3 d, showing approximate equivalence in time to resolution compared with ciprofloxacin. However, the ciprofloxacin group was significantly superior in having fewer treatment failures, and having a higher rate of microbiological eradication of pathogens (92.5% for ciprofloxacin, 76.7% for rifaximin), and having a more favorable response to invasive pathogens (e.g., Shigella) (Taylor et al. 2006).

More recently a meta-analysis summarized the results of four double-blind trials, conducted to test for prevention of TD. Subjects were dosed with rifaximin prophylactically during a trip. Subjects of the rifaximin test group had a lower rate of contracting TD than did the placebo group. The results taken together showed that for every four subjects who took rifaximin, one case of TD was prevented (Hu et al. 2012).

In summary rifaximin showed clear benefit in both treating and preventing TD, and equivalence or near equivalence to ciprofloxacin in efficacy. An important consideration is the safety and tolerability of treatment, and here rifaximin has the advantage of confinement to the GI tract, devoid of systemic safety concerns and drug–drug interactions, and being a well-tolerated drug. Despite these apparently favorable factors, the CDC states that "fluoroquinolones are the drugs of choice" if TD patients seek chemotherapy, and makes no mention of rifaximin despite its approval for this indication (see wwwnc.cdc.gov/travel/page/travelers-diarrhea).

RIFAXIMIN IS PRONE TO RESISTANCE DEVELOPMENT, DESPITE THE PREVAILING SENTIMENT TO THE CONTRARY

Resistance to rifaximin is mediated in three ways. In the most frequently encountered mechanism, mutations arise during DNA replication, unavoidable mistakes that encode RNA polymerase that binds rifamycins less tightly. The mutations mediating strong resistance to one rifamycin are cross-resistant to all approved rifamycins (Tupin et al. 2010; Goldstein 2014). Hence, whenever the bioburden exceeds 10^8 bacteria, a mutant strongly resistant to all approved rifamycins has been created by an error in DNA replication. (For more details, see the above section on rifamycin resistance.) The other two mechanisms are efflux pumps, which expel rifamycins from Gram-negative bacteria, and modifying enzymes, which are rarely encountered.

It has been suggested that rifaximin is less prone to resistance development compared with other rifamycins (for one of many examples, see Rivkin and Gim 2011). It is conceivable that the extremely high concentrations of rifaximin in the GI tract, nominally at least 8000 µg/mL (Jiang et al. 2000), might exceed the MIC of every rifamycin-resistant mutant, neutralizing or nullifying their selective advantage. However, the isolation of clinical strains of *E. coli* from IBD patients treated with rifaximin for 12 wk revealed strains having *rpoB* mutations, an efflux pump, or both (Kothary et al. 2013). Rifamycin-resistant strains of Enterobacteriaceae isolated from TD patients were found to be mediated by mutations in *rpoB*, by efflux, and in one case by rifaximin inactivation (Hopkins et al. 2014). Finally, starting with four clinical isolates from TD patients, rifaximin-resistant strains were selected that contained *rpoB* mutations, and that showed enhanced expression of efflux activity (Pons et al. 2012)

A dynamic demonstration that rifaximin selects for resistance with facility in vivo is shown in Figure 2 (De Leo et al. 1986). Human volunteers donated fecal samples, and analysis of the microbiome showed no detectable rifaximin-resistant mutants before dosing, whereas after 5 d of rifaximin treatment, Enterobacteriaceae Enterococcus, Bacteroides, Clostridium, and anaerobic cocci contained from 30% to 90% rifaximin-resistant strains. After dosing, rifaximin selection unwound, resulting in a return of the population to the original rifamycin-sensitive status. A similar elasticity of the microbiome with regard to rifamycin susceptibility was observed when fecal samples were collected from UC patients following three treatment periods of 10 d each, interspersed by 25 d of washout (Brigidi et al. 2002). Rifamycin-resistant mutants became more abundant during rifaximin treatment, and diminished in frequency during washout periods.

The conservation of the rifamycin-binding site in the β subunit of RNA polymerase from

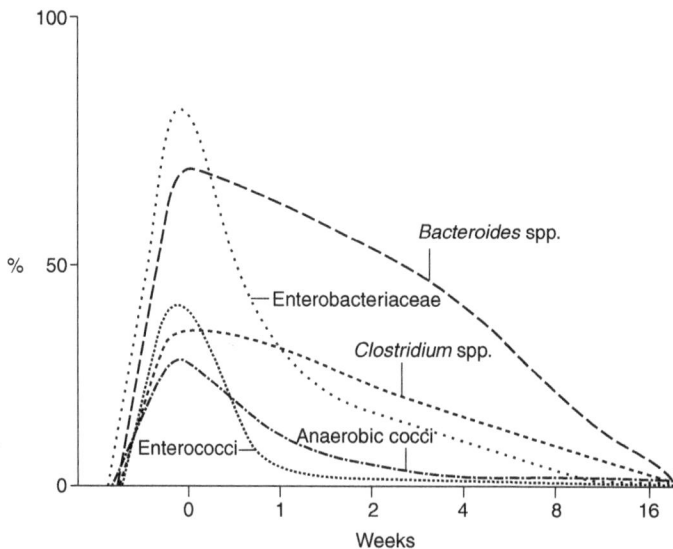

Figure 2. Percentage of bacteria resistant to rifamycins diminishes in the human gastrointestinal tract after discontinuing rifaximin treatment at week 0. (Data based on De Leo et al. 1986.)

Cite this article as *Cold Spring Harb Perspect Med* doi: 10.1101/cshperspect.a027011

multiple bacterial genera suggests the importance of retaining the binding site, and a mutant that has lost the conserved site would probably be less competitively fit. Thus, once rifaximin was washed out of the GI tract, the less fit rifaximin-resistant bacteria were probably displaced by immigrant rifaximin-sensitive bacteria that were ingested. This transition to a rifaximin-sensitive microbiome may be an essential condition for future successful cycles of rifaximin treatment.

A second resistance concern, beyond the efficacy of rifaximin, is the possibility that rifaximin treatment might endanger patients by selecting for rifamycin resistance outside of the GI tract. It has already been documented that rifaximin treatment can select rifamycin-resistant strains of *S. aureus* on the skin (Valentin et al. 2011, 2014), perhaps not so remote a danger for a patient with a prosthetic joint who was at risk for infection. The selection of resistance probably occurred because of the interface of rifaximin and *S. aureus* in the perianal area during elimination of the drug. The fact that in relatively healthy patients, the systemic exposure to rifaximin is very low mitigates the risk of selection of rifamycin resistance outside of the GI tract. However, it is important to consider that in less healthy patients, for example, cirrhotic patients, bioavailability of rifaximin can increase 10- or 20-fold (Rivkin and Gim 2011). Thus, a physician may have difficult decisions with regard to therapeutic options and risk of resistance selection outside of the GI tract, during 6 mo of rifaximin treatment with less healthy patients. Finally, the physician would hopefully be aware that branded rifaximin α is the crystal form that maximally confines rifaximin to the GI tract; it has the lowest systemic absorption, whereas generic rifaximin in the amorphous form may have five times the systemic exposure (Blandizzi et al. 2014).

RIFAXIMIN AND *C. Difficile*—A BATTLE BETWEEN CURE AND RESISTANCE SELECTION

It has been suggested that rifaximin could have utility in treating *C. difficile* infections (Rivkin and Gim 2011), as has been reported anecdotally. However, treating an invasive pathogen, *C. difficile*, with rifaximin monotherapy can lead to the failure of therapy because of the selection of a highly rifamycin-resistant *C. difficile* strain (Johnson et al. 2009). Therefore, rifaximin does not appear to be a suitable first-line monotherapeutic agent for treating *C. difficile* infections, in which the invading bacterium has to be eradicated. The mission is to kill the invader, not to rebalance the genome, as is the goal of rifaximin treatment for HE, IBD, or IBS. Perhaps a more appealing protocol is serial combination therapy, in which oral vancomycin would be administered for a few days, and then, when the bioburden has diminished, follow up with rifaximin to eradicate the offending organism with little chance of rifaximin-resistance development (Johnson et al. 2009). This protocol could be used in stubborn cases, in which vancomycin therapy in the past has failed.

Rifaximin-resistant *C. difficile* has been observed occasionally (O'Connor et al. 2008; Johnson et al. 2009; Carman et al. 2012). It would be wise to test the pathogen before initiating therapy, because of the instances of increases in rifamycin-resistant *C. difficile* pathogens, which have been reported to be >30%, an astonishing increase in resistance reported in specific medical centers (Curry et al. 2009; Huang et al. 2013). It is likely that rifaximin would not be successful in curing patients afflicted with a highly rifamycin-resistant strain, despite the high nominal concentration of 8000 μg/mL reported in fecal samples following rifaximin therapy (Jiang et al. 2000). Perhaps the best strategy is to use other options for *C. difficile* therapy if possible.

SUMMARY

Rifampicin continues to be very effective in combination therapies with careful selection of partner antibiotics, particularly as a mainstay therapeutic for treating active and latent TB, as well as selective, serious Gram-positive infections. Rifampicin is well established as the most essential agent to treat biofilms in PJI infections, and it is possible that its role could expand to other serious Gram-positive infections, provided that optimal combinations can be established and agreed to. One obstacle to clinical research is the lack of entrepreneurial incentive for an older drug off patent. We all benefit, however, if an infusion of money to finance clinical research is increased in the future, so that the proper evidence can lead the way to improved therapies for stubborn Gram-positive infections. Rifabutin is especially useful in treating TB in AIDS patients, and in preventing MAC infections. Rifabutin also has a role in *H. pylori* eradication as part of second-line combination therapy. Rifapentine, with its increased half-life, has potential to shorten active and latent TB therapy. Rifaximin, an oral drug confined to the GI tract, is used to treat GI disorders that involve dysfunction of the GI microbiome. Its efficacy for at least some indications probably has an additional (non-antibacterial) component as a regulatory effector of the pregnane X receptor. The continued exploration of rifaximin's uses is fostered in part by its excellent safety profile, as a drug confined to the GI tract, with both potent and broad-spectrum antibacterial activity and a soothing effect of toning down the anti-inflammatory response.

ACKNOWLEDGMENTS

I thank Andrej Trampuz (Charite–University Medicine Berlin) for his attention and advice for the section on Gram-positive systemic infections, Michael Cynamon (Syracuse Veterans Affairs Medical Center) for lending his expertise on mycobacterial infections, Simon Hirota (University of Calgary) for his interesting insights on rifamycins as effectors of the pregnane receptor, and A.L. Sonenshein (Tufts University School of Medicine) for his critical reading of the manuscript.

REFERENCES

Achermann Y, Eigenmann K, Ledergerber B, Derksen L, Rafeiner P, Clauss M, Nüesch R, Zellweger C, Vogt M, Zimmerli W. 2013. Factors associated with rifampin resistance in staphylococcal periprosthetic joint infections (PJI): A matched case-control study. *Infection* **41:** 431–437.

Adachi JA, DuPont HL. 2006. Rifaximin: A novel nonabsorbed rifamycin for gastrointestinal disorders. *Clin Infect Dis* **42:** 541–547.

Amorena B, Gracia E, Monzón M, Leiva J, Oteiza C, Pérez M, Alabart JL, Hernández-Yago J. 1999. Antibiotic susceptibility assay for *Staphylococcus aureus* in biofilms developed in vitro. *J Antimicrob Chemother* **44:** 43–55.

Aubry-Damon H, Soussy CJ, Courvalin P. 1998. Characterization of mutations in the rpoB gene that confer rifampin resistance in *Staphylococcus aureus*. *Antimicrob Agents Chemother* **42:** 2590–2594.

Baldoni D, Haschke M, Rajacic Z, Zimmerli W, Trampuz A. 2009. Linezolid alone or combined with rifampin against methicillin-resistant *Staphylococcus aureus* in experimental foreign-body infection. *Antimicrob Agents Chemother* **53:** 1142–1148.

Bass NM, Mullen KD, Sanyal A, Poordad F, Neff G, Leevy CB, Sigal S, Sheikh MY, Beavers K, Frederick T, et al. 2010. Rifaximin treatment in hepatic encephalopathy. *N Engl J Med.* **362:** 1071–1081.

Billington OJ, McHugh TD, Gillespie SH. 1999. Physiological cost of rifampin resistance induced in vitro in *Mycobacterium tuberculosis*. *Antimicrob Agents Chemother* **43:** 1866–1869.

Blandizzi C, Viscomi GC, Scarpignato C. 2014. Impact of crystal polymorphism on the systemic bioavailability of rifaximin, an antibiotic acting locally in the gastrointestinal tract, in healthy volunteers. *Drug Des Devel Ther* **9:** 1–11

Boeree MJ, Diacon AH, Dawson R, Narunsky K, du Bois J, Venter A, Phillips PP, Gillespie SH, McHugh TD, Hoelscher M, et al. 2015. A dose-ranging trial to optimize the dose of rifampin in the treatment of tuberculosis. *Am J Respir Crit Care Med* **191:** 1058–1065.

Brigidi P, Swennen E, Rizzello F, Bozzolasco M, Matteuzzi D. 2002. Effects of rifaximin administration on the intestinal microbiota in patients with ulcerative colitis. *J Chemother* **14:** 290–295.

Burman WJ, Gallicano K, Peloquin C. 2001. Comparative pharmacokinetics and pharmacodynamics of the rifamycin antibacterials. *Clin Pharmacokinet* **40:** 327–341.

Burman W, Benator D, Vernon A, Khan A, Jones B, Silva C, Lahart C, Weis S, King B, Mangura B, et al. 2006. Acquired rifamycin resistance with twice-weekly treatment of HIV-related tuberculosis. *Am J Respir Crit Care Med* **173:** 350–356.

Carman RJ, Boone JH, Grover H, Wickham KN, Chen L. 2012. In vivo selection of rifamycin-resistant *Clostridium difficile* during rifaximin therapy. *Antimicrob Agents Chemother* **56:** 6019–6020.

Chen LF, Kaye D. 2009. Current use for old antibacterial agents: Polymyxins, rifamycins, and aminoglycosides. *Infect Dis Clin North Am* **7523:** 1053–1075.

Cheng J, Shah YM, Ma X, Pang X, Tanaka T, Kodama T, Krausz KW, Gonzalez FJ. 2010. Therapeutic role of rifaximin in inflammatory bowel disease: Clinical implication of human pregnane X receptor activation. *J Pharmacol Exp Ther* **335:** 32–41.

Cho JH, Brant SR. 2011. Recent insights into the genetics of inflammatory bowel disease. *Gastroenterology* **140:** 1704–1712.

Costerton JW, Stewart PS, Greenberg EP. 1999. Bacterial biofilms: A common cause of persistent infections. *Science* **284:** 1318–1322.

Curry SR, Marsh JW, Shutt KA, Muto CA, O'Leary MM, Saul MI, Pasculle AW, Harrison LH. 2009. High frequency of rifampin resistance identified in an epidemic *Clostridium difficile* clone from a large teaching hospital. *Clin Infect Dis* **48:** 425–429.

Darkoh C, Lichtenberger LM, Ajami N, Dial EJ, Jiang ZD, DuPont HL. 2010. Bile acids improve the antimicrobial effect of rifaximin. *Antimicrob Agents Chemother* **54:** 3618–3624.

De Leo C, Eftimiadi C, Schito GC. 1986. Rapid disappearance from the intestinal tract of bacteria resistant to rifaximin. *Drugs Exp Clin Res* **12:** 979– 981.

Dring MM, Goulding CA, Trimble VI, Keegan D, Ryan AW, Brophy KM, Smyth CM, Keeling PW, O'Donoghue D, O'Sullivan M, et al. 2006. The pregnane X receptor locus is associated with susceptibility to inflammatory bowel disease. *Gastroenterology* **130:** 341–348.

Dworkin RJ, Lee BL, Sande MA, Chambers HF. 1989. Treatment of right-sided *Staphylococcus aureus* endocarditis in intravenous drug users with ciprofloxacin and rifampicin. *Lancet* **334:** 1071–1073.

Field SK, Fisher D, Jarand JM, Cowie RL. 2012. New treatment options for multidrug-resistant tuberculosis. *Ther Adv Respir Dis* **6:** 255–268.

Forrest GN, Tamura K. 2010. Rifampin combination therapy for nonmycobacterial infections. *Clin Microbiol Rev* **23:** 14–34.

Fox W, Ellard GA, Mitchison DA. 1999. Studies on the treatment of tuberculosis undertaken by the British Medical Research Council tuberculosis units, 1946–1986, with relevant subsequent publications. *Int J Tuberc Lung Dis* **3:** S231–S279.

Furustrand Tafin U, Corvec S, Betrisey B, Zimmerli W, Trampuz A. 2012. Role of rifampin against *Propionibacterium acnes* biofilm in vitro and in an experimental foreign-body infection model. *Antimicrob Agents Chemother* **56:** 1885–1891.

Getahun H, Matteelli A, Chaisson RE, Raviglione M. 2015. Latent *Mycobacterium tuberculosis* infection. *N Engl J Med* **372:** 2127–2135.

Gisbert JP, Calvet X. 2012. Review article: Rifabutin in the treatment of refractory *Helicobacter pylori* infection. *Aliment Pharmacol Ther* **35:** 209–221.

Goldstein BP. 2014. Resistance to rifampicin: A review. *J Antibiot* **67:** 625–630.

Gomes F, Teixeira P, Ceri H, Oliveira R. 2012. Evaluation of antimicrobial activity of certain combinations of antibiotics against in vitro *Staphylococcus epidermidis* biofilms. *Indian J Med Res* **135:** 542–547.

Guslandi M. 2011. Rifaximin in the treatment of inflammatory bowel disease. *World J Gastroenterol* **17:** 4643–4646.

Habib G, Lancellotti P, Antunes MJ, Bongiorni MG, Casalta JP, Del Zotti F, Dulgheru R, El Khoury G, Erba PA, Iung B, et al. 2015. 2015 ESC Guidelines for the management of infective endocarditis: The Task Force for the Management of Infective Endocarditis of the European Society of Cardiology (ESC) Endorsed by: European Association for Cardio-Thoracic Surgery (EACTS), the European Association of Nuclear Medicine (EANM). *Eur Heart J* **36:** 3075–3128.

Heldman AW, Hartert TV, Ray SC, Daoud EG, Kowalski TE, Pompili VJ, Sisson SD, Tidmore WC, vom Eigen KA, Goodman SN, et al. 1996. Oral antibiotic treatment of right-sided staphylococcal endocarditis in injection drug users: Prospective randomized comparison with parenteral therapy. *Am J Med* **101:** 68–76.

Hirota SA. 2015. Understanding the molecular mechanisms of rifaximin in the treatment of gastrointestinal disorders—A focus on the modulation of host tissue function. *Mini Rev Med Chem* **16:** 206–217.

Hoover WW, Gerlach EH, Hoban DJ, Eliopoulos GM, Pfaller MA, Jones RN. 1993. Antimicrobial activity and spectrum of rifaximin, a new topical rifamycin derivative. *Diagn Microbiol Infect Dis* **16:** 111–118.

Hopkins KL, Mushtaq S, Richardson JF, Doumith M, de Pinna E, Cheasty T, Wain J, Livermore DM, Woodford N. 2014. In vitro activity of rifaximin against clinical isolates of *Escherichia coli* and other enteropathogenic bacteria isolated from travellers returning to the UK. *Int J Antimicrob Agents* **43:** 431–437.

Horsburgh CR, Rubin EJ. 2011. Clinical practice. Latent tuberculosis infection in the United States. *N Engl J Med* **364:** 1441–1448.

Hu Y, Ren J, Zhan M, Li W, Dai H. 2012. Efficacy of rifaximin in prevention of travelers' diarrhea: A meta-analysis of randomized, double-blind, placebo-controlled trials. *J Travel Med* **19:** 352–356.

Huang JS, Jiang ZD, Garey KW, Lasco T, Dupont HL. 2013. Use of rifamycin drugs and development of infection by rifamycin-resistant strains of *Clostridium difficile*. *Antimicrob Agents Chemother* **57:** 2690–2693.

Iorio N, Malik Z, Schey R. 2015. Profile of rifaximin and its potential in the treatment of irritable bowel syndrome. *Clin Exp Gastroenterol* **8:** 159–167.

Jacqueline C, Caillon J. 2014. Impact of bacterial biofilm on the treatment of prosthetic joint infections. *J Antimicrob Chemother* **69:** 37–40.

Jiang ZD, Ke S, Palazzini E, Riopel L, Dupont H. 2000. In vitro activity and fecal concentration of rifaximin after oral administration. *Antimicrob Agents Chemother* **44:** 2205–2206.

Jiang ZD, DuPont HL, La Rocco M, Garey KW. 2010. In vitro susceptibility of *Clostridium difficile* to rifaximin and rifampin in 359 consecutive isolates at a university hospital in Houston, Texas. *J Clin Pathol* **63:** 355–358.

Jindani A, Doré CJ, Mitchison DA. 2003. Bactericidal and sterilizing activities of antituberculosis drugs during the first 14 days. *Am J Respir Crit Care Med* **167:** 1348–1354.

Jindani A, Harrison TS, Nunn AJ, Phillips PP, Churchyard GJ, Charalambous S, Hatherill M, Geldenhuys H, McIlleron HM, Zvada SP, et al. 2014. High-dose rifapentine with moxifloxacin for pulmonary tuberculosis. *N Engl J Med* **371:** 1599–1608.

John AK, Baldoni D, Haschke M, Rentsch K, Schaerli P, Zimmerli W, Trampuz A. 2009. Efficacy of daptomycin in implant-associated infection due to methicillin-resistant *Staphylococcus aureus*: Importance of combination with rifampin. *Antimicrob Agents Chemother* **53:** 2719–2724.

Johnson S, Schriever C, Patel U, Patel T, Hecht DW, Gerding DN. 2009. Rifaximin redux: Treatment of recurrent *Clostridium difficile* infections with rifaximin immediately post-vancomycin treatment. *Anaerobe* **15:** 290–291.

Kassinen A, Krogius-Kurikka L, Mäkivuokko H, Rinttilä T, Paulin L, Corander J, Malinen E, Apajalahti J, Palva A. 2007. The fecal microbiota of irritable bowel syndrome patients differs significantly from that of healthy subjects. *Gastroenterology* **133:** 24–33.

Klemens SP, Grossi MA, Cynamon MH. 1994. Comparative in vivo activities of rifabutin and rifapentine against *Mycobacterium avium complex*. *Antimicrob Agents Chemother* **38:** 234–237.

Kothary V, Scherl EJ, Bosworth B, Jiang ZD, Dupont HL, Harel J, Simpson KW, Dogan B. 2013. Rifaximin resistance in *Escherichia coli* associated with inflammatory bowel disease correlates with prior rifaximin use, mutations in *rpoB*, and activity of Phe-Arg-β-naphthylamide-inhibitable efflux pumps. *Antimicrob Agents Chemother* **57:** 811–817.

Kunin CM. 1996. Antimicrobial activity of rifabutin. *Clin Infect Dis* **22:** S3–S14.

Langmann T, Moehle C, Mauerer R, Scharl M, Liebisch G, Zahn A, Stremmel W, Schmitz G. 2004. Loss of detoxification in inflammatory bowel disease: Dysregulation of pregnane X receptor target genes. *Gastroenterology* **127:** 26–40.

Levine DP, Fromm BS, Reddy BR. 1991. Slow response to vancomycin or vancomycin plus rifampin in methicillin-resistant *Staphylococcus aureus* endocarditis. *Ann Intern Med* **115:** 674–680.

Ma X, Shah YM, Guo GL, Wang T, Krausz KW, Idle JR, Gonzalez FJ. 2007. Rifaximin is a gut-specific human pregnane X receptor activator. *J Pharmacol Exp Ther* **322:** 391–398.

Mariam DH, Mengistu Y, Hoffner SE, Andersson DI. 2004. Effect of *rpoB* mutations conferring rifampin resistance on fitness of *Mycobacterium tuberculosis*. *Antimicrob Agents Chemother* **48:** 1289–1294.

Mihailescu R, Furustrand Tafin U, Corvec S, Oliva A, Betrisey B, Borens O, Trampuz A. 2014. High activity of fosfomycin and rifampin against methicillin-resistant *Staphylococcus aureus* biofilm in vitro and in an experimental foreign-body infection model. *Antimicrob Agents Chemother* **58:** 2547–2553.

Mitchison D, Davies G. 2012. The chemotherapy of tuberculosis: Past, present and future. *Tuberc Lung Dis* **16:** 724–732.

Mitnick CD, McGee B, Peloquin CA. 2009. Tuberculosis pharmacotherapy: Strategies to optimize patient care. *Expert Opin Pharmacother* **10:** 381–401.

Mullen KD, Sanyal AJ, Bass NM, Poordad FF, Sheikh MY, Frederick RT, Bortey E, Forbes WP. 2014. Rifaximin is safe and well tolerated for long-term maintenance of remission from overt hepatic encephalopathy. *Clin Gastroenterol Hepatol* **12:** 1390–1397.

Munsiff SS, Kambili C, Ahuja SD. 2006. Rifapentine for the treatment of pulmonary tuberculosis. *Clin Infect Dis* **43:** 1468–1475.

Murphy CK, Mullin S, Osburne MS, van Duzer J, Siedlecki J, Yu X, Kerstein K, Cynamon M, Rothstein DM. 2006. In vitro activity of novel rifamycins against rifamycin-resis-

tant *Staphylococcus aureus*. *Antimicrob Agents Chemother* **50:** 827–834.

Neu HC. 1983. Antibacterial activity of DL 473, a C3-substituted rifamycin derivative. *Antimicrob Agents Chemother* **24:** 457–460.

Nuermberger EL, Spigelman MK, Yew WW. 2010. Current development and future prospects in chemotherapy of tuberculosis. *Respirology* **15:** 764–778.

O'Brien RJ, Vernon AA. 1998. New tuberculosis drug development. How can we do better? *Am J Respir Crit Care Med* **157:** 1705–1707.

O'Connor JR, Galang MA, Sambol SP, Hecht DW, Vedantam G, Gerding DN, Johnson S. 2008. Rifampin and rifaximin resistance in clinical isolates of *Clostridium difficile*. *Antimicrob Agents Chemother* **52:** 2813–2817.

O'Neill AJ, Huovinen T, Fishwick CW, Chopra I. 2006. Molecular genetic and structural modeling studies of *Staphylococcus aureus* RNA polymerase and the fitness of rifampin resistance genotypes in relation to clinical prevalence. *Antimicrob Agents Chemother* **50:** 298–309.

Osmon DR, Berbari EF, Berendt AR, Lew D, Zimmerli W, Steckelberg JM, Rao N, Hanssen A, Wilson WR; Infectious Diseases Society of America. 2013. Diagnosis and management of prosthetic joint infection: Clinical practice guidelines by the Infectious Diseases Society of America. *Clin Infect Dis* **56:** e1–e25.

Patidar KR, Bajaj JS. 2013. Antibiotics for the treatment of hepatic encephalopathy. *Metab Brain Dis* **28:** 307–312.

Peel TN, Buising KL, Dowsey MM, Aboltins CA, Daffy JR, Stanley PA, Choong PF. 2013. Outcome of debridement and retention in prosthetic joint infections by methicillin-resistant *Staphylococci*, with special reference to rifampin and fusidic acid combination therapy. *Antimicrob Agents Chemother* **57:** 350–355.

Perlroth J1, Kuo M, Tan J, Bayer AS, Miller LG. 2008. Adjunctive use of rifampin for the treatment of *Staphylococcus aureus* infections: A systematic review of the literature. *Arch Intern Med* **168:** 805–819.

Pimentel M, Lembo A, Chey WD, Zakko S, Ringel Y, Yu J, Mareya SM, Shaw AL, Bortey E, Forbes WP; TARGET Study Group. 2011a. Rifaximin therapy for patients with irritable bowel syndrome without constipation. *N Engl J Med* **364:** 22–32.

Pimentel M, Morales W, Chua K, Barlow G, Weitsman S, Kim G, Amichai MM, Pokkunuri V, Rook E, Mathur R, et al. 2011b. Effects of rifaximin treatment and retreatment in nonconstipated IBS subjects. *Dig Dis Sci* **56:** 2067–2072.

Pistiki A, Galani I, Pyleris E, Barbatzas C, Pimentel M, Giamarellos-Bourboulis EJ. 2014. In vitro activity of rifaximin against isolates from patients with small intestinal bacterial overgrowth. *Int J Antimicrob Agents* **43:** 236–241.

Pons MJ, Mensa L, Gascón J, Ruiz J. 2012. Fitness and molecular mechanisms of resistance to rifaximin in in vitro selected *Escherichia coli* mutants. *Microb Drug Resist* **18:** 376–379.

Prantera C, Lochs H, Campieri M, Scribano ML, Sturniolo GC, Castiglione F, Cottone M. 2006. Antibiotic treatment of Crohn's disease: Results of a multicentre, double blind, randomized, placebo-controlled trial with rifaximin. *Aliment Pharmacol Ther* **23:** 1117–1125.

Prantera C, Lochs H, Grimaldi M, Danese S, Scribano ML, Gionchetti P; Retic Study Group (Rifaximin-EIR Treatment in Crohn's Disease). 2012. Rifaximin-extended intestinal release induces remission in patients with moderately active Crohn's disease. *Gastroenterology* **142:** 473–481.

Rastogi N, Goh KS, Berchel M, Bryskier A. 2000. Activity of rifapentine and its metabolite 25-O-desacetylrifapentine compared with rifampin and rifabutin against *Mycobacterium tuberculosis, Mycobacterium africanum, Mycobacterium bovis* and *M. bovis* BCG. *J Antimicrob Chemother* **46:** 565–570.

Regazzi M, Carvalho AC, Villani P, Matteelli A. 2014. Treatment optimization in patients co-infected with HIV and Mycobacterium tuberculosis infections: Focus on drug–drug interactions with rifamycins. *Clin Pharmacokinet* **53:** 489–507.

Rivkin A, Gim S. 2011. Rifaximin: New therapeutic indication and future directions. *Clin Ther* **33:** 812–827.

Saadi M, McCallum RW. 2013. Rifaximin in irritable bowel syndrome: Rationale, evidence and clinical use. *Ther Adv Chronic Dis* **4:** 71–75.

Saginur R, Stdenis M, Ferris W, Aaron SD, Chan F, Lee C, Ramotar K. 2006. Multiple combination bactericidal testing of staphylococcal biofilms from implant-associated infections. *Antimicrob Agents Chemother* **50:** 55–61.

Saleh-Mghir A, Muller-Serieys C, Dinh A, Massias L, Crémieux AC. 2011. Adjunctive rifampin is crucial to optimizing daptomycin efficacy against rabbit prosthetic joint infection due to methicillin-resistant *Staphylococcus aureus*. *Antimicrob Agents Chemother* **55:** 4589–4893.

Sanders WE. 1976. Rifampin. *Ann Intern Med* **85:** 82–86.

Schrenzel J, Harbarth S, Schockmel G, Genné D, Bregenzer T, Flueckiger U, Petignat C, Jacobs F, Francioli P, Zimmerli W, et al. 2004. A randomized clinical trial to compare fleroxacin-rifampicin with flucloxacillin or vancomycin for the treatment of staphylococcal infection. *Clin Infect Dis* **39:** 1285–1292.

Scott LJ. 2014. Rifaximin: A review of its use in reducing recurrence of overt hepatic encephalopathy episodes. *Drugs* **74:** 2153–2160.

Silver LL. 2011. Challenges of antibacterial discovery. *Clin Microbiol Rev* **24:** 71–109.

Stagg HR, Zenner D, Harris RJ, Muñoz L, Lipman MC, Abubakar I. 2014. Treatment of latent tuberculosis infection: A network meta-analysis. *Ann Intern Med* **161:** 419–428.

Sussman NL. 2015. Treatment of overt hepatic encephalopathy. *Clin Liver Dis* **19:** 551–563.

Taylor DN, Bourgeois AL, Ericsson CD, Steffen R, Jiang ZD, Halpern J, Haake R, Dupont HL. 2006. A randomized, double-blind, multicenter study of rifaximin compared with placebo and with ciprofloxacin in the treatment of travelers' diarrhea. *Am J Trop Med Hyg* **74:** 1060–1066.

Thornsberry C, Hill BC, Swenson JM, McDougal LK. 1983. Rifampin: Spectrum of antibacterial activity. *Rev Infect Dis* **3:** S412–S417.

Tupin A, Gualtieri M, Roquet-Banères F, Morichaud Z, Brodolin K, Leonetti JP. 2010. Resistance to rifampicin: At the crossroads between ecological, genomic and medical concerns. *Int J Antimicrob Agents* **35:** 519–523.

Valentin T, Leitner E, Rohn A, Zollner-Schwetz I, Hoenigl M, Salzer HJ, Krause R. 2011. Rifaximin intake leads to emergence of rifampin-resistant Staphylococci. *J Infect* **62:** 34–38.

Valentin T, Hoenigl M, Wagner J, Krause R, Zollner-Schwetz I. 2014. Bacteraemia with rifampin-resistant *Staphylococcus aureus* and the potential role of cross-resistance between rifampin and rifaximin. *J Infect* **69:** 295–297.

Van der Auwera P, Klastersky J, Thys JP, Meunier-Carpentier F, Legrand JC. 1985. Double-blind, placebo-controlled study of oxacillin combined with rifampin in the treatment of staphylococcal infections. *Antimicrob Agents Chemother* **28:** 467–472.

Vergidis P, Rouse MS, Euba G, Karau MJ, Schmidt SM, Mandrekar JN, Steckelberg JM, Patel R. 2011. Treatment with linezolid or vancomycin in combination with rifampin is effective in an animal model of methicillin-resistant *Staphylococcus aureus* foreign body osteomyelitis. *Antimicrob Agents Chemother* **55:** 1182–1186.

Wichelhaus TA, Schäfer V, Brade V, Böddinghaus B. 1999. Molecular characterization of rpoB mutations conferring cross-resistance to rifamycins on methicillin-resistant *Staphylococcus aureus*. *Antimicrob Agents Chemother* **43:** 2813–2816.

Wichelhaus TA, Böddinghaus B, Besier S, Schäfer V, Brade V, Ludwig A. 2002. Biological cost of rifampin resistance from the perspective of *Staphylococcus aureus*. *Antimicrob Agents Chemother* **46:** 3381–3385.

Widmer AF, Frei R, Rajacic Z, Zimmerli W. 1990. Correlation between in vivo and in vitro efficacy of antimicrobial agents against foreign body infections. Correlation between in vivo and in vitro efficacy of antimicrobial agents against foreign body infections. *J Infect Dis* **162:** 96–102.

Widmer AF, Gaechter A, Ochsner PE, Zimmerli W. 1992. Antimicrobial treatment of orthopedic implant-related infections with rifampin combinations. *Clin Infect Dis* **14:** 1251–1253.

Williams DL, Spring L, Gillis TP, Salfinger M, Persing DH. 1998. Evaluation of a polymerase chain reaction-based universal heteroduplex generator assay for direct detection of rifampin susceptibility of *Mycobacterium tuberculosis* from sputum specimens. *Clin Infect Dis* **26:** 446–450.

Zhang J, Zhu L, Stonier M, Coumbis J, Xu X, Wu Y, Arikan D, Farajallah A, Bertz R. 2011. Determination of rifabutin dosing regimen when administered in combination with ritonavir-boosted atazanavir. *J Antimicrob Chemother* **66:** 2075–2082.

Zimmerli W, Waldvogel FA, Vaudaux P, Nydegger UE. 1982. Pathogenesis of foreign body infection: Description and characteristics of an animal model. *J Infect Dis* **146:** 487–497.

Zimmerli W, Widmer AF, Blatter M, Frei R, Ochsner PE. 1998. Role of rifampin for treatment of orthopedic implant-related staphylococcal infections: A randomized controlled trial. Foreign-Body Infection (FBI) Study Group. *JAMA* **279:** 1537–1541.

Zimmerli W, Trampuz A, Ochsner PE. 2004. Prosthetic-joint infections. *N Engl J Med* **351:** 1645–1654.

Bacterial Protein Synthesis as a Target for Antibiotic Inhibition

Stefan Arenz[1] and Daniel N. Wilson[1,2]

[1]Center for Integrated Protein Science Munich (CiPSM), University of Munich, 81377 Munich, Germany

[2]Gene Center and Department for Biochemistry, University of Munich, 81377 Munich, Germany

Correspondence: wilson@lmb.uni-muenchen.de

Protein synthesis occurs on macromolecular machines, called ribosomes. Bacterial ribosomes and the translational machinery represent one of the major targets for antibiotics in the cell. Therefore, structural and biochemical investigations into ribosome-targeting antibiotics provide not only insight into the mechanism of action and resistance of antibiotics, but also insight into the fundamental process of protein synthesis. This review summarizes the recent advances in our understanding of protein synthesis, particularly with respect to X-ray and cryoelectron microscopy (cryo-EM) structures of ribosome complexes, and highlights the different steps of translation that are targeted by the diverse array of known antibiotics. Such findings will be important for the ongoing development of novel and improved antimicrobial agents to combat the rapid emergence of multidrug resistant pathogenic bacteria.

THE BACTERIAL RIBOSOME AND PROTEIN SYNTHESIS

The ribosome is the macromolecular machine that converts the genetic information encoded in the messenger RNA (mRNA) into the polypeptide sequence that comprises the proteins and enzymes of the cell (Schmeing and Ramakrishnan 2009; Voorhees and Ramakrishnan 2013). Bacterial 70S ribosomes are comprised of two subunits, a small 30S subunit and a large 50S subunit (Fig. 1A), both of which are ribonucleoprotein particles. In the bacterium *Escherichia coli*, the small subunit (SSU) is assembled from 21 ribosomal proteins and a single 16S ribosomal RNA (rRNA) of 1541 nucleotides, whereas the large subunit (LSU) is assembled from 33 ribosomal proteins and two rRNAs, a 5S rRNA of 115 nucleotides, and a 23S rRNA of 2904 nucleotides (Fig. 1B). The ribosome provides the platform for the binding of transfer RNAs (tRNAs), which are adaptor molecules that contain at one end the anticodon to recognize a specific codon of the mRNA, and at the other end the CCA-end that is covalently linked to the amino acid specific for the mRNA codon. There are three tRNA binding sites in the 70S ribosome, termed the A-site, P-site, and E-site (Fig. 1C). The P-site is the peptidyl-tRNA binding site, which is occupied by the tRNA carrying the polypeptide chain during elongation (or the initiator

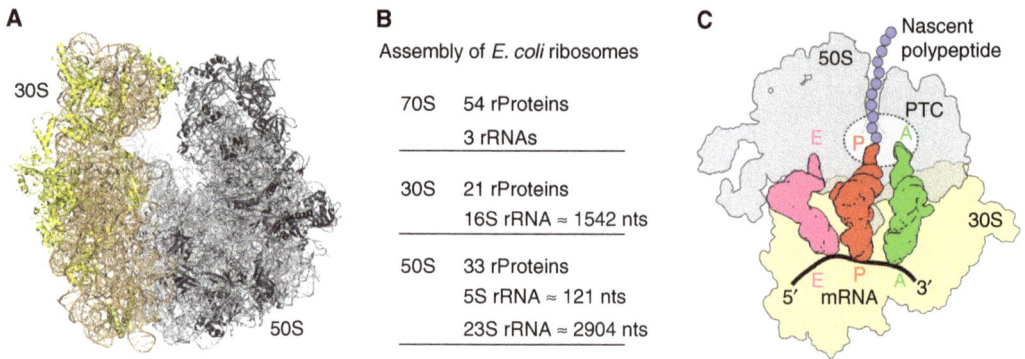

Figure 1. The prokaryotic ribosome. (A) Overview of the *Escherichia coli* 70S ribosome (Dunkle et al. 2011) with 30S subunit colored in yellow and 50S subunit in gray. (B) Table of assembly components of the 70S ribosome as well as the 30S and 50S subunits. (C) Schematic representation of the prokaryotic ribosome bound with three transfer RNAs (tRNAs) showing the 30S subunit (yellow), 50S subunit (gray), A-tRNA (green), P-tRNA (red), E-tRNA (pink), nascent polypeptide chain (violet), and mRNA (black). The peptidyltransferase center (PTC) on the 50S subunit is depicted as a dashed sphere.

tRNA during initiation). The A-site binds the incoming aminoacylated or charged tRNA, whereas the E-site is the exit site and binds only outgoing deacylated or uncharged tRNA. Thus, during translation elongation, the tRNAs pass consecutively through the A-, P-, and then E-sites before dissociating from the ribosome. There are two main functional centers on the ribosome, namely, the decoding center (DC) and peptidyltransferase center (PTC). The DC is located on the SSU in the A-site and monitors the correctness of the interaction between the codon of the mRNA and the anticodon of the tRNA and thereby ensures that the correct amino acid is delivered, that is, the amino acid corresponding to the codon of the mRNA. The PTC is located on the LSU and catalyzes the process of peptide bond formation. Protein synthesis can be divided into four main steps: initiation, elongation, termination, and ribosome recycling, each of which is targeted by a plethora of different antibiotics (Fig. 2) (Wilson 2009, 2014). Therefore, studies into antibiotic action during translation provide not only insight into the inhibitory action of the antibiotics, but also insight into the fundamental process of protein synthesis. This review provides an overview of the individual steps of protein synthesis and provides a brief overview of how these steps are inhibited by antibiotics.

INITIATION OF TRANSLATION

Initiation of translation is the rate-limiting step during translation of mRNA molecules into proteins (Laursen et al. 2005; Simonetti et al. 2009). Prokaryotic translation initiation requires formation of the 30S preinitiation complex (PIC), which in bacteria involves the three initiation factors (IFs), IF1, IF2, and IF3, as well as mRNA and the formylated initiator fMet-tRNA^{fMet} (Fig. 3A). The methionyl-tRNA transformylase-mediated formylation of the initiator tRNA distinguishes initiator fMet-tRNA^{fMet} from elongation Met-tRNA^{Met}. The main goal of 30S-PIC formation is to position the initiator fMet-tRNA^{fMet} in a peptidyl/initiation (P/I) state, such that it is bound to the start codon of the mRNA in the P-site of the SSU (Fig. 3A). Subsequently, the LSU joins to form the 70S preinitiation complex (70S-PIC) (Fig. 3B). 70S-IC formation is accompanied by the dissociation of initiation factors IF1–IF3 from the ribosome, leaving the fMet-tRNA^{fMet} positioned in the P/P state, thereby priming the ribosome for initiation of translation elongation (Fig. 3C).

During initiation, IF1 and IF3 ensure fidelity of the process, whereas IF2 recruits fMet-tRNA^{fMet}. IF3 binds to the E-site of the 30S subunit to form the IF3–30S complex (Dallas

Cite this article as *Cold Spring Harb Perspect Med* doi: 10.1101/cshperspect.a025361

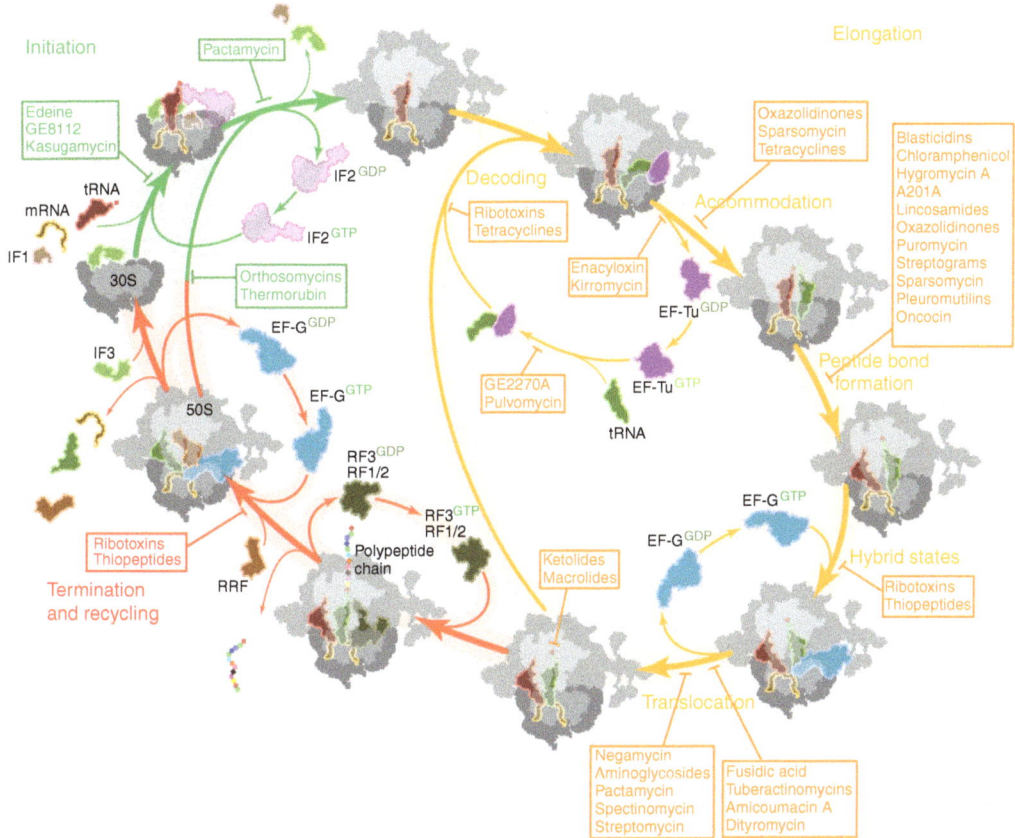

Figure 2. Overview of antibiotics inhibiting the prokaryotic translation cycle. Overview of antibiotics, inhibiting translation initiation (green), translation elongation (yellow), and translation termination/recycling (red) of the prokaryotic translation cycle (modified from Sohmen et al. 2009). tRNA, Transfer RNA; mRNA, messenger RNA; IF, initiation factor; EF-Tu, elongation factor Tu; GDP, guanosine diphosphate; EF-G, elongation factor G; GTP, guanosine triphosphate; RRF, ribosome recycling factor.

and Noller 2001) and, thereby, prevents premature 50S subunit joining before association with IF1, IF2, mRNA, and initiator tRNA (Karimi et al. 1999). The interaction of the Shine–Dalgarno (SD) sequence of canonical mRNAs with the anti-SD at the 3′ end of the 16S rRNA (Fig. 3D) places the start codon in the P-site and allows for subsequent association of fMet-tRNA^{fMet}, IF1, and IF2. Discrimination of the initiator tRNA is performed by IF3 through monitoring of three unique G:C base pairs in fMet-tRNA^{fMet}. Furthermore, the presence of IF3 is required to ensure the fidelity of the codon–anticodon interaction in the P-site of the SSU. IF1 binds at the A-site of the SSU (Fig. 3E)

(Carter et al. 2001) where it stabilizes IF2 binding (Julian et al. 2011; Simonetti et al. 2013) and accelerates IF2-dependent fMet-tRNA^{fMet} recruitment (Laursen et al. 2005). Binding of IFs to the SSU stabilizes a swiveled conformation of the head with respect to the body (Julian et al. 2011). The LSU joins the 30S-PIC in ratcheted conformation to form the 70S-IC (Allen et al. 2005). Formation of the 70S-IC activates guanosine triphosphate (GTP) hydrolysis by IF2, which leads to unratcheting of the ribosome, allowing the conformational transition of the fMet-tRNA^{fMet} from the P/I state to the accommodated P/P-state (Fig. 3C). At the same time, the IFs dissociate from the complex, thus

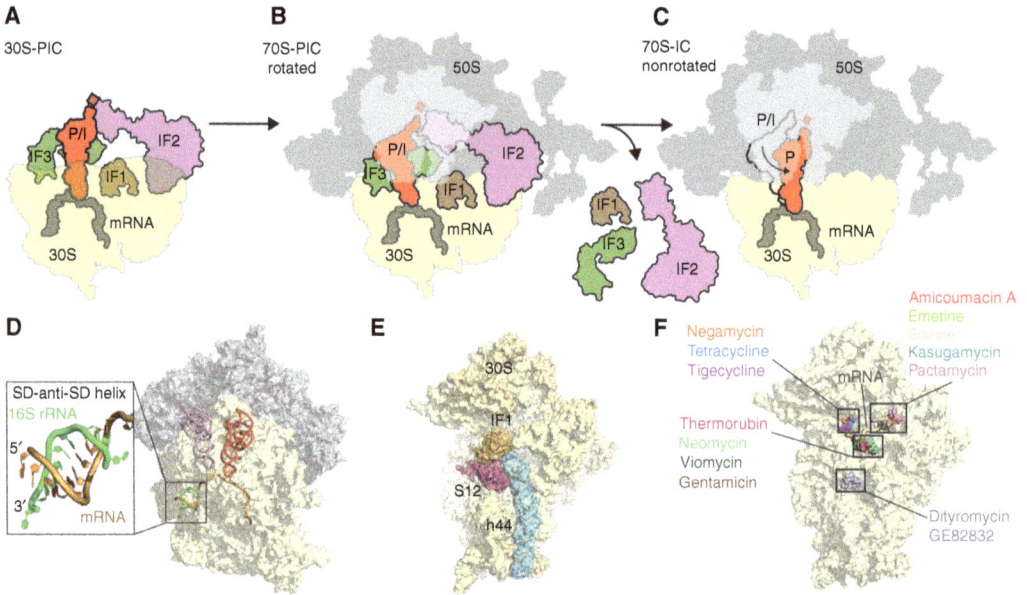

Figure 3. Initiation of translation. Schematic assembly of the (*A*) 30S preinitiation complex (PIC), (*B*) 70S-PIC, and (*C*) 70S-IC during translation initiation with 30S subunit (yellow), 50S subunit (gray), messenger RNA (mRNA) (dark gray), initiator transfer RNA (tRNA) (red), initiation factors (IFs): IF1 (brown), IF2 (purple), and IF3 (green). (*D*) Crystal structure of the prokaryotic ribosome with zoom onto the interaction of the Shine–Dalgarno (SD) sequence of canonical mRNAs (orange) with the anti-SD at the 3′ end of the 16S ribosomal RNA (rRNA) (green), including P-site tRNA (red) and E-site tRNA (pink) (Yusupova et al. 2006). (*E*) Crystal structure of IF1 (brown) bound to the 30S subunit (yellow) with highlighted h44 (blue) of the 16S rRNA and ribosomal protein S12 (red) (Carter et al. 2001). (*F*) Binding sites of dityromycin (PDB 4NVU) (Bulkley et al. 2014), gentamicin (PDB 4V53) (Borovinskaya et al. 2007a), thermorubin (PDB 3UXT) (Bulkley et al. 2012), viomycin (PDB 3KNH) (Stanley et al. 2010), neomycin (PDB 2QAL) (Borovinskaya et al. 2007a), negamycin (PDB 4RBH) (Polikanov et al. 2014c), tetracycline (PDB 4G5K) (Jenner et al. 2013), tigecycline (PDB 4G5T) (Jenner et al. 2013), amicoumacin A (PDB 4RB5) (Polikanov et al. 2014a), edeine (PDB 1I95) (Pioletti et al. 2001), kasugamycin (PDB 1VS5) (Schuwirth et al. 2006), pactamycin (PDB 4RBB) (Polikanov et al. 2014c), and emetine (PDB 3J7A) (Wong et al. 2014), on the small subunit (SSU) (yellow). P/I, Peptidyl/initiation.

readying the ribosome to enter into the elongation cycle.

There are a number of antibiotics that are commonly referred to as translation initiation inhibitors, namely, kasugamycin, pactamycin, edeine, GE81112, which interact with the SSU (Fig. 3F), the orthosomycins evernimicin and avilamycin, which interact with the LSU, as well as thermorubin, the binding site of which comprises components of both the SSU and LSU (Brandi et al. 2008; Wilson 2009; Bulkley et al. 2012). Kasugamycin binds within the E-site of the SSU in a position that overlaps with the mRNA (Fig. 3F) (Schluenzen et al. 2006; Schuwirth et al. 2006). By disturbing the path of the mRNA, kasugamycin prevent the initiator fMet-tRNA^{fMet} binding to the 30S-PIC (Schluenzen et al. 2006; Schuwirth et al. 2006). Similarly, pactamycin has also been reported to prevent 30S-PIC formation by perturbing the path of the mRNA through the E-site (Fig. 3F) (Brodersen et al. 2000); however, subsequent studies suggested that inhibition of translocation, rather than initiation, was the mechanism of action (Dinos et al. 2004). In contrast, edeine and GE81112 are suggested to inhibit 30S-PIC formation by directly blocking the binding of fMet-tRNA^{fMet} (Pioletti et al. 2001; Dinos et al. 2004; Brandi et al. 2006). The binding site of thermorubin is located within a pocket formed by h44 of the SSU and H69 of the LSU (Bulkley et al. 2012) and, therefore, does not directly overlap with the mRNA or initiator-tRNA, but rather inhibits 30S-PIC formation by inducing conformational changes that perturb IF binding (Bulkley et al. 2012). Evernimicin and avilamycin are suggested to block the association of the 30S-PIC with the LSU by preventing the accommodation of IF2 on the LSU during subunit joining (Belova et al. 2001).

ELONGATION PHASE

After initiation, the ribosomal P-site is occupied by the initiator fMet-tRNA^{fMet}, whereas the A-site remains empty. The next aminoacyl-tRNA (aa-tRNA) is delivered to the empty A-site as a ternary complex with elongation factor Tu (EF-Tu) and GTP (Schmeing and Ramakrishnan 2009; Voorhees and Ramakrishnan 2013). The ternary complex initially binds with a bent or kinked conformation of the tRNA that allows the anticodon stem loop (ASL) to interact with the codon of the mRNA in the DC, whereas the aminoacylated 3'-acceptor stem remains bounds to EF-Tu and is, hence, termed the A/T state (Fig. 4A) (Stark et al. 1997; Valle et al. 2002; Blanchard et al. 2004a; Schmeing et al. 2009; Schuette et al. 2009; Fischer et al. 2015). To ensure translational fidelity, the ribosome discriminates between cognate and noncognate tRNA binding by monitoring the interaction between the A-site codon of the mRNA and the anticodon of the tRNA (Ogle and Ramakrishnan 2005). During codon recognition, nucleotides A1492 and A1493 (*E. coli* numbering) adopt a conformation flipped-out of helix 44 (h44) of the 16S rRNA and together with G530 monitor the correct Watson–Crick geometry of the first two base pairs of the codon–anticodon interaction in the form of A-minor motifs (Fig. 4B) (Ogle and Ramakrishnan 2005). At the third nucleotide position of the codon a wobble pair (e.g., G·U) is tolerated. This allows a single tRNA to decode multiple codons that only differ in the third anticodon position, thus providing an explanation for the degeneracy of the genetic code. Recent crystal structures of complete cognate or near-cognate tRNAs bound to the full 70S ribosome have challenged the hypothesis of how the ribosome distinguishes cognate tRNAs from near-cognate or noncognate tRNAs (Demeshkina et al. 2012). Surprisingly, in these structures, even the near-cognate codon–anticodon interactions are observed to form Watson–Crick-like interactions, suggesting that A1492, A1493, and G530 do not discriminate cognate and near-cognate tRNAs (Demeshkina et al. 2012). Rather, the ribosome stabilizes an energetically unfavorable tautomer of the nucleotide to achieve Watson–Crick geometry, leading the investigators to suggest that this energetic penalty is then used by the ribosome to discriminate cognate from near-cognate tRNAs (Demeshkina et al. 2013; Rozov et al. 2015).

Nevertheless, recognition of a correct codon–anticodon interaction triggers large-scale conformational changes in the ribosome that induce a domain closure of the 30S, involving movement of the SSU shoulder toward EF-Tu (Ogle and Ramakrishnan 2005; Schmeing et al. 2009). These structural rearrangements are propagated to EF-Tu, which ultimately leads to stimulation of GTP hydrolysis (Schmeing and Ramakrishnan 2009; Voorhees and Ramakrishnan 2013). The GTPase activity of EF-Tu is controlled by positioning the catalytic histidine 84 (H84) of EF-Tu in proximity to the phosphate of A2662 of the sarcin–ricin loop (SRL) in H95 of the 23S rRNA. This enables H84 to coordinate a water molecule for nucleophilic attack on the γ-phosphate of GTP, which is then hydrolyzed (Voorhees et al. 2010). GTP hydrolysis and inorganic phosphate (P$_i$) release cause structural rearrangements in EF-Tu, leading to dissociation of EF-Tu from the ribosome and, thus, allowing the tRNA to transition from the A/T-state into the A/A-state. During accommodation, the acceptor stem of the tRNA moves into the A-site of the PTC in the large ribosomal subunit (Blanchard et al. 2004a; Sanbonmatsu et al. 2005).

There are many antibiotics that inhibit delivery and accommodation of the aminoacyl-tRNA on the ribosome (Wilson 2009). These range from antibiotics that interact with

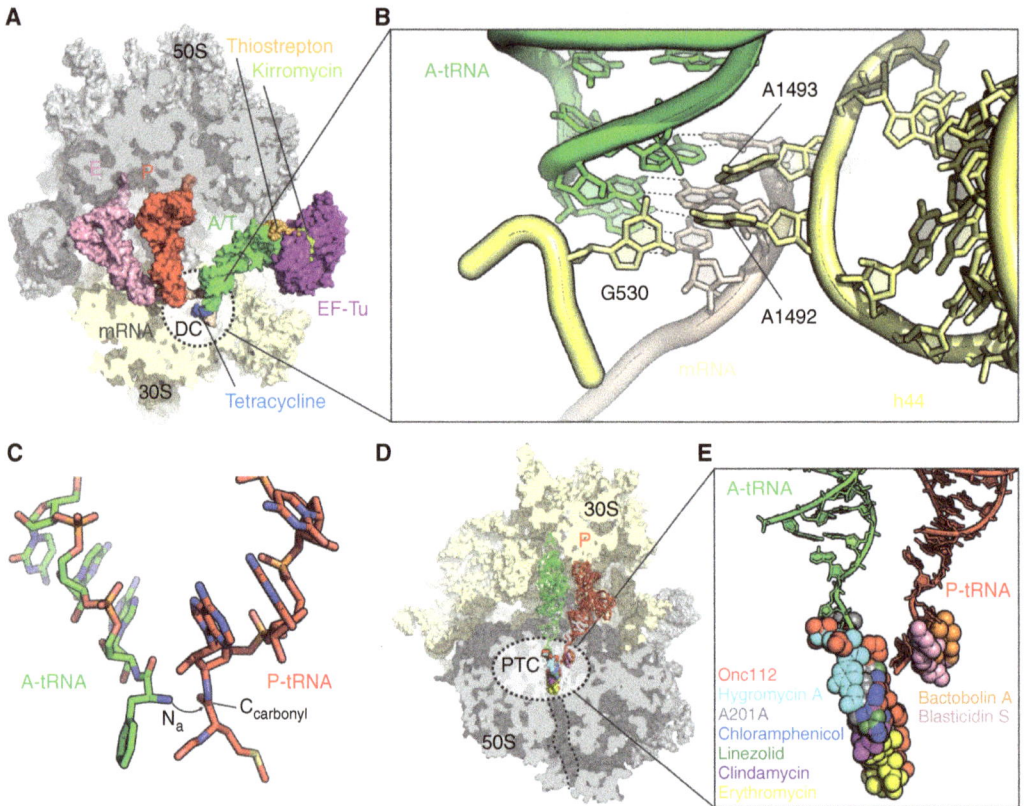

Figure 4. Decoding and peptide bond formation on the ribosome. (*A*) Crystal structure of the ribosome in complex with elongation factor Tu (EF-Tu) (purple), A/T-transfer RNA (tRNA) (green), P-tRNA (red), E-tRNA (pink), kirromycin (light green), messenger RNA (mRNA) (brown), 30S subunit (yellow), and 50S subunit (gray) (Voorhees et al. 2010) with superimposed binding positions of thiostrepton (orange) (Harms et al. 2008) and tetracycline (blue) (Jenner et al. 2013). The decoding center (DC) on the 30S subunit is depicted as dashed-lined sphere. (*B*) DC within the small ribosomal subunit with A-tRNA (green), mRNA (brown), and 16S ribosomal RNA (rRNA) nucleotides G530, A1492, and A1493 (yellow) (Demeshkina et al. 2012). (*C*) Positions of the A-tRNA (green) and P-tRNA (red) within the peptidyltransferase center (PTC) of the 50S subunit. The nucleophilic attack of the A-tRNA α-amino group (Nα) onto the P-tRNA carbonyl-carbon ($C_{carbonyl}$) is indicated with an arrow. (*D*) Overview and (*E*) zoom onto the binding sites of Onc112 (PDB 4ZER) (Seefeldt et al. 2015), hygromycin A (PDB 4Z3R) (Polikanov et al. 2015), A201A (PDB 4Z3S) (Polikanov et al. 2015), chloramphenicol (PDB 3OFC) (Dunkle et al. 2010), linezolid (PDB 3DLL) (Wilson et al. 2008), clindamycin (PDB 3OFZ) (Dunkle et al. 2010), erythromycin (PDB 3OFR) (Dunkle et al. 2010), bactobolin A (PDB 4WWE) (Amunts et al. 2015), and blasticidin S (PDB 4L6J) (Svidritskiy et al. 2013) within the PTC (dashed lines in *D*) of the 50S subunit (gray) with A-tRNA (green) and P-tRNA (red).

EF-Tu, such as GE2270A and kirromycin, to those that interact directly with the ribosome, such as thiostreptons and tetracyclines (Fig. 4A). In addition, antibiotics of the negamycin and aminoglycoside classes also bind to the ribosome and interfere with the decoding process. Thiopeptide antibiotics, such as GE2270A and related derivatives, such as LFF571, bind to EF-Tu and prevent formation of the complex with the aminoacyl-tRNA, whereas in contrast kirromycin binds to the EF-Tu-aa-tRNA complex and traps it on the ribosome (Fig. 4A) (Wilson 2009). The thiostrepton-like antibiotics interact with the large ribosomal subunit and interfere with the binding of translational GTPases, including EF-Tu and elongation factor

Cite this article as *Cold Spring Harb Perspect Med* doi: 10.1101/cshperspect.a025361

G (EF-G) to the ribosome (Fig. 4A) (Wilson 2009). The tetracycline family of antibiotics, including the third-generation glycylcyclines, such as tigecycline, bind to the SSU and sterically block the recognition of the codon of the mRNA by the anticodon of the aa-tRNA (Nguyen et al. 2014). Although negamycin and aminoglycoside antibiotics have distinct binding sites on the SSU, both classes stabilize the binding of aa-tRNAs, including near-cognate aa-tRNAs, thus leading to misreading and stop-codon suppression (Wilson 2009; Olivier et al. 2014; Polikanov et al. 2014c).

PEPTIDE BOND FORMATION AND NASCENT CHAIN PROLONGATION

Peptide bond formation between the incoming amino acid attached to the A-site tRNA and the nascent polypeptide attached to the P-site tRNA represents the main function of the ribosome (Simonovic and Steitz 2009; Rodnina 2013). On accommodation of the A-site tRNA, the peptidyl transfer reaction occurs rapidly, with a rate that is $\sim 2 \times 10^7$-fold faster than the rate of spontaneous peptide bond formation in solution (Sievers et al. 2004). The ribosome acts as an entropy trap by precisely positioning the aminoacylated tRNA CCA-end substrates for *trans*-esterification and formation of the new peptide bond (Sievers et al. 2004). Positioning of the tRNAs is facilitated by stabilizing interactions between 23S rRNA nucleotides and the tRNA CCA-ends. The P-site CCA end is stabilized by Watson–Crick base pairs of nucleotides C74 and C75 with P-loop nucleotides G2251 and G2252, respectively, whereas the A-tRNA C74 stacks up on U2555, C75 forms a Watson–Crick base pair with G2553 and A76 interacts with G2583 in the form of a class I A-minor motif (Kim and Green 1999; Nissen et al. 2000; Hansen et al. 2002; Voorhees et al. 2009; Polikanov et al. 2014b). Proper tRNA accommodation into the A-site leads to conformational changes within the PTC, namely, 23S rRNA nucleotides G2583-U2585 undergo a shift of 1–2 Å, while U2506 rotates by 90° to provide space for A-tRNA accommodation into the PTC (Schmeing et al. 2005). These confor-

mational changes convert the PTC into its induced state by exposing the peptidyl-tRNA ester for peptide bond formation, which occurs through nucleophilic attack of the α-amino group of the A-tRNA onto the carbonyl-carbon of the aminoacyl ester of the peptidyl-tRNA in the P-site (Fig. 4C). The most recent model for peptide bond formation is based on high-resolution crystal structures of a *Thermus thermophilus* 70S ribosome in both pre-attack and postcatalysis states (Polikanov et al. 2014b). Three water molecules trapped in the PTC before catalysis allowed the investigators to suggest a proton wire mechanism that couples aa-tRNA accommodation and peptide bond formation. Both tRNAs, 23S rRNA nucleotides A2451, U2584, C2063, and A2602, as well as the amino terminus of ribosomal protein L27 contribute to the coordination of the water molecules. L27 and N6 of A2602 activate a water molecule (W1) to initiate the proton wire via the 2′ OH of A2451 to the 2′ OH of the P-site A76, which deprotonates the α-amino group for concerted nucleophilic attack onto the ester carbonyl carbon. The tetrahedral intermediate state is stabilized by a second water molecule (W2), which donates a proton to the negatively charged ester carbonyl carbon. Breakdown of the intermediate state occurs via protonation of the 3′ ester oxygen of the leaving group via a third water (W3) and a partially reversed proton wire via the 2′ OH of P-site A76, the 2′ OH of A2451 back to W1 (Polikanov et al. 2014b).

Antibiotics that directly interfere with peptide bond formation generally do so by preventing the accurate placement of the aminoacylated-CCA-end of the A-tRNA at the PTC (Fig. 4D). Therefore, these antibiotics can also be thought of as inhibiting a final stage in the accommodation of the aminoacyl-tRNA at the A-site. Well-known examples include the phenicols (chloramphenicol), oxazolidinones (linezolid), pleuromutilins (tiamulin), and lincosamides (clindamycin) (Wilson 2009, 2014), but was also shown more recently for hygromycin A, the nucleoside antibiotic A201A (Polikanov et al. 2015), and the antimicrobial peptide oncocin (Fig. 4E) (Roy et al. 2015; Seefeldt et al. 2015). Some of the larger macrolide antibiotics, such as josamycin, tylosin, and spiramycin, also interfere with peptide bond formation by perturbing A-tRNA accommodation at the PTC (Wilson 2009), but generally the binding of macrolide antibiotics within the ribosomal tunnel is thought to interfere with prolongation of the nascent polypeptide chain (Kannan et al. 2014). However, recent studies have revealed that some polypeptides manage to bypass the drug in the tunnel and can even become fully synthesized in the presence of the drug (Kannan and Mankin 2012; Kannan et al. 2012).

TRANSLOCATION AND EF-G

After peptide bond formation, the A-site is bound by the peptidyl-tRNA lengthened by one amino acid and the P-site is bound by deacylated tRNA. To allow the next round of elongation, the tRNAs together with the mRNA need to move with respect to the ribosome, namely, to shift the peptidyl-tRNA from the A-site to the P-site and the deacylated tRNA from the P-site to the E-site (Yamamoto et al. 2014). The mRNA shifts precisely by one codon, placing the next codon in the A-site. Translocation is catalyzed by the GTPase elongation factor G (EF-G) and provides an empty A-site, which in turn allows binding and accommodation of the next cognate aa-tRNA and, thus, the elongation cycle to proceed (Fig. 2). However, the ribosome has an intrinsic capability to translocate tRNAs both forward and backward, and therefore, the function of EF-G is to accelerate and direct the process in forward direction (Yamamoto et al. 2014).

In the pretranslocation state, the 3′-ends of A- and P-site tRNA spontaneously move back and forth between the P- and E-sites on the LSU, respectively, while their ASLs remain anchored within the A- and P-sites on the SSU, which creates A/P and P/E hybrid tRNA binding states (Fig. 5A) (Moazed and Noller, 1989; Blanchard et al. 2004b). The spontaneous formation of tRNA hybrid states is driven by decreased affinity for peptidyl-tRNA in the A-site and deacylated tRNA in the P-site by ~1000-fold (Semenkov et al. 2000). Moreover, the E-site on the 50S subunit sterically occludes

binding of peptidyl-tRNA (Rheinberger et al. 1981; Schmeing et al. 2003) and, thus, ensures that translocation occurs only after peptide bond formation is completed. Formation of tRNA hybrid states is coupled to a large-scale conformational rotation (ratcheting) of the SSU ~3–10° counterclockwise relative to the LSU (Frank and Agrawal 2001; Valle et al. 2003).

It has been shown that EF-G binds to both the ratcheted and nonratcheted states of the ribosome (Chen et al. 2013; Pulk and Cate 2013), but translocation occurs via hybrid state formation for both scenarios. Rotation of the SSU head relative to the body (head swiveling) opens a constriction that allows the passage of the mRNA and ASLs through the SSU (Zhang et al. 2009; Ratje et al. 2010; Yamamoto et al. 2014). Domain IV of EF-G, overlapping the A-site on the SSU, is crucial to facilitate GTP-dependent translocation by disrupting interactions of the codon–anticodon duplex with the DC (Liu et al. 2014). Translocation of tRNAs through the ribosome proceeds via a series of intermediate states including intrasubunit hybrid states on the SSU, where the mRNA and ASLs simultaneously bind to the A- and P-site (ap/P) and to the P- and the E-site (pe/E) (Ratje et al. 2010; Ramrath et al. 2013). Following GTP hydrolysis and P_i release, the 30S head swivel and the ratcheting is reversed, EF-G dissociates and leaves the ribosome in the posttranslocation (POST) state with tRNAs bound in the classical P/P and E/E sites (Fig. 5B) (Ratje et al. 2010; Ermolenko and Noller 2011).

There are a plethora of antibiotics that interfere with the process of translocation, including aminoglycosides, tuberactinomycins (viomycin, capreomycin), and spectinomycin (Wilson 2009), as well as more recently characterized translocation inhibitors, such as negamycin, amicoumacin A, and dityromycin (Figs. 2 and 3F) (Bulkley et al. 2014; Olivier et al. 2014; Polikanov et al. 2014a,c). Aminoglycosides, such as kanamycin and gentamycin, bind within h44 of the SSU (Fig. 3F) and stabilize the pretranslocation state (Wilson 2009), whereas aminoglycosides, such as neomycin, have an additional binding site located in H69 of the LSU, and appear to inhibit translocation

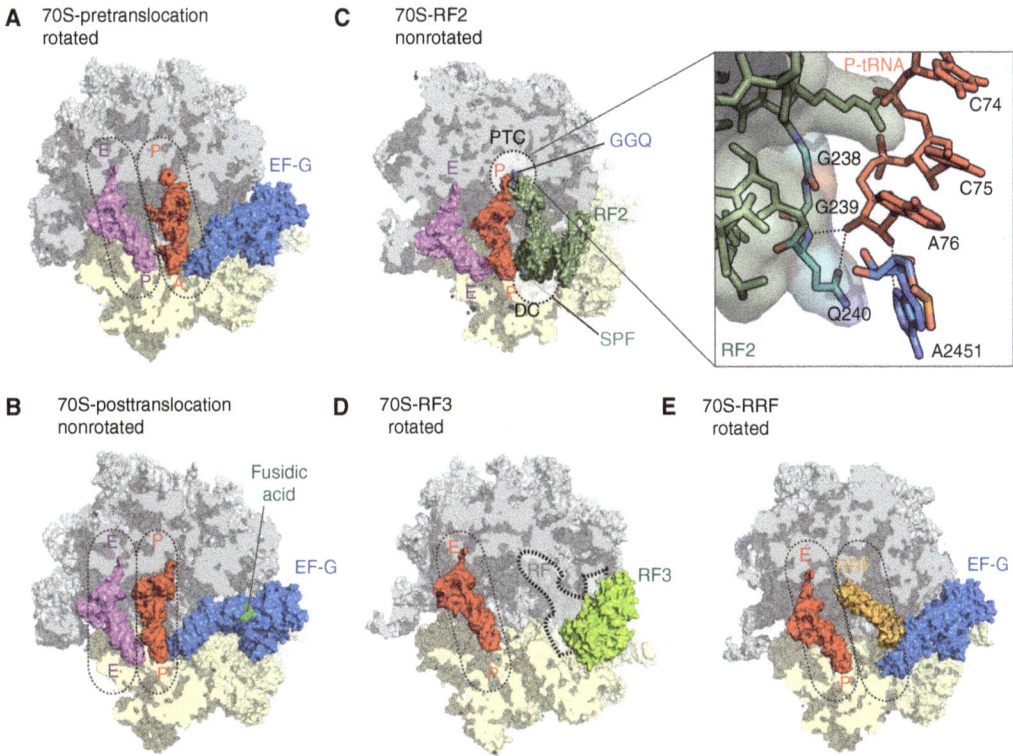

Figure 5. The prokaryotic ribosome. Crystal structures of the ribosome bound with elongation factor G (EF-G) in the (A) pre- (Brilot et al. 2013), and (B) posttranslocation state (Lin et al. 2015), with 30S subunit (yellow), 50S subunit (gray), EF-G (blue), P-transfer RNA (tRNA) (red), E-tRNA (pink), and fusidic acid (green). (C) Crystal structure of RF2 (dark green) bound to the poststate ribosome with P-tRNA (red) and E-tRNA (pink) (Weixlbaumer et al. 2008). The peptidyltransferase center (PTC) on the 50S (gray) and the DC on the 30S subunit (yellow) are indicated as dashed lines. The insert zooms onto the PTC showing the GGQ motif of RF2 interacting with the CCA-end of the P-tRNA. (D) Crystal structure of RF3 (pale green) bound to the rotated 70S ribosome (Zhou et al. 2012b) with superimposed position of the P/E hybrid tRNA (red) (Brilot et al. 2013). The position of class I release factors is indicated with dashed lines. (E) Crystal structure of ribosome recycling factor (RRF, orange) bound to the rotated ribosome (Borovinskaya et al. 2007a) with superimposed positions of P/E hybrid tRNA (red) and prestate EF-G (blue) (Brilot et al. 2013).

by trapping intermediate hybrid states (Wang et al. 2012). Similarly, the binding site of viomycin and capreomycin span the ribosomal interface between h44 (Fig. 3F) and H69 (Stanley et al. 2010), and inhibit translocation by stabilizing a distinct intermediate hybrid state (Ermolenko et al. 2007b). In contrast, spectinomycin binds to the neck region of the SSU (Carter et al. 2000), locking a rotated conformation of the SSU head (Borovinskaya et al. 2007b) and, thus, trapping an intermediate state during translocation (Pan et al. 2007). Negamycin inhibits translocation by interact-

ing with the A-tRNA and stabilizing it in the A-site (Olivier et al. 2014; Polikanov et al. 2014c), whereas amicoumacin A interacts with the mRNA in the E-site (Fig. 3F) and, thus, prevents translocation by preventing movement of the mRNA (Polikanov et al. 2014a). In contrast, dityromycin and the related antibiotic GE82832 interact exclusively with ribosomal protein S12 on the SSU (Fig. 3F) and block translocation by preventing EF-G from adopting the final state necessary for translocation of the tRNAs and mRNA on the SSU (Bulkley et al. 2014).

TERMINATION AND RECYCLING

Class I termination release factors 1 (RF1) or 2 (RF2) recognize mRNA stop codons in the A-site of the SSU and trigger translation termination by mediating hydrolysis of the P-site peptidyl-tRNA and subsequent peptide release (Fig. 2) (Klaholz 2011; Rodnina 2013). In a following step, the class I release factors are removed from the ribosome with the help of the GTPase class II release factor RF3 in a GTPase-dependent manner. RF1 and RF2 differ with respect to stop-codon recognition (RF1 recognizes UAG/UAA; RF2 recognizes UGA/UAA), but both mediate peptidyl-tRNA hydrolysis using their universally conserved GGQ-motif (Fig. 5C). Early structures revealed that ribosome binding induces conformational changes in RF1 and RF2 that stabilize an open conformation, which allows the concurrent insertion of the GGQ motif into the PTC on stop-codon recognition (Fig. 5C) (Klaholz 2011; Zhou et al. 2012a). Subsequent X-ray structures of RF1 and RF2 bound to the ribosome provided molecular insight into the structural determinants of specificity of stop-codon recognition (Korostelev et al. 2008; Laurberg et al. 2008; Weixlbaumer et al. 2008; Jin et al. 2010). The exact mechanism by which the GGQ motif coordinates peptidyl-tRNA hydrolysis remains unclear. The amino acid side chain of Q of the GGQ motif is not essential to mediate hydrolysis (Seit-Nebi et al. 2001), but rather the backbone nitrogen of Q240 of RF2 (or Q230 for RF1) that accounts for the catalytic activity by interacting with the 3' OH of A76 (Fig. 5C) (Laurberg et al. 2008). However, loss of the posttranslational N5-methylation of Q230/Q240 reduces the efficiency of peptide release (Dincbas-Renqvist et al. 2000) and, therefore, favors a model in which the glutamine side chain, together with the backbone amine, the 2' OH of A76, and A2451, directly coordinate a water molecule for nucleophilic attack (Weixlbaumer et al. 2008; Jin et al. 2010; Klaholz 2011). After hydrolysis of the nascent chain, the class II release factor RF3 binds to the ribosome and promotes dissociation of RF1 and RF2 from the ribosome in a GTP-dependent manner (Freistroffer et al.

1997; Koutmou et al. 2014; Peske et al. 2014). RF3·GTP binding to RF1/RF2·ribosome complexes stabilizes the ratcheted state of the ribosome with tRNAs in hybrid states (Fig. 5D) (Ermolenko et al. 2007a; Gao et al. 2007; Jin et al. 2011; Zhou et al. 2012b). The ratcheted conformation of the ribosome creates a series of steric clashes between RF domains I and IV with the L11 stalk and the 30S head, respectively, which are supposed to contribute to the destabilization of the RFs (Fig. 5D) (Gao et al. 2007).

Following termination, the ribosome still contains mRNA and deacylated tRNA in the P-site, which needs to be recycled to allow a new round of translation. During recycling, the ribosome is split into subunits by the combined action of ribosome recycling factor (RRF) and EF-G in a GTP-dependent manner (Fig. 2) (Kaji et al. 2001). Recent crystal structures of RRF bound to the ribosome show RRF bound in the ribosomal P-site stabilizing the ratcheted conformation of the ribosome and would, therefore, presumably also stabilize a deacylated tRNA in a P/E hybrid state (Fig. 5E) (Weixlbaumer et al. 2007; Dunkle et al. 2011). Binding of EF-G to the RRF·mRNA·tRNA·70S complex subsequently splits the ribosome into subunits on GTP hydrolysis (Peske et al. 2005; Zavialov et al. 2005; Barat et al. 2007). Binding of IF3 to the 30S subunit leads to dissociation of mRNA and deacylated tRNA and simultaneously prevents re-association with the 50S subunit (Zavialov et al. 2005) and thereby links the last steps in translation termination to the first steps of translation initiation (Fig. 2).

To date, there are no antibiotics that specifically target the termination and recycling phases of translation. Many antibiotics that act during elongation to inhibit factor binding, for example, thiostrepton, or prevent peptide bond formation, for example, chloramphenicol, also inhibit RF binding or peptidyl-tRNA hydrolysis (Wilson 2009). However, a few antibiotics have been suggested to act preferentially during termination, rather than elongation, namely, blasticidin S (Svidritskiy et al. 2013) and fusidic acid (Savelsbergh et al. 2009). Fusidic acid does not bind to the free form of EF-G, but rather when EF-G−GTP is in complex with the ribosome

Cite this article as *Cold Spring Harb Perspect Med* doi: 10.1101/cshperspect.a025361

(Fig. 5B). Fusidic acid allows GTP hydrolysis but prevents the associated changes in EF-G that are necessary for dissociation and, thus, traps EF-G on the ribosome. Inhibition of EF-G's function during ribosome recycling, rather than during elongation, has been reported to be more sensitive to the inhibitory action of fusidic acid (Savelsbergh et al. 2009). Blasticidin S binds to the P-site of the LSU and overlaps the binding position of C75 of the CCA-end of the P-tRNA (Fig. 4E) (Hansen et al. 2003; Svidritskiy et al. 2013). As expected, the presence of blasticidin S reduces the rate of peptide bond formation, but is even more effective at inhibiting peptidyl-tRNA hydrolysis by RF1 (Svidritskiy et al. 2013).

ACKNOWLEDGMENTS

The Wilson laboratory is supported by grants from the Deutsche Forschungsgemeinschaft (FOR1805, WI3285/4-1, and GRK1721 to D.N.W.).

REFERENCES

Allen G, Zavialov A, Gursky R, Ehrenberg M, Frank J. 2005. The cryo-EM structure of a translation initiation complex from *Escherichia coli*. *Cell* 121: 703–712.

Amunts A, Fiedorczuk K, Truong TT, Chandler J, Peter Greenberg E, Ramakrishnan V. 2015. Bactobolin A binds to a site on the 70S ribosome distinct from previously seen antibiotics. *J Mol Biol* 427: 753–755.

Barat C, Datta PP, Raj VS, Sharma MR, Kaji H, Kaji A, Agrawal RK. 2007. Progression of the ribosome recycling factor through the ribosome dissociates the two ribosomal subunits. *Mol Cell* 27: 250–261.

Belova L, Tenson T, Xiong LQ, McNicholas PM, Mankin AS. 2001. A novel site of antibiotic action in the ribosome: Interaction of evernimicin with the large ribosomal subunit. *Proc Natl Acad Sci* 98: 3726–3731.

Blanchard SC, Gonzalez RL, Kim HD, Chu S, Puglisi JD. 2004a. tRNA selection and kinetic proofreading in translation. *Nat Struct Mol Biol* 11: 1008–1014.

Blanchard SC, Kim HD, Gonzalez RL Jr, Puglisi JD, Chu S. 2004b. tRNA dynamics on the ribosome during translation. *Proc Natl Acad Sci* 101: 12893–12898.

Borovinskaya MA, Pai RD, Zhang W, Schuwirth BS, Holton JM, Hirokawa G, Kaji H, Kaji A, Cate JH. 2007a. Structural basis for aminoglycoside inhibition of bacterial ribosome recycling. *Nat Struct Mol Biol* 14: 727–732.

Borovinskaya MA, Shoji S, Holton JM, Fredrick K, Cate JH. 2007b. A steric block in translation caused by the antibiotic spectinomycin. *ACS Chem Biol* 2: 545–552.

Brandi L, Fabbretti A, La Teana A, Abbondi M, Losi D, Donadio S, Gualerzi C. 2006. Specific, efficient, and selective inhibition of prokaryotic translation initiation by a novel peptide antibiotic. *Proc Natl Acad Sci* 103: 39–44.

Brandi L, Fabbretti A, Pon CL, Dahlberg AE, Gualerzi CO. 2008. Initiation of protein synthesis: A target for antimicrobials. *Expert Opin Ther Targets* 12: 519–534.

Brilot AF, Korostelev AA, Ermolenko DN, Grigorieff N. 2013. Structure of the ribosome with elongation factor G trapped in the pretranslocation state. *Proc Natl Acad Sci* 110: 20994–20999.

Brodersen DE, Clemons WM, Carter AP, Morgan-Warren RJ, Wimberly BT, Ramakrishnan V. 2000. The structural basis for the action of the antibiotics tetracycline, pactamycin, and hygromycin B on the 30S ribosomal subunit. *Cell* 103: 1143–1154.

Bulkley D, Johnson F, Steitz TA. 2012. The antibiotic thermorubin inhibits protein synthesis by binding to intersubunit bridge B2a of the ribosome. *J Mol Biol* 416: 571–578.

Bulkley D, Brandi L, Polikanov YS, Fabbretti A, O'Connor M, Gualerzi CO, Steitz TA. 2014. The antibiotics dityromycin and GE82832 bind protein S12 and block EF-G-catalyzed translocation. *Cell Rep* 6: 357–365.

Carter AP, Clemons WM, Brodersen DE, Morgan-Warren RJ, Wimberly BT, Ramakrishnan V. 2000. Functional insights from the structure of the 30S ribosomal subunit and its interactions with antibiotics. *Nature* 407: 340–348.

Carter AP, Clemons WM Jr, Brodersen DE, Morgan-Warren RJ, Hartsch T, Wimberly BT, Ramakrishnan V. 2001. Crystal structure of an initiation factor bound to the 30S ribosomal subunit. *Science* 291: 498–501.

Chen Y, Feng S, Kumar V, Ero R, Gao YG. 2013. Structure of EF-G-ribosome complex in a pretranslocation state. *Nat Struct Mol Biol* 20: 1077–1084.

Dallas A, Noller HF. 2001. Interaction of translation initiation factor 3 with the 30S ribosomal subunit. *Mol Cell* 8: 855–864.

Demeshkina N, Jenner L, Westhof E, Yusupov M, Yusupova G. 2012. A new understanding of the decoding principle on the ribosome. *Nature* 484: 256–259.

Demeshkina N, Jenner L, Westhof E, Yusupov M, Yusupova G. 2013. New structural insights into the decoding mechanism: Translation infidelity via a G·U pair with Watson–Crick geometry. *FEBS Lett* 587: 1848–1857.

Dincbas-Renqvist V, Engstrom A, Mora L, Heurgue-Hamard V, Buckingham R, Ehrenberg M. 2000. A posttranslational modification in the GGQ motif of RF2 from *Escherichia coli* stimulates termination of translation. *EMBO J* 19: 6900–6907.

Dinos G, Wilson DN, Teraoka Y, Szaflarski W, Fucini P, Kalpaxis D, Nierhaus KH. 2004. Dissecting the ribosomal inhibition mechanisms of edeine and pactamycin: The universally conserved residues G693 and C795 regulate P-site tRNA binding. *Mol Cell* 13: 113–124.

Dunkle JA, Xiong L, Mankin AS, Cate JH. 2010. Structures of the *Escherichia coli* ribosome with antibiotics bound near the peptidyl transferase center explain spectra of drug action. *Proc Natl Acad Sci* 107: 17152–17157.

Dunkle JA, Wang L, Feldman MB, Pulk A, Chen VB, Kapral GJ, Noeske J, Richardson JS, Blanchard SC, Cate JH. 2011. Structures of the bacterial ribosome in classical and hybrid states of tRNA binding. *Science* **332**: 981–984.

Ermolenko DN, Noller HF. 2011. mRNA translocation occurs during the second step of ribosomal intersubunit rotation. *Nat Struct Mol Biol* **18**: 457–462.

Ermolenko DN, Majumdar ZK, Hickerson RP, Spiegel PC, Clegg RM, Noller HF. 2007a. Observation of intersubunit movement of the ribosome in solution using FRET. *J Mol Biol* **370**: 530–540.

Ermolenko DN, Spiegel PC, Majumdar ZK, Hickerson RP, Clegg RM, Noller HF. 2007b. The antibiotic viomycin traps the ribosome in an intermediate state of translocation. *Nat Struct Mol Biol* **14**: 493–497.

Fischer N, Neumann P, Konevega AL, Bock LV, Ficner R, Rodnina MV, Stark H. 2015. Structure of the *E. coli* ribosome–EF-Tu complex at <3 Å resolution by C_s-corrected cryo-EM. *Nature* **520**: 567–570.

Frank J, Agrawal R. 2001. Ratchet-like movements between the two ribosomal subunits: Their implications in elongation factor recognition and tRNA translocation. *Cold Spring Harb Symp Quant Biol* **66**: 67–75.

Freistroffer DV, Pavlov MY, MacDougall J, Buckingham RH, Ehrenberg M. 1997. Release factor RF3 in *E. coli* accelerates the dissociation of release factors RF1 and RF2 from the ribosome in a GTP-dependent manner. *EMBO J* **16**: 4126–4133.

Gao H, Zhou Z, Rawat U, Huang C, Bouakaz L, Wang C, Cheng Z, Liu Y, Zavialov A, Gursky R, et al. 2007. RF3 induces ribosomal conformational changes responsible for dissociation of class I release factors. *Cell* **129**: 929–941.

Hansen JL, Schmeing TM, Moore PB, Steitz TA. 2002. Structural insights into peptide bond formation. *Proc Natl Acad Sci* **99**: 11670–11675.

Hansen JL, Moore PB, Steitz TA. 2003. Structures of five antibiotics bound at the peptidyl transferase center of the large ribosomal subunit. *J Mol Biol* **330**: 1061–1075.

Harms JM, Wilson DN, Schluenzen F, Connell SR, Stachelhaus T, Zaborowska Z, Spahn CM, Fucini P. 2008. Translational regulation via L11: Molecular switches on the ribosome turned on and off by thiostrepton and micrococcin. *Mol Cell* **30**: 26–38.

Jenner L, Starosta AL, Terry DS, Mikolajka A, Filonava L, Yusupov M, Blanchard SC, Wilson DN, Yusupova G. 2013. Structural basis for potent inhibitory activity of the antibiotic tigecycline during protein synthesis. *Proc Natl Acad Sci* **110**: 3812–3816.

Jin H, Kelley AC, Loakes D, Ramakrishnan V. 2010. Structure of the 70S ribosome bound to release factor 2 and a substrate analog provides insights into catalysis of peptide release. *Proc Natl Acad Sci* **107**: 8593–8598.

Jin H, Kelley AC, Ramakrishnan V. 2011. Crystal structure of the hybrid state of ribosome in complex with the guanosine triphosphatase release factor 3. *Proc Natl Acad Sci* **108**: 15798–15803.

Julian P, Milon P, Agirrezabala X, Lasso G, Gil D, Rodnina MV, Valle M. 2011. The Cryo-EM structure of a complete 30S translation initiation complex from *Escherichia coli*. *PLoS Biol* **9**: e1001095.

Kaji A, Kiel M, Hirokawa G, Muto A, Inokuchi Y, Kaji H. 2001. The fourth step of protein synthesis: Disassembly of the posttermination complex is catalyzed by elongation factor G and ribosome recycling factor, a near-perfect mimic of tRNA. *Cold Spring Harb Symp Quant Biol* **66**: 515–529.

Kannan K, Mankin AS. 2012. Macrolide antibiotics in the ribosome exit tunnel: Species-specific binding and action. *Ann NY Acad Sci* **1241**: 33–47.

Kannan K, Vazquez-Laslop N, Mankin AS. 2012. Selective protein synthesis by ribosomes with a drug-obstructed exit tunnel. *Cell* **151**: 508–520.

Kannan K, Kanabar P, Schryer D, Florin T, Oh E, Bahroos N, Tenson T, Weissman JS, Mankin AS. 2014. The general mode of translation inhibition by macrolide antibiotics. *Proc Natl Acad Sci* **111**: 15958–15963.

Karimi R, Pavlov M, Buckingham R, Ehrenberg M. 1999. Novel roles for classical factors at the interface between translation termination and initiation. *Mol Cell* **3**: 601–609.

Kim D, Green R. 1999. Base-pairing between 23S rRNA and tRNA in the ribosomal A site. *Mol Cell* **4**: 859–864.

Klaholz BP. 2011. Molecular recognition and catalysis in translation termination complexes. *Trends Biochem Sci* **36**: 282–292.

Korostelev A, Asahara H, Lancaster L, Laurberg M, Hirschi A, Zhu J, Trakhanov S, Scott WG, Noller HF. 2008. Crystal structure of a translation termination complex formed with release factor RF2. *Proc Natl Acad Sci* **105**: 19684–19689.

Koutmou KS, McDonald ME, Brunelle JL, Green R. 2014. RF3:GTP promotes rapid dissociation of the class 1 termination factor. *RNA* **20**: 609–620.

Laurberg M, Asahara H, Korostelev A, Zhu J, Trakhanov S, Noller HF. 2008. Structural basis for translation termination on the 70S ribosome. *Nature* **454**: 852–857.

Laursen BS, Sorensen HP, Mortensen KK, Sperling-Petersen HU. 2005. Initiation of protein synthesis in bacteria. *Microbiol Mol Biol Rev* **69**: 101–123.

Lin J, Gagnon MG, Bulkley D, Steitz TA. 2015. Conformational changes of elongation factor G on the ribosome during tRNA translocation. *Cell* **160**: 219–227.

Liu G, Song G, Zhang D, Zhang D, Li Z, Lyu Z, Dong J, Achenbach J, Gong W, Zhao XS, et al. 2014. EF-G catalyzes tRNA translocation by disrupting interactions between decoding center and codon–anticodon duplex. *Nat Struct Mol Biol* **21**: 817–824.

Moazed D, Noller HF. 1989. Intermediate states in the movement of transfer RNA in the ribosome. *Nature* **342**: 142–148.

Nguyen F, Starosta AL, Arenz S, Sohmen D, Dönhöfer A, Wilson DN. 2014. Tetracycline antibiotics and resistance mechanisms. *Biol Chem* **395**: 559–575.

Nissen P, Hansen J, Ban N, Moore PB, Steitz TA. 2000. The structural basis of ribosome activity in peptide bond synthesis. *Science* **289**: 920–930.

Ogle JM, Ramakrishnan V. 2005. Structural insights into translational fidelity. *Annu Rev Biochem* **74**: 129–177.

Olivier NB, Altman RB, Noeske J, Basarab GS, Code E, Ferguson AD, Gao N, Huang J, Juette MF, Livchak S, et al. 2014. Negamycin induces translational stalling and

Cite this article as *Cold Spring Harb Perspect Med* doi: 10.1101/cshperspect.a025361

miscoding by binding to the small subunit head domain of the *Escherichia coli* ribosome. *Proc Natl Acad Sci* **111:** 16274–16279.

Pan D, Kirillov SV, Cooperman BS. 2007. Kinetically competent intermediates in the translocation step of protein synthesis. *Mol Cell* **25:** 519–529.

Peske F, Rodnina M, Wintermeyer W. 2005. Sequence of steps in ribosome recycling as defined by kinetic analysis. *Mol Cell* **18:** 403–412.

Peske F, Kuhlenkoetter S, Rodnina MV, Wintermeyer W. 2014. Timing of GTP binding and hydrolysis by translation termination factor RF3. *Nucleic Acids Res* **42:** 1812–1820.

Pioletti M, Schlunzen F, Harms J, Zarivach R, Gluhmann M, Avila H, Bashan A, Bartels H, Auerbach T, Jacobi C, et al. 2001. Crystal structures of complexes of the small ribosomal subunit with tetracycline, edeine and IF3. *EMBO J* **20:** 1829–1839.

Polikanov YS, Osterman IA, Szal T, Tashlitsky VN, Serebryakova MV, Kusochek P, Bulkley D, Malanicheva IA, Efimenko TA, Efremenkova OV, et al. 2014a. Amicoumacin A inhibits translation by targeting mRNA interaction with the ribosome. *Mol Cell* **56:** 531–540.

Polikanov YS, Steitz TA, Innis CA. 2014b. A proton wire to couple aminoacyl-tRNA accommodation and peptide-bond formation on the ribosome. *Nat Struct Mol Biol* **21:** 787–793.

Polikanov YS, Szal T, Jiang F, Gupta P, Matsuda R, Shiozuka M, Steitz TA, Vazquez-Laslop N, Mankin AS. 2014c. Negamycin interferes with decoding and translocation by simultaneous interaction with rRNA and tRNA. *Mol Cell* **56:** 541–550.

Polikanov YS, Starosta AL, Juette MF, Altman RB, Terry DS, Lu W, Burnett BJ, Dinos G, Reynolds KA, Blanchard SC, et al. 2015. Distinct tRNA Accommodation intermediates observed on the ribosome with the antibiotics Hygromycin A and A201A. *Mol Cell* **58:** 832–844.

Pulk A, Cate JH. 2013. Control of ribosomal subunit rotation by elongation factor G. *Science* **340:** 1235970.

Ramrath DJ, Lancaster L, Sprink T, Mielke T, Loerke J, Noller HF, Spahn CM. 2013. Visualization of two transfer RNAs trapped in transit during elongation factor G–mediated translocation. *Proc Natl Acad Sci* **110:** 20964–20969.

Ratje AH, Loerke J, Mikolajka A, Brunner M, Hildebrand PW, Starosta AL, Donhofer A, Connell SR, Fucini P, Mielke T, et al. 2010. Head swivel on the ribosome facilitates translocation by means of intra-subunit tRNA hybrid sites. *Nature* **468:** 713–716.

Rheinberger HJ, Sternbach H, Nierhaus KH. 1981. Three tRNA binding sites on *Escherichia coli* ribosomes. *Proc Natl Acad Sci* **78:** 5310–5314.

Rodnina MV. 2013. The ribosome as a versatile catalyst: Reactions at the peptidyl transferase center. *Curr Opin Struct Biol* **23:** 595–602.

Roy RN, Lomakin IB, Gagnon MG, Steitz TA. 2015. The mechanism of inhibition of protein synthesis by the proline-rich peptide oncocin. *Nat Struct Mol Biol* **22:** 466–469.

Rozov A, Demeshkina N, Westhof E, Yusupov M, Yusupova G. 2015. Structural insights into the translational infidelity mechanism. *Nat Commun* **6:** 7251.

Sanbonmatsu KY, Joseph S, Tung CS. 2005. Simulating movement of tRNA into the ribosome during decoding. *Proc Natl Acad Sci* **102:** 15854–15859.

Savelsbergh A, Rodnina MV, Wintermeyer W. 2009. Distinct functions of elongation factor G in ribosome recycling and translocation. *RNA* **15:** 772–780.

Schluenzen F, Takemoto C, Wilson DN, Kaminishi T, Harms JM, Hanawa-Suetsugu K, Szaflarski W, Kawazoe M, Shirouzu M, Nierhaus KH, et al. 2006. The antibiotic kasugamycin mimics mRNA nucleotides to destabilize tRNA binding and inhibit canonical translation initiation. *Nat Struct Mol Biol* **13:** 871–878.

Schmeing TM, Ramakrishnan V. 2009. What recent ribosome structures have revealed about the mechanism of translation. *Nature* **461:** 1234–1242.

Schmeing TM, Moore PB, Steitz TA. 2003. Structures of deacylated tRNA mimics bound to the E site of the large ribosomal subunit. *RNA* **9:** 1345–1352.

Schmeing TM, Huang KS, Strobel SA, Steitz TA. 2005. An induced-fit mechanism to promote peptide bond formation and exclude hydrolysis of peptidyl-tRNA. *Nature* **438:** 520–524.

Schmeing TM, Voorhees RM, Kelley AC, Gao YG, Murphy FVT, Weir JR, Ramakrishnan V. 2009. The crystal structure of the ribosome bound to EF-Tu and aminoacyl-tRNA. *Science* **326:** 688–694.

Schuette JC, Murphy FVt, Kelley AC, Weir JR, Giesebrecht J, Connell SR, Loerke J, Mielke T, Zhang W, Penczek PA, et al. 2009. GTPase activation of elongation factor EF-Tu by the ribosome during decoding. *EMBO J* **28:** 755–765.

Schuwirth BS, Day JM, Hau CW, Janssen GR, Dahlberg AE, Cate JH, Vila-Sanjurjo A. 2006. Structural analysis of kasugamycin inhibition of translation. *Nat Struct Mol Biol* **13:** 879–886.

Seefeldt AC, Nguyen F, Antunes S, Perebaskine N, Graf M, Arenz S, Inampudi KK, Douat C, Guichard G, Wilson DN, et al. 2015. The proline-rich antimicrobial peptide Onc112 inhibits translation by blocking and destabilizing the initiation complex. *Nat Struct Mol Biol* **22:** 470–475.

Seit-Nebi A, Frolova L, Justesen J, Kisselev L. 2001. Class-1 translation termination factors: Invariant GGQ minidomain is essential for release activity and ribosome binding but not for stop codon recognition. *Nucleic Acids Res* **29:** 3982–3987.

Semenkov YP, Rodnina MV, Wintermeyer W. 2000. Energetic contribution of tRNA hybrid state formation to translocation catalysis on the ribosome. *Nat Struct Biol* **7:** 1027–1031.

Sievers A, Beringer M, Rodnina MV, Wolfenden R. 2004. The ribosome as an entropy trap. *Proc Natl Acad Sci* **101:** 7897–7901.

Simonetti A, Marzi S, Jenner L, Myasnikov A, Romby P, Yusupova G, Klaholz BP, Yusupov M. 2009. A structural view of translation initiation in bacteria. *Cell Mol Life Sci* **66:** 423–436.

Simonetti A, Marzi S, Billas IM, Tsai A, Fabbretti A, Myasnikov AG, Roblin P, Vaiana AC, Hazemann I, Eiler D, et al. 2013. Involvement of protein IF2 N domain in ribosomal subunit joining revealed from architecture and function of the full-length initiation factor. *Proc Natl Acad Sci* **110:** 15656–15661.

Simonovic M, Steitz TA. 2009. A structural view on the mechanism of the ribosome-catalyzed peptide bond formation. *Biochim Biophys Acta* **1789:** 612–623.

Sohmen D, Harms JM, Schlunzen F, Wilson DN. 2009. Enhanced SnapShot: Antibiotic inhibition of protein synthesis II. *Cell* **139:** 211e–212e.

Stanley RE, Blaha G, Grodzicki RL, Strickler MD, Steitz TA. 2010. The structures of the anti-tuberculosis antibiotics viomycin and capreomycin bound to the 70S ribosome. *Nat Struct Mol Biol* **17:** 289–293.

Stark H, Rodnina MV, Rinkeappel J, Brimacombe R, Wintermeyer W, Vanheel M. 1997. Visualization of elongation factor Tu on the *Escherichia coli* ribosome. *Nature* **389:** 403–406.

Svidritskiy E, Ling C, Ermolenko DN, Korostelev AA. 2013. Blasticidin S inhibits translation by trapping deformed tRNA on the ribosome. *Proc Natl Acad Sci* **110:** 12283–12288.

Valle M, Sengupta J, Swami NK, Grassucci RA, Burkhardt N, Nierhaus KH, Agrawal RK, Frank J. 2002. Cryo-EM reveals an active role for aminoacyl-tRNA in the accommodation process. *EMBO J* **21:** 3557–3567.

Valle M, Zavialov A, Sengupta J, Rawat U, Ehrenberg M, Frank J. 2003. Locking and unlocking of ribosomal motions. *Cell* **114:** 123–134.

Voorhees RM, Ramakrishnan V. 2013. Structural basis of the translational elongation cycle. *Annu Rev Biochem* **82:** 203–236.

Voorhees RM, Weixlbaumer A, Loakes D, Kelley AC, Ramakrishnan V. 2009. Insights into substrate stabilization from snapshots of the peptidyl transferase center of the intact 70S ribosome. *Nat Struct Mol Biol* **16:** 528–533.

Voorhees RM, Schmeing TM, Kelley AC, Ramakrishnan V. 2010. The mechanism for activation of GTP hydrolysis on the ribosome. *Science* **330:** 835–838.

Wang L, Pulk A, Wasserman MR, Feldman MB, Altman RB, Cate JH, Blanchard SC. 2012. Allosteric control of the ribosome by small-molecule antibiotics. *Nat Struct Mol Biol* **19:** 957–963.

Weixlbaumer A, Petry S, Dunham CM, Selmer M, Kelley AC, Ramakrishnan V. 2007. Crystal structure of the ribosome recycling factor bound to the ribosome. *Nat Struct Mol Biol* **14:** 733–737.

Weixlbaumer A, Jin H, Neubauer C, Voorhees R, Petry S, Kelley A, Ramakrishnan V. 2008. Insights into translational termination from the structure of RF2 bound to the ribosome. *Science* **322:** 953–956.

Wilson DN. 2009. The A–Z of bacterial translation inhibitors. *Crit Rev Biochem Mol Biol* **44:** 393–433.

Wilson DN. 2014. Ribosome-targeting antibiotics and bacterial resistance mechanisms. *Nat Rev Microbiol* **12:** 35–48.

Wilson DN, Schluenzen F, Harms JM, Starosta AL, Connell SR, Fucini P. 2008. The oxazolidinone antibiotics perturb the ribosomal peptidyl-transferase center and effect tRNA positioning. *Proc Natl Acad Sci* **105:** 13339–13344.

Wong W, Bai XC, Brown A, Fernandez IS, Hanssen E, Condron M, Tan YH, Baum J, Scheres SH. 2014. Cryo-EM structure of the *Plasmodium falciparum* 80S ribosome bound to the anti-protozoan drug emetine. *eLife* **3:** e03080.

Yamamoto H, Qin Y, Achenbach J, Li C, Kijek J, Spahn CM, Nierhaus KH. 2014. EF-G and EF4: Translocation and back-translocation on the bacterial ribosome. *Nat Rev Microbiol* **12:** 89–100.

Yusupova G, Jenner L, Rees B, Moras D, Yusupov M. 2006. Structural basis for messenger RNA movement on the ribosome. *Nature* **444:** 391–394.

Zavialov AV, Hauryliuk VV, Ehrenberg M. 2005. Splitting of the posttermination ribosome into subunits by the concerted action of RRF and EF-G. *Mol Cell* **18:** 675–686.

Zhang W, Dunkle JA, Cate JH. 2009. Structures of the ribosome in intermediate states of ratcheting. *Science* **325:** 1014–1017.

Zhou J, Korostelev A, Lancaster L, Noller HF. 2012a. Crystal structures of 70S ribosomes bound to release factors RF1, RF2 and RF3. *Curr Opin Struct Biol* **22:** 733–742.

Zhou J, Lancaster L, Trakhanov S, Noller HF. 2012b. Crystal structure of release factor RF3 trapped in the GTP state on a rotated conformation of the ribosome. *RNA* **18:** 230–240.

Aminoglycosides: An Overview

Kevin M. Krause,[1] Alisa W. Serio,[1] Timothy R. Kane,[1] and Lynn E. Connolly[1,2]

[1]Achaogen, South San Francisco, California 94080

[2]Department of Medicine, Division of Infectious Diseases, University of California, San Francisco, San Francisco, California 94143

Correspondence: kevinmichaelkrause@gmail.com

Aminoglycosides are natural or semisynthetic antibiotics derived from actinomycetes. They were among the first antibiotics to be introduced for routine clinical use and several examples have been approved for use in humans. They found widespread use as first-line agents in the early days of antimicrobial chemotherapy, but were eventually replaced in the 1980s with cephalosporins, carbapenems, and fluoroquinolones. Aminoglycosides synergize with a variety of other antibacterial classes, which, in combination with the continued increase in the rise of multidrug-resistant bacteria and the potential to improve the safety and efficacy of the class through optimized dosing regimens, has led to a renewed interest in these broad-spectrum and rapidly bactericidal antibacterials.

Aminoglycosides are potent, broad-spectrum antibiotics that act through inhibition of protein synthesis. The class has been a cornerstone of antibacterial chemotherapy since streptomycin (Fig. 1) was first isolated from *Streptomyces griseus* and introduced into clinical use in 1944. Several other members of the class (Fig. 1) were introduced over the intervening years including neomycin (1949, *S. fradiae*), kanamycin (1957, *S. kanamyceticus*), gentamicin (1963, *Micromonospora purpurea*), netilmicin (1967, derived from sisomicin), tobramycin (1967, *S. tenebrarius*), and amikacin (1972, derived from kanamycin). A shift away from systemic use of the class began in the 1980s with the availability of the third-generation cephalosporins, carbapenems, and fluoroquinolones, which were perceived to be less toxic and/or provide broader coverage than the aminoglycosides. However, increasing resistance to these classes of drugs, combined with more extensive knowledge of the basis of aminoglycoside resistance, has led to renewed interest in the legacy aminoglycosides and the development of novel aminoglycosides such as arbekacin and plazomicin (Fig. 4). These latter agents were designed to overcome common aminoglycoside resistance mechanisms thereby maintaining potency against multidrug-resistant (MDR) pathogens.

Additionally, improved dosing schemes have been developed that attempt to reduce aminoglycoside toxicity while maintaining efficacy. Specifically, clinical studies have reported a lower incidence of nephrotoxicity with once-daily dosing (Nicolau et al. 1995). These data

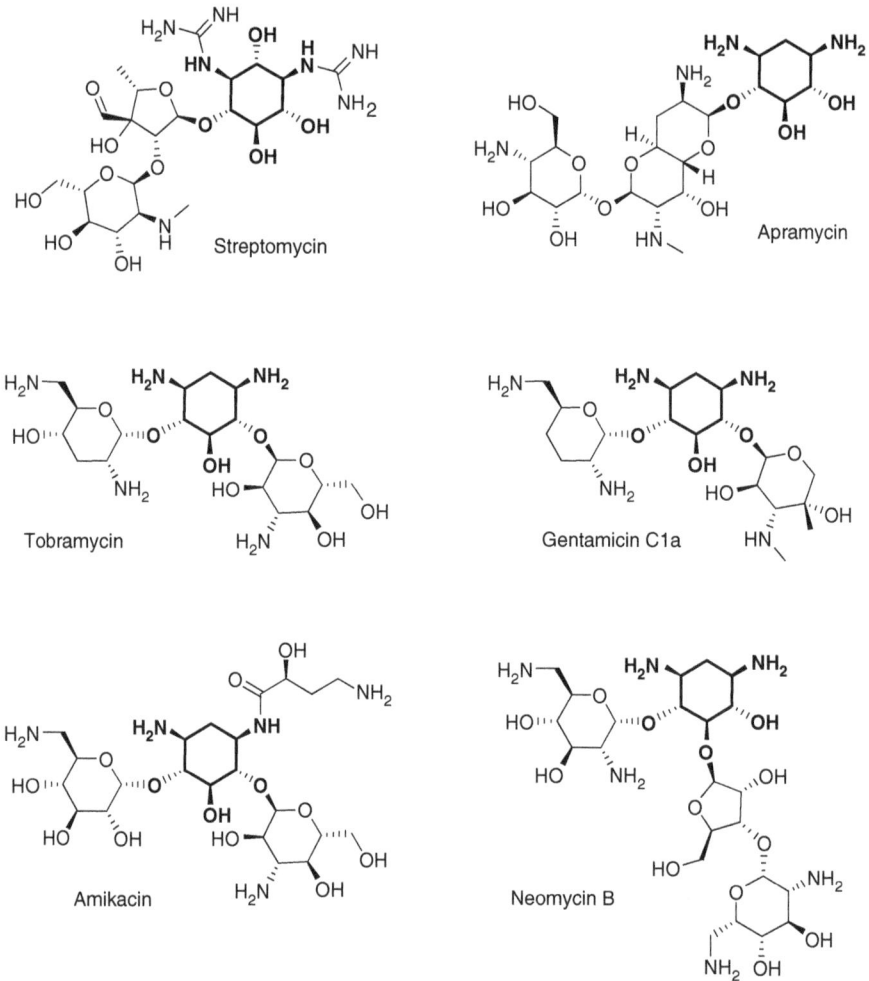

Figure 1. Structures of representative aminoglycosides, including the atypical aminoglycosides streptomycin and apramycin, 4,6-substituted AGs tobramycin, gentamcin, and amikacin, and the 4,5-substituted AG neomycin. The deoxystreptamine or streptidine rings are in bold.

are consistent with the concept that higher doses given less often should reduce the risk of toxicity while maintaining and possibly enhancing efficacy (Drusano et al. 2007). The advantages of once-daily dosing of aminoglycosides are now widely accepted and, for many infection types, this dosing schedule has become the standard of care (Avent et al. 2011). Additionally, inhaled delivery of aminoglycosides has become an area of renewed interest because it allows for greater local exposure within the lungs with reduced systemic toxicity. Inhaled tobramycin is

available in the European Union (EU) and the United States for the treatment of patients with chronic *Pseudomonas aeruginosa* lung infection associated with cystic fibrosis (CF), and inhaled amikacin and arbekacin are in development for potential use in CF and acute respiratory tract infections.

SPECTRUM OF ACTIVITY

Aminoglycosides are active against various Gram-positive and Gram-negative organisms.

Cite this article as *Cold Spring Harb Perspect Med* doi: 10.1101/cshperspect.a027029

Aminoglycosides are particularly potent against members of the Enterobacteriaceae family, including *Escherichia coli, Klebsiella pneumoniae* and *K. oxytoca, Enterobacter cloacae* and *E. aerogenes, Providencia* spp., *Proteus* spp., *Morganella* spp., and *Serratia* spp. (Ristuccia and Cunha 1985; Aggen et al. 2010; Landman et al. 2010). Furthermore, aminoglycosides are active against *Yersinia pestis* (Heine 2015) and *Francisella tularensis* (Ikäheimo et al. 2000), the causative agents of plague and tularemia, respectively. The class also has good activity against *Staphylococcus aureus*, including methicillin-resistant and vancomycin-intermediate and -resistant isolates, *P. aeruginosa* and to a lesser extent *Acinetobacter baumannii* (Ristuccia and Cunha 1985; Karlowsky et al. 2003; Aggen et al. 2010; Landman et al. 2011). Many *Mycobacterium* spp. are also susceptible to aminoglycosides including *Mycobacterium tuberculosis, M. fortuitum, M. chelonae,* and *M. avium* (Swenson et al. 1985; Ho et al. 1997).

Active electron transport is required for aminoglycoside uptake into cells, so the class inherently lacks activity against anaerobic bacteria (Kislak 1972; Martin et al. 1972; Ramirez and Tolmasky 2010). Aminoglycosides are also inactive against most *Burkholderia* spp. and *Stenotrophomonas* spp. as well as *Streptococcus* spp. and *Enterococcus* spp. (Brogden et al. 1976; Vakulenko and Mobashery 2003; Brooke 2012; Podnecky et al. 2015).

Contemporary large-scale surveillance programs provide an understanding of current aminoglycoside susceptibility among important pathogens associated with common infection types. A recent surveillance study of Gram-negative organisms isolated from patients hospitalized in intensive care units (ICUs) in the United States and the EU found that amikacin and gentamicin showed good activity against key Gram-negative pathogens (Sader et al. 2014). In the United States, 99.5% and 87.9% of *E. coli* isolates were susceptible to amikacin and gentamicin, respectively, according to the Clinical and Laboratory Standards Institute (CLSI) criteria. Likewise, 97.3% and 87.2% of *E. coli* isolates from the EU were susceptible to

amikacin and gentamicin, respectively, according to the European Committee on Antimicrobial Susceptibility Testing (EUCAST) criteria. Among *Klebsiella* spp., 94.8% and 92.7% of U.S. isolates and 90.5% and 83.3% of EU isolates were susceptible to amikacin and gentamicin, respectively. Amikacin was one of the few agents that retained activity against *P. aeruginosa* (97.3% susceptibility in the United States and 84.9% in the EU according to CLSI and EUCAST criteria, respectively). Similarly, aminoglycoside susceptibility rates were high among isolates collected from U.S. medical centers in 2012 (Sader et al. 2015). In this study, CLSI-based susceptibility rates were 99.0%, 88.2%, and 86.3% among *E. coli* and 88.2%, 89.2%, and 82.4% against *K. pneumoniae* for amikacin, gentamicin, and tobramycin, respectively. However, activity was reduced among *K. pneumoniae* carbapenemase (KPC)-producing *K. pneumoniae*, which are often resistant to multiple classes of drugs, for amikacin (42.5% susceptible), gentamicin (50.0% susceptible), and tobramycin (25.0% susceptible). Amikacin was the most active agent tested against *P. aeruginosa* (98% susceptible), followed closely by tobramycin (90% susceptible) and gentamicin (88.0% susceptible). Broad potency was also observed against *S. aureus* (96.0%, 95.0%, and 76.0% for amikacin, gentamicin, and tobramycin, respectively), but these compounds were less active against *A. baumannii* ($\leq$58.0% susceptible).

The broad-spectrum activity of aminoglycosides is enhanced in vitro through synergy with other classes of antimicrobials. This phenomenon, in which the combined effect of two antimicrobial agents is greater than the sum of their individual effects, is particularly well characterized between aminoglycosides and cell-wall-active agents such as β-lactams. In vitro synergy between aminoglycosides and β-lactams has been observed in both Gram-negative and -positive organisms, including wild-type and MDR isolates, using a variety of methodologies (Eliopoulos and Eliopoulos 1988). These in vitro observations have helped to stimulate the use of aminoglycoside-containing combination therapy in the treatment of a number of infection types (see below).

MECHANISM OF ACTION

Aminoglycosides inhibit protein synthesis by binding, with high affinity, to the A-site on the 16S ribosomal RNA of the 30S ribosome (Kotra et al. 2000). Although aminoglycoside class members have a different specificity for different regions on the A-site, all alter its conformation. As a result of this interaction, the antibiotic promotes mistranslation by inducing codon misreading on delivery of the aminoacyl transfer RNA. This results in error prone protein synthesis, allowing for incorrect amino acids to assemble into a polypeptide that is subsequently released to cause damage to the cell membrane and elsewhere (Davis et al. 1986; Mingeot-Leclercq et al. 1999; Ramirez and Tolmasky 2010; Wilson 2014). Some aminoglycosides can also impact protein synthesis by blocking elongation or by directly inhibiting initiation (Davis 1987; Kotra et al. 2000; Wilson 2014). The exact mechanism of binding and the subsequent downstream effects varies by chemical structure, but all aminoglycosides are rapidly bactericidal (Davis 1987; Mingeot-Leclercq et al. 1999) and typically produce a prolonged postantibiotic effect (PAE) (Zhanel et al. 1991) The PAE has been shown to be directly related to the length of time that the bacteria take to recover from the inhibition of protein synthesis (Stubbings et al. 2006). It is hypothesized that this is dependent on the eventual disassociation of the antibiotic from its target and exit from the cell.

Aminoglycosides are characterized by a core structure of amino sugars connected via glycosidic linkages to a dibasic aminocyclitol, which is most commonly 2-deoxystreptamine (Mingeot-Leclercq et al. 1999). Aminoglycosides are broadly classified into four subclasses based on the identity of the aminocyclitol moiety: (1) no deoxystreptamine (e.g., streptomycin, which has a streptidine ring); (2) a mono-substituted deoxystreptamine ring (e.g., apramycin); (3) a 4,5-di-substituted deoxystreptamine ring (e.g., neomycin, ribostamycin); or (4) a 4,6-di-substituted deoxystreptamine ring (e.g., gentamicin, amikacin, tobramycin, and plazomicin) (Magnet and Blanchard 2005; Wachino and Arakawa 2012). Examples of each subclass are shown in Figure 1. The core structure is decorated with a variety of amino and hydroxyl substitutions that have a direct influence on the mechanisms of action and susceptibility to various aminoglycoside-modifying enzymes (AMEs) associated with each of the aminoglycosides.

Aminoglycoside entry into bacterial cells is comprised of three distinct stages, the first of which increases permeability of the bacterial membrane, whereas the second and third are energy-dependent. The first stage involves electrostatic binding of the polycationic aminoglycoside to the negatively charged components of the bacterial membrane, such as the phospholipids and teichoic acids of Gram-positive organisms and the phospholipids and lipopolysaccharide (LPS) of Gram-negative organisms, followed by displacement of magnesium ions (Davis 1987; Taber et al. 1987; Ramirez and Tolmasky 2010). These cations are responsible for cross bridging and stabilization of the lipid components of the bacterial membrane and their removal leads to disruption of the outer membrane, enhanced permeability, and initiation of aminoglycoside uptake (Hancock et al. 1981, 1991; Hancock 1984; Ramirez and Tolmasky 2010). This phenomenon facilitates entry into the cytoplasm via a slow, energy-dependent, electron-transport-mediated process (Kislak 1972; Martin et al. 1972; Davis 1987; Taber et al. 1987; Ramirez and Tolmasky 2010). Inhibition of protein synthesis and mistranslation of proteins occurs once aminoglycoside molecules access the cytoplasm. These mistranslated proteins insert into and cause damage to the cytoplasmic membrane itself and facilitate subsequent aminoglycoside entry (Nichols and Young 1985; Davis et al. 1986). This then leads to rapid uptake of additional aminoglycoside molecules into the cytoplasm, increased inhibition of protein synthesis, mistranslation, and accelerated cell death (Davis et al. 1986; Davis 1987; Taber et al. 1987; Ramirez and Tolmasky 2010).

Cite this article as *Cold Spring Harb Perspect Med* doi: 10.1101/cshperspect.a027029

MECHANISMS OF AMINOGLYCOSIDE RESISTANCE

Aminoglycoside resistance takes many different forms including enzymatic modification, target site modification via an enzyme or chromosomal mutation, and efflux. Each of these mechanisms has varying effects on different members of the class and often multiple mechanisms are involved in any given resistant isolate. Resistance to aminoglycosides via target site mutations has not been observed because nearly all prokaryotes, with the exception of *Mycobacterium* spp. (Bercovier et al. 1986) and *Borrelia* spp. (Schwartz et al. 1992), encode multiple copies of rRNA. Although contemporary large-scale surveillance programs provide an understanding of phenotypic aminoglycoside resistance among important pathogens, these studies have generally not focused on the epidemiology of specific resistance mechanisms (Jones et al. 2014; Sader et al. 2015).

Enzymatic Drug Modification

AMEs are often found on plasmids containing multiple resistance elements, including other AMEs or β-lactamases. The mobility of these enzymes might be tied to their origins, which has been hypothesized to be via horizontal gene transfer from the actinomycetes responsible for the natural production of aminoglycosides (Shaw et al. 1993; Ramirez and Tolmasky 2010). More than 100 AMEs have been described and are broadly categorized into three groups based on their ability to acetylate, phosphorylate, or adenylate amino or hydroxyl groups found at various positions around the aminoglycoside core scaffold (Ramirez and Tolmasky 2010). These modifications decrease the binding affinity of the drug for its target and lead to a loss in antibacterial potency (Llano-Sotelo et al. 2002). These three families of AMEs include aminoglycoside *N*-acetyltransferases (abbreviated AACs), aminoglycoside *O*-nucleotidyltransferases (ANTs), and aminoglycoside *O*-phosphotransferases (APHs). The classes are further divided into subtypes according to the position on the aminoglycoside that the enzyme modifies followed by a Roman numeral and, in some cases, a letter when multiple enzymes exist that modify the same position (Shaw et al. 1993; Ramirez and Tolmasky 2010). The major sites of aminoglycoside modification for the most common AMEs are shown for kanamycin A in Figure 2.

Aminoglycoside Acetyltransferases

The aminoglycoside acetyltransferases or AACs comprise the largest group of AMEs. They are part of the GCN5-related *N*-acetyltransferase (GNAT) superfamily of ~10,000 described proteins (Ramirez and Tolmasky 2010). These enzymes acetylate amino groups found at various positions on the aminoglycoside scaffold in an acetyl-CoA-dependent reaction (Fig. 3A). There are four main subclasses of this group

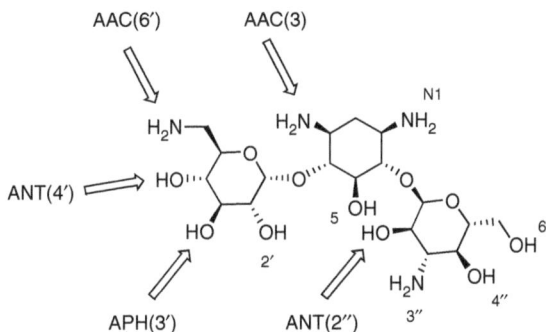

Figure 2. Sites of chemical modification by representative aminoglycoside-modifying enzymes (AMEs) on kanamycin A.

A

B

C

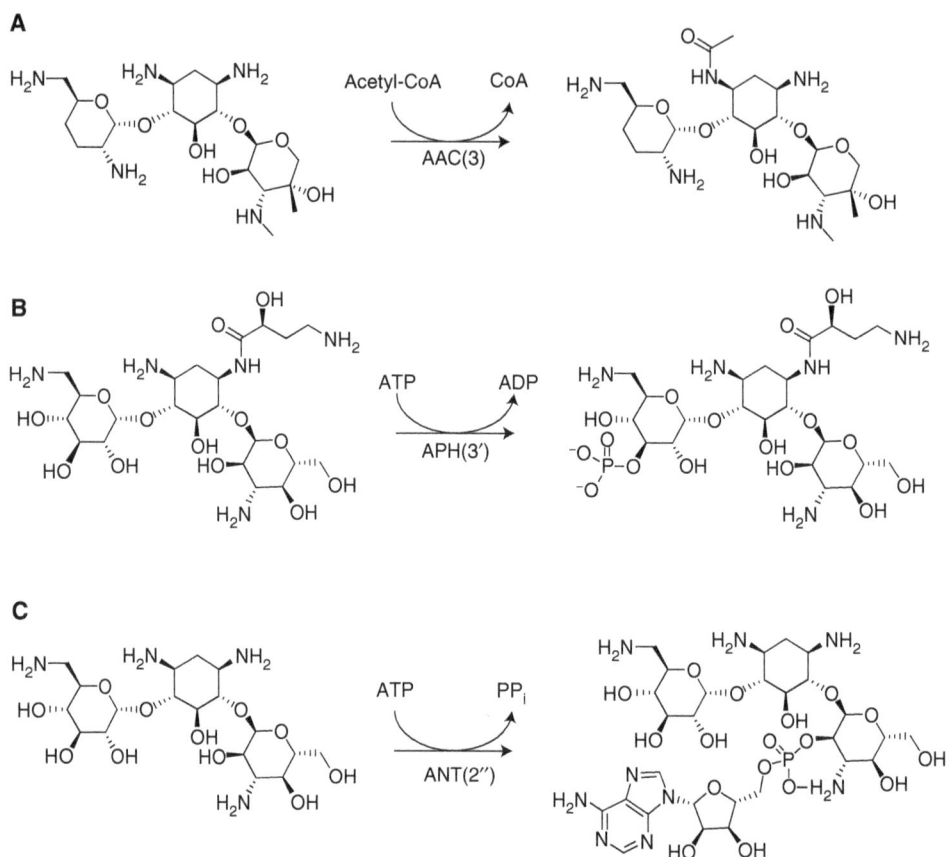

Figure 3. Examples of aminoglycoside modification by AMEs. (*A*) An example of chemical modification of gentamicin catalyzed by the aminoglycoside acetyltransferase AAC(3). (*B*) An example of chemical modification catalyzed by the aminoglycoside phosphotransferase APH(3′) on amikacin. (*C*) Adenylation of the 2″ hydroxyl of kanamycin A catalyzed by the aminoglycoside nucleotidyltransferase ANT(2″).

of enzymes whose nomenclature is derived from the specific amino group that is modified. Specifically, AAC(1) and AAC(3) target the amino groups found at position 1 and 3, respectively, of the 2-deoxystreptamine ring, whereas AAC(2′) and AAC(6′) target amino groups found at the 2′ and 6′ position of the 2,6-dideoxy-2,6-diaminoglucose ring (Mingeot-Leclercq et al. 1999; Wright and Thompson 1999; Azucena and Mobashery 2001; Magnet and Blanchard 2005).

A representative of the AAC family (AAC(6′)-IV) was the first AME found in bacteria to be described in the literature (Okamoto and Suzuki 1965). Since then, >70 members of the AAC family have been described. Fre-

quently observed class members found among Gram-negative bacteria include the AAC(6′)-1 enzyme that leads to amikacin, netilmicin, and tobramycin resistance, AAC(3)-IIa, which is responsible for resistance to gentamicin, tobramycin, and netilmicin, and AAC(3)-I, which modifies gentamicin (Shaw et al. 1993; Castanheira et al. 2015). Less common are the AAC(6′)-APH(2″) hybrid enzyme responsible for high-level aminoglycoside resistance in *Enterococcus faecalis* (Culebras and Martínez 1999), the chromosomally encoded AAC(6′)-Ii enzyme responsible for intrinsic resistance to aminoglycosides among *E. faecium* (Costa et al. 1993) and the chromosomal AAC(2′) enzyme found in *Mycobacterium* spp. and *Provi-*

Cite this article as *Cold Spring Harb Perspect Med* doi: 10.1101/cshperspect.a027029

dencia stuartii (Rather et al. 1993; Aínsa et al. 1997; Macinga and Rather 1999; Ramirez and Tolmasky 2010). Last, a variant of AAC(6′)-Ib that has acquired the ability to modify fluoroquinolones without significantly altering its activity against aminoglycosides has been identified in clinical isolates of Gram-negative bacteria (Robicsek et al. 2006; Strahilevitz et al. 2009).

Aminoglycoside Phosphotransferases

The second largest group of AMEs is the APHs. This structurally diverse group of enzymes acts like kinases in that they catalyze the ATP-dependent phosphorylation of hydroxyl groups found on aminoglycosides (Fig. 3B). APH enzymes are functionally and structurally similar to the serine–threonine and tyrosine kinases found in eukaryotes (Wright and Thompson 1999). The modifications made by these enzymes lower binding affinity to the target by decreasing the hydrogen bonding potential of aminoglycoside hydroxyl groups with important rRNA residues. Most of the >30 described APH enzymes belong to the APH(3′) subfamily (Kim and Mobashery 2005), although variants that target the 2″ hydroxyl also exist (Ramirez and Tolmasky 2010). These enzymes are found in diverse groups of Gram-negative bacteria, although APH(3′)-IIIa was discovered in *S. aureus* and *Enterococcus* spp. All members of the family lead to kanamycin and neomycin resistance with various members of the family also able to modify a variety of other aminoglycosides, including amikacin and gentamicin B (Shaw et al. 1993).

Aminoglycoside Nucleotidyltransferases

The final group of AMEs is the ANTs. These enzymes act by adding AMP from an ATP donor to hydroxyl groups at the 2″, 3″, 4′, 6, and 9 positions (Fig. 3C). The most clinically relevant members of the class include ANT(2″) and ANT(4′) (Kotra et al. 2000; Magnet and Blanchard 2005), which were first described in *K. pneumoniae* and *S. aureus*, respectively (Benveniste and Davies 1971; Le Goffic et al. 1976).

ANT(2″) broadly effects the activity of 4,6-di-substituted aminoglycosides (Gates 1988), whereas ANT(4′) targets kanamycin A, B, and C, gentamicin A, amikacin, tobramycin, and neomycin B and C. Other members of this class include ANT(3″), ANT(6), and ANT(9), which confer resistance to streptomycin and spectinomycin (Hollingshead and Vapnek 1985; Murphy 1985; Ounissi et al. 1990).

16S rRNA Methylation

Target site modification leading to aminoglycoside resistance occurs via the action of 16S rRNA methyltransferases (RMTs). These enzymes modify specific rRNA nucleotide residues in a manner that blocks aminoglycosides from effectively binding to their target (Beauclerk and Cundliffe 1987; Cundliffe 1989; Wachino and Arakawa 2012). There are two general classes of RMTs that are characterized by the specific nucleotide residues that they modify. These include enzymes that render bacteria resistant to 4,6-di-substituted aminoglycosides via methylation of the N7 position of nucleotide G^{1405} (Thompson et al. 1985; Beauclerk and Cundliffe 1987) and those that affect both 4,6- and 4,5-di-substituted aminoglycosides through methylation of the N1 position of nucleotide A^{1408} (Skeggs et al. 1985; Beauclerk and Cundliffe 1987; Mingeot-Leclercq et al. 1999).

The first clinical case of a pathogen with an RMT as a mechanism of aminoglycoside resistance was reported in a *P. aeruginosa* isolate from Japan in 2003 (Yokoyama et al. 2003). This isolate contained a plasmid-encoded RMT, named RmtA for ribosomal methyltransferase A. Subsequently, several additional plasmid-borne RMTs, encoded by the genes *armA*, *rmtB1*, *rmtB2*, *rmtC*, *rmtD*, *rmtD2*, *rmtE*, *rmtF*, *rmtG*, and *rmtH*, have emerged in clinical isolates that show high-level resistance to multiple aminoglycosides (Yokoyama et al. 2003; Doi et al. 2004; Yan et al. 2004; Yamane et al. 2005; Wachino et al. 2006; Wachino and Arakawa 2012). These enzymes modify the G^{1405} nucleotide and, thus, impact the activity of all 4,6-di-substituted aminoglycosides (i.e., amikacin, gentamicin, and tobramycin). In 2007, the enzyme NpmA was discovered encoded on a plasmid in an aminoglycoside-resistant *E. coli* clinical isolate, also from Japan (Wachino et al. 2007). This methyltransferase modifies the A^{1408} nucleotide and, thus, impacts 4,6- and 4,5-di-substituted as well as monosubstituted aminoglycosides, conferring pan-aminoglycoside resistance. To date, there has been only one additional report of this enzyme in a clinical isolate (Al Sheikh et al. 2014).

Efflux-Mediated Resistance

Several members of the resistance–nodulation–division (RND) family of efflux systems have been shown to be involved in intrinsic aminoglycoside resistance in various pathogens (Aires et al. 1999; Westbrock-Wadman et al. 1999; Rosenberg et al. 2000; Magnet et al. 2001; Hocquet et al. 2003; Islam et al. 2004). In the opportunistic pathogen *P. aeruginosa*, intrinsic low-level resistance to aminoglycosides, tetracycline and erythromycin, is mediated by the expression of the multiple efflux (Mex) XY-OprM system. MexXY orthologs are found in other species of bacteria. For example, members of the *Burkholderia cenopacia* complex are often intrinsically resistant to aminoglycosides via RND efflux pumps (Buroni et al. 2009). A homologous transporter in *E. coli*, AcrD, participates in efflux of aminoglycosides (Rosenberg et al. 2000) as does the Acinetobacter drug efflux (Ade) ABC efflux system in *A. baumannii* (Magnet et al. 2001) and the major facilitator superfamily (MFS) in *Mycobacteria* spp.

Molecular Epidemiology of Aminoglycoside Resistance Mechanisms

Comprehensive molecular data regarding the current prevalence of specific aminoglycoside resistance mechanisms among common pathogens is sparse. One recent study evaluated the aminoglycoside resistance mechanisms found among 200 Gram-negative bacilli isolates selected at random from an extensive culture collection (Castanheira et al. 2015). Ninety-nine Enterobacteriaceae (91.9% AME-positive), 49 *A. baumannii* (79.6% AME-positive), and 52

P. aeruginosa (63.5% AME positive) with a variety of aminoglycoside resistance profiles were included. A diversity of AME genes were identified with the most prevalent being *aac(6′)-Ib* ($n = 75$), *ant(3′)-Ia* ($n = 51$), and *aac(3)-IIa* ($n = 45$), and with 26 isolates harboring more than one AME gene. In addition, 21 of the isolates were found to carry an RMT including 12 Enterobacteriaceae, seven *A. baumannii*, and two *P. aeruginosa*. Many of these isolates were also resistant to other common antibiotics used to treat Gram-negative infections, highlighting the ability of these resistance mechanisms to spread through conjugation of plasmids and nonreplicative transposons among bacteria (Courvalin 1994; Waters 1999; Dzidic and Bedeković 2003; Feizabadi et al. 2004).

AMEs are commonly found in association with other key resistance elements, such as carbapenemases and extended spectrum β-lactamases (ESBLs). Among 50 carbapenem-resistant *K. pneumoniae* clinical isolates from two U.S. medical centers (80% possessing KPC-2, 10% possessing KPC-3 with all KPC$^+$ isolates also expressing TEM-1 and SHV-12), 98% of isolates expressed at least one AME (Almaghrabi et al. 2014). Specifically, 98% were positive for *aac(6′)-Ib*, 56% positive for *aph(3′)-Ia*, 38% positive for *aac(3′)-IV*, and 2% positive for *ant(2″)-Ia*. Overall, 40%, 98%, and 16% of the strains were nonsusceptible to gentamicin, tobramycin, and amikacin, respectively. Plazomicin, a novel aminoglycoside in clinical development that was designed to evade AME-based resistance, had MICs that ranged from 0.25 to 1 μg/mL against this set of isolates. AME characterization in 330 aminoglycoside-resistant clinical Enterobacteriaceae isolates from Spain revealed the presence of *aph(3″)-Ib* and *ant(3″)-Ia* genes in 65.4% and 37.5% of the isolates, consistent with 92% phenotypic resistance to streptomycin found in this strain collection (Miró et al. 2013). These isolates were resistant to other aminoglycosides to varying degrees, including gentamicin (18.4%), tobramycin (16.9%), and amikacin (1.5%), indicating the presence of other AMEs; *aph(3′)-Ia* was found in 13.9% of isolates, *aac(3)-IIa* in 12.4%, *aac(6′)-Ib* in 4.2%, *ant(2″)-Ia* in 3.6%, and *aph(3″)-IIa* 1.2% (Miró et al. 2013). Underscoring the association of AMEs with resistance elements to other key antibiotic classes, many of these isolates also produced ESBLs and were resistant to the fluoroquinolones.

Similar to the AMEs, comprehensive data describing the prevalence and molecular epidemiology of RMTs is lacking. A recent literature review described reports of widespread, global dissemination of genes encoding RMTs among isolates from both human and livestock sources (Wachino and Arakawa 2012). Despite the widespread detection of RMTs across many regions, the prevalence appears to vary by region with the highest rates reported in Asia. A SENTRY surveillance study from the Asia-Pacific region (2007–2008) detected genes encoding RMTs in 6.9% (China), 10.5% (India), 1.5% (Hong Kong), 6.1% (Korea), and 5.0% (Taiwan) of Enterobacteriaceae isolates (Bell et al. 2010). Lower rates ($\leq$1.3%) of RMTs among Enterobacteriaceae have been reported from single institution or local studies from European medical centers (Wachino and Arakawa 2012). Similar to the association between AMEs and other key resistance mechanisms, an association between RMTs and specific β-lactamases has been described. Seventy-six percent of the RMT containing isolates described in the SENTRY surveillance study above also possessed a CTX-M ESBL (Bell et al. 2010), and RMTs are frequently found in association with the New Delhi metallo-β-lactamase (NDM) (Berçot et al. 2011; Livermore et al. 2011; Mushtaq et al. 2011; Poirel et al. 2014).

AGENTS IN DEVELOPMENT

Plazomicin is a new aminoglycoside that was specifically engineered to be resistant to the action of the AMEs that are prevalent in key Gram-negative pathogens (Armstrong and Miller 2010). It is synthesized from a sisomicin scaffold that is intrinsically refractory to modification by APH(3′)-III, -VI, and -VII and ANT(4′), which confer amikacin resistance because of an absence of the 3′- and 4′-OH groups. Modification at the N-1 position via addition of a hydroxylaminobutyric acid substituent sterically hinders the action of the AAC(3), ANT(2″), and APH(2″) enzymes, which confer resistance to gentamicin and tobramycin. Finally, addition of a hydroxyethyl substituent at the 6′ position inhibits the action of the AAC(6′) enzymes, which confer resistance to a broad range of agents, including amikacin, tobramycin, and gentamicin (Fig. 4). Importantly, these modifications to sisomicin do not reduce intrinsic potency as has been associated with previous efforts to protect the 6′ position and, as predicted, lead to improved activity against Enterobacteriaceae (MIC$_{90}$ $\leq$2 mg/L) that are resistant to currently available aminoglycosides (Nagabhushan et al. 1982; Aggen et al. 2010). Plazomicin retains vulnerability to modification by AAC(2′)-I, a chromosomal AME found in *P. stuartii* and some mycobacterial species. However, this enzyme is rare, has not been found on a mobile element and has not been shown to have clinical relevance in *Mycobacterium* spp. In addition, plazomicin, like all

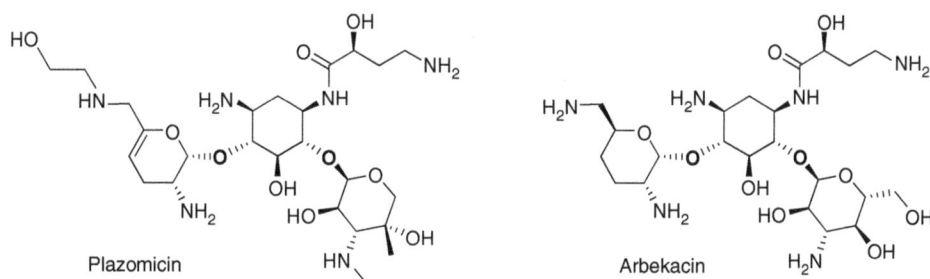

Figure 4. Structures of plazomicin and arbekacin.

4,6-linked aminoglycosides, is inactive against isolates that produce RMTs. As described above, this mechanism of resistance frequently travels on mobile genetic elements with NDM-positive Enterobacteriaceae, and, thus, plazomicin is not active against many isolates harboring this enzyme (Berçot et al. 2011; Livermore et al. 2011; Mushtaq et al. 2011; Poirel et al. 2014).

Similar to plazomicin, arbekacin, a semisynthetic derivative of dibekacin, was specifically engineered to overcome the action of a subset of clinically important AMEs (Fig. 4). Arbekacin is stable to the action of AMEs commonly found in methicillin-resistant *S. aureus* (MRSA), such as APH, ANT, and AAC and possesses potent activity against this clinically important pathogen (Matsumoto 2014). Although arbekacin has been approved for use in Japan since 1990 for the treatment of sepsis and pneumonia caused by MRSA, this molecule also retains potent activity against key Gram-negative pathogens, including MDR strains, and has more recently gained attention as a potential therapy for infections caused by these organisms. In a recent surveillance study of isolates from hospitalized patients with pneumonia, arbekacin was the most potent aminoglycoside tested against ESBL-producing *E. coli* and was slightly more active than amikacin and tobramycin against ESBL- and KPC-expressing *K. pneumoniae*. Similarly, arbekacin possessed greater activity against *P. aeruginosa* and *Acinetobacter* spp., including MDR isolates, than the other aminoglycosides tested (amikacin, gentamicin, and tobramcin) (Sader et al. 2015). Because of its broad-spectrum in vitro activity against both MDR Gram-positive and Gram-negative pathogens, arbekacin is currently under development as an inhalational agent for the treatment of mechanically ventilated patients with bacterial pneumonia (clinicaltrials.gov/ct2/show/NCT02459158) and is under study at Walter Reed Army Medical Hospital for the treatment of patients with infections caused by MDR organisms that have limited treatment options (clinicaltrials.gov/ct2/show/NCT01659515).

PHARMACOKINETICS AND PHARMACODYNAMICS

Aminoglycosides are poorly absorbed via the gastrointestinal (GI) tract and are, thus, administered via the intravenous or intramuscular route (Ramirez and Tolmasky 2010; Craig 2011). The volume of distribution for members of the aminoglycoside class approaches total body volume, indicating a broad distribution into tissues, including the lung (Simon et al. 1973). This feature has led to extensive use of aminoglycosides as part of combination regimens for the treatment of pneumonia. Aminoglycosides are rapidly cleared through the urinary tract, which also makes these drugs ideal for the treatment of urinary tract infections (Lode et al. 1976; Ramirez and Tolmasky 2010; Craig 2011).

The relationship between aminoglycoside pharmacokinetics (PKs) and pharmacodynamics (PDs) has been studied extensively in mice. The PK/PD variable that is most often correlated with efficacy of aminoglycosides is the ratio of area under the concentration–time curve (AUC) to MIC, although peak concentration also appears to play a role. The magnitude of the PK/PD target for aminoglycosides is not as well defined as it is for other antibiotic classes because of, until recently, the lack of development of new aminoglycosides. Available data suggests that significant variations in the PK/PD target exist between species and body site of infection. For example, an AUC/MIC target of 100 was reportedly associated with a 1- to 2-$\log_{10}$ kill in the mouse neutropenic thigh infection model with amikacin and *K. pneumoniae* (Craig 2011), whereas this same group reported better efficacy at the same dose level and with the same strain in the mouse lung infection model (Craig et al. 1991). These results were potentially a result of a longer measured PAE in the lung compared with the thigh (Craig et al. 1991).

AUC/MIC thresholds have been correlated to efficacy in patients in whom extensive PK sampling was also conducted. These case reports and reviews describe a variety of different ways that the AUC/MIC ratio might be used to predict outcomes in patients. Smith et al. (2001)

reviewed 23 patients with intra-abdominal infections ($n = 16$) or lower respiratory infections ($n = 7$) treated with tobramycin infused >30 min to achieve a C_{max} of between 4 and 10 mg/L. A PK/PD model constructed from these patients data showed that improved efficacy was observed when an AUC/MIC ratio of ≥110 was achieved compared with AUC/MIC ratios of <110 (80% clinical cure rate compared with 47%). Other investigators have found similar correlations between exposure and efficacy. Jacobs (2001) describes the need to achieve an AUC/MIC ratio of 25 for less severe infections or in immune competent patients and a ratio of at least 100 in patients with severe infections and/or those that are immune-compromised. Zelenitsky et al. (2003) found significantly improved outcomes in patients with bacteremia treated with gentamicin when AUC/MIC ratios exceeded 70 compared with AUC/MIC ratios <70 (90% cure vs. 45.5%, respectively). However, a much stronger correlation with outcome was noted for C_{max}/MIC with 84% and 90% clinical cure rates for C_{max}/MIC values of 4.8 or 8 compared with 0% clinical cure for C_{max}/ MIC < 2.9.

There is substantial evidence that administering larger doses of aminoglycosides less frequently may be associated with improved outcomes compared with providing the same total daily dose over more frequent dosing schedules (Barclay et al. 1999). This dosing approach, known as once daily or extended interval dosing, takes advantage of three features of aminoglycosides—concentration-dependent killing, rapid elimination, and a prolonged PAE. Larger doses are thought to increase target body site concentrations to improve PD while minimizing potential toxicity through allowance for a period of time in which there is little or no drug in circulation. The PAE properties of the class allow for an extended period of killing after the drug is cleared from the body and before the next dose is administered. A number of meta-analyses of results from clinical trials comparing once versus multiple daily administration of aminoglycosides have been published (Barclay et al. 1999). Overall, the results of these studies suggest that once daily dosing is associated with

reduced nephrotoxicity and equal, if not slightly improved, efficacy.

Therapeutic drug management (TDM), the clinical practice of measuring drug concentrations at designated intervals for use in optimizing individualized dosage regimens, has further improved the safety profile of aminoglycosides. Older studies using multiple daily doses of these drugs have shown nephrotoxicity rates of 10% to 20% (Humes et al. 1982; Moore et al. 1984), whereas lower rates of nephrotoxicity ranging from 0% to 14% have been reported with once-daily dosing regimens with dose adjustment guided by the use of TDM (Prins et al. 1993; Nicolau et al. 1995; Murry et al. 1999; Rybak et al. 1999; Buijk et al. 2002). The advantages of once-daily or extended interval dosing of aminoglycosides combined with the use of TDM are now widely accepted and, for many infection types, this dosing approach has become the standard of care (Avent et al. 2011).

CLINICAL USES OF AMINOGLYCOSIDES

The spectrum of activity, rapid bactericidal activity, and favorable chemical and pharmacokinetic properties of aminoglycosides make them a clinically useful class of drugs. Aminoglycosides are used as single agents and in combination with other antibiotics in both empirical and definitive therapy for a broad range of indications (Avent et al. 2011; Jackson et al. 2013).

In patients with serious infections caused by Gram-negative pathogens, the receipt of empiric combination therapy containing at least one antimicrobial agent to which the pathogen is susceptible leads to lower mortality and improved outcomes (Tamma et al. 2012). In addition to helping ensure that the pathogen is adequately covered by at least one active drug, the use of empiric aminoglycoside-β-lactam combination therapy has also been theorized to contribute to improved outcomes by taking advantage of the in vitro synergy observed between these classes and to prevent the emergence of resistance (Pankuch et al. 2010; Le et al. 2011). Although clinical data to support the latter two theoretical benefits of combination therapy are conflicting (Tamma et al. 2012), aminoglycosides are often combined with β-lactams for the empirical treatment of severe sepsis and certain nosocomial infections in patients with a high risk of mortality or when there is concern that the causative pathogen may be resistant to more commonly used agents (American Thoracic Society; Infectious Diseases Society of America 2005; Dellinger et al. 2013). More recently, aminoglycosides have increasingly become important components of therapy for patients infected with MDR pathogens, such as carbapenem-resistant Enterobacteriaceae (CRE), for whom few treatment options remain. In a retrospective cohort study of 50 cases of sepsis caused by carbapenem- and colistin-resistant *K. pneumoniae*, the receipt of gentamicin as part of definitive therapy was associated with lower mortality compared with the receipt of non-gentamicin-containing regimens (Gonzalez-Padilla et al. 2015). The novel aminoglycoside plazomicin (see section on Agents in Development) is currently under phase 3 study for the treatment of serious infections caused by CRE (clinicaltrials .gov/ct2/show/NCT01970371).

Aminoglycosides are also an important component of combination therapy for multidrug-resistant tuberculosis (MDR-TB) and certain non-tuberculous mycobacterial (NTM) infections. Current MDR-TB treatment guidelines recommend inclusion of one of the following agents during the intensive phase of therapy: amikacin, kanamycin, streptomycin, or capreomycin, a cyclic peptide antibiotic that is often considered as an aminoglycoside because of its mechanism of action. Each of these agents possesses potent bactericidal activity against *M. tuberculosis* (Ho et al. 1997) and the choice of agent depends on previous injectable use (if any) and the likelihood of resistance. A meta-analysis including 32 studies with >9000 treatment episodes did not reveal any clear differences in efficacy among the available agents (World Health Organization 2011). Similar to treatment of MDR-TB, combination therapy for patients with fibrocavitary, severe nodular/bronchiectatic or macrolide-resistant lung disease because of the *M. avium* complex generally includes amikacin or streptomycin (Griffith et al. 2007). Among the rapidly growing myco-

bacteria, amikacin is the preferred agent for infections because of *M. fortuitum* or *M. abscessus*, whereas tobramycin is the most active agent against *M. chelonae* (Griffith et al. 2007).

Aminoglycosides remain the preferred therapy for certain zoonotic infections such as plague and tularemia. Although streptomycin has traditionally been the agent of choice for these infection types, gentamicin is now widely used because of the broader availability of this agent as well as data suggesting similar efficacy to streptomycin (Mwengee et al. 2006; Snowden and Stovall 2011). Aminoglycosides are bactericidal against these organisms and the use of bacteriostatic agents, such as doxycycline or chloramphenicol has led to treatment failures (Dennis et al. 2001; Snowden and Stovall 2011).

Inhaled tobramycin therapy in CF patients with chronic lung infection caused by *P. aeruginosa* has been shown to improve respiratory function, decrease hospitalizations, and reduce systemic antibiotic use (Ramsey et al. 1999), and has contributed to a significant increase in survival for these patients (Sawicki et al. 2012). Inhaled aminoglycosides, alone or as part of combination therapy, are currently under evaluation as adjunctive agents for treatment of additional respiratory infection types including chronic lung infections associated with non-CF bronchiectasis and refractory NTM infections of the lung, and for the prevention and/or treatment of ventilator associated infections, tracheobronchitis (VAT), and pneumonia (VAP). Regarding the latter indication, although definitive data from large randomized trials is not available, a number of small studies focused on the prevention or treatment of VAP have provided encouraging results. A meta-analysis of eight comparative trials of prophylactic aerosolized antibiotics (four of which used inhaled tobramycin or gentamicin) found that ICU-acquired pneumonia was less common in the group of patients that received antibiotic prophylaxis compared with those who received no prophylaxis (Falagas et al. 2006). Similarly, a meta-analysis of five comparative trials of adjunctive aerosolized antibiotics in the treatment of VAP, each of which evaluated an inhaled aminoglycoside versus placebo or no therapy, revealed that administration of aerosolized antibiotics was associated with better treatment success compared with control in both the intent-to-treat and clinically evaluable populations (Ioannidou et al. 2007). The relative inefficiency of drug delivery through the ventilator circuit is a significant challenge with the use of aerosolized antibiotics for the treatment of ventilator-associated infections. To overcome this issue, BAY41-6551, an investigational drug–device combination of amikacin specifically formulated for inhalation, has been developed. This drug–device combination is currently under phase 3 study as adjunctive therapy in intubated and mechanically ventilated patients with Gram-negative pneumonia (clinicaltrials .gov/ct2/show/NCT01799993 and clinicaltrials .gov/ct2/show/results/NCT00805168).

Because of their poor oral bioavailability, aminoglycosides are a key element of oropharyngeal or gut decolonization/decontamination regimens, including those targeting MDR pathogens. The purpose of these decontamination regimens is to eradicate potential pathogens from the oropharynx and digestive tract of patients at risk for nosocomial or postoperative infections. Selective digestive decontamination (SDD) consists of the oropharyngeal and gastric administration of non-absorbable antibiotics lacking anaerobic activity (often a polymyxin, an aminoglycoside, and amphotericin) along with a short course of systemic antibiotic therapy, whereas selective oropharyngeal decontamination (SOD) consists of application of non-absorbable antibiotics to the oropharynx alone. More than 50 randomized studies and 10 meta-analyses of SDD/SOD have been published. Overall, these data suggest that SDD/SOD are associated with improved survival in ICU patients (Price et al. 2014) and SDD is associated with a reduction in the rate of postoperative infection, including anastomotic leakage, in patients undergoing elective GI surgery (Abis et al. 2013; Roos et al. 2013). Despite these successes in the use of SOD and SDD to improve patient outcomes in the setting of low levels of antibiotic resistance, controversy remains regarding their effectiveness in the setting of high levels of antibiotic resistance as well as their im-

pact on antibiotic resistance. No relationship between the use of SDD or SOD and the development of antimicrobial resistance has been shown in individual studies or meta-analyses in the setting of low antibiotic resistance (Daneman et al. 2013; Plantinga and Bonten 2015) and an international multicenter study of the effects of SDD and SOD on ICU-level antibiotic resistance in countries with higher levels of resistance is currently ongoing (clinicaltrials.gov/ct2/show/NCT02208154).

The ability of paromomycin to bind to eukaryotic ribosomes has led to the use of this agent in the treatment of protozoal infections. Like other aminoglycosides, oral paromomycin is poorly absorbed and may be used for the treatment of noninvasive amebiasis, cryptosporidiosis, trichomoniasis, and giardiasis in patients in whom other agents are contraindicated (Stover et al. 2012). More recently, this agent has been used to treat both cutaneous and visceral leishmaniasis. Topical paromomycin yielded a significantly higher cure rate compared with control therapy in patients with cutaneous leishmaniasis caused by *Leishmania major* (Ben Salah et al. 2013). Intramuscular paromomycin monotherapy was noninferior to standard amphotericin B therapy in a randomized control trial in patients with visceral disease in India (Sundar et al. 2007) and short-course combination regimens containing this agent were also noninferior to standard therapy with fewer adverse events (Sundar et al. 2011).

CONCLUSIONS

The aminoglycosides are a critical component of the current antibacterial arsenal. Their broad spectrum of activity, rapid bactericidal action, and favorable chemical and pharmacokinetic properties make them a clinically useful class of drugs across numerous infection types, including certain protozoal infections. The use of aminoglycosides waned as a result of the emergence of other classes of broad-spectrum agents with improved safety profiles, but the emergence of MDR pathogens has led to renewed interest in this class of drugs. Improved understanding of the drivers of toxicity and efficacy has led to the implementation of optimized dosing regimens that improve safety while maintaining efficacy. Increased understanding of key aminoglycoside resistance mechanisms combined with innovative medicinal chemistry approaches have led to the synthesis and development of novel agents specifically designed to evade resistance while maintaining potency against fully susceptible isolates. Given the dearth of new agents in the antibiotic pipeline and the ever-increasing specter of resistance, further optimization of the aminoglycoside scaffold to generate new agents with superior potency against MDR pathogens as well as an improved safety profile is warranted.

REFERENCES

Abis GS, Stockmann HB, van Egmond M, Bonjer HJ, Vandenbroucke-Grauls CM, Oosterling SJ. 2013. Selective decontamination of the digestive tract in gastrointestinal surgery: Useful in infection prevention? A systematic review. *J Gastrointest Surg* 17: 2172–2178.

Aggen JB, Armstrong ES, Goldblum AA, Dozzo P, Linsell MS, Gliedt MJ, Hildebrandt DJ, Feeney LA, Kubo A, Matias RD, et al. 2010. Synthesis and spectrum of the neoglycoside ACHN-490. *Antimicrob Agents Chemother* 54: 4636–4642.

Aínsa JA, Pérez E, Pelicic V, Berthet FX, Gicquel B, Martín C. 1997. Aminoglycoside 2′-N-acetyltransferase genes are universally present in mycobacteria: Characterization of the aac(2′)-Ic gene from *Mycobacterium tuberculosis* and the aac(2′)-Id gene from *Mycobacterium smegmatis*. *Mol Microbiol* 24: 431–441.

Aires JR, Köhler T, Nikaido H, Plésiat P. 1999. Involvement of an active efflux system in the natural resistance of *Pseudomonas aeruginosa* to aminoglycosides. *Antimicrob Agents Chemother* 43: 2624–2628.

Almaghrabi R, Clancy CJ, Doi Y, Hao B, Chen L, Shields RK, Press EG, Iovine NM, Townsend BM, Wagener MM, et al. 2014. Carbapenem-resistant *Klebsiella pneumoniae* strains exhibit diversity in aminoglycoside-modifying enzymes, which exert differing effects on plazomicin and other agents. *Antimicrob Agents Chemother* 58: 4443–4451.

Al Sheikh YA, Marie MA, John J, Krishnappa LG, Dabwab KH. 2014. Prevalence of 16S rRNA methylase genes among β-lactamase-producing Enterobacteriaceae clinical isolates in Saudi Arabia. *Libyan J Med* 9: 24432.

American Thoracic Society; Infectious Diseases Society of America. 2005. Guidelines for the management of adults with hospital-acquired, ventilator-associated, and healthcare-associated pneumonia. *Am J Respir Crit Care Med* 171: 388–416.

Armstrong ES, Miller GH. 2010. Combating evolution with intelligent design: The neoglycoside ACHN-490. *Curr Opin Microbiol* 13: 565–573.

Avent ML, Rogers BA, Cheng AC, Paterson DL. 2011. Current use of aminoglycosides: Indications, pharmacokinetics and monitoring for toxicity. *Intern Med J* 41: 441–449.

Azucena E, Mobashery S. 2001. Aminoglycoside-modifying enzymes: Mechanisms of catalytic processes and inhibition. *Drug Resist Updat* 4: 106–117.

Barclay ML, Kirkpatrick CM, Begg EJ. 1999. Once daily aminoglycoside therapy. Is it less toxic than multiple daily doses and how should it be monitored? *Clin Pharmacokinet* 36: 89–98.

Beauclerk AA, Cundliffe E. 1987. Sites of action of two ribosomal RNA methylases responsible for resistance to aminoglycosides. *J Mol Biol* 193: 661–671.

Bell J, Andersson P, Jones R, Turnidge J. 2010. 16S rRNA methylase containing Enterobacteriaceae in the SENTRY Asia-Pacific region frequently harbour plasmid-mediated quinolone resistance CTXM types. *20th European Congress of Clinical Microbiology and Infectious Diseases (ECCMID)*, Abstract O559. Vienna, April 10–13.

Ben Salah A, Ben Messaoud N, Guedri E, Zaatour A, Ben Alaya N, Bettaieb J, Gharbi A, Belhadj Hamida N, Boukthir A, Chlif S, et al. 2013. Topical paromomycin with or without gentamicin for cutaneous leishmaniasis. *N Engl J Med* 368: 524–532.

Benveniste R, Davies J. 1971. R-factor mediated gentamicin resistance: A new enzyme which modifies aminoglycoside antibiotics. *FEBS Lett* 14: 293–296.

Berçot B, Poirel L, Nordmann P. 2011. Updated multiplex polymerase chain reaction for detection of 16S rRNA methylases: High prevalence among NDM-1 producers. *Diagn Microbiol Infect Dis* 71: 442–445.

Bercovier H, Kafri O, Sela S. 1986. Mycobacteria possess a surprisingly small number of ribosomal RNA genes in relation to the size of their genome. *Biochem Biophys Res Commun* 136: 1136–1141.

Brogden RN, Pinder RM, Sawyer PR, Speight TM, Avery GS. 1976. Tobramycin: A review of its antibacterial and pharmacokinetic properties and therapeutic use. *Drugs* 12: 166–200.

Brooke JS. 2012. *Stenotrophomonas maltophilia*: An emerging global opportunistic pathogen. *Clin Microbiol Rev* 25: 2–41.

Buijk SE, Mouton JW, Gyssens IC, Verbrugh HA, Bruining HA. 2002. Experience with a once-daily dosing program of aminoglycosides in critically ill patients. *Intensive Care Med* 28: 936–942.

Buroni S, Pasca MR, Flannagan RS, Bazzini S, Milano A, Bertani I, Venturi V, Valvano MA, Riccardi G. 2009. Assessment of three resistance-nodulation-cell division drug efflux transporters of *Burkholderia cenocepacia* in intrinsic antibiotic resistance. *BMC Microbiol* 9: 200.

Castanheira M, Costello SE, Jones RN, Mendes RE. 2015. Prevalence of aminoglycoside resistance genes among contemporary Gram-negative resistant isolates collected worldwide. *25th European Congress of Clinical Microbiology and Infectious Diseases (ECCMID)*, Abstract O011. Copenhagen, April 25–28.

Cattoir V, Nordmann P. 2009. Plasmid-mediated quinolone resistance in Gram-negative bacterial species: An update. *Curr Med Chem* 16: 1028–1046.

Costa Y, Galimand M, Leclercq R, Duval J, Courvalin P. 1993. Characterization of the chromosomal *aac(6′)-Ii* gene specific for *Enterococcus faecium*. *Antimicrob Agents Chemother* **37:** 1896–1903.

Courvalin P. 1994. Transfer of antibiotic resistance genes between Gram-positive and Gram-negative bacteria. *Antimicrob Agents Chemother* **38:** 1447–1451.

Craig WA. 2011. Optimizing aminoglycoside use. *Crit Care Clin* **27:** 107–121.

Craig WA, Redington J, Ebert SC. 1991. Pharmacodynamics of amikacin in vitro and in mouse thigh and lung infections. *J Antimicrob Chemother* **27:** 29–40.

Culebras E, Martínez JL. 1999. Aminoglycoside resistance mediated by the bifunctional enzyme 6′-N-aminoglycoside acetyltransferase-2″-O-aminoglycoside phosphotransferase. *Front Biosci* **4:** D1–D8.

Cundliffe E. 1989. How antibiotic-producing organisms avoid suicide. *Annu Rev Microbiol* **43:** 207–233.

Daneman N, Sarwar S, Fowler RA, Cuthbertson BH, Group SCS. 2013. Effect of selective decontamination on antimicrobial resistance in intensive care units: A systematic review and meta-analysis. *Lancet Infect Dis* **13:** 328–341.

Davis BD. 1987. Mechanism of bactericidal action of aminoglycosides. *Microbiol Rev* **51:** 341–350.

Davis BD, Chen LL, Tai PC. 1986. Misread protein creates membrane channels: An essential step in the bactericidal action of aminoglycosides. *Proc Natl Acad Sci* **83:** 6164–6168.

Dellinger RP, Levy MM, Rhodes A, Annane D, Gerlach H, Opal SM, Sevransky JE, Sprung CL, Douglas IS, Jaeschke R, et al. 2013. Surviving sepsis campaign: International guidelines for management of severe sepsis and septic shock: 2012. *Crit Care Med* **41:** 580–637.

Dennis DT, Inglesby TV, Henderson DA, Bartlett JG, Ascher MS, Eitzen E, Fine AD, Friedlander AM, Hauer J, Layton M, et al. 2001. Tularemia as a biological weapon: Medical and public health management. *JAMA* **285:** 2763–2773.

Doi Y, Yokoyama K, Yamane K, Wachino J, Shibata N, Yagi T, Shibayama K, Kato H, Arakawa Y. 2004. Plasmid-mediated 16S rRNA methylase in *Serratia marcescens* conferring high-level resistance to aminoglycosides. *Antimicrob Agents Chemother* **48:** 491–496.

Drusano GL, Ambrose PG, Bhavnani SM, Bertino JS, Nafziger AN, Louie A. 2007. Back to the future: Using aminoglycosides again and how to dose them optimally. *Clin Infect Dis* **45:** 753–760.

Dzidic S, Bedeković V. 2003. Horizontal gene transfer-emerging multidrug resistance in hospital bacteria. *Acta Pharmacol Sin* **24:** 519–526.

Eliopoulos GM, Eliopoulos CT. 1988. Antibiotic combinations: Should they be tested? *Clin Microbiol Rev* **1:** 139–156.

Falagas ME, Siempos II, Bliziotis IA, Michalopoulos A. 2006. Administration of antibiotics via the respiratory tract for the prevention of ICU-acquired pneumonia: A meta-analysis of comparative trials. *Crit Care* **10:** R123.

Feizabadi MM, Asadi S, Zohari M, Gharavi S, Etemadi G. 2004. Genetic characterization of high-level gentamicin-resistant strains of *Enterococcus faecalis* in Iran. *Can J Microbiol* **50:** 869–872.

Gonzalez-Padilla M, Torre-Cisneros J, Rivera-Espinar F, Pontes-Moreno A, López-Cerero L, Pascual A, Natera C, Rodríguez M, Salcedo I, Rodríguez-López F, et al. 2015. Gentamicin therapy for sepsis due to carbapenem-resistant and colistin-resistant *Klebsiella pneumoniae*. *J Antimicrob Chemother* **70:** 905–913.

Griffith DE, Aksamit T, Brown-Elliott BA, Catanzaro A, Daley C, Gordin F, Holland SM, Horsburgh R, Huitt G, Iademarco MF, et al. 2007. An official ATS/IDSA statement: Diagnosis, treatment, and prevention of nontuberculous mycobacterial diseases. *Am J Respir Crit Care Med* **175:** 367–416.

Hancock RE. 1984. Alterations in outer membrane permeability. *Annu Rev Microbiol* **38:** 237–264.

Hancock RE, Raffle VJ, Nicas TI. 1981. Involvement of the outer membrane in gentamicin and streptomycin uptake and killing in *Pseudomonas aeruginosa*. *Antimicrob Agents Chemother* **19:** 777–785.

Hancock RE, Farmer SW, Li ZS, Poole K. 1991. Interaction of aminoglycosides with the outer membranes and purified lipopolysaccharide and OmpF porin of *Escherichia coli*. *Antimicrob Agents Chemother* **35:** 1309–1314.

Ho YI, Chan CY, Cheng AF. 1997. In-vitro activities of aminoglycoside-aminocyclitols against mycobacteria. *J Antimicrob Chemother* **40:** 27–32.

Hocquet D, Vogne C, El Garch F, Vejux A, Gotoh N, Lee A, Lomovskaya O, Plésiat P. 2003. MexXY-OprM efflux pump is necessary for adaptive resistance of *Pseudomonas aeruginosa* to aminoglycosides. *Antimicrob Agents Chemother* **47:** 1371–1375.

Hollingshead S, Vapnek D. 1985. Nucleotide sequence analysis of a gene encoding a streptomycin/spectinomycin adenylyltransferase. *Plasmid* **13:** 17–30.

Humes HD, Weinberg JM, Knauss TC. 1982. Clinical and pathophysiologic aspects of aminoglycoside nephrotoxicity. *Am J Kidney Dis* **2:** 5–29.

Ikäheimo I, Syrjälä H, Karhukorpi J, Schildt R, Koskela M. 2000. In vitro antibiotic susceptibility of *Francisella tularensis* isolated from humans and animals. *J Antimicrob Chemother* **46:** 287–290.

Ioannidou E, Siempos II, Falagas ME. 2007. Administration of antimicrobials via the respiratory tract for the treatment of patients with nosocomial pneumonia: A meta-analysis. *J Antimicrob Chemother* **60:** 1216–1226.

Islam S, Jalal S, Wretlind B. 2004. Expression of the MexXY efflux pump in amikacin-resistant isolates of *Pseudomonas aeruginosa*. *Clin Microbiol Infect* **10:** 877–883.

Jackson J, Chen C, Buising K. 2013. Aminoglycosides: How should we use them in the 21st century? *Curr Opin Infect Dis* **26:** 516–525.

Jones RN, Flonta M, Gurler N, Cepparulo M, Mendes RE, Castanheira M. 2014. Resistance surveillance program report for selected European nations (2011). *Diagn Microbiol Infect Dis* **78:** 429–436.

Karlowsky JA, Draghi DC, Jones ME, Thornsberry C, Friedland IR, Sahm DF. 2003. Surveillance for antimicrobial susceptibility among clinical isolates of *Pseudomonas aeruginosa* and *Acinetobacter baumannii* from hospitalized patients in the United States, 1998 to 2001. *Antimicrob Agents Chemother* **47:** 1681–1688.

Kim C, Mobashery S. 2005. Phosphoryl transfer by amino-glycoside 3′-phosphotransferases and manifestation of antibiotic resistance. *Bioorg Chem* **33:** 149–158.

Kislak JW. 1972. The susceptibility of *Bacteroides fragilis* to 24 antibiotics. *J Infect Dis* **125:** 295–299.

Kotra LP, Haddad J, Mobashery S. 2000. Aminoglycosides: Perspectives on mechanisms of action and resistance and strategies to counter resistance. *Antimicrob Agents Chemother* **44:** 3249–3256.

Landman D, Babu E, Shah N, Kelly P, Bäcker M, Bratu S, Quale J. 2010. Activity of a novel aminoglycoside, ACHN-490, against clinical isolates of *Escherichia coli* and *Klebsiella pneumoniae* from New York City. *J Antimicrob Chemother* **65:** 2123–2127.

Landman D, Kelly P, Bäcker M, Babu E, Shah N, Bratu S, Quale J. 2011. Antimicrobial activity of a novel amino-glycoside, ACHN-490, against *Acinetobacter baumannii* and *Pseudomonas aeruginosa* from New York City. *J Antimicrob Chemother* **66:** 332–334.

Le J, McKee B, Srisupha-Olarn W, Burgess DS. 2011. In vitro activity of carbapenems alone and in combination with amikacin against KPC-producing *Klebsiella pneumoniae*. *J Clin Med Res* **3:** 106–110.

Le Goffic F, Baca D, Soussy CJ, Dublanchet A, Duval J. 1976. ANT(4′)I: A new aminoglycoside nucleotidyltransferase found in "*staphylococcus aureus*." *Ann Microbiol (Paris)* **127:** 391–399 (author's transl.).

Livermore DM, Mushtaq S, Warner M, Zhang JC, Maharjan S, Doumith M, Woodford N. 2011. Activity of amino-glycosides, including ACHN-490, against carbapenem-resistant Enterobacteriaceae isolates. *J Antimicrob Chemother* **66:** 48–53.

Llano-Sotelo B, Azucena EF, Kotra LP, Mobashery S, Chow CS. 2002. Aminoglycosides modified by resistance enzymes display diminished binding to the bacterial ribosomal aminoacyl-tRNA site. *Chem Biol* **9:** 455–463.

Lode H, Grunert K, Koeppe P, Langmaack H. 1976. Pharmacokinetic and clinical studies with amikacin, a new aminoglycoside antibiotic. *J Infect Dis* **134:** S316–S322.

Macinga DR, Rather PN. 1999. The chromosomal 2′-N-acetyltransferase of *Providencia stuartii*: Physiological functions and genetic regulation. *Front Biosci* **4:** D132–D140.

Magnet S, Blanchard JS. 2005. Molecular insights into aminoglycoside action and resistance. *Chem Rev* **105:** 477–498.

Magnet S, Courvalin P, Lambert T. 2001. Resistance-nodulation-cell division-type efflux pump involved in aminoglycoside resistance in *Acinetobacter baumannii* strain BM4454. *Antimicrob Agents Chemother* **45:** 3375–3380.

Martin WJ, Gardner M, Washington JA. 1972. In vitro antimicrobial susceptibility of anaerobic bacteria isolated from clinical specimens. *Antimicrob Agents Chemother* **1:** 148–158.

Matsumoto T. 2014. Arbekacin: Another novel agent for treating infections due to methicillin-resistant *Staphylococcus aureus* and multidrug-resistant Gram-negative pathogens. *Clin Pharmacol* **6:** 139–148.

Mingeot-Leclercq MP, Glupczynski Y, Tulkens PM. 1999. Aminoglycosides: Activity and resistance. *Antimicrob Agents Chemother* **43:** 727–737.

Miró E, Grünbaum F, Gómez L, Rivera A, Mirelis B, Coll P, Navarro F. 2013. Characterization of aminoglycoside-modifying enzymes in enterobacteriaceae clinical strains and characterization of the plasmids implicated in their diffusion. *Microb Drug Resist* **19:** 94–99.

Moore RD, Smith CR, Lipsky JJ, Mellits ED, Lietman PS. 1984. Risk factors for nephrotoxicity in patients treated with aminoglycosides. *Ann Intern Med* **100:** 352–357.

Murphy E. 1985. Nucleotide sequence of a spectinomycin adenyltransferase AAD(9) determinant from *Staphylococcus aureus* and its relationship to AAD(3″)(9). *Mol Gen Genet* **200:** 33–39.

Murry KR, McKinnon PS, Mitrzyk B, Rybak MJ. 1999. Pharmacodynamic characterization of nephrotoxicity associated with once-daily aminoglycoside. *Pharmacotherapy* **19:** 1252–1260.

Mushtaq S, Irfan S, Sarma JB, Doumith M, Pike R, Pitout J, Livermore DM, Woodford N. 2011. Phylogenetic diversity of *Escherichia coli* strains producing NDM-type carbapenemases. *J Antimicrob Chemother* **66:** 2002–2005.

Mwengee W, Butler T, Mgema S, Mhina G, Almasi Y, Bradley C, Formanik JB, Rochester CG. 2006. Treatment of plague with gentamicin or doxycycline in a randomized clinical trial in Tanzania. *Clin Infect Dis* **42:** 614–621.

Nagabhushan T, Miller G, Weinstein M. 1982. Structure–activity relationships in aminoglycoside-aminocyclitol antibiotics. In *The aminoglycosides: Microbiology, clinical use and toxicology* (ed. Whelton A, Neu HC), pp. 3–27. Marcel Dekker, New York.

Nichols WW, Young SN. 1985. Respiration-dependent uptake of dihydrostreptomycin by *Escherichia coli*. Its irreversible nature and lack of evidence for a uniport process. *Biochem J* **228:** 505–512.

Nicolau DP, Belliveau PP, Nightingale CH, Quintiliani R, Freeman CD. 1995. Implementation of a once-daily aminoglycoside program in a large community-teaching hospital. *Hosp Pharm* **30:** 674–676, 679–680.

Okamoto S, Suzuki Y. 1965. Chloramphenicol-, dihydrostreptomycin-, and kanamycin-inactivating enzymes from multiple drug-resistant *Escherichia coli* carrying episome "R". *Nature* **208:** 1301–1303.

Ounissi H, Derlot E, Carlier C, Courvalin P. 1990. Gene homogeneity for aminoglycoside-modifying enzymes in Gram-positive cocci. *Antimicrob Agents Chemother* **34:** 2164–2168.

Pankuch GA, Seifert H, Appelbaum PC. 2010. Activity of doripenem with and without levofloxacin, amikacin, and colistin against *Pseudomonas aeruginosa* and *Acinetobacter baumannii*. *Diagn Microbiol Infect Dis* **67:** 191–197.

Plantinga NL, Bonten MJ. 2015. Selective decontamination and antibiotic resistance in ICUs. *Crit Care* **19:** 259.

Podnecky NL, Rhodes KA, Schweizer HP. 2015. Efflux pump-mediated drug resistance in *Burkholderia*. *Front Microbiol* **6:** 305.

Poirel L, Savov E, Nazli A, Trifonova A, Todorova I, Gergova I, Nordmann P. 2014. Outbreak caused by NDM-1- and RmtB-producing *Escherichia coli* in Bulgaria. *Antimicrob Agents Chemother* **58:** 2472–2474.

Price R, MacLennan G, Glen J; SuDDICU Collaboration. 2014. Selective digestive or oropharyngeal decontamination and topical oropharyngeal chlorhexidine for prevention of death in general intensive care: Systematic review and network meta-analysis. *BMJ* **348:** g2197.

Prins JM, Büller HR, Kuijper EJ, Tange RA, Speelman P. 1993. Once versus thrice daily gentamicin in patients with serious infections. *Lancet* **341:** 335–339.

Ramirez MS, Tolmasky ME. 2010. Aminoglycoside modifying enzymes. *Drug Resist Updat* **13:** 151–171.

Ramsey BW, Pepe MS, Quan JM, Otto KL, Montgomery AB, Williams-Warren J, Vasiljev-K M, Borowitz D, Bowman CM, Marshall BC, et al. 1999. Intermittent administration of inhaled tobramycin in patients with cystic fibrosis. Cystic Fibrosis Inhaled Tobramycin Study Group. *N Engl J Med* **340:** 23–30.

Rather PN, Orosz E, Shaw KJ, Hare R, Miller G. 1993. Characterization and transcriptional regulation of the 2'-*N*-acetyltransferase gene from *Providencia stuartii*. *J Bacteriol* **175:** 6492–6498.

Ristuccia AM, Cunha BA. 1985. An overview of amikacin. *Ther Drug Monit* **7:** 12–25.

Robicsek A, Strahilevitz J, Jacoby GA, Macielag M, Abbanat D, Park CH, Bush K, Hooper DC. 2006. Fluoroquinolone-modifying enzyme: A new adaptation of a common aminoglycoside acetyltransferase. *Nat Med* **12:** 83–88.

Roos D, Dijksman LM, Tijssen JG, Gouma DJ, Gerhards MF, Oudemans-van Straaten HM. 2013. Systematic review of perioperative selective decontamination of the digestive tract in elective gastrointestinal surgery. *Br J Surg* **100:** 1579–1588.

Rosenberg EY, Ma D, Nikaido H. 2000. AcrD of *Escherichia coli* is an aminoglycoside efflux pump. *J Bacteriol* **182:** 1754–1756.

Rybak MJ, Abate BJ, Kang SL, Ruffing MJ, Lerner SA, Drusano GL. 1999. Prospective evaluation of the effect of an aminoglycoside dosing regimen on rates of observed nephrotoxicity and ototoxicity. *Antimicrob Agents Chemother* **43:** 1549–1555.

Sader HS, Farrell DJ, Flamm RK, Jones RN. 2014. Antimicrobial susceptibility of Gram-negative organisms isolated from patients hospitalised with pneumonia in U.S. and European hospitals: Results from the SENTRY Antimicrobial Surveillance Program, 2009–2012. *Int J Antimicrob Agents* **43:** 328–334.

Sader HS, Rhomberg PR, Farrell DJ, Jones RN. 2015. Arbekacin activity against contemporary clinical bacteria isolated from patients hospitalized with pneumonia. *Antimicrob Agents Chemother* **59:** 3263–3270.

Sawicki GS, Signorovitch JE, Zhang J, Latremouille-Viau D, von Wartburg M, Wu EQ, Shi L. 2012. Reduced mortality in cystic fibrosis patients treated with tobramycin inhalation solution. *Pediatr Pulmonol* **47:** 44–52.

Schwartz JJ, Gazumyan A, Schwartz I. 1992. rRNA gene organization in the Lyme disease spirochete, *Borrelia burgdorferi*. *J Bacteriol* **174:** 3757–3765.

Shaw KJ, Rather PN, Hare RS, Miller GH. 1993. Molecular genetics of aminoglycoside resistance genes and familial relationships of the aminoglycoside-modifying enzymes. *Microbiol Rev* **57:** 138–163.

Simon VK, Mösinger EU, Malerczy V. 1973. Pharmacokinetic studies of tobramycin and gentamicin. *Antimicrob Agents Chemother* **3:** 445–450.

Skeggs PA, Thompson J, Cundliffe E. 1985. Methylation of 16S ribosomal RNA and resistance to aminoglycoside antibiotics in clones of *Streptomyces lividans* carrying DNA from *Streptomyces tenjimariensis*. *Mol Gen Genet* **200:** 415–421.

Snowden J, Stovall S. 2011. Tularemia: Retrospective review of 10 years' experience in Arkansas. *Clin Pediatr (Phila)* **50:** 64–68.

Stover KR, Riche DM, Gandy CL, Henderson H. 2012. What would we do without metronidazole? *Am J Med Sci* **343:** 316–319.

Strahilevitz J, Jacoby GA, Hooper DC, Robicsek A. 2009. Plasmid-mediated quinolone resistance: A multifaceted threat. *Clin Microbiol Rev* **22:** 664–689.

Stubbings W, Bostock J, Ingham E, Chopra I. 2006. Mechanisms of the post-antibiotic effects induced by rifampicin and gentamicin in *Escherichia coli*. *J Antimicrob Chemother* **58:** 444–448.

Sundar S, Jha TK, Thakur CP, Sinha PK, Bhattacharya SK. 2007. Injectable paromomycin for visceral leishmaniasis in India. *N Engl J Med* **356:** 2571–2581.

Sundar S, Sinha PK, Rai M, Verma DK, Nawin K, Alam S, Chakravarty J, Vaillant M, Verma N, Pandey K, et al. 2011. Comparison of short-course multidrug treatment with standard therapy for visceral leishmaniasis in India: An open-label, non-inferiority, randomised controlled trial. *Lancet* **377:** 477–486.

Swenson JM, Wallace RJ, Silcox VA, Thornsberry C. 1985. Antimicrobial susceptibility of five subgroups of *Mycobacterium fortuitum* and *Mycobacterium chelonae*. *Antimicrob Agents Chemother* **28:** 807–811.

Taber HW, Mueller JP, Miller PF, Arrow AS. 1987. Bacterial uptake of aminoglycoside antibiotics. *Microbiol Rev* **51:** 439–457.

Tamma PD, Cosgrove SE, Maragakis LL. 2012. Combination therapy for treatment of infections with Gram-negative bacteria. *Clin Microbiol Rev* **25:** 450–470.

Thompson J, Skeggs PA, Cundliffe E. 1985. Methylation of 16S ribosomal RNA and resistance to the aminoglycoside antibiotics gentamicin and kanamycin determined by DNA from the gentamicin-producer, *Micromonospora purpurea*. *Mol Gen Genet* **201:** 168–173.

Vakulenko SB, Mobashery S. 2003. Versatility of aminoglycosides and prospects for their future. *Clin Microbiol Rev* **16:** 430–450.

Wachino J, Arakawa Y. 2012. Exogenously acquired 16S rRNA methyltransferases found in aminoglycoside-resistant pathogenic Gram-negative bacteria: An update. *Drug Resist Updat* **15:** 133–148.

Wachino J, Yamane K, Shibayama K, Kurokawa H, Shibata N, Suzuki S, Doi Y, Kimura K, Ike Y, Arakawa Y. 2006. Novel plasmid-mediated 16S rRNA methylase, RmtC, found in a *Proteus mirabilis* isolate demonstrating extraordinary high-level resistance against various aminoglycosides. *Antimicrob Agents Chemother* **50:** 178–184.

Wachino J, Shibayama K, Kurokawa H, Kimura K, Yamane K, Suzuki S, Shibata N, Ike Y, Arakawa Y. 2007. Novel plasmid-mediated 16S rRNA m1A1408 methyltransferase, NpmA, found in a clinically isolated *Escherichia coli* strain resistant to structurally diverse aminoglycosides. *Antimicrob Agents Chemother* **51:** 4401–4409.

Waters VL. 1999. Conjugative transfer in the dissemination of β-lactam and aminoglycoside resistance. *Front Biosci* **4:** D433–D456.

Westbrock-Wadman S, Sherman DR, Hickey MJ, Coulter SN, Zhu YQ, Warrener P, Nguyen LY, Shawar RM, Folger KR, Stover CK. 1999. Characterization of a *Pseudomonas aeruginosa* efflux pump contributing to aminoglycoside impermeability. *Antimicrob Agents Chemother* **43:** 2975–2983.

Wilson DN. 2014. Ribosome-targeting antibiotics and mechanisms of bacterial resistance. *Nat Rev Microbiol* **12:** 35–48.

World Health Organization. 2011. *Guidelines for the programmatic management of drug-resistant tuberculosis: 2011 update.* World Health Organization, Geneva.

Wright GD, Thompson PR. 1999. Aminoglycoside phosphotransferases: Proteins, structure, and mechanism. *Front Biosci* **4:** D9–D21.

Yamane K, Wachino J, Doi Y, Kurokawa H, Arakawa Y. 2005. Global spread of multiple aminoglycoside resistance genes. *Emerg Infect Dis* **11:** 951–953.

Yan JJ, Wu JJ, Ko WC, Tsai SH, Chuang CL, Wu HM, Lu YJ, Li JD. 2004. Plasmid-mediated 16S rRNA methylases conferring high-level aminoglycoside resistance in *Escherichia coli* and *Klebsiella pneumoniae* isolates from two Taiwanese hospitals. *J Antimicrob Chemother* **54:** 1007–1012.

Yokoyama K, Doi Y, Yamane K, Kurokawa H, Shibata N, Shibayama K, Yagi T, Kato H, Arakawa Y. 2003. Acquisition of 16S rRNA methylase gene in *Pseudomonas aeruginosa*. *Lancet* **362:** 1888–1893.

Zhanel GG, Hoban DJ, Harding GK. 1991. The postantibiotic effect: A review of in vitro and in vivo data. *DICP* **25:** 153–163.

Tetracycline Antibiotics and Resistance

Trudy H. Grossman

Tetraphase Pharmaceuticals, Watertown, Massachusetts 02472

Correspondence: tgrossman@tphase.com

Tetracyclines possess many properties considered ideal for antibiotic drugs, including activity against Gram-positive and -negative pathogens, proven clinical safety, acceptable tolerability, and the availability of intravenous (IV) and oral formulations for most members of the class. As with all antibiotic classes, the antimicrobial activities of tetracyclines are subject to both class-specific and intrinsic antibiotic-resistance mechanisms. Since the discovery of the first tetracyclines more than 60 years ago, ongoing optimization of the core scaffold has produced tetracyclines in clinical use and development that are capable of thwarting many of these resistance mechanisms. New chemistry approaches have enabled the creation of synthetic derivatives with improved in vitro potency and in vivo efficacy, ensuring that the full potential of the class can be explored for use against current and emerging multidrug-resistant (MDR) pathogens, including carbapenem-resistant Enterobacteriaceae, MDR *Acinetobacter* species, and *Pseudomonas aeruginosa*.

Tetracycline antibiotics are well known for their broad spectrum of activity, spanning a wide range of Gram-positive and -negative bacteria, spirochetes, obligate intracellular bacteria, as well as protozoan parasites. The first tetracyclines were natural products derived from the fermentations of actinomycetes. Chlortetracycline, produced by *Streptomyces aureofaciens*, and marketed as Aureomycin, was first reported by Benjamin Duggar at Lederle Laboratories in 1948 and approved for clinical use that same year (Duggar 1948). Soon after, Pfizer (New York) scientists isolated oxytetracycline, approved by the U.S. Food and Drug Administration (FDA) in 1950 and marketed as Terramycin (Finlay et al. 1950). Other tetracyclines that followed over the next two decades were also natural products produced by streptomycetes (tetracycline, demethylchlortetracycline) or semisynthetic derivatives with improved antibacterial potency, spectrum, resistance coverage, solubility, and/or oral bioavailability (methacycline, rolitetracycline, lymecycline, doxycycline, and minocycline) (Jarolmen et al. 1970; Cunha et al. 1982; Nelson and Levy 2011). Several of these "legacy" tetracyclines remain in clinical use for the treatment of uncomplicated respiratory, urogenital, gastrointestinal, and other rare and serious infections; however, the dissemination of tetracycline-resistant mechanisms has narrowed their utility, limiting use to only infections with confirmed susceptibility (Fig. 1).

After a long pause in the advancement of the tetracycline class, renewed interest in optimization of tetracyclines during the late 1980s led to the discovery of semisynthetic derivatives with improved potency against difficult-to-treat emerging multidrug-resistant (MDR) Gram-

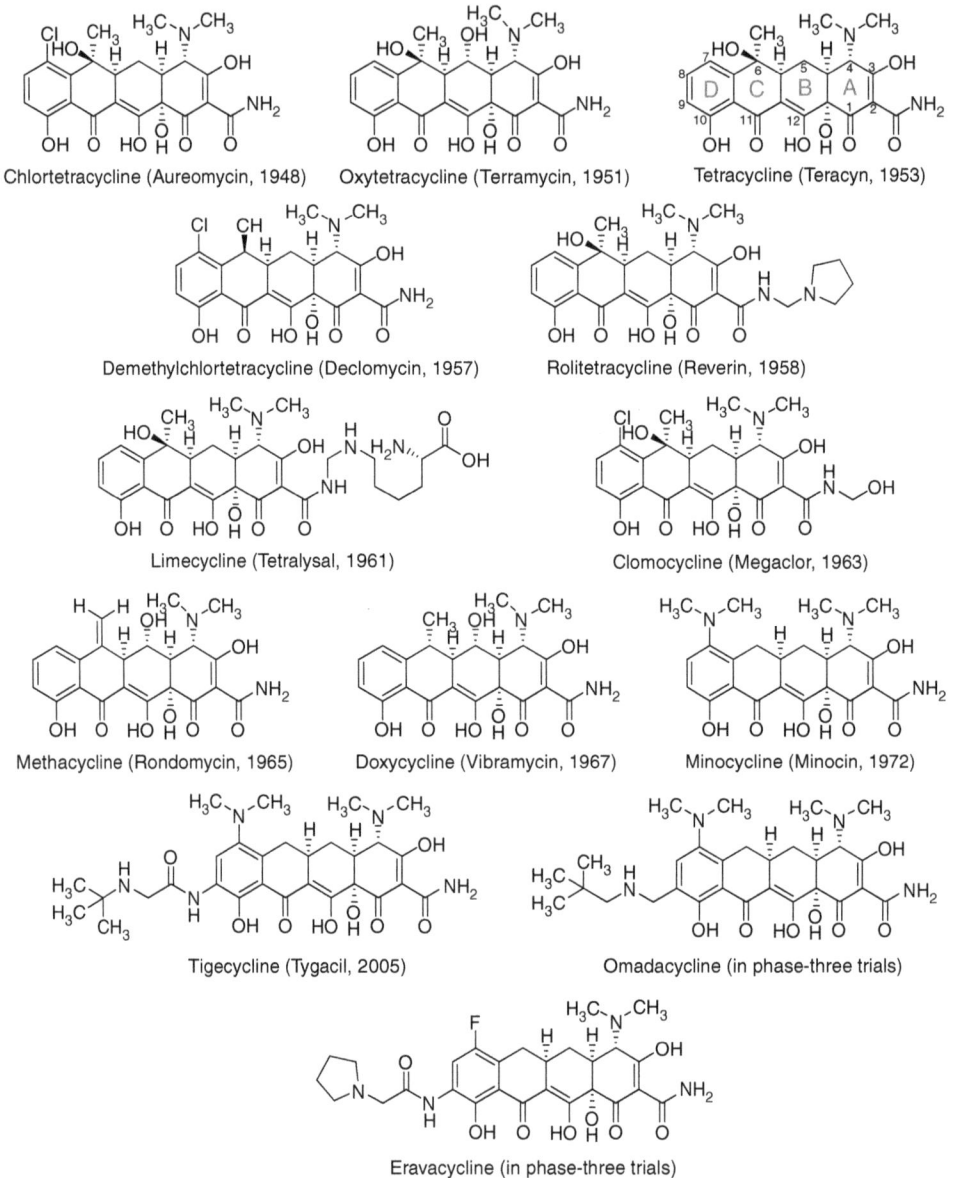

Figure 1. Chemical structures of clinically used tetracyclines and development candidates. Tetracycline structures are labeled with generic names; trade names and year of discovery are indicated within parentheses. The core structure rings (A–D) and carbons (1–12) are labeled in the chemical structure of tetracycline using the convention for tetracycline carbon numbering and ring letter assignments.

negative and -positive pathogens, including bacteria with tetracycline-specific resistance mechanisms. Tigecycline, a semisynthetic parenteral glycylcycline, was discovered in 1993 by scientists at Lederle (which later became Wyeth, New York), and introduced into clinical use in 2005 (Sum and Petersen 1999; Zhanel et al. 2004). Tigecycline continues to be an important treatment option for serious infections caused by pathogens resistant to other antibiotic classes. In recent years, two new tetracyclines have entered clinical development: omadacycline, a semisynthetic aminomethylcycline derivative of minocycline discovered at Paratek Pharmaceuticals (Boston, MA) (Draper et al. 2014), and eravacycline, a fully synthetic fluorocycline discovered at Tetraphase Pharmaceuticals (Watertown, MA) (Clark et al. 2012; Xiao et al. 2012). In addition to efficacy against MDR infections, an important feature of these two new antibiotics is their oral formulations. This review will focus on recent developments in the understanding of tetracycline-resistance mechanisms and their potential impact on the clinical utility of tetracycline-class antibiotics.

MECHANISM, UPTAKE, AND TETRACYCLINE-SPECIFIC RESISTANCE

In recent surveillance studies, the prevalence of tetracycline resistance in selected European countries was found to be 66.9% and 44.9% for extended-spectrum β-lactamase (ESBL)-producing *Escherichia coli* and *Klebsiella* species (spp.), respectively (Jones et al. 2014), and global tetracycline-resistance percentages were 8.7% and 24.3% for methicillin-resistant *Staphylococcus aureus* (MRSA) and *Streptococcus pneumoniae*, respectively (Mendes et al. 2015). Resistance to tetracyclines is usually attributed to one or more of the following: the acquisition of mobile genetic elements carrying tetracycline-specific resistance genes, mutations within the ribosomal binding site, and/or chromosomal mutations leading to increased expression of intrinsic resistance mechanisms. Three general class-specific mechanisms have been well described: efflux, ribosomal protection, and enzymatic inactivation of tetracycline drugs. As there are several recent reviews on the topics of tetracycline-specific resistance determinants and their prevalence in clinical and environmental settings (Roberts 2005, 2011; Jones et al. 2008; Thaker et al. 2010), only a limited discussion of these areas will be covered here.

Uptake and Mechanism of Action

Tetracyclines preferentially bind to bacterial ribosomes and interact with a highly conserved 16S ribosomal RNA (rRNA) target in the 30S ribosomal subunit, arresting translation by sterically interfering with the docking of amino-acyl-transfer RNA (tRNA) during elongation (Maxwell 1967; Brodersen et al. 2000; Pioletti et al. 2001). Tetracyclines are usually considered bacteriostatic antibiotics; however, organism- and isolate-specific bactericidal activity in vitro has been described (Norcia et al. 1999; Petersen et al. 2007; Bantar et al. 2008; Noviello et al. 2008), and, recently, the bactericidal activity of tigecycline against *E. coli* and *K. pneumoniae* in a mouse model suggests that in vitro bactericidal assessments may not necessarily predict in vivo outcomes (Tessier and Nicolau 2013).

The mechanism of tetracycline uptake has been reviewed by Nikaido and Thanassi (1993). Briefly, in Gram-negative cells such as *E. coli*, tetracycline passively diffuses through the outer membrane porins OmpF and OmpC (Mortimer and Piddock 1993; Thanassi et al. 1995), most likely as a Mg^{2+} chelate, and this is consistent with the finding that outer membrane porin mutants show decreased susceptibility to tetracyclines (Pugsley and Schnaitman 1978). Accumulation of tetracycline in the periplasm is driven by the Donnan potential across the outer membrane. The dissociation of tetracycline from Mg^{2+} enables the weakly lipophilic, uncharged form to diffuse through the inner membrane to the cytoplasm where it may be complexed with magnesium and reach its ribosomal target. Uptake into the cytoplasm is partially energy dependent, involving passive diffusion, proton motive force, and phosphate bond hydrolysis (McMurry and Levy 1978; Smith and Chopra 1984; Yamaguchi et al. 1991).

Ribosomal Interactions

Crystallographic studies with the *Thermus thermophilus* 30S ribosomal subunit have revealed at least one high-occupancy tetracycline-binding site (Tet-1) and five other minor binding sites in 16S rRNA (Brodersen et al. 2000; Pioletti et al. 2001). Tetracycline most likely binds complexed with two Mg^{2+} ions at the Tet-1 site located in a pocket formed between helices h34 and h31, near the A-site where aminoacyl-tRNA docks onto the 30S subunit, consistent with the known mechanism of action (Jenner et al. 2013). The significance of the other five tetracycline-binding sites located elsewhere within the 30S subunit is unclear, and recent crystallographic studies with tigecycline and tetracycline binding to the *T. thermophilus* 70S ribosome (Jenner et al. 2013) and tigecycline binding to the 30S ribosome (Schedlbauer et al. 2015) showed that tigecycline was bound only to the Tet-1 site, and secondary binding sites were not observed (Fig. 2). Additional interactions made between the 9-*tert*-butylglycylamido moiety of tigecycline and C1054 in h34 are consistent with the higher binding affinity and greater antitranslational potency of tigecycline compared with tetracycline (Olson et al. 2006). Interestingly, a different orientation of this tigecycline side chain was observed in the 30S versus the 70S structure (Fig. 2), suggesting that tigecycline must accommodate conformation changes in the primary binding site that occur during decoding (Schedlbauer et al. 2015). Consistent with this recent finding, earlier

work by Bauer et al. (2004) showed that tigecycline and tetracycline produced slightly different patterns of Fe^{2+}-mediated RNA cleavage and dimethylsulfate modification, suggesting that both antibiotics bind at the same binding site, but in somewhat different orientations. Ribosome-binding competition experiments with [^{3}H]tetracycline show relative IC_{50} values as follows: eravacycline, 0.22 μM; tigecycline, 0.22 μM; minocycline, 1.63 μM; omadacycline, 1.96 μM; and tetracycline, 4 μM; and results were consistent with these and other novel tetracycline derivatives binding at a single major site (Olson et al. 2006; Grossman et al. 2012; Jenner et al. 2013; Draper et al. 2014).

Binding-Site Mutations

Because most bacteria have multiple rRNA copies, target-based mutations in rRNA conferring tetracycline resistance are usually found in bacteria with low rRNA gene copy numbers. Mutations in 16S rRNA have been reported in *Propionibacterium acnes* (2–3 16S rRNA

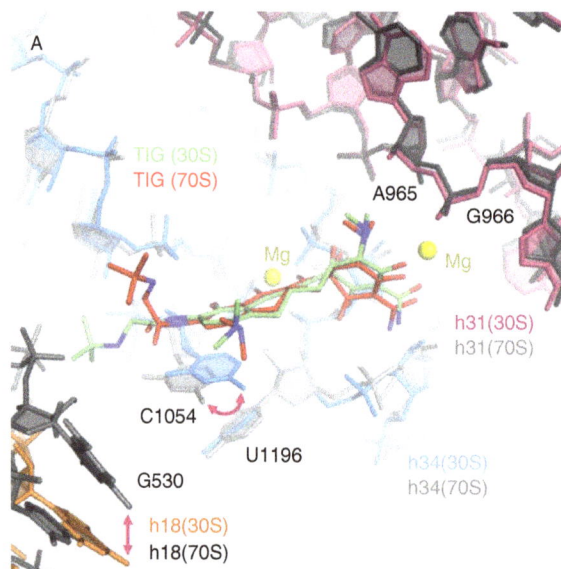

Figure 2. Alternative binding modes of tigecycline at the primary ribosomal-binding site. Alternative tigecycline-binding modes in the 30S (green) and 70S (red) structures are shown, superimposed within the primary tetracycline-binding site. Key nucleotides (G530, A965, G966, C1054, U1196) and helices (h18, h31, h34) are shown in both structures. (From Schedlbauer et al. 2015; reprinted, with permission, from the American Society for Microbiology © 2015.)

copies), *Helicobacter pylori* (1–2 16S rRNA copies), *Mycoplasma bovis* (1–2 16S rRNA copies), and *S. pneumoniae* (4 16S rRNA copies), and the effects of these mutations on tetracycline binding can generally be explained by crystallographic or biophysical data. In *H. pylori*, a triple mutation AGA 965-967 TTC in the h31 loop, and a deletion of G942 (*E. coli* numbering), each conferred tetracycline resistance (Trieber and Taylor 2002). Residues 965–967 are located in the primary, or Tet-1, binding site, whereas G942 is located in the Tet-4 secondary binding site (Brodersen et al. 2000; Pioletti et al. 2001). Mutations in h34 of 16S rRNA were associated with increased tetracycline resistance in *P. acnes* (G1058C) and *M. bovis* (G1058A/C), and tetracycline-resistance mutations A965T, A967T/C, and U1199C (which base pairs with G1058 in h34) were also found in *M. bovis* (Ross et al. 1998; Amram et al. 2015). Although G1058 does not directly interact with tetracycline, mutation to cytosine likely causes a conformational change in the binding site, reducing the affinity of tetracycline for the 30S ribosomal subunit. Preexisting G1058C mutations in *P. acnes* reduced the antibacterial activities of tetracycline, doxycycline, eravacycline, and tigecycline, consistent with all of these tetracycline antibiotics having common interactions with rRNA in bacteria (Grossman et al. 2012). In *S. pneumoniae*, mutations in 16S RNA C1054T and T1062G/A conferred tigecycline resistance when present in the four genomic copies of 16S rRNA (Lupien et al. 2015). Whereas resistance caused by a mutation in C1054 can be explained by the interaction of this residue with tigecycline, a more indirect effect on tigecycline binding may be conferred by mutations in T1062. Nonsense mutations in a gene encoding a 16S rRNA methyltransferase in *S. pneumoniae* were also found to confer reduced tigecycline susceptibility in the study by Lupien et al. (2015). This enzyme methylates position N(2) of G966 in h31 of 16S rRNA in *E. coli* and the alterations in this activity may reduce tigecycline binding to the ribosome.

Unlike rRNA genes, genes encoding ribosomal proteins are single copy and mutations in these genes can confer antibiotic resistance.

Mutations in the *rpsJ*, encoding changes or deletions in residues 53–60 in the 30S ribosomal subunit protein S10, have been linked to tetracycline or tigecycline resistance in in vitro studies with Gram-positive bacteria *Bacillus subtilis*, *Enterococcus faecium*, *E. faecalis*, and *S. aureus* (Williams and Smith 1979; Wei and Bechhofer 2002; Beabout et al. 2015a; Cattoir et al. 2015), in clinical isolates of Gram-negative bacteria *Neisseria gonorrhoeae* and *K. pneumoniae* (Hu et al. 2005; Villa et al. 2014), and in in vitro studies with *E. coli* and *Acinetobacter baumannii* (Beabout et al. 2015a). Identification of a tigecycline-resistant *K. pneumoniae* strain with an *rpsJ* mutation encoding Val57Leu in S10 was the first description of tetracycline resistance attributable, at least in part, to a target site mutation in Enterobacteriaeceae (Villa et al. 2014). In the *T. thermophilus* crystal structure, these S10 residues map to a loop projecting toward the aminoacyl-tRNA-binding site in the 30S structure (Brodersen et al. 2000; Carter et al. 2000). Although located ~8.5 Å from tetracycline in the structure, it has been proposed that this region of the S10 protein may alter the interaction of tetracyclines and 16S rRNA in this region (Hu et al. 2005). Mutations in *rpsC* encoding Lys4Arg and His175Asp variations in ribosomal protein S3 were associated with reduced tigecycline susceptibility in *S. pneumoniae* (Lupien et al. 2015). Ribosomal protein S3 has been shown to be important for tetracycline binding to the ribosome (Buck and Cooperman 1990).

Tetracycline-Specific Ribosomal Protection

Tetracycline ribosomal protection proteins (RPPs), originally described in *Campylobacter jejuni* and *Streptococcus* spp., are GTPases with significant sequence and structural similarity to elongation factors EF-G and EF-Tu (Burdett 1986; Taylor et al. 1987; Sanchez-Pescador et al. 1988; Kobayashi et al. 2007). According to a nomenclature list maintained at the University of Washington (faculty.washington.edu/marilynr), there are currently 12 reported ribosomal protection genes. These genes are disseminated through bacterial populations on mobile genetic elements, and many of the genes are found in both Gram-negative and Gram-positive organisms (Roberts 2011). The most common and best characterized RPPs are Tet(O) and Tet(M), with 75% sequence similarity to each other. These proteins catalyze the GTP-dependent release of tetracycline from the ribosome (Connell et al. 2003a,b). Cryoelectron microscopic structural studies indicate that RPPs compete with EF-G for an overlapping binding site, and it is thought that RPPs dissociate tetracycline from its binding site by directly interfering with the stacking interaction of the tetracycline D-ring and 16S rRNA base C1054 within h34 (Donhofer et al. 2012; Li et al. 2013). Conformational changes induced by RPPs promote rapid binding of the EF-Tu • GTP • aa-tRNA ternary complex, enabling translation to continue in the presence of tetracycline (Donhofer et al. 2012). RPP mechanisms confer resistance to tetracycline, minocycline, and doxycycline; however, other tetracyclines containing side chains at the C-9 position of the D-ring, such as tigecycline and other glycylcyclines, eravacycline and other fluorocyclines, and omadacycline, generally retain translational inhibitory and antibacterial activities in the presence of RPPs (Table 1) (Rasmussen et al. 1994; Bergeron et al. 1996; Grossman et al. 2012; Jenner et al. 2013). The 9-*t*-butylglycylamido moiety at the C-9 position in tigecycline was shown to improve binding affinity and translational inhibition by >100-fold and 20-fold, respectively, over that of tetracycline;

however, the mechanism of RPP evasion could not be fully explained (Olson et al. 2006). Recently, using a set of novel synthetic tetracycline derivatives containing C-9 side chains with different degrees of bulkiness, Jenner et al. (2013) showed that, in addition to conferring enhanced interactions with C1054 (Schedlbauer et al. 2015), steric interference by the bulk of the C-9 side chain is also a significant factor in maintaining ribosome binding in the presence of RPPs. Although earlier reports have shown the relative immunity of tigecycline to RPP mechanisms, a recent study by Beabout et al. (2015b) has linked *Tn*916-associated constituitive overexpression and increased copy number of *tet*(M) to tigecycline resistance in *E. faecalis*.

Tetracycline-Specific Efflux

The most common tetracycline-specific efflux pumps are members of the major facilitator superfamily (MFS) of transporters (Chopra and Roberts 2001); however, there have been rare reports of non-MFS pumps (Teo et al. 2002; Warburton et al. 2013). The latest tally shows that 30 distinct tetracycline-specific efflux pumps reported in bacteria (faculty .washington.edu/marilynr; updated August 6, 2015). These pumps extrude tetracycline antibiotics from the inside of cells at the expense of a proton, and have been assigned to seven different groups according to amino acid sequence

Table 1. The activities of tetracyclines against recombinant *E. coli* expressing major tetracycline-specific resistance mechanisms

	MIC (µg/mL)					
	E. coli lacZ	*E. coli tet*(M)	*E. coli tet*(K)	*E. coli tet*(A)	*E. coli tet*(B)	*E. coli tet*(X)
Eravacycline	0.063	0.063	0.031	0.25	0.063	4
Tigecycline	0.063	0.13	0.063	1	0.063	2
Doxycycline	2	64	4	32	32	16
Minocycline	0.5	64	1	8	16	4
Tetracycline	2	128	128	>128	>128	128
Ceftriaxone	0.063	0.13	0.063	0.13	0.13	0.13

Genes were overexpressed in *E. coli* DH10B from a recombinant expression vector under the control of an arabinose promoter. Standardized MIC assays were performed according to CLSI methodology as previously described.

MIC, minimal inhibitory concentration; *tet*(M), ribosomal protection; *tet*(K), Gram-positive tetracycline efflux; *tet*(A) and *tet*(B), Gram-negative efflux; *tet*(X), flavin-dependent monooxygenase.

Data is reprinted, with permission, from Grossman et al. (2012).

similarities and the number of times they traverse the inner membrane (9–14 times) (Guillaume et al. 2004; Thaker et al. 2010). The most clinically prevalent pumps are members of either group 1 or group 2. The group 1 drug−H⁺ antiporters contain 12 transmembrane segments organized into α and β domains connected by a large interdomain cytoplasmic loop. This group includes Tet(A) and Tet(B), the most commonly found tetracycline pumps in Gram-negative clinical isolates. The group 2 pumps possess 14 transmembrane segments and include Tet(K) and Tet(L), the most common tetracycline-specific efflux pumps in Gram-positive clinical isolates. In addition to their role in conferring tetracycline resistance, group 2 pumps are also monovalent cation−H⁺ antiporters, and may play a role in coping with sodium stress, alkali stress, and potassium insufficiency (Guay et al. 1993; Krulwich et al. 2001). Pumps assigned to group 3–7 include pumps that are less prevalent clinically (Guillaume et al. 2004).

The order of substrate preference across all tetracycline efflux pump types can be shown in recombinant *E. coli* strains overexpressing representative pumps in an isogenic background: tetracycline > minocycline, doxycycline > tigecycline, eravacycline (Table 1) (Grossman et al. 2012). It should be noted, however, that it is likely that multiple strain-specific factors in clinical isolates affecting uptake and intrinsic efflux systems, in addition to the level of expression of tetracycline-specific pumps, play a coordinated role in the overall susceptibility to tetracyclines. Tet(A), Tet(B), and Tet(K) pumps are all able to recognize tetracycline, minocycline, and doxycycline. Whereas Tet(B) and Tet(K) overexpression had no effect on tigecycline and eravacycline, overexpression of Tet(A) produced a fourfold increase in eravacycline minimal inhibitory concentration (MIC) and a 16-fold increase in tigecycline MIC versus the negative control strain, indicating that these newer tetracyclines are recognized to differing extents by the Tet(A) pump (Table 1) (Grossman et al. 2012). Earlier characterizations of the substrate specificity of Tet pumps in nonisogenic strain backgrounds led to the conclusion

that tigecycline was not a substrate for Tet(A) (Petersen et al. 1999), and that a naturally occurring amino acid sequence variation (Ser-Phe-Val→Ala-Ser-Phe) in the interdomain loop sequence at residues 201–203 affected recognition of tigecycline and minocycline (Tuckman et al. 2000). More recent work has shown that recombinant expression of either Tet(A) pump variation in *E. coli* produced similar tigecycline and minocycline susceptibility, confirming that these amino acid residues do not appear to be involved in substrate recognition (Fyfe et al. 2013). The notion that mutations in tetracycline pumps can alter substrate specificity is, however, supported by studies with *tet*(B) in which mutations encoding residues in transmembrane domains had opposing effects on tetracycline versus glycylcycline susceptibility (Guay et al. 1994), and site-directed mutations in the interdomain loop had opposing effects on tetracycline versus minocycline and doxycycline susceptibility (Sapunaric and Levy 2005). These studies suggest the possibility that tetracyclines could select for resistance mutations within tetracycline pump genes during clinical use; however, this has not yet been reported in clinical isolates.

Enzymatic Inactivation of Tetracyclines

Evidence of a tetracycline-modifying enzyme mechanism was first described as an activity encoded by a *Bacteroides* plasmid expressed in *E. coli* (Speer and Salyers 1988, 1989). This activity was subsequently characterized as a flavin-dependent monooxygenase, encoded by an expanding family of *tet*(X) orthologs, capable of covalently inactivating all tetracyclines with the addition of a hydroxyl group to the C-11a position located between the C and B rings of the tetracycline core (Fig. 1) (Speer et al. 1991; Yang et al. 2004; Moore et al. 2005; Grossman et al. 2012; Aminov 2013). Because *Bacteroides* species are obligate aerobes, it is not surprising that the oxidoreductases encoded by *tet*(X) and its orthologs *tetX1* and *tetX2* do not confer resistance in the isolates in which they were originally found (Whittle et al. 2001). The environmental origin of *tet*(X) is suggested by its identification in *Sphinogbacterium* spp., a Gram-negative soil bacterium that expresses a functional Tet(X) (Ghosh et al. 2009). Further, the presence of *tet*(X) and genes encoding similar tetracycline-inactivating activities, also known as "tetracycline destructases," in agricultural and aquacultural bacteria ensures the persistence of this resistance mechanism in the food chain, facilitating crossover into human pathogens (Aminov 2013; Forsberg et al. 2015). Because of the conjugative nature of *tet*(X)-containing plasmids and transposons, recent reports of *tet*(X) in Enterobacteriaceae and Pseudomonadaceae hospital urinary tract infection (UTI) isolates in Sierra Leone, and *tet*(X) in *A. baumannii* in a Chinese hospital, are of concern with regard to the spread of this mechanism (Leski et al. 2013; Deng et al. 2014).

Other less well-characterized tetracycline-modifying mechanisms have also been described. An NADP-requiring tetracycline-modifying activity similar to that of Tet(X) was expressed from the metagenomic DNA of uncultivatable oral microflora; however, there is no homology between the deduced amino acid sequence of Tet(37) from the oral metagenome and Tet(X) from *Bacteroides* (Diaz-Torres et al. 2003). Another gene, *tet*(34), has been cloned from the chromosome of *Vibrio* spp. and encodes a xanthine-guanine phosphoribosyl-transferase capable of conferring resistance to oxytetracycline (Nonaka and Suzuki 2002). The clinical relevance of these two tetracycline-modifying enzymes remains to be determined.

INTRINSIC MULTIDRUG-RESISTANCE MECHANISMS AFFECTING TETRACYCLINES

Complex intrinsic regulatory networks in bacteria modulate the uptake and intracellular accumulation of most antibiotics, including tetracyclines. Mutations affecting expression and/or function of one or more key repressor, activator, pump, or porin can simultaneously impact the susceptibility to a broad range of antibiotic classes (Fig. 2).

AraC Transcriptional Activators in Gram-Negative Bacteria

MarA, RamA, SoxS, RobA, and the newly described RarA are members of the "AraC-family" of bacterial transcriptional activators that enable Gram-negative bacteria to respond to different types of environmental stress, including antibiotic exposure (Fig. 3) (Martin and Rosner 2001; Grkovic et al. 2002; De Majumdar et al. 2013). Each activator regulates a set of genes in response to a specific type of stress (Martin et al. 2008; Martin and Rosner 2011); for instance, MarA regulates more than 60 genes collectively referred to as the "*mar* regulon," for multiple antibiotic resistance (Barbosa and Levy 2000). AraC-family activators bind to a consensus 20 base pair sequence via two helix–turn–helix motifs that comprise the DNA-binding domain. The DNA-binding site is known as the "box" and is located in the promoter region of stress-responsive genes (i.e., "marbox" for MarA, etc.). Further, AraC-family activators can bind their own promoters and autoactivate their own expression (Alekshun and Levy 1997; Rosenblum et al. 2011). Mutations promoting constituitive expression of AraC-family regulons are now known to be common mechanisms contributing to multidrug resistance.

The first description of *mar* in *E. coli* by George and Levy in 1983 showed that amplifiable resistance to tetracyclines, as well as structurally and mechanistically unrelated antibiotics, including chloramphenicol, penicillins, cephalosporins, puromycin, nalidixic acid, and rifampin, was caused by an energy-dependent efflux system (George and Levy 1983a,b). The Mar regulon is now known to be widespread in enteric Gram-negative species, including *E. coli*, *K. pneumoniae*, *Salmonella* spp., *Shigella* spp., *Citrobacter* spp., *Enterobacter* spp., and *Yersinia* spp. (Cohen et al. 1993; Alekshun and Levy 1997). The *mar* locus encodes two divergent operons regulated by a repressor MarR that binds to an operator MarO (Martin and Rosner 1995; Seoane and Levy 1995). Induction of MarR triggers the expression of *marC* in one direction and *marRAB* in the other direction. Whereas tetracycline has been shown to be an

Cite this article as *Cold Spring Harb Perspect Med* doi: 10.1101/cshperspect.a025387

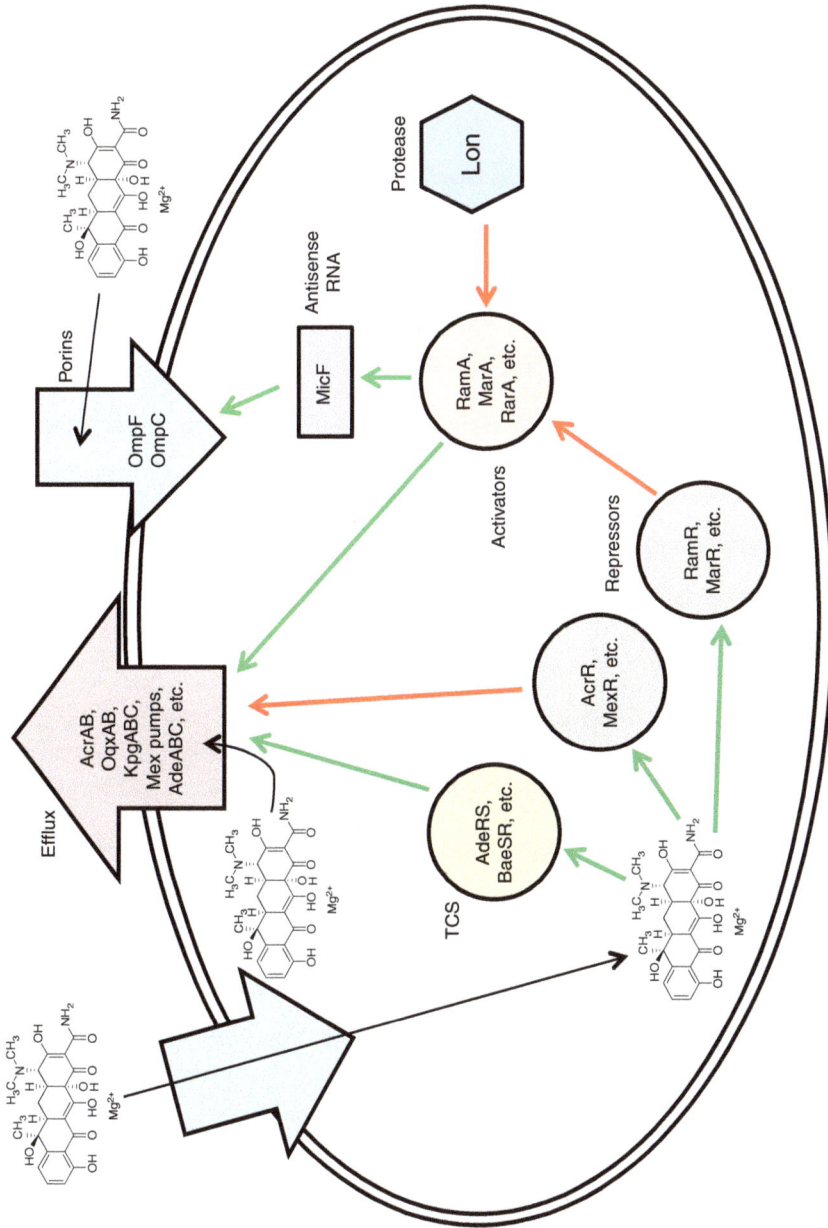

Figure 3. Regulation of expression of Gram-negative intrinsic multidrug-resistance mechanisms affecting tetracyclines. A summary of known regulatory mechanisms affecting tetracycline susceptibility are shown. Green arrows indicate interactions in which tetracycline resistance is "increased," and red arrows indicate interactions in which tetracycline resistance is "reduced." See the text for details. TCS, Two-component signal transduction system.

inducer of *marA* expression (Hachler et al. 1991), induction appears to be via an indirect mechanism because direct binding of tetracycline to MarR could not be shown (Martin and Rosner 1995). MarA is a key activator of stress-responsive genes, and its role in promoting overexpression of the major multidrug efflux pump, AcrAB (Li and Nikaido 2009), is central to conferring an MDR phentoype in enteric bacteria (Gambino et al. 1993; Alekshun and Levy 1997). MarA also controls the expression of the major Gram-negative porin OmpF through the up-regulation of *micF*. MicF is an antisense RNA regulator of *ompF* expression, acting by reducing the levels of *ompF* mRNA (Cohen et al. 1988; Andersen and Delihas 1990; Gambino et al. 1993). Reduction in *ompF* expression contributes to reduced accumulation of tetracycline and other antibiotics (Mortimer and Piddock 1993; Thanassi et al. 1995). The roles of MarC and MarB in multidrug resistance are less well defined; MarC has been shown to encode a periplasmic protein, which appears to indirectly affect the transcription of *marA* (Vinue et al. 2013). Whereas first-step *mar* mutants may not confer clinically relevant resistance to some classes of antibiotics, it is possible that first-step mutants can achieve clinically relevant resistance to tetracycline (George and Levy 1983b); however, this has not yet been shown in clinical isolates. Reduced susceptibility to tigecycline in *E. coli* clinical isolates has been attributed to increased overexpression of AcrAB correlating with mutations in *marR* and increased transcription of *marA* (Keeney et al. 2008). In a study by Linkevicius et al. (2013), targeted sequencing of loci suspected to be involved in tigecycline resistance found a deletion in *marR* in one of eight *E. coli* clinical isolates with reduced tigecycline susceptibility; however, MICs were still well below the resistance breakpoint (MIC = 0.19 μg/mL).

RamA, another AraC-family activator, was first identified in *K. pneumoniae* showing reduced susceptibility to a range of unrelated antibiotics, including tetracycline (George et al. 1995). Expression of the *ramA* gene is repressed by RamR, encoded by the *ramR* gene that is

divergently transcribed from the nearby *ramA* gene. Similar to MarR, tetracycline does not directly bind RamR; thus, induction of *ramA* appears indirect (Yamasaki et al. 2013). Analogous to regulation by *marA* in *E. coli*, overexpression of *ramA* was shown to also reduce porin expression and up-regulate AcrAB efflux in *K. pneumoniae* (George et al. 1995; Ruzin et al. 2005b), *S. enterica* (van der Straaten et al. 2004a,b; Nikaido et al. 2008), and *Enterobacter cloacae* (Keeney et al. 2007). RamA function appears to be independent of MarA, as RamA-mediated increases in AcrAB expression were not associated with increases in MarA expression (Ruzin et al. 2005b). Although, heterologously expressed *ramA* from bacteria, including *K. pneumoniae, Salmonella, Citrobacter,* and *Enterobacter* spp., is functional in *E. coli* (Chollet et al. 2004; van der Straaten et al. 2004b; Ruzin et al. 2005b; Reinhardt 2014), a *ramA* gene has not been identified in several enteric species, notably *E. coli* and *Shigella* spp.

In *K. pneumoniae*, a survey of recent literature suggests an emerging theme that AraC-family activators, especially *ramA*, play a prominent role in clinically relevant resistance to tetracycline antibiotics. A study by Bratu et al. (2009) showed that reduced tigecycline susceptibility in *K. pneumoniae* clinical isolates from New York City correlated with *ramA* and *soxS* expression, but not with *marA* or *acrAB* expression (Bratu et al. 2009). However, in the same study, *K. pneumoniae* mutants selected in vitro for reduced tigecycline susceptibility showed increases in *marA* and *acrB* expression, but not *soxS* and *ramA*, so the interplay of regulators appears complicated. In another study, analysis of 72 demographically and geographically diverse *K. pneumoniae* clinical isolates from the tigecycline phase 3 clinical trials showed that isolates with tigecycline MIC values >2 μg/mL had a statistically significant correlation with elevated expression of *ramA* and a less significant trend with *acrA* expression (Ruzin et al. 2008). The association of tigecycline resistance with mutations in genes encoding MDR repressors (*ramR, acrR*) and/or increased expression of genes encoding AraC-family activators (*rarA, marA, ramA*) and efflux pump subunits (*acrB, oqxB*) has been described in several studies with geographically diverse isolates from Germany (Hentschke et al. 2010); Turkey, Singapore, Chile, and Pakistan (Rosenblum et al. 2011); Italy (Villa et al. 2014); and China (Bratu et al. 2009; Sheng et al. 2014; Zhong et al. 2014; He et al. 2015).

Additional pathways to tigecycline resistance in *K. pneumoniae* are suggested by the identification of tigecycline-resistant *K. pneumoniae* clinical isolates that do not overexpress *ramA* (Rosenblum et al. 2011) and by the isolation of low-level tigecycline-resistant strains from a *K. pneumoniae ramA* deletion mutant (Veleba and Schneiders 2012). Bioinformatic scanning of the *K. pneumoniae* genome for new AraC-family regulators identified *rarA* (Veleba et al. 2012). Expression of *rarA* and the nearby operon *oqxAB* encoding an MDR efflux pump were found to be elevated in geographically diverse *K. pneumoniae* MDR clinical isolates (Veleba et al. 2012) and *E. cloacae* isolates (Veleba et al. 2013) with reduced tigecycline susceptibility. The presence of *rarA* was confirmed in the genomes of *Enterobacter* and *Serratia* spp., and similar to other AraC-family regulators, overexpression of *rarA* produced a low-level MDR phenotype, including tigecycline resistance in *K. pneumoniae* and *E. cloacae* (Veleba et al. 2012, 2013). RarA is thought to be the activator of *oqxAB* (Kim et al. 2009), and has been linked to the regulation of *acrAB* and *ompF* expression (De Majumdar et al. 2013).

Two-Component Systems

Two-component signal transduction systems (TCSs) in bacteria are the most common form of bacterial signal transduction, and are generally composed of a membrane-bound histidine kinase and a response regulator, to which a phoshoryl group is transferred, allowing it to function as a transcription factor affecting the expression of responsive genes (Bem et al. 2015). Several TCSs have been implicated in modulating susceptibility to tetracycline-class antibiotics in Gram-negative and -positive bacteria, presumably by affecting permeability and/or expression of intrinsic multidrug efflux systems.

In *A. baumannii*, the AdeRS TCS controls expression of the major multidrug efflux pump AdeABC (Marchand et al. 2004). Mutations in the AdeR regulator and/or AdeS sensor affecting the normal phosphotransfer process can lead to the constituitive expression of the AdeABC efflux pump (Marchand et al. 2004). Ruzin et al. (2007) were the first to show that elevated tigecycline MICs (4 μg/mL) were associated with constituitive overexpression of AdeABC in two clinical isolates, and coincided with an insertion element in the *adeS* gene in both isolates. In more recent studies, characterizing 81 genetically diverse XDR and 38 carbapenem-resistant MDR *A. baumannii* clinical isolates from Taiwan (Sun et al. 2012, 2014b) reported that tigecycline resistance (MIC ≥8 μg/mL) correlated with overexpression of AdeABC in the majority of isolates, likely resulting from mutations in *adeR* and *adeS* encoding changes in conserved amino acid residues, or an insertion sequence (IS) in *adeS* producing a truncated constituitively "on" form of AdeS. Although a more detailed understanding of how these mutations impact AdeRS signaling and AdeABC expression remains to be elucidated, the recurrence of genetic alterations in *adeR* and *adeS* genes strongly implicates the involvement of AdeRS and the AdeABC efflux system in tigecycline resistance. The existence of multiple mechanisms affecting tigecycline susceptibility in *A. baumannii* is suggested by cases of tigecycline-resistant isolates in which either no mutations in *adeR* and *adeS* were found (Hornsey et al. 2010c; Yoon et al. 2013; Sun et al. 2014b) or additional mutations in *rpsJ*, *rrf*, *msbA*, and *gna* were associated with increasing the level of tigecycline resistance in *adeS* mutants; the possible roles of *rrf*, *msbA*, and *gna* in tigecycline resistance remains to be shown (Hammerstrom et al. 2015). Interestingly, reduced susceptibility to tigecycline, minocycline, and doxycycline was associated with a deletion in the *trm* (tigecycline-related methyltransferase) gene encoding an *S*-adenosyl-L-methionine-dependent methyltransferase in an *A. baumannii* isolate; this newly identified mechanism may be responsible for some resistance not attributable to AdeABC (Chen et al. 2014; Lomovskaya et al. 2015).

Other TCSs that have been associated with resistance to tetracyclines include BaeSR in *A. baumannii* (Lin et al. 2014), PhoBR in *K. pneumoniae* (Srinivasan et al. 2012), and RprXY in *Bacteroides fragilis* (Rasmussen and Kovacs 1993); however, the relevance of these systems in conferring clinical resistance to tetracyclines is not yet understood.

Lon Protease

Induction of multidrug resistance through AraC-family regulators in Gram-negative bacteria is posttranslationally regulated by the cytoplasmic ATP-dependent serine protease, Lon, which is involved in the degradation of unstable or misfolded proteins (Tsilibaris et al. 2006). In the absence of environmental stress, Lon promotes rapid reversion of stress phenotypes by binding at amino-terminal residues of activators MarA, RamA, SoxS and proteolytically degrading them (Griffith et al. 2004; Nicoloff et al. 2006; Ricci et al. 2014). It follows that mutations in *lon* may prolong the stability of these stress-responsive activators, increasing expression of *acrAB* and other resistance genes, leading to antibiotic resistance or reduced susceptibility.

The involvement of Lon protease in the development of antibiotic resistance was shown in a series of 13 *E. coli* cultures derived from a single inoculum in which a significant subpopulation ($\sim 3.7 \times 10^{-4}$) contained a *lon*::IS*186* mutation, or deletion in *lon*, and was capable of growing in low-level tetracycline and chloramphenicol (Nicoloff et al. 2007). Most mutants characterized in this study also contained IS elements in *marR* or *acrR*, or tandem amplifications of the *acrAB* region, and the antibiotic-resistance phenotype, at least in part, could be attributed to these mutations. Because Lon protease is also involved in the stability of transposase enzymes from IS elements and transposons, *E. coli lon* mutant strains show higher transposition rates and greater genome instability (Derbyshire et al. 1990; Nagy and Chandler 2004; Rouquette et al. 2004). Further, the *lon* gene itself is a hotspot for IS insertions (SaiSree et al. 2001). Thus, the potential to select for early steps in drug resistance in vitro appears to be much higher in *lon* mutants, and this is supported by the finding that genomic duplications of the region encoding the major efflux pump, AcrAB, can be readily isolated in an *E. coli lon* mutant (Nicoloff et al. 2006; Nicoloff and Andersson 2013).

Whether *lon* mutations increase the potential to select for resistance to tetracyclines, or any other antibiotic class, in clinical isolates during infection is not entirely clear (Butler et al. 2006). There has been at least one report of an *E. coli acrR* (A191V), *lon*::IS*186* mutant isolated from a UTI, and this mutant showed reduced susceptibility to tigecycline (MIC = 0.25 µg/mL), but still did not reach a level considered clinically significant (Linkevicius et al. 2013). A tigecycline-resistant *K. pneumoniae* clinical isolate containing a frameshift within the coding region of *lon* was reported by Fyfe et al. (2015); however, this mutant also had a deletion in *ramR*, which presumably also contributed to the tigecycline-resistant phenotype (MIC = 8 µg/mL). In this same study, *K. pneumoniae lon* mutants generated by transposon mutagenesis showed 8- to 32-fold increases in the parental tigecycline MIC (0.5 µg/mL), suggesting that mutation in *lon* can contribute to clinically significant resistance levels. Additional studies are needed to clarify the contribution of *lon* to the development of resistance to tetracyclines and other antibiotics in clinical isolates.

Intrinsic Efflux of Tetracyclines

A large component of the intrinsic antibiotic-resistance response in bacteria is due to increased expression of intrinsic efflux pumps (Piddock 2006; Li et al. 2015). As described earlier, expression of these genes can be modulated by locally or distally encoded negative and positive regulators, and mutations up-regulating or down-regulating expression of the regulators themselves, or the efflux pumps they regulate, can impact antibiotic susceptibility. Susceptibility to tetracycline-class antibiotics has been

Cite this article as *Cold Spring Harb Perspect Med* doi: 10.1101/cshperspect.a025387

linked to a wide variety of intrinsic efflux systems in Gram-negative and -positive bacteria summarized in Table 2.

Overexpression of AcrAB, the major pump found in Enterobacteriaceae, and a member of the resistance-nodulation-division (RND) superfamily, has been implicated in resistance to tigecycline, in *E. coli* (Hirata et al. 2004), *Enterobacter* spp. (Keeney et al. 2007), *K. pneumoniae* (Ruzin et al. 2005b), *Morganella morganii* (Ruzin et al. 2005a), and *Proteus mirabilis* (Visalli et al. 2003). Two recently identified pumps, OqxAB and KpgABC, in *K. pneumoniae* also appear to have some association with tigecycline resistance, but their clinical significance is uncertain (Nielsen et al. 2014; Bialek-Davenet et al. 2015; He et al. 2015). In *Serratia marcescens*, pumps with specificity for tetracycline and/or tigecycline include SdeXY-HasF and SmdAB (Chen et al. 2003; Matsuo et al. 2008; Hornsey et al. 2010a).

RND-type pumps are also implicated in conferring reduced susceptibility in nonfermenter and anaerobic Gram-negative bacteria. In *A. baumannii* clinical isolates, as discussed earlier, reduced susceptibility to tigecycline and eravacycline has been correlated with AdeABC pump expression (Ruzin et al. 2010; Abdallah et al. 2015). Interestingly, the AdeABC pump appears to show some selectivity among the tetracyclines, as minocycline is reported to be a weaker substrate than other tetracyclines (Coyne et al. 2011; Lomovskaya et al. 2014). *Pseudomonas aeruginosa* strains are generally less susceptible to tetracycline antibiotics, including tigecycline, and this is largely because of expression of the MexAB-OprM, MexCD-OprJ, MexXY-OprM pumps (Dean et al. 2003). In *Stenotrophomonas maltophilia*, SmrA and SmeDEF (Alonso and Martinez 2001; Zhang et al. 2001; Al-Hamad et al. 2009), and in *B. fragilis*, BmeABC (Pumbwe et al. 2006), have been reported to recognize tetracycline, but their clinical significance remains to be shown.

Much less is known about the regulation of intrinsic resistance to tetracyclines in Gram-positive bacteria. The best-characterized intrinsic Gram-positive pump with demonstrated specificity for tetracyclines is the multidrug and toxic compound extrusion (MATE)-family pump, MepA, in *S. aureus*. Although this pump does not appear to recognize tetracycline as a substrate, fourfold and 64-fold increases in MIC for eravacycline and tigecycline, respectively, were observed for a MepA overexpressing strain versus the isogenic parent, indicating a distinct tetracycline substrate specificity for this pump (McAleese et al. 2005; Sutcliffe et al. 2013). The NorB pump, negatively regulated by MgrA in *S. aureus*, has also been reported to recognize tetracycline (Truong-Bolduc et al. 2005).

THE PRESENT AND FUTURE FOR TETRACYCLINES

Minocycline

Historically, minocycline has been available in both oral and intravenous dosage formulations. As options for the treatment of MDR *A. baumannii* are limited, the recent approval of a new intravenous (IV) formulation, Minocin IV, for treatment of *Acinetobacter* spp. and other difficult-to-treat Gram-positive and -negative pathogens, is a valuable repurposing of an old antibiotic for targeted use (The Medicines Company 2015). In the 2004–2013 Tigecycline Evaluation and Surveillance Trial (TEST) report, the highest level of in vitro susceptibility against *A. baumannii* isolates was reported for minocycline (84.5%), and 70.3% susceptibility was observed against MDR *A. baumannii* (Hoban et al. 2015). In the global 2007–2011 SENTRY surveillance program, minocycline was the second most active antibiotic against *A. baumannii* (79.1% susceptible) (Castanheira et al. 2014). This might be explained to some extent by the ability of minocycline to thwart AdeABC efflux, and a lower rate of minocycline-resistance development in *A. baumannii* (Lomovskaya et al. 2014). Clinical responses to Minocin IV used as a monotherapy or in combination for the treatment of MDR *A. baumannii* infections appear encouraging (Goff et al. 2014; Ritchie and Garavaglia-Wilson 2014; Falagas et al. 2015), but this therapy will likely be a stop-gap as the spread of RPPs should increase minocycline resistance.

Table 2. Intrinsic bacterial multidrug efflux mechanisms conferring resistance to tetracycline drugs

Pathogen	Efflux pump family	Known tetracycline specificity	References
A. baumannii	RND	AdeABC: tetracycline, tigecycline, minocycline,[a] doxycycline[a] AdeDE: tetracycline AdeFGH: tetracycline, minocycline, tigecycline AdeIJK: tetracycline, minocycline, doxycycline, tigecycline	Chau et al. 2004; Ruzin et al. 2007; Damier-Piolle et al. 2008; Coyne et al. 2010, 2011; Ruzin et al. 2010; Lomovskaya et al. 2014, 2015
B. fragilis	RND	BmeABC: tetracycline	Pumbwe et al. 2006
E. coli	RND	AcrAB: tetracycline, tigecycline, minocycline, doxycycline AcrEF: tetracycline, tigecycline, minocycline, doxycycline	Hirata et al. 2004
Enterobacter spp.	RND	AcrAB: tetracycline, tigecycline, minocycline OqxAB: tigecycline	Keeney et al. 2007; Hornsey et al. 2010b; Veleba et al. 2013
E. faecalis	ABC	EfrAB: doxycycline (not tetracycline)	Lee et al. 2003
K. pneumoniae	RND	AcrAB: tetracycline, tigecycline, minocycline OqxAB: tigecycline, tetracycline KpgABC: tigecycline	Ruzin et al. 2005b; Ruzin et al. 2008; Veleba and Schneiders 2012; Nielsen et al. 2014; Zhong et al. 2014; He et al. 2015
P. aeruginosa	RND	MexAB-OprM: tetracycline, minocycline, doxycycline, chlortetracycline, oxytetracycline, tigecycline MexCD-OprJ: tetracycline, chlortetracycline, oxytetracycline, tigecycline MexJK: tetracycline MexXY-OprM: tetracycline, minocycline, doxycycline, chlortetracycline, oxytetracycline, tigecycline	Masuda et al. 2000; Morita et al. 2001; Chuanchuen et al. 2002; Dean et al. 2003; Schweizer 2003
P. mirabilis	RND	AcrAB: tigecycline, minocycline	Visalli et al. 2003
S. aureus	MATE	MepA: tigecycline, eravacycline (not tetracycline) NorB: tetracycline (not minocycline)	McAleese et al. 2005; Truong-Bolduc et al. 2005; Sutcliffe et al. 2013
Stenotrophomonas maltophilia	ABC RND	SmrA: tetracycline SmeDEF: tetracycline, doxycycline, minocycline, tigecycline	Alonso and Martinez 2001; Zhang et al. 2001; Al-Hamad et al. 2009
Serratia marcescens	RND ABC	SdeXY-HasF: tetracycline, tigecycline SmdAB: tetracycline	Chen et al. 2003; Matsuo et al. 2008; Hornsey et al. 2010a

[a]Poorer substrates for AdeABC as compared with other tetracycline drugs.

RND, Resistance-nodulation-division superfamily; MATE, multidrug and toxic compound extrusion family; ABC, ATP-binding cassette transporter family.

Tigecycline

Tigecycline has a broad spectrum of coverage, including activity against MRSA, vancomycin-resistant *Enterococcus* spp. (VRE), MDR *A. baumannii*, and ESBL-producing and carbapenem-resistant Enterobacteriaceae (CRE), supporting the currently approved indications of complicated skin and skin structure infections, complicated intra-abdominal infections (cIAI), and community-acquired bacterial pneumonia (CABP) (Stein and Babinchak 2013; Wyeth Pharmaceuticals 2016). The administration of tigecycline is limited to IV only. Given its activity against MDR pathogens, tigecycline was evaluated for use in hospital-acquired pneumonia, ventilator-associated pneumonia, and diabetic foot infections; however, clinical studies showed lower cure rates versus comparator drugs (Wyeth Pharmaceuticals 2016). Broader usage and alternative dosing regimens for serious infections continue to be explored in clinical studies (Ramirez et al. 2013; Stein and Babinchak 2013). Based on meta-analyses of clinical trial data, the FDA issued a safety alert in 2010 and a black box warning in 2013 because of an observed increase in mortality risk in patients treated with tigecycline, as compared with other drugs (U.S. Food and Drug Administration 2010, 2013). Whereas the cause of death during tigecycline treatment remains uncertain, mortality appeared to occur in patients with complicated worsening infections or underlying medical conditions.

Since its approval in 2005, tigecycline maintains high levels of susceptibility in global surveillance studies despite sporadic reports of resistance during use: *E. coli* and *K. pneumoniae* in the United Kingdom (Stone et al. 2011; Spanu et al. 2012); *K. pneumoniae* in Greece (Neonakis et al. 2011), Saudi Arabia (Al-Qadheeb et al. 2010), Spain (Rodriguez-Avial et al. 2012), the United States (Nigo et al. 2013); *E. hormaechei* in France (Daurel et al. 2009); *A. baumannii* in the United States (Peleg et al. 2007; Reid et al. 2007; Anthony et al. 2008); *E. faecalis* in the United Kingdom (Cordina et al. 2012) and Germany (Werner et al. 2008); and *B. fragilis* in the United States (Sherwood et al. 2011). There are

also reports in which tigecycline resistance actually predated the use of tigecycline in institutions in which resistance was detected (Rosenblum et al. 2011; Zhong et al. 2014) or arose during treatment with another antibiotic (Hornsey et al. 2010b).

In the 2004–2013 TEST report, among the Enterobacteriaceae ($n = 118,899$), enterococci ($n = 20,782$), methicillin-resistant *S. aureus* ($n = 14,647$), and *S. pneumoniae* ($n = 14,562$), susceptibility to tigecycline was 97%, >99%, $\geq 99.9\%$, and $\geq 99.9\%$, respectively (Hoban et al. 2015). Against MDR Enterobacteriaceae ($n = 9372$), defined as resistant to more than three different classes of antibiotics, susceptibility to tigecycline was 83.2%. And, among the carbapenem-resistant *Enterobacter* spp. ($n = 578$), *E. coli* ($n = 181$), and *K. pneumoniae* ($n = 1330$), susceptibility to tigecycline was 83%, 97.2%, and 92%, respectively. Although there are no breakpoints available against *A. baumannii*, tigecycline maintained an MIC_{90} (MIC inhibiting 90% of the isolates) of 2 μg/mL against all *A. baumannii* ($n = 16,778$), as well as an MDR subset ($n = 6743$). The gap in coverage of *P. aeruginosa* is evident by an MIC_{90} of 16 μg/mL against all collected isolates ($n = 28,413$) and ≥ 32 μg/mL against the MDR subset ($n = 3496$). Similar findings were observed in recent reports from the SENTRY (Sader et al. 2013, 2014), Regional Resistance Surveillance (Jones et al. 2014), and CANWARD (Zhanel et al. 2013) programs.

Omadacycline

The aminomethylcycline derivative of minocycline, omadacycline, has completed a phase 2 trial for safety and efficacy in skin and skin structure infections (SSSI) and is being developed for use in SSSI, CABP, and UTIs with IV and oral formulations (Noel et al. 2012). Omadacycline was shown to have MIC_{90} values against MRSA ($n = 39$), VRE ($n = 19$), *Streptococcus pyogenes* ($n = 30$), penicillin-resistant *S. pneumoniae* ($n = 23$), and *Haemophilus influenzae* ($n = 53$) of 0.5, 0.5, 0.25, ≤ 0.06, and 2 μg/mL, respectively (Macone et al. 2014). Against *E. coli* ($n = 23$) and *K. pneumo-*

niae ($n = 14$), MIC_{90} values were 2 μg/mL and 4 μg/mL, respectively. The 9-alkylamino-methyl modification of minocycline endows omadacycline with activity against ribosomal protection mechanisms (Draper et al. 2014).

Eravacycline

Eravacycline, a broad spectrum, fully synthetic fluorocycline with novel C-9 pyrrolidino-acetamido and C-7 fluoro modifications, completed a phase 2 trial for cIAI and has completed pivotal phase 3 trials for cIAI and complicated UTI (Solomkin et al. 2014), with future indications expected to include other serious infections. Both IV and oral formulations are in development. In evaluations against large panels of aerobic and anaerobic Gram-negative and -positive bacteria, eravacycline showed MIC_{90} values ranging from ≤ 0.008 to 2 μg/mL for all species, except *P. aeruginosa* and *Burkholderia cenocepacia* (MIC_{90} values of 16–32 μg/mL) (Sutcliffe et al. 2013; McDermott et al. 2015; Morrissey et al. 2015a,b,c). In the study by Sutcliffe et al. (2013), eravacycline showed activity against tetracycline-resistant *E. coli* ($MIC_{50/90} = 0.25/0.5$ μg/mL; $n = 157$), *E. cloacae* ($MIC_{50/90} = 2/4$ μg/mL; $n = 25$), and *P. mirabilis* ($MIC_{50/90} = 1/2$ μg/mL; $n = 109$). In a recent study with more than 4000 contemporary Gram-negative pathogens from New York City hospitals, eravacycline $MIC_{50/90}$ values were 0.12/0.5 μg/mL for *E. coli*, 0.25/ 1 μg/mL for *K. pneumoniae*, 0.25/1 μg/mL for *Enterobacter aerogenes*, 0.5/1 μg/mL for *E. cloacae*, and 0.5/1 μg/mL for *A. baumannii* (Abdallah et al. 2015). Eravacycline also shows good activity against MDR bacteria, including Enterobacteriaceae and *A. baumannii* expressing extended spectrum β-lactamases, carbapenem resistance, and mechanisms conferring resistance to other antibiotic classes (Sutcliffe et al. 2013; Grossman et al. 2014b; Abdallah et al. 2015).

CONCLUSION

Tetracycline-class antibiotics have treated serious life-threatening infections for more than 60 years; however, as with every other antibiotic class, use has led to resistance development. Historically, potency, spectrum, and tetracycline-resistance hurdles have been addressed semisynthetically with chemical modifications of earlier natural product derivatives. The most successful examples of this approach include minocycline, doxycycline, tigecycline, and omadacycline. More recently, a fully synthetic chemistry approach has led to the discovery of eravacycline, which shows promise in the treatment of serious infections caused by a broad range of bacterial pathogens. Ongoing exploration of synthetic tetracycline derivatives has enabled improvements in potency against *P. aeruginosa* and other difficult-to-treat MDR Gram-negative pathogens (Deng et al. 2012; O'Brien et al. 2012; Xiao et al. 2013; Grossman et al. 2014a; Sun et al. 2014a, 2015). The ability to synthesize completely novel, "unnatural" tetracyclines opens new opportunities to more fully explore the potential of this familiar and clinically validated antibiotic class.

ACKNOWLEDGMENTS

I thank Joyce Sutcliffe, Patricia Bradford, Kathy Kerstein, and Corey Fyfe for reading this manuscript, and Charlie Xiao for helping to prepare Figure 1.

REFERENCES

Abdallah M, Olafisoye O, Cortes C, Urban C, Landman D, Quale J. 2015. Activity of eravacycline against Enterobacteriaceae and *Acinetobacter baumannii*, including multi-drug-resistant isolates, from New York City. *Antimicrob Agents Chemother* **59:** 1802–1805.

Alekshun MN, Levy SB. 1997. Regulation of chromosomally mediated multiple antibiotic resistance: The mar regulon. *Antimicrob Agents Chemother* **41:** 2067–2075.

Al-Hamad A, Upton M, Burnie J. 2009. Molecular cloning and characterization of SmrA, a novel ABC multidrug efflux pump from *Stenotrophomonas maltophilia*. *J Antimicrob Chemother* **64:** 731–734.

Alonso A, Martinez JL. 2001. Expression of multidrug efflux pump SmeDEF by clinical isolates of *Stenotrophomonas maltophilia*. *Antimicrob Agents Chemother* **45:** 1879–1881.

Al-Qadheeb NS, Althawadi S, Alkhalaf A, Hosaini S, Alrajhi AA. 2010. Evolution of tigecycline resistance in *Klebsiella pneumoniae* in a single patient. *Ann Saudi Med* **30:** 404–407.

Cite this article as *Cold Spring Harb Perspect Med* doi: 10.1101/cshperspect.a025387

Aminov RI. 2013. Evolution in action: Dissemination of tet(X) into pathogenic microbiota. *Front Microbiol* **4:** 192.

Amram E, Mikula I, Schnee C, Ayling RD, Nicholas RA, Rosales RS, Harrus S, Lysnyansky I. 2015. 16S rRNA gene mutations associated with decreased susceptibility to tetracycline in *Mycoplasma bovis*. *Antimicrob Agents Chemother* **59:** 796–802.

Andersen J, Delihas N. 1990. micF RNA binds to the 5′ end of ompF mRNA and to a protein from *Escherichia coli*. *Biochemistry* **29:** 9249–9256.

Anthony KB, Fishman NO, Linkin DR, Gasink LB, Edelstein PH, Lautenbach E. 2008. Clinical and microbiological outcomes of serious infections with multidrug-resistant Gram-negative organisms treated with tigecycline. *Clin Infect Dis* **46:** 567–570.

Bantar C, Schell C, Posse G, Limansky A, Ballerini V, Mobilia L. 2008. Comparative time-kill study of doxycycline, tigecycline, sulbactam, and imipenem against several clones of *Acinetobacter baumannii*. *Diagn Microbiol Infect Dis* **61:** 309–314.

Barbosa TM, Levy SB. 2000. Differential expression of over 60 chromosomal genes in *Escherichia coli* by constitutive expression of MarA. *J Bacteriol* **182:** 3467–3474.

Bauer G, Berens C, Projan SJ, Hillen W. 2004. Comparison of tetracycline and tigecycline binding to ribosomes mapped by dimethylsulphate and drug-directed Fe^{2+} cleavage of 16S rRNA. *J Antimicrob Chemother* **53:** 592–599.

Beabout K, Hammerstrom TG, Perez AM, Magalhaes BF, Prater AG, Clements TP, Arias CA, Saxer G, Shamoo Y. 2015a. The ribosomal S10 protein is a general target for decreased tigecycline susceptibility. *Antimicrob Agents Chemother* **59:** 5561–5566.

Beabout K, Hammerstrom TG, Wang TT, Bhatty M, Christie PJ, Saxer G, Shamoo Y. 2015b. Rampant parasexuality evolves in a hospital pathogen during antibiotic selection. *Mol Biol Evol* **32:** 2585–2597.

Bem AE, Velikova N, Pellicer MT, Baarlen P, Marina A, Wells JM. 2015. Bacterial histidine kinases as novel antibacterial drug targets. *ACS Chem Biol* **10:** 213–224.

Bergeron J, Ammirati M, Danley D, James L, Norcia M, Retsema J, Strick CA, Su WG, Sutcliffe J, Wondrack L. 1996. Glycylcyclines bind to the high-affinity tetracycline ribosomal binding site and evade Tet(M)- and Tet(O)-mediated ribosomal protection. *Antimicrob Agents Chemother* **40:** 2226–2228.

Bialek-Davenet S, Lavigne JP, Guyot K, Mayer N, Tournebize R, Brisse S, Leflon-Guibout V, Nicolas-Chanoine MH. 2015. Differential contribution of AcrAB and OqxAB efflux pumps to multidrug resistance and virulence in *Klebsiella pneumoniae*. *J Antimicrob Chemother* **70:** 81–88.

Bratu S, Landman D, George A, Salvani J, Quale J. 2009. Correlation of the expression of acrB and the regulatory genes marA, soxS and ramA with antimicrobial resistance in clinical isolates of *Klebsiella pneumoniae* endemic to New York City. *J Antimicrob Chemother* **64:** 278–283.

Brodersen DE, Clemons WM Jr, Carter AP, Morgan-Warren RJ, Wimberly BT, Ramakrishnan V. 2000. The structural basis for the action of the antibiotics tetracycline, pacta-

mycin, and hygromycin B on the 30S ribosomal subunit. *Cell* **103:** 1143–1154.

Buck MA, Cooperman BS. 1990. Single protein omission reconstitution studies of tetracycline binding to the 30S subunit of *Escherichia coli* ribosomes. *Biochemistry* **29:** 5374–5379.

Burdett V. 1986. Streptococcal tetracycline resistance mediated at the level of protein synthesis. *J Bacteriol* **165:** 564–569.

Butler SM, Festa RA, Pearce MJ, Darwin KH. 2006. Self-compartmentalized bacterial proteases and pathogenesis. *Mol Microbiol* **60:** 553–562.

Carter AP, Clemons WM, Brodersen DE, Morgan-Warren RJ, Wimberly BT, Ramakrishnan V. 2000. Functional insights from the structure of the 30S ribosomal subunit and its interactions with antibiotics. *Nature* **407:** 340–348.

Castanheira M, Mendes RE, Jones RN. 2014. Update on *Acinetobacter* species: Mechanisms of antimicrobial resistance and contemporary in vitro activity of minocycline and other treatment options. *Clin Infect Dis* **59:** S367–S373.

Cattoir V, Isnard C, Cosquer T, Odhiambo A, Bucquet F, Guerin F, Giard JC. 2015. Genomic analysis of reduced susceptibility to tigecycline in *Enterococcus faecium*. *Antimicrob Agents Chemother* **59:** 239–244.

Chau SL, Chu YW, Houang ET. 2004. Novel resistance-nodulation-cell division efflux system AdeDE in *Acinetobacter* genomic DNA group 3. *Antimicrob Agents Chemother* **48:** 4054–4055.

Chen J, Kuroda T, Huda MN, Mizushima T, Tsuchiya T. 2003. An RND-type multidrug efflux pump SdeXY from *Serratia marcescens*. *J Antimicrob Chemother* **52:** 176–179.

Chen Q, Li X, Zhou H, Jiang Y, Chen Y, Hua X, Yu Y. 2014. Decreased susceptibility to tigecycline in *Acinetobacter baumannii* mediated by a mutation in trm encoding SAM-dependent methyltransferase. *J Antimicrob Chemother* **69:** 72–76.

Chollet R, Chevalier J, Bollet C, Pages JM, Davin-Regli A. 2004. RamA is an alternate activator of the multidrug resistance cascade in *Enterobacter aerogenes*. *Antimicrob Agents Chemother* **48:** 2518–2523.

Chopra I, Roberts M. 2001. Tetracycline antibiotics: Mode of action, applications, molecular biology, and epidemiology of bacterial resistance. *Microbiol Mol Biol Rev* **65:** 232–260.

Chuanchuen R, Narasaki CT, Schweizer HP. 2002. The MexJK efflux pump of *Pseudomonas aeruginosa* requires OprM for antibiotic efflux but not for efflux of triclosan. *J Bacteriol* **184:** 5036–5044.

Clark RB, Hunt DK, He M, Achorn C, Chen CL, Deng Y, Fyfe C, Grossman TH, Hogan PC, O'Brien WJ, et al. 2012. Fluorocyclines. 2: Optimization of the C-9 side-chain for antibacterial activity and oral efficacy. *J Med Chem* **55:** 606–622.

Cohen SP, McMurry LM, Levy SB. 1988. marA locus causes decreased expression of OmpF porin in multiple-antibiotic-resistant (Mar) mutants of *Escherichia coli*. *J Bacteriol* **170:** 5416–5422.

Cohen SP, Yan W, Levy SB. 1993. A multidrug resistance regulatory chromosomal locus is widespread among enteric bacteria. *J Infect Dis* **168:** 484–488.

Connell SR, Tracz DM, Nierhaus KH, Taylor DE. 2003a. Ribosomal protection proteins and their mechanism of tetracycline resistance. *Antimicrob Agents Chemother* **47:** 3675–3681.

Connell SR, Trieber CA, Dinos GP, Einfeldt E, Taylor DE, Nierhaus KH. 2003b. Mechanism of Tet(O)-mediated tetracycline resistance. *EMBO J* **22:** 945–953.

Cordina C, Hill R, Deshpande A, Hood J, Inkster T. 2012. Tigecycline-resistant *Enterococcus faecalis* associated with omeprazole use in a surgical patient. *J Antimicrob Chemother* **67:** 1806–1807.

Coyne S, Rosenfeld N, Lambert T, Courvalin P, Perichon B. 2010. Overexpression of resistance-nodulation-cell division pump AdeFGH confers multidrug resistance in *Acinetobacter baumannii. Antimicrob Agents Chemother* **54:** 4389–4393.

Coyne S, Courvalin P, Perichon B. 2011. Efflux-mediated antibiotic resistance in *Acinetobacter* spp. *Antimicrob Agents Chemother* **55:** 947–953.

Cunha BA, Sibley CM, Ristuccia AM. 1982. Doxycycline. *Ther Drug Monit* **4:** 115–135.

Damier-Piolle L, Magnet S, Bremont S, Lambert T, Courvalin P. 2008. AdeIJK, a resistance-nodulation-cell division pump effluxing multiple antibiotics in *Acinetobacter baumannii. Antimicrob Agents Chemother* **52:** 557–562.

Daurel C, Fiant AL, Bremont S, Courvalin P, Leclercq R. 2009. Emergence of an *Enterobacter hormaechei* strain with reduced susceptibility to tigecycline under tigecycline therapy. *Antimicrob Agents Chemother* **53:** 4953–4954.

Dean CR, Visalli MA, Projan SJ, Sum PE, Bradford PA. 2003. Efflux-mediated resistance to tigecycline (GAR-936) in *Pseudomonas aeruginosa* PAO1. *Antimicrob Agents Chemother* **47:** 972–978.

De Majumdar S, Veleba M, Finn S, Fanning S, Schneiders T. 2013. Elucidating the regulon of multidrug resistance regulator RarA in *Klebsiella pneumoniae. Antimicrob Agents Chemother* **57:** 1603–1609.

Deng Y, Grossman T, Clark RB, Xiao XY, Sutcliffe J. 2012. The intravenous pharmacokinetics (PK) and efficacy of TP-433 in murine infection models with *Pseudomonas aeruginosa.* In *Abstr 25th ECCMID*, Abstract AbsP1426. London.

Deng M, Zhu MH, Li JJ, Bi S, Sheng ZK, Hu FS, Zhang JJ, Chen W, Xue XW, Sheng JF, et al. 2014. Molecular epidemiology and mechanisms of tigecycline resistance in clinical isolates of *Acinetobacter baumannii* from a Chinese university hospital. *Antimicrob Agents Chemother* **58:** 297–303.

Derbyshire KM, Kramer M, Grindley ND. 1990. Role of instability in the cis action of the insertion sequence IS903 transposase. *Proc Natl Acad Sci* **87:** 4048–4052.

Diaz-Torres ML, McNab R, Spratt DA, Villedieu A, Hunt N, Wilson M, Mullany P. 2003. Novel tetracycline resistance determinant from the oral metagenome. *Antimicrob Agents Chemother* **47:** 1430–1432.

Donhofer A, Franckenberg S, Wickles S, Berninghausen O, Beckmann R, Wilson DN. 2012. Structural basis for TetM-mediated tetracycline resistance. *Proc Natl Acad Sci* **109:** 16900–16905.

Draper MP, Weir S, Macone A, Donatelli J, Trieber CA, Tanaka SK, Levy SB. 2014. Mechanism of action of the novel aminomethylcycline antibiotic omadacycline. *Antimicrob Agents Chemother* **58:** 1279–1283.

Duggar BM. 1948. Aureomycin; a product of the continuing search for new antibiotics. *Ann NY Acad Sci* **51:** 177–181.

Falagas ME, Vardakas KZ, Kapaskelis A, Triarides NA, Roussos NS. 2015. Tetracyclines for multidrug-resistant *Acinetobacter baumannii* infections. *Int J Antimicrob Agents* **45:** 455–460.

Finlay AC, Hobby GL, P'an SY, Regna PP, Routien JB, Seeley DB, Shull GM, Sobin BA, Solomons IA, Vinson JW, et al. 1950. Terramycin, a new antibiotic. *Science* **111:** 85.

Forsberg KJ, Patel S, Wencewicz TA, Dantas G. 2015. The tetracycline destructases: A novel family of tetracycline-inactivating enzymes. *Chem Biol* **22:** 888–897.

Fyfe C, Sutcliffe JA, Grossman TH. 2013. Susceptibility of tetracyclines to Tet(A) resistance is independent of interdomain loop sequence. *Antimicrob Agents Chemother* **57:** 2430–2431.

Fyfe C, Norris D, Sutcliffe JA, Grossman TH. 2015. Identification of *Klebsiella pneumoniae* genes involved in tigecycline-resistance using transposon mutagenesis. In *Abstr 25th ECCMID*, Abstract P1022. Copenhagen, Denmark.

Gambino L, Gracheck SJ, Miller PF. 1993. Overexpression of the MarA positive regulator is sufficient to confer multiple antibiotic resistance in *Escherichia coli. J Bacteriol* **175:** 2888–2894.

George AM, Levy SB. 1983a. Amplifiable resistance to tetracycline, chloramphenicol, and other antibiotics in *Escherichia coli*: Involvement of a non-plasmid-determined efflux of tetracycline. *J Bacteriol* **155:** 531–540.

George AM, Levy SB. 1983b. Gene in the major cotransduction gap of the *Escherichia coli* K-12 linkage map required for the expression of chromosomal resistance to tetracycline and other antibiotics. *J Bacteriol* **155:** 541–548.

George AM, Hall RM, Stokes HW. 1995. Multidrug resistance in *Klebsiella pneumoniae*: A novel gene, ramA, confers a multidrug resistance phenotype in *Escherichia coli. Microbiology* **141:** 1909–1920.

Ghosh S, Sadowsky MJ, Roberts MC, Gralnick JA, LaPara TM. 2009. *Sphingobacterium* sp. strain PM2-P1-29 harbours a functional tet(X) gene encoding for the degradation of tetracycline. *J Appl Microbiol* **106:** 1336–1342.

Goff DA, Bauer KA, Mangino JE. 2014. Bad bugs need old drugs: A stewardship program's evaluation of minocycline for multidrug-resistant *Acinetobacter baumannii* infections. *Clin Infect Dis* **1:** S381–S387.

Griffith KL, Shah IM, Wolf RE Jr, 2004. Proteolytic degradation of *Escherichia coli* transcription activators SoxS and MarA as the mechanism for reversing the induction of the superoxide (SoxRS) and multiple antibiotic resistance (Mar) regulons. *Mol Microbiol* **51:** 1801–1816.

Grkovic S, Brown MH, Skurray RA. 2002. Regulation of bacterial drug export systems. *Microbiol Mol Biol Rev* **66:** 671–701.

Grossman TH, Starosta AL, Fyfe C, O'Brien W, Rothstein DM, Mikolajka A, Wilson DN, Sutcliffe JA. 2012. Target-

and resistance-based mechanistic studies with TP-434, a novel fluorocycline antibiotic. *Antimicrob Agents Chemother* **56:** 2559–2564.

Grossman T, Fyfe C, Kerstein K, Sun C, Clark R, Xiao XY, Sutcliffe J. 2014a. In vivo efficacy of novel, fully synthetic tetracyclines in a murine lung infection model challenged with KPC-producing *Klebsiella pneumonia*, Abstract P0300. In *Abstr 24th ECCMID*, Barcelona, Spain.

Grossman T, O'Brien W, Fyfe C, Sutcliffe J. 2014b. Eravacycline is potent against third generation cephalosporin- and carbapenem-resistant Enterobacteriaceae, carbapenem-resistant *Acinetobacter baumannii* and has isolate-specific bactericidal activity. In *Abstr 54th Intersci Conf Antimicrob Agents Chemother*, Abstract C-1374. Washington, DC.

Guay GG, Tuckman M, McNicholas P, Rothstein DM. 1993. The tet(K) gene from *Staphylococcus aureus* mediates the transport of potassium in *Escherichia coli*. *J Bacteriol* **175:** 4927–4929.

Guay GG, Tuckman M, Rothstein DM. 1994. Mutations in the tetA(B) gene that cause a change in substrate specificity of the tetracycline efflux pump. *Antimicrob Agents Chemother* **38:** 857–860.

Guillaume G, Ledent V, Moens W, Collard JM. 2004. Phylogeny of efflux-mediated tetracycline resistance genes and related proteins revisited. *Microb Drug Resist* **10:** 11–26.

Hachler H, Cohen SP, Levy SB. 1991. marA, a regulated locus which controls expression of chromosomal multiple antibiotic resistance in *Escherichia coli*. *J Bacteriol* **173:** 5532–5538.

Hammerstrom TG, Beabout K, Clements TP, Saxer G, Shamoo Y. 2015. *Acinetobacter baumannii* repeatedly evolves a hypermutator phenotype in response to tigecycline that effectively surveys evolutionary trajectories to resistance. *PLoS ONE* **10:** e0140489.

He F, Fu Y, Chen Q, Ruan Z, Hua X, Zhou H, Yu Y. 2015. Tigecycline susceptibility and the role of efflux pumps in tigecycline resistance in KPC-producing *Klebsiella pneumoniae*. *PLoS ONE* **10:** e0119064.

Hentschke M, Wolters M, Sobottka I, Rohde H, Aepfelbacher M. 2010. ramR mutations in clinical isolates of *Klebsiella pneumoniae* with reduced susceptibility to tigecycline. *Antimicrob Agents Chemother* **54:** 2720–2723.

Hirata T, Saito A, Nishino K, Tamura N, Yamaguchi A. 2004. Effects of efflux transporter genes on susceptibility of *Escherichia coli* to tigecycline (GAR-936). *Antimicrob Agents Chemother* **48:** 2179–2184.

Hoban DJ, Reinert RR, Bouchillon SK, Dowzicky MJ. 2015. Global in vitro activity of tigecycline and comparator agents: Tigecycline Evaluation and Surveillance Trial 2004–2013. *Ann Clin Microbiol Antimicrob* **14:** 27.

Hornsey M, Ellington MJ, Doumith M, Hudson S, Livermore DM, Woodford N. 2010a. Tigecycline resistance in *Serratia marcescens* associated with up-regulation of the SdeXY-HasF efflux system also active against ciprofloxacin and cefpirome. *J Antimicrob Chemother* **65:** 479–482.

Hornsey M, Ellington MJ, Doumith M, Scott G, Livermore DM, Woodford N. 2010b. Emergence of AcrAB-mediated tigecycline resistance in a clinical isolate of *Enterobacter cloacae* during ciprofloxacin treatment. *Int J Antimicrob Agents* **35:** 478–481.

Hornsey M, Ellington MJ, Doumith M, Thomas CP, Gordon NC, Wareham DW, Quinn J, Lolans K, Livermore DM, Woodford N. 2010c. AdeABC-mediated efflux and tigecycline MICs for epidemic clones of *Acinetobacter baumannii*. *J Antimicrob Chemother* **65:** 1589–1593.

Hu M, Nandi S, Davies C, Nicholas RA. 2005. High-level chromosomally mediated tetracycline resistance in *Neisseria gonorrhoeae* results from a point mutation in the rpsJ gene encoding ribosomal protein S10 in combination with the mtrR and penB resistance determinants. *Antimicrob Agents Chemother* **49:** 4327–4334.

Jarolmen H, Hewel D, Kain E. 1970. Activity of minocycline against R factor-carrying Enterobacteriaceae. *Infect Immun* **1:** 321–326.

Jenner L, Starosta AL, Terry DS, Mikolajka A, Filonava L, Yusupov M, Blanchard SC, Wilson DN, Yusupova G. 2013. Structural basis for potent inhibitory activity of the antibiotic tigecycline during protein synthesis. *Proc Natl Acad Sci* **110:** 3812–3816.

Jones CH, Murphy E, Bradford PA. 2008. Genetic determinants of tetracycline resistance and their effect on tetracycline and glycylcycline antibiotics. *Anti-Infect Agents Med Chem* **7:** 84–96.

Jones RN, Flonta M, Gurler N, Cepparulo M, Mendes RE, Castanheira M. 2014. Resistance surveillance program report for selected European nations (2011). *Diagn Microbiol Infect Dis* **78:** 429–436.

Keeney D, Ruzin A, Bradford PA. 2007. RamA, a transcriptional regulator, and AcrAB, an RND-type efflux pump, are associated with decreased susceptibility to tigecycline in *Enterobacter cloacae*. *Microb Drug Resist* **13:** 1–6.

Keeney D, Ruzin A, McAleese F, Murphy E, Bradford PA. 2008. MarA-mediated overexpression of the AcrAB efflux pump results in decreased susceptibility to tigecycline in *Escherichia coli*. *J Antimicrob Chemother* **61:** 46–53.

Kim HB, Wang M, Park CH, Kim EC, Jacoby GA, Hooper DC. 2009. oqxAB encoding a multidrug efflux pump in human clinical isolates of Enterobacteriaceae. *Antimicrob Agents Chemother* **53:** 3582–3584.

Kobayashi T, Nonaka L, Maruyama F, Suzuki S. 2007. Molecular evidence for the ancient origin of the ribosomal protection protein that mediates tetracycline resistance in bacteria. *J Mol Evol* **65:** 228–235.

Krulwich TA, Jin J, Guffanti AA, Bechhofer H. 2001. Functions of tetracycline efflux proteins that do not involve tetracycline. *J Mol Microbiol Biotechnol* **3:** 237–246.

Lee EW, Huda MN, Kuroda T, Mizushima T, Tsuchiya T. 2003. EfrAB, an ABC multidrug efflux pump in *Enterococcus faecalis*. *Antimicrob Agents Chemother* **47:** 3733–3738.

Leski TA, Bangura U, Jimmy DH, Ansumana R, Lizewski SE, Stenger DA, Taitt CR, Vora GJ. 2013. Multidrug-resistant tet(X)-containing hospital isolates in Sierra Leone. *Int J Antimicrob Agents* **42:** 83–86.

Li XZ, Nikaido H. 2009. Efflux-mediated drug resistance in bacteria: An update. *Drugs* **69:** 1555–1623.

Li W, Atkinson GC, Thakor NS, Allas U, Lu CC, Chan KY, Tenson T, Schulten K, Wilson KS, Hauryliuk V, et al. 2013. Mechanism of tetracycline resistance by ribosomal protection protein Tet(O). *Nat Commun* **4:** 1477.

Li XZ, Plesiat P, Nikaido H. 2015. The challenge of efflux-mediated antibiotic resistance in Gram-negative bacteria. *Clin Microbiol Rev* **28:** 337–418.

Lin MF, Lin YY, Yeh HW, Lan CY. 2014. Role of the BaeSR two-component system in the regulation of *Acinetobacter baumannii* adeAB genes and its correlation with tigecycline susceptibility. *BMC Microbiol* **14:** 1471–2180.

Linkevicius M, Sandegren L, Andersson DI. 2013. Mechanisms and fitness costs of tigecycline resistance in *Escherichia coli*. *J Antimicrob Chemother* **68:** 2809–2819.

Lomovskaya O, Sun D, King P, Dudley MN. 2014. Tigecycline (TIG) but not minocycline (MINO) selects for clinically relevant efflux-mediated resistance (R) in *Acinetobacter* spp. (ACB). In *Abstr 54th Intersci Conf Antimicrob Agents Chemother*, Abstract C1-1087.

Lomovskaya O, Sun D, Rubio-Aparicio D, Dudley MN. 2015. Accumulation of several chromosomal mutations have limited impact on the sensitivity of *Acinetobacter baumannii* (ACB) to minocycline (MINO). In *Abstr Intersci Conf Antimicrob Agents Chemother/Internat Cong Chemother Infect*, Abstract C-1009.

Lupien A, Gingras H, Leprohon P, Ouellette M. 2015. Induced tigecycline resistance in *Streptococcus pneumoniae* mutants reveals mutations in ribosomal proteins and rRNA. *J Antimicrob Chemother* **70:** 2973–2980.

Macone AB, Caruso BK, Leahy RG, Donatelli J, Weir S, Draper MP, Tanaka SK, Levy SB. 2014. In vitro and in vivo antibacterial activities of omadacycline, a novel aminomethylcycline. *Antimicrob Agents Chemother* **58:** 1127–1135.

Marchand I, Damier-Piolle L, Courvalin P, Lambert T. 2004. Expression of the RND-type efflux pump AdeABC in *Acinetobacter baumannii* is regulated by the AdeRS two-component system. *Antimicrob Agents Chemother* **48:** 3298–3304.

Martin RG, Rosner JL. 1995. Binding of purified multiple antibiotic-resistance repressor protein (MarR) to mar operator sequences. *Proc Natl Acad Sci* **92:** 5456–5460.

Martin RG, Rosner JL. 2001. The AraC transcriptional activators. *Curr Opin Microbiol* **4:** 132–137.

Martin RG, Rosner JL. 2011. Promoter discrimination at class I MarA regulon promoters mediated by glutamic acid 89 of the MarA transcriptional activator of *Escherichia coli*. *J Bacteriol* **193:** 506–515.

Martin RG, Bartlett ES, Rosner JL, Wall ME. 2008. Activation of the *Escherichia coli* marA/soxS/rob regulon in response to transcriptional activator concentration. *J Mol Biol* **380:** 278–284.

Masuda N, Sakagawa E, Ohya S, Gotoh N, Tsujimoto H, Nishino T. 2000. Substrate specificities of MexAB-OprM, MexCD-OprJ, and MexXY-oprM efflux pumps in *Pseudomonas aeruginosa*. *Antimicrob Agents Chemother* **44:** 3322–3327.

Matsuo T, Chen J, Minato Y, Ogawa W, Mizushima T, Kuroda T, Tsuchiya T. 2008. SmdAB, a heterodimeric ABC-type multidrug efflux pump, in *Serratia marcescens*. *J Bacteriol* **190:** 648–654.

Maxwell IH. 1967. Partial removal of bound transfer RNA from polysomes engaged in protein synthesis in vitro after addition of tetracycline. *Biochim Biophys Acta* **138:** 337–346.

McAleese F, Petersen P, Ruzin A, Dunman PM, Murphy E, Projan SJ, Bradford PA. 2005. A novel MATE family efflux pump contributes to the reduced susceptibility of laboratory-derived *Staphylococcus aureus* mutants to tigecycline. *Antimicrob Agents Chemother* **49:** 1865–1871.

McDermott L, Jacobus NV, Snydman DR, Kerstein K, Grossman TH, Sutcliffe JA. 2015. Evaluation of the in vitro activity of eravacycline against a broad spectrum of recent clinical anaerobic isolates. In *Abstr Intersci Conf Antimicrob Agents Chemother/Internat Cong Chemother Infect*, Abstract C-547.

McMurry L, Levy SB. 1978. Two transport systems for tetracycline in sensitive *Escherichia coli*: Critical role for an initial rapid uptake system insensitive to energy inhibitors. *Antimicrob Agents Chemother* **14:** 201–209.

Mendes RE, Farrell DJ, Sader HS, Streit JM, Jones RN. 2015. Update of the telavancin activity in vitro tested against a worldwide collection of Gram-positive clinical isolates (2013), when applying the revised susceptibility testing method. *Diagn Microbiol Infect Dis* **81:** 275–279.

Moore IF, Hughes DW, Wright GD. 2005. Tigecycline is modified by the flavin-dependent monooxygenase TetX. *Biochemistry* **44:** 11829–11835.

Morita Y, Komori Y, Mima T, Kuroda T, Mizushima T, Tsuchiya T. 2001. Construction of a series of mutants lacking all of the four major mex operons for multidrug efflux pumps or possessing each one of the operons from *Pseudomonas aeruginosa* PAO1: MexCD-OprJ is an inducible pump. *FEMS Microbiol Lett* **202:** 139–143.

Morrissey I, Sutcliffe J, Hackel M, Hawser S. 2015a. Assessment of eravacycline against 3,467 recent gram-positive bacteria, including multidrug-resistant isolates collected from 2013–2014. In *Abstr Intersci Conf Antimicrob Agents Chemother/Internat Cong Chemother Infect*, Abstract C-563.

Morrissey I, Sutcliffe J, Hackel M, Hawser S. 2015b. Assessment of Eravacycline against a recent global collection of 4,462 Enterobacteriaceae clinical isolates (2013–2014). In *Abstr Intersci Conf Antimicrob Agents Chemother/Internat Cong Chemother Infect*, Abstract C-619.

Morrissey I, Sutcliffe J, Hackel M, Hawser S. 2015c. Assessment of eravacycline against non-fermenting gram-negative clinical isolates isolated in 2013–2014. *Abstr Intersci Conf Antimicrob Agents Chemother/Internat Cong Chemother Infect*, Abstract C-095.

Mortimer PG, Piddock LJ. 1993. The accumulation of five antibacterial agents in porin-deficient mutants of *Escherichia coli*. *J Antimicrob Chemother* **32:** 195–213.

Nagy Z, Chandler M. 2004. Regulation of transposition in bacteria. *Res Microbiol* **155:** 387–398.

Nelson ML, Levy SB. 2011. The history of the tetracyclines. *Ann NY Acad Sci* **1241:** 17–32.

Neonakis IK, Stylianou K, Daphnis E, Maraki S. 2011. First case of resistance to tigecycline by *Klebsiella pneumoniae* in a European University Hospital. *Indian J Med Microbiol* **29:** 78–79.

Nicoloff H, Andersson DI. 2013. Lon protease inactivation, or translocation of the *lon* gene, potentiate bacterial evolution to antibiotic resistance. *Mol Microbiol* **90:** 1233–1248.

Nicoloff H, Perreten V, McMurry LM, Levy SB. 2006. Role for tandem duplication and lon protease in AcrAB-TolC-dependent multiple antibiotic resistance (Mar) in an *Escherichia coli* mutant without mutations in marRAB or acrRAB. *J Bacteriol* **188:** 4413–4423.

Nicoloff H, Perreten V, Levy SB. 2007. Increased genome instability in *Escherichia coli lon* mutants: Relation to emergence of multiple-antibiotic-resistant (Mar) mutants caused by insertion sequence elements and large tandem genomic amplifications. *Antimicrob Agents Chemother* **51:** 1293–1303.

Nielsen LE, Snesrud EC, Onmus-Leone F, Kwak YI, Aviles R, Steele ED, Sutter DE, Waterman PE, Lesho EP. 2014. IS5 element integration, a novel mechanism for rapid in vivo emergence of tigecycline nonsusceptibility in *Klebsiella pneumoniae*. *Antimicrob Agents Chemother* **58:** 6151–6156.

Nigo M, Cevallos CS, Woods K, Flores VM, Francis G, Perlman DC, Revuelta M, Mildvan D, Waldron M, Gomez T, et al. 2013. Nested case-control study of the emergence of tigecycline resistance in multidrug-resistant *Klebsiella pneumoniae*. *Antimicrob Agents Chemother* **57:** 5743–5746.

Nikaido H, Thanassi DG. 1993. Penetration of lipophilic agents with multiple protonation sites into bacterial cells: Tetracyclines and fluoroquinolones as examples. *Antimicrob Agents Chemother* **37:** 1393–1399.

Nikaido E, Yamaguchi A, Nishino K. 2008. AcrAB multidrug efflux pump regulation in *Salmonella enterica* serovar *Typhimurium* by RamA in response to environmental signals. *J Biol Chem* **283:** 24245–24253.

Noel GJ, Draper MP, Hait H, Tanaka SK, Arbeit RD. 2012. A randomized, evaluator-blind, phase 2 study comparing the safety and efficacy of omadacycline to those of linezolid for treatment of complicated skin and skin structure infections. *Antimicrob Agents Chemother* **56:** 5650–5654.

Nonaka L, Suzuki S. 2002. New Mg^{2+}-dependent oxytetracycline resistance determinant tet 34 in Vibrio isolates from marine fish intestinal contents. *Antimicrob Agents Chemother* **46:** 1550–1552.

Norcia LJ, Silvia AM, Hayashi SF. 1999. Studies on time-kill kinetics of different classes of antibiotics against veterinary pathogenic bacteria including *Pasteurella*, *Actinobacillus* and *Escherichia coli*. *J Antibiot* **52:** 52–60.

Noviello S, Ianniello F, Leone S, Fiore M, Esposito S. 2008. In vitro activity of tigecycline: MICs, MBCs, time-kill curves and post-antibiotic effect. *J Chemother* **20:** 577–580.

O'Brien W, Fyfe C, Grossman T, Chen CL, Clark R, Deng Y, He M, Hunt D, Sun C, Xiao XY, et al. 2012. In vitro potency of novel tetracyclines against *Pseudomonas aeruginosa* and other major Gram-negative pathogens. In *Abstr 22th ECCMID*, Abstract P1448. London, United Kingdom.

Olson MW, Ruzin A, Feyfant E, Rush TS III, O'Connell J, Bradford PA. 2006. Functional, biophysical, and structural bases for antibacterial activity of tigecycline. *Antimicrob Agents Chemother* **50:** 2156–2166.

Peleg AY, Potoski BA, Rea R, Adams J, Sethi J, Capitano B, Husain S, Kwak EJ, Bhat SV, Paterson DL. 2007. *Acinetobacter baumannii* bloodstream infection while receiving tigecycline: A cautionary report. *J Antimicrob Chemother* **59:** 128–131.

Petersen PJ, Jacobus NV, Weiss WJ, Sum PE, Testa RT. 1999. In vitro and in vivo antibacterial activities of a novel glycylcycline, the 9-*t*-butylglycylamido derivative of minocycline (GAR-936). *Antimicrob Agents Chemother* **43:** 738–744.

Petersen PJ, Jones CH, Bradford PA. 2007. In vitro antibacterial activities of tigecycline and comparative agents by time-kill kinetic studies in fresh Mueller–Hinton broth. *Diagn Microbiol Infect Dis* **59:** 347–349.

Piddock LJ. 2006. Clinically relevant chromosomally encoded multidrug resistance efflux pumps in bacteria. *Clin Microbiol Rev* **19:** 382–402.

Pioletti M, Schlunzen F, Harms J, Zarivach R, Gluhmann M, Avila H, Bashan A, Bartels H, Auerbach T, Jacobi C, et al. 2001. Crystal structures of complexes of the small ribosomal subunit with tetracycline, edeine and IF3. *EMBO J* **20:** 1829–1839.

Pugsley AP, Schnaitman CA. 1978. Outer membrane proteins of *Escherichia coli*. VII: Evidence that bacteriophage-directed protein 2 functions as a pore. *J Bacteriol* **133:** 1181–1189.

Pumbwe L, Ueda O, Yoshimura F, Chang A, Smith RL, Wexler HM. 2006. *Bacteroides fragilis* BmeABC efflux systems additively confer intrinsic antimicrobial resistance. *J Antimicrob Chemother* **58:** 37–46.

Ramirez J, Dartois N, Gandjini H, Yan JL, Korth-Bradley J, McGovern PC. 2013. Randomized phase 2 trial to evaluate the clinical efficacy of two high-dosage tigecycline regimens versus imipenem-cilastatin for treatment of hospital-acquired pneumonia. *Antimicrob Agents Chemother* **57:** 1756–1762.

Rasmussen BA, Kovacs E. 1993. Cloning and identification of a two-component signal-transducing regulatory system from *Bacteroides fragilis*. *Mol Microbiol* **7:** 765–776.

Rasmussen BA, Gluzman Y, Tally FP. 1994. Inhibition of protein synthesis occurring on tetracycline-resistant, TetM-protected ribosomes by a novel class of tetracyclines, the glycylcyclines. *Antimicrob Agents Chemother* **38:** 1658–1660.

Reid GE, Grim SA, Aldeza CA, Janda WM, Clark NM. 2007. Rapid development of *Acinetobacter baumannii* resistance to tigecycline. *Pharmacotherapy* **27:** 1198–1201.

Reinhardt A, Kuhn F, Heisig A, Heisig P. 2014. Overexpression of *ramA* from *Citrobacter freundii* mediates MDR phenotype in various *Citrobacter* species and also in *Escherichia coli*. In *Abstr 24th ECCMID*, Abstract P1102. Barcelona, Spain.

Ricci V, Blair JM, Piddock LJ. 2014. RamA, which controls expression of the MDR efflux pump AcrAB-TolC, is regulated by the Lon protease. *J Antimicrob Chemother* **69:** 643–650.

Ritchie DJ, Garavaglia-Wilson A. 2014. A review of intravenous minocycline for treatment of multidrug-resistant *Acinetobacter* infections. *Clin Infect Dis* **59:** S374–S380.

Roberts MC. 2005. Update on acquired tetracycline resistance genes. *FEMS Microbiol Lett* **245:** 195–203.

Roberts MC. 2011. Mechanisms of bacterial antibiotic resistance and lessons learned from environmental tetracycline-resistant bacteria. In *Antimicrobial Resistance in the Environment* (ed. Keen P, Montforts MHMM). Wiley, Hoboken, NJ.

Rodriguez-Avial C, Rodriguez-Avial I, Merino P, Picazo JJ. 2012. *Klebsiella pneumoniae*: Development of a mixed population of carbapenem and tigecycline resistance during antimicrobial therapy in a kidney transplant patient. *Clin Microbiol Infect* **18**: 61–66.

Rosenblum R, Khan E, Gonzalez G, Hasan R, Schneiders T. 2011. Genetic regulation of the ramA locus and its expression in clinical isolates of *Klebsiella pneumoniae*. *Int J Antimicrob Agents* **38**: 39–45.

Ross JI, Eady EA, Cove JH, Cunliffe WJ. 1998. 16S rRNA mutation associated with tetracycline resistance in a gram-positive bacterium. *Antimicrob Agents Chemother* **42**: 1702–1705.

Rouquette C, Serre MC, Lane D. 2004. Protective role for H-NS protein in IS1 transposition. *J Bacteriol* **186**: 2091–2098.

Ruzin A, Keeney D, Bradford PA. 2005a. AcrAB efflux pump plays a role in decreased susceptibility to tigecycline in *Morganella morganii*. *Antimicrob Agents Chemother* **49**: 791–793.

Ruzin A, Visalli MA, Keeney D, Bradford PA. 2005b. Influence of transcriptional activator RamA on expression of multidrug efflux pump AcrAB and tigecycline susceptibility in *Klebsiella pneumoniae*. *Antimicrob Agents Chemother* **49**: 1017–1022.

Ruzin A, Keeney D, Bradford PA. 2007. AdeABC multidrug efflux pump is associated with decreased susceptibility to tigecycline in *Acinetobacter calcoaceticus–Acinetobacter baumannii* complex. *J Antimicrob Chemother* **59**: 1001–1004.

Ruzin A, Immermann FW, Bradford PA. 2008. Real-time PCR and statistical analyses of acrAB and ramA expression in clinical isolates of *Klebsiella pneumoniae*. *Antimicrob Agents Chemother* **52**: 3430–3432.

Ruzin A, Immermann FW, Bradford PA. 2010. RT-PCR and statistical analyses of adeABC expression in clinical isolates of *Acinetobacter calcoaceticus–Acinetobacter baumannii* complex. *Microb Drug Resist* **16**: 87–89.

Sader HS, Flamm RK, Jones RN. 2013. Tigecycline activity tested against antimicrobial resistant surveillance subsets of clinical bacteria collected worldwide (2011). *Diagn Microbiol Infect Dis* **76**: 217–221.

Sader HS, Farrell DJ, Flamm RK, Jones RN. 2014. Variation in potency and spectrum of tigecycline activity against bacterial strains from U.S. medical centers since its approval for clinical use (2006 to 2012). *Antimicrob Agents Chemother* **58**: 2274–2280.

SaiSree L, Reddy M, Gowrishankar J. 2001. IS*186* insertion at a hot spot in the *lon* promoter as a basis for lon protease deficiency of *Escherichia coli* B: Identification of a consensus target sequence for IS*186* transposition. *J Bacteriol* **183**: 6943–6946.

Sanchez-Pescador R, Brown JT, Roberts M, Urdea MS. 1988. Homology of the TetM with translational elongation factors: Implications for potential modes of tetM-conferred tetracycline resistance. *Nucleic Acids Res* **16**: 1218.

Sapunaric FM, Levy SB. 2005. Substitutions in the interdomain loop of the Tn10 TetA efflux transporter alter tetracycline resistance and substrate specificity. *Microbiology* **151**: 2315–2322.

Schedlbauer A, Kaminishi T, Ochoa-Lizarralde B, Dhimole N, Zhou S, Lopez-Alonso JP, Connell SR, Fucini P. 2015. Structural characterization of an alternative mode of

tigecycline binding to the bacterial ribosome. *Antimicrob Agents Chemother* **59**: 2849–2854.

Schweizer HP. 2003. Efflux as a mechanism of resistance to antimicrobials in *Pseudomonas aeruginosa* and related bacteria: Unanswered questions. *Genet Mol Res* **2**: 48–62.

Seoane AS, Levy SB. 1995. Characterization of MarR, the repressor of the multiple antibiotic resistance (mar) operon in *Escherichia coli*. *J Bacteriol* **177**: 3414–3419.

Sheng ZK, Hu F, Wang W, Guo Q, Chen Z, Xu X, Zhu D, Wang M. 2014. Mechanisms of tigecycline resistance among *Klebsiella pneumoniae* clinical isolates. *Antimicrob Agents Chemother* **58**: 6982–6985.

Sherwood JE, Fraser S, Citron DM, Wexler H, Blakely G, Jobling K, Patrick S. 2011. Multi-drug resistant *Bacteroides fragilis* recovered from blood and severe leg wounds caused by an improvised explosive device (IED) in Afghanistan. *Anaerobe* **17**: 152–155.

Smith MC, Chopra I. 1984. Energetics of tetracycline transport into *Escherichia coli*. *Antimicrob Agents Chemother* **25**: 446–449.

Solomkin JS, Ramesh MK, Cesnauskas G, Novikovs N, Stefanova P, Sutcliffe JA, Walpole SM, Horn PT. 2014. Phase 2, randomized, double-blind study of the efficacy and safety of two dose regimens of eravacycline versus ertapenem for adult community-acquired complicated intra-abdominal infections. *Antimicrob Agents Chemother* **58**: 1847–1854.

Spanu T, De Angelis G, Cipriani M, Pedruzzi B, D'Inzeo T, Cataldo MA, Sganga G, Tacconelli E. 2012. In vivo emergence of tigecycline resistance in multidrug-resistant *Klebsiella pneumoniae* and *Escherichia coli*. *Antimicrob Agents Chemother* **56**: 4516–4518.

Speer BS, Salyers AA. 1988. Characterization of a novel tetracycline resistance that functions only in aerobically grown *Escherichia coli*. *J Bacteriol* **170**: 1423–1429.

Speer BS, Salyers AA. 1989. Novel aerobic tetracycline resistance gene that chemically modifies tetracycline. *J Bacteriol* **171**: 148–153.

Speer BS, Bedzyk L, Salyers AA. 1991. Evidence that a novel tetracycline resistance gene found on two *Bacteroides* transposons encodes an NADP-requiring oxidoreductase. *J Bacteriol* **173**: 176–183.

Srinivasan VB, Venkataramaiah M, Mondal A, Vaidyanathan V, Govil T, Rajamohan G. 2012. Functional characterization of a novel outer membrane porin KpnO, regulated by PhoBR two-component system in *Klebsiella pneumoniae* NTUH-K2044. *PLoS ONE* **7**: 25.

Stein GE, Babinchak T. 2013. Tigecycline: An update. *Diagn Microbiol Infect Dis* **75**: 331–336.

Stone NR, Woodford N, Livermore DM, Howard J, Pike R, Mushtaq S, Perry C, Hopkins S. 2011. Breakthrough bacteraemia due to tigecycline-resistant *Escherichia coli* with New Delhi metallo-β-lactamase (NDM)-1 successfully treated with colistin in a patient with calciphylaxis. *J Antimicrob Chemother* **66**: 2677–2678.

Sum PE, Petersen P. 1999. Synthesis and structure-activity relationship of novel glycylcycline derivatives leading to the discovery of GAR-936. *Bioorg Med Chem Lett* **9**: 1459–1462.

Sun JR, Perng CL, Chan MC, Morita Y, Lin JC, Su CM, Wang WY, Chang TY, Chiueh TS. 2012. A truncated AdeS kinase protein generated by ISAba1 insertion correlates with tigecycline resistance in *Acinetobacter baumannii*. *PLoS ONE* **7:** 14.

Sun C, Hunt D, Clark R, Fyfe C, Kerstein K, Grossman T, Sutcliffe J, Xiao XY. 2014a. In vitro potency of novel, fully synthetic tetracyclines against MDR Gram-negative pathogens including carbapenem-resistant Enterobacteriaceae. In *24th ECCMID*, Abstract P0299. Barcelona, Spain.

Sun JR, Perng CL, Lin JC, Yang YS, Chan MC, Chang TY, Lin FM, Chiueh TS. 2014b. AdeRS combination codes differentiate the response to efflux pump inhibitors in tigecycline-resistant isolates of extensively drug-resistant *Acinetobacter baumannii*. *Eur J Clin Microbiol Infect Dis* **33:** 2141–2147.

Sun C, Hunt DK, Chen CL, Deng Y, He M, Clark RB, Fyfe C, Grossman TH, Sutcliffe JA, Xiao XY. 2015. Design, synthesis, and biological evaluation of hexacyclic tetracyclines as potent, broad spectrum antibacterial agents. *J Med Chem* **20:** 20.

Sutcliffe JA, O'Brien W, Fyfe C, Grossman TH. 2013. Antibacterial activity of eravacycline (TP-434), a novel fluorocycline, against hospital and community pathogens. *Antimicrob Agents Chemother* **57:** 5548–5558.

Taylor DE, Hiratsuka K, Ray H, Manavathu EK. 1987. Characterization and expression of a cloned tetracycline resistance determinant from *Campylobacter jejuni* plasmid pUA466. *J Bacteriol* **169:** 2984–2989.

Teo JW, Tan TM, Poh CL. 2002. Genetic determinants of tetracycline resistance in *Vibrio harveyi*. *Antimicrob Agents Chemother* **46:** 1038–1045.

Tessier PR, Nicolau DP. 2013. Tigecycline displays in vivo bactericidal activity against extended-spectrum-β-lactamase-producing Enterobacteriaceae after 72-hour exposure period. *Antimicrob Agents Chemother* **57:** 640–642.

Thaker M, Spanogiannopoulos P, Wright GD. 2010. The tetracycline resistome. *Cell Mol Life Sci* **67:** 419–431.

Thanassi DG, Suh GS, Nikaido H. 1995. Role of outer membrane barrier in efflux-mediated tetracycline resistance of *Escherichia coli*. *J Bacteriol* **177:** 998–1007.

The Medicines Company. 2015. Prescribing Information, Minocin. Parsippany, NJ.

Trieber CA, Taylor DE. 2002. Mutations in the 16S rRNA genes of *Helicobacter pylori* mediate resistance to tetracycline. *J Bacteriol* **184:** 2131–2140.

Truong-Bolduc QC, Dunman PM, Strahilevitz J, Projan SJ, Hooper DC. 2005. MgrA is a multiple regulator of two new efflux pumps in *Staphylococcus aureus*. *J Bacteriol* **187:** 2395–2405.

Tsilibaris V, Maenhaut-Michel G, Van Melderen L. 2006. Biological roles of the Lon ATP-dependent protease. *Res Microbiol* **157:** 701–713.

Tuckman M, Petersen PJ, Projan SJ. 2000. Mutations in the interdomain loop region of the tetA(A) tetracycline resistance gene increase efflux of minocycline and glycylcyclines. *Microb Drug Resist* **6:** 277–282.

U.S. Food and Drug Administration. 2010. FDA Drug Safety Communication: Increased risk of death with Tygacil (tigecycline) compared to other antibiotics used to treat similar infections. FDA, Silver Spring, MD.

U.S. Food and Drug Administration. 2013. FDA Drug Safety Communication: FDA warns of increased risk of death with IV antibacterial Tygacil (tigecycline) and approves new Boxed Warning. FDA, Silver Spring, MD.

van der Straaten T, Janssen R, Mevius DJ, van Dissel JT. 2004a. *Salmonella* gene *rma* (*ramA*) and multiple-drug-resistant *Salmonella enterica* serovar typhimurium. *Antimicrob Agents Chemother* **48:** 2292–2294.

van der Straaten T, Zulianello L, van Diepen A, Granger DL, Janssen R, van Dissel JT. 2004b. *Salmonella enterica* serovar Typhimurium RamA, intracellular oxidative stress response, and bacterial virulence. *Infect Immun* **72:** 996–1003.

Veleba M, Schneiders T. 2012. Tigecycline resistance can occur independently of the ramA gene in *Klebsiella pneumoniae*. *Antimicrob Agents Chemother* **56:** 4466–4467.

Veleba M, Higgins PG, Gonzalez G, Seifert H, Schneiders T. 2012. Characterization of RarA, a novel AraC family multidrug resistance regulator in *Klebsiella pneumoniae*. *Antimicrob Agents Chemother* **56:** 4450–4458.

Veleba M, De Majumdar S, Hornsey M, Woodford N, Schneiders T. 2013. Genetic characterization of tigecycline resistance in clinical isolates of *Enterobacter cloacae* and *Enterobacter aerogenes*. *J Antimicrob Chemother* **68:** 1011–1018.

Villa L, Feudi C, Fortini D, Garcia-Fernandez A, Carattoli A. 2014. Genomics of KPC-producing *Klebsiella pneumoniae* sequence type 512 clone highlights the role of RamR and ribosomal S10 protein mutations in conferring tigecycline resistance. *Antimicrob Agents Chemother* **58:** 1707–1712.

Vinue L, McMurry LM, Levy SB. 2013. The 216-bp marB gene of the marRAB operon in *Escherichia coli* encodes a periplasmic protein which reduces the transcription rate of marA. *FEMS Microbiol Lett* **345:** 49–55.

Visalli MA, Murphy E, Projan SJ, Bradford PA. 2003. AcrAB multidrug efflux pump is associated with reduced levels of susceptibility to tigecycline (GAR-936) in *Proteus mirabilis*. *Antimicrob Agents Chemother* **47:** 665–669.

Warburton PJ, Ciric L, Lerner A, Seville LA, Roberts AP, Mullany P, Allan E. 2013. TetAB46, a predicted heterodimeric ABC transporter conferring tetracycline resistance in *Streptococcus australis* isolated from the oral cavity. *J Antimicrob Chemother* **68:** 17–22.

Wei Y, Bechhofer DH. 2002. Tetracycline induces stabilization of mRNA in *Bacillus subtilis*. *J Bacteriol* **184:** 889–894.

Werner G, Gfrorer S, Fleige C, Witte W, Klare I. 2008. Tigecycline-resistant *Enterococcus faecalis* strain isolated from a German intensive care unit patient. *J Antimicrob Chemother* **61:** 1182–1183.

Whittle G, Hund BD, Shoemaker NB, Salyers AA. 2001. Characterization of the 13-kilobase ermF region of the *Bacteroides* conjugative transposon CTnDOT. *Appl Environ Microbiol* **67:** 3488–3495.

Williams G, Smith I. 1979. Chromosomal mutations causing resistance to tetracycline in *Bacillus subtilis*. *Mol Gen Genet* **177:** 23–29.

Wyeth Pharmaceuticals. 2016. Prescribing information, Tygacil. Philadelphia, PA.

Xiao XY, Hunt DK, Zhou J, Clark RB, Dunwoody N, Fyfe C, Grossman TH, O'Brien WJ, Plamondon L, Ronn M, et al. 2012. Fluorocyclines. 1: 7-fluoro-9-pyrrolidinoaceta-mido-6-demethyl-6-deoxytetracycline: A potent, broad spectrum antibacterial agent. *J Med Chem* **55:** 597–605.

Xiao XY, Deng Y, Sun C, Hunt D, Clark R, Fyfe C, Grossman T, Sutcliffe J. 2013. Novel 7-CF$_3$-8-heterocyclyl tetracy-clines with promising antibacterial activity against Gram-positive and Gram-negative pathogens, including *Pseudomonas aeruginosa*. In *Abstr 53th Intersci Conf Anti-microb Agents Chemother*, Abstract F-632. Denver, CO.

Yamaguchi A, Ohmori H, Kaneko-Ohdera M, Nomura T, Sawai T. 1991. ΔpH-dependent accumulation of tetracy-cline in *Escherichia coli*. *Antimicrob Agents Chemother* **35:** 53–56.

Yamasaki S, Nikaido E, Nakashima R, Sakurai K, Fujiwara D, Fujii I, Nishino K. 2013. The crystal structure of multi-drug-resistance regulator RamR with multiple drugs. *Nat Commun* **4:** 2078.

Yang W, Moore IF, Koteva KP, Bareich DC, Hughes DW, Wright GD. 2004. TetX is a flavin-dependent monooxy-genase conferring resistance to tetracycline antibiotics. *J Biol Chem* **279:** 52346–52352.

Yoon EJ, Courvalin P, Grillot-Courvalin C. 2013. RND-type efflux pumps in multidrug-resistant clinical isolates of *Acinetobacter baumannii*: Major role for AdeABC over-expression and AdeRS mutations. *Antimicrob Agents Chemother* **57:** 2989–2995.

Zhanel GG, Homenuik K, Nichol K, Noreddin A, Vercaigne L, Embil J, Gin A, Karlowsky JA, Hoban DJ. 2004. The glycylcyclines: A comparative review with the tetracy-clines. *Drugs* **64:** 63–88.

Zhanel GG, Adam HJ, Baxter MR, Fuller J, Nichol KA, De-nisuik AJ, Lagace-Wiens PR, Walkty A, Karlowsky JA, Schweizer F, et al. 2013. Antimicrobial susceptibility of 22746 pathogens from Canadian hospitals: Results of the CANWARD 2007-11 study. *J Antimicrob Chemother* **68:** 7–22.

Zhang L, Li XZ, Poole K. 2001. SmeDEF multidrug efflux pump contributes to intrinsic multidrug resistance in *Stenotrophomonas maltophilia*. *Antimicrob Agents Che-mother* **45:** 3497–3503.

Zhong X, Xu H, Chen D, Zhou H, Hu X, Cheng G. 2014. First emergence of acrAB and oqxAB mediated tigecy-cline resistance in clinical isolates of *Klebsiella pneumo-niae* pre-dating the use of tigecycline in a Chinese hospital. *PLoS ONE* **9:** e115185.

Resistance to Macrolide Antibiotics in Public Health Pathogens

Corey Fyfe, Trudy H. Grossman, Kathy Kerstein, and Joyce Sutcliffe

Tetraphase Pharmaceuticals, Watertown, Massachusetts 02472

Correspondence: jsutcliffe@tphase.com

Macrolide resistance mechanisms can be target-based with a change in a 23S ribosomal RNA (rRNA) residue or a mutation in ribosomal protein L4 or L22 affecting the ribosome's interaction with the antibiotic. Alternatively, mono- or dimethylation of A2058 in domain V of the 23S rRNA by an acquired rRNA methyltransferase, the product of an *erm* (erythromycin ribosome methylation) gene, can interfere with antibiotic binding. Acquired genes encoding efflux pumps, most predominantly *mef*(A) + *msr*(D) in pneumococci/streptococci and *msr*(A/B) in staphylococci, also mediate resistance. Drug-inactivating mechanisms include phosphorylation of the 2′-hydroxyl of the amino sugar found at position C5 by phosphotransferases and hydrolysis of the macrocyclic lactone by esterases. These acquired genes are regulated by either translation or transcription attenuation, largely because cells are less fit when these genes, especially the rRNA methyltransferases, are highly induced or constitutively expressed. The induction of gene expression is cleverly tied to the mechanism of action of macrolides, relying on antibiotic-bound ribosomes stalled at specific sequences of nascent polypeptides to promote transcription or translation of downstream sequences.

Macrolide antibiotics are polyketides composed of a 14-, 15-, or 16-membered macrocyclic lactone ring (14-, 15-, and 16-membered) to which several sugars and/or side chains have been attached by the producing organism or as modifications during semisynthesis in the laboratory (Figs. 1 and 2). Newer semisynthetic derivations, like ketolides telithromycin, and solithromycin, have a C3-keto group in place of the C3 cladinose (akin to naturally occurring pikromycin) (Brockmann and Henkel 1950) and an 11,12-cyclic carbamate with an extended alkyl–aryl side chain that increases the affinity of the antibiotic for the ribosome by 10- to 100-fold (Hansen et al. 1999;

Dunkle et al. 2010); in the case of solithromycin, a fluorine substituent at C2 provides an additional ribosomal interaction (Llano-Sotelo et al. 2010). Macrolides continue to be important in the therapeutic treatment of community-acquired pneumonia (*Streptococcus pneumoniae*, *Haemophilus influenzae*, *Moraxella catarrhalis*, and atypicals *Legionella pneumophila*, *Mycoplasma pneumoniae*, *Chlamydia pneumoniae*), sexually transmitted diseases (*Neiserria gonorhoeae*, *Chlamydia trachomatis*, *Mycoplasma genitalium*), shigellosis, and salmonellosis. With solithromycin heading for a new drug application (NDA) filing in 2016 and having the in vitro potency to treat erythromycin-resistant

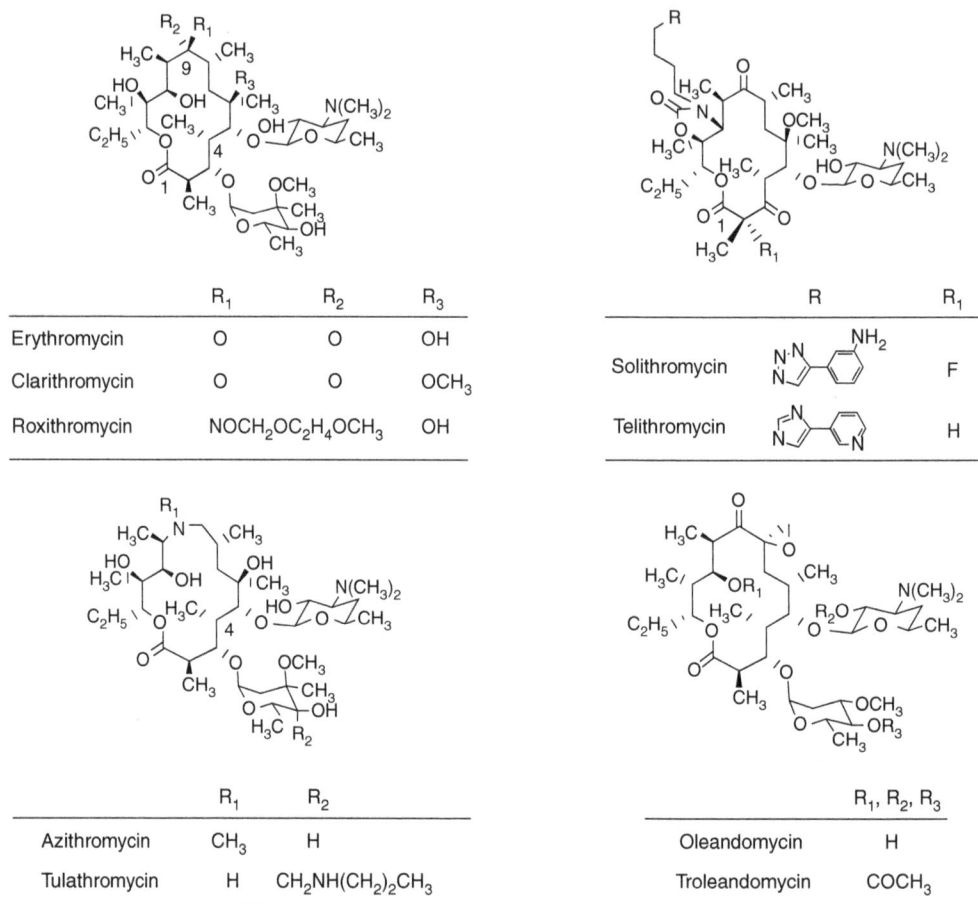

Figure 1. Structures of 14- and 15-membered macrolides.

	R_1	R_2	R_3
Erythromycin	O	O	OH
Clarithromycin	O	O	OCH_3
Roxithromycin	$NOCH_2OC_2H_4OCH_3$	OH	

	R	R_1
Solithromycin		F
Telithromycin		H

	R_1	R_2
Azithromycin	CH_3	H
Tulathromycin	H	$CH_2NH(CH_2)_2CH_3$

	R_1, R_2, R_3
Oleandomycin	H
Troleandomycin	$COCH_3$

pneumococci and gonococci (Farrell et al. 2015; Hook et al. 2015), macrolides/ketolides will continue as an important part of the antibiotic armamentarium.

The mechanism of action of macrolides has been further refined through a combination of genetic, biochemical, crystallographic, and ribosome profiling studies (Tu et al. 2005; Dunkle et al. 2010; Kannan et al. 2012, 2014; Gupta et al. 2016). Macrolides/ketolides are sensed by the ribosome and, in the presence of certain macrolide-stalling nascent amino acid chain-dependent motifs, selectively inhibit protein synthesis. Further, and to different extents, ketolides and macrolides cause frameshifting, leading to aberrant protein synthesis.

Shortly after its clinical debut in 1953, resistance to erythromycin in staphylococci was described and was likely mediated by methylation of the 23S ribosomal RNA (rRNA) at nucleotide A2058 (*Escherichia coli* numbering) encoded by an erythromycin ribosomal methyltransferase (*erm*) gene (Weisblum 1995a). Erm methyltransferases add one or two methyl groups to the N-6 exocyclic amino group of A2058, disrupting the key hydrogen bond between A2058 and the desosamine sugar at C5 (Fig. 3). Ribosomal methylation by methyltransferases encoded by *erm* genes remains the most widespread macrolide resistance in pathogenic bacteria, with certain *erm* genes more predominantly found in some species. Streptococci

Cite this article as *Cold Spring Harb Perspect Med* doi: 10.1101/cshperspect.a025395

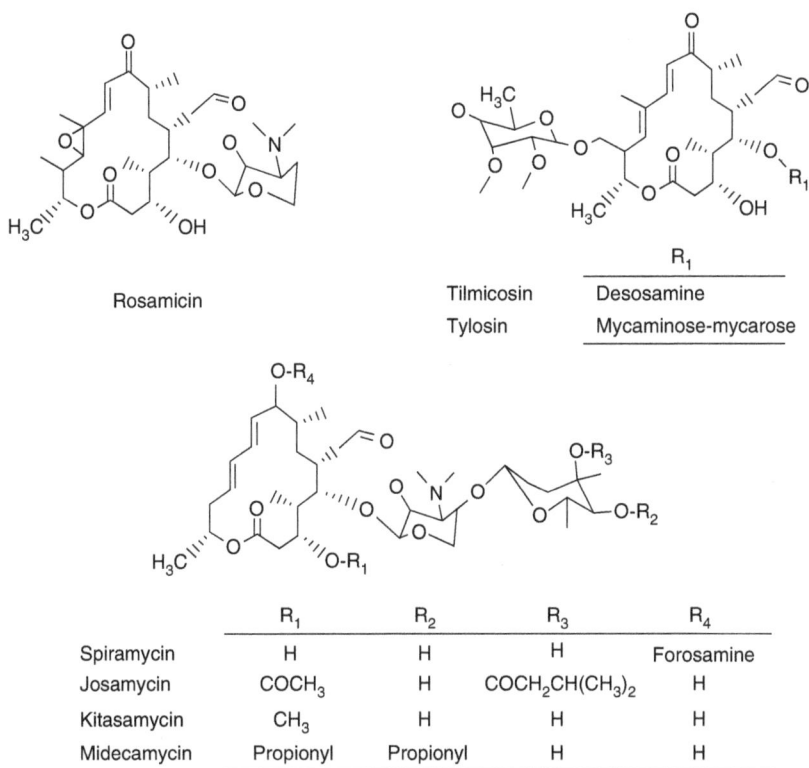

Figure 2. Structures of 16-membered macrolides.

generally have *erm*(B) or *erm*(A), subclass *erm*(TR), whereas *erm*(A), *erm*(B), or *erm*(C) are found in staphylococci and *erm*(F) in anaerobes and *H. influenzae* (Table 2, and references therein; see also faculty.washington.edu/marilynr). There can be either mono- or dimethylation of A2058 and the degree of rRNA dimethylation can determine ketolide resistance (Douthwaite et al. 2005). Most *erm* genes are inducible by 14- and 15-membered macrolides, whereby translation repression of the *erm* methyltransferase gene, because of the sequestration of its ribosome-binding site (RBS) by messenger RNA (mRNA) secondary structure, is relieved by binding of the inducer to the ribosome (Horinouchi and Weisblum 1980; Depardieu et al. 2007; Subramaniam et al. 2011). Upstream of the start codon of a methyltransferase gene is an open reading frame (ORF) that produces leader peptides of different lengths (8–38 amino acids), each containing a macrolide stalling motif; when the macrolide-bound ribosome pauses, the attenuator, a stem and loop structure that encompasses the RBS, is disrupted, resulting in ribosomal binding and synthesis of the methyltransferase (Subramaniam et al. 2011; Arenz et al. 2014a,b). Although most *erm* genes are regulated by translation attenuation, a few genes (e.g., *erm*(K)) are regulated by transcription attenuation (Kwak et al. 1991; Choi et al. 1997) or through inducible transcription factors (Morris et al. 2005). Ketolide induction has been described for *erm*(C) and involves promotion of frameshifting in the *erm*(C) leader (*ermCL*) mRNA, leading to bypass of the *ermCL* stop codon, via rearrangement of the secondary mRNA structure, allowing expression of the downstream resistance gene (Gupta et al. 2013a).

There are two families of macrolide efflux pumps with regulation that is at least in part, transcriptionally mediated—*mef*, a major-facil-

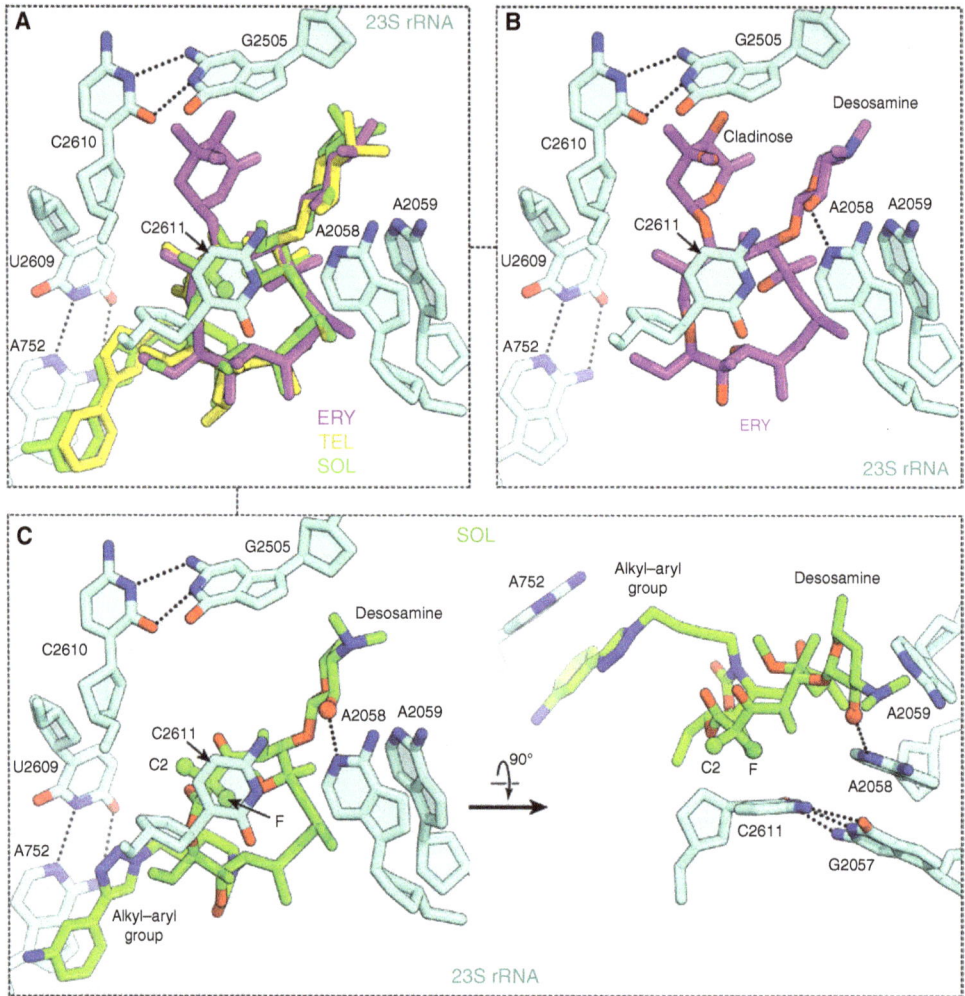

Figure 3. A model based on the crystal structure of the 70S *Escherichia coli* ribosome bound to erythromycin (PDB ID codes 3OFO, 3OFP, 3OFR, 3OFQ), telithromycin (PDB ID codes 3OAQ, 3OAR, 3OAS, 3OAT), and solithromycin (PDB ID code 4WWW) (Dunkle et al. 2010; Llano-Sotelo et al. 2010). (*A*) A comparison of the conformations of erythromycin (ERY, magenta), telithromycin (TEL, gold), and solithromycin (SOL, green) in their binding sites at the top of the nascent peptide exit tunnel (PET) comprised of 23S ribosomal RNA (rRNA). 23S rRNA residues are marked, with nitrogen in dark blue and oxygen in red. Hydrogen bonds are indicated between residues by dotted lines, including between residues U2609 in domain V and A752 in domain II of 23S rRNA. The alkyl–aryl arm of telithromycin and solithromycin is shown stacking with A752. (*B*) Erythromycin-only view. The key hydrogen bond between the 2′ hydroxyl of the desosamine and the N1 of A2058 is indicated. The exocyclic N6 amino group that is methylated by Erm methyltransferases is notable next to the N1 of A2058. (*C*) Solithromycin-only view. The *left* side of the figure displays solithromycin in the same conformation as macrolides in *A* and *B*. The C2-F is visible through the ring of C2611, but a better view of its interaction with C2611 is displayed when the view is rotated by 90°, with the C2-F stacking with the hydrophobic side of C2611. C2611 is paired through three hydrogen bonds to G2057.

itator-superfamily pump that confers resistance to most 14- and 15-membered macrolides (Leclercq and Courvalin 2002; Sutcliffe and Leclercq 2002; Chancey et al. 2011) and *msr*, a member of the ATP-binding cassette (ABC) superfamily that generally confers resistance to 14- and 15-membered macrolides and streptogramin B and low-level resistance to ketolides (Sutcliffe and Leclercq 2002; Chancey et al. 2011).

Intrinsic efflux pumps that are not specific to macrolides exist in different species. These pumps are often responsible for limiting macrolide spectrum in Gram-negative species and overexpression of multidrug efflux pumps is associated with clinically relevant drug resistance in both Gram-negative and Gram-positive species. Interested readers are referred to recent reviews (Costa et al. 2013; Blair et al. 2014; Delmar et al. 2014; Sun et al. 2014).

Mutations in 23S rRNA, L4, and/or L22 ribosomal proteins can confer macrolide resistance because the mutation is technically in the 23S rRNA gene. In addition, macrolides can be inactivated by esterases or phosphotransferases (in public health pathogens and macrolide producers) or by glycosyltransferases (decribed in many strains of *Streptomyces* producing polyketides or polyether antibiotics; *Micromonospora purpurea; Nocardia asteroides*), deacylases (*N. asteroides*), or formyl reductases (*N. asteroides, Nocardia brasiliensis, Nocardia otitidiscaviarum*) (Sutcliffe and Leclercq 2002; Roberts 2008; Shakya and Wright 2010; Morar et al. 2012). Many strains carry more than one macrolide resistance mechanism, sometimes on the same mobile element.

This review will focus on antimicrobial resistance mechanisms to macrolides primarily in public health pathogens. Recent reviews on the mechanisms of macrolide resistance are recommended (Leclercq and Courvalin 2002; Sutcliffe and Leclercq 2002; Franceschi et al. 2004; Depardieu et al. 2007; Roberts 2008; Kannan and Mankin 2011; Wilson 2014) as well as the website for macrolide–lincosamide–streptogramin resistances maintained by Marilyn Roberts (see faculty.washington.edu/marilynr). Based on the paper published in 1999 that set out to prevent duplicate genes being renamed when discovered

in a new species or as part of a novel mobile element (Roberts et al. 1999), macrolide-resistant genes are considered in the same family if they have $\geq 80\%$ amino acid identity from the original gene identified for that family.

MACROLIDE MECHANISM OF ACTION

All members of the macrolide class inhibit bacterial protein synthesis by binding to the 23S rRNA in the large ribosomal subunit (50S) downstream from the peptidyltransferase center (PTC), the catalytic site for peptide bond formation (for overview of protein synthesis, see Arenz and Wilson 2016) (Wilson 2009, 2014; Dunkle et al. 2010; Kannan et al. 2014). Macrolides/ketolides bind at the entrance of the peptide exit tunnel (PET) just above the constriction formed by extended loops of ribosomal proteins L4 and L22 (Yusupov et al. 2001; Davydova et al. 2002; Hansen et al. 2002; Schlunzen et al. 2003; Tu et al. 2005; Dunkle et al. 2010), further restricting the effective diameter of the PET. The macrocyclic lactone and the C5 sugars overlap (Fig. 3A). The sugar at C5 (often desosamine) is positioned toward the PTC, and macrolides like tylosin that have a disaccharide at the C5 position, reach deeper into the PTC. The 2′ hydroxyl of desosamine sugar at C5 makes a key hydrogen bond contact with the N1 atom of A2058 and modification at this position by either mutation or methylation of the N6 exocyclic amine results in macrolide resistance (see Fig. 3B) (Sutcliffe and Leclercq 2002; Franceschi et al. 2004; Tu et al. 2005; Dunkle et al. 2010). Other residues help define a local binding conformation for macrolides, including G2057 and C2611 that form a Watson–Crick base pair with each other and to which the hydrophobic face of the lactone ring is packed (seen best in Fig. 3C). For ketolides, telithromycin, and solithromycin, the extended alkyl–aryl arm of each drug is oriented down the tunnel and makes a stacking interaction with a base pair formed by A752 and U2609 in the 23S rRNA (Fig. 3A,C); these side chains align closely in the crystal structure of each drug complexed to *E. coli* 70S ribosome, but are positioned differently from the crystal structures of telithromycin complexed with either *Deinococcus radiodurans* or *Haloarcula marismortui* (Schlunzen et al. 2003; Tu et al. 2005; Dunkle et al. 2010; Llano-Sotelo et al. 2010), likely a result of the absence of the A752-U2609 base pair seen in most pathogenic bacteria. Solithromycin has an additional stacking interaction with the hydrophobic portion of C2611 via its C2-fluorine, thereby conferring eightfold better activity against *S. pneumoniae* constitutive *erm*(B) isolates than a corresponding structure with hydrogen at C2 (Fig. 3C) (Llano-Sotelo et al. 2010).

Until recently, macrolides were thought to inhibit protein synthesis by sterically blocking nascent peptides as they transversed the PET (Hansen et al. 2002; Voss et al. 2006). However, despite the constriction formed from L4 and L22 loops and bound macrolide, there is still room in the PET for nascent, unfolded peptides to successfully negotiate the tunnel (Tu et al. 2005; Dunkle et al. 2010; Kannan et al. 2012). Further, genome-wide ribosome profiling analyses in *E. coli* have shown that ribosomes with bound erythromycin or telithromycin allow a compound-dependent subset of proteins to be synthesized, rather than act as general translation inhibitors (Kannan et al. 2012, 2014). Further work has shown that ribosomes bound with a macrolide or ketolide are impaired in the efficient catalysis of peptide bond formation and that this impairment is sequence- and context-specific (i.e., dependent on macrolide stalling motifs) (Arenz et al. 2014a; Kannan et al. 2014; Sothiselvam et al. 2014). If a macrolide-stalling motif is encountered near the amino terminus of an ORF, then ribosomal stalling of a short nascent peptide likely leads to premature release of peptidyl-tRNA, consistent with early biochemical studies with erythromycin (Otaka and Kaji 1975; Menninger 1985; Tenson et al. 2003). However, macrolide stalling motifs can be hundreds of codons away from the start codon, resulting in the synthesis of large peptides and, for some proteins that have no translation arrest sequences, synthesis of full length proteins (Kannan et al. 2012). Translation arrest can occur because specific sequences (macrolide stalling motifs) of the nascent leader peptide in the ribosomal tunnel sense the ribosome-bound antibiotic and, through interactions with it and with elements of the tunnel wall, induce conformational rearrangements that are communicated to the PTC so as to stop translation (Arenz et al. 2014a,b; Sothiselvam et al. 2014). Alternatively, macrolide- and peptide-dependent programmed translation arrest is also defined by the nature of the amino acid residues in the PTC (Kannan et al. 2014). Amazingly, the "nose" of the ribosome can sense small structural changes in the macrolide as well as discriminate a single amino acid difference in the nascent peptide (Gupta et al. 2016). Further, the discriminating properties of the PET allow for regulation of *cis*-located target gene expression, protein targeting and folding, and response to additional cellular factors (Ito et al. 2010; Kannan et al. 2014; Gupta et al. 2016).

Investigation into the mechanism by which ketolides induce *erm*(C) uncovered another mechanism of action of ketolides and macrolides—promotion of frameshifting (Gupta et al. 2013a). Intriguingly, the extent of reduction in translational fidelity is compound-dependent and is reliant on the antibiotic allosterically influencing the reading-frame maintenance in the 30S ribosomal subunit, some 90 Å away from the macrolide/ketolide binding site in the 50S ribosomal subunit (Gupta et al. 2013a). Because 25% of the entire *E. coli* proteome continues to be synthesized in the presence of telithromycin (Kannan et al. 2012), production of aberrant cellular proteins may also be important to its antibacterial action.

MACROLIDE-RESISTANT MECHANISMS

Ribosomal Modifications

23S rRNA Mutations

Mutants that are resistant to one or more of the MLS_B antibiotics, because of base substitutions in either domain V or helix 35 in domain II of 23S rRNA or in ribosomal proteins L4 or L22, provide genetic evidence that these antibiotics interact with the ribosome (Figs. 3 and 4) (Vester and Douthwaite 2001; Sutcliffe and Leclercq 2002; Franceschi et al. 2004). Macrolides primarily interact with A2058 and A2059 of the

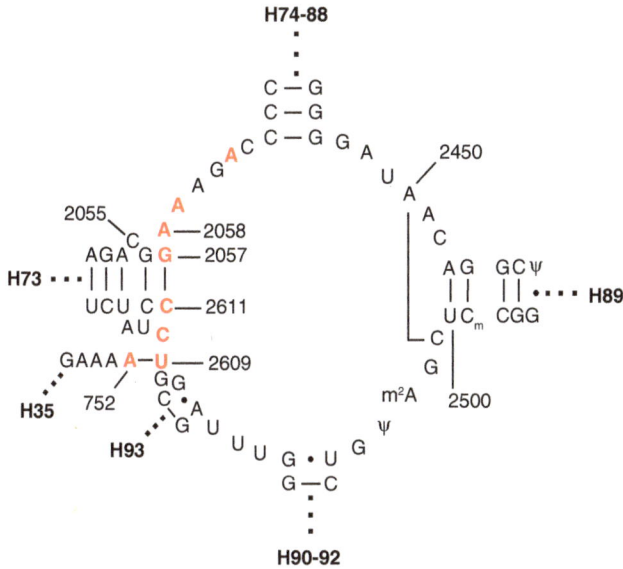

Figure 4. The secondary structure of the 3′ region of 23S ribosomal RNA (rRNA) domain V. Nucleotides in red indicate mutations that can yield 14/15-membered macrolide, 16-membered macrolide, and/or ketolide resistance (see Schwarz et al. 2016 for mutations in this region that alter lincosamides, streptogramins, phenicols, and pleuromutilins). rRNA helices that stem from this region are designated with dotted lines. Residue A752 is in the hairpin 35 loop of domain II.

23S rRNA and mutations in these nucleotides have been found in many macrolide-resistant bacterial strains, generally in pathogens (*Mycobacterium, Brachyspira, Helicobacter, Treponema*) with just one or two copies of *rrl*, the gene that codes for 23S rRNA or, in pathogens with three or more rRNA genes, may develop during chronic treatment of macrolides (Table 1). However, mutations in these positions as well as at G2057 in combination with A2059, and at C2611, have been found in clinical isolates and laboratory mutants of *S. pneumoniae* (Tait-Kamradt et al. 2000a,b; Canu et al. 2002), at A2058 and C2611 in clinical isolates of *Streptococcus pyogenes* (Malbruny et al. 2002; Jalava et al. 2004), and at A2058, A2059, or both A2059 and G2160 in clinical isolates of *H. influenzae* (Peric et al. 2003). Consistently, mutations at A2058 and A2059 are the most frequently observed and have a strong phenotype in all species, generally conferring macrolide–lincosamide–streptogramin B-ketolide (MLS$_B$K) resistance in most isolates. A comprehensive listing of ribosomal 23S rRNA mutations isolated

in *S. pneumoniae, S. pyogenes,* and *H. influenzae* before 2005 has been assembled by Franceschi et al. (2004). Table 1 updates the base substitutions and extends the citings to other species.

With the increase in use of the macrolide azithromycin as a maintenance treatment for cystic fibrosis (CF) patients, there has been an increase in the levels of MLS$_B$-resistant *Staphylococcus aureus* isolated from CF patients. Six azithromycin- and erythromycin-resistant isolates of *S. aureus* from CF patients after treatment with azithromycin (Prunier et al. 2002), that did not carry resistance determinant *erm* or *msr*(A) genes, were found to carry mutations A2058G, A2058T, or A2059G with copy numbers of mutant alleles ranging from three of five and four of five to four of six *rrl* genes (*S. aureus* can have five or six *rrl* genes). A more recent characterization of *S. aureus* strains isolated from Czech CF patients showed high rates (29%) of strains with ribosomal mutations conferring resistance to MLS$_B$ antibiotics with the majority in 23S rRNA (23%) (Tkadlec et al. 2015).

Table 1. Mutations in 23S rRNA (*E. coli* numbering) conferring macrolide/ketolide resistance

Nucleotide	Organism	Wild-type	Mutant	References
752	*Mycoplasma genitalium*	A	C	Jensen et al. 2014
	Streptococcus pneumoniae	A	Deletion	Canu et al. 2000
754	*Escherichia coli*	U	A	Xiong et al. 1999
2038	*M. genitalium*	C	T	Chrisment et al. 2012
2057	*E. coli*	G	A	Ettayebi et al. 1985
	Mycoplasma fermentans	G	A	Pereyre et al. 2002
	Mycoplasma hominis	G	A	Pereyre et al. 2002
	Propionibacterium spp.	G	A	Ross et al. 1997
2057 + 2032	*E. coli*	G/G	A/A	Douthwaite 1992
2057 + 2032	*Helicobacter pylori*	A/G	G/A	Hulten et al. 1997
2057 + 2059	*S. pneumoniae*	G/A	A/C	Fu et al. 2000
	S. pneumoniae	G/A	A/G	Farrell et al. 2004
	Streptococcus pyogenes	G/A	A/G	Doktor et al. 2001
2058 + 2160	*Haemophilus influenzae*	A/G	G/T	Peric et al. 2003
2058 + 2166	*Streptococcus pyogenes*	A/U	G/C	Farrell et al. 2006
2058 + 2160	*Haemophilus influenzae*	A/G	G/U	Peric et al. 2003
2059 + 2059	*S. pneumoniae*	A/A	G/C	Farrell et al. 2004
2058	*Brachyspira hyodysenteriae*	A	G, T	Karlsson et al. 1999
	E. coli	A	G	Vester and Garrett 1987; Douthwaite 1992
	E. coli	A	T	Sigmund et al. 1984
	H. influenzae	A	G	Clark et al. 2002
	H. pylori	A	C, G	Stone et al. 1996; Hulten et al. 1997; Occhialini et al. 1997; Versalovic et al. 1997; Debets-Ossenkopp et al. 1998; Wang and Taylor 1998
	Moraxella catarrhalis	A	T	Saito et al. 2012; Iwata et al. 2015
	Mycobacterium abscessus	A	G	Wallace et al. 1996
	Mycobacterium avium	A	C, G, T	Nash and Inderlied 1995
	Mycobacterium chelonae	A	C, G	Wallace et al. 1996
	Mycobacterium intracellulare	A	C, G, T	Meier et al. 1994
	Mycobacterium kansasii	A	T	Burman et al. 1998
	Mycobacterium smegmatis	A	G	Sander et al. 1997
	M. genitalium	A	C, G	Jensen et al. 2008; Ito et al. 2011; Chrisment et al. 2012; Gesink et al. 2012; Twin et al. 2012; Touati et al. 2014; Bissessor et al. 2015
	M. hominis	A	G	Pereyre et al. 2002
	Mycoplasma pneumoniae	A	C, G, T	Lucier et al. 1995; Matsuoka et al. 2004; Liu et al. 2009b; Peuchant et al. 2009; Xin et al. 2009; Cao et al. 2010; Kawai et al. 2013; Ye et al. 2013; Zhou et al. 2015
	Propionibacterium spp.	A	G	Ross et al. 1997
	Staphylococcus aureus	A	G, T	Prunier et al. 2002
	S. pneumoniae	A	G, T	Tait-Kamradt et al. 2000a,b; Canu et al. 2002; Farrell et al. 2004
	S. pyogenes	A	G	Jalava et al. 2004
	Treponema pallidum	A	G	Stamm and Bergen 2000

Continued

Table 1. *Continued*

Nucleotide	Organism	Wild-type	Mutant	References
2059	*H. influenzae*	A	G	Clark et al. 2002
	H. pylori	A	C, G, T	Hulten et al. 1997; Occhialini et al. 1997; Versalovic et al. 1997; Debets-Ossenkopp et al. 1998; Wang and Taylor 1998
	Propionibacterium spp.	A	G	Ross et al. 1997
	M. abscessus	A	C, G	Wallace et al. 1996
	M. avium	A	C	Nash and Inderlied 1995
	M. chelonae	A	G	Wallace et al. 1996
	M. intracellulare	A	C	Meier et al. 1994
	M. genitalium	A	C, G	Jensen et al. 2008; Ito et al. 2011; Chrisment et al. 2012; Gesink et al. 2012; Twin et al. 2012; Touati et al. 2014; Bissessor et al. 2015
	M. pneumoniae	A	G	Matsuoka et al. 2004; Peuchant et al. 2009; Xin et al. 2009; Cao et al. 2010; Kawai et al. 2013
	S. aureus	A	G	Prunier et al. 2002
	S. pneumoniae	A	C, G	Tait-Kamradt et al. 2000a; Farrell et al. 2004; Rantala et al. 2005
2062	*M. genitalium*	A	T	Chrisment et al. 2012
	M. pneumoniae	A	G	Bebear and Pereyre 2005; Peuchant et al. 2009
	S. pneumoniae	A	C	Depardieu and Courvalin 2001
2098	*H. pylori*	T	C	Kim et al. 2008; Rimbara et al. 2008[a]
2160	*H. influenzae*	G	U	Peric et al. 2003
2160–2162	*H. influenzae*	GGA	UAU	Peric et al. 2003
2164	*H. influenzae*	C	G	Peric et al. 2003
2185	*M. genitalium*	T	G	Shimada et al. 2011
2609	*E. coli*	U	C	Garza-Ramos et al. 2001
2610	*M. hominis*	C	U	Pereyre et al. 2002
2611	*Chlamydia trachomatis*	C	T	Misyurina et al. 2004
	E. coli	C	T	Vannuffel et al. 1992
	H. pylori	C	A	Rimbara et al. 2008
	M. hominis	C	T	Pereyre et al. 2002
	M. pneumoniae	C	A, G	Matsuoka et al. 2004; Peuchant et al. 2009; Kawai et al. 2013; Ye et al. 2013
	Neisseria gonorrhoeae	C		Ng et al. 2002
	S. pneumoniae	C	A, G	Tait-Kamradt et al. 2000b; Pihlajamaki et al. 2002; Farrell et al. 2003; Farrell and Felmingham 2004
	S. pneumoniae	C	T	Rantala et al. 2005
	S. pyogenes	C	T	Malbruny et al. 2002

Not all mutations have been shown to solely cause macrolide/ketolide resistance; base substitutions can occur in the background of ribosomal protein changes.

[a]This mutation was not required for clarithromycin resistance in the background of C2611A mutation in (Rimbara et al. 2008).

An analysis of 14 of 217 erythromycin-resistant clinical *S. pneumoniae* isolates collected in Finland in 2002 (Rantala et al. 2005) characterized by polymerase chain reaction (PCR) did not harbor an efflux mechanism, *mef*(E) or *mef*(A), or a target modification mechanism, *erm*(B) or *erm*(A) subclass *erm*(TR), but did have previously identified A2059G mutations in one or more 23S rRNA genes. As had been shown previously, erythromycin minimum inhibitory concentrations (MICs) increased with an increasing number of *rrl* alleles containing A2059G (Tait-Kamradt et al. 2000a), whereas a C2611T mutation was present in all four alleles (Rantala et al. 2005). In *Helicobacter pylori*, only one of the two alleles needs to contain a 23S rRNA mutation to result in macrolide resistance (Hulten et al. 1997). Mutations in *rrl* are also found in combination with mutations in genes encoding L4 or L22 in many species.

Ribosomal Protein Mutations

Mutations in genes encoding ribosomal proteins L4 and L22 in laboratory isolates of *E. coli* and clinical isolates of *S. pneumoniae* can confer erythromycin resistance and reduced telithromycin susceptibility (Tait-Kamradt et al. 2000a; Pihlajamaki et al. 2002). In addition to the changes detailed below, a list of L4 and L22 mutations can be found in Franceschi et al. (2004).

Changes within a highly conserved sequence of *S. pneumoniae* L4 ($_{63}$KPWRQKGTG RAR$_{74}$) can result in decreased susceptibility to macrolides or ketolides (a 500-fold increase to a telithromycin MIC of 3.12 μg/mL for one variation) as well as alter fitness or confer temperature sensitivity of growth (Tait-Kamradt et al. 2000a,b; Farrell et al. 2004). This sequence forms the loop that extends into the PET. Mutations in L4 that have been identified within the conserved sequence encode $_{68}$E$_{69}$, $_{68}$KEG$_{69}$, or $_{68}$GQK$_{69}$ insertions; T$_{94}$I, E$_{30}$K, S$_{20}$N, G$_{71}$R, I$_{78}$V, K$_{68}$S, K$_{68}$Q, $_{69}$VP$_{70}$, $_{69}$TPS$_{71}$, or V$_{88}$I substitutions; $_{69}$GTGR$_{72}$ or $_{64}$P—Q$_{67}$ deletions. In *S. pyogenes* isolates from children treated with azithromycin, amino acid variations in L4 ($_{64}$WR$_{65}$ or $_{69}$TG$_{70}$ deletion; insertion of RA

after position 73, $_{73}$RA) were uncovered (Bingen et al. 2002). In *H. influenzae*, L4 amino acid variations (insertion $_{65}$GT, K$_{61}$Q, T$_{64}$K, G$_{65}$D; deletion $_{65}$GR, G$_{53}$A; deletion of $_{66}$RA, A$_{69}$S, T$_{82}$I, D$_{94}$E, D$_{139}$G), some outside the loop region, could provide high-level resistance of up to 128 μg/mL for 14- and 15-membered macrolides (Clark et al. 2002; Peric et al. 2003). *S. aureus* isolates with L4 amino acid changes R$_{168}$S, G$_{69}$A, and T$_{70}$P have been described in CF patients (Prunier et al. 2005). Amino acid variations in L4 from *M. genitalium* (N$_{21}$K, H$_{69}$R, V$_{84}$G, E$_{128}$G, P$_{81}$S, Y$_{135}$P, N$_{172}$S, N$_{172}$S, A$_{114}$V, A$_{116}$V, A$_{114}$S, R$_{45}$K) were found encoded in DNA from the urine of men with nongonococcal urethritis (Shimada et al. 2011), and in the chromosomal DNA isolated from a collection of *M. genitalium* isolates (Jensen et al. 2014). The L4 A$_{209}$T variation was found in chromosomal DNA of *Mycoplasma pneumoniae* isolates from patients (Cao et al. 2010).

Mutations encoding amino acid changes in the carboxy-terminal region of ribosomal protein L22 (e.g., G$_{95}$D, P$_{99}$Q, A$_{93}$E, P$_{91}$S, G$_{83}$E, A$_{101}$P, $_{109}$RTAHIT$_{114}$ tandem duplication) resulted in decreased susceptibility to macrolides and ketolides, although the MICs were not greater than 1 μg/mL in *S. pneumoniae* (Canu et al. 2002; Farrell et al. 2003). A mutation encoding K$_{94}$Q along with a large deletion in the *erm*(B) upstream region was selected by telithromycin in a *S. pneumoniae* isolate with *erm*(B) (Walsh et al. 2003). In *H. influenzae* clinical isolates, MICs increased 4- to 16-fold with insertions or deletions in L22 (G$_{91}$D; insertions of $_{77}$DEGPSM, $_{88}$RAKG, $_{91}$KG, $_{91}$RAG, or $_{91}$RADR; deletions of $_{81}$S, $_{82}$M, $_{91}$KG, $_{95}$R, $_{95}$RI, or $_{96}$ILKR) (Clark et al. 2002; Peric et al. 2003). A deletion of three amino acids in L22 associated with an A2058 mutation has also been reported in a *S. aureus* isolate from CF patients (Prunier et al. 2002). Amino acid changes in L22 from *M. genitalium* (A$_{43}$V, G$_{93}$E + D$_{109}$E, S$_{81}$T, S$_{81}$N, M$_{82}$L, N$_{112}$D, R$_{114}$K, E$_{123}$K) were found in men with nongonococcal urethritis (Ito et al. 2011; Shimada et al. 2011). In *M. pneumoniae*, all 14-membered macrolide-resistant isolates harbored a T$_{508}$C mutation in L22 and, for most, either an

   Cite this article as *Cold Spring Harb Perspect Med* doi: 10.1101/cshperspect.a025395

A2058G or A2059G mutation in 23S rRNA (Cao et al. 2010; Jensen et al. 2014).

Resistance to telithromycin in *S. pneumoniae* significantly increases when 23S rRNA methylation/mutations are combined with ribosomal protein mutations. For example, a combination of a truncated leader peptide leading to constitutive synthesis of *erm*(B) conferred a telithromycin MIC of 16 μg/mL (Wolter et al. 2008a), whereas clinical isolates with both a constitutive *erm*(B) and a $_{69}GTG_{71}$ to TPS substitution in L4 (Wolter et al. 2007) or a combined A2058T mutation and a three-amino acid deletion in L22 (Faccone et al. 2005), provided high-level telithromycin resistance (≥256 μg/mL). A patient with a *S. pneumoniae* isolate harboring an A2058G mutation in 23S rRNA and an RTAHIT insertion in L22 between amino acid T_{108} and V_{109} resulted in a telithromycin MIC of 16 μg/mL (Perez-Trallero et al. 2003). In addition, a telithromycin-resistant isolate with a MIC of 8 μg/mL was found to contain an *erm*(B) gene, an $S_{20}N$ variation in L4, and a number of mutations in 23S rRNA (Reinert et al. 2005). A highly resistant laboratory-generated *S. pneumoniae* strain (MIC, 32 μg/mL) contained a 210-bp deletion in the *erm*(B) upstream region together with a $K_{94}Q$ mutation in L22 (Walsh et al. 2003).

erm Genes

A major and widespread mechanism of resistance to the macrolide class of antibiotics is mediated by *erm* genes that encode rRNA methyltransferases that add one or two methyl groups to the exocyclic amino group of A2058 (Figs. 3 and 4) located in the PET of 23S rRNA (Horinouchi and Weisblum 1980; Weisblum 1995a). In addition to conferring resistance to 14-, 15-, and 16-membered macrolides and ketolides, resistance to two other classes of antibiotics, lincosamides and streptogramin B, is imparted, giving the host a MLS_BK phenotype (Sutcliffe and Leclercq 2002; Roberts 2008; Schwarz et al. 2016).

As of January 2016, 38 *erm* genes have been reported (see faculty.washington.edu/marilynr). Among the *erm* genes, the most commonly carried is *erm*(B) (36 genera), followed by *erm*(C) (32 genera), *erm*(F) (25 genera), *erm*(X) (15 genera), *erm*(V) (11 genera), *erm*(A) (nine genera), *erm*(G) and *erm*(E) (seven genera each), *erm*(Q) (six genera), *erm*(T) (four genera), *erm*(42) (three genera), *erm*(D) and *erm*(R) (two genera each). The remaining 25 *erm* genes are found in a single genus. Sixteen (46%) of the *erm* genes (*erm*(H), *erm*(I), *erm*(N), *erm*(O), *erm*(R), *erm*(S), *erm*(U), *erm*(W), *erm*(Z), *erm*(30), *erm*(31), *erm*(32), *erm*(34), *erm*(36), *erm*(37), *erm*(38), *erm*(39), *erm*(40), *erm*(41), and *erm*(46)) are unique to environmental bacteria, defined as those species and genera that are primarily found outside of humans and animals.

Inducible or Constitutive MLS_B Phenotype

Depending on the nature of leader sequences upstream of the translational start site, *erm* genes are either inducible by antibiotics or constitutively expressed; examples include *erm*(A) (Murphy 1985), *erm*(B) (Min et al. 2008), *erm*(C) (Gryczan et al. 1980; Horinouchi and Weisblum 1980; Weisblum 1995b), and *erm*(D) (Hue and Bechhofer 1992). For inducible *erm* genes, there are leader sequences upstream of the translational start site that form at least two stem and loop structures, one of which sequesters the ribosomal start site for the resistance gene, and the other upstream stem-loop structure that overlaps ORFs for one (*erm*(C), *erm*(B), *erm*(D)) or two (*erm*(A)) short peptides. Thus, in the absence of an inducing antibiotic, the upstream leader sequence and attending peptide is synthesized, but there is no synthesis of the *erm* gene because of sequestration of its ribosome-binding site. In the macrolide-bound ribosome, a macrolide-stalling motif in the nascent leader peptide is encountered and translation is stalled. The stalled ribosome allows an alternative messenger RNA (mRNA) secondary structure to form, such that the ribosome-binding site for the *erm* gene is exposed and available for translation by a ribosome not bound by erythromycin (Min et al. 2008; Arenz et al. 2014b). The programmed arrest of translation is both inducer (small molecule)- and leader peptide-specific (Mayford and Weisblum 1990; Vazquez-Laslop et al. 2011; Kannan et al. 2014).

The high degree of variability among the regulatory leader regions of the mRNA transcripts for the different classes of *erm* genes, despite the highly conserved nature of the genes themselves, allows for a variety of phenotypes pertaining to induction by particular antibiotics (Subramaniam et al. 2011). Although most *erm* genes are induced by the 14- or 15-membered macrolides and not by 16-membered macrolides or ketolides, exceptions have been noted. The *erm*(B) subgroup *erm*(AMR) from a clinical strain of *Enterococcus faecalis* (Oh et al. 1998), *erm*(S) subgroup *erm*(SF) (Kelemen et al. 1994), and *erm*(V) subgroup *erm*(SV) (Fujisawa and Weisblum 1981) in 16-membered macrolide-producing *Streptomyces* spp. have been shown to be induced by tylosin and, in the latter, other 16-membered macrolides as well (Kamimiya and Weisblum 1997). Inducible resistance in *Streptomyces* spp. is the most diverse, with induction by lincomycin and streptogramin B in corresponding producers resulting in N^6 dimethylation of 23S rRNA and a MLS$_B$-resistant phenotype (Fujisawa and Weisblum 1981).

The length of the leader peptide can vary. The Erm(C) leader peptide (ErmCL) is 19 amino acids, with amino acids IFVI (I6–I9) constituting an important macrolide-stalling motif that triggers ribosome pausing. Cryoelectron microscopy (cryo-EM) of the erythromycin-dependent ErmCL-stalled ribosome complex (SRC) (Arenz et al. 2014b) revealed the path of the ErmCL nascent polypeptide chain, its contact with erythromycin, and its interactions with 23S rRNA nucleotides U2506, U2586, and A2062 within the ribosomal tunnel. Interactions of ErmCL amino acids V8 and F7 with U2506 and I6 with U2586 are consistent with experiments that show that mutations in the conserved I6–I9 motif severely reduce ribosome stalling (Vazquez-Laslop et al. 2008; Johansson et al. 2014). An interaction of the amino terminus (minimally I3) of ErmCL with A2062 stabilizes an unusual conformation such that this nucleotide is forced to lie flat against the tunnel wall instead of protruding into the tunnel lumen, thereby allowing an interaction with A2503, consistent with the findings that mutations A2062U/C or A2503G dramatically alleviate ErmCL stalling (Vazquez-Laslop et al. 2008, 2010). In addition, ErmCL was observed to directly interact with the cladinose sugar of erythromycin, providing a structural explanation for how the nascent chain monitors the presence of erythromycin in PET. In the ErmCL-SRC, the PTC is remodeled because of interactions of the ErmCL nascent chain with U2586, U2506, and A2062, promoting a flipped conformation of U2585, which makes it unfavorable for the A-tRNA to fully accommodate, leading to dissociation and translation arrest (Arenz et al. 2014b). Previous crystallographic studies have shown that accommodation of the CCA-end of the A-site tRNA requires movement of nucleotides U2584 and U2585 (Schmeing et al. 2005; Simonovic and Steitz 2009).

Although ketolides lack the C3 cladinose and do not induce *erm*(C) by the mechanism described above, they can promote its expression by inducing ribosomal frame-shifting errors within the *erm*(C) leader ORF (Gupta et al. 2013a). Telithromycin induces a (-1) frameshift within a string of four adenine residues in the last two sense codons of *ermCL*, resulting in a read-through of the stop codon and unmasking of the *erm*(C) start codon because of the subsequent change in secondary mRNA structure. When other macrolides were tested on a model leader construct, it was found that frameshifting was also an intrinsic property (although to different degrees) of 14-membered macrolides as well as other ketolides.

In contrast to ErmCL, the ErmBL is 36 amino acids in length and *erm*(B) is induced by a wider range of 14- and 15-membered macrolides, including those that lack the C3 cladinose or have modifications of this sugar (Arenz et al. 2014a). Cryo-EM of ErmBL-SRC with erythromycin shows that nascent ErmBL travels a unique path in the PET and does not come into contact with the antibiotic, thereby defining a paradigm distinctly different than the one used for *erm*(C) induction (Arenz et al. 2014b). Stalling occurs after 10 amino acids have been polymerized and cryo-EM of the ErmBL-SRC

Cite this article as *Cold Spring Harb Perspect Med* doi: 10.1101/cshperspect.a025395

shows that the P-site and the A-site are filled with the ErmBL-tRNA and the Lys-tRNA (K_{11}), respectively (Min et al. 2008; Vazquez-Laslop et al. 2010). No further peptide bond formation occurs because interactions of amino acids $_9VD_{10}$ and R_7 with U2585 and U2586, respectively, stabilize U2585 in a position that precludes Lys-tRNA from being properly accommodated in the A-site. Notably missing in the ErmBL-SRC was any interaction of the nascent peptide with A2062 that was so critical for ErmCL-mediated ribosome stalling.

Gene regulation by nascent-peptide-dependent ribosome stalling expands beyond antibiotic resistance genes. Other examples in bacteria include translation arrest at the *secM* ORF activating the expression of *secA* (Nakatogawa and Ito 2002; Bhushan et al. 2011) and ribosome stalling of the *tnaC* ORF regulating the expression of the tryptophanase operon (Gong and Yanofsky 2002; Seidelt et al. 2009).

The *erm*(D) subclass *erm*(K) is regulated by both transcription and translation attenuation (Kwak et al. 1991; Choi et al. 1997), whereas *erm*(37) in *Mycobacterium tuberculosis* (Buriankova et al. 2004) has been shown to be activated by the erythromycin-inducible transcription activator WhiB7 (Morris et al. 2005). The *erm*(41) gene intrinsic in *Mycobacterium abscessus* and *M. bolletii* has been shown to be inducible by macrolides and ketolides; however, sequence analysis of the upstream region does not provide compelling evidence for regulation of this gene by either transcription or translation attenuation or by the inducible transcription factor WhiB (Nash et al. 2009).

Constitutive MLS$_B$ resistance can be conferred by a variety of mutations in the leader sequence (Sutcliffe and Leclercq 2002; Subramaniam et al. 2011), and includes deletions of the entire attenuator region for *erm*(C) in clinical isolates of *S. epidermidis* and *S. aureus* (Lampson and Parisi 1986) and for *erm*(B) in *E. faecalis*, *S. agalactiae*, and *S. pneumoniae* (Martin et al. 1987; Rosato et al. 1999; Wolter et al. 2008b) as well as tandem duplications in the attenuator of *erm*(C) of *S. aureus* and *S. equorum* (Oliveira et al. 1993; Lodder et al. 1997), which either destabilize the hairpin

structure sequestering the initiation sequences for the methyltransferase or duplicate the initiation sequences, leaving one unsequestered and available for translation. Notably, constitutive *erm*(B)-containing pneumococcal isolates with a higher percentage of 23S rRNA methylation were telithromycin-resistant (Douthwaite et al. 2005; Wolter et al. 2008a).

Clindamycin and 16-membered macrolides with more than an amino sugar at C5 do not induce *erm* expression in most species. This is now understood because both clindamycin and 16-membered macrolides directly interact with the PTC, inhibiting peptide bond synthesis; thus, is it unlikely that the synthesis of nascent peptide longer than a few amino acids could be synthesized, too short for the ribosome to sense or the nascent peptide to interact with the antibiotic (Tenson et al. 2003). Clinical isolates of *S. pyogenes* or *S. agalactiae* with variations in the leader sequence, including point mutations, insertions, deletions, and duplications, were shown to be resistant to both erythromycin and clindamycin, showing that constitutive resistance yields an MLS$_B$ phenotype (Culebras et al. 2005; Doktor and Shortridge 2005). In this study, three isolates with a 44-base duplication/insertion corresponding to bases 188 to 231, duplicating the *erm*(A) ribosomal binding site and start site, and one isolate with a 68-bp deletion of the entire leader peptide 2 region, were also resistant to the ketolide telithromycin.

Why are most *erm* genes inducible rather than constitutively expressed? Ribosome methylation at A2058 exerts a fitness cost because of the change in the ribosome's ability to sense/respond to nascent peptides, thereby changing the expression of a number of cellular polypeptides (Ramu et al. 2009; Gupta et al. 2013b; Wilson 2014). Thus, deregulation of translation may well explain why bacteria prefer to retain the ability to become conditionally resistant.

cis-Acting Peptides

Translation of a pentapeptide encoded in *E. coli* 23S rRNA can cause macrolides to dissociate from the ribosome, thereby conferring macrolide resistance (Tenson et al. 1996). Other *cis*-

acting peptides (resistance conferred only to a ribosome on which the peptide is synthesized) have been identified using a random-library approach, providing a consensus sequence, fMet-(bulky/hydrophobic)-(Leu/Ile)-(hydrophobic)-Val for erythromycin resistance; other consensus peptides specific for different macrolides (e.g., oleandomycin, ketolides, 15-membered macrolides) were also identified, consistent with the ability of the PET to distinguish small changes in antibiotic/nascent peptide interactions (Tenson et al. 1997; Vimberg et al. 2004). When key amino acids are synthesized in specific short peptides, the affinity of the macrolide/ketolide for its binding site is weakened, but removal of the antibiotic from the ribosome is most likely when the pentapeptide is removed from the peptidyl-tRNA by class I release factor (Lovmar et al. 2006).

Rlm Methyltransferases

The importance of modifications to 23S rRNA is not completely understood. However, for tylosin producer *Streptomyces fradiae*, monomethylation of G748 and A2058 by rRNA methyltransferase RlmA^{II} and Erm(N), respectively, is needed for self-preservation (Liu and Douthwaite 2002b; Takaya et al. 2013). Certain Gram-positive bacteria like *S. pneumoniae* have an intrinsic chromosomal *rlmA^{II}* gene (Liu and Douthwaite 2002a). Molecular modeling shows that the methyl group of G748 stabilizes the binding of telithromycin to the ribosome by moving the alkyl–aryl arm of telithromycin toward the aromatic rings of A752 in helix 35 (Fig. 3 shows the positions of A752 and A2058) (Takaya et al. 2013). Mutations in *rlmA^{II}* preventing methylation of G748 in *S. pneumoniae* isolates that also harbor a constitutive *erm*(B) result in 16- to 32-fold greater resistance to telithromycin (Takaya et al. 2013). Just to keep things interesting, *S. pneumoniae* has another methyltransferase, RlmCD, which mediates a methyl transfer to both U747 (in helix 35) and U1939. Recent data have shown that RlmA^{II} prefers U747-methylated 23S rRNA as a substrate; thus, when these two methyltransferases work sequentially, the binding of telithromycin to the ribosome is facilitated and, in their absence, telithromycin resistance occurs when there is dimethylation of A2058 by an *erm* methylase (Shoji et al. 2015).

Acquired Macrolide Efflux Mechanisms

Mef Family. Mef pumps are members of the major facilitator superfamily and have 12-transmembrane domains connected by hydrophilic loops, with both the amino and carboxyl termini located in the cytoplasm (Paulsen et al. 1996; Pao et al. 1998). They are antiporters with a binding site for macrolide antibiotics and have a rocker-switch type of movement, in which conformational changes in the protein are elicited to efflux a macrolide in exchange for a proton (Law et al. 2008). The *mef* genes are largely found among Gram-positive bacteria, but have also been reported in Gram-negative species (see faculty.washington.edu/marilynr; Ojo et al. 2004). The two major subclasses, *mef*(A) (Clancy et al. 1996), first identified in *S. pyogenes* isolates, and *mef*(E) (Tait-Kamradt et al. 1997), initially identified in *S. pneumoniae*, were characterized as providing resistance to 14- and 15-membered macrolides but not 16-membered macrolides, lincosamides, or streptogramin B, thereby affording the M (macrolide-resistant only) phenotype (distinguishable from the MLS$_B$ phenotype). Both pumps are collectively categorized as *mef*(A) because of their >80% amino acid sequence identity (Roberts et al. 1999). However, the two genes are carried on distinct genetic elements; Tn*1207.1*, Tn*1207.3*, φ10394.4, or φm46.1 for *mef*(A) (Santagati et al. 2000, 2003; Giovanetti et al. 2003; Varaldo et al. 2009) and a macrolide efflux genetic assembly (megacomplex and derivative transposons Tn*2009* and Tn*2010* for *mef*(E)) (Gay and Stephens 2001; Del Grosso et al. 2002, 2004, 2006; Varaldo et al. 2009). Interestingly, although the megacomplex does not carry a transposase or recombinase as in Tn*1207.1*, the related *mef*-family complexes carry an adjacent ATP-binding cassette (ABC)-type transporter gene with sequences encoding two fused nucleotide-binding domains but no membrane-spanning domains, known as *mel* or *msr*(D) (because of its similarity to the *S. aureus* *msr*(A) gene) (Gay and Stephens 2001; Del Grosso et al. 2002). The *msr*(D) gene is adjacent to and cotranscribed with *mef*(E) in the presence of inducers like erythromycin. The exact function of Msr(D) in streptococci has yet to be fully elucidated, but coexpression of *msr*(D) and *mef*(E) is required for high-level macrolide efflux in *S. pneumoniae*, and both proteins interact synergistically to increase macrolide resistance in *E. coli* (Ambrose et al. 2005; Nunez-Samudio and Chesneau 2013). In *E. coli*, a physical association of Msr(D) and Mef(E) was shown using a mef(E)-green fluorescent protein (GFP) fusion, in which it appeared that Msr(D) directed Mef(E)-GFP to the cell poles, possibly assisting in the assembly of Mef(E) in the membrane and/or enhancing macrolide efflux as part of a composite transporter (Nunez-Samudio and Chesneau 2013).

Owing, in part, to the genomic plasticity and natural competency of *S. pneumoniae*, mega has been found in multiple chromosomal locations as well as inserted into other composite mobile elements carrying genetic markers of multiple *Streptococcus* species (Chancey et al. 2015a). Conjugal transfer rates of Tn*1703.1* carrying *mef*(A) have been found to be highly variable between different *S. pyogenes emm*-types, ranging in frequencies of 1.13×10^{-6} to 7.2×10^{-8} in various isolates, with higher frequencies in *emm1* and *emm4* isolates (Hadjirin et al. 2013). The *mef*(A) gene has also been found on a large chimeric chromosomal element that also carries the *tet*(O) gene, an element that lends a distinct *Sma*I-typable pulse-field gel electrophoresis profile to isolates carrying the construct (Brenciani et al. 2004; Bacciaglia et al. 2007).

A novel *mef*(B) gene found in *E. coli* porcine isolates located near *sul3* on plasmids has been described with 38% protein identity (62% similarity) to Mef(A) (Liu et al. 2009a). When *mef*(B) was cloned into a plasmid and transformed into *E. coli* JM109, transformants had the M phenotype. The plasmid location as well as the genetic organization of the *mef*(B) gene were distinct from its organization in conjugative transposons. The GC content (44.95%) was lower than that of *E. coli*, suggesting horizontal transition from another organism.

More recently, a novel subclass *mef*(C) gene was identified, along with a macrolide phosphotransferase *mph*(G), on plasmid pAQU1 isolated from marine bacteria including *Vibrio* and *Photobacterium* (Nonaka et al. 2015). Another subclass, termed *mef*(I) and first identified in isolates of *S. pseudopneumoniae* (Cochetti et al. 2005), has since been isolated from *S. pneumoniae* located on a novel IQ element inserted into defective Tn*5252* and Tn*916* sequences along with a unique *msr*(D) gene variant and *catQ*, a chloramphenicol acetyltransferase (Mingoia et al. 2007). The *mef*(O) subclass, which has a high degree of similarity to *mef* genes from *S. dysgalactiae*, was identified in *S. pyogenes* isolates from Norway (Sangvik 2005; Blackman Northwood et al. 2009). In addition to these, *mef*(B) and *mef*(G) genes were identified in *S. agalactiae* and group G streptococci, respectively, conferring an M phenotype and showing high degrees of sequence identity to each other, although <90% sequence identity to *mef*(A) or *mef*(E) (Amezaga and McKenzie 2006; Cai et al. 2007). However, *mef*(C), *mef*(G), *mef*(I), and *mef*(O) have >80% amino acid sequence homology with class *mef*(A), so none are recognized as a separate class at the website maintained by Marilyn Roberts (see faculty.washington.edu/marilynr; Roberts et al. 1999).

Derivatives of *S. pyogenes* sequence type 39 have been found carrying multiple mosaic *mef* gene variants encompassing the 5′ and terminal portions of *mef*(A) combined with a region of *mef*(E) spanning the majority of bases 570–1100 (*mef*(A) sequence numbering used), often on a φm46.1-like element and in conjunction with the tetracycline resistance gene *tet*(O) (Blackman Northwood et al. 2009; Del Grosso et al. 2011). Some composite resistance genes containing *mef*-family sequences have been detected in which the *msr*(D) gene was not readily amplified by PCR; however, it was not clear whether this was because of sequence variance or absence of the genes (Cerda Zolezzi et al. 2004).

Induction of the *mef*(E)-*msr*(D) operon has been linked to the presence of substrate macrolides (Daly et al. 2004; Ambrose et al. 2005;

Wierzbowski et al. 2005), with induction occurring by most 14- and 15-membered macrolides, including ketolides (Chancey et al. 2011). The expression of the efflux operon in response to drug exposure appears to be correlated with the presence of a free hydroxyl at the 2′ position of a monosaccharide amino sugar like desosamine at position C5, rather than correlated to macrocyclic ring size or C3 sugar composition (see Fig. 1 for macrolide structures). For example, troleandomycin with an acetate substitution at the 2′ hydroxyl of desosamine, does not induce mef(E)-msr(D) (Chancey et al. 2011) nor do the majority of 16-membered macrolides with a disaccharide at C5 (see Fig. 2 for 16-membered macrolide structures). mef(E)-msr(D)-inducing 16-membered macrolides, such as tilmicosin and rosamicin, and the 14-membered ketolide telithromycin have C5 monosaccharides that bind in the ribosome in locations similar to efflux substrates erythromycin and azithromycin, but distinct from weakly/noninducing macrolides with C5 disaccharides or modifications to C3 or C5 sugars (Chancey et al. 2011). Although both rosamicin and tilmicosin induced expression of mef(E)/msr(D), only tilmicosin appears to be a substrate for the Mef(E)/Msr(D) pumps (Chancey et al. 2011).

mef genes are regulated by transcription attenuation, with the induction of the mef(E)/msr(D) operon occurring by anti-attenuation of transcription in the presence of inducing macrolides; however, there is also evidence that other regulatory mechanisms influence the control of mef(E)/msr(D). There is a leader peptide encoded 34 bp upstream of the mef(E) start codon that is required for full expression of mef(E)/msr(D) (Subramaniam et al. 2011; Chancey et al. 2015b). Macrolide-bound ribosomes stall in the leader peptide, causing a shift in mRNA conformation, similar to induction of activity of erm methyltransferase genes (Chancey et al. 2015a). The mef(E) gene is also induced by unrelated structures such as LL-37, a cationic antimicrobial peptide produced in human macrophages, and two related murine homologs (Zahner et al. 2010). LL-37 may induce mef(E) by a different mechanism, but induction

by either LL-37 or erythromycin confers resistance to both.

Msr Family. There are four macrolide efflux msr types, with each class having ≤80% amino acid homology with any member of any other type (see faculty.washington.edu/marilynr). All Msr classes have ATP-binding motif sequence homology with the ATP-binding transport superfamily (Ross et al. 1990). msr(A) or msr(B) genes (encoding a polypeptide homologous to the carboxyl terminus of Msr(A)) were first identified in *S. epidermidis* and *S. xylosus*, respectively (Ross et al. 1990; Milton et al. 1992), but are now characterized as a single class. These genes have also been described in clinical *S. aureus* isolates (Wondrack et al. 1996; Matsuoka et al. 1999, 2003). The msr(A) family genes confer resistance to 14- and 15-membered macrolides and streptogramin B (MS phenotype), and low-level resistance to ketolides (Ross et al. 1995, 1996; Wondrack et al. 1996; Canton et al. 2005; Reynolds and Cove 2005; Vimberg et al. 2015). Related msr-family efflux genes have been isolated from other genera, including *Enterococcus* (msr(C), msr(A)), *Streptococcus* (msr(D), also known as mel), *Pseudomonas* (msr(A)), *Corynebacterium* (msr(A)), and in various environmental isolates (msr(E)) (Portillo et al. 2000; Ojo et al. 2006; Varaldo et al. 2009; Desmolaize et al. 2011; Roberts 2011).

The structure of the Msr(A) protein is classified as a class 2 ABC-transporter, containing two ATP-binding domains and a long Q-linker region, but not a typical membrane-spanning region. Thus, there exists more than one hypothesized mechanism for Msr(A) function (Ross et al. 1995, 1996; Reynolds et al. 2003). Theories describing the mechanism of action for Msr(A)-linked resistance involve interaction of Msr(A) with the ribosome, blocking binding to the 23S rRNA target site that overlaps macrolides and streptogramin B, or an ATP-dependent efflux pump activity mediated by interaction with membrane-spanning binding proteins (possibly Mef(E)) (Kerr et al. 2005; Nunez-Samudio and Chesneau 2013), or as a structure that helps with the localization and/or assembly of Mef(E) into the membrane (Nunez-Samudio and Chesneau 2013). Studies

Cite this article as *Cold Spring Harb Perspect Med* doi: 10.1101/cshperspect.a025395

with efflux pump inhibitors show Msr(A) function is uninhibited by reserpine, a common Gram-positive efflux pump inhibitor, but efflux activity was inhibited by arsenate, dinitrophenol, or CCCP, supporting the ATP-dependent function of the pump. To date, there has been no direct evidence to support the hypothesis of ribosome protection and recent evidence suggests that Mef(E) and Msr(D) may form a composite efflux pump (Nunez-Samudio and Chesneau 2013).

A recent study that examined the nature of telithromycin resistance in mutants selected in *S. aureus* RN4220 recombinantly expressing *msr*(A) found that mutations mapped to *clpX*, a protein that functions as both the substrate-recognizing component of the ClpXP proteolytic system and as a ClpP-independent chaperone for protein–DNA and protein–protein complexes (Burton et al. 2001). The decreased susceptibility of telithromycin (and erythromycin) was Msr(A)-mediated and related to loss-of-function mutations in ClpX only (Vimberg et al. 2015).

The *msr*(A) gene was initially isolated on a *S. epidermidis* plasmid designated pUL5050, along with a single-domain ATP-binding protein (*stpA*) and a hydrophobic protein (*smpA*), which are similar to *S. aureus* chromosomal genes stpC and smpC, but it was shown that these genes played no role in conferring macrolide resistance (Ross et al. 1996). A variety of hybrid resistance plasmids have been found carrying *msr*(A) and similar genes along with other resistance elements; pMS97 carrying *msr*(A) and the macrolide-inactivating phosphotransferase *mph*(C) (Matsuoka et al. 2003) and hybrid plasmids mediating combinations of penicillinase, tetracycline-efflux, and ribosomal methylation functions (Argudin et al. 2014) as examples. Expression of *msr*(A) is mediated in a similar manner to *erm* genes, via translation attenuation mechanisms, but requires higher amounts of inducer (Ross et al. 1990, 1996; Subramaniam et al. 2011). If the 320-bp control region upstream of *msr*(A) is deleted, the strain is constitutively macrolide-streptogramin B–resistant, analogous to deletions or mutations that destroy the secondary structure that is at-

tenuated by drug-dependent stalling within the leader peptide of *erm* genes.

MACROLIDE INACTIVATION

Macrolide Esterases

Inactivation of erythromycin by hydrolysis was first shown to be widespread in Enterobacteriaceae isolated from the human fecal flora and was usually associated with erythromycin therapy (Barthelemy et al. 1984; Andremont et al. 1985, 1986; Arthur and Courvalin 1986). Hydrolytic inactivation of macrolides by esterases specifically involves 14- and 15-membered macrolides; josamycin, midecamycin, rosaramycin, and spiramycin are not substrates (Arthur and Courvalin 1986; Arthur et al. 1987; Morar et al. 2012).

Two plasmid-encoded esterases, *ere*(A) and *ere*(B), conferring high-level erythromycin resistance (MIC ≥ 1 mg/mL), have been isolated from *E. coli*. The *ere*(A) gene on the self-transmissible plasmid pIP1100 encodes a product with a molecular weight of 37,765. The *ere*(B) gene, encoding an enzyme with a molecular weight of 51,000, was first identified on the self-transmissible plasmid pIP1527, which also contained the *erm*(B) gene, formerly known as *erxA* and *ermAM*, encoding an rRNA-methylating enzyme commonly found in streptococci (Arthur and Courvalin 1986). Based on GC content and codon usage, *ere*(A) (GC content 50%) is thought to have originated in Gram-negative bacteria, whereas *ere*(B) (GC content 36%), although originally discovered in *E. coli*, is thought to have originated from Gram-positive bacteria (Arthur et al. 1987). The gene for *erm*(B) is linked to *ere*(B) on plasmid pIP1527, and their physical linkage may be responsible for codissemination of the genes (Arthur et al. 1987). Similar codon usage in *ere*(B) and *erm*(B) suggests a similar Gram-positive bacterial origin; however, the separation of both genes on pIP1527 by GC-rich sequences suggests that both genes were integrated into plasmid pIP1527 by separate genetic events (Arthur and Courvalin 1986). It has been shown that coexpression of *ere*(B) and *erm*(B) more than additively contributes to erythromycin resistance in *E. coli* (Arthur and Courvalin 1986).

Ere(A), a type I esterase, and Ere(B), a type II esterase, both hydrolyze the lactone ring in 14-membered macrolides (Morar et al. 2012); however, the two enzymes are only weakly related with 25% protein sequence identity. Using a genomic enzymology approach, the catalytic mechanisms of the "erythromycin esterase superfamily" enzymes were compared (Morar et al. 2012). Ere(A), Ere(B), and two related enzymes from *Bacillus cereus*, Bcr135 and Bcr136, whose three-dimensional structures had previously been determined, were studied. Ere(A), Ere(B), and Bcr136 were found to be distinct, with only Ere(A) inhibited by chelating agents and hypothesized to contain a noncatalytic metal. Data from kinetic, mutagenesis, and modeling studies are consistent with all of the erythromycin esterases sharing a common catalytic mechanism, and efforts to detect a tightly bound metal in Ere(B) and Bcr136 were unsuccessful, leaving the hypothesis that Ere enzymes do not require a metal ion for their catalytic mechanism. Thus, the metal dependence of Ere(A) may be structural. A histidine residue, H_{46} (Ere(B) numbering), was found to be essential for catalytic function and proposed to serve as a general base in the activation of a nucleophilic water molecule. Ere(A) and Ere(B) substrate profiles differed. Ere(B) inactivated erythromycin, clarithromycin, roxithromycin, and azithromycin, but was inactive against the ketolide telithromycin. Ere(A) was unable to inactivate either telithromycin or azithromycin.

In recent years, the *ere*(A2) gene, a variant of *ere*(A) located in a class 1 integron cassette, has been found in *Enterobacter aerogenes*, *E. cloacae*, *E. coli*, *Klebsiella oxytoca*, *K. pneumoniae*, *Providencia stuartii*, *Pseudomonas* spp., *Salmonella enterica*, and *Vibrio cholera* (Chang et al. 2000; Peters et al. 2001; Kim et al. 2002; Thungapathra et al. 2002; Plante et al. 2003; Verdet et al. 2006; Abbassi et al. 2008; Chen et al. 2009; Krauland et al. 2010). Although macrolide antibiotics are generally not used in the treatment of nongastrointestinal infections caused by enteric bacteria, the spread of *ere*(A2) in Enterobacteriaceae is concerning because macrolides are often used in the treatment of traveler's diarrhea, and erythromycin is a common treatment of cholera in children and pregnant women (see cdc.gov/cholera/doc/recommend-anitbiotics-treatment.docx).

Phosphotransferases

Macrolide phosphotransferases are macrolide-inactivating enzymes widespread in Gram-negative and Gram-positive bacteria (Sutcliffe and Leclercq 2002; Roberts 2008) that, by in silico analysis, are in the same family as aminoglycoside and protein kinases (Shakya and Wright 2010). The first reported purifications of macrolide-2′-phosphotransferases were from macrolide-resistant *E. coli*, and this mechanism was soon shown to be prevalent in *E. coli* clinical isolates in Japan (O'Hara et al. 1989; Kono et al. 1992; Taniguchi et al. 2004). Macrolide 2′-phosphotransferases, commonly found on mobile genetic elements, are inducible (e.g., *mph*(A)) or constitutively expressed (e.g., *mph*(B)) intracellular enzymes capable of transferring the γ-phosphate of nucleotide triphosphate to the 2′-OH group of 14-, 15-, and 16-membered-ring macrolide antibiotics, thereby disrupting the macrolide's key interaction with A2058. Although early studies showed Mph enzymes could use ATP, more recent work with Mph(A) has shown a preference for GTP under physiologically relevant in vitro assay conditions (Shakya and Wright 2010). Expression of *mph*(A) is induced by erythromycin, and, recently, the structure of the MphR(A) repressor protein, a negative regulator of *mph*(A) expression, has been solved uncomplexed and complexed with erythromycin to 2.00 Å and 1.76 Å resolutions, respectively (Zheng et al. 2009). Erythromycin binds with a stoichiometry of 1:1 to each monomer of the functional MphR(A) dimer in a large hydrophobic cavern composed of residues from α helices of one monomer and the dimeric interface of the other monomer that appears to close around the ligand as it binds (Zheng et al. 2009).

Seven distinct macrolide phosphotransferases have been identified to date (see faculty.washington.edu/marilynr). The first identified

Cite this article as *Cold Spring Harb Perspect Med* doi: 10.1101/cshperspect.a025395

phosphotransferases, Mph(A) and Mph(B), share 37% amino acid identity (O'Hara et al. 1989; Kono et al. 1992). The G+C contents of the mph(A) gene (66%) (Noguchi et al. 1995) and the mph(B) gene (38%) (Noguchi et al. 1996) differ significantly from each other and the G+C content of $E.$ $coli$ chromosome (50%) (Muto and Osawa 1987), suggesting an exogenous nature of their origins. The mph(B) has been functionally expressed in both $E.$ $coli$ and $S.$ $aureus$, providing a first indication of the potential promiscuity of this macrolide resistance mechanism (Noguchi et al. 1998). In contrast, the mph(A) gene could be expressed in $E.$ $coli$ but not in $S.$ $aureus$, presumably because of its relatively higher G+C content relative to that of the $S.$ $aureus$ chromosome (33%) (Muto and Osawa 1987). The mph(C) gene, formerly $mphBM$, along with genes encoding the Msr(A) efflux pump, and Erm methyltransferase, were first identified as naturally occurring on a transmissible plasmid in $S.$ $aureus$ clinical isolate (Matsuoka et al. 1998). The sequence of Mph(C) showed 67% amino acid similarity to Mph(B) from $E.$ $coli$. In this study, expression of mph(C) in $S.$ $aureus$ was shown to be highly dependent on the presence of a portion of the gene encoding the Msr(A) efflux pump; however, the nature of this dependence is not fully understood (Matsuoka et al. 2003).

Of the two enzymes originally found in $E.$ $coli$, Mph(A) preferentially phosphorylates 14- and 15-membered ring versus 16-membered macrolides, whereas Mph(B) phosphorylates 14- and 16-membered macrolides efficiently (Kono et al. 1992; O'Hara and Yamamoto 1996). Clear substrate specificity of these enzymes was shown by recombinant overexpression of mph(A), mph(B), and mph(C) in an isogenic strain background using an efflux-deficient laboratory $E.$ $coli$ strain (Chesneau et al. 2007). This study found that Mph(A) conferred resistance to erythromycin, telithromycin, azithromycin, and spiramycin. The closely related Mph(B) and Mph(C) enzymes both conferred resistance to erythromycin, spiramycin, and telithromycin, but no activity against azithromycin was observed. Further, functional expression of mph(C) in $E.$ $coli$ showed phos-

photransferase activity in the absence of msr(A) (Chesneau et al. 2007); however, the same plasmid did not confer macrolide resistance in $S.$ $aureus$ for reasons unknown, similar to previous observations (Matsuoka et al. 2003).

Residues shown to be important for enzymatic activity were found in the same relative positions in an alignment of Mph(A), Mph(B), and Mph(C), making it difficult to attribute substrate specificities to specific sequence variations (Chesneau et al. 2007). Site-directed mutagenesis of five aspartic acid residues (D_{200}, D_{209}, D_{219}, D_{227}, and D_{231}) thought to be located in the active site of Mph(B) based on alignments with the aminoglycoside phosphotransferase APH(3')-IIa (Wright and Thompson 1999), showed that replacements of all aspartic acid residues with alanine, except for D_{227}, completely inactivated Mph(B). The D_{227}A mutant retained 7% of the wild-type activity and showed altered substrate specificity with regard to 16-membered ring macrolides, suggesting a role for D_{227} in substrate recognition (Taniguchi et al. 1999). A similar site-directed mutagenesis study investigated conserved histidines H_{198} and H_{205} located in the active site of Mph(B) (Taniguchi et al. 2004). In this study, an H_{198}A mutant retained 50% of the specific enzymatic activity, suggesting that H_{198} was not a catalytically essential residue. In contrast, the H_{205}A mutant retained only 0.7% of wild-type levels of activity, and an H_{205}N mutant retained greater than half of wild-type levels, suggesting that H_{205} was essential for catalysis. Based on alignments with the active site in the structure of an aminoglycoside phosphotransferase, H_{205} was proposed to contact the γ-phosphate of ATP through magnesium and aid in the transfer of phosphate from ATP to the 2'-hydroxyl of the desosamine.

Macrolide phosphotransferases are widespread in bacteria of clinical, veterinary, agricultural, and environmental origins. Genes encoding Mph enzymes are usually found on mobile genetic elements containing other macrolide resistance genes and genes conferring resistance to other antibiotic classes. The mph(A) gene has been found on plasmids that encode CTX-M extended-spectrum β-lactamases originating in *E. coli* ST131 (Woodford et al. 2009; Sandegren et al. 2012) and 16S rRNA methyltransferases (i.e., *armA*) that encode aminoglycoside resistance. The *mph*(A) gene has also been detected in multidrug-resistant (MDR) and KPC carbapenemase-producing *K. pneumoniae* (Soge et al. 2006; Sandegren et al. 2012; Lee et al. 2014), *Shigella* spp. isolates (Boumghar-Bourtchai et al. 2008; Howie et al. 2010; Gaudreau et al. 2014), in globally collected MDR and susceptible *E. coli* isolates (Phuc Nguyen et al. 2009), as well as other Gram-negative pathogens.

The *mph*(C) gene appears to be widespread in staphylococci and has been found in isolates from horse skin (Schnellmann et al. 2006), bovine mastitis (Luthje 2006; Li et al. 2015), and dogs, cats, and pigs (Luthje and Schwarz 2007). *mph*(C) has also been identified in corynebacteria from healthy human skin (Szemraj et al. 2014) and, interestingly, in *Stenotrophomonas maltophilia* (Alonso et al. 2000).

A partial sequence of *mph*(D) (AB048591) has been described from *Pseudomonas aeruginosa* clinical isolate M398 from Japan (Nakamura et al. 2000). Inactivation of oleandomycin was dependent on either ATP or GTP addition to crude extracts and the inactivated product chromatographed with the standard oleadomycin 2′-phosphate. Although the strain was resistant to 14-, 15-, and 16-membered macrolides, crude extracts only inactivated 14-membered macrolides, with some activity (15% inactivation) toward azithromycin. Because this PCR product had only 53% identity with other *mph* genes, it was given a separate designation. Variants of this gene have also been described in *E. coli*, *Klebsiella*, *Pantoeae*, *Proteus*, and *Stenotrophomonas* (see faculty.washington.edu/marilynr).

Macrolide phosphotransferase genes designated as *mph*(E) have been found in the chromosomes of *Acinetobacter baumannii* (Poirel et al. 2008) and bovine respiratory *Pasturella multocida* and *Mannheimia haemolytica* isolates (Desmolaize et al. 2011; Kadlec et al. 2011). They are also transferable on plasmids, including mobile, broad-host range IncP-1β plasmids, and have been described in *Serratia marscescens* (Bae et al. 2009), *K. pneumoniae* (Shen et al. 2009; Jiang et al. 2010), *A. baumannii* (Poirel et al. 2008; Zarrilli et al. 2008), *E. coli* (GenBank #FJ187822, partial sequence) (Gonzalez-Zorn et al. 2005; Bercot et al. 2008), *Citrobacter freundii* (Golebiewski et al. 2007), and in plasmid DNA from uncultured bacterium from wastewater treatment facilities (Schluter et al. 2007; Szczepanowski et al. 2007) (note that although the investigators designate the *mph* gene as *mph*(E), it is listed at macrolide nomenclature center as *mph*(F)). *mph*(A) and *mph*(E) genes are often found in the context of a macrolide resistance operon, either *mph*(A)-*mrx*-*mphR*(A) or *mphR*(E)-*mph*(E)-*mrx*(E), and the operons are bordered by inverted repeat motifs of IS elements, suggesting that the latter could play an important role in the acquisition and spread of these resistance genes (Noguchi et al. 1995; Poole et al. 2006; Szczepanowski et al. 2007). The deduced gene product of *mphR* is a transcriptional regulator (where studied, a negative regulator of *mph*(A) gene expression; Noguchi et al. 2000) of the TetR/AcrR family, whereas the *mrx* genes encode a putative transmembrane transport protein; both are needed for high-level expression of macrolide resistance.

The most recently identified macrolide phosphotransferase, *mph*(G), has been found in *Vibrio* spp. and photobacteria in the seawater of fish farms (Nonaka et al. 2015).

SURVEILLANCE OF MACROLIDE RESISTANCE AND CHARACTERIZATION OF MOLECULAR MECHANISMS

For treatment of community-acquired pneumonia, a 14- or 15-membered macrolide plus a β-lactam is part of the regimen for patients with risk factors and is recommended as a single agent in patients without risk factors (Mandell et al. 2007). In a study that assessed the macrolide failures in patients with pneumococcal bacteremia, *mef*(A) and *erm*(B) were equally overrepresented, but MIC increases >1 µg/mL were not associated with any greater failure rate (Daneman et al. 2006), thus showing that the lower level resistance generally seen in pneumococci harboring *mef*(A) is clinically signif-

Cite this article as *Cold Spring Harb Perspect Med* doi: 10.1101/cshperspect.a025395

icant. Mutations in 23S rRNA are more frequently found where there are chronic or prolonged treatment regimens, such as for CF patients and those with *M. pneumoniae* or *H. pylori* infections (Table 2). The studies can be difficult to compare as investigators choose different genes to monitor their surveillance population.

Surveillance studies (published largely in 2006–2015) find macrolide resistance rates ranging from <10% (Columbia, Hidalgo et al. 2011; Alaska, Rudolph et al. 2013) to >60% (Asia, Song et al. 2004; Lebanon, Taha et al. 2012) in pneumococci. For group A streptococci, there was also a wide range in macrolide-resistant rates, varying from 2% (Utah, Rowe et al. 2009; Scotland, Amezaga and McKenzie 2006; The Netherlands, Buter et al. 2010) to 98% in *S. pyogenes* (Chengdu, China, Zhou et al. 2014). It has been shown that macrolide resistance rates can increase with erythromycin usage (Seppala et al. 1997) and intermediate/long-acting macrolide consumption (Italy, Cornaglia et al. 1996; Spain, Perez-Trallero et al. 1998; Slovenia, Cizman et al. 2001), but not all countries have increasing macrolide resistance paralleling increase in macrolide consumption (Portugal, Silva-Costa et al. 2015). Clonality can also play a role as was seen in erythromycin-resistant *S. pyogenes* in Pittsburgh, where all of the macrolide resistance (48%) was the result of a single strain of *S. pyogenes* (Martin et al. 2002), showing resistance rates in one city or small region are possibly not representative for an entire country. Other factors yet determined also play a role. The rates of macrolide resistance in group B streptococci, including *S. agalactiae*, range from 4% to 5% (Scotland, Amezaga and McKenzie 2006; The Netherlands, Buter et al. 2010) to 40% (France, Bergal et al. 2015; Tunisia, Hraoui et al. 2012) and viridans streptococci generally have higher rates, ranging from 27% (Turkey, Ergin et al. 2006) to 63% (Canada, Thornton et al. 2015). Thus, it is important to continue surveillance and monitor resistance rates locally and globally.

A high rate of azithromycin-resistant streptococci was resident and characterized in adults

with CF, with half of the isolates harboring A2058G or A2059G mutations in 23S rRNA (Thornton et al. 2015). In *S. aureus* isolates from adult and children patients with CF (Tkadlec et al. 2015), 52% of the macrolide-resistant isolates had 23S rRNA or L4 ribosomal mutations. Mechanisms of macrolide resistance appear to be different in Gram-negative isolates from children with CF (Roberts et al. 2011). In patients participating in a randomized placebo-controlled trial with azithromycin, there was 25.5% frank macrolide resistance in *H. influenzae* with all but one of the remaining isolates intermediate to azithromycin. Rather than ribosomal protein mutations, *erm*(B) and *erm*(F) were frequently identified, usually in combination with *mef*(A); 23S rRNA mutations were not interrogated.

The rates of macrolide resistance in methicillin-resistant *S. aureus* (MRSA) and co-agulase-negative staphylococci remain high (44%–100%), with *erm* genes as the most predominant mechanisms. Macrolide resistance in MRSA is significantly higher than in methicillin-susceptible *S. aureus* (MSSA) and, in a recent study in Turkey, there was a sevenfold difference in macrolide resistance between MRSA and MSSA (Gul et al. 2008; Yildiz et al. 2014; Aydeniz Ozansoy et al. 2015). The rates of macrolide resistance in coagulase-negative staphylococci vary from 44% (*Staphylococcus saprophyticus*, France, Le Bouter et al. 2011) to 100% (*S. haemolyticus*, Poland, Brzychczy-Wloch et al. 2013; *S. hominis*, Poland, Szczuka et al. 2015) in surveillance studies (Table 2). Msr-mediated efflux appears to be increasing in staphylococci, often in conjunction with an *erm* gene. A study of isolates collected from European hospitals in 1997–1998 found *msr* genes in only 13% of MSSA isolates and did not detect the gene in MRSA (Schmitz et al. 2000). Similarly, a surveillance of isolates from French hospitals published in 1999 showed only 2.1% of MRSA/MSSA isolates carrying *msr*(A) resistance genes (Lina et al. 1999). More recent studies have shown *msr* genes present at rates ranging from 1.6% (Iran, Shahsavan et al. 2012) to 79% (Spain, Argudin et al. 2014; Aydeniz Ozansoy et al. 2015) of *S. aureus* and 15% (Tunisia, Bou-

Table 2. Surveillance and mechanisms of macrolide resistance in clinical isolates

Species	Percent macrolide resistance	Number of isolates	Macrolide-resistant mechanism						Country or region	Isolation year(s)	References
			erm^a	mef(A)	$erm^a +$ mef(A)	Target[b]	msr(A)	mph(A)[c]			
Streptococcus pneumoniae	36.5%	20,142	16.3%	9.8%	3.6%	0.4%	ND	ND	Global	2001–2005	Felmingham et al. 2007
S. pneumoniae	9%–14%	15,982	57%	27%	15%	1%	ND	ND	South Africa	2000–2005	Wolter et al. 2008a
S. pneumoniae	8%–22%	12,759	36.9%	51.0%	8.0%	ND	ND	ND	Canada	1998–2008	Wierzbowski et al. 2014
S. pneumoniae	37.2%	7083	55%	30.6%	12.0%	ND	ND	ND	Global	2003–2004	Farrell et al. 2008
S. pneumoniae	35.3%	6747	18.8%	53.8%	24.1%	1.7%	ND	ND	United States	2005–2006	Jenkins and Farrell 2009
S. pneumoniae	9.2%	2923	12.0%	77.0%	7.0%	ND	ND	ND	Alaska	1986–2010	Rudolph et al. 2013
S. pneumoniae	6.2%	3571	30%	56%	1.0%	ND	ND	ND	Finland	2002–2006	Siira et al. 2009
S. pneumoniae	21.5%	1007	41%	50%	2%	6%	ND	ND	Finland	2002	Rantala et al. 2005
S. pneumoniae	26%	863	54%	13%	31%	ND	ND	ND	Russia	2009–2013	Mayanskiy et al. 2014
S. pneumoniae	59.3%	555	47.7%	30.7%	21.6%	ND	ND	ND	Asia	1998–2001	Song et al. 2004
S. pneumoniae	9.5%	410	43.5%	56.5%	ND	ND	ND	ND	Chile	1997–1999	Palavecino et al. 2002
S. pneumoniae	67.7%	65	36.0%	18.0%	32%	ND	ND	ND	Lebanon	2008–2010	Taha et al. 2012
S. pneumoniae	2.4%–6.9%	3241	61.0%	33.1%	ND	ND	ND	ND	Columbia	1994–2008	Hidalgo et al. 2011
S. pneumoniae	26.4%	151	95.0%	5.0%	ND	ND	ND	ND	Turkey	1998–2002	Gulay et al. 2008
S. pneumoniae	21.4%	2045	27.8%	68.9%	4.1%	ND	ND	ND	Germany	2005–2006	Bley et al. 2011
Streptococcus pyogenes	8.2%	352	48.3%	31.0%	0%	ND	ND	ND	Germany		
S. pyogenes	12.5%	3893	28.1%	71.5%	ND[e]	ND	ND	ND	Serbia	2007–2008	Opavski et al. 2015
S. pyogenes	98.4%	127	100%	0.0%	0.0%	ND	ND	ND	China	2004–2011	Zhou et al. 2014
S. pyogenes	2.4%	739	29.7%	48.1%	22.8%	ND	ND	ND	United States (Utah)	2007–2008	Rowe et al. 2009
β-Hemolytic streptococci group											
A	1.9%	1625	57.7%	42.3%	0%	ND	ND	ND	Scotland	2000–2001	Amezaga and McKenzie 2006
B	4.3%	1233	88.1%	11.9%	0%	ND	ND	ND	Scotland	2000–2001	Amezaga and McKenzie 2006
C	3.8%	479	11.1%	66.7%	11.1%	ND	ND	ND	Scotland	2000–2001	Amezaga and McKenzie 2006
G	6.2%	1034	90.4%	9.6%	0%	ND	ND	ND	Scotland	2000–2001	Amezaga and McKenzie 2006

Cite this article as *Cold Spring Harb Perspect Med* doi: 10.1101/cshperspect.a025395

Continued

β-Hemolytic streptococci group

Organism	%	n							Country	Years	Reference
A	1.4%	219	66.7%	33.3%	0%	ND	ND	ND	The Netherlands	2005–2006	Buter et al. 2010
B	5.3%	562	93.3%	0%	0%	ND	ND	ND	The Netherlands	2005–2006	Buter et al. 2010
C	6.9%	58	100%	0%	0%	ND	ND	ND	The Netherlands	2005–2006	Buter et al. 2010
G	4.6%	237	45.5%	18.2%	9.1%	ND	ND	ND	The Netherlands	2005–2006	Buter et al. 2010
Group B streptococci	12.6%	143	77.8%	0%	22.2%	0%	0%	ND	Kuwait	2007	Boswihi et al. 2012
S. agalactiae	38.1%	93	75.7%	24.3%	ND	ND	ND	ND	France	2011–2012	Bergal et al. 2015
S. agalactiae	40%	226	97.8%	2.2%	0%	ND	ND	ND	Tunisia	2007–2009	Hraoui et al. 2012
Viridans group streptococci	27%	85	12.9%	12.9%	4.7%	ND	ND	ND	Turkey	1996–2004	Ergin et al. 2006
Viridans group streptococci	28.2%	85	66.7%	33.3%	ND	ND	ND	ND	Gran Canaria, Spain	2004–2006	Artiles Campelo et al. 2007
Streptococci	63.0%	413	22.6%	29.6%	0.4%	47.0%	ND	ND	Canada[f]	2006–2011	Thornton et al. 2015
S. aureus	49.7%	656	100%	ND	ND	ND	0%	ND	Greece	2003–2008	Vallianou et al. 2015
S. aureus	58.0%	106	95.1%	ND	ND	ND	1.6%	ND	Iran	2010–2011	Shahsavan et al. 2012
S. aureus	NR	97[e]	73.1%	ND	ND	ND	13.4%	ND	Poland	NR	Piatkowska et al. 2012
S. aureus	56.0%	100	41.1%	ND	ND	51.8%	7.1%	ND	Czech Republic[f]	2011–2013	Tkadlec et al. 2015
MRSA	72.8%	397	88.9%	ND	ND	ND	4.4%	ND	Turkey	2006–2008	Yildiz et al. 2014
MRSA	84.9%	265	83.1%	ND	ND	ND	35.6%	ND	Turkey	2003–2006	Gul et al. 2008
MRSA	79.1	158	89.6%	ND	ND	ND	10.4%	ND	Turkey	2012–2013	Aydeniz Ozansoy et al. 2015
MSSA	9.8%	246	75.0%	ND	ND	ND	25.0%	ND	Turkey	2012–2013	Aydeniz Ozansoy et al. 2015
S. aureus[d]	25.2%	111	82.1%	ND	ND	ND	75.0%	ND	Spain	1997–2006	Argudin et al. 2014
S. aureus	45.3%	203	51.1%	ND	ND	ND	34.8%	ND	Spain	2006–2007	Perez-Vazquez et al. 2009
S. aureus	34.1%	91	96.7%	ND	ND	ND	3.3%	ND	Italy	2005–2006	Gherardi et al. 2009
CoNS[g]	62.7%	89	47.1%	ND	ND	ND	52.9%	ND	Italy	2005–2006	Gherardi et al. 2009
CoNS	61.7%	494	73.1%	ND	ND	ND	27.5%	24.9%	Germany	2004–2006	Gatermann et al. 2007
S. epidermidis	57.4%	47	70.4%	ND	ND	ND	33.3%	ND	Mexico	2002–2004	Castro-Alarcon et al. 2011
S. epidermidis	62.8%	77	85.3%	0%	0%	ND	14.7%	ND	Tunisia	2002	Bouchami et al. 2007
S. epidermidis	90%	63	60%	ND	ND	ND	40.4%	ND	Poland	2009	Brzychczy-Wloch et al. 2013
S. haemolyticus	100%	28	7%	ND	ND	ND	92.9%	ND	Poland	2009	Brzychczy-Wloch et al. 2013
Staphylococcus hominis	100%	55	67.5%	ND	ND	ND	18.2%	ND	Poland	ND	Szczuka et al. 2015
Staphylococcus saprophyticus	44.4%	72	15.6%	ND	ND	0.0%	81.3%	ND	France	2005–2009	Le Bouter et al. 2011

Table 2. *Continued*

Species	Percent macrolide resistance	Number of isolates	Macrolide-resistant mechanism						Country or region	Isolation year(s)	References
			erm^a	mef(A)	$erm^a +$ mef(A)	Target[b]	msr(A)	mph(A)[c]			
Haemophilus influenzae	25.5%	106	18.5%	29.6%	51.9%	0.0%	ND	ND	United States + Canada[f]	2007–2008	Roberts et al. 2011
Corynebacterium species	89.7%	140	87.8%	ND	ND	ND	ND	ND	Poland	2008–2011	Olender 2013
Mycoplasma genitalium	36.1%	155	ND	ND	ND	100%	ND	ND	Australia	2012–2013	Bissessor et al. 2015
M. genitalium	9.8%	297	ND	ND	ND	100%	ND	ND	Greenland	2008–2009	Gesink et al. 2012
M. genitalium	40%	1121	ND	ND	ND	100%	ND	ND	Denmark	2006–2010	Salado-Rasmussen and Jensen 2014
Mycoplasma pneumoniae	68.7%	67	ND	ND	ND	100%	ND	ND	China	2008–2009	Cao et al. 2010
M. pneumoniae	67.8%	1655	ND	ND	ND	100%	ND	ND	Japan	2008–2012	Kawai et al. 2013
M. pneumoniae	83%	53	ND	ND	ND	100%	ND	ND	China	2005–2008	Liu et al. 2009b
M. pneumoniae	40.4%	27	ND	ND	ND	100%	ND	ND	Korea	2011	Yoo et al. 2012
M. pneumoniae	88.0%	309	ND	ND	ND	100%	ND	ND	China	2008–2011	Zhao et al. 2013
M. pneumoniae	13.2%	91	ND	ND	ND	100%	ND	ND	United States	2012–2014	Zheng et al. 2015
M. pneumoniae	100%	71	ND	ND	ND	100%	ND	ND	China	2012–2014	Zhou et al. 2015
Ureaplasma urealyticum	80.6%	72	36.2%	ND	ND	ND	63.8%	ND	China	2008	Lu et al. 2010

Macrolide resistance is defined by resistance to erythromycin or clarithromycin; in the case of sexually transmitted pathogens, the species and macrolide resistance mechanisms were usually determined by real-time hydrolysis probe PCR targeting directly on urine samples/vaginal swabs. *M. pneumoniae* isolates were identified by colony morphology and PCR assay, real-time PCR melt curve analysis, and/or real-time PCR targeting of conserved genes and 23S rRNA mutations.

ND, Not determined; NR, not reported; MRSA, methicillin-resistant *S. aureus*; MSSA, methicillin-susceptible *S. aureus*.

[a] Generally *erm*(B) or *erm*(A), including subclass *erm*(TR), in streptococci; *erm*(A) *erm*(B), and/or *erm*(C) in staphylococci; *erm*(A), *erm*(B), *erm*(C), and/or *erm*(F) in *H. influenza*.

[b] Mutation(s) in ribosomal protein L4 or L22 or in domain V of 23S rRNA.

[c] *mph*(C) in Gatermann et al. (2007).

[d] Majority of isolates had multiple macrolide resistance genes and the most frequently found was *msr*(B)-*erm*(C).

[e] All erythromycin-resistant isolates characterized, no total number provided.

[f] Patients with cystic fibrosis.

[g] CoNS, coagulase-negative staphylococci.

Cite this article as *Cold Spring Harb Perspect Med* doi: 10.1101/cshperspect.a025395

chami et al. 2007) to 81% (France, Le Bouter et al. 2011) of coagulase-negative staphylococci (Gatermann et al. 2007; Perez-Vazquez et al. 2009; Le Bouter et al. 2011; Zmantar et al. 2011; Argudin et al. 2014).

Erythromycin, clarithromycin, and azithromycin are the therapeutic agents of choice for *M. pneumoniae* infections in children, and the first macrolide-resistant strain was isolated in Japan in 2000 (Okazaki et al. 2001). By 2003, 13% (13/76) of *M. pneumoniae* isolates in Japan were resistant to erythromycin (Matsuoka et al. 2004) with the majority ($n = 10$) associated with A2063G (A2058 *E. coli* numbering) and one each of A2063C, A2064G, and C2617G (C2611 *E. coli* numbering), with the latter only expressing weak resistant to erythromycin (MIC = 8 μg/mL) (Tables 1 and 2). By 2008, the prevalence had reached 30.6%. Prevalence increased from 2008 to 2012 (Kawai et al. 2013) with regional differences of macrolide resistance, varying from 50% to 93%, and with resistance rates higher in patients that had received macrolides before the surveillance study. The majority of the 561 isolates from 769 patients had mutations of A2063G or A2063T; less commonly, A2063C, A2064G, and C2617G were found. Rates of macrolide-resistant *M. pneumoniae* (MRMP) have exceeded 90% in Beijing, China (Zhao et al. 2013), with the majority of MRMP carrying the A2063G followed by A2064G and a single isolate with A2063T mutation. In Zhejiang, China, 100% of *M. pneumoniae* strains isolated from adults with community-acquired pneumonia carried the resistance determinant, A2063G mutation (Zhou et al. 2015).

Macrolides are often used for first-line therapies of *Ureaplasma urealyticum* infections. Interestingly, 80.6% of *U. urealyticum* in a 2008 study in China were macrolide-resistant (Lu et al. 2010). About 64% of the isolates harbored *msr*(A) $\pm$ *msr*(D), whereas 36% carried *erm*(B), perhaps reflecting that this species can host plasmids and transposons. For *M. genitalium*, another causative agent of sexually transmitted infections, only 23S rRNA mutations have been identified in macrolide-resistant isolates.

CONCLUDING REMARKS

Cryo-EM and X-ray crystallography have provided structural insights into how the ribosome interacts with and responds to small molecules like antibiotics. These studies help to explain how different target-based mutations or methylation of A2058 confer resistance, as well as provide an understanding into how regulation of *erm* methyltransferases and efflux genes has evolved. Along with the ribosome-macrolide X-ray structures, we now have a much clearer understanding of how macrolides inhibit protein synthesis (i.e., the link between the ribosomal tunnel and the PTC) and the data show us that macrolide action is specific, targeting a subset of proteins. Assays could be developed to ensure that the synthesis of certain vital proteins is impacted and/or to monitor the mechanism(s) of action, potentially enriching for compounds that promote frameshifting, for example. The route to total synthesis will allow the exploration of structure–activity relationships, overcoming any limitations of semisynthesis (see macrolide.com; Zhang et al. 2016) and potentially extending spectrum beyond the community-acquired respiratory pathogens. With our present understanding, perhaps it is prime time to rethink macrolide drug discovery and use the existing and expanding tool sets to find molecules with more refined, targeted actions.

ACKNOWLEDGMENTS

The authors thank Philip C. Hogan, Macrolide Pharmaceuticals, for his chemistry prowess and review of macrolide structures, and Yury S. Polikanov, University of Illinois at Chicago, for his enthusiasm and expertise in producing Figure 3.

REFERENCES

*Reference is also in this collection.

Abbassi MS, Torres C, Achour W, Vinue L, Saenz Y, Costa D, Bouchami O, Ben Hassen A. 2008. Genetic characterisation of CTX-M-15-producing *Klebsiella pneumoniae* and *Escherichia coli* strains isolated from stem cell transplant patients in Tunisia. *Int J Antimicrob Agents* **32:** 308–314.

Alonso A, Sanchez P, Martinez JL. 2000. *Stenotrophomonas maltophilia* D457R contains a cluster of genes from Gram-positive bacteria involved in antibiotic and heavy metal resistance. *Antimicrob Agents Chemother* **44:** 1778–1782.

Ambrose KD, Nisbet R, Stephens DS. 2005. Macrolide efflux in *Streptococcus pneumoniae* is mediated by a dual efflux pump (*mel* and *mef*) and is erythromycin inducible. *Antimicrob Agents Chemother* **49:** 4203–4209.

Amezaga MR, McKenzie H. 2006. Molecular epidemiology of macrolide resistance in β-haemolytic streptococci of Lancefield groups A, B, C and G and evidence for a new *mef* element in group G streptococci that carries allelic variants of *mef* and *msr*(D). *J Antimicrob Chemother* **57:** 443–449.

Andremont A, Gerbaud G, Tancrede C, Courvalin P. 1985. Plasmid-mediated susceptibility to intestinal microbial antagonisms in *Escherichia coli*. *Infect Immun* **49:** 751–755.

Andremont A, Sancho-Garnier H, Tancrede C. 1986. Epidemiology of intestinal colonization by members of the family Enterobacteriaceae highly resistant to erythromycin in a hematology–oncology unit. *Antimicrob Agents Chemother* **29:** 1104–1107.

* Arenz S, Wilson DN. 2016. Bacterial protein synthesis as a target for antibiotic inhibition. *Cold Spring Harb Perspect Med* doi: 10.1101/cshperspect.a025361.

Arenz S, Meydan S, Starosta AL, Berninghausen O, Beckmann R, Vazquez-Laslop N, Wilson DN. 2014a. Drug sensing by the ribosome induces translational arrest via active site perturbation. *Mol Cell* **56:** 446–452.

Arenz S, Ramu H, Gupta P, Berninghausen O, Beckmann R, Vázquez-Laslop N, Mankin AS, Wilson DN. 2014b. Molecular basis for erythromycin-dependent ribosome stalling during translation of the ErmBL leader peptide. *Nat Commun* **5:** 3501.

Argudin MA, Mendoza MC, Martin MC, Rodicio MR. 2014. Molecular basis of antimicrobial drug resistance in *Staphylococcus aureus* isolates recovered from young healthy carriers in Spain. *Microb Pathog* **74:** 8–14.

Arthur M, Courvalin P. 1986. Contribution of two different mechanisms to erythromycin resistance in *Escherichia coli*. *Antimicrob Agents Chemother* **30:** 694–700.

Arthur M, Brisson-Noel A, Courvalin P. 1987. Origin and evolution of genes specifying resistance to macrolide, lincosamide and streptogramin antibiotics: Data and hypotheses. *J Antimicrob Chemother* **20:** 783–802.

Artiles Campelo F, Horcajada Herrera I, Alamo Antunez I, Canas Pedrosa A, Lafarga Capuz B. 2007. Phenotypes and genetic mechanisms of resistance to macrolides and lincosamides in viridans group streptococci. *Rev Esp Quimioter* **20:** 317–322.

Aydeniz Ozansoy F, Cevahir N, Kaleli I. 2015. Investigation of macrolide, lincosamide and streptogramin B resistance in *Staphylococcus aureus* strains isolated from clinical samples by phenotypical and genotypical methods. *Mikrobiyol Bul* **49:** 1–14.

Bacciaglia A, Brenciani A, Varaldo PE, Giovanetti E. 2007. SmaI typeability and tetracycline susceptibility and resistance in *Streptococcus pyogenes* isolates with efflux-mediated erythromycin resistance. *Antimicrob Agents Chemother* **51:** 3042–3043.

Bae IK, Woo G-J, Park I, Jeong SH, Lee SH. 2009. Genetic environment of plasmid-mediated *armA* gene in *Serratia marcescens* clinical isolate. GenBank #FJ917355.1.

Barthelemy P, Autissier D, Gerbaud G, Courvalin P. 1984. Enzymic hydrolysis of erythromycin by a strain of *Escherichia coli*. A new mechanism of resistance. *J Antibiot (Tokyo)* **37:** 1692–1696.

Bebear CM, Pereyre S. 2005. Mechanisms of drug resistance in *Mycoplasma pneumoniae*. *Curr Drug Targets Infect Disord* **5:** 263–271.

Bercot B, Poirel L, Nordmann P. 2008. Plasmid-mediated 16S rRNA methylases among extended-spectrum β-lactamase-producing Enterobacteriaceae isolates. *Antimicrob Agents Chemother* **52:** 4526–4527.

Bergal A, Loucif L, Benouareth DE, Bentorki AA, Abat C, Rolain JM. 2015. Molecular epidemiology and distribution of serotypes, genotypes, and antibiotic resistance genes of *Streptococcus agalactiae* clinical isolates from Guelma, Algeria and Marseille, France. *Eur J Clin Microbiol Infect Dis* **34:** 2339–2348.

Bhushan S, Hoffmann T, Seidelt B, Frauenfeld J, Mielke T, Berninghausen O, Wilson DN, Beckmann R. 2011. SecM-stalled ribosomes adopt an altered geometry at the peptidyl transferase center. *PLoS Biol* **9:** e1000581.

Bingen E, Leclercq R, Fitoussi F, Brahimi N, Malbruny B, Deforche D, Cohen R. 2002. Emergence of group A streptococcus strains with different mechanisms of macrolide resistance. *Antimicrob Agents Chemother* **46:** 1199–1203.

Bissessor M, Tabrizi SN, Twin J, Abdo H, Fairley CK, Chen MY, Vodstrcil LA, Jensen JS, Hocking JS, Garland SM, et al. 2015. Macrolide resistance and azithromycin failure in a *Mycoplasma genitalium*-infected cohort and response of azithromycin failures to alternative antibiotic regimens. *Clin Infect Dis* **60:** 1228–1236.

Blackman Northwood J, Del Grosso M, Cossins LR, Coley MD, Creti R, Pantosti A, Farrell DJ. 2009. Characterization of macrolide efflux pump *mef* subclasses detected in clinical isolates of *Streptococcus pyogenes* isolated between 1999 and 2005. *Antimicrob Agents Chemother* **53:** 1921–1925.

Blair JM, Richmond GE, Piddock LJ. 2014. Multidrug efflux pumps in Gram-negative bacteria and their role in antibiotic resistance. *Future Microbiol* **9:** 1165–1177.

Bley C, van der Linden M, Reinert RR. 2011. *mef*(A) is the predominant macrolide resistance determinant in *Streptococcus pneumoniae* and *Streptococcus pyogenes* in Germany. *Int J Antimicrob Agents* **37:** 425–431.

Boswihi SS, Udo EE, Al-Sweih N. 2012. Serotypes and antibiotic resistance in Group B streptococcus isolated from patients at the Maternity Hospital, Kuwait. *J Med Microbiol* **61:** 126–131.

Bouchami O, Achour W, Ben Hassen A. 2007. Prevalence and mechanisms of macrolide resistance among *Staphylococcus epidermidis* isolates from neutropenic patients in Tunisia. *Clin Microbiol Infect* **13:** 103–106.

Boumghar-Bourtchai L, Mariani-Kurkdjian P, Bingen E, Filliol I, Dhalluin A, Ifrane SA, Weill FX, Leclercq R. 2008. Macrolide-resistant *Shigella sonnei*. *Emerg Infect Dis* **14:** 1297–1299.

Brenciani A, Ojo KK, Monachetti A, Menzo S, Roberts MC, Varaldo PE, Giovanetti E. 2004. Distribution and molecular analysis of *mef*(A)-containing elements in tetracycline-susceptible and -resistant *Streptococcus pyogenes* clinical isolates with efflux-mediated erythromycin resistance. *J Antimicrob Chemother* **54**: 991–998.

Brockmann H, Henkel W. 1950. Pikromycin, ein neues Antibiotikum aus Actinomyceten. *Naturwissenschaften* **37**: 138–139.

Brzychczy-Wloch M, Borszewska-Kornacka M, Gulczynska E, Wojkowska-Mach J, Sulik M, Grzebyk M, Luchter M, Heczko PB, Bulanda M. 2013. Prevalence of antibiotic resistance in multi-drug resistant coagulase-negative staphylococci isolated from invasive infection in very low birth weight neonates in two Polish NICUs. *Ann Clin Microbiol Antimicrob* **12**: 41.

Buriankova K, Doucet-Populaire F, Dorson O, Gondran A, Ghnassia JC, Weiser J, Pernodet JL. 2004. Molecular basis of intrinsic macrolide resistance in the *Mycobacterium tuberculosis* complex. *Antimicrob Agents Chemother* **48**: 143–150.

Burman WJ, Stone BL, Brown BA, Wallace RJ Jr, Bottger EC. 1998. AIDS-related *Mycobacterium kansasii* infection with initial resistance to clarithromycin. *Diagn Microbiol Infect Dis* **31**: 369–371.

Burton BM, Williams TL, Baker TA. 2001. ClpX-mediated remodeling of Mu transpososomes: Selective unfolding of subunits destabilizes the entire complex. *Mol Cell* **8**: 449–454.

Buter CC, Mouton JW, Klaassen CH, Handgraaf CM, Sunnen S, Melchers WJ, Sturm PD. 2010. Prevalence and molecular mechanism of macrolide resistance in β-haemolytic streptococci in The Netherlands. *Int J Antimicrob Agents* **35**: 590–592.

Cai Y, Kong F, Gilbert GL. 2007. Three new macrolide efflux (*mef*) gene variants in *Streptococcus agalactiae*. *J Clin Microbiol* **45**: 2754–2755.

Canton R, Mazzariol A, Morosini MI, Baquero F, Cornaglia G. 2005. Telithromycin activity is reduced by efflux in *Streptococcus pyogenes*. *J Antimicrob Chemother* **55**: 489–495.

Canu A, Malbruny B, Coquemont M, Davies TA, Appelbaum PC, Leclercq R. 2000. Diversity of mutations in L22, L4 ribosomal proteins and 23S ribosomal RNA in pneumococcal mutants resistant to macrolides, telithromycin, and clindamycin selected in vitro. *40th Interscience Conference on Antimicrobial Agents and Chemotherapy (ICAAC)*, Abstract 1927. Toronto, September 17–20.

Canu A, Malbruny B, Coquemont M, Davies TA, Appelbaum PC, Leclercq R. 2002. Diversity of ribosomal mutations conferring resistance to macrolides, clindamycin, streptogramin, and telithromycin in *Streptococcus pneumoniae*. *Antimicrob Agents Chemother* **46**: 125–131.

Cao B, Zhao CJ, Yin YD, Zhao F, Song SF, Bai L, Zhang JZ, Liu YM, Zhang YY, Wang H, et al. 2010. High prevalence of macrolide resistance in *Mycoplasma pneumoniae* isolates from adult and adolescent patients with respiratory tract infection in China. *Clin Infect Dis* **51**: 189–194.

Castro-Alarcon N, Ribas-Aparicio RM, Silva-Sanchez J, Calderon-Navarro A, Sanchez-Perez A, Parra-Rojas I, Aparicio-Ozores G. 2011. Molecular typing and characterization of macrolide, lincosamide and streptogramin resistance in *Staphylococcus epidermidis* strains isolated in a Mexican hospital. *J Med Microbiol* **60**: 730–736.

Cerda Zolezzi P, Rubio Calvo MC, Millan L, Goni P, Canales M, Capilla S, Duran E, Gomez-Lus R. 2004. Macrolide resistance phenotypes of commensal viridans group streptococci and *Gemella* spp. and PCR detection of resistance genes. *Int J Antimicrob Agents* **23**: 582–589.

Chancey ST, Zhou X, Zahner D, Stephens DS. 2011. Induction of efflux-mediated macrolide resistance in *Streptococcus pneumoniae*. *Antimicrob Agents Chemother* **55**: 3413–3422.

Chancey ST, Agrawal S, Schroeder MR, Farley MM, Tettelin H, Stephens DS. 2015a. Composite mobile genetic elements disseminating macrolide resistance in *Streptococcus pneumoniae*. *Front Microbiol* **6**: 26.

Chancey ST, Bai X, Kumar N, Drabek EF, Daugherty SC, Colon T, Ott S, Sengamalay N, Sadzewicz L, Tallon LJ, et al. 2015b. Transcriptional attenuation controls macrolide inducible efflux and resistance in *Streptococcus pneumoniae* and in other Gram-positive bacteria containing *mef*/*mel*(*msr*(D)) elements. *PLoS ONE* **10**: e0116254.

Chang CY, Chang LL, Chang YH, Lee TM, Chang SF. 2000. Characterisation of drug resistance gene cassettes associated with class 1 integrons in clinical isolates of *Escherichia coli* from Taiwan, ROC. *J Med Microbiol* **49**: 1097–1102.

Chen YT, Liao TL, Liu YM, Lauderdale TL, Yan JJ, Tsai SF. 2009. Mobilization of *qnrB2* and *ISCR1* in plasmids. *Antimicrob Agents Chemother* **53**: 1235–1237.

Chesneau O, Tsvetkova K, Courvalin P. 2007. Resistance phenotypes conferred by macrolide phosphotransferases. *FEMS Microbiol Lett* **269**: 317–322.

Choi SS, Kim SK, Oh TG, Choi EC. 1997. Role of mRNA termination in regulation of *ermK*. *J Bacteriol* **179**: 2065–2067.

Chrisment D, Charron A, Cazanave C, Pereyre S, Bebear C. 2012. Detection of macrolide resistance in *Mycoplasma genitalium* in France. *J Antimicrob Chemother* **67**: 2598–2601.

Cizman M, Pokorn M, Seme K, Orazem A, Paragi M. 2001. The relationship between trends in macrolide use and resistance to macrolides of common respiratory pathogens. *J Antimicrob Chemother* **47**: 475–477.

Clancy J, Petitpas J, Dib-Hajj F, Yuan W, Cronan M, Kamath AV, Bergeron J, Retsema JA. 1996. Molecular cloning and functional analysis of a novel macrolide-resistance determinant, *mefA*, from *Streptococcus pyogenes*. *Mol Microbiol* **22**: 867–879.

Clark C, Bozdogan B, Peric M, Dewasse B, Jacobs MR, Appelbaum PC. 2002. In vitro selection of resistance in *Haemophilus influenzae* by amoxicillin-clavulanate, cefpodoxime, cefprozil, azithromycin, and clarithromycin. *Antimicrob Agents Chemother* **46**: 2956–2962.

Cochetti I, Vecchi M, Mingoia M, Tili E, Catania MR, Manzin A, Varaldo PE, Montanari MP. 2005. Molecular characterization of pneumococci with efflux-mediated erythromycin resistance and identification of a novel *mef* gene subclass, *mef*(I). *Antimicrob Agents Chemother* **49**: 4999–5006.

Cornaglia G, Ligozzi M, Mazzariol A, Valentini M, Orefici G, Fontana R. 1996. Rapid increase of resistance to erythromycin and clindamycin in *Streptococcus pyogenes* in Italy, 1993–1995. The Italian Surveillance Group for Antimicrobial Resistance. *Emerg Infect Dis* **2**: 339–342.

Costa SS, Viveiros M, Amaral L, Couto I. 2013. Multidrug efflux pumps in *Staphylococcus aureus*: An update. *Open Microbiol J* 7: 59–71.

Culebras E, Rodriguez-Avial I, Betriu C, Picazo JJ. 2005. Differences in the DNA sequence of the translational attenuator of several constitutively expressed *erm*(A) genes from clinical isolates of *Streptococcus agalactiae*. *J Antimicrob Chemother* 56: 836–840.

Daly MM, Doktor S, Flamm R, Shortridge D. 2004. Characterization and prevalence of MefA, MefE, and the associated *msr*(D) gene in *Streptococcus pneumoniae* clinical isolates. *J Clin Microbiol* 42: 3570–3574.

Daneman N, McGeer A, Green K, Low DE. 2006. Macrolide resistance in bacteremic pneumococcal disease: implications for patient management. *Clin Infect Dis* 43: 432–438.

Davydova N, Streltsov V, Wilce M, Liljas A, Garber M. 2002. L22 ribosomal protein and effect of its mutation on ribosome resistance to erythromycin. *J Mol Biol* 322: 635–644.

Debets-Ossenkopp YJ, Brinkman AB, Kuipers EJ, Vandenbroucke-Grauls CM, Kusters JG. 1998. Explaining the bias in the 23S rRNA gene mutations associated with clarithromycin resistance in clinical isolates of *Helicobacter pylori*. *Antimicrob Agents Chemother* 42: 2749–2751.

Del Grosso M, Iannelli F, Messina C, Santagati M, Petrosillo N, Stefani S, Pozzi G, Pantosti A. 2002. Macrolide efflux genes *mef*(A) and *mef*(E) are carried by different genetic elements in *Streptococcus pneumoniae*. *J Clin Microbiol* 40: 774–778.

Del Grosso M, Scotto d'Abusco A, Iannelli F, Pozzi G, Pantosti A. 2004. Tn*2009*, a Tn*916*-like element containing *mef*(E) in *Streptococcus pneumoniae*. *Antimicrob Agents Chemother* 48: 2037–2042.

Del Grosso M, Camilli R, Iannelli F, Pozzi G, Pantosti A. 2006. The *mef*(E)-carrying genetic element (mega) of *Streptococcus pneumoniae*: Insertion sites and association with other genetic elements. *Antimicrob Agents Chemother* 50: 3361–3366.

Del Grosso M, Camilli R, Barbabella G, Blackman Northwood J, Farrell DJ, Pantosti A. 2011. Genetic resistance elements carrying *mef* subclasses other than *mef*(A) in *Streptococcus pyogenes*. *Antimicrob Agents Chemother* 55: 3226–3230.

Delmar JA, Su CC, Yu EW. 2014. Bacterial multidrug efflux transporters. *Annu Rev Biophys* 43: 93–117.

Depardieu F, Courvalin P. 2001. Mutation in 23S rRNA responsible for resistance to 16-membered macrolides and streptogramins in *Streptococcus pneumoniae*. *Antimicrob Agents Chemother* 45: 319–323.

Depardieu F, Podglajen I, Leclercq R, Collatz E, Courvalin P. 2007. Modes and modulations of antibiotic resistance gene expression. *Clin Microbiol Rev* 20: 79–114.

Desmolaize B, Rose S, Wilhelm C, Warrass R, Douthwaite S. 2011. Combinations of macrolide resistance determinants in field isolates of *Mannheimia haemolytica* and *Pasteurella multocida*. *Antimicrob Agents Chemother* 55: 4128–4133.

Doktor SZ, Shortridge V. 2005. Differences in the DNA sequences in the upstream attenuator region of *erm*(A) in clinical isolates of *Streptococcus pyogenes* and their correlation with macrolide/lincosamide resistance. *Antimicrob Agents Chemother* 49: 3070–3072.

Doktor S, Shortridge V, Zhong P, Flamm R. 2001. Ribosomal mutations and macrolide-lincosamide resistance in *Streptococcus pneumoniae*. *41st Interscience Conference on Antimicrobial Agents and Chemotherapy (ICAAC)*, Abstract C1-1812. Chicago, December 16–19.

Douthwaite S. 1992. Functional interactions within 23S rRNA involving the peptidyltransferase center. *J Bacteriol* 174: 1333–1338.

Douthwaite S, Jalava J, Jakobsen L. 2005. Ketolide resistance in *Streptococcus pyogenes* correlates with the degree of rRNA dimethylation by Erm. *Mol Microbiol* 58: 613–622.

Dunkle JA, Xiong L, Mankin AS, Cate JH. 2010. Structures of the *Escherichia coli* ribosome with antibiotics bound near the peptidyl transferase center explain spectra of drug action. *Proc Natl Acad Sci* 107: 17152–17157.

Ergin A, Ercis S, Hascelik G. 2006. Macrolide resistance mechanisms and in vitro susceptibility patterns of viridans group streptococci isolated from blood cultures. *J Antimicrob Chemother* 57: 139–141.

Ettayebi M, Prasad SM, Morgan EA. 1985. Chloramphenicol-erythromycin resistance mutations in a 23S rRNA gene of *Escherichia coli*. *J Bacteriol* 162: 551–557.

Faccone D, Andres P, Galas M, Tokumoto M, Rosato A, Corso A. 2005. Emergence of a *Streptococcus pneumoniae* clinical isolate highly resistant to telithromycin and fluoroquinolones. *J Clin Microbiol* 43: 5800–5803.

Farrell DJ, Felmingham D. 2004. Activities of telithromycin against 13,874 *Streptococcus pneumoniae* isolates collected between 1999 and 2003. *Antimicrob Agents Chemother* 48: 1882–1884.

Farrell DJ, Douthwaite S, Morrissey I, Bakker S, Poehlsgaard J, Jakobsen L, Felmingham D. 2003. Macrolide resistance by ribosomal mutation in clinical isolates of *Streptococcus pneumoniae* from the PROTEKT 1999–2000 study. *Antimicrob Agents Chemother* 47: 1777–1783.

Farrell DJ, Morrissey I, Bakker S, Buckridge S, Felmingham D. 2004. In vitro activities of telithromycin, linezolid, and quinupristin-dalfopristin against *Streptococcus pneumoniae* with macrolide resistance due to ribosomal mutations. *Antimicrob Agents Chemother* 48: 3169–3171.

Farrell DJ, Shackcloth J, Barbadora KA, Green MD. 2006. *Streptococcus pyogenes* isolates with high-level macrolide resistance and reduced susceptibility to telithromycin associated with 23S rRNA mutations. *Antimicrob Agents Chemother* 50: 817–818.

Farrell DJ, Couturier C, Hryniewicz W. 2008. Distribution and antibacterial susceptibility of macrolide resistance genotypes in *Streptococcus pneumoniae*: PROTEKT Year 5 (2003–2004). *Int J Antimicrob Agents* 31: 245–249.

Farrell DJ, Mendes RE, Jones RN. 2015. Antimicrobial activity of solithromycin against serotyped macrolide-resistant *Streptococcus pneumoniae* isolates collected from U.S. medical centers in 2012. *Antimicrob Agents Chemother* 59: 2432–2434.

Felmingham D, Canton R, Jenkins SG. 2007. Regional trends in β-lactam, macrolide, fluoroquinolone and telithromycin resistance among *Streptococcus pneumoniae* isolates 2001–2004. *J Infect* 55: 111–118.

Cite this article as *Cold Spring Harb Perspect Med* doi: 10.1101/cshperspect.a025395

Franceschi F, Kanyo Z, Sherer EC, Sutcliffe J. 2004. Macrolide resistance from the ribosome perspective. *Curr Drug Targets Infect Disord* 4: 177–191.

Fu W, Anderson M, Williams S, Tait-Kamradt A, Sutcliffe J, Retsema J. 2000. In vitro derived macrolide-resistant *Streptococcus pneumoniae* strains have ribosomal mechanisms of resistance. *40th Interscience Conference on Antimicrobial Agents and Chemotherapy (ICAAC)*, Abstract 07–10. Toronto, September 17–20.

Fujisawa Y, Weisblum B. 1981. A family of r-determinants in *Streptomyces* spp. that specifies inducible resistance to macrolide, lincosamide, and streptogramin type B antibiotics. *J Bacteriol* 146: 621–631.

Garza-Ramos G, Xiong L, Zhong P, Mankin A. 2001. Binding site of macrolide antibiotics on the ribosome: New resistance mutation identifies a specific interaction of ketolides with rRNA. *J Bacteriol* 183: 6898–6907.

Gatermann SG, Koschinski T, Friedrich S. 2007. Distribution and expression of macrolide resistance genes in coagulase-negative staphylococci. *Clin Microbiol Infect* 13: 777–781.

Gaudreau C, Barkati S, Leduc JM, Pilon PA, Favreau J, Bekal S. 2014. *Shigella* spp. with reduced azithromycin susceptibility, Quebec, Canada, 2012–2013. *Emerg Infect Dis* 20: 854–856.

Gay K, Stephens DS. 2001. Structure and dissemination of a chromosomal insertion element encoding macrolide efflux in *Streptococcus pneumoniae*. *J Infect Dis* 184: 56–65.

Gesink DC, Mulvad G, Montgomery-Andersen R, Poppel U, Montgomery-Andersen S, Binzer A, Vernich L, Frosst G, Stenz F, Rink E, et al. 2012. *Mycoplasma genitalium* presence, resistance and epidemiology in Greenland. *Int J Circumpolar Health* 71: 1–8.

Gherardi G, De Florio L, Lorino G, Fico L, Dicuonzo G. 2009. Macrolide resistance genotypes and phenotypes among erythromycin-resistant clinical isolates of *Staphylococcus aureus* and coagulase-negative staphylococci, Italy. *FEMS Immunol Med Microbiol* 55: 62–67.

Gibbons S, Oluwatuyi M, Kaatz GW. 2003. A novel inhibitor of multidrug efflux pumps in *Staphylococcus aureus*. *J Antimicrob Chemother* 51: 13–17.

Giovanetti E, Brenciani A, Lupidi R, Roberts MC, Varaldo PE. 2003. Presence of the *tet*(O) gene in erythromycin- and tetracycline-resistant strains of *Streptococcus pyogenes* and linkage with either the *mef*(A) or the *erm*(A) gene. *Antimicrob Agents Chemother* 47: 2844–2849.

Golebiewski M, Kern-Zdanowicz I, Zienkiewicz M, Adamczyk M, Zylinska J, Baraniak A, Gniadkowski M, Bardowski J, Ceglowski P. 2007. Complete nucleotide sequence of the pCTX-M3 plasmid and its involvement in spread of the extended-spectrum β-lactamase gene bla$_{CTX-M-3}$. *Antimicrob Agents Chemother* 51: 3789–3795.

Gong F, Yanofsky C. 2002. Instruction of translating ribosome by nascent peptide. *Science* 297: 1864–1867.

Gonzalez-Zorn B, Catalan A, Escudero JA, Dominguez L, Teshager T, Porrero C, Moreno MA. 2005. Genetic basis for dissemination of *armA*. *J Antimicrob Chemother* 56: 583–585.

Gryczan TJ, Grandi G, Hahn J, Grandi R, Dubnau D. 1980. Conformational alteration of mRNA structure and the posttranscriptional regulation of erythromycin-induced drug resistance. *Nucleic Acids Res* 8: 6081–6097.

Gul HC, Kilic A, Guclu AU, Bedir O, Orhon M, Basustaoglu AC. 2008. Macrolide-lincosamide-streptogramin B resistant phenotypes and genotypes for methicillin-resistant *Staphylococcus aureus* in Turkey, from 2003 to 2006. *Pol J Microbiol* 57: 307–312.

Gulay Z, Ozbek OA, Bicmen M, Gur D. 2008. Macrolide resistance determinants in erythromycin-resistant *Streptococcus pneumoniae* in Turkey. *Jpn J Infect Dis* 61: 490–493.

Gupta P, Kannan K, Mankin AS, Vazquez-Laslop N. 2013a. Regulation of gene expression by macrolide-induced ribosomal frameshifting. *Mol Cell* 52: 629–642.

Gupta P, Sothiselvam S, Vazquez-Laslop N, Mankin AS. 2013b. Deregulation of translation due to post-transcriptional modification of rRNA explains why *erm* genes are inducible. *Nat Commun* 4: 1984.

Gupta P, Liu B, Klepacki D, Gupta V, Schulten K, Mankin AS, Vazquez-Laslop N. 2016. Nascent peptide assists the ribosome in recognizing chemically distinct small molecules. *Nat Chem Biol* 12: 153–158.

Hadjirin NF, Harrison EM, Holmes MA, Paterson GK. 2013. Conjugative transfer frequencies of *mef*(A)-containing Tn*1207.3* to macrolide-susceptible *Streptococcus pyogenes* belonging to different *emm* types. *Lett Appl Microbiol* 58: 299–302.

Hansen LH, Mauvais P, Douthwaite S. 1999. The macrolide-ketolide antibiotic binding site is formed by structures in domains II and V of 23S ribosomal RNA. *Mol Microbiol* 31: 623–631.

Hansen JL, Ippolito JA, Ban N, Nissen P, Moore PB, Steitz TA. 2002. The structures of four macrolide antibiotics bound to the large ribosomal subunit. *Mol Cell* 10: 11–128.

Hidalgo M, Santos C, Duarte C, Castaneda E, Agudelo CI. 2011. Increase in erythromycin-resistant *Streptococcus pneumoniae* in Colombia, 1994–2008. *Biomedica* 31: 124–131.

Hook EW III, Golden M, Jamieson BD, Dixon PB, Harbison HS, Lowens S, Fernandes P. 2015. A phase 2 trial of oral solithromycin 1200 mg or 1000 mg as single-dose oral therapy for uncomplicated gonorrhea. *Clin Infect Dis* 61: 1043–1048.

Horinouchi S, Weisblum B. 1980. Posttranscriptional modification of mRNA conformation: Mechanism that regulates erythromycin-induced resistance. *Proc Natl Acad Sci* 77: 7079–7083.

Howie RL, Folster JP, Bowen A, Barzilay EJ, Whichard JM. 2010. Reduced azithromycin susceptibility in *Shigella sonnei*, United States. *Microb Drug Resist* 16: 245–248.

Hraoui M, Boutiba-Ben Boubaker I, Rachdi M, Slim A, Ben Redjeb S. 2012. Macrolide and tetracycline resistance in clinical strains of *Streptococcus agalactiae* isolated in Tunisia. *J Med Microbiol* 61: 1109–1113.

Hue KK, Bechhofer DH. 1992. Regulation of the macrolide-lincosamide-streptogramin B resistance gene *ermD*. *J Bacteriol* 174: 5860–5868.

Hulten K, Gibreel A, Skold O, Engstrand L. 1997. Macrolide resistance in *Helicobacter pylori*: Mechanism and stability in strains from clarithromycin-treated patients. *Antimicrob Agents Chemother* 41: 2550–2553.

Ito K, Chiba S, Pogliano K. 2010. Divergent stalling sequences sense and control cellular physiology. *Biochem Biophys Res Commun* **393**: 1–5.

Ito S, Shimada Y, Yamaguchi Y, Yasuda M, Yokoi S, Nakano M, Ishiko H, Deguchi T. 2011. Selection of *Mycoplasma genitalium* strains harbouring macrolide resistance-associated 23S rRNA mutations by treatment with a single 1 g dose of azithromycin. *Sex Transm Infect* **87**: 412–414.

Iwata S, Sato Y, Toyonaga Y, Hanaki H, Sunakawa K. 2015. Genetic analysis of a pediatric clinical isolate of *Moraxella catarrhalis* with resistance to macrolides and quinolones. *J Infect Chemother* **21**: 308–311.

Jalava J, Vaara M, Huovinen P. 2004. Mutation at the position 2058 of the 23S rRNA as a cause of macrolide resistance in *Streptococcus pyogenes*. *Ann Clin Microbiol Antimicrob* **3**: 5.

Jenkins SG, Farrell DJ. 2009. Increase in pneumococcus macrolide resistance, United States. *Emerg Infect Dis* **15**: 1260–1264.

Jensen JS, Bradshaw CS, Tabrizi SN, Fairley CK, Hamasuna R. 2008. Azithromycin treatment failure in *Mycoplasma genitalium*-positive patients with nongonococcal urethritis is associated with induced macrolide resistance. *Clin Infect Dis* **47**: 1546–1553.

Jensen JS, Fernandes P, Unemo M. 2014. In vitro activity of the new fluoroketolide solithromycin (CEM-101) against macrolide-resistant and -susceptible *Mycoplasma genitalium* strains. *Antimicrob Agents Chemother* **58**: 3151–3156.

Jiang Y, Yu D, Wei Z, Shen P, Zhou Z, Yu Y. 2010. Complete nucleotide sequence of *Klebsiella pneumoniae* multidrug resistance plasmid pKP048, carrying bla_{KPC-2}, bla_{DHA-1}, *qnrB4*, and *armA*. *Antimicrob Agents Chemother* **54**: 3967–3969.

Johansson M, Chen J, Tsai A, Kornberg G, Puglisi JD. 2014. Sequence-dependent elongation dynamics on macrolide-bound ribosomes. *Cell Rep* **7**: 1534–1546.

Kadlec K, Brenner MG, Sweeney MT, Brzuszkiewicz E, Liesegang H, Daniel R, Watts JL, Schwarz S. 2011. Molecular basis of macrolide, triamilide, and lincosamide resistance in *Pasteurella multocida* from bovine respiratory disease. *Antimicrob Agents Chemother* **55**: 2475–2477.

Kamimiya S, Weisblum B. 1997. Induction of *ermSV* by 16-membered-ring macrolide antibiotics. *Antimicrob Agents Chemother* **41**: 530–534.

Kannan K, Mankin AS. 2011. Macrolide antibiotics in the ribosome exit tunnel: Species-specific binding and action. *Ann NY Acad Sci* **1241**: 33–47.

Kannan K, Vázquez-Laslop N, Mankin Alexander S. 2012. Selective protein synthesis by ribosomes with a drug-obstructed exit tunnel. *Cell* **151**: 508–520.

Kannan K, Kanabar P, Schryer D, Florin T, Oh E, Bahroos N, Tenson T, Weissman JS, Mankin AS. 2014. The general mode of translation inhibition by macrolide antibiotics. *Proc Natl Acad Sci* **111**: 15958–15963.

Karlsson M, Fellstrom C, Heldtander MU, Johansson KE, Franklin A. 1999. Genetic basis of macrolide and lincosamide resistance in *Brachyspira (Serpulina) hyodysenteriae*. *FEMS Microbiol Lett* **172**: 255–260.

Kawai Y, Miyashita N, Kubo M, Akaike H, Kato A, Nishizawa Y, Saito A, Kondo E, Teranishi H, Wakabayashi T, et al.

2013. Nationwide surveillance of macrolide-resistant *Mycoplasma pneumoniae* infection in pediatric patients. *Antimicrob Agents Chemother* **57**: 4046–4049.

Kelemen GH, Zalacain M, Culebras E, Seno ET, Cundliffe E. 1994. Transcriptional attenuation control of the tylosin-resistance gene *tlrA* in *Streptomyces fradiae*. *Mol Microbiol* **14**: 833–842.

Kerr ID, Reynolds ED, Cove JH. 2005. ABC proteins and antibiotic drug resistance: Is it all about transport? *Biochem Soc Trans* **33**: 1000–1002.

Kim YH, Cha CJ, Cerniglia CE. 2002. Purification and characterization of an erythromycin esterase from an erythromycin-resistant *Pseudomonas* sp. *FEMS Microbiol Lett* **210**: 239–244.

Kim JM, Kim JS, Kim N, Kim YJ, Kim IY, Chee YJ, Lee CH, Jung HC. 2008. Gene mutations of 23S rRNA associated with clarithromycin resistance in *Helicobacter pylori* strains isolated from Korean patients. *J Microbiol Biotechnol* **18**: 1584–1589.

Kono M, O'Hara K, Ebisu T. 1992. Purification and characterization of macrolide 2′-phosphotransferase type II from a strain of *Escherichia coli* highly resistant to macrolide antibiotics. *FEMS Microbiol Lett* **76**: 89–94.

Krauland M, Harrison L, Paterson D, Marsh J. 2010. Novel integron gene cassette arrays identified in a global collection of multi-drug resistant non-typhoidal *Salmonella enterica*. *Curr Microbiol* **60**: 217–223.

Kwak JH, Choi EC, Weisblum B. 1991. Transcriptional attenuation control of *ermK*, a macrolide-lincosamide-streptogramin B resistance determinant from *Bacillus licheniformis*. *J Bacteriol* **173**: 4725–4735.

Lampson BC, Parisi JT. 1986. Naturally occurring *Staphylococcus epidermidis* plasmid expressing constitutive macrolide-lincosamide-streptogramin B resistance contains a deleted attenuator. *J Bacteriol* **166**: 479–483.

Law CJ, Maloney PC, Wang DN. 2008. Ins and outs of major facilitator superfamily antiporters. *Annu Rev Microbiol* **62**: 289–305.

Le Bouter A, Leclercq R, Cattoir V. 2011. Molecular basis of resistance to macrolides, lincosamides and streptogramins in *Staphylococcus saprophyticus* clinical isolates. *Int J Antimicrob Agents* **37**: 118–123.

Leclercq R, Courvalin P. 2002. Resistance to macrolides and related antibiotics in *Streptococcus pneumoniae*. *Antimicrob Agents Chemother* **46**: 2727–2734.

Lee Y, Kim BS, Chun J, Yong JH, Lee YS, Yoo JS, Yong D, Hong SG, D'Souza R, Thomson KS, et al. 2014. Clonality and resistome analysis of KPC-producing *Klebsiella pneumoniae* strain isolated in Korea using whole genome sequencing. *Biomed Res Int* **2014**: 352862.

Li L, Feng W, Zhang Z, Xue H, Zhao X. 2015. Macrolide-lincosamide-streptogramin resistance phenotypes and genotypes of coagulase-positive *Staphylococcus aureus* and coagulase-negative staphylococcal isolates from bovine mastitis. *BMC Vet Res* **11**: 168.

Lina G, Quaglia A, Reverdy ME, Leclercq R, Vandenesch F, Etienne J. 1999. Distribution of genes encoding resistance to macrolides, lincosamides, and streptogramins among staphylococci. *Antimicrob Agents Chemother* **43**: 1062–1066.

Liu M, Douthwaite S. 2002a. Methylation at nucleotide G745 or G748 in 23S rRNA distinguishes Gram-negative from Gram-positive bacteria. *Mol Microbiol* **44:** 195–204.

Liu M, Douthwaite S. 2002b. Resistance to the macrolide antibiotic tylosin is conferred by single methylations at 23S rRNA nucleotides G748 and A2058 acting in synergy. *Proc Natl Acad Sci* **99:** 14658–14663.

Liu J, Keelan P, Bennett PM, Enne VI. 2009a. Characterization of a novel macrolide efflux gene, *mef*(B), found linked to *sul3* in porcine *Escherichia coli*. *J Antimicrob Chemother* **63:** 423–426.

Liu Y, Ye X, Zhang H, Xu X, Li W, Zhu D, Wang M. 2009b. Antimicrobial susceptibility of *Mycoplasma pneumoniae* isolates and molecular analysis of macrolide-resistant strains from Shanghai, China. *Antimicrob Agents Chemother* **53:** 2160–2162.

Llano-Sotelo B, Dunkle J, Klepacki D, Zhang W, Fernandes P, Cate JH, Mankin AS. 2010. Binding and action of CEM-101, a new fluoroketolide antibiotic that inhibits protein synthesis. *Antimicrob Agents Chemother* **54:** 4961–4970.

Lodder G, Werckenthin C, Schwarz S, Dyke K. 1997. Molecular analysis of naturally occurring *ermC*-encoding plasmids in staphylococci isolated from animals with and without previous contact with macrolide/lincosamide antibiotics. *FEMS Immunol Med Microbiol* **18:** 7–15.

Lovmar M, Nilsson K, Vimberg V, Tenson T, Nervall M, Ehrenberg M. 2006. The molecular mechanism of peptide-mediated erythromycin resistance. *J Biol Chem* **281:** 6742–6750.

Lu C, Ye T, Zhu G, Feng P, Ma H, Lu R, Lai W. 2010. Phenotypic and genetic characteristics of macrolide and lincosamide resistant *Ureaplasma urealyticum* isolated in Guangzhou, China. *Curr Microbiol* **61:** 44–49.

Lucier TS, Heitzman K, Liu SK, Hu PC. 1995. Transition mutations in the 23S rRNA of erythromycin-resistant isolates of *Mycoplasma pneumoniae*. *Antimicrob Agents Chemother* **39:** 2770–2773.

Luthje P. 2006. Antimicrobial resistance of coagulase-negative staphylococci from bovine subclinical mastitis with particular reference to macrolide-lincosamide resistance phenotypes and genotypes. *J Antimicrob Chemother* **57:** 966–969.

Luthje P, Schwarz S. 2007. Molecular basis of resistance to macrolides and lincosamides among staphylococci and streptococci from various animal sources collected in the resistance monitoring program BfT-GermVet. *Int J Antimicrob Agents* **29:** 528–535.

Malbruny B, Nagai K, Coquemont M, Bozdogan B, Andrasevic AT, Hupkova H, Leclercq R, Appelbaum PC. 2002. Resistance to macrolides in clinical isolates of *Streptococcus pyogenes* due to ribosomal mutations. *J Antimicrob Chemother* **49:** 935–939.

Mandell LA, Wunderink RG, Anzueto A, Bartlett JG, Campbell GD, Dean NC, Dowell SF, File TM Jr, Musher DM, Niederman MS, et al. 2007. Infectious Diseases Society of America/American Thoracic Society consensus guidelines on the management of community-acquired pneumonia in adults. *Clin Infect Dis* **44:** S27–72.

Martin B, Alloing G, Mejean V, Claverys JP. 1987. Constitutive expression of erythromycin resistance mediated by the *ermAM* determinant of plasmid pAM β1 results from deletion of 5' leader peptide sequences. *Plasmid* **18:** 250–253.

Martin JM, Green M, Barbadora KA, Wald ER. 2002. Erythromycin-resistant group A streptococci in schoolchildren in Pittsburgh. *N Engl J Med* **346:** 1200–1206.

Matsuoka M, Endou K, Kobayashi H, Inoue M, Nakajima Y. 1998. A plasmid that encodes three genes for resistance to macrolide antibiotics in *Staphylococcus aureus*. *FEMS Microbiol Lett* **167:** 221–227.

Matsuoka M, Janosi L, Endou K, Nakajima Y. 1999. Cloning and sequences of inducible and constitutive macrolide resistance genes in *Staphylococcus aureus* that correspond to an ABC transporter. *FEMS Microbiol Lett* **181:** 91–100.

Matsuoka M, Inoue M, Endo Y, Nakajima Y. 2003. Characteristic expression of three genes, *msr*(A), *mph*(C) and *erm*(Y), that confer resistance to macrolide antibiotics on *Staphylococcus aureus*. *FEMS Microbiol Lett* **220:** 287–293.

Matsuoka M, Narita M, Okazaki N, Ohya H, Yamazaki T, Ouchi K, Suzuki I, Andoh T, Kenri T, Sasaki Y, et al. 2004. Characterization and molecular analysis of macrolide-resistant *Mycoplasma pneumoniae* clinical isolates obtained in Japan. *Antimicrob Agents Chemother* **48:** 4624–4630.

Mayanskiy N, Alyabieva N, Ponomarenko O, Lazareva A, Katosova L, Ivanenko A, Kulichenko T, Namazova-Baranova L, Baranov A. 2014. Serotypes and antibiotic resistance of non-invasive *Streptococcus pneumoniae* circulating in pediatric hospitals in Moscow, Russia. *Int J Infect Dis* **20:** 58–62.

Mayford M, Weisblum B. 1990. The *ermC* leader peptide: Amino acid alterations leading to differential efficiency of induction by macrolide-lincosamide-streptogramin B antibiotics. *J Bacteriol* **172:** 3772–3779.

Meier A, Kirschner P, Springer B, Steingrube VA, Brown BA, Wallace RJ Jr, Bottger EC. 1994. Identification of mutations in 23S rRNA gene of clarithromycin-resistant *Mycobacterium intracellulare*. *Antimicrob Agents Chemother* **38:** 381–384.

Menninger JR. 1985. Functional consequences of binding macrolides to ribosomes. *J Antimicrob Chemother* **16:** 23–34.

Milton ID, Hewitt CL, Harwood CR. 1992. Cloning and sequencing of a plasmid-mediated erythromycin resistance determinant from *Staphylococcus xylosus*. *FEMS Microbiol Lett* **76:** 141–147.

Min YH, Kwon AR, Yoon EJ, Shim MJ, Choi EC. 2008. Translational attenuation and mRNA stabilization as mechanisms of *erm*(B) induction by erythromycin. *Antimicrob Agents Chemother* **52:** 1782–1789.

Mingoia M, Vecchi M, Cochetti I, Tili E, Vitali LA, Manzin A, Varaldo PE, Montanari MP. 2007. Composite structure of *Streptococcus pneumoniae* containing the erythromycin efflux resistance gene *mef*(I) and the chloramphenicol resistance gene *catQ*. *Antimicrob Agents Chemother* **51:** 3983–3987.

Misyurina OY, Chipitsyna EV, Finashutina YP, Lazarev VN, Akopian TA, Savicheva AM, Govorun VM. 2004. Mutations in a 23S rRNA gene of *Chlamydia trachomatis* associated with resistance to macrolides. *Antimicrob Agents Chemother* **48:** 1347–1349.

Morar M, Pengelly K, Koteva K, Wright GD. 2012. Mechanism and diversity of the erythromycin esterase family of enzymes. *Biochemistry* **51:** 1740–1751.

Morris RP, Nguyen L, Gatfield J, Visconti K, Nguyen K, Schnappinger D, Ehrt S, Liu Y, Heifets L, Pieters J, et al. 2005. Ancestral antibiotic resistance in *Mycobacterium tuberculosis*. *Proc Natl Acad Sci* **102:** 12200–12205.

Murphy E. 1985. Nucleotide sequence of *ermA*, a macrolide-lincosamide-streptogramin B determinant in *Staphylococcus aureus*. *J Bacteriol* **162:** 633–640.

Muto A, Osawa S. 1987. The guanine and cytosine content of genomic DNA and bacterial evolution. *Proc Natl Acad Sci* **84:** 166–169.

Nakamura A, Miyakozawa I, Nakazawa K, K OH, Sawai T. 2000. Detection and characterization of a macrolide 2′-phosphotransferase from a *Pseudomonas aeruginosa* clinical isolate. *Antimicrob Agents Chemother* **44:** 3241–3242.

Nakatogawa H, Ito K. 2002. The ribosomal exit tunnel functions as a discriminating gate. *Cell* **108:** 629–636.

Nash KA, Inderlied CB. 1995. Genetic basis of macrolide resistance in *Mycobacterium avium* isolated from patients with disseminated disease. *Antimicrob Agents Chemother* **39:** 2625–2630.

Nash KA, Brown-Elliott BA, Wallace RJ Jr. 2009. A novel gene, *erm*(41), confers inducible macrolide resistance to clinical isolates of *Mycobacterium abscessus* but is absent from *Mycobacterium chelonae*. *Antimicrob Agents Chemother* **53:** 1367–1376.

Ng LK, Martin I, Liu G, Bryden L. 2002. Mutation in 23S rRNA associated with macrolide resistance in *Neisseria gonorrhoeae*. *Antimicrob Agents Chemother* **46:** 3020–3025.

Noguchi N, Emura A, Matsuyama H, O'Hara K, Sasatsu M, Kono M. 1995. Nucleotide sequence and characterization of erythromycin resistance determinant that encodes macrolide 2′-phosphotransferase I in *Escherichia coli*. *Antimicrob Agents Chemother* **39:** 2359–2363.

Noguchi N, Katayama J, O'Hara K. 1996. Cloning and nucleotide sequence of the *mphB* gene for macrolide 2′-phosphotransferase II in *Escherichia coli*. *FEMS Microbiol Lett* **144:** 197–202.

Noguchi N, Tamura Y, Katayama J, Narui K. 1998. Expression of the *mphB* gene for macrolide 2′-phosphotransferase II from *Escherichia coli* in *Staphylococcus aureus*. *FEMS Microbiol Lett* **159:** 337–342.

Noguchi N, Takada K, Katayama J, Emura A, Sasatsu M. 2000. Regulation of transcription of the *mph*(A) gene for macrolide 2′-phosphotransferase I in *Escherichia coli*: Characterization of the regulatory gene *mphR*(A). *J Bacteriol* **182:** 5052–5058.

Nonaka L, Maruyama F, Suzuki S, Masuda M. 2015. Novel macrolide-resistance genes, *mef*(C) and *mph*(G), carried by plasmids from *Vibrio* and *Photobacterium* isolated from sediment and seawater of a coastal aquaculture site. *Lett Appl Microbiol* **61:** 1–6.

Nunez-Samudio V, Chesneau O. 2013. Functional interplay between the ATP binding cassette Msr(D) protein and the membrane facilitator superfamily Mef(E) transporter for macrolide resistance in *Escherichia coli*. *Res Microbiol* **164:** 226–235.

Occhialini A, Urdaci M, Doucet-Populaire F, Bebear CM, Lamouliatte H, Megraud F. 1997. Macrolide resistance in *Helicobacter pylori*: Rapid detection of point mutations and assays of macrolide binding to ribosomes. *Antimicrob Agents Chemother* **41:** 2724–2728.

Oh TG, Kwon AR, Choi EC. 1998. Induction of *ermAMR* from a clinical strain of *Enterococcus faecalis* by 16-membered-ring macrolide antibiotics. *J Bacteriol* **180:** 5788–5791.

O'Hara K, Yamamoto K. 1996. Reaction of roxithromycin and clarithromycin with macrolide-inactivating enzymes from highly erythromycin-resistant *Escherichia coli*. *Antimicrob Agents Chemother* **40:** 1036–1038.

O'Hara K, Kanda T, Ohmiya K, Ebisu T, Kono M. 1989. Purification and characterization of macrolide 2′-phosphotransferase from a strain of *Escherichia coli* that is highly resistant to erythromycin. *Antimicrob Agents Chemother* **33:** 1354–1357.

Ojo KK, Ulep C, Van Kirk N, Luis H, Bernardo M, Leitao J, Roberts MC. 2004. The *mef*(A) gene predominates among seven macrolide resistance genes identified in Gram-negative strains representing 13 genera, isolated from healthy Portuguese children. *Antimicrob Agents Chemother* **48:** 3451–3456.

Ojo KK, Striplin MJ, Ulep CC, Close NS, Zittle J, Luis H, Bernardo M, Leitao J, Roberts MC. 2006. Staphylococcus efflux *msr*(A) gene characterized in *Streptococcus, Enterococcus, Corynebacterium,* and *Pseudomonas* isolates. *Antimicrob Agents Chemother* **50:** 1089–1091.

Okazaki N, Narita M, Yamada S, Izumikawa K, Umetsu M, Kenri T, Sasaki Y, Arakawa Y, Sasaki T. 2001. Characteristics of macrolide-resistant Mycoplasma pneumoniae strains isolated from patients and induced with erythromycin in vitro. *Microbiol Immunol* **45:** 617–620.

Olender A. 2013. Antibiotic resistance and detection of the most common mechanism of resistance (MLS$_B$) of opportunistic *Corynebacterium*. *Chemotherapy* **59:** 294–306.

Oliveira SS, Murphy E, Gamon MR, Bastos MC. 1993. pRJ5: A naturally occurring *Staphylococcus aureus* plasmid expressing constitutive macrolide-lincosamide-streptogramin B resistance contains a tandem duplication in the leader region of the *ermC* gene. *J Gen Microbiol* **139:** 1461–1467.

Oliveira CS, Moura A, Henriques I, Brown CJ, Rogers LM, Top EM, Correia A. 2013. Comparative genomics of IncP-1ε plasmids from water environments reveals diverse and unique accessory genetic elements. *Plasmid* **70:** 412–419.

Opavski N, Gajic I, Borek AL, Obszanska K, Stanojevic M, Lazarevic I, Ranin L, Sitkiewicz I, Mijac V. 2015. Molecular characterization of macrolide resistant *Streptococcus pyogenes* isolates from pharyngitis patients in Serbia. *Infect Genet Evol* **33:** 246–252.

Otaka T, Kaji A. 1975. Release of (oligo) peptidyl-tRNA from ribosomes by erythromycin A. *Proc Natl Acad Sci* **72:** 2649–2652.

Palavecino EL, Riedel I, Duran C, Bajaksouzian S, Joloba M, Davies T, Appelbaum PC, Jacobs MR. 2002. Macrolide resistance phenotypes in *Streptococcus pneumoniae* in Santiago, Chile. *Int J Antimicrob Agents* **20:** 108–112.

Pao SS, Paulsen IT, Saier MH Jr. 1998. Major facilitator superfamily. *Microbiol Mol Biol Rev* **62:** 1–34.

Cite this article as *Cold Spring Harb Perspect Med* doi: 10.1101/cshperspect.a025395

Paulsen IT, Brown MH, Skurray RA. 1996. Proton-dependent multidrug efflux systems. *Microbiol Rev* **60:** 575–608.

Pereyre S, Gonzalez P, De Barbeyrac B, Darnige A, Renaudin H, Charron A, Raherison S, Bebear C, Bebear CM. 2002. Mutations in 23S rRNA account for intrinsic resistance to macrolides in *Mycoplasma hominis* and *Mycoplasma fermentans* and for acquired resistance to macrolides in *M. hominis. Antimicrob Agents Chemother* **46:** 3142–3150.

Perez-Trallero E, Urbieta M, Montes M, Ayestaran I, Marimon JM. 1998. Emergence of *Streptococcus pyogenes* strains resistant to erythromycin in Gipuzkoa, Spain. *Eur J Clin Microbiol Infect Dis* **17:** 25–31.

Perez-Trallero E, Marimon JM, Iglesias L, Larruskain J. 2003. Fluoroquinolone and macrolide treatment failure in pneumococcal pneumonia and selection of multidrug-resistant isolates. *Emerg Infect Dis* **9:** 1159–1162.

Perez-Vazquez M, Vindel A, Marcos C, Oteo J, Cuevas O, Trincado P, Bautista V, Grundmann H, Campos J, Group ESs-t. 2009. Spread of invasive Spanish *Staphylococcus aureus* spa-type t067 associated with a high prevalence of the aminoglycoside-modifying enzyme gene *ant(4')-Ia* and the efflux pump genes *msrA/msrB. J Antimicrob Chemother* **63:** 21–31.

Peric M, Bozdogan B, Jacobs MR, Appelbaum PC. 2003. Effects of an efflux mechanism and ribosomal mutations on macrolide susceptibility of *Haemophilus influenzae* clinical isolates. *Antimicrob Agents Chemother* **47:** 1017–1022.

Peters ED, Leverstein-van Hall MA, Box AT, Verhoef J, Fluit AC. 2001. Novel gene cassettes and integrons. *Antimicrob Agents Chemother* **45:** 2961–2964.

Peuchant O, Menard A, Renaudin H, Morozumi M, Ubukata K, Bebear CM, Pereyre S. 2009. Increased macrolide resistance of *Mycoplasma pneumoniae* in France directly detected in clinical specimens by real-time PCR and melting curve analysis. *J Antimicrob Chemother* **64:** 52–58.

Phuc Nguyen MC, Woerther PL, Bouvet M, Andremont A, Leclercq R, Canu A. 2009. *Escherichia coli* as reservoir for macrolide resistance genes. *Emerg Infect Dis* **15:** 1648–1650.

Piatkowska E, Piatkowski J, Przondo-Mordarska A. 2012. The strongest resistance of *Staphylococcus aureus* to erythromycin is caused by decreasing uptake of the antibiotic into the cells. *Cell Mol Biol Lett* **17:** 633–645.

Pihlajamaki M, Kataja J, Seppala H, Elliot J, Leinonen M, Huovinen P, Jalava J. 2002. Ribosomal mutations in *Streptococcus pneumoniae* clinical isolates. *Antimicrob Agents Chemother* **46:** 654–658.

Plante I, Centron D, Roy PH. 2003. An integron cassette encoding erythromycin esterase, *ere*(A), from *Providencia stuartii. J Antimicrob Chemother* **51:** 787–790.

Poirel L, Mansour W, Bouallegue O, Nordmann P. 2008. Carbapenem-resistant *Acinetobacter baumannii* isolates from Tunisia producing the OXA-58-like carbapenem-hydrolyzing oxacillinase OXA-97. *Antimicrob Agents Chemother* **52:** 1613–1617.

Poole TL, Callaway TR, Bischoff KM, Warnes CE, Nisbet DJ. 2006. Macrolide inactivation gene cluster *mphA-mrx-mphR* adjacent to a class 1 integron in *Aeromonas hydro-*

phila isolated from a diarrhoeic pig in Oklahoma. *J Antimicrob Chemother* **57:** 31–38.

Portillo A, Ruiz-Larrea F, Zarazaga M, Alonso A, Martinez JL, Torres C. 2000. Macrolide resistance genes in *Enterococcus* spp. *Antimicrob Agents Chemother* **44:** 967–971.

Prunier AL, Malbruny B, Tande D, Picard B, Leclercq R. 2002. Clinical isolates of *Staphylococcus aureus* with ribosomal mutations conferring resistance to macrolides. *Antimicrob Agents Chemother* **46:** 3054–3056.

Prunier AL, Trong HN, Tande D, Segond C, Leclercq R. 2005. Mutation of L4 ribosomal protein conferring unusual macrolide resistance in two independent clinical isolates of *Staphylococcus aureus. Microb Drug Resist* **11:** 18–20.

Ramu H, Mankin A, Vazquez-Laslop N. 2009. Programmed drug-dependent ribosome stalling. *Mol Microbiol* **71:** 811–824.

Rantala M, Huikko S, Huovinen P, Jalava J. 2005. Prevalence and molecular genetics of macrolide resistance among *Streptococcus pneumoniae* isolates collected in Finland in 2002. *Antimicrob Agents Chemother* **49:** 4180–4184.

Reinert RR, van der Linden M, Al-Lahham A. 2005. Molecular characterization of the first telithromycin-resistant *Streptococcus pneumoniae* isolate in Germany. *Antimicrob Agents Chemother* **49:** 3520–3522.

Reynolds ED, Cove JH. 2005. Resistance to telithromycin is conferred by *msr*(A), *msr*C and *msr*(D) in *Staphylococcus aureus. J Antimicrob Chemother* **56:** 1179–1180.

Reynolds E, Ross JI, Cove JH. 2003. *Msr*(A) and related macrolide/streptogramin resistance determinants: Incomplete transporters? *Int J Antimicrob Agents* **22:** 228–236.

Rimbara E, Noguchi N, Kawai T, Sasatsu M. 2008. Novel mutation in 23S rRNA that confers low-level resistance to clarithromycin in *Helicobacter pylori. Antimicrob Agents Chemother* **52:** 3465–3466.

Roberts MC. 2008. Update on macrolide-lincosamide-streptogramin, ketolide, and oxazolidinone resistance genes. *FEMS Microbiol Lett* **282:** 147–159.

Roberts MC. 2011. Environmental macrolide-lincosamide-streptogramin and tetracycline resistant bacteria. *Front Microbiol* **2:** 40.

Roberts MC, Sutcliffe J, Courvalin P, Jensen LB, Rood J, Seppala H. 1999. Nomenclature for macrolide and macrolide-lincosamide-streptogramin B resistance determinants. *Antimicrob Agents Chemother* **43:** 2823–2830.

Roberts MC, Soge OO, No DB. 2011. Characterization of macrolide resistance genes in *Haemophilus influenzae* isolated from children with cystic fibrosis. *J Antimicrob Chemother* **66:** 100–104.

Rosato A, Vicarini H, Leclercq R. 1999. Inducible or constitutive expression of resistance in clinical isolates of streptococci and enterococci cross-resistant to erythromycin and lincomycin. *J Antimicrob Chemother* **43:** 559–562.

Ross JI, Eady EA, Cove JH, Cunliffe WJ, Baumberg S, Wootton JC. 1990. Inducible erythromycin resistance in staphylococci is encoded by a member of the ATP-binding transport super-gene family. *Mol Microbiol* **4:** 1207–1214.

Ross JI, Eady EA, Cove JH, Baumberg S. 1995. Identification of a chromosomally encoded ABC-transport system with which the staphylococcal erythromycin exporter MsrA may interact. *Gene* **153:** 93–98.

Ross JI, Eady EA, Cove JH, Baumberg S. 1996. Minimal functional system required for expression of erythromycin resistance by *msrA* in *Staphylococcus aureus* RN4220. *Gene* **183:** 143–148.

Ross JI, Eady EA, Cove JH, Jones CE, Ratyal AH, Miller YW, Vyakrnam S, Cunliffe WJ. 1997. Clinical resistance to erythromycin and clindamycin in cutaneous propionibacteria isolated from acne patients is associated with mutations in 23S rRNA. *Antimicrob Agents Chemother* **41:** 1162–1165.

Rowe RA, Stephenson RM, East DL, Wright S. 2009. Mechanisms of resistance for *Streptococcus pyogenes* in northern Utah. *Clin Lab Sci* **22:** 39–44.

Rudolph K, Bulkow L, Bruce M, Zulz T, Reasonover A, Harker-Jones M, Hurlburt D, Hennessy T. 2013. Molecular resistance mechanisms of macrolide-resistant invasive *Streptococcus pneumoniae* isolates from Alaska, 1986 to 2010. *Antimicrob Agents Chemother* **57:** 5415–5422.

Saito R, Nonaka S, Nishiyama H, Okamura N. 2012. Molecular mechanism of macrolide-lincosamide resistance in *Moraxella catarrhalis*. *J Med Microbiol* **61:** 1435–1438.

Salado-Rasmussen K, Jensen JS. 2014. *Mycoplasma genitalium* testing pattern and macrolide resistance: A Danish nationwide retrospective survey. *Clin Infect Dis* **59:** 24–30.

Sandegren L, Linkevicius M, Lytsy B, Melhus A, Andersson DI. 2012. Transfer of an *Escherichia coli* ST131 multiresistance cassette has created a *Klebsiella* pneumoniae-specific plasmid associated with a major nosocomial outbreak. *J Antimicrob Chemother* **67:** 74–83.

Sander P, Prammananan T, Meier A, Frischkorn K, Bottger EC. 1997. The role of ribosomal RNAs in macrolide resistance. *Mol Microbiol* **26:** 469–480.

Sangvik M. 2005. *mef*(A), *mef*(E) and a new *mef* allele in macrolide-resistant *Streptococcus* spp isolates from Norway. *J Antimicrob Chemother* **56:** 841–846.

Santagati M, Iannelli F, Oggioni MR, Stefani S, Pozzi G. 2000. Characterization of a genetic element carrying the macrolide efflux gene *mef*(A) in *Streptococcus pneumoniae*. *Antimicrob Agents Chemother* **44:** 2585–2587.

Santagati M, Iannelli F, Cascone C, Campanile F, Oggioni MR, Stefani S, Pozzi G. 2003. The novel conjugative transposon Tn*1207.3* carries the macrolide efflux gene *mef*(A) in *Streptococcus pyogenes*. *Microb Drug Resist* **9:** 243–247.

Schlunzen F, Harms JM, Franceschi F, Hansen HA, Bartels H, Zarivach R, Yonath A. 2003. Structural basis for the antibiotic activity of ketolides and azalides. *Structure* **11:** 329–338.

Schluter A, Szczepanowski R, Kurz N, Schneiker S, Krahn I, Puhler A. 2007. Erythromycin resistance-conferring plasmid prsb105, isolated from a sewage treatment plant, harbors a new macrolide resistance determinant, an integron-containing Tn*402*-like element, and a large region of unknown function. *Appl Environ Microbiol* **73:** 1952–1960.

Schmeing TM, Huang KS, Strobel SA, Steitz TA. 2005. An induced-fit mechanism to promote peptide bond formation and exclude hydrolysis of peptidyl-tRNA. *Nature* **438:** 520–524.

Schmitz FJ, Sadurski R, Kray A, Boos M, Geisel R, Kohrer K, Verhoef J, Fluit AC. 2000. Prevalence of macrolide-resistance genes in *Staphylococcus aureus* and *Enterococcus faecium* isolates from 24 European university hospitals. *J Antimicrob Chemother* **45:** 891–894.

Schnellmann C, Gerber V, Rossano A, Jaquier V, Panchaud Y, Doherr MG, Thomann A, Straub R, Perreten V. 2006. Presence of new *mecA* and *mph*(C) variants conferring antibiotic resistance in *Staphylococcus* spp. isolated from the skin of horses before and after clinic admission. *J Clin Microbiol* **44:** 4444–4454.

* Schwarz S, Shen J, Kadlec K, Wang Y, Brenner Michael G, Feßler AT, Vester B. 2016. Lincosamides, streptogramins, phenicols, and pleuromutilins: Mode of action and mechanisms of resistance. *Cold Spring Harb Perspect Med* doi: 10.1101/cshperspect.a027037.

Seidelt B, Innis CA, Wilson DN, Gartmann M, Armache JP, Villa E, Trabuco LG, Becker T, Mielke T, Schulten K, et al. 2009. Structural insight into nascent polypeptide chain-mediated translational stalling. *Science* **326:** 1412–1415.

Seppala H, Klaukka T, Vuopio-Varkila J, Muotiala A, Helenius H, Lager K, Huovinen P. 1997. The effect of changes in the consumption of macrolide antibiotics on erythromycin resistance in group A streptococci in Finland. Finnish Study Group for antimicrobial resistance. *N Engl J Med* **337:** 441–446.

Shahsavan S, Emaneini M, Noorazar Khoshgnab B, Khoramian B, Asadollahi P, Aligholi M, Jabalameli F, Eslampour MA, Taherikalani M. 2012. A high prevalence of mupirocin and macrolide resistance determinant among *Staphylococcus aureus* strains isolated from burnt patients. *Burns* **38:** 378–382.

Shakya T, Wright GD. 2010. Nucleotide selectivity of antibiotic kinases. *Antimicrob Agents Chemother* **54:** 1909–1913.

Shen P, Wei Z, Jiang Y, Du X, Ji S, Yu Y, Li L. 2009. Novel genetic environment of the carbapenem-hydrolyzing β-lactamase KPC-2 among Enterobacteriaceae in China. *Antimicrob Agents Chemother* **53:** 4333–4338.

Shimada Y, Deguchi T, Nakane K, Yasuda M, Yokoi S, Ito S, Nakano M, Ishiko H. 2011. Macrolide resistance-associated 23S rRNA mutation in *Mycoplasma genitalium*, Japan. *Emerg Infect Dis* **17:** 1148–1150.

Shoji T, Takaya A, Sato Y, Kimura S, Suzuki T, Yamamoto T. 2015. RlmCD-mediated U747 methylation promotes efficient G748 methylation by methyltransferase RlmAII in 23S rRNA in *Streptococcus pneumoniae*; interplay between two rRNA methylations responsible for telithromycin susceptibility. *Nucleic Acids Res* **43:** 8964–8972.

Sigmund CD, Ettayebi M, Morgan EA. 1984. Antibiotic resistance mutations in 16S and 23S ribosomal RNA genes of *Escherichia coli*. *Nucleic Acids Res* **12:** 4653–4663.

Siira L, Rantala M, Jalava J, Hakanen AJ, Huovinen P, Kaijalainen T, Lyytikainen O, Virolainen A. 2009. Temporal trends of antimicrobial resistance and clonality of invasive *Streptococcus pneumoniae* isolates in Finland, 2002 to 2006. *Antimicrob Agents Chemother* **53:** 2066–2073.

Silva-Costa C, Friaes A, Ramirez M, Melo-Cristino J. 2015. Macrolide-resistant *Streptococcus pyogenes*: Prevalence and treatment strategies. *Expert Rev Anti Infect Ther* **13:** 615–628.

Simonovic M, Steitz TA. 2009. A structural view on the mechanism of the ribosome-catalyzed peptide bond formation. *Biochim Biophys Acta* **1789:** 612–623.

Soge OO, Queenan AM, Ojo KK, Adeniyi BA, Roberts MC. 2006. CTX-M-15 extended-spectrum β-lactamase from Nigerian *Klebsiella pneumoniae*. *J Antimicrob Chemother* **57:** 24–30.

Song JH, Chang HH, Suh JY, Ko KS, Jung SI, Oh WS, Peck KR, Lee NY, Yang Y, Chongthaleong A, et al. 2004. Macrolide resistance and genotypic characterization of *Streptococcus pneumoniae* in Asian countries: A study of the Asian Network for Surveillance of Resistant Pathogens (ANSORP). *J Antimicrob Chemother* **53:** 457–463.

Sothiselvam S, Liu B, Han W, Ramu H, Klepacki D, Atkinson GC, Brauer A, Remm M, Tenson T, Schulten K, et al. 2014. Macrolide antibiotics allosterically predispose the ribosome for translation arrest. *Proc Natl Acad Sci* **111:** 9804–9809.

Stamm LV, Bergen HL. 2000. A point mutation associated with bacterial macrolide resistance is present in both 23S rRNA genes of an erythromycin-resistant *Treponema pallidum* clinical isolate. *Antimicrob Agents Chemother* **44:** 806–807.

Stone GG, Shortridge D, Flamm RK, Versalovic J, Beyer J, Idler K, Zulawinski L, Tanaka SK. 1996. Identification of a 23S rRNA gene mutation in clarithromycin-resistant *Helicobacter pylori*. *Helicobacter* **1:** 227–228.

Subramaniam SL, Ramu H, Mankin AS. 2011. Inducible resistance to macrolide antibiotics. In *Antibiotic drug discovery and development* (ed. Dougherty TJ, Pucci MJ), pp 445–484. Springer, New York.

Sun J, Deng Z, Yan A. 2014. Bacterial multidrug efflux pumps: Mechanisms, physiology and pharmacological exploitations. *Biochem Biophys Res Commun* **453:** 254–267.

Sutcliffe J, Leclercq R, ed. 2002. Mechanisms of resistance to macrolides, lincosamides, and ketolides. In *Macrolide antibiotics* (ed. Schonfeld W, Kirst HA), pp. 281–317. Birkhauser Verlag, Basel, Switzerland.

Szczepanowski R, Krahn I, Bohn N, Puhler A, Schluter A. 2007. Novel macrolide resistance module carried by the IncP-1β resistance plasmid pRSB111, isolated from a wastewater treatment plant. *Antimicrob Agents Chemother* **51:** 673–678.

Szczuka E, Makowska N, Bosacka K, Slotwinska A, Kaznowski A. 2015. Molecular basis of resistance to macrolides, lincosamides and streptogramins in *Staphylococcus hominis* strains isolated from clinical specimens. *Folia Microbiol (Praha)* **61:** 143–147.

Szemraj M, Kwaszewska A, Pawlak R, Szewczyk EM. 2014. Macrolide, lincosamide, and streptogramin B resistance in lipophilic corynebacteria inhabiting healthy human skin. *Microb Drug Resist* **20:** 404–409.

Taha N, Araj GF, Wakim RH, Kanj SS, Kanafani ZA, Sabra A, Khairallah MT, Nassar FJ, Shehab M, Baroud M, et al. 2012. Genotypes and serotype distribution of macrolide resistant invasive and non-invasive *Streptococcus pneumoniae* isolates from Lebanon. *Ann Clin Microbiol Antimicrob* **11:** 2.

Tait-Kamradt A, Clancy J, Cronan M, Dib-Hajj F, Wondrack L, Yuan W, Sutcliffe J. 1997. *mefE* is necessary for the erythromycin-resistant M phenotype in *Streptococcus*

pneumoniae. *Antimicrob Agents Chemother* **41:** 2251–2255.

Tait-Kamradt A, Davies T, Appelbaum PC, Depardieu F, Courvalin P, Petitpas J, Wondrack L, Walker A, Jacobs MR, Sutcliffe J. 2000a. Two new mechanisms of macrolide resistance in clinical strains of *Streptococcus pneumoniae* from Eastern Europe and North America. *Antimicrob Agents Chemother* **44:** 3395–3401.

Tait-Kamradt A, Davies T, Cronan M, Jacobs MR, Appelbaum PC, Sutcliffe J. 2000b. Mutations in 23S rRNA and ribosomal protein L4 account for resistance in pneumococcal strains selected in vitro by macrolide passage. *Antimicrob Agents Chemother* **44:** 2118–2125.

Takaya A, Sato Y, Shoji T, Yamamoto T. 2013. Methylation of 23S rRNA nucleotide G748 by RlmAII methyltransferase renders *Streptococcus pneumoniae* telithromycin susceptible. *Antimicrob Agents Chemother* **57:** 3789–3796.

Taniguchi K, Nakamura A, Tsurubuchi K, Ishii A, O'Hara K, Sawai T. 1999. Identification of functional amino acids in the macrolide 2′-phosphotransferase II. *Antimicrob Agents Chemother* **43:** 2063–2065.

Taniguchi K, Nakamura A, Tsurubuchi K, O'Hara K, Sawai T. 2004. The role of histidine residues conserved in the putative ATP-binding region of macrolide 2′-phosphotransferase II. *FEMS Microbiol Lett* **232:** 123–126.

Tenson T, DeBlasio A, Mankin A. 1996. A functional peptide encoded in the *Escherichia coli* 23S rRNA. *Proc Natl Acad Sci* **93:** 5641–5646.

Tenson T, Xiong L, Kloss P, Mankin AS. 1997. Erythromycin resistance peptides selected from random peptide libraries. *J Biol Chem* **272:** 17425–17430.

Tenson T, Lovmar M, Ehrenberg M. 2003. The mechanism of action of macrolides, lincosamides and streptogramin B reveals the nascent peptide exit path in the ribosome. *J Mol Biol* **330:** 1005–1014.

Thornton CS, Grinwis ME, Sibley CD, Parkins MD, Rabin HR, Surette MG. 2015. Antibiotic susceptibility and molecular mechanisms of macrolide resistance in streptococci isolated from adult cystic fibrosis patients. *J Med Microbiol* **64:** 1375–1386.

Thungapathra M, Amita, Sinha KK, Chaudhuri SR, Garg P, Ramamurthy T, Nair GB, Ghosh A. 2002. Occurrence of antibiotic resistance gene cassettes *aac(6′)-Ib*, *dfrA5*, *dfrA12*, and *ereA2* in class I integrons in non-O1, non-O139 *Vibrio cholerae* strains in India. *Antimicrob Agents Chemother* **46:** 2948–2955.

Tkadlec J, Varekova E, Pantucek R, Doskar J, Ruzickova V, Botka T, Fila L, Melter O. 2015. Characterization of *Staphylococcus aureus* strains isolated from Czech cystic fibrosis patients: High rate of ribosomal mutation conferring resistance to MLS$_B$ antibiotics as a result of long-term and low-dose azithromycin treatment. *Microb Drug Resist* **21:** 416–423.

Touati A, Peuchant O, Jensen JS, Bebear C, Pereyre S. 2014. Direct detection of macrolide resistance in *Mycoplasma genitalium* isolates from clinical specimens from France by use of real-time PCR and melting curve analysis. *J Clin Microbiol* **52:** 1549–1555.

Tu D, Blaha G, Moore PB, Steitz TA. 2005. Structures of MLS$_B$K antibiotics bound to mutated large ribosomal subunits provide a structural explanation for resistance. *Cell* **121:** 257–270.

Twin J, Jensen JS, Bradshaw CS, Garland SM, Fairley CK, Min LY, Tabrizi SN. 2012. Transmission and selection of macrolide resistant *Mycoplasma genitalium* infections detected by rapid high resolution melt analysis. *PLoS ONE* 7: 20.

Vallianou N, Evangelopoulos A, Hadjisoteriou M, Avlami A, Petrikkos G. 2015. Prevalence of macrolide, lincosamide, and streptogramin resistance among staphylococci in a tertiary care hospital in Athens, Greece. *J Chemother* 27: 319–323.

Vannuffel P, Di Giambattista M, Morgan EA, Cocito C. 1992. Identification of a single base change in ribosomal RNA leading to erythromycin resistance. *J Biol Chem* 267: 8377–8382.

Varaldo PE, Montanari MP, Giovanetti E. 2009. Genetic elements responsible for erythromycin resistance in streptococci. *Antimicrob Agents Chemother* 53: 343–353.

Vazquez-Laslop N, Thum C, Mankin AS. 2008. Molecular mechanism of drug-dependent ribosome stalling. *Mol Cell* 30: 190–202.

Vazquez-Laslop N, Ramu H, Klepacki D, Kannan K, Mankin AS. 2010. The key function of a conserved and modified rRNA residue in the ribosomal response to the nascent peptide. *EMBO J* 29: 3108–3117.

Vazquez-Laslop N, Klepacki D, Mulhearn DC, Ramu H, Krasnykh O, Franzblau S, Mankin AS. 2011. Role of antibiotic ligand in nascent peptide-dependent ribosome stalling. *Proc Natl Acad Sci* 108: 10496–10501.

Verdet C, Benzerara Y, Gautier V, Adam O, Ould-Hocine Z, Arlet G. 2006. Emergence of DHA-1-producing *Klebsiella* spp. in the Parisian region: Genetic organization of the *ampC* and *ampR* genes originating from *Morganella morganii*. *Antimicrob Agents Chemother* 50: 607–617.

Versalovic J, Osato MS, Spakovsky K, Dore MP, Reddy R, Stone GG, Shortridge D, Flamm RK, Tanaka SK, Graham DY. 1997. Point mutations in the 23S rRNA gene of *Helicobacter pylori* associated with different levels of clarithromycin resistance. *J Antimicrob Chemother* 40: 283–286.

Vester B, Douthwaite S. 2001. Macrolide resistance conferred by base substitutions in 23S rRNA. *Antimicrob Agents Chemother* 45: 1–12.

Vester B, Garrett RA. 1987. A plasmid-coded and site-directed mutation in *Escherichia coli* 23S RNA that confers resistance to erythromycin: Implications for the mechanism of action of erythromycin. *Biochimie* 69: 891–900.

Vimberg V, Xiong L, Bailey M, Tenson T, Mankin A. 2004. Peptide-mediated macrolide resistance reveals possible specific interactions in the nascent peptide exit tunnel. *Mol Microbiol* 54: 376–385.

Vimberg V, Lenart J, Janata J, Balikova Novotna G. 2015. ClpP-independent function of ClpX interferes with telithromycin resistance conferred by Msr(A) in *Staphylococcus aureus*. *Antimicrob Agents Chemother* 59: 3611–3614.

Voss NR, Gerstein M, Steitz TA, Moore PB. 2006. The geometry of the ribosomal polypeptide exit tunnel. *J Mol Biol* 360: 893–906.

Wallace RJ Jr., Meier A, Brown BA, Zhang Y, Sander P, Onyi GO, Bottger EC. 1996. Genetic basis for clarithromycin resistance among isolates of *Mycobacterium chelonae* and *Mycobacterium abscessus*. *Antimicrob Agents Chemother* 40: 1676–1681.

Walsh F, Willcock J, Amyes S. 2003. High-level telithromycin resistance in laboratory-generated mutants of *Streptococcus pneumoniae*. *J Antimicrob Chemother* 52: 345–353.

Wang G, Taylor DE. 1998. Site-specific mutations in the 23S rRNA gene of *Helicobacter pylori* confer two types of resistance to macrolide-lincosamide-streptogramin B antibiotics. *Antimicrob Agents Chemother* 42: 1952–1958.

Weisblum B. 1995a. Erythromycin resistance by ribosome modification. *Antimicrob Agents Chemother* 39: 577–585.

Weisblum B. 1995b. Insights into erythromycin action from studies of its activity as inducer of resistance. *Antimicrob Agents Chemother* 39: 797–805.

Wierzbowski AK, Boyd D, Mulvey M, Hoban DJ, Zhanel GG. 2005. Expression of the *mef*(E) gene encoding the macrolide efflux pump protein increases in *Streptococcus pneumoniae* with increasing resistance to macrolides. *Antimicrob Agents Chemother* 49: 4635–4640.

Wierzbowski AK, Karlowsky JA, Adam HJ, Nichol KA, Hoban DJ, Zhanel GG. 2014. Evolution and molecular characterization of macrolide-resistant *Streptococcus pneumoniae* in Canada between 1998 and 2008. *J Antimicrob Chemother* 69: 59–66.

Wilson DN. 2009. The A-Z of bacterial translation inhibitors. *Crit Rev Biochem Mol Biol* 44: 393–433.

Wilson DN. 2014. Ribosome-targeting antibiotics and mechanisms of bacterial resistance. *Nat Rev Microbiol* 12: 35–48.

Wolter N, Smith AM, Farrell DJ, Klugman KP. 2006. Heterogeneous macrolide resistance and gene conversion in the pneumococcus. *Antimicrob Agents Chemother* 50: 359–361.

Wolter N, Smith AM, Low DE, Klugman KP. 2007. High-level telithromycin resistance in a clinical isolate of *Streptococcus pneumoniae*. *Antimicrob Agents Chemother* 51: 1092–1095.

Wolter N, Smith AM, Farrell DJ, Northwood JB, Douthwaite S, Klugman KP. 2008a. Telithromycin resistance in *Streptococcus pneumoniae* is conferred by a deletion in the leader sequence of *erm*(B) that increases rRNA methylation. *Antimicrob Agents Chemother* 52: 435–440.

Wolter N, von Gottberg A, du Plessis M, de Gouveia L, Klugman KP. 2008b. Molecular basis and clonal nature of increasing pneumococcal macrolide resistance in South Africa, 2000–2005. *Int J Antimicrob Agents* 32: 62–67.

Wondrack L, Massa M, Yang BV, Sutcliffe J. 1996. Clinical strain of *Staphylococcus aureus* inactivates and causes efflux of macrolides. *Antimicrob Agents Chemother* 40: 992–998.

Woodford N, Carattoli A, Karisik E, Underwood A, Ellington MJ, Livermore DM. 2009. Complete nucleotide sequences of plasmids pEK204, pEK499, and pEK516, encoding CTX-M enzymes in three major *Escherichia coli* lineages from the United Kingdom, all belonging to the international O25:H4-ST131 clone. *Antimicrob Agents Chemother* 53: 4472–4482.

Wright GD, Thompson PR. 1999. Aminoglycoside phosphotransferases: Proteins, structure, and mechanism. *Front Biosci* **4:** D9–D21.

Xin D, Mi Z, Han X, Qin L, Li J, Wei T, Chen X, Ma S, Hou A, Li G, et al. 2009. Molecular mechanisms of macrolide resistance in clinical isolates of *Mycoplasma pneumoniae* from China. *Antimicrob Agents Chemother* **53:** 2158–2159.

Xiong L, Shah S, Mauvais P, Mankin AS. 1999. A ketolide resistance mutation in domain II of 23S rRNA reveals the proximity of hairpin 35 to the peptidyl transferase centre. *Mol Microbiol* **31:** 633–639.

Ye Y, Li S, Li Y, Ren T, Liu K. 2013. *Mycoplasma pneumoniae* 23S rRNA gene mutations and mechanisms of macrolide resistance. *Lab Med* **44:** 63–68.

Yildiz O, Coban AY, Sener AG, Coskuner SA, Bayramoglu G, Guducuoglu H, Ozyurt M, Tatman-Otkun M, Karabiber N, Ozkutuk N, et al. 2014. Antimicrobial susceptibility and resistance mechanisms of methicillin resistant *Staphylococcus aureus* isolated from 12 Hospitals in Turkey. *Ann Clin Microbiol Antimicrob* **13:** 44.

Yoo SJ, Kim HB, Choi SH, Lee SO, Kim SH, Hong SB, Sung H, Kim MN. 2012. Differences in the frequency of 23S rRNA gene mutations in *Mycoplasma pneumoniae* between children and adults with community-acquired pneumonia: Clinical impact of mutations conferring macrolide resistance. *Antimicrob Agents Chemother* **56:** 6393–6396.

Yusupov MM, Yusupova GZ, Baucom A, Lieberman K, Earnest TN, Cate JH, Noller HF. 2001. Crystal structure of the ribosome at 5.5 Å resolution. *Science* **292:** 883–896.

Zahner D, Zhou X, Chancey ST, Pohl J, Shafer WM, Stephens DS. 2010. Human antimicrobial peptide LL-37 induces MefE/Mel-mediated macrolide resistance in *Streptococcus pneumoniae*. *Antimicrob Agents Chemother* **54:** 3516–3519.

Zarrilli R, Vitale D, Di Popolo A, Bagattini M, Daoud Z, Khan AU, Afif C, Triassi M. 2008. A plasmid-borne bla_{OXA-58} gene confers imipenem resistance to *Acinetobacter baumannii* isolates from a Lebanese hospital. *Antimicrob Agents Chemother* **52:** 4115–4120.

Zhang Z, Fukuzaki T, Myers AG. 2016. Synthesis of D-desosamine and analogs by rapid assembly of 3-amino sugars. *Angew Chem Int Ed Engl* **55:** 523–527.

Zhao F, Liu G, Wu J, Cao B, Tao X, He L, Meng F, Zhu L, Lv M, Yin Y, et al. 2013. Surveillance of macrolide-resistant *Mycoplasma pneumoniae* in Beijing, China, from 2008 to 2012. *Antimicrob Agents Chemother* **57:** 1521–1523.

Zheng J, Sagar V, Smolinsky A, Bourke C, LaRonde-LeBlanc N, Cropp TA. 2009. Structure and function of the macrolide biosensor protein, MphR(A), with and without erythromycin. *J Mol Biol* **387:** 1250–1260.

Zheng X, Lee S, Selvarangan R, Qin X, Tang YW, Stiles J, Hong T, Todd K, Ratliff AE, Crabb DM, et al. 2015. Macrolide-resistant *Mycoplasma pneumoniae*, United States. *Emerg Infect Dis* **21:** 1470–1472.

Zhou W, Jiang YM, Wang HJ, Kuang LH, Hu ZQ, Shi H, Shu M, Wa CM. 2014. Erythromycin-resistant genes in group A β-haemolytic *Streptococci* in Chengdu, Southwestern China. *Indian J Med Microbiol* **32:** 290–293.

Zhou Z, Li X, Chen X, Luo F, Pan C, Zheng X, Tan F. 2015. Macrolide-resistant *Mycoplasma pneumoniae* in adults in Zhejiang, China. *Antimicrob Agents Chemother* **59:** 1048–1051.

Zmantar T, Kouidhi B, Miladi H, Bakhrouf A. 2011. Detection of macrolide and disinfectant resistance genes in clinical *Staphylococcus aureus* and coagulase-negative staphylococci. *BMC Res Notes* **4:** 453.

Lincosamides, Streptogramins, Phenicols, and Pleuromutilins: Mode of Action and Mechanisms of Resistance

Stefan Schwarz,[1,2] Jianzhong Shen,[2] Kristina Kadlec,[1] Yang Wang,[2] Geovana Brenner Michael,[1] Andrea T. Feßler,[1] and Birte Vester[3]

[1]Institute of Farm Animal Genetics, Friedrich-Loeffler-Institut (FLI), 31535 Neustadt-Mariensee, Germany

[2]Beijing Key Laboratory of Detection Technology for Animal-Derived Food Safety, College of Veterinary Medicine, China Agricultural University, Beijing, P.R. China

[3]Department of Biochemistry and Molecular Biology, University of Southern Denmark, 5230 Odense M, Denmark

Correspondence: stefan.schwarz@fli.bund.de

Lincosamides, streptogramins, phenicols, and pleuromutilins (LSPPs) represent four structurally different classes of antimicrobial agents that inhibit bacterial protein synthesis by binding to particular sites on the 50S ribosomal subunit of the ribosomes. Members of all four classes are used for different purposes in human and veterinary medicine in various countries worldwide. Bacteria have developed ways and means to escape the inhibitory effects of LSPP antimicrobial agents by enzymatic inactivation, active export, or modification of the target sites of the agents. This review provides a comprehensive overview of the mode of action of LSPP antimicrobial agents as well as of the mutations and resistance genes known to confer resistance to these agents in various bacteria of human and animal origin.

For more than 70 years, antimicrobial agents have been indispensable for the control of bacterial infections in human and veterinary medicine. They inhibit bacteria mainly by interfering with cell-wall synthesis, nucleic acid synthesis, or protein synthesis. Their efficacy is, however, hampered by the continuous development of resistance not only by the target bacteria but also by members of the physiological microbiota in both humans and animals. Lincosamides, streptogramins, phenicols, and pleuromutilins (LSPPs) represent four classes of antimicrobial agents that inhibit protein synthesis by interacting with the 50S subunit of bacterial

ribosomes. Bacterial resistance to LSPP antimicrobial agents can be a result of the acquisition of endogenous mutations or horizontally transmitted resistance genes (Schwarz et al. 2006). The mechanisms known so far, associated with resistance to LSPP antimicrobial agents, commonly fall into three categories: enzymatic inactivation, active efflux, and/or structural changes at the ribosomal target site. Many of the resistance genes known so far that confer LSPP resistance are located on mobile genetic elements, which facilitate their dissemination across strain, species, and even genus boundaries. The understanding of the mode of action of antimicrobial agents has in the past led to the development of derivatives of known antimicrobial agents, designed to escape preexisting bacterial resistance mechanisms (Schwarz and Kehrenberg 2006). As such, the knowledge of not only the mode of action but also of the mode of resistance is a key parameter in the understanding of the complex interaction of antimicrobial agents and bacteria.

In the present review, we summarize the current knowledge of the mode of action of LSPP antimicrobial agents and provide an overview of the genetic basis of resistance to these agents in bacteria of human and animal origin.

USE OF LSPP ANTIMICROBIAL AGENTS IN HUMAN AND VETERINARY MEDICINE

Lincosamides

Lincosamides consist of three components: an amino acid (aa) (L-proline substituted by a 4′-alkyl chain) and a sugar (lincosamine), connected by an amide bond (Fig. 1A) (Bryskier 2005a). The first lincosamide, lincomycin, was isolated in 1962 from *Streptomyces lincolnensis* ssp. *lincolnensis* found in a soil sample from Lincoln, NE (MacLeod et al. 1964; Bryskier 2005a). In 1967, lincomycin was licensed in the United States for the treatment of infections caused by Gram-positive bacteria. Although approved for use in human medicine, lincomycin is rarely used nowadays. In veterinary medicine, lincomycin is approved for use in various infections in swine, dogs, and cats. In combination with spectinomycin, it is not only approved for ruminants, pigs, and poultry (Giguère 2013) but also for dogs, cats, and carrier pigeons.

As lincomycin has only a limited spectrum of activity, various chemical modifications were introduced to improve the pharmacokinetics of lincomycin and to expand its antibacterial spectrum. The 7-chloro-7-deoxylincomycin derivative, clindamycin, proved to be the most effective. Clindamycin was approved by the Food and Drug Administration (FDA) in 1970 in the United States. The antimicrobial spectrum of clindamycin includes staphylococci, group A and B streptococci, *Streptococcus pneumoniae*,

most anaerobic bacteria, and *Chlamydia trachomatis* (Smieja 1998). Moreover, clindamycin also shows activity against several protozoa, such as *Plasmodium* spp. and *Toxoplasma* spp. (Bryskier 2005a). However, it shows little, if no, activity against most aerobic Gram-negative bacilli, *Nocardia* spp., *Mycobacterium* spp., as well as *Enterococcus faecalis* and *Enterococcus faecium* (Giguère 2013). In veterinary medicine, clindamycin must not be used in food-producing animals, but may be used in dogs and cats under the Animal Medicinal Drug Use Clarification Act of 1994 (AMDUCA) in the United States and under similar regulations in other countries. Clindamycin is often used in human medicine for the treatment of infections caused by anaerobic bacteria. Because of its activity against anaerobes, it can lead to disruption of the intestinal microbiota and *Clostridium difficile* overgrowth causing diarrhea and colitis (Gerding et al. 1995). A similar situation has been observed in horses in which lincosamide administration can cause *C. difficile*–associated disease (CDAD), that is, a severe to fatal enterocolitis (Diab et al. 2013; Giguère 2013). It should be noted that lincosamides are also highly toxic to rabbits, guinea pigs, and hamsters, where toxins produced by *Clostridium perfringens* and other clostridia have been implicated in lincomycin- and clindamycin-induced enteritis (Morris 1995).

A new lincosamide, pirlimycin, was approved in 2000 in the United States and in 2001 in the European Union (EU). Pirlimycin represents a *cis*-4-ethyl-L-picecolic acid amide of clindamycin (Ahonkhai et al. 1982). It is exclusively approved for veterinary applications, that is, as an intramammary infusion for the control of staphylococci and streptococci associated with bovine subclinical mastitis.

Streptogramins

Streptogramins (pristinamycin, virginiamycin, mikamycin, and quinupristin–dalfopristin) consist of two structurally different components, A and B (Fig. 1B). The A components, such as pristinamycin IIA, virginiamycin M, mikamycin A, or dalfopristin, are polyunsaturated macro-

Cite this article as *Cold Spring Harb Perspect Med* doi: 10.1101/cshperspect.a027037

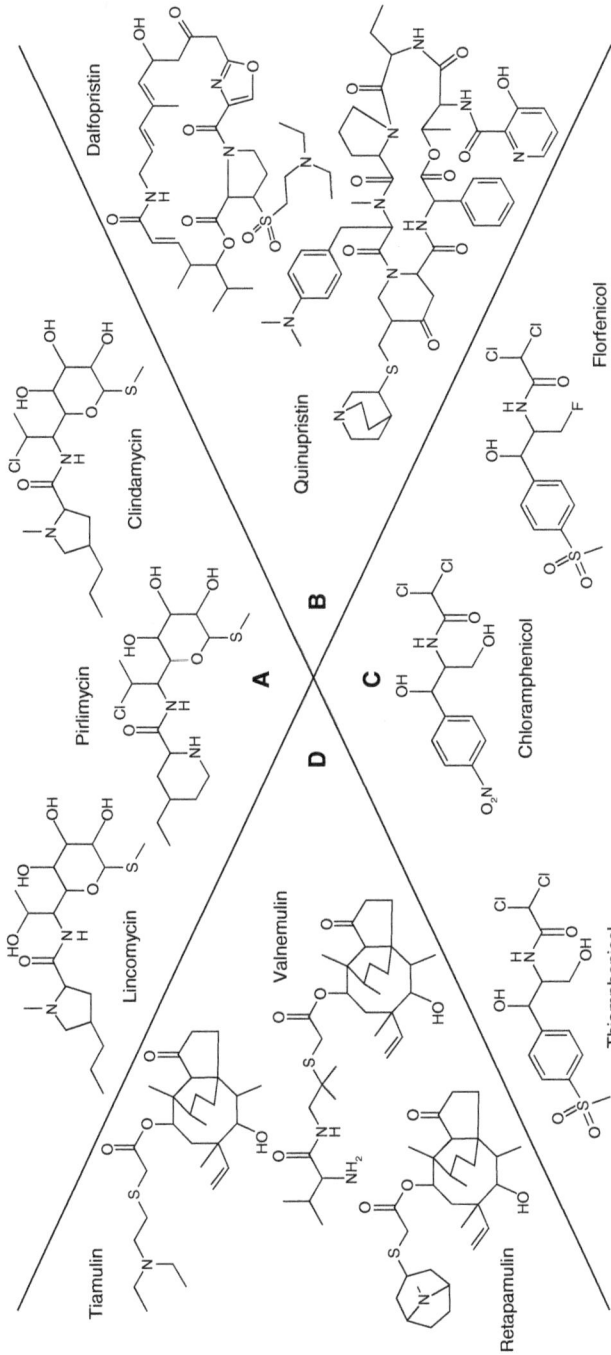

Figure 1. Structural formulas of the lincosamides, streptogramins, phenicols, and pleuromutilins (LSPP) antimicrobial agents. (*A*) The lincosamides lincomycin, clindamycin, and pirlimycin, (*B*) the streptogramin A dalfopristin and the streptogramin B quinupristin, (*C*) the phenicols chloramphenicol, thiamphenicol, and florfenicol, and (*D*) the pleuromutilins tiamulin, valnemulin, and retapamulin.

lactones. The B components, such as pristinamycin IB, virginiamycin S, mikamycin B, or quinupristin, are cyclic hexadepsipeptides (Allignet et al. 1996; Giguère 2013).

Pristinamycin was identified in culture filtrates of *Streptomyces pristinaespiralis* in 1962, virginiamycin in *Streptomyces virginiae* in 1955, and mikamycin in *Streptomyces mitakaensis* in 1956 (Vazquez 1966). Natural mixtures of streptogramins have been available for clinical use in Europe since the mid-1950s. Streptogramins cannot cross the outer membrane of most Gram-negative bacteria and are primarily effective against Gram-positive bacteria (Wright 2007). They have been used topically or orally in the treatment of skin, bone, and respiratory infections, mainly caused by staphylococci (Allignet et al. 1996). The first semisynthetic injectable streptogramin compound, quinupristin–dalfopristin, was approved in 1999. Quinupristin– dalfopristin represents one of the few potential antimicrobial agents for the treatment of infections in humans caused by multiresistant *E. faecium*, especially in cases of vancomycin- and/or linezolid-resistant isolates and also for multiresistant staphylococci, including methicillin-resistant *Staphylococcus aureus* (MRSA) (see who .int/iris/bitstream/10665/43765/1/978924159 5742_eng.pdf). Therefore, it is considered as a last resort drug for human use (see ema.euro pa.eu/docs/en_GB/document_library/Scien tific_guideline/2011/11/WC500118230.pdf).

Virginiamycin has been developed largely as a growth promoter (Giguère 2013) but its use for that purpose was phased out in the EU by the end of 2005. Since January 1, 2006, antimicrobial agents are not allowed to be used as growth promoters in the EU anymore. Streptogramins are currently not approved for therapeutic purposes in veterinary medicine in the EU, but virginiamycin may be used for specific indications in swine and horses in North America (Giguère 2013).

Phenicols

Chloramphenicol was isolated from *Streptomyces venezuelae* in 1947. It was the first phenicol antimicrobial agent and also the first natural product found to contain a nitro group (Fig. 1C). Because of the relative simplicity of its structure, chloramphenicol has been produced synthetically since 1950 (Schwarz et al. 2004). Thiamphenicol is a derivative of chloramphenicol, in which the *p*-nitro group has been replaced by a sulfomethyl group. Florfenicol is a fluorinated derivative of thiamphenicol in which the hydroxyl group at C3 has been replaced with fluorine (Dowling 2013).

Based on its activity against a wide range of Gram-positive and Gram-negative bacteria, chloramphenicol was initially considered a promising antimicrobial agent (Shaw 1983). However, serious adverse effects have been observed since the mid-1960s. These included a dose-unrelated irreversible aplastic anemia, a dose-related reversible bone marrow suppression, or the Gray syndrome in neonates and infants. In addition, hypersensitivity to chloramphenicol has been observed occasionally (Schwarz et al. 2004). Based on these adverse effects and the availability of less toxic antimicrobial agents with a similar spectrum of activity, chloramphenicol is now used in human medicine only for the treatment of a small number of life-threatening infections or for topical applications, for example, in eye infections. In veterinary medicine, chloramphenicol is still used in pets and non-food-producing animals. It was banned from use in food-producing animals in the EU in 1994 and in many other countries soon thereafter.

Thiamphenicol shows lower antimicrobial activity than chloramphenicol and, therefore, has rarely been used in human and veterinary medicine. Both, thiamphenicol and florfenicol do not cause dose-unrelated irreversible aplastic anemia, but may cause dose-dependent bone marrow suppression in animals (Dowling 2013). Florfenicol is exclusively approved for use in food-producing animals. Since 1995, florfenicol has been approved in numerous countries for the treatment of respiratory disease, pododermatitis, and keratoconjunctivitis in cattle, swine respiratory disease, air sacculitis in broiler chickens, but also for various diseases in fish. The application of florfenicol in horses is not recommended (Dowling 2013).

Pleuromutilins

Pleuromutilins are diterpene antimicrobial agents (Fig. 1D). They were discovered as natural antimicrobial agents in 1950/1951 (Bryskier 2005b; Novak and Shlaes 2010). The first pleuromutilin was found in the basidiomycete *Pleurotus mutilus*, later renamed as *Clitopilus scyphoides* (Bryskier 2005b). Tiamulin and valnemulin are semisynthetic pleuromutilins. Tiamulin was the first pleuromutilin to be approved for veterinary use in 1979, followed by valnemulin in 1999 (Novak and Shlaes 2010). Both tiamulin and valnemulin are exclusively used in veterinary medicine, mainly for the control of bacterial infections in pigs and poultry. They are particularly active against anaerobic bacteria, *Mycoplasma* spp., and selected Gram-positive and Gram-negative bacteria (Giguère 2013). In swine, tiamulin is commonly used for the treatment of swine dysentery caused by *Brachyspira hyodysenteriae*, swine pneumonia caused by *Actinobacillus pleuropneumoniae* and *Mycoplasma hyodysenteriae*, colonic spirochaetosis caused by *Brachyspira pilosicoli*, and proliferative enteropathy caused by *Lawsonia intracellularis*. In poultry, both pleuromutilins are commonly used for the treatment of *Mycoplasma gallisepticum* infections. Pleuromutilins are not approved for use in ruminants and should also not be administered to horses (Giguère 2013).

In 2007, the first pleuromutilin antimicrobial agent, retapamulin, was approved for use in humans (Novak and Shlaes 2010). However, the use of retapamulin is limited to topical treatment of impetigo caused by methicillin-susceptible *S. aureus* (MSSA) or *Streptococcus pyogenes* in patients of 9 months or older.

MODE OF ACTION OF LSPP ANTIMICROBIAL AGENTS

Early studies of the mechanisms of action of various antimicrobial agents relied on their effects in various functional assays. The interpretations of these studies were influenced by the lack of knowledge of their binding sites, and whether or not the effect assayed for was a major/minor or a direct/indirect effect. Details

and summaries of these studies for translational inhibitors have been reviewed by Wilson (2009). After solution of the bacterial ribosome structure in the year 2000 (Ban et al. 2000), the unraveling of mechanisms of action for protein synthesis inhibitors increased tremendously because the binding sites of the antimicrobial agents could now be determined. The ribosome crystals were coincubated with the antimicrobial agents and thereafter the detailed position of binding could be determined for each antimicrobial agent. From the position in the ribosomes, the overall mode of action could then be deduced.

LSPP antimicrobial agents have very diverse chemical structures as shown by the examples presented in Figure 1. Despite their differences, they have all been found to bind at the peptidyl-transferase center (PTC) of bacterial ribosomes. The ribosome structures with the bound antimicrobial agents have been obtained from various bacteria (*Escherichia coli*, *Deinococcus radiodurans*, *Thermus thermophilus*, and *S. aureus*) and one archaeon *Haloarcula marismortui*, and they show some minor differences in antibiotic binding. Details and summaries of these structures have been described by Wilson (2011, 2014) and Eyal et al. (2015). It is possible to transfer the binding positions from one structure to another to compare their binding sites as presented in Figure 2, which shows binding of one example from each antimicrobial class placed together. It is thus obvious that LSPP antimicrobial agents have overlapping binding sites and that each of them occupies specific sites in the ribosome in the area where the ribosome extends the nascent peptide chain. The binding site is in the bottom of the cleft of the 50S ribosomal subunit where the 3′-ends of aminoacyl-tRNA and peptidyl-tRNA are positioned for peptide transfer. The site is composed exclusively of RNA, and mainly RNA from the central part of domain V of 23S RNA (Nissen et al. 2000), and it is highly conserved in all bacteria. The same site in the ribosome also binds the oxazolidinones including linezolid.

Streptogramin B antimicrobial agents bind at a site adjacent to the PTC in the beginning of the nascent peptide exit tunnel (Fig. 2) and over-

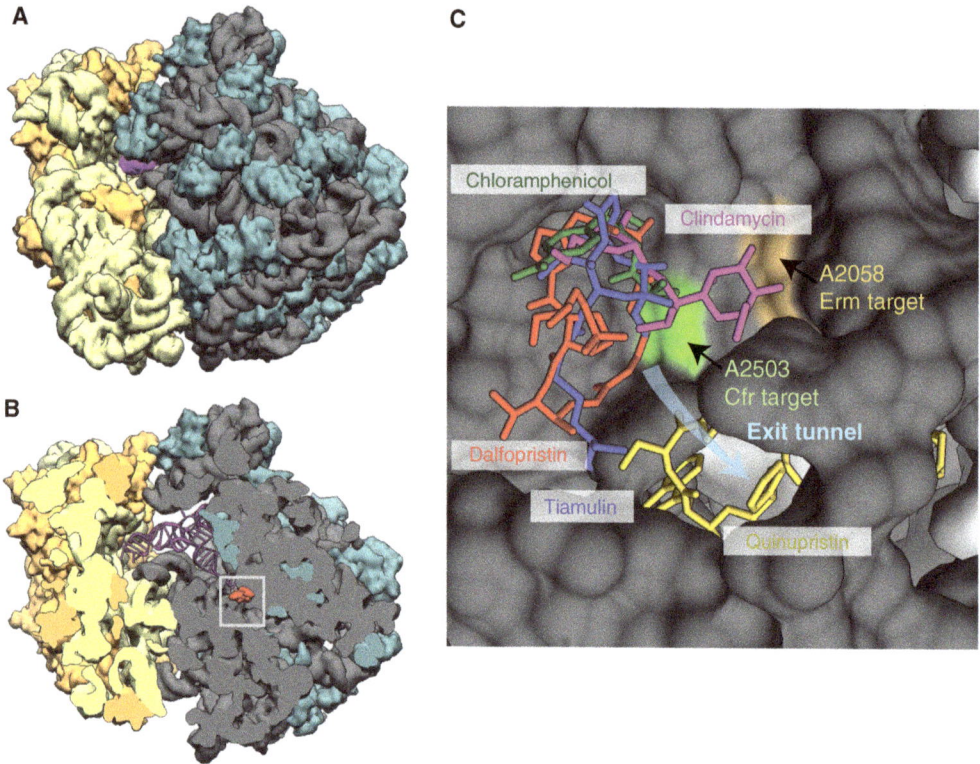

Figure 2. Lincosamides, streptogramins, phenicols, and pleuromutilins (LSPP) binding to the ribosome. (*A*) A model of the *Escherichia coli* bacterial ribosome with the two ribosomal subunits (based on RCSB Protein Data Bank [PDB] 4V9D; see rcsb.org, last accessed August 9, 2015), 30S with RNA in light yellow and proteins in darker yellow, 50S with RNA in grey and proteins in greenish. The magenta is the anticodon tip of tRNA in P-site. (*B*) A cut-view of the 50S subunit from *A* with a streptogramin A molecule (dalfopristin from PDB 4U26) in red to mark the peptidyltransferase center (PTC) area. The square marks the approximate area shown as a blow-up in *C*. (*C*) The overlapping binding sites of lincosamides, streptogramins, phenicols, and pleuromutilins exemplified by clindamycin (in pink) from PDB 4V7T, quinupristin (in yellow), and dalfopristin (in red) from PDB 4U26, chloramphenicol (in green) from PDB 4V7V, and tiamulin (in blue) from PDB 1XBP, respectively, and all placed in the *E. coli* structure. All PDB structures are from rcsb.org/pdb/home/home.do (Berman et al. 2000). The colored nucleotide bases indicate the positions methylated by Erm and Cfr methyltransferases. Nucleotides "in front" of the antimicrobial agents have been removed to be able to see the agents in their binding site.

lap the binding site of macrolides (reviewed in Poehlsgaard and Douthwaite 2005). The environment of the PTC seems to facilitate binding of a range of antimicrobial agents, which in this way interfere with the peptide transfer process. They can thus either disturb the positioning of aminoacyl-tRNA or peptidyl-tRNA for peptide transfer or directly block some movements required during peptide transfer. The overall mechanism of action is the inhibition of peptide transfer by sterically blocking the transfer process. The exact effect depends on the access of the

antimicrobial agents to the PTC (during the initiation or the nascent chain elongation), "on" and "off" rates of each agent, and their binding to the ribosomal A- and/or P-site.

The phenicols are small molecules and at least one of them, chloramphenicol, seems to have an alternative binding site in archael ribosomes as discussed by Dunkle et al. (2010), but still with its active site in the A-site of the PTC. The structural data on phenicol binding to the ribosome have been obtained with chloramphenicol, but florfenicol and thiamphenicol probably bind similarly. Lincosamides bind adjacent to and a bit overlapping with chloramphenicol and also at the A-site of PTC (Fig. 2) (Dunkle et al. 2010). The streptogramins appear naturally in pairs and bind together and work synergistically. The A component binds in PTC and the B component binds right beside it at the entrance of the tunnel (Fig. 2). The binding and effect on translation has recently been revisited by Noeske et al. (2014). All pleuromutilins have a conserved tricyclic core that fits nicely in a "cave" in the A-site PTC of the ribosomes (Davidovich et al. 2007). They vary by extensions pointing toward the P-site and occupying different positions (Davidovich et al 2007).

The older literature focused on the effect of various assays such as the puromycin reaction, fMet-tRNA binding, A- and P-site tRNA binding, translocation, etc.; however, some of these assays were under unnatural conditions and sometimes contradicting results were obtained. Nowadays, focus is more on defining the binding sites and using this knowledge: (1) to develop derivatives that bind more strongly to the ribosome and/or that bind despite target site modifications providing resistance; (2) to avoid other resistance mechanisms; and (3) to improve solubility and pharmacokinetics of the antimicrobial agents without compromising their effect.

PREVALENCE OF RESISTANCE TO LSPP ANTIMICROBIAL AGENTS

To classify a bacterial isolate as resistant or susceptible to an antimicrobial agent, clinical breakpoints are necessary. The Clinical and Laboratory Standards Institute (CLSI) has published a wide range of clinical breakpoints applicable to bacteria from humans and from animals (CLSI 2015, 2016). Among the lincosamides, no clinical breakpoints are available for lincomycin, whereas those for pirlimycin are applicable only to *S. aureus* and certain streptococcal species from bovine mastitis. For clindamycin, clinical breakpoints are only available for human staphylococci, streptococci, and anaerobes as well as for canine staphylococci and β-hemolytic streptococci. For the streptogramin

combination quinupristin–dalfopristin, only clinical breakpoints for staphylococci, streptococci, and enterococci are available. Among the pleuromutilins, the only clinical breakpoint available is for tiamulin and applicable to porcine *A. pleuropneumoniae*. In contrast, clinical breakpoints for chloramphenicol are available for a wide variety of Gram-positive and Gram-negative bacteria of human origin, whereas those for florfenicol are only applicable to bovine and porcine respiratory tract pathogens. In addition to CLSI, the European Committee on Antimicrobial Suceptibility Testing (EUCAST) provides slightly divergent clinical breakpoints (see eucast.org/fileadmin/src/media/PDFs/EUCAST_files/Breakpoint_tables/v_6.0_Breakpoint_table.pdf). Although there are numerous national resistance monitoring programs, those in human medicine often cover only pathogens from specific infections, mainly food-borne diarrheal diseases, whereas those in veterinary medicine usually monitor indicator bacteria and commensal bacteria, and only a very limited number of pathogenic bacteria (Silley et al. 2012).

Table 1 shows some examples of resistance prevalences for LSPP antimicrobial agents among pathogenic bacteria from the national resistance monitoring programs DANMAP (see danmap.org), GE*RM*-Vet (see www.bvl.bund.de/DE/09_Untersuchungen/untersuchungen_node.html;jsessionid=76F5E0F5BC375CA81F9CEF5BE44991E1.2_cid322), NARMS (see cdc.gov/narms/reports), NORM/NORM-VET (see vetinst.no/eng/Publications/NORM-NORM-VET-Report), as well as some long-term surveillance studies (Portis et al. 2012; Lindeman et al. 2013).

MECHANISMS OF RESISTANCE

In general, antimicrobial resistance can be based on two different mechanisms: the acquisition of mutations or resistance genes. Resistance-mediating mutations usually occur in genes or regions that represent the target site of the antimicrobial agents and prevent efficient binding of the antimicrobial agents to these sites. However, mutations may also enhance expression of

Table 1. Resistance rates to LSPP antimicrobial agents from selected national resistance monitoring programs and surveillance studies

LSPP antimicrobial agent	Bacteria	Origin	Year	Isolates tested	Resistant isolates (%)	MIC (mg/L)	Country	References[a]
Clindamycin	*Staphylococcus aureus* (MSSA)	Human—blood culture	2013	1155	1.5	≥ 1	Norway	NORM/NORM-VET
	S. aureus (MRSA)	Human	2013	1528	18.9	≥ 1	Norway	NORM/NORM-VET
	S. aureus (MSSA)	Human—bacteraemia	2014	381	8.4	≥ 1	Denmark	DANMAP
	S. aureus (MRSA)	Human	2014	1932	33.0	≥ 1	Denmark	DANMAP
	Staphylococcus pseudintermedius	Dog—clinical	2013	201	18.0	≥ 0.5	Norway	NORM/NORM-VET
	S. pseudintermedius	Dog—skin infections	2011	54	38.9	≥ 4	Germany	GERM-Vet
Pirlimycin	*S. aureus*	Cattle—mastitis	2009	210	1.4	≥ 4	Germany	GERM-Vet
	S. aureus	Cattle—mastitis	2010	342	3.0	≥ 4	United States, Canada	Lindeman et al. 2013
	Streptococcus dysgalactiae	Cattle—mastitis	2009	158	17.7	≥ 4	Germany	GERM-Vet
	S. dysgalactiae	Cattle—mastitis	2010	257	7.0	≥ 4	United States, Canada	Lindeman et al. 2013
	Streptococcus uberis	Cattle—mastitis	2009	289	27.0	≥ 4	Germany	GERM-Vet
	S. uberis	Cattle—mastitis	2010	289	25.0	≥ 4	United States, Canada	Lindeman et al. 2013
Quinupristin–dalfopristin	*S. aureus*	Poultry—clinical	2011	43	27.9	≥ 2	Germany	GERM-Vet
	S. aureus	Horse—clinical	2011	33	0.0	≥ 2	Germany	GERM-Vet
	S. pseudintermedius	Dog—skin infections	2011	54	0.0	≥ 2	Germany	GERM-Vet
	Enterococcus faecium	Broiler meat	2014	177	2.8	≥ 8	Denmark	DANMAP
	E. faecium	Beef	2014	56	0.0	≥ 8	Denmark	DANMAP
	E. faecium	Pork	2014	23	0.0	≥ 8	Denmark	DANMAP
Tiamulin	*Actinobacillus pleuropneumoniae*	Swine—respiratory tract	2012	41	2.4	≥ 32	Germany	GERM-Vet
Chloramphenicol	Nontyphoidal *Salmonella*	Human	2013	2178	3.9	≥ 32	United States	NARMS
	Salmonella Typhi	Human	2013	279	9.3	≥ 32	United States	NARMS
	Escherichia coli O157	Human	2013	177	2.8	≥ 32	United States	NARMS

Continued

Cite this article as *Cold Spring Harb Perspect Med* doi: 10.1101/cshperspect.a027037

Table 1. *Continued*

LSPP antimicrobial agent	Bacteria	Origin	Year	Isolates tested	Resistant isolates (%)	MIC (mg/L)	Country	References[a]
	E. coli	Turkey—clinical	2012	159	18.9	≥32	Germany	GERM-Vet
	S. pseudintermedius	Dog—skin infections	2011	54	24.1	≥32	Germany	GERM-Vet
Florfenicol	*Pasteurella multocida*	Cattle—respiratory tract	2012	77	1.3	≥8	Germany	GERM-Vet
	P. multocida	Cattle—respiratory tract	2009	328	11.6	≥8	United States, Canada	Portis et al. 2012
	Mannheimia haemolytica	Cattle—respiratory tract	2009	304	8.6	≥8	United States, Canada	Portis et al. 2012
	Bordetella bronchiseptica	Swine—respiratory tract	2012	90	2.2	≥8	Germany	GERM-Vet
	A. pleuropneumoniae	Swine—respiratory tract	2012	41	0.0	≥8	Germany	GERM-Vet

LSPP, Lincosamides, streptogramins, phenicols, and pleuromutilins.

[a]Websites of the national resistance monitoring programs are given in the section "Prevalence of Resistance to LSPP Antimicrobial Agents."

efflux genes or alter the substrate spectrum of transporters and thereby cause resistance or decreased susceptibility. Resistance genes may confer antimicrobial resistance by either enzymatic inactivation, active efflux, or modifications at the target sites of the antimicrobial agents. Although mutations occur spontaneously and are vertically transferred during division of a cell that harbors the mutation, resistance genes are often associated with mobile genetic elements that are disseminated vertically during cell division, but also horizontally by the gene transfer processes.

Ribosomal Mutations Associated with Resistance to LSPP Antimicrobial Agents

All LSPP antimicrobial agents bind to a well-conserved area of the ribosome consisting mainly of RNA. Although the RNA sequence in the area is conserved and therefore not expected to allow many mutations, a number of mutations in domain V of 23S RNA causing

resistance has been observed. It is also noteworthy that most bacteria have multiple 23S rRNA (ribosomal RNA) copies meaning that recombinations have to occur to obtain a full effect of such mutations. It has been shown that, for example, *S. aureus* under antimicrobial pressure will increase the number of mutated ribosomal RNA operons over time (Besier et al. 2008). Mutations occur constantly but maintaining 23S RNA resistance mutations depends on the number of copies of 23S RNA genes, how dominant the mutation is, and how "expensive" it is to contain a mutated RNA nucleotide in the exact position. In addition, it differs from species to species which mutations show up, and sometimes it is hard to prove that the resistance effect is because of the mutations observed as other mutations may contribute to or be the main effector. In addition, the identification of resistance mutations in clinical or animal isolates might be complicated by the lack of a wild-type comparator or the bacterial species might be hard to assay for resistance. The published literature thus contains more- or less-proven examples of resistance mutations, and only the most proven and solid data has been included here. All 23S RNA mutations causing resistance to the LPPS antimicrobial agents have been found in the part of 23S RNA shown in the secondary structure model shown in Figure 3. This region encompasses the RNA located at the PTC and contains the nucleotides known to be involved in binding of antimicrobial agents to the PTC.

As can be seen, some of the RNA mutations cause resistance to multiple antimicrobial agents and some also to agents not mentioned in this review. Especially the adenine residues at positions 2058 and 2059 are well-studied nucleotides for macrolide and lincosamide resistance (Vester and Douthwaite 2001). Antibiotics with partly overlapping binding sites may also bind to some of the same nucleotides and, thus,

cross-resistance can be seen. However, the predictions of such a cross-resistance do not follow an easy recognizable pattern (Long et al. 2010) and, furthermore, multiple mutations can show enhanced effects (Douthwaite 1992; Long et al. 2010).

In addition to 23S rRNA, there are also a few ribosomal proteins located close to PTC, including L3 and L4 (Nissen et al. 2000), in which mutations have been correlated with antimicrobial resistance. L3 mutations associated with tiamulin resistance have been summarized by Klitgaard et al. (2015), and most data relate to the aa 144–151 region of L3 (*E. coli* numbering), which is also the region closest to PTC. The mutations have been found in *E. coli*, *Staphylococcus* spp., and *Brachyspira* spp., and especially mutations at positions 149–150 have been proven to confer pleuromutilin resistance (see Klitgaard et al. 2015, and references therein).

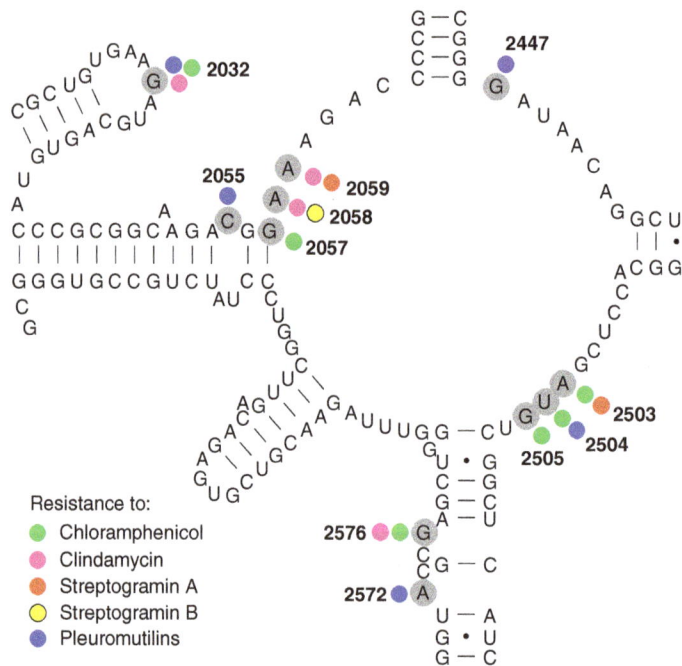

Figure 3. A secondary structure model of the peptidyl transferase loop of domain V of 23S rRNA (*Escherichia coli* sequence and numbering) with nucleotides providing antibiotic resistance marked with gray circles. Data are all from bacteria except the streptogramin A data that come from an archae (Porse and Garrett 1999). The bacterial data are from the following studies: Douthwaite (1992), Vester and Douthwaite (2001), Miller et al. (2008), Long et al. (2009, 2010), and Li et al. (2011), and references herein. The smaller circles indicate resistance to the various antibiotics and with the same color code as in Figure 2.

Cite this article as *Cold Spring Harb Perspect Med* doi: 10.1101/cshperspect.a027037

The same L3 region contains mutations associated with linezolid resistance and also examples of cross-resistance to linezolid and tiamulin (Klitgaard et al. 2015). Ribosomal protein L22 is located further down the tunnel, and a six aa insertion is reported to confer resistance to streptogramin B compounds in *S. pneumoniae* (Cattoir et al. 2007). As for the 23S RNA mutations, it is a cost-benefit matter whether a resistance-mediating mutation in a ribosomal protein "survives," and it is probably often followed by other "helper mutations" (also called compensatory mutations) as seen by Gentry et al. (2007). Increasing genome-sequencing capability and decreasing costs should facilitate more investigations on these contexts in the near future. In the current situation, investigators are often looking for what others have seen and might, therefore, ignore other positions that may be important for resistance.

Genes Conferring Resistance to LSPP Antimicrobial Agents by Enzymatic Inactivation

Enzymatic Inactivation of Lincosamides

Lincosamides are commonly inactivated by lincosamide nucleotidyltransferases, and their genes are found in various organisms. The known lincosamide nucleotidyltransferases show highest activity, as measured by the corresponding minimum inhibitory concentrations (MICs), against lincomycin, moderate activity against pirlimycin, and lowest activity against clindamycin (Lüthje and Schwarz 2006; Zhao et al. 2014).

The gene *lnu*(A) encodes a lincosamide nucleotidyltransferase of 161 aa (Brisson-Noël and Courvalin 1986). This gene is often located on small plasmids that usually contain only the *lnu*(A) gene and a plasmid replication gene. Several different types of *lnu*(A)-carrying plasmids have been described, many of them in *S. aureus* and coagulase-negative staphylococci (CoNS) of animal origin (Loeza-Lara et al. 2004; Lüthje and Schwarz 2007b; Lozano et al. 2012a).

The gene *lnu*(B) codes for a lincosamide nucleotidyltransferase of 267 aa and was first described in *E. faecium* of human origin (Boz-

dogan et al. 1999), and later identified in porcine *S. dysgalactiae* ssp. *equisimilis* (Lüthje and Schwarz 2007b). Recently, it has been detected as part of multiresistance gene clusters on plasmids or in the chromosomal DNA of *S. aureus*, *Staphylococcus hyicus*, *E. faecium*, *E. faecalis*, *Streptococcus agalactiae*, and *Erysipelothrix rhusiopathiae* of human and animal origin (Fig. 4) (Lozano et al. 2012a; Li et al. 2013, 2014b; Montilla et al. 2014; Silva et al. 2014; Wendlandt et al. 2014, 2015a; Zhang et al. 2015a).

The gene *lnu*(C) was first found in a human *S. agalactiae* isolate and codes for a 164-aa nucleotidyltransferase that inactivates lincomycin and clindamycin (Achard et al. 2005). It has also been discovered in the swine pathogen *Haemophilus parasuis* (Chen et al. 2010). The gene *lnu*(D) was detected in the chromosomal DNA of a *Streptococcus uberis* isolate from a case of bovine mastitis and codes for a nucleotidyltransferase of 164 aa (Petinaki et al. 2008). The gene *lnu*(E), which codes for a 173 aa protein, was found on a plasmid in *Streptococcus suis* in which it was interrupted by the integration of an IS*Enfa5-cfr-*IS*Enfa5* segment (Zhao et al. 2014). This gene was de novo synthesized, and after cloning and expression in *S. aureus* RN4220, shown to confer resistance to lincomycin. The gene *lnu*(F), originally referred to as *linF* codes for a 273-aa nucleotidyltransferase and was first identified on a plasmid from an *E. coli* isolate of human origin. Further analysis showed that the *lnu*(F) gene was part of a gene cassette located in a class 1 integron (Heir et al. 2004). A *lnu*(F)-related gene, originally reported as *linG*, which also encodes a 273-aa nucleotidyltransferase that shows 93% identity to Lnu(F), has been found as part of a gene cassette in a class 1 integron from a *Salmonella enterica* serovar Stanley isolate of human origin (Levings et al. 2006). Finally, a gene, described as *linA*$_{N2}$, has been detected on the mobilizable 11-kb transposon NBU2 in human clinical isolates of *Bacteroides fragilis* and *Bacteroides thetaiotaomicron*. It codes for a nucleotidyltransferase of 171 aa, which shares only about 50% identity to Lnu(A) proteins (Wang et al. 2000).

From an evolutionary point of view, at least two groups of lincosamide nucleotidyltrans-

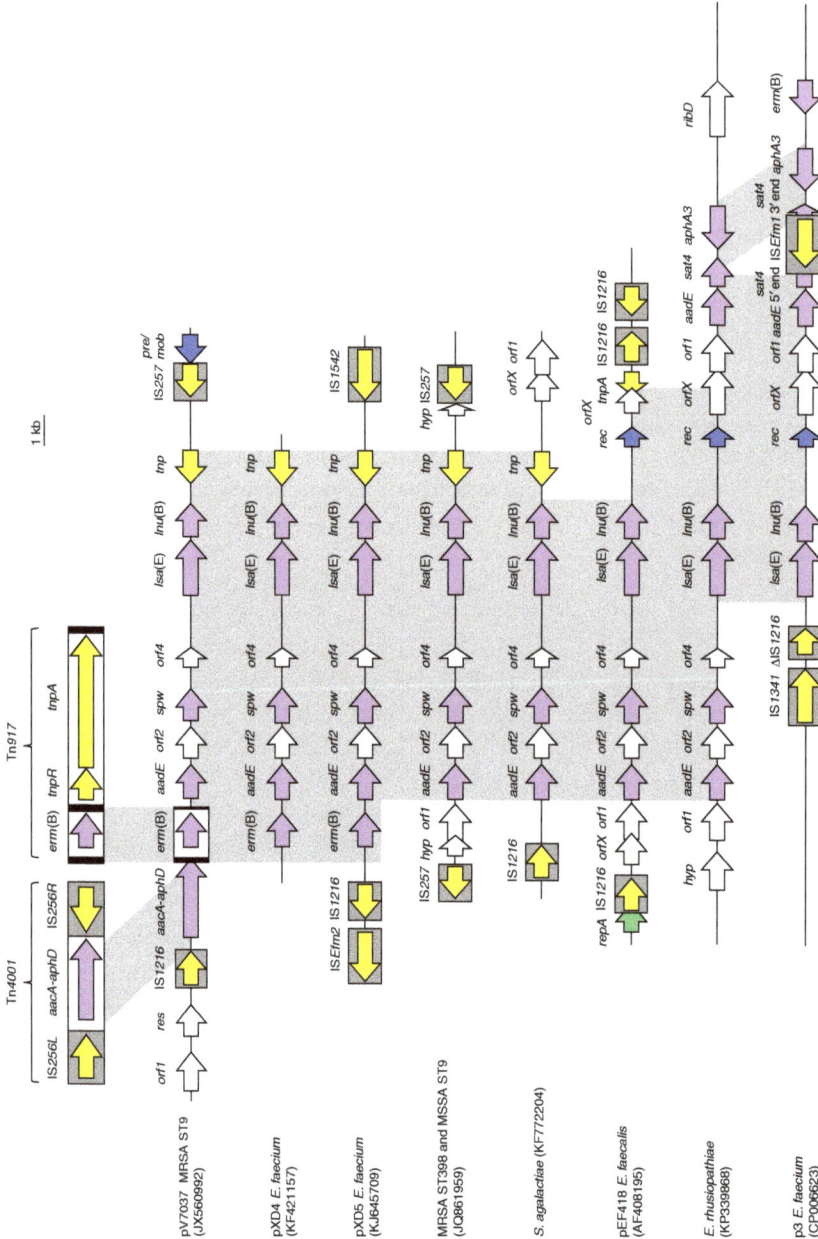

Figure 4. lsa(E)- and lnu(B)-carrying multiresistance gene clusters. Schematic presentation of the structural variability among *lsa*(E) and *lnu*(B)-carrying multiresistance gene clusters found on plasmids or in the chromosomal DNA of *Enterococcus faecalis*, *E. faecium*, *Streptococcus agalactiae*, *Erysipelothrix rhusiopathiae*, MRSA (methicillin-resistant *Staphylococcus aureus*), and MSSA (methicillin-susceptible *S. aureus*). All genes are indicated by arrows with the arrowhead showing the direction of transcription. Insertion sequences are presented as gray boxes with the arrow inside indicating the transposase gene. All antimicrobial resistance genes are depicted as violet arrows. *aacA-aphD*, gentamicin/kanamycin/tobramycin resistance; *aadE*, streptomycin resistance; *aphA3*, kanamycin/neomycin resistance; *erm*(B), MLS$_B$ resistance; *lnu*(B), lincosamide resistance; *lsa*(E), lincosamide/pleuromutilin/streptogramin A resistance; *spw*, spectinomycin resistance; *sat4*, streptothricin resistance. Genes involved in transposition are shown in yellow, whereas the genes involved in plasmid replication and plasmid recombination/mobilization are shown in green and blue, respectively. Genes with other functions are displayed in white. The Δ symbol indicates a truncated gene. A 1-kb scale is shown in the *upper right* corner. The gray-shaded regions show >99% sequence identity.

ferases can be differentiated. One group includes the proteins Lnu(A), Lnu(C), Lnu(D), and Lnu(E) that harbor a conserved domain at their amino terminus, which shows similarity to the aminoglycoside nucleotidyltransferase ANT(2″)-Ia (Petinaki et al. 2008). This observation points toward a divergent evolution from a common ancestor. The other group includes the proteins Lnu(B) and Lnu(F) with sequence similarity to the β-subunit of DNA polymerases. It is thus likely that they have developed from nucleotide polymerases, which are ubiquitous in bacteria (Morar et al. 2009).

Enzymatic Inactivation of Streptogramins

Inactivation of streptogramin A antimicrobial agents is commonly mediated by acetyltransferases. In staphylococci, three genes, *vat*(A) (Allignet et al. 1993), *vat*(B) (Allignet and El Solh 1995), or *vat*(C) (Allignet et al. 1998), which code for acetyltransferases of 219, 212, or 212 aa, respectively, have been described. These three *vat* genes have been identified on plasmids of different sizes, which occasionally also harbor additional resistance genes. In contrast to staphylococci from human sources, *vat* genes have rarely been detected in animal staphylococci. Solely, the *vat*(B) gene has been identified in two *Staphylococcus xylosus* isolates of poultry origin (Aarestrup et al. 2000). In enterococci, the plasmid-borne *vat*(D) gene from *E. faecium*, initially described as *satA*, was the first *vat* gene detected in this genus (Rende-Fournier et al. 1993). In another study, it was shown that the *vat*(D) gene, which codes for a 209-aa protein, is linked to the macrolide–lincosamide–streptogramin B (MLS_B)-resistance gene *erm*(B) and that both genes are cotransferred (Hammerum et al. 2001). The *vat*(E) gene, coding for an acetyltransferase of 214 aa and originally published as *satG*, was first identified in an *E. faecium* isolate from a sewage treatment plant in Germany (Werner and Witte 1999). A survey of streptogramin-resistant *E. faecium* from retail meat samples revealed some variability in the *vat*(E) genes. However, no correlation was seen between the number of aa substitutions and the MICs of quinupristin–dalfopristin (Simjee et al. 2001).

Besides Gram-positive bacteria, a novel type of *vat* gene, *vat*(F), which codes for a 221-aa protein, was detected in the chromosomal DNA of *Yersinia enterocolitica* (Seoane and García Lobo 2000). A *vat*(H) gene coding for a 216-aa acetyltransferase was detected together with the *vga*(D) gene on a plasmid in an *E. faecium* isolate of human origin (Jung et al. 2010). The origin of the VAT enzymes remains to be elucidated. Homologs and orthologs of the Vat enzymes, which are present in clinically resistant bacteria, are widely distributed in chromosomes of numerous environmental bacterial species, including species, such as *Y. enterocolitica*, which are intrinsically resistant to streptogramins (Wright 2007). It should be noted that streptogramin A–acetylating Vat proteins are related to CatB chloramphenicol acetyltransferases in their sizes, aa sequence, presumable active center, and tertiary structure (Murray and Shaw 1997), which points toward a common evolutionary origin.

So far, two genes, *vgb*(A) and *vgb*(B), have been described in staphylococci, which code for streptogramin B–specific lactone hydrolases of 298 and 295 aa, respectively (Allignet et al. 1988, 1998). Both genes are plasmid-borne. Plasmid pIP1714, which carries *vgb*(B), also harbors *vat*(C). This plasmid was isolated from a *Staphylococcus cohnii* ssp. *cohnii* found in the environment of a hospital where pristinamycin was extensively used (Allignet et al. 1998). There is little information about *vgb* genes in staphylococci from animals. The same two *vat*(B)-positive *S. xylosus* from poultry also carried a *vgb*(B) gene (Aarestrup et al. 2000). Orthologous and homologous *vgb* genes are present in the genomes of many environmental bacteria, including *Streptomyces coelicolor* and other *Streptomyces* spp., which belong to the same genus as the streptogramin producers (Wright 2007). Such genes may have served as precursors from which the *vgb* genes found in pathogenic bacteria have evolved.

Enzymatic Inactivation of Phenicols

Nonfluorinated phenicols, such as chloramphenicol and thiamphenicol, are commonly inactivated by chloramphenicol *O*-acetyltrans-ferases (CATs) (Shaw 1983; Schwarz et al. 2004). All CATs transfer an acetyl group from a donor molecule (e.g., acetyl-CoA) to the hydroxyl group at C3 of the phenicol molecule. This ace-tyl group is then shifted to the hydroxyl group at C1, and the hydroxyl group at C3 is available again for a second acetylation step. Neither mono- nor diacetylated phenicol molecules have antimicrobial activity (Murray and Shaw 1997). None of the CAT enzymes are able to inactivate florfenicol, because the hydroxyl group at C3 is replaced by fluorine and cannot act as an acceptor site for acetyl groups. There are two types of CATs, both of which have a trimeric structure composed of three identical monomers, and the respective *cat* gene codes for the monomer (Murray and Shaw 1997). The sizes of the classical CAT monomers vary between 207 and 238 aa, whereas CATs of the second type vary in their sizes between 209 and 219 aa (Schwarz et al. 2004).

The classical CATs, encoded by *catA* genes (Schwarz et al. 2004; Roberts and Schwarz 2009), represent a highly diverse group of en-zymes whose members show an overall identity of approximately 44%. These enzymes have been detected in Gram-positive and Gram-neg-ative aerobic and anaerobic bacteria and can be subdivided into at least 22 different groups when using a threshold value of ≥80% aa iden-tity to define a group (Schwarz et al. 2004). The genes *catA1*, *catA2*, and *catA3* (also known as *catI*, *catII*, and *catIII*) are exclusively found in Gram-negative bacteria and are expressed con-stitutively. The CatA3 (CATIII) enzyme was the first to be crystallized, and the analysis of its crystal structure provided insight into the fold-ing of the CAT monomers and helped to identify the aa that were important for the structure and function of the CAT enzyme (Leslie et al. 1988). Another three groups of classical CATs were named according to the plasmids (pC221, pC223/pSCS7, and pC194), on which their genes were first detected. Although initially found in staphylococci, the corresponding *cat* genes have in the meantime been identified in a number of Gram-positive genera. The ex-pression of these *cat* genes is inducible by chlor-amphenicol and is regulated by translational attenuation. The regulatory region comprises a reading frame for a short peptide and two in-verted repeated sequences, and is located im-mediately upstream of the respective *cat* gene (Lovett 1990). Closely related CATP and CATD proteins were first identified in the Gram-positive anaerobe *Clostridium* spp. where they are located on transposons (Lyras and Rood 2000). However, these genes have also been identified in Gram-negative *Neisseria meningi-tidis* (Galimand et al. 1998; Shultz et al. 2003). Both genes are expressed constitutively. The re-maining 15 groups of classical CATs are repre-sented by individual enzymes whose genes have so far been detected in only a single species of either Gram-positive or Gram-negative bacteria (Schwarz et al. 2004).

The second type of CAT enzymes, encoded by *catB* genes (Murray and Shaw 1997; Schwarz et al. 2004; Roberts and Schwarz 2009), is only distantly related to the classical CATs. Their members are structurally similar to acetyltrans-ferases involved in streptogramin A resistance (Murray and Shaw 1997). In general, these CAT enzymes confer lower MICs to chloram-phenicol than the classical CATs. Hence, it was speculated that members of this second type of CAT might have a physiological role other than chloramphenicol resistance in their host bacte-ria (Murray and Shaw 1997). Using the same threshold value as for the classical CATs, at least five different groups can be distinguished, al-though all enzymes have approximately 77% identity with each other (Schwarz et al. 2004). Many of these *cat* genes are part of gene cassettes in class 1 and class 2 integrons in Gram-negative bacteria (Recchia and Hall 1995), whereas oth-ers have been identified on transposons.

In addition to CATs, inactivation of chlor-amphenicol by *O*-phosphorylation has been observed in the chloramphenicol producer *S. venezuelae* and is believed to contribute to self-defense of the host (Mosher et al. 1995). It is not known whether fluorinated chloramphenicol analogues can be inactivated by phosphoryla-tion. Moreover, a chloramphenicol hydrolase gene has been found in the chloramphenicol producer *S. venezuelae* and considered to con-

Cite this article as *Cold Spring Harb Perspect Med* doi: 10.1101/cshperspect.a027037

tribute to self-defense of the host (Mosher et al. 1990). Moreover, the gene *estDL136* from a soil metagenome library was found to specify a hydrolase, which, when cloned in *E. coli*, inactivated both chloramphenicol and florfenicol (Tao et al. 2012).

Enzymatic Inactivation of Pleuromutilins

To the best of our knowledge, no pleuromutilin-inactivating enzymes have been described so far.

Genes Conferring Resistance to LSPP Antimicrobial Agents by Active Efflux

Active efflux can be based on multidrug transporters or specific transporters. In the following sections, multidrug transporters and specific transporters are described whose substrate spectrum includes one or more of the LSPP antimicrobial agents.

Resistance to LSPP Antimicrobial Agents by Multidrug Transporters

Usually, multidrug transporters confer an increase in the MICs of their substrates, but not necessarily to levels that are indicative of clinical resistance. Multidrug transporter systems assigned to the resistance/nodulation/cell division (RND) family have been reported to export phenicols from the bacterial cell. They include the AcrAB-TolC system in *E. coli* (McMurray et al. 1994), the MexAB-OprM and MexCD-OprJ systems in *P. aeruginosa* (Paulsen et al. 1996), as well as the OqxAB system in Enterobacteriaceae (Hansen et al. 2007), among others. Initially, intrinsic resistance of Enterobacteriaceae and other Gram-negative bacteria to macrolides, lincosamides, and streptogramins was considered to be based on the relative impermeability of the outer membrane to these compounds (Leclercq and Courvalin 1991). However, the observation that *E. coli* strains, such as CS1562 or AS19, are hypersusceptible to these antimicrobial agents caused by a deficiency of the TolC porin, suggests the involvement of RND pumps, which use TolC as

an outer membrane component, in intrinsic resistance to macrolides, lincosamides, and streptogramins.

In Gram-positive bacteria, several 12-trans-membrane segments (TMS) multidrug transporters of the major facilitator superfamily (MFS), such as Blt and Bmr proteins from *Bacillus subtilis* and NorA from *S. aureus*, have been reported to have a substrate spectrum that includes chloramphenicol (Paulsen et al. 1996). Another two closely related 12-TMS multidrug efflux proteins, MdfA and Cmr, which are able to export chloramphenicol, have been identified in *E. coli* (Nilsen et al. 1996; Edgar and Bibi 1997). Overexpression of the chromosomal multidrug exporter MdeA of *S. aureus* conferred a 16-fold increase in the MIC of virginiamycin, while no significant increase in the MICs of other LSPP antimicrobial agents tested was seen (Huang et al. 2004). In addition, the gene *lmr*(B) from *B. subtilis* coding for a 479-aa MFS protein (Kumano et al. 2003; Murata et al. 2003) was shown to confer resistance to multiple antimicrobial agents, including lincomycin, when overexpressed. The Lmr(P) protein of *Lactococcus lactis* specifies a 408-aa multidrug transporter of the MFS, which confers resistance to lincosamides, macrolides, streptogramins, and tetracyclines (Putman et al. 2001; Poelarends et al. 2002). A similar substrate spectrum was determined for the ATP-binding cassette (ABC) transporter encoded by the gene *lmr*(A) from *L. lactis* when expressed in a hypersusceptible *E. coli* strain (Putman et al. 2000; Poelarends et al. 2002). The 480-aa MFS protein LmrS from *S. aureus* has been reported to confer 4- to 16-fold increases in the MICs of numerous antimicrobial agents (including lincomycin, chloramphenicol, florfenicol, and linezolid) and other substances when expressed in *E. coli* (Floyd et al. 2010).

Active Efflux of Lincosamides by Specific Exporters

The 492-aa ABC transporter Lsa(B) conferred, in contrast to other Lsa proteins, only elevated MICs of lincosamides, which, however, are below the clinical breakpoint for resistance (Kehrenberg et al. 2004). The *lsa*(B) gene was detected in close proximity to the multiresistance gene *cfr* on plasmids in *Staphylococcus sciuri* and *Staphylococcus warneri* (Kehrenberg et al. 2004, 2007). The *lmrA* gene of the lincomycin producer *S. lincolnensis* specifies an MFS protein of 481 aa, which exports lincomycin and it is believed to be part of a self-defense system of this organism (Peschke et al. 1995). A gene, designated *lmrB*, which codes for a 481-aa MFS exporter in *Corynebacterium glutamicum*, was shown to confer resistance to lincosamides but not to other antimicrobial agents tested (Kim et al. 2001). This Lmr(B) protein is only distantly related to the Lmr(B) protein in *B. subtilis* described above.

Active Efflux of Lincosamides, Pleuromutilins, and Streptogramin A Antimicrobial Agents by Specific Exporters

During recent years, several ABC transporters have been identified in staphylococci, streptococci, and enterococci, which confer combined resistance to lincosamides, pleuromutilins, and streptogramin A antimicrobial agents. A recent review provided detailed information about most of the corresponding multiresistance genes, such as *vga*(A), *vga*(A)$_V$, *vga*(A)$_{LC}$, *vga*(C), *vga*(E), *vga*(E)$_V$, *lsa*(A), *lsa*(C), *lsa*(E), *eat*(A)v, and *sal*(A) (Wendlandt et al. 2015b).

The ABC transporters specified by the genes *vga*(A) (Allignet et al. 1992), *vga*(A)$_V$ (Haroche et al. 2000), *vga*(A)$_{LC}$ (Novotná and Janata 2006; Gentry et al. 2008), *vga*(B) (Allignet and El Solh 1997), *vga*(C) (Kadlec and Schwarz 2009), *vga*(E) (Schwendener and Perreten 2011), and *vga*(E)$_V$ (Li et al. 2014a) show sizes of 522, 524, 522, 552, 522, 524, and 524 aa, respectively. All of these can export streptogramin A antimicrobial agents, whereas the Vga(A), Vga(C), and Vga(E) proteins also export lincosamides and pleuromutilins. The *vga*(A) genes are most widespread and can be located on the 5.5-kb transposon Tn5406 (Haroche et al. 2002). Studies on clinical *S. aureus* isolates from France identified Tn5406 inserted into the chromosomal att554 site, a site that resembles the integration site in a type III SCC*mec*

cassette and on plasmids. The *vga*(A) genes can also be located on plasmids, ranging from small plasmids of 5.7 kb that harbor only the *vga*(A) gene (Kadlec et al. 2010) to large plasmids of 25–46 kb that carry additional resistance genes (Allignet and El Solh 1999; Weiß et al. 2014). The *vga*(C) gene was initially identified on a 14-kb multiresistance plasmid from a porcine livestock–associated (LA)-MRSA isolate (Kadlec and Schwarz 2009) but has also been identified on a small plasmid of 5.3 kb (Kadlec et al. 2010).

The *vga*(E) was shown to be part of the 11.5-kb transposon Tn6133, which was detected initially in a porcine LA-MRSA from Switzerland (Schwendener and Perreten 2011) and soon thereafter in LA-MRSA from cattle, chickens, and turkeys in Germany (Hauschild et al. 2012; Monecke et al. 2013). In 2014, a variant of the Vga(E) protein was described, which shared only 86% aa identity with the original Vga(E) protein (Li et al. 2014a). The Vga(E) variant also confers pleuromutilin–lincosamide–streptogramin A resistance, and the corresponding gene *vga*(E)$_V$ was located on a 5.6-kb plasmid in porcine *Staphylococcus simulans* and *S. cohnii* isolates. The *vga*(D) gene codes for a 525-aa protein and was found in *E. faecium* of human origin (Jung et al. 2010). Vga(D) was the first ABC transporter in *E. faecium*, which conferred resistance to streptogramin A antimicrobial agents. However, it is not known whether this ABC transporter also exports lincosamides and pleuromutilins.

The species-specific chromosomal gene *lsa*(A) from *E. faecalis* codes for an ABC transporter of 496 aa, which is believed to play a role in the intrinsic resistance of *E. faecalis* to lincosamides and streptogramins. Disruption of the *lsa*(A) gene was associated with an at least 40-fold decrease in MICs of quinupristin–dalfopristin, clindamycin, and dalfopristin, whereas complementation of the disruption mutant with an intact *lsa*(A) resulted in restoration of the MICs to wild-type levels (Singh et al. 2002). Another study reported that Lsa(A) not only confers resistance to lincosamide and streptogramin A, but also to pleuromutilins (Malbruny et al. 2011). The same resistance pattern is seen

Cite this article as *Cold Spring Harb Perspect Med* doi: 10.1101/cshperspect.a027037

for the *lsa*(C) gene from *S. agalactiae* that codes for an ABC transporter of 492 aa (Malbruny et al. 2011). The first *lsa* gene in staphylococci, *lsa*(E), which mediates combined resistance to lincosamides, pleuromutilins, and streptogramin A, has been described in human MRSA ST398 and MSSA ST9 (Wendlandt et al. 2013b). This gene codes for an ABC transporter protein of 494 aa and is part of a multiresistance gene cluster that is most likely of enterococcal origin (Fig. 4) (Wendlandt et al. 2013b). Several variants of this multiresistance gene cluster have been identified in MRSA ST398 of human origin and in MRSA/MSSA ST9 of human and pig origin in Europe and Asia (Lozano et al. 2012a; Li et al. 2013; Wendlandt et al. 2013a, 2014, 2015b). Moreover, as summarized in Figure 4, variants of the *lsa*(E)-containing multiresistance gene cluster have also been found in *E. faecalis* and *E. faecium* of human and swine origin in China (Li et al. 2014b; Si et al. 2015), in *S. agalactiae* of human origin in Argentina (Montilla et al. 2014), and also in *E. rhusiopathiae* of swine origin in China (Zhang et al. 2015a).

Although *E. faecalis* is intrinsically resistant to lincosamides, pleuromutilins, and streptogramin A by production of the ABC transporter Lsa(A), *E. faecium* is naturally susceptible. *E. faecium* harbors a gene *eat*(A) for an ABC transporter of 500 aa that shows only 66% aa identity to Lsa(A) and has no function in antimicrobial resistance. Despite this, an aa substitution (Thr450Ile) was found in the Eat(A) protein of isolates that were resistant to lincosamides, pleuromutilins, and streptogramin A antimicrobial agents. Single-nucleotide replacement and transfer of the mutated gene *eat*(A)$_V$ into a susceptible *E. faecium* isolate confirmed a role of this mutation in resistance to lincosamides, pleuromutilins, and streptogramin A (Isnard et al. 2013).

The gene *sal*(A) from *S. sciuri* codes for an ABC transporter of 541 aa, which was reported to confer resistance to streptogramin A and lincosamides (Hot et al. 2014). The *sal*(A) gene has exclusively been found in the chromosomal DNA of *S. sciuri* isolates inserted between the two housekeeping genes, *iscS* and *mnmA*, of the staphylococcal core genome. A recent study confirmed that Sal(A) also confers pleuromutilin resistance (Wendlandt et al. 2015a).

Active Efflux of Macrolides and Streptogramin B by Specific Exporters

The gene *msr*(A) codes for a 488-aa ABC transporter protein that confers resistance to macrolides and streptogramin B (Ross et al. 1990) and is often found together with the gene *mph*(C) that codes for a macrolide phosphotransferase (Lüthje and Schwarz 2006). These two genes have been detected among CoNS and MRSA CC398 from humans in Germany and Spain, respectively (Gatermann et al. 2007; Lozano et al. 2012b), and on plasmids of the *S. aureus* clone USA300 (Kennedy et al. 2010). The 33-kb plasmid pMS97 has been reported to carry *msr*(A) and *mph*(C) together with the MLS$_B$-resistance gene *erm*(Y) (Matsuoka et al. 2003). In veterinary medicine, the *msr*(A) gene has been detected in *S. aureus* isolates of poultry (Nawaz et al. 2000) and dog origin (Lüthje and Schwarz 2007b), in canine *S. pseudintermedius* (Lüthje and Schwarz 2007b), and in various species of CoNS from cases of bovine mastitis (Lüthje and Schwarz 2006). Moreover, Jaglic et al. (2012) identified the gene *msr*(A) in not-further-specified CoNS isolates from turkeys, pigs, and cattle.

The species-specific and chromosomally located gene *msr*(C) of *E. faecium* codes for a 492-aa ABC transporter. Functional deletion of *msr*(C) resulted in a two- to eightfold decrease in MICs of the macrolides erythromycin, azithromycin, and tylosin, as well as of the streptogramin B quinupristin. This *E. faecium*–specific gene shows only 53% identity to *msr*(A) (Singh et al. 2001). The *msr*(D) from *S. pyogenes* (Gay and Stephens 2001) has been reported to confer macrolide and ketolide, but not streptogramin resistance (Poole 2005). The *msr*(E) gene found in various Gram-negative bacteria is also known to be involved in the efflux of macrolides (Kadlec et al. 2011), but it is not known whether this gene also confers resistance to streptogramin B antimicrobial agents.

Active Efflux of Phenicols by Specific Exporters

Specific exporters that transport only chloramphenicol or chloramphenicol and florfenicol out of the bacterial cell are mainly members of the MFS efflux proteins (Paulsen et al. 1996; Poole 2005). They commonly show 10 to 14 TMS (Butaye et al. 2003). Based on an 80% threshold value, at least 11 genetic groups of specific phenicol exporters can be distinguished, including four groups from soil and environmental bacteria (Schwarz et al. 2004; Roberts and Schwarz 2009). The *cmr* and *cmx* genes found in *Corynebacterium* spp. are located on plasmids, whereas the *cmx* gene is associated with transposon Tn*5564* (Schwarz et al. 2004). The genes *cmr* and *cmrA* from *Rhodococcus* spp. are also found on plasmids, with the *cmrA* being part of transposon Tn*5561* (Nagy et al. 1997). The *S. venezuelae cmlv* gene is believed to play a role in self-defense of the chloramphenicol producer (Mosher et al. 1990).

In clinically important bacteria, such as *S. enterica, E. coli, Klebsiella pneumoniae*, or *Pseudomonas aeruginosa*, several closely related *cmlA* genes have been identified on gene cassettes. Unlike other cassette-borne genes, *cmlA* is inducibly expressed via translational attenuation (Stokes and Hall 1991; Recchia and Hall 1995). The chloramphenicol exporter CmlB1, which shares 74%–77% identity with CmlA proteins, was identified on a plasmid from *Bordetella bronchiseptica* (Kadlec et al. 2007), a bacterium involved in respiratory tract infections in swine. The *cmlB1* gene is also inducibly expressed via translational attenuation. Neither CmlA nor CmlB1 proteins can efficiently export florfenicol from the bacterial cell, and bacteria carrying the corresponding genes are classified as florfenicol susceptible (Schwarz et al. 2004).

In contrast to the aforementioned genes, the *floR* gene codes for a MFS protein that can export both chloramphenicol and florfenicol (Schwarz et al. 2004). The *floR* gene (also referred to as *flo* or *pp-flo*) can be found in the chromosomal DNA of multiresistant *S. enterica* serovars, including Typhimurium DT104 and Newport, *Vibrio cholerae, E. coli, B. bronchisep-*tica, and *Acinetobacter baumannii*, or on plasmids of *E. coli, K. pneumoniae, Pasteurella multocida, Pasteurella trehalosi, A. pleuropneumoniae*, and *Stenotrophomonas maltophilia* (Schwarz et al. 2004). Recently, the *floR* gene has been identified as part of the chromosomally located integrative and conjugative element ICE*Pmu1* from *P. multocida*, which harbors a total of 12 antimicrobial resistance genes (Michael et al. 2012). A new *floR* gene variant, *floRv*, whose product showed only 84%–92% aa identity to the so-far known FloR proteins, has recently been identifed in a multidrug resistance genomic island of a porcine *S. maltophilia* isolate (He et al. 2015).

There are also a few chloramphenicol/florfenicol exporters in Gram-positive bacteria. The gene *fexA* is part of transposon Tn*558* and was first identified on a plasmid from *Staphylococcus lentus* (Kehrenberg and Schwarz 2005). Expression of *fexA* is inducible with either chloramphenicol or florfenicol. A translational attenuator, similar to those of *cat* genes from *Staphylococcus* spp. and *Bacillus pumilus*, is located immediately upstream of the *fexA* gene (Kehrenberg and Schwarz 2004). A *fexA* gene variant, *fexAv*, which conferred only chloramphenicol resistance, was detected in a canine *S. pseudintermedius* (Gómez-Sanz et al. 2013). In comparison to FexA, FexAv showed two aa substitutions Gly33Ala and Ala37Val. Both substitutions appear to be important for substrate recognition as site-directed mutagenesis to the original *fexA* gene, restored the chloramphenicol/florfenicol resistance phenotype. In 2010, a novel chloramphenicol/florfenicol resistance gene, designated *pexA*, was identified from an Alaskan soil sample (Lang et al. 2010). The gene *fexB*, which also confers resistance to chloramphenicol and florfenicol, was found on nonconjugative plasmids of *E. faecium* and *Enterococcus hirae* (Liu et al. 2012a).

Active Efflux of Phenicols and Oxazolidinones by a Specific Exporter

Most recently, a novel gene, designated *optrA* and coding for an ABC transporter of 655 aa, has been identified on a conjugative plasmid in

E. faecalis (Wang et al. 2015). In contrast to the MFS transporters described above, the *optrA* gene confers resistance, not only to chloramphenicol and florfenicol, but also to the oxazolidinones linezolid and tedizolid. The *optrA* gene was shown to be functionally active in *E. faecalis*, *E. faecium*, and *S. aureus*. A first survey conducted in China revealed that the *optrA* gene was more frequently found in enterococci of animal than of human origin (Wang et al. 2015). In a study including 1159 enterococcal isolates from five hospitals in China, the *optrA* gene was detected at a prevalence of 2.9%, and a distinct increase in *optrA* carriers was seen from 2010 to 2014 (Cai et al. 2015a). The analysis of the genetic environment of *optrA* in *E. faecalis* revealed a substantial heterogeneity (He et al. 2016). Besides in isolates from China, the *optrA* gene has also been identified in two clinical *E. faecalis* isolates from Italy (Brenciani et al. 2016). Finally, the analysis of 50 porcine staphylococci from China identified the *optrA* gene in a single *S. sciuri* isolate where it was located together with the genes *cfr*, *fexA*, *aadD*, *ble*, and *aacA-aphD* on a 60.5-kb nonconjugative multiresistance plasmid (Li et al. 2016).

Genes Conferring Resistance to LSPP Antimicrobial Agents by Target Site Modification

Target Site Modifications that Confer Resistance to Macrolides, Lincosamides, and Streptogramin B Antimicrobial Agents

Combined resistance to macrolides, lincosamides, and streptogramin B antimicrobial agents is commonly mediated by rRNA methylases that target the adenine residue at position 2058 in the domain V of the 23S rRNA. The adenine residue at position 2058 (Fig. 2) is located in the overlapping binding region of these three classes of antimicrobial agents. Methylation of this residue prevents macrolides, lincosamides, and streptogramin B antimicrobial agents from binding to their ribosomal target sites (reviewed in Weisblum 1995a; Roberts 2008). The corresponding *erm* genes, which code for these methylases, are often located on plasmids, transpo-

sons, or integrative and conjugative elements, which facilitate their dissemination across strain, species, and sometimes even genus boundaries. Based on a threshold value of <80% aa identity (Roberts et al. 1999), 46 classes of rRNA methylases are currently distinguished (see faculty .washington.edu/marilynr/ermweb1.pdf). Many *erm* genes are inducibly expressed via translational attenuation with 14- and 15-membered macrolides acting as inducers (Weisblum 1995b). Lincosamides and streptogramins, but also 16-membered macrolides usually do not act as inducers. Studies on *erm*(A) and *erm*(C) genes in staphylococci showed that constitutive expression can develop rapidly in the presence of such noninducers and is commonly a result of deletions, duplications, or point mutations in the translational attenuator (Schmitz et al. 2002a,b; Lüthje and Schwarz 2007a). Detailed information on the *erm* genes and their occurrence in the various bacteria can be seen in a review on macrolides and ketolides (Fyfe et al. 2016).

Target Site Modifications that Confer Resistance to Phenicols, Lincosamides, Oxazolidinones, Pleuromutilins, and Streptogramin A Antimicrobial Agents

The gene *cfr* was the first gene that conferred combined resistance to phenicols, lincosamides, oxazolidinones, pleuromutilins, and streptogramin A (PhLOPS$_A$) antimicrobial agents. This gene was identified on a plasmid in a bovine *S. sciuri* and initially referred to as a chloramphenicol/florfenicol resistance gene (Schwarz et al. 2000). The clarification of the resistance mechanism, however, showed that *cfr* is a multiresistance gene, which confers resistance to the aforementioned classes of antimicrobial agents (Kehrenberg et al. 2005; Long et al. 2006). In addition, Cfr also provides decreased susceptibility to some large 16-membered macrolides, such as spiramycin and josamycin (Smith and Mankin 2008). The gene *cfr* encodes an rRNA methylase that targets the adenine residue at position 2503 in the domain V of the 23S rRNA (Fig. 2). This target site is located in the overlapping binding site of the PhLOPS$_A$ agents and the additional Cfr-mediated methylation is believed to prevent these antimicrobial agents from binding to the ribosome. Further studies identified the *cfr* gene on a number of plasmids in *S. aureus*, *Staphylococcus hyicus*, and various CoNS species from pigs, cattle, horses, chickens, and ducks (Kehrenberg and Schwarz 2006; Kehrenberg et al. 2007, 2009; Wang et al. 2012d,e, 2013a; He et al. 2014). The gene *cfr* was also detected in *S. aureus* and CoNS from infections in humans (Toh et al. 2007; Mendes et al. 2008, 2013; Bonilla et al. 2010; Shore et al. 2010; Gopegui et al. 2012; Locke et al. 2012; Cui et al. 2013; LaMarre et al 2013; Feßler et al. 2014; Bender et al. 2015, Cai et al. 2015b). Most recently, the *cfr* gene was found to be integrated in a type IVb SCC*mec* cassette of an MRSA CC9 isolate of swine origin (Fig. 5) (Li et al. 2015).

Screening studies conducted in China revealed the presence of the gene *cfr* also in other Gram-positive bacteria such as *Bacillus* spp. (Dai et al. 2010; Zhang et al. 2011; Wang et al. 2012b), *Enterococcus* spp. (Liu et al. 2012b, 2013, 2014), *Macrococcus caseolyticus* and *Jeotgalicoccus pinnipedialis* (Wang et al. 2012c), and *S. suis* (Wang et al. 2013b). In most of these cases, the *cfr* gene was located on plasmids of variable sizes. Moreover, the *cfr* gene was also found in a small number of Gram-negative bacteria, for example, in the chromosomal DNA of a *Proteus vulgaris* isolate (Wang et al. 2011) or on plasmids in *E. coli* (Fig. 5) (Wang et al. 2012a; Zhang et al. 2014, 2015b).

A comparative analysis of the plasmids on which the *cfr* gene was found in the different bacteria revealed a high structural variability in the regions surrounding the *cfr* gene (Shen et al. 2013). Moreover, numerous insertion sequences, such as IS*21-558*, IS*256*, IS*1216*, IS*Enfa4*, IS*Enfa5*, and IS*26*, were found in close proximity to the *cfr* gene (Fig. 5). These insertion sequences may play a role in the transfer of *cfr* between different plasmids but also in the chromosomal integration of *cfr*-carrying segments. In addition, many *cfr*-carrying plasmids harbor additional resistance genes, which enable the coselection and persistence of the *cfr*

gene under a selection pressure imposed by non-PhLOPS$_A$ antimicrobial agents.

Recently, a *cfr*-like gene, whose product showed only 75% aa identity to the original Cfr protein from *S. sciuri*, was detected in *C. difficile*, and claimed to be responsible for linezolid resistance (Marín et al. 2015; Schwarz and Wang 2015). In a different study, this gene, meanwhile designated *cfr*(B), has been shown to confer multiple antimicrobial resistance by the same mechanism as the original *cfr* gene (Hansen and Vester 2015). Most recently, the *cfr*(B) gene was also detected in *E. faecium* recovered from human specimens in the United States (Deshpande et al. 2015).

CONCLUDING REMARKS

For a number of years, fewer and fewer new antimicrobial agents have been approved for use in human and veterinary medicine despite rising trends in antimicrobial resistance in bacterial pathogens of human, veterinary, and zoonotic relevance. This is particularly true for veterinary medicine as all new antimicrobial agents approved during the last 25 years represent derivatives of already known substances. For the development of antimicrobial agents with improved binding to the ribosome, the exact mode of action of the antimicrobial agents, but also the knowledge about their binding sites, are indispensable prerequisites. Moreover, the knowledge about resistance-mediating mutations and the mechanisms specified by resistance genes are important aspects that need to be taken into account when developing new antimicrobial agents that may even overcome existing resistance mechanisms. Florfenicol is a good example to illustrate how knowledge about the most common phenicol resistance mechanism, namely, the enzymatic inactivation of chloramphenicol by chloramphenicol acetyltransferases, was used to generate an antimicrobial agent that was resistant to inactivation by the most widespread resistance mechanism and, in addition, did not show the adverse side effects of the parental substance. However, bacteria are versatile organisms that are able to quickly develop or

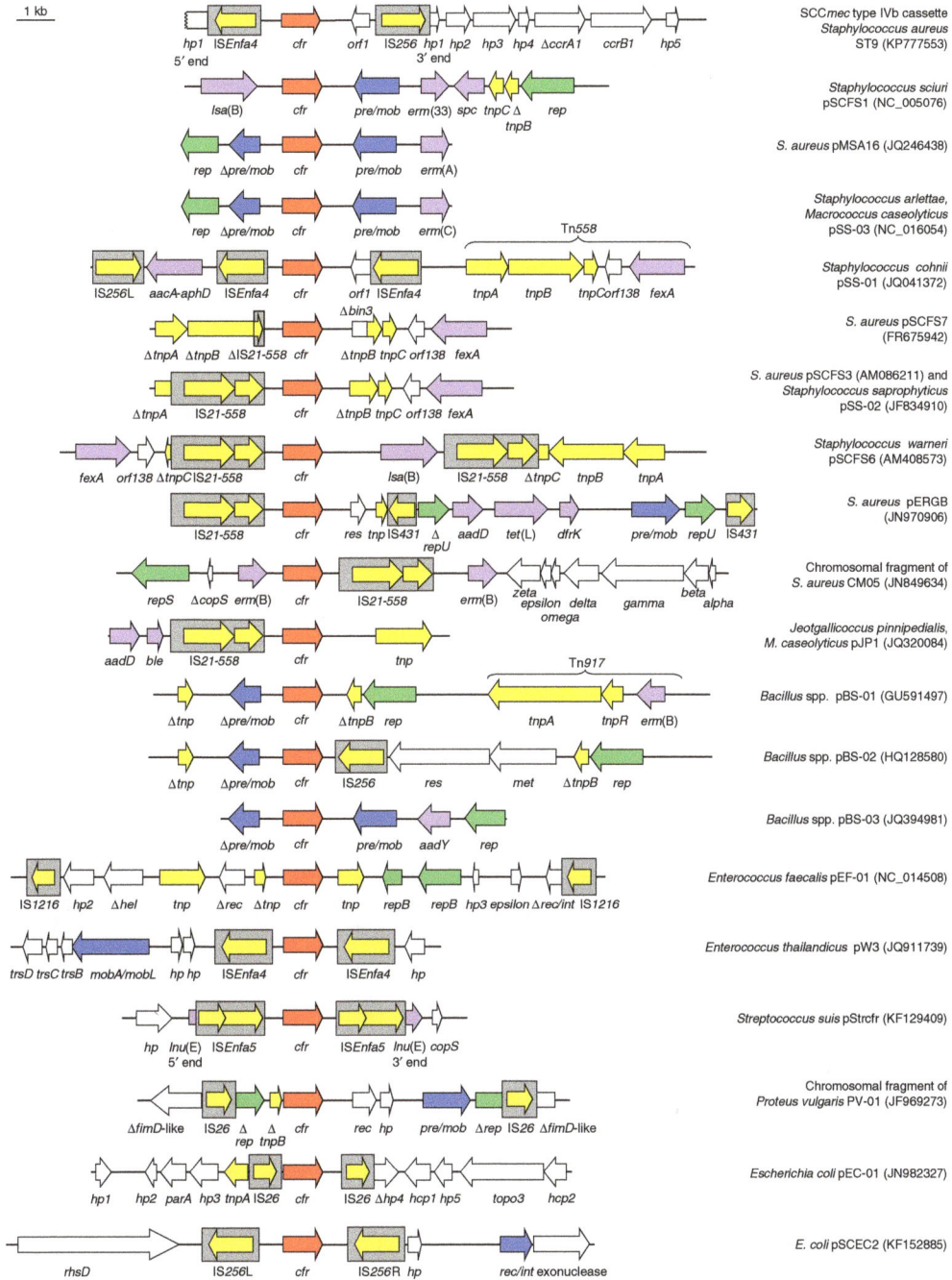

Figure 5. Schematic presentation of the structural variability in the regions surrounding the *cfr* gene on plasmids or in the chromosomal DNA of various Gram-positive and Gram-negative bacteria. All genes are indicated by arrows with the arrowhead showing the direction of transcription. Insertion sequences are presented as gray boxes with the arrow inside indicating the transposase gene. The *cfr* gene is shown as a red arrow, whereas all other antimicrobial resistance genes are depicted as violet arrows. Genes involved in transposition are shown in yellow, whereas the genes involved in plasmid replication and plasmid recombination/mobilization are shown in green and blue, respectively. Genes with other functions are displayed in white. The Δ symbol indicates a truncated gene. A 1-kb scale is shown in the *upper left* corner. *aadD*, kanamycin/neomycin resistance; *aadY*, streptomycin resistance; *aacA-aphD*, gentamicin/kanamycin/tobramycin resistance; *ble*, bleomycin resistance; *dfrK*, trimethoprim resistance, *erm*(A), *erm*(B), *erm*(C), *erm*(33), MLS$_B$ resistance; *fexA*, chloramphenicol/florfenicol resistance; *lsa*(B), elevated minimum inhibitory concentrations (MICs) of lincosamides; *spc*, specinomycin resistance; *tet*(L), tetracycline resistance.

acquire new resistance mechanisms. In the case of the synthetic and new agent florfenicol, the first phenicol exporters and target site–modifying enzymes, that conferred florfenicol resistance, were identified only a few years after the introduction of florfenicol into use in animals (Kim and Aoki 1996; Arcangioli et al. 1999; Schwarz et al. 2000).

The introduction of any new antimicrobial agent into clinical use will create a new selection pressure under which bacteria develop and/or acquire sooner or later new resistance genes or resistance-mediating mutations. This will reduce the efficacy of the antimicrobial agents and create a demand for newer and better antimicrobial agents. Changes in the current practice of how we use antimicrobial agents not only in human and veterinary medicine, but also in horticulture and aquaculture, are unavoidable, and good antimicrobial stewardship practice should be adopted by everyone involved in antimicrobial use and application (Prescott 2014). The future will show whether new technologies, such as whole-genome sequencing, which have become widely available during the past few years, can successfully be used for the identification of new bacterial targets that serve for the development of future antimicrobial agents for specific bacterial pathogens and disease conditions (Prescott 2014).

ACKNOWLEDGMENTS

We thank our technical staff, Kerstin Meyer, Roswitha Becker, Vivian Hensel, Ute Beermann, Marita Meurer, and Regina Ronge for their invaluable help and support and our colleagues and cooperation partners for the fruitful collaboration in the various research projects that formed the basis of this article. Work in our laboratories is supported by the German Federal Ministry of Education and Research (BMBF) through the German Aerospace Center (DLR) Grant number 01KI1313D (RESET II) and Grant number 01KI1301D (MedVet-Staph II), respectively, as well as by the German Research Foundation (DFG) Grant numbers SCHW382/10-1 and SCHW382/10-2; Grants from the National Basic Research Program of China (2013CB127200) and the National Natural Science Foundation of China (31370046); and a Grant from the Danish National Research Foundation (12-125943).

REFERENCES

Reference is also in this collection.

Aarestrup FM, Agersø Y, Ahrens P, Jørgensen JC, Madsen M, Jensen LB. 2000. Antimicrobial susceptibility and presence of resistance genes in staphylococci from poultry. *Vet Microbiol* **74:** 353–364.

Achard A, Villers C, Pichereau V, Leclercq R. 2005. New *lnu*(C) gene conferring resistance to lincomycin by nucleotidylation in *Streptococcus agalactiae* UCN36. *Antimicrob Agents Chemother* **49:** 2716–2719.

Ahonkhai VI, Cherubin CE, Shulman MA, Jhagroo M, Bancroft U. 1982. In vitro activity of U-57930E, a new clindamycin analog, against aerobic Gram-positive bacteria. *Antimicrob Agents Chemother* **21:** 902–905.

Allignet J, El Solh N. 1995. Diversity among the Gram-positive acetyltransferases inactivating streptogramin A and structurally related compounds and characterization of a new staphylococcal determinant, *vatB*. *Antimicrob Agents Chemother* **39:** 2027–2036.

Allignet J, El Solh N. 1997. Characterization of a new staphylococcal gene, *vgaB*, encoding a putative ABC transporter conferring resistance to streptogramin A and related compounds. *Gene* **202:** 133–138.

Allignet J, El Solh N. 1999. Comparative analysis of staphylococcal plasmids carrying three streptogramin-resistance genes: *vat-vgb-vga*. *Plasmid* **42:** 134–138.

Allignet J, Loncle V, Mazodier P, El Solh N. 1988. Nucleotide sequence of a staphylococcal plasmid gene, *vgb*, encoding a hydrolase inactivating the B components of virginiamycin-like antibiotics. *Plasmid* **20:** 271–275.

Allignet J, Loncle V, El Solh N. 1992. Sequence of a staphylococcal plasmid gene, *vga*, encoding a putative ATP-binding protein involved in resistance to virginiamycin A-like antibiotics. *Gene* **117:** 45–51.

Allignet J, Loncle V, Simenel C, Delepierre M, el Solh N. 1993. Sequence of a staphylococcal gene, *vat*, encoding an acetyltransferase inactivating the A-type compounds of virginiamycin-like antibiotics. *Gene* **130:** 91–98.

Allignet J, Aubert S, Morvan A, El Solh N. 1996. Distribution of genes encoding resistance to streptogramin A and related compounds among staphylococci resistant to these antibiotics. *Antimicrob Agents Chemother* **40:** 2523–2528.

Allignet J, Liassine N, El Solh N. 1998. Characterization of a staphylococcal plasmid related to pUB110 and carrying two novel genes, *vatC* and *vgbB*, encoding resistance to streptogramins A and B and similar antibiotics. *Antimicrob Agents Chemother* **42:** 1794–1798.

Arcangioli MA, Leroy-Sétrin S, Martel JL, Chaslus-Dancla E. 1999. A new chloramphenicol and florfenicol resistance gene flanked by two integron structures in *Salmonella typhimurium* DT104. *FEMS Microbiol Lett* **174:** 327–332.

Ban N, Nissen P, Hansen J, Moore PB, Steitz TA. 2000. The complete atomic structure of the large ribosomal subunit at 2.4 Å resolution. *Science* **289:** 905–920.

Bender J, Strommenger B, Steglich M, Zimmermann O, Fenner I, Lensing C, Dagwadordsch U, Kekulé AS, Werner G, Layer F. 2015. Linezolid resistance in clinical isolates of *Staphylococcus epidermidis* from German hospitals and characterization of two *cfr*-carrying plasmids. *J Antimicrob Chemother* **70:** 1630–1638.

Berman HM, Westbrook J, Feng Z, Gilliland G, Bhat TN, Weissig H, Shindyalov IN, Bourne PE. 2000. The protein data bank. *Nucl Acids Res* **28:** 235–242.

Besier S, Ludwig A, Zander J, Brade V, Wichelhaus TA. 2008. Linezolid resistance in *Staphylococcus aureus*: Gene dosage effect, stability, fitness costs, and cross-resistances. *Antimicrob Agents Chemother* **52:** 1570–1572.

Bonilla H, Huband MD, Seidel J, Schmidt H, Lescoe M, McCurdy SP, Lemmon MM, Brennan LA, Tait-Kamradt A, Puzniak L, et al. 2010. Multicity outbreak of linezolid-resistant *Staphylococcus epidermidis* associated with clonal spread of a *cfr*-containing strain. *Clin Infect Dis* **51:** 796–800.

Bozdogan B, Berrezouga L, Kuo MS, Yurek DA, Farley KA, Stockman BJ, Leclercq R. 1999. A new resistance gene, *linB*, conferring resistance to lincosamides by nucleotidylation in *Enterococcus faecium* HM1025. *Antimicrob Agents Chemother* **43:** 925–929.

Brenciani A, Morroni G, Vincenzi C, Manso E, Mingoia M, Giovanetti E, Varaldo PE. 2016. Detection in Italy of two clinical *Enterococcus faecium* isolates carrying both the oxazolidinone and phenicol resistance gene *optrA* and a silent multiresistance gene *cfr*. *J Antimicrob Chemother* **71:** 307–313.

Brisson-Noël A, Courvalin P. 1986. Nucleotide sequence of gene *linA* encoding resistance to lincosamides in *Staphylococcus haemolyticus*. *Gene* **43:** 247–253.

Bryskier A. 2005a. Lincosamines. In *Antimicrobial agents: Antibacterials antifungals* (ed. Bryskier A), pp. 592–603. ASM, Washington, DC.

Bryskier A. 2005b. Mutilins. In *Antimicrobial agents: Antibacterials antifungals* (ed. Bryskier A), pp. 1239–1241. ASM, Washington, DC.

Butaye P, Cloeckaert A, Schwarz S. 2003. Mobile genes coding for efflux-mediated antimicrobial resistance in Gram-positive and Gram-negative bacteria. *Int J Antimicrob Agents* **22:** 205–210.

Cai J, Wang Y, Schwarz S, Lv H, Li Y, Liao K, Yu S, Zhao K, Gu D, Wang X, et al. 2015a. Enterococcal isolates carrying the novel oxazolidinone resistance gene *optrA* from hospitals in Zhejiang, Guangdong, and Henan, China, 2010–2014. *Clin Microbiol Infect* **21:** 1095.e1–1095.e4.

Cai JC, Hu YY, Chen GX, Zhou HW, Zhang R. 2015b. Dissemination of the same *cfr*-carrying plasmid among methicillin-resistant *Staphylococcus aureus* and coagulase-negative staphylococcal isolates in China. *Antimicrob Agents Chemother* **59:** 3669–3671.

Cattoir V, Merabet L, Legrand P, Soussy CJ, Leclercq R. 2007. Emergence of a *Streptococcus pneumoniae* isolate resistant to streptogramins by mutation in ribosomal protein L22 during pristinamycin therapy of pneumococcal pneumonia. *J Antimicrob Chemother* **59:** 1010–1012.

Chen LP, Cai XW, Wang XR, Zhou XL, Wu DF, Xu XJ, Chen HC. 2010. Characterization of plasmid-mediated lincosamide resistance in a field isolate of *Haemophilus parasuis*. *J Antimicrob Chemother* **65:** 2256–2258.

Clinical and Laboratory Standards Institute (CLSI). 2015. *Performance standards for antimicrobial disk and dilution susceptibility tests for bacteria isolated from animals*, 3rd ed. CLSI supplement VET01S, CLSI, Wayne, PA.

Clinical and Laboratory Standards Institute (CLSI). 2016. *Performance standards for antimicrobial susceptibility testing*, 26th ed. CLSI supplement M100S, CLSI, Wayne, PA.

Cui L, Wang Y, Li Y, He T, Schwarz S, Ding Y, Shen J, Lv Y. 2013. Cfr-mediated linezolid-resistance among methicillin-resistant coagulase-negative staphylococci from infections of humans. *PLoS ONE* **8:** e57096.

Dai L, Wu CM, Wang MG, Wang Y, Wang Y, Huang SY, Xia LN, Li BB, Shen JZ. 2010. First report of the multidrug resistance gene *cfr* and the phenicol resistance gene *fexA* in a *Bacillus* strain from swine feces. *Antimicrob Agents Chemother* **54:** 3953–3955.

Davidovich C, Bashan A, Auerbach-Nevo T, Yaggie RD, Gontarek RR, Yonath A. 2007. Induced-fit tightens pleuromutilins binding to ribosomes and remote interactions enable their selectivity. *Proc Natl Acad Sci* **104:** 4291–4296.

Deshpande L, Ashcraft D, Kahn H, Pankey G, Jones R, Farrell D, Mendes R. 2015. Detection of a new *cfr*-like gene, *cfr*(B), in *Enterococcus faecium* recovered from human specimens in the United States: Report from The SENTRY Antimicrobial Surveillance Program. *Antimicrob Agents Chemother* **59:** 6256–6261.

Diab SS, Songer G, Uzal FA. 2013. *Clostridium difficile* infection in horses: A review. *Vet Microbiol* **167:** 42–49.

Douthwaite S. 1992. Functional interactions within 23S rRNA involving the peptidyltransferase center. *J Bacteriol* **174:** 1333–1338.

Dowling PM. 2013. Chloramphenicol, thiamphenicol, and florfenicol. In *Antimicrobial therapy in veterinary medicine*, 5th ed. (ed. Giguère S, Prescott JF, Dowling PM), pp. 269–277. John Wiley, Hoboken, NJ.

Dunkle JA, Xiong L, Mankin AS, Cate JH. 2010. Structures of the *Escherichia coli* ribosome with antibiotics bound near the peptidyl transferase center explain spectra of drug action. *Proc Natl Acad Sci* **107:** 17152–17157.

Edgar R, Bibi E. 1997. MdfA, an *Escherichia coli* multidrug resistance protein with an extraordinarily broad spectrum of drug recognition. *J Bacteriol* **179:** 2274–2280.

Eyal Z, Matzov D, Krupkin M, Wekselman I, Paukner S, Zimmerman E, Rozenberg H, Bashan A, Yonath A. 2015. Structural insights into species-specific features of the ribosome from the pathogen *Staphylococcus aureus*. *Proc Natl Acad Sci* **112:** E5805–E5814.

Feßler AT, Calvo N, Gutiérrez N, Muñoz Bellido JL, Fajardo M, Garduño E, Monecke S, Ehricht R, Kadlec K, Schwarz S. 2014. Cfr-mediated linezolid resistance in methicillin-resistant *Staphylococcus aureus* and *Staphylococcus haemolyticus* associated with clinical infections in humans: Two case reports. *J Antimicrob Chemother* **69:** 268–270.

Floyd JL, Smith KP, Kumar SH, Floyd JT, Varela MF. 2010. LmrS is a multidrug efflux pump of the major facilitator superfamily from *Staphylococcus aureus*. *Antimicrob Agents Chemother* **54:** 5406–5412.

* Fyfe C, Grossman TH, Kerstein K, Sutcliffe J. 2016. Resistance to macrolide antibiotics in public health pathogens. *Cold Spring Harb Perspect Med* doi: 10.1101/cshperspect.a025395.

Galimand M, Gerbaud G, Guibourdenche M, Riou JY, Courvalin P. 1998. High-level chloramphenicol resistance in *Neisseria meningitidis*. *N Engl J Med* **339:** 868–874.

Gatermann SG, Koschinski T, Friedrich S. 2007. Distribution and expression of macrolide resistance genes in coagulase-negative staphylococci. *Clin Microbiol Infect* **13:** 777–781.

Gay K, Stephens DS. 2001. Structure and dissemination of a chromosomal insertion element encoding macrolide efflux in *Streptococcus pneumoniae*. *J Infect Dis* **184:** 56–65.

Gentry DR, Rittenhouse SF, McCloskey L, Holmes DJ. 2007. Stepwise exposure of *Staphylococcus aureus* to pleuromutilins is associated with stepwise acquisition of mutations in *rplC* and minimally affects susceptibility to retapamulin. *Antimicrob Agents Chemother* **51:** 2048–2052.

Gentry DR, McCloskey L, Gwynn MN, Rittenhouse SF, Scangarella N, Shawar R, Holmes DJ. 2008. Genetic characterization of Vga ABC proteins conferring reduced susceptibility of *Staphylococcus aureus* to pleuromutilins. *Antimicrob Agents Chemother* **52:** 4507–4509.

Gerding DN, Johnson S, Peterson LR, Mulligan ME, Silva J Jr. 1995. *Clostridium difficile*-associated diarrhea and colitis. *Infect Control Hosp Epidemiol* **16:** 459–477.

Giguère S. 2013. Lincosamides, pleuromutilins, and streptogramins. In *Antimicrobial therapy in veterinary medicine*, 5th ed. (ed. Giguère S, Prescott JF, Dowling PM), pp. 199–210. Wiley, Hoboken, NJ.

Gómez-Sanz E, Kadlec K, Feßler AT, Zarazaga M, Torres C, Schwarz S. 2013. A novel *fexA* variant from a canine *Staphylococcus pseudintermedius* isolate that does not confer florfenicol resistance. *Antimicrob Agents Chemother* **57:** 5763–5766.

Gopegui ER, Juan C, Zamorano L, Pérez JL, Oliver A. 2012. Transferable multidrug resistance plasmid carrying *cfr* associated with *tet*(L), *ant(4′)-Ia*, and *dfrK* genes from a clinical methicillin-resistant *Staphylococcus aureus* ST125 strain. *Antimicrob Agents Chemother* **56:** 2139–2142.

Hammerum AM, Flannagan SE, Clewell DB, Jensen LB. 2001. Indication of transposition of a mobile DNA element containing the *vat*(D) and *erm*(B) genes in *Enterococcus faecium*. *Antimicrob Agents Chemother* **45:** 3223–3225.

Hansen LH, Vester B. 2015. A *cfr*-like gene from *Clostridium difficile* confers multiple antibiotic resistance by the same mechanism as the *cfr* gene. *Antimicrob Agents Chemother* **59:** 5841–5843.

Hansen LH, Jensen LB, Sørensen HI, Sørensen SJ. 2007. Substrate specificity of the OqxAB multidrug resistance pump in *Escherichia coli* and selected enteric bacteria. *J Antimicrob Chemother* **60:** 145–147.

Haroche J, Allignet J, Buchrieser C, El Solh N. 2000. Characterization of a variant of *vga*(A) conferring resistance to streptogramin A and related compounds. *Antimicrob Agents Chemother* **44:** 2271–2275.

Haroche J, Allignet J, El Solh N. 2002. Tn*5406*, a new staphylococcal transposon conferring resistance to streptogramin A and related compounds including dalfopristin. *Antimicrob Agents Chemother* **46:** 2337–2343.

Hauschild T, Feßler AT, Kadlec K, Billerbeck C, Schwarz S. 2012. Detection of the novel *vga*(E) gene in methicillin-resistant *Staphylococcus aureus* CC398 isolates from cattle and poultry. *J Antimicrob Chemother* **67:** 503–504.

He T, Wang Y, Schwarz S, Zhao Q, Shen J, Wu C. 2014. Genetic environment of the multi-resistance gene *cfr* in methicillin-resistant coagulase-negative staphylococci from chickens, ducks, and pigs in China. *Int J Med Microbiol* **304:** 257–261.

He T, Shen J, Schwarz S, Wu C, Wang Y. 2015. Characterization of a genomic island in *Stenotrophomonas maltophilia* that carries a novel *floR* gene variant. *J Antimicrob Chemother* **70:** 1031–1036.

He T, Shen Y, Schwarz S, Cai J, Lv Y, Li J, Feßler AT, Zhang R, Wu C, Shen J, et al. 2016. Genetic environment of the transferable oxazolidinone/phenicol resistance gene *optrA* in *Enterococcus faecalis* isolates of human and animal origin. *J Antimicrob Chemother* doi: 10.1093/jac/dkw016.

Heir E, Lindstedt BA, Leegaard TM, Gjernes E, Kapperud G. 2004. Prevalence and characterization of integrons in blood culture Enterobacteriaceae and gastrointestinal *Escherichia coli* in Norway and reporting of a novel class 1 integron–located lincosamide resistance gene. *Ann Clin Microbiol Antimicrob* **3:** 12.

Hot C, Berthet N, Chesneau O. 2014. Characterization of *sal*(A), a novel gene responsible for lincosamide and streptogramin A resistance in *Staphylococcus sciuri*. *Antimicrob Agents Chemother* **58:** 3335–3341.

Huang J, O'Toole PW, Shen W, Amrine-Madsen H, Jiang X, Lobo N, Palmer LM, Voelker L, Fan F, Gwynn MN, et al. 2004. Novel chromosomally encoded multidrug efflux transporter MdeA in *Staphylococcus aureus*. *Antimicrob Agents Chemother* **48:** 909–917.

Isnard C, Malbruny B, Leclercq R, Cattoir V. 2013. Genetic basis for in vitro and in vivo resistance to lincosamides, streptogramins A, and pleuromutilins (LSAP phenotype) in *Enterococcus faecium*. *Antimicrob Agents Chemother* **57:** 4463–4469.

Jaglic Z, Vlkova H, Bardon J, Michu E, Cervinkova D, Babak V. 2012. Distribution, characterization and genetic bases of erythromycin resistance in staphylococci and enterococci originating from livestock. *Zoonoses Public Health* **59:** 202–211.

Jung YH, Shin ES, Kim O, Yoo JS, Lee KM, Yoo JI, Chung GT, Lee YS. 2010. Characterization of two newly identified genes, *vgaD* and *vatG*, conferring resistance to streptogramin A in *Enterococcus faecium*. *Antimicrob Agents Chemother* **54:** 4744–4749.

Kadlec K, Schwarz S. 2009. Identification of a novel ABC transporter gene, *vga*(C), located on a multiresistance plasmid from a porcine methicillin-resistant *Staphylococcus aureus* ST398 strain. *Antimicrob Agents Chemother* **53:** 3589–3591.

Kadlec K, Kehrenberg C, Schwarz S. 2007. Efflux-mediated resistance to florfenicol and/or chloramphenicol in *Bordetella bronchiseptica*: Identification of a novel chloramphenicol exporter. *J Antimicrob Chemother* **59:** 191–196.

Kadlec K, Pomba CF, Couto N, Schwarz S. 2010. Small plasmids carrying *vga*(A) or *vga*(C) genes mediate resistance to lincosamides, pleuromutilins and streptogramin antibiotics in methicillin-resistant *Staphylococcus aureus* ST398 from swine. *J Antimicrob Chemother* **65:** 2692–2693.

Kadlec K, Brenner Michael G, Sweeney MT, Brzuszkiewicz E, Liesegang H, Daniel R, Watts JL, Schwarz S. 2011. Molecular basis of macrolide, triamilide, and lincosamide resistance in *Pasteurella multocida* from bovine respiratory disease. *Antimicrob Agents Chemother* **55:** 2475–2477.

Kehrenberg C, Schwarz S. 2004. *fexA*, a novel *Staphylococcus lentus* gene encoding resistance to florfenicol and chloramphenicol. *Antimicrob Agents Chemother* **48:** 615–618.

Kehrenberg C, Schwarz S. 2005. Florfenicol–chloramphenicol exporter gene *fexA* is part of the novel transposon Tn*558*. *Antimicrob Agents Chemother* **49:** 813–815.

Kehrenberg C, Schwarz S. 2006. Distribution of florfenicol resistance genes *fexA* and *cfr* among chloramphenicol-resistant *Staphylococcus* isolates. *Antimicrob Agents Chemother* **50:** 1156–1163.

Kehrenberg C, Ojo KK, Schwarz S. 2004. Nucleotide sequence and organization of the multiresistance plasmid pSCFS1 from *Staphylococcus sciuri*. *J Antimicrob Chemother* **54:** 936–939.

Kehrenberg C, Schwarz S, Jacobsen L, Hansen LH, Vester B. 2005. A new mechanism for chloramphenicol, florfenicol and clindamycin resistance: Methylation of 23S ribosomal RNA at A2503. *Mol Microbiol* **57:** 1064–1073.

Kehrenberg C, Aarestrup FM, Schwarz S. 2007. IS*21-558* insertion sequences are involved in the mobility of the multiresistance gene *cfr*. *Antimicrob Agents Chemother* **51:** 483–487.

Kehrenberg C, Cuny C, Strommenger B, Schwarz S, Witte W. 2009. Methicillin-resistant and -susceptible *Staphylococcus aureus* strains of clonal lineages ST398 and ST9 from swine carry the multidrug resistance gene *cfr*. *Antimicrob Agents Chemother* **53:** 779–781.

Kennedy AD, Porcella SF, Martens C, Whitney AR, Braughton KR, Chen L, Craig CT, Tenover FC, Kreiswirth BN, Musser JM, et al. 2010. Complete nucleotide sequence analysis of plasmids in strains of *Staphylococcus aureus* clone USA300 reveals a high level of identity among isolates with closely related core genome sequences. *J Clin Microbiol* **48:** 4504–4511.

Kim EH, Aoki T. 1996. Sequence analysis of the florfenicol resistance gene encoded in the transferable R-plasmid of a fish pathogen, *Pasteurella piscicida*. *Microbiol Immunol* **40:** 665–669.

Kim HJ, Kim Y, Lee MS, Lee HS. 2001. Gene *lmrB* of *Corynebacterium glutamicum* confers efflux-mediated resistance to lincomycin. *Mol Cells* **12:** 112–116.

Klitgaard RN, Ntokou E, Norgaard K, Biltoft D, Hansen LH, Traedholm NM, Kongsted J, Vester B. 2015. Mutations in the bacterial ribosomal protein L3 and their association with antibiotic resistance. *Antimicrob Agents Chemother* **59:** 3518–3528.

Kumano M, Fujita M, Nakamura K, Murata M, Ohki R, Yamane K. 2003. Lincomycin resistance mutations in two regions immediately downstream of the −10 region of *lmr* promoter cause overexpression of a putative multidrug efflux pump in *Bacillus subtilis* mutants. *Antimicrob Agents Chemother* **47:** 432–435.

LaMarre J, Mendes RE, Szal T, Schwarz S, Jones RN, Mankin AS. 2013. The genetic environment of the *cfr* gene and the presence of other mechanisms account for the very high linezolid resistance of *Staphylococcus epidermidis* isolate 426-3147L. *Antimicrob Agents Chemother* **57:** 1173–1179.

Lang KS, Anderson JM, Schwarz S, Williamson L, Handelsman J, Singer RS. 2010. Novel florfenicol and chloramphenicol resistance gene discovered in Alaskan soil by using functional metagenomics. *Appl Environ Microbiol* **76:** 5321–5326.

Leclercq R, Courvalin P. 1991. Intrinsic and unusual resistance to macrolide, lincosamide, and streptogramin antibiotics in bacteria. *Antimicrob Agents Chemother* **35:** 1273–1276.

Leslie AGW, Moody PCE, Shaw WV. 1988. Structure of chloramphenicol acetyltransferase at 1.75 Å resolution. *Proc Natl Acad Sci* **85:** 4133–4137.

Levings RS, Hall RM, Lightfoot D, Djordjevic SP. 2006. *linG*, a new integron-associated gene cassette encoding a lincosamide nucleotidyltransferase. *Antimicrob Agents Chemother* **50:** 3514–3515.

Li BB, Wu CM, Wang Y, Shen JZ. 2011. Single and dual mutations at positions 2058, 2503 and 2504 of 23S rRNA and their relationship to resistance to antibiotics that target the large ribosomal subunit. *J Antimicrob Chemother* **66:** 1983–1986.

Li B, Wendlandt S, Yao J, Liu Y, Zhang Q, Shi Z, Wei J, Shao D, Schwarz S, Wang S, et al. 2013. Detection and new genetic environment of the pleuromutilin-lincosamide-streptogramin A resistance gene *lsa*(E) in methicillin-resistant *Staphylococcus aureus* of swine origin. *J Antimicrob Chemother* **68:** 1251–1255.

Li J, Li B, Wendlandt S, Schwarz S, Wang Y, Wu C, Ma Z, Shen J. 2014a. Identification of a novel *vga*(E) gene variant that confers resistance to pleuromutilins, lincosamides and streptogramin A antibiotics in staphylococci of porcine origin. *J Antimicrob Chemother* **69:** 919–923.

Li XS, Dong WC, Wang XM, Hu GZ, Wang YB, Cai BY, Wu CM, Wang Y, Du XD. 2014b. Presence and genetic environment of pleuromutilin–lincosamide–streptogramin A resistance gene *lsa*(E) in enterococci of human and swine origin. *J Antimicrob Chemother* **69:** 1424–1426.

Li D, Wu C, Wang Y, Fan R, Schwarz S, Zhang S. 2015. Identification of multiresistance gene *cfr* in methicillin-resistant *Staphylococcus aureus* from pigs: Plasmid location and integration into a staphylococcal cassette chromosome *mec* complex. *Antimicrob Agents Chemother* **59:** 3641–3644.

Li D, Wang Y, Schwarz S, Cai J, Fan R, Li J, Feßler AT, Zhang R, Wu C, Shen J. 2016. Co-location of the oxazolidinone resistance genes *optrA* and *cfr* on a multi-resistance plasmid from *Staphylococcus sciuri*. *J Antimicrobial Chemother* doi: 10.1093/jac/dkw040.

Lindeman CJ, Portis E, Johansen L, Mullins LM, Stoltman GA. 2013. Susceptibility to antimicrobial agents among bovine mastitis pathogens isolated from North American dairy cattle, 2002–2010. *J Vet Diagn Invest* **25:** 581–591.

Liu H, Wang Y, Wu C, Schwarz S, Shen Z, Jeon B, Ding S, Zhang Q, Shen J. 2012a. A novel phenicol exporter gene, *fexB*, found in enterococci of animal origin. *J Antimicrob Chemother* **67:** 322–325.

Liu Y, Wang Y, Wu C, Shen Z, Schwarz S, Du XD, Dai L, Zhang W, Zhang Q, Shen J. 2012b. First report of the multidrug resistance gene *cfr* in *Enterococcus faecalis* of animal origin. *Antimicrob Agents Chemother* **56:** 1650–1654.

Liu Y, Wang Y, Schwarz S, Li Y, Shen Z, Zhang Q, Wu C, Shen J. 2013. Transferable multiresistance plasmids carrying *cfr* in *Enterococcus* spp. from swine and farm environment. *Antimicrob Agents Chemother* **57:** 42–48.

Liu Y, Wang Y, Dai L, Wu C, Shen J. 2014. First report of multiresistance gene *cfr* in *Enterococcus* species *casseliflavus* and *gallinarum* of swine origin. *Vet Microbiol* **170:** 352–357.

Locke JB, Rahawi S, LaMarre J, Mankin AS, Shaw KJ. 2012. Genetic environment and stability of *cfr* in methicillin-resistant *Staphylococcus aureus* CM05. *Antimicrob Agents Chemother* **56:** 332–340.

Loeza-Lara PD, Soto-Huipe M, Baizabal-Aguirre VM, Ochoa-Zarzosa A, Valdez-Alarcón JJ, Cano-Camacho H, López-Meza JE. 2004. pBMSa1, a plasmid from a dairy cow isolate of *Staphylococcus aureus*, encodes a lincomycin resistance determinant and replicates by the rolling-circle mechanism. *Plasmid* **52:** 48–56.

Long KS, Poehlsgaard J, Kehrenberg C, Schwarz S, Vester B. 2006. The Cfr rRNA methyltransferase confers resistance to phenicols, lincosamides, oxazolidinones, pleuromutilins, and streptogramin A antibiotics. *Antimicrob Agents Chemother* **50:** 2500–2505.

Long KS, Poehlsgaard J, Hansen LH, Hobbie SN, Bottger EC, Vester B. 2009. Single 23S rRNA mutations at the ribosomal peptidyl transferase centre confer resistance to valnemulin and other antibiotics in *Mycobacterium smegmatis* by perturbation of the drug binding pocket. *Mol Microbiol* **71:** 1218–1227.

Long KS, Munck C, Andersen TM, Schaub MA, Hobbie SN, Böttger EC, Vester B. 2010. Mutations in 23S rRNA at the peptidyl transferase center and their relationship to linezolid binding and cross-resistance. *Antimicrob Agents Chemother* **54:** 4705–4713.

Lovett PS. 1990. Translational attenuation as the regulator of inducible *cat* genes. *J Bacteriol* **172:** 1–6.

Lozano C, Aspiroz C, Sáenz Y, Ruiz-García M, Royo-García G, Gómez-Sanz E, Ruiz-Larrea F, Zarazaga M, Torres C. 2012a. Genetic environment and location of the *lnu*(A) and *lnu*(B) genes in methicillin-resistant *Staphylococcus aureus* and other staphylococci of animal and human origin. *J Antimicrob Chemother* **67:** 2804–2808.

Lozano C, Rezusta A, Gómez P, Gómez-Sanz E, Báez N, Martin-Saco G, Zarazaga M, Torres C. 2012b. High prevalence of *spa* types associated with the clonal lineage CC398 among tetracycline-resistant methicillin-resistant *Staphylococcus aureus* strains in a Spanish hospital. *J Antimicrob Chemother* **67:** 330–334.

Lüthje P, Schwarz S. 2006. Antimicrobial resistance of coagulase-negative staphylococci from bovine subclinical mastitis with particular reference to macrolide-lincosamide resistance phenotypes and genotypes. *J Antimicrob Chemother* **57:** 966–969.

Lüthje P, Schwarz S. 2007a. Molecular analysis of constitutively expressed *erm*(C) genes selected in vitro in the presence of the non-inducers pirlimycin, spiramycin and tylosin. *J Antimicrob Chemother* **59:** 97–101.

Lüthje P, Schwarz S. 2007b. Molecular basis of resistance to macrolides and lincosamides among staphylococci and streptococci from various animal sources collected in the resistance monitoring program BfT-GermVet. *Int J Antimicrob Agents* **29:** 528–535.

Lyras D, Rood JI. 2000. Transposition of Tn*4451* and Tn*4453* involves a circular intermediate that forms a promoter for the large resolvase, TnpX. *Mol Microbiol* **38:** 588–601.

MacLeod AJ, Ross HB, Ozere RL, Digout G, van Rooyen CE. 1964. Lincomycin: A new antibiotic active against staphylococci and other Gram-positive cocci: Clinical and laboratory studies. *Can Med Assoc J* **91:** 1056–1060.

Malbruny B, Werno AM, Murdoch DR, Leclercq R, Cattoir V. 2011. Cross-resistance to lincosamides, streptogramins A, and pleuromutilins due to the *lsa*(C) gene in *Streptococcus agalactiae* UCN70. *Antimicrob Agents Chemother* **55:** 1470–1474.

Marín M, Martín A, Alcalá L, Cercenado E, Iglesias C, Reigadas E, Bouza E. 2015. *Clostridium difficile* with high linezolid MICs harbor the multiresistance gene *cfr*. *Antimicrob Agents Chemother* **59:** 586–589.

Matsuoka M, Inoue M, Endo Y, Nakajima Y. 2003. Characteristic expression of three genes, *msr*(A), *mph*(C) and *erm*(Y), that confer resistance to macrolide antibiotics on *Staphylococcus aureus*. *FEMS Microbiol Lett* **220:** 287–293.

McMurray LM, George AM, Levy SB. 1994. Active efflux of chloramphenicol in susceptible *Escherichia coli* strains and in multiple-antibiotic-resistant (Mar) mutants. *Antimicrob Agents Chemother* **38:** 542–546.

Mendes RE, Deshpande LM, Castanheira M, DiPersio J, Saubolle MA, Jones RN. 2008. First report of *cfr*-mediated resistance to linezolid in human staphylococcal clinical isolates recovered in the United States. *Antimicrob Agents Chemother* **52:** 2244–2246.

Mendes RE, Deshpande LM, Bonilla HF, Schwarz S, Huband MD, Jones RN, Quinn JP. 2013. Dissemination of a pSCFS3-like *cfr*-carrying plasmid in *Staphylococcus aureus* and *Staphylococcus epidermidis* clinical isolates recovered from hospitals in Ohio. *Antimicrob Agents Chemother* **57:** 2923–2928.

Michael GB, Kadlec K, Sweeney MT, Brzuszkiewicz E, Liesegang H, Daniel R, Murray RW, Watts JL, Schwarz S. 2012. ICE*Pmu1*, an integrative conjugative element (ICE) of *Pasteurella multocida*: Analysis of the regions that comprise 12 antimicrobial resistance genes. *J Antimicrob Chemother* **67:** 84–90.

Miller K, Dunsmore CJ, Fishwick CW, Chopra I. 2008. Linezolid and tiamulin cross-resistance in *Staphylococcus aureus* mediated by point mutations in the peptidyl transferase center. *Antimicrob Agents Chemother* **52:** 1737–1742.

Monecke S, Ruppelt A, Wendlandt S, Schwarz S, Slickers P, Ehricht R, Jäckel SC. 2013. Genotyping of *Staphylococcus aureus* isolates from diseased poultry. *Vet Microbiol* **162:** 806–812.

Montilla A, Zavala A, Cáceres Cáceres R, Cittadini R, Vay C, Gutkind G, Famiglietti A, Bonofiglio L, Mollerach M. 2014. Genetic environment of the *lnu*(B) gene in a *Streptococcus agalactiae* clinical isolate. *Antimicrob Agents Chemother* **58:** 5636–5637.

Morar M, Bhullar K, Hughes DW, Junop M, Wright GD. 2009. Structure and mechanism of the lincosamide antibiotic adenylyltransferase LinB. *Structure* **17:** 1649–1659.

Morris TH. 1995. Antibiotic therapeutics in laboratory animals. *Lab Anim* **29:** 16–36.

Mosher RH, Ranade NP, Schrempf H, Vining LC. 1990. Chloramphenicol resistance in *Streptomyces*: Cloning and characterization of a chloramphenicol hydrolase gene from *Streptomyces venezuelae*. *J Gen Microbiol* **136:** 293–301.

Mosher RH, Camp DJ, Yang K, Brown MP, Shaw WV, Vining LC. 1995. Inactivation of chloramphenicol by O-phosphorylation. *J Biol Chem* **27:** 27000–27006.

Murata M, Ohno S, Kumano M, Yamamane K, Ohki R. 2003. Multidrug resistant phenotype of *Bacillus subtilis* spontaneous mutants isolated in the presence of puromycin and lincomycin. *Can J Microbiol* **49:** 71–77.

Murray IA, Shaw WV. 1997. O-Acetyltransferases for chloramphenicol and other natural products. *Antimicrob Agents Chemother* **41:** 1–6.

Nagy I, Schoofs G, Vanderleyden J, De Mot R. 1997. Transposition of the IS*v21*-related element IS*v1415* in *Rhodococcus erythropolis*. *J Bacteriol* **179:** 4635–4638.

Nawaz MS, Khan SA, Khan AA, Khambaty FM, Cerniglia CE. 2000. Comparative molecular analysis of erythromycin-resistance determinants in staphylococcal isolates of poultry and human origin. *Mol Cell Probes* **14:** 311–319.

Nilsen IW, Bakke I, Vader A, Olsvik O, El-Gewely MR. 1996. Isolation of *cmr*, a novel chloramphenicol resistance gene encoding a putative efflux pump. *J Bacteriol* **178:** 3188–3193.

Nissen P, Hansen J, Ban N, Moore PB, Steitz TA. 2000. The structural basis of ribosome activity in peptide bond synthesis. *Science* **289:** 920–930.

Noeske J, Huang J, Olivier NB, Giacobbe RA, Zambrowski M, Cate JH. 2014. Synergy of streptogramin antibiotics occurs independently of their effects on translation. *Antimicrob Agents Chemother* **58:** 5269–5279.

Novak R, Shlaes DM. 2010. The pleuromutilin antibiotics: A new class for human use. *Curr Opin Investig Drugs* **11:** 182–191.

Novotná G, Janata J. 2006. A new evolutionary variant of the streptogramin A resistance protein, Vga(A)$_{LC}$, from *Staphylococcus haemolyticus* with shifted substrate specificity towards lincosamides. *Antimicrob Agents Chemother* **50:** 4070–4076.

Paulsen IT, Brown MH, Skurray RA. 1996. Proton-dependent multidrug efflux systems. *Microbiol Rev* **60:** 575–608.

Peschke U, Schmidt H, Zhang HZ, Piepersberg W. 1995. Molecular characterization of the lincomycin-production gene cluster of *Streptomyces lincolnensis* 78-11. *Mol Microbiol* **16:** 1137–1156.

Petinaki E, Guérin-Faublée V, Pichereau V, Villers C, Achard A, Malbruny B, Leclercq R. 2008. Lincomycin resistance gene *lnu*(D) in *Streptococcus uberis*. *Antimicrob Agents Chemother* **52:** 626–630.

Poehlsgaard J, Douthwaite S. 2005. The bacterial ribosome as a target for antibiotics. *Nat Rev Microbiol* **3:** 870–881.

Poelarends G, Mazurkiewicz P, Konings W. 2002. Multidrug transporters and antibiotic resistance in *Lactococcus lactis*. *Biochim Biophys Acta* **1555:** 1–7.

Poole K. 2005. Efflux-mediated antimicrobial resistance. *J Antimicrob Chemother* **56:** 20–51.

Porse BT, Garrett RA. 1999. Sites of interaction of streptogramin A and B antibiotics in the peptidyl transferase loop of 23 S rRNA and the synergism of their inhibitory mechanisms. *J Mol Biol* **286:** 375–387.

Portis E, Lindeman C, Johansen L, Stoltman G. 2012. A ten-year (2000–2009) study of antimicrobial susceptibility of bacteria that cause bovine respiratory disease complex— *Mannheimia haemolytica*, *Pasteurella multocida*, and *Histophilus somni*—in the United States and Canada. *J Vet Diagn Invest* **24:** 932–944.

Prescott JF. 2014. The resistance tsunami, antimicrobial stewardship, and the golden age of microbiology. *Vet Microbiol* **171:** 273–278.

Putman M, van Veen HW, Degener JE, Konings WN. 2000. Antibiotic resistance: Era of the multidrug pump. *Mol Microbiol* **36:** 772–774.

Putman M, van Veen HW, Degener JE, Konings WN. 2001. The lactococcal secondary multidrug transporter LmrP confers resistance to lincosamides, macrolides, streptogramins and tetracyclines. *Microbiology* **147:** 2873–2880.

Recchia GD, Hall RM. 1995. Gene cassettes: A new class of mobile element. *Microbiology* **141:** 3015–3027.

Rende-Fournier R, Leclercq R, Galimand M, Duval J, Courvalin P. 1993. Identification of the *satA* gene encoding a streptogramin A acetyltransferase in *Enterococcus faecium* BM4145. *Antimicrob Agents Chemother* **37:** 2119–2125.

Roberts MC. 2008. Update on macrolide–lincosamide–streptogramin, ketolide, and oxazolidinone resistance genes. *FEMS Microbiol Lett* **282:** 147–159.

Roberts MC, Schwarz S. 2009. Tetracycline and chloramphenicol resistance mechanisms. In *Antimicrobial drug resistance, Vol. 1, Mechanisms of drug resistance* (ed. Mayers DJ, et al.), pp. 183–193. Springer, New York.

Roberts MC, Sutcliffe J, Courvalin P, Jensen LB, Rood J, Seppala H. 1999. Nomenclature for macrolide and macrolide–lincosamide–streptogramin B resistance determinants. *Antimicrob Agents Chemother* **43:** 2823–2830.

Ross JI, Eady EA, Cove JH, Cunliffe WJ, Baumberg S, Wootton JC. 1990. Inducible erythromycin resistance in staphylococci is encoded by a member of the ATP-binding transport super-gene family. *Mol Microbiol* **4:** 1207–1214.

Schmitz F-J, Petridou J, Astfalk N, Köhrer K, Scheuring S, Schwarz S. 2002a. Molecular analysis of constitutively expressed *erm*(C) genes selected in vitro by incubation in the presence of the noninducers quinupristin, telithromycin, or ABT-773. *Microb Drug Resist* **8:** 171–177.

Schmitz F-J, Petridou J, Jagusch H, Astfalk N, Scheuring S, Schwarz S. 2002b. Molecular characterization of ketolide-resistant *erm*(A)-carrying *Staphylococcus aureus* isolates selected in vitro by telithromycin, ABT-773, quinupristin and clindamycin. *J Antimicrob Chemother* **49:** 611–617.

Schwarz S, Kehrenberg C. 2006. Old dogs that learn new tricks: Modified antimicrobial agents that escape pre-existing resistance mechanisms. *Int J Med Microbiol* **296:** 45–49.

Schwarz S, Wang Y. 2015. Nomenclature and functionality of the so-called *cfr* gene from *Clostridium difficile*. *Antimicrob Agents Chemother* **59:** 2476–2477.

Schwarz S, Werckenthin C, Kehrenberg C. 2000. Identification of a plasmid-borne chloramphenicol–florfenicol resistance gene in *Staphylococcus sciuri*. *Antimicrob Agents Chemother* **44:** 2530–2533.

Schwarz S, Kehrenberg C, Doublet B, Cloeckaert A. 2004. Molecular basis of bacterial resistance to chloramphenicol and florfenicol. *FEMS Microbiol Rev* **28:** 519–542.

Schwarz S, Cloeckaert A, Roberts MC. 2006. Mechanisms and spread of bacterial resistance to antimicrobial agents. In *Antimicrobial resistance in bacteria of animal origin* (ed. Aarestrup FM), pp. 73–98. ASM, Washington, DC.

Schwendener S, Perreten V. 2011. New transposon Tn*6133* in methicillin-resistant *Staphylococcus aureus* ST398 contains *vga*(E), a novel streptogramin A, pleuromutilin, and lincosamide resistance gene. *Antimicrob Agents Chemother* **55:** 4900–4904.

Seoane A, García Lobo JM. 2000. Identification of a streptogramin A acetyltransferase gene in the chromosome of *Yersinia enterocolitica*. *Antimicrob Agents Chemother* **44:** 905–909.

Shaw WV. 1983. Chloramphenicol acetyltransferase, enzymology and molecular biology. *Crit Rev Biochem* **14:** 1–46.

Shen J, Wang Y, Schwarz S. 2013. Presence and dissemination of the multiresistance gene *cfr* in Gram-positive and Gram-negative bacteria. *J Antimicrob Chemother* **68:** 1697–1706.

Shore AC, Brennan OM, Ehricht R, Monecke S, Schwarz S, Slickers P, Coleman DC. 2010. Identification and characterization of the multidrug resistance gene *cfr* in a Panton-Valentine leukocidin-positive sequence type 8 methicillin-resistant *Staphylococcus aureus* IVa (USA300) isolate. *Antimicrob Agents Chemother* **54:** 4978–4984.

Shultz TR, Tapsall JW, White PA, Ryan CS, Lyras D, Rood JI. 2003. Chloramphenicol-resistant *Neisseria meningitidis* containing *catP* isolated in Australia. *J Antimicrob Chemother* **52:** 856–859.

Si H, Zhang WJ, Chu S, Wang XM, Dai L, Hua X, Dong Z, Schwarz S, Liu S. 2015. Novel plasmid-borne multidrug resistance gene cluster including *lsa*(E) from a linezolid-resistant *Enterococcus faecium* isolate of swine origin. *Antimicrob Agents Chemother* **59:** 7113–7116.

Silley P, Simjee S, Schwarz S. 2012. Surveillance and monitoring of antimicrobial resistance and antibiotic consumption in humans and animals. *Rev Sci Tech* **31:** 105–120.

Silva NC, Guimarães FF, Manzi MP, Júnior AF, Gómez-Sanz E, Gómez P, Langoni H, Rall VL, Torres C. 2014. Methicillin-resistant *Staphylococcus aureus* of lineage ST398 as cause of mastitis in cows. *Lett Appl Microbiol* **59:** 665–669.

Simjee S, McDermott PF, Wagner DD, White DG. 2001. Variation within the *vat*(E) allele of *Enterococcus faecium* isolates from retail poultry samples. *Antimicrob Agents Chemother* **45:** 2931–2932.

Singh KV, Malathum K, Murray BE. 2001. Disruption of an *Enterococcus faecium* species-specific gene, a homologue of acquired macrolide resistance genes of staphylococci, is associated with an increase in macrolide susceptibility. *Antimicrob Agents Chemother* **45:** 263–266.

Singh KV, Weinstock GM, Murray BE. 2002. An *Enterococcus faecalis* ABC homologue (Lsa) is required for the resistance of this species to clindamycin and quinupristin-dalfopristin. *Antimicrob Agents Chemother* **46:** 1845–1850.

Smieja M. 1998. Current indications for the use of clindamycin: A critical review. *Can J Infect Dis* **9:** 22–28.

Smith LK, Mankin AS. 2008. Transcriptional and translational control of the *mlr* operon, which confers resistance to seven classes of protein synthesis inhibitors. *Antimicrob Agents Chemother* **52:** 1703–1712.

Stokes HW, Hall RM. 1991. Sequence analysis of the inducible chloramphenicol resistance determinant in the Tn*1696* integron suggests regulation by translational attenuation. *Plasmid* **26:** 10–19.

Tao W, Lee MH, Wu J, Kim H, Kim JC, Chung E, Hwang EC, Lee SW. 2012. Inactivation of chloramphenicol and florfenicol by a novel chloramphenicol hydrolase. *Appl Environ Microbiol* **78:** 6295–6301.

Toh SM, Xiong L, Arias CA, Villegas MV, Lolans K, Quinn J, Mankin AS. 2007. Acquisition of a natural resistance gene renders a clinical strain of methicillin-resistant *Staphylococcus aureus* resistant to the synthetic antibiotic linezolid. *Mol Microbiol* **64:** 1506–1514.

Vazquez D. 1966. Studies on the mode of action of the streptomycin antibiotics. *J Gen Microbiol* **42:** 93–106.

Vester B, Douthwaite S. 2001. Macrolide resistance conferred by base substitutions in 23S rRNA. *Antimicrob Agents Chemother* **45:** 1–12.

Wang J, Shoemaker NB, Wang G-R, Salyers AA. 2000. Characterization of a *Bacteroides* mobilizable transposon, NBU2, which carries a functional lincomycin resistance gene. *J Bacteriol* **182:** 3559–3571.

Wang Y, Wang Y, Wu CM, Schwarz S, Shen Z, Zhang W, Zhang Q, Shen JZ. 2011. Detection of the staphylococcal multiresistance gene *cfr* in *Proteus vulgaris* of food animal origin. *J Antimicrob Chemother* **66:** 2521–2526.

Wang Y, He T, Schwarz S, Zhou D, Shen Z, Wu C, Wang Y, Ma L, Zhang Q, Shen J. 2012a. Detection of the staphylococcal multiresistance gene *cfr* in *Escherichia coli* of domestic-animal origin. *J Antimicrob Chemother* **67:** 1094–1098.

Wang Y, Schwarz S, Shen Z, Zhang W, Qi J, Liu Y, He T, Shen J, Wu C. 2012b. Co-location of the multiresistance gene *cfr* and the novel streptomycin resistance gene *aadY* on a small plasmid in a porcine *Bacillus* strain. *J Antimicrob Chemother* **67:** 1547–1549.

Wang Y, Wang Y, Schwarz S, Shen Z, Zhou N, Lin J, Wu C, Shen J. 2012c. Detection of the staphylococcal multiresistance gene *cfr* in *Macrococcus caseolyticus* and *Jeotgalicoccus pinnipedialis*. *J Antimicrob Chemother* **67:** 1824–1827.

Wang Y, Zhang W, Wang J, Wu C, Shen Z, Fu X, Yan Y, Zhang Q, Schwarz S, Shen J. 2012d. Distribution of the multidrug resistance gene *cfr* in *Staphylococcus* species isolates from swine farms in China. *Antimicrob Agents Chemother* **56:** 1485–1490.

Wang XM, Zhang WJ, Schwarz S, Yu SY, Liu H, Si W, Zhang RM, Liu S. 2012e. Methicillin-resistant *Staphylococcus aureus* ST9 from a case of bovine mastitis carries the genes *cfr* and *erm*(A) on a small plasmid. *J Antimicrob Chemother* **67:** 1287–1289.

Wang Y, He T, Schwarz S, Zhao Q, Shen Z, Wu C, Shen J. 2013a. Multidrug resistance gene *cfr* in methicillin-resistant coagulase-negative staphylococci from chickens, ducks, and pigs in China. *Int J Med Microbiol* **303:** 84–87.

Wang Y, Li D, Song L, Liu Y, He T, Liu H, Wu C, Schwarz S, Shen J. 2013b. First report of the multiresistance gene *cfr* in *Streptococcus suis*. *Antimicrob Agents Chemother* **57:** 4061–4063.

Wang Y, Lv Y, Cai J, Schwarz S, Cui L, Hu Z, Zhang R, Li J, Zhao Q, He T, et al. 2015. A novel gene, *optrA*, that confers transferable resistance to oxazolidinones and phenicols and its presence in *Enterococcus faecalis* and *Enterococcus faecium* of human and animal origin. *J Antimicrob Chemother* **70:** 2182–2190.

Weisblum B. 1995a. Erythromycin resistance by ribosome modification. *Antimicrob Agents Chemother* **39:** 577–585.

Weisblum B. 1995b. Insights into erythromycin action from studies of its activity as inducer of resistance. *Antimicrob Agents Chemother* **39:** 797–805.

Weiß S, Kadlec K, Feßler AT, Schwarz S. 2014. Complete sequence of a multiresistance plasmid from a methicillin-resistant *Staphylococcus epidermidis* ST5 isolated in a small animal clinic. *J Antimicrob Chemother* **69:** 847–859.

Wendlandt S, Li B, Ma Z, Schwarz S. 2013a. Complete sequence of the multi-resistance plasmid pV7037 from a porcine methicillin-resistant *Staphylococcus aureus*. *Vet Microbiol* **166:** 650–654.

Wendlandt S, Lozano C, Kadlec K, Gómez-Sanz E, Zarazaga M, Torres C, Schwarz S. 2013b. The enterococcal ABC transporter gene *lsa*(E) confers combined resistance to lincosamides, pleuromutilins and streptogramin A antibiotics in methicillin-susceptible and methicillin-resistant *Staphylococcus aureus*. *J Antimicrob Chemother* **68:** 473–475.

Wendlandt S, Li J, Ho J, Porta MA, Feßler AT, Wang Y, Kadlec K, Monecke S, Ehricht R, Boost M, et al. 2014. Enterococcal multiresistance gene cluster in methicillin-resistant *Staphylococcus aureus* from various origins and geographical locations. *J Antimicrob Chemother* **69:** 2573–2575.

Wendlandt S, Kadlec K, Feßler AT, Schwarz S. 2015a. Identification of ABC transporter genes conferring combined pleuromutilin–lincosamide–streptogramin A resistance in bovine methicillin-resistant *Staphylococcus aureus* and

coagulase-negative staphylococci. *Vet Microbiol* **177:** 353–358.

Wendlandt S, Shen J, Kadlec K, Wang Y, Li B, Zhang WJ, Feßler AT, Wu C, Schwarz S. 2015b. Multidrug resistance genes in staphylococci from animals that confer resistance to critically and highly important antimicrobial agents in human medicine. *Trends Microbiol* **23:** 44–54.

Werner G, Witte W. 1999. Characterization of a new enterococcal gene, *satG*, encoding a putative acetyltransferase conferring resistance to streptogramin A compounds. *Antimicrob Agents Chemother* **43:** 1813–1814.

Wilson DN. 2009. The A–Z of bacterial translation inhibitors. *Crit Rev Biochem Mol Biol* **44:** 393–433.

Wilson DN. 2011. On the specificity of antibiotics targeting the large ribosomal subunit. *Ann NY Acad Sci* **1241:** 1–16.

Wilson DN. 2014. Ribosome-targeting antibiotics and mechanisms of bacterial resistance. *Nat Rev Microbiol* **12:** 35–48.

Wright GW. 2007. The antibiotic resistome: The nexus of chemical and genetic diversity. *Nature Rev Microbiol* **5:** 175–186.

Zhang WJ, Wu CM, Wang Y, Shen ZQ, Dai L, Han J, Foley SL, Shen JZ, Zhang Q. 2011. The new genetic environment of *cfr* on plasmid pBS-02 in a *Bacillus* strain. *J Antimicrob Chemother* **66:** 1174–1175.

Zhang WJ, Xu XR, Schwarz S, Wang XM, Dai L, Zheng HJ, Liu S. 2014. Characterization of the IncA/C plasmid pSCEC2 from *Escherichia coli* of swine origin that harbours the multiresistance gene *cfr*. *J Antimicrob Chemother* **69:** 385–389.

Zhang A, Xu C, Wang H, Lei C, Liu B, Guan Z, Yang C, Yang Y, Peng L. 2015a. Presence and new genetic environment of pleuromutilin–lincosamide–streptogramin A resistance gene *lsa*(E) in *Erysipelothrix rhusiopathiae* of swine origin. *Vet Microbiol* **177:** 162–167.

Zhang WJ, Wang XM, Dai L, Hua X, Dong Z, Schwarz S, Liu S. 2015b. Novel conjugative plasmid from *Escherichia coli* of swine origin that coharbors the multiresistance gene *cfr* and the extended-spectrum-β-lactamase gene *bla*CTX-M-14b. *Antimicrob Agents Chemother* **59:** 1337–1340.

Zhao Q, Wendlandt S, Li H, Li J, Wu C, Shen J, Schwarz S, Wang Y. 2014. Identification of the novel lincosamide resistance gene *lnu*(E) truncated by IS*Enfa5-cfr-ISEnfa5* insertion in *Streptococcus suis*: De novo synthesis and confirmation of functional activity in *Staphylococcus aureus*. *Antimicrob Agents Chemother* **58:** 1785–1788.

Pleuromutilins: Potent Drugs for Resistant Bugs—Mode of Action and Resistance

Susanne Paukner and Rosemarie Riedl

Nabriva Therapeutics AG, A-1110 Vienna, Austria

Correspondence: susanne.paukner@nabriva.com; rosemarie.riedl@nabriva.com

Pleuromutilins are antibiotics that selectively inhibit bacterial translation and are semisynthetic derivatives of the naturally occurring tricyclic diterpenoid pleuromutilin, which received its name from the pleuromutilin-producing fungus *Pleurotus mutilus*. Tiamulin and valnemulin are two established derivatives in veterinary medicine for oral and intramuscular administration. As these early pleuromutilin drugs were developed at a time when companies focused on major antibacterial classes, such as the β-lactams, and resistance was not regarded as an issue, interest in antibiotic research including pleuromutilins was limited. Over the last decade or so, there has been a resurgence in interest to develop this class for human use. This has resulted in a topical derivative, retapamulin, and additional derivatives in clinical development. The most advanced compound is lefamulin, which is in late-stage development for the intravenous and oral treatment of community-acquired bacterial pneumonia and acute bacterial skin infections. Overall, pleuromutilins and, in particular, lefamulin are characterized by potent activity against Gram-positive and fastidious Gram-negative pathogens as well as against mycoplasmas and intracellular organisms, such as *Chlamydia* spp. and *Legionella pneumophila*. Pleuromutilins are unaffected by resistance to other major antibiotic classes, such as macrolides, fluoroquinolones, tetracyclines, β-lactam antibiotics, and others. Furthermore, pleuromutilins display very low spontaneous mutation frequencies and slow, stepwise resistance development at sub-MIC in vitro. The potential for resistance development in clinic is predicted to be slow as confirmed by extremely low resistance rates to this class despite the use of pleuromutilins in veterinary medicine for >30 years. Although rare, resistant strains have been identified in human- and livestock-associated environments and as with any antibiotic class, require close monitoring as well as prudent use in veterinary medicine. This review focuses on the structural characteristics, mode of action, antibacterial activity, and resistance development of this potent and novel antibacterial class for systemic use in humans.

Pleuromutilins are a well-known class of antibiotics discovered in the 1950s by the isolation of the naturally occurring pleuromutilin from *Pleurotus mutilus* (now renamed *Clitophilus scyphoides*), an edible mushroom (Fig. 1)

(Kavanagh et al. 1951). Semisynthetic derivatizations have led to tiamulin and valnemulin, which were introduced to veterinary medicine in 1979 and 1999, respectively, for the treatment of pulmonary and intestinal infections caused

Pleurotus mutilus
(Clitopilus scyphoides)

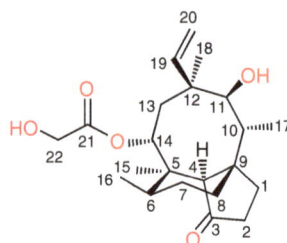

Figure 1. Structure of pleuromutilin with numbering system (Arigoni 1968). (*Left* panel from Lindsey 2006.)

by *Mycoplasma* spp., *Brachyspira* spp., and *Lawsonia intracellularis* in pigs, poultry, and, to some extent, in rabbits. Despite the use of pleuromutilins for treatment in veterinary medicine for more than three decades, resistance development has been uncommon. This can likely be attributed to several factors including the unique and highly specific mode of action of the pleuromutilins. Further, this class has not been used for enhancement of food-producing animal production (e.g., as growth promoters or for enhancement of feed efficiency) unlike the tetracyclines, penicillins, or sulfonamides (EMA 2014a,b). Even though oral valnemulin has been reported to be efficacious in the treatment of persistent or life-threatening mycoplasma infection in humans (Heilmann et al. 2001), no pleuromutilin had received marketing authorization by the end of the last century.

In the new millennium, interest in the pleuromutilin class significantly increased as evidenced by the development of new derivatives for human use. Retapamulin, a topical agent, was the first to be approved for the treatment of impetigo and infected small lacerations, abrasion or sutured wounds caused by *Staphylococcus aureus* and *Streptococcus pyogenes* (FDA 2007; EMA 2008). More recently, lefamulin, the first pleuromutilin for intravenous and oral administration, has entered into late-stage clinical development for the treatment of community-acquired bacterial pneumonia (CABP) and acute bacterial skin and skin structure infections (ABSSSIs). BC-7013 and BC-3205, which have a similar antibacterial profile but differ in ADMET properties, are in early-stage clinical

development for topical and oral administration, respectively. In addition, recent research has been directed at further extending the antibacterial spectrum to include the ESKAPE pathogens (Boucher et al. 2009; Paukner et al. 2014a,b, 2015a,b,c; Strickmann et al. 2014; Wicha and Ivezic-Schoenfeld 2014; Wicha et al. 2015b).

To the question "Are pleuromutilins finally fit for human use?" (Novak 2011), which has been raised because of unjustified anecdotal concerns regarding metabolic stability, gastrointestinal side effects, cardiac safety, or intravenous tolerability, a clear response can be given: Yes. In a phase 2 study, lefamulin was well-tolerated and showed comparable efficacy to IV vancomycin in patients with ABSSSI (Prince et al. 2013). Despite challenging medicinal chemistry, a number of compounds are in the pipeline and further developments in this antibiotic class are anticipated.

PLEUROMUTILINS—MODE OF ACTION, ACTIVITY, AND RESISTANCE

Structure

The diterpenoid pleuromutilin comprises a tricyclic scaffold with unique annelation of a five-, six-, and eight-membered ring and eight stable chiral centers, as well as a glycolic ester moiety forming the side chain also regarded as an extension at position C14 (Fig. 1) (Anchel 1952; Arigoni 1962, 1968; Birch et al. 1963, 1966). Remarkable efforts have been made to achieve chemical modifications at several positions of

Cite this article as *Cold Spring Harb Perspect Med* doi: 10.1101/cshperspect.a027110

the tricyclic core (Naegeli 1961; Berner et al. 1980, 1981, 1983, 1987; Brooks and Hunt 2000; Bacqué et al. 2002; Springer et al. 2003, 2008; Takadoi et al. 2007; Wang et al. 2009; Paukner et al. 2015b), as well as biotransformation (Hanson et al. 2002), and the total synthesis of this unique scaffold (Gibbons 1982; Paquette and Bullman-Page 1985; Paquette and Wiedemann 1985; Bacqué et al. 2003; Liu et al. 2011; Lotesta et al. 2011; Ruscoe et al. 2015). Most modifications, however, are primarily performed at the glycolic side chain of pleuromutilin with replacement of the terminal hydroxyl group or the entire side chain resulting in semisynthetic pleuromutilin analogs of two main types: (1) the flexible sulfanylacetyl, and (2) the rigid acylcarbamate linker type. Despite significant efforts in the field of acylcarbamate pleuromutilins by the GlaxoSmithKline group (Hunt 2000; Brooks et al. 2001; Andemichael et al. 2009), only sulfanylacetyl derivatives, mostly with one basic function at the side chain, have progressed beyond phase 1 clinical studies. The early work of the Sandoz group resulted in lipophilic orally available veterinary products, tiamulin (Egger and Reinshagen 1978) and valnemulin (Fig. 2A) (Berner and Vyplel 1987), whereas work from GlaxoSmithKline and Sandoz/Nabriva led to the lipophilic topical products retapamulin (Berry et al. 1999) and BC-7013 (Fig. 2B) (Thirring et al. 2007). Extensive modification of the C14 side chain culminated in lefamulin (Fig. 2C) (Mang et al. 2008), the first pleuromutilin with optimized physicochemical characteristics, including exceptional solubility, potent antimicrobial activity, and excellent ADMET properties including metabolic stability enabling administration by both the intravenous and oral routes.

Mode of Action

Pleuromutilins inhibit bacterial protein synthesis by binding to the central part of domain V of the 50S ribosomal subunit at the peptidyl transferase center (PTC), which prevents the correct positioning of the CCA ends of tRNAs for peptide transfer in the A- and P-site, thereby inhibiting peptide bond formation (Hogenauer 1975; Hogenauer and Ruf 1981; Hogenauer et al. 1981; Poulsen et al. 2001; Schlunzen et al. 2004; Long et al. 2006; Davidovich et al. 2007). Figure 3 shows the positioning of lefamulin in the PTC of the bacterial ribosome in relation to the positions of A- and P-site tRNA. Positioning is similar for various pleuromutilin derivatives in that the tricyclic core is located in a pocket close to the A-site, whereas the C14 side chain extends toward the P-site hindering the 3'-end tRNA A- to P-site rotary motion, as shown by various footprinting and crystallographic studies for tiamulin, valnemulin (Poulsen et al. 2001; Schlunzen et al. 2004; Long et al. 2006; Davidovich et al. 2007), retapamulin (Yan et al. 2006), lefamulin (Nabriva, unpubl.), and BC-3205 (Eyal et al. 2015). Crystallography data using ribosomal preparations from *Deinococcus radiodurans* and *S. aureus* show that the tricyclic

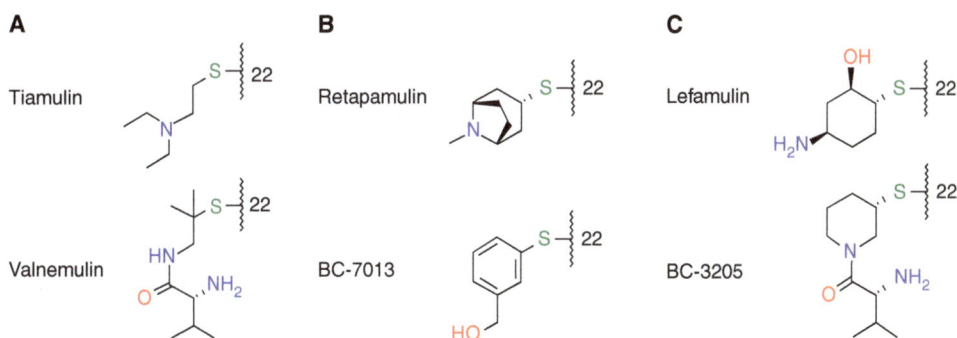

Figure 2. Structures of side chains at C22. Side chains of various pleuromutilin derivatives: (*A*) veterinary, (*B*) topical human, and (*C*) systemical human.

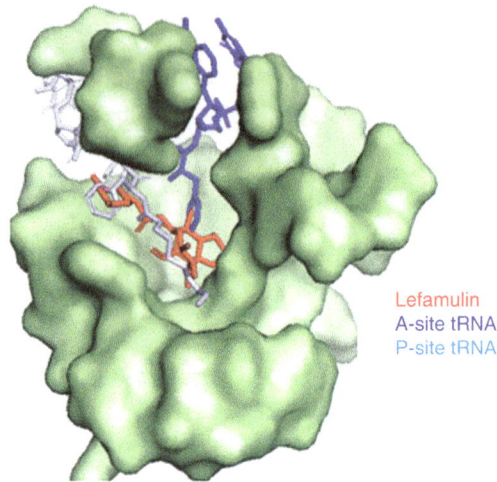

Lefamulin
A-site tRNA
P-site tRNA

Figure 3. Lefamulin positioning in the peptidyl transferase center (PTC). PTC of the bacterial ribosome in relation to the positions of A- and P-site tRNA. Red, Lefamulin; blue, A-site tRNA; teal, P-site tRNA.

pleuromutilin core interacts with the ribosomal nucleotides mainly through hydrophobic interactions, van der Waal forces, and hydrogen bonds with nucleotides of domain V of 23S rRNA, namely, A2503, U2504, G2505, U2506, C2452, and U2585 (Fig. 4). For the tricyclic pleuromutilin core, specific hydrogen bonds have been reported for the hydroxyl group at C11 with nucleotides G2505 or A2503 (Davidovich et al. 2007) and for the hydroxyl group at C2 (present in only few selected derivatives) with G2505. Further, hydrogen bonds have been reported for the C21 carboxyl group and the sulfur in sulfanylacetyl or acyl in acylcarbamate of the C14 extension (linker) with the nucleotide G2061, which are similar for both linker types (Schlunzen et al. 2004; Davidovich et al. 2007). In previous studies using *D. radiodurans* ribosomes, it was concluded that the rest of the C14 extension is involved only in minor hydrophobic interactions (Davidovich et al. 2007). Recent studies using *S. aureus* ribosomes, however, clearly showed additional hydrogen bonds of the C14 extensions, specifically the amino groups of BC-3205 and lefamulin, with the nucleotides U2506 (Eyal et al. 2015) and A2062 (A Yonath, Z Eyal, E Zimmerman, et al., unpubl.), respectively. Most notably, the C14 extensions of all pleuromutilins sterically interfere

with the highly flexible nucleotides U2585 and U2506 causing rotational movements of these nucleotides, which consequently interact with each other by the formation of one or more additional hydrogen bonds or at least van der Waal or similar interactions. This closing of the binding pocket around pleuromutilins, also regarded as the induced-fit mechanism, tightens the binding of pleuromutilins to the ribosome (Davidovich et al. 2007; Eyal et al. 2015) and leads to the protection of these nucleotides in footprinting experiments (Schlunzen et al. 2004). Interestingly, the amino group of the C14 extension of BC-3205 forms a hydrogen bond with U2506 in *S. aureus* causing a larger shift of U2506, consequently further stabilizing BC-3205 in the binding pocket and indicating a better fit of this molecule in the pocket (Eyal et al. 2015).

It has been further hypothesized that pleuromutilins might also interfere with translation initiation or at an early point of the elongation cycle with particular sensitivity of the first peptide bond formation (Hunt 2000; Novak 2011). This is based on the fact that (1) radiolabeled tiamulin did not bind to the 50S subunit of the ribosome once elongation has begun, (2) addition of tiamulin to intact cells led to the formation of defective initiation complexes reflected

Cite this article as *Cold Spring Harb Perspect Med* doi: 10.1101/cshperspect.a027110

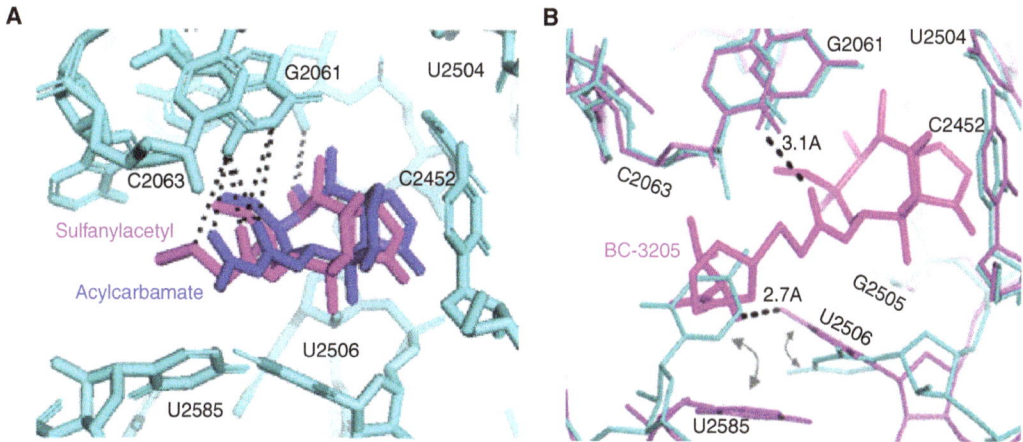

Figure 4. Interaction network. (*A*) The tricyclic pleuromutilin core and sulfanylacetyl as well as acylcarbamate linker (side chains omitted for clarity), and (*B*) the C14 side chain extension with nucleotides of the peptidyl transferase center (PTC). Hydrogen bonds are shown as dotted lines (Schlunzen et al. 2004; Davidovich et al. 2007; Eyal et al. 2015).

by the depletion of the polysome pool through blockage of reinitiation (Dornhelm and Hogenauer 1978), and (3) retapamulin partially inhibited binding of fMet-tRNA during initiation complex formation (Yan et al. 2006).

Coupled in vitro transcription/translation (TT) assays with bacterial ribosomes have shown high specificity of pleuromutilins for the inhibition of bacterial protein translation, whereas no effect on eukaryotic nonorganelle protein synthesis was observed for tiamulin and retapamulin (Yan et al. 2006). The specificity for bacterial protein synthesis was also confirmed for lefamulin with IC_{50} values of 0.51 µM and 0.31 µM in *Escherichia coli* and *S. aureus* TT-assays, respectively, whereas the IC_{50} in the eukaryotic TT-assay was with 952 µM >2000-fold higher than the IC_{50} in the bacterial system. Cycloheximide, a TT inhibitor of eukaryotic protein synthesis, and puromycin, a nonspecific inhibitor of bacterial and eukaryotic TT, were used as controls. Comparison of IC_{50} values of various pleuromutilin derivatives showed a slight trend of higher IC_{50} for *E. coli* than for *S. aureus* (Table 1; Nabriva, unpubl.). It further showed that the IC_{50} does not necessarily correlate with high antibacterial activity; for example, BC-7013, which has the highest IC_{50}, was one of the most active com-

pounds in vitro (Biedenbach et al. 2009). Additional important factors other than translation inhibition, such as intracellular concentration, uptake, or efflux, contribute to in vitro antimicrobial activity as well (Paukner et al. 2014b).

Antibacterial Activity

The antibacterial spectrum of the pleuromutilins is characterized by potent activity against Gram-positive organisms including staphylococcal species (e.g., community-acquired methicillin-resistant *S. aureus* [CA-MRSA], hospital-acquired MRSA [HA-MRSA], vancomycin-resistant *S. aureus* [VRSA], vancomycin-intermediate *S. aureus* [VISA], heteroresistant VISA [hVISA]), streptococcal species (e.g., penicillin-resistant *Streptococcus pneumoniae* [PRSP], multidrug-resistant *S. pneumoniae* [MDR-SP]), and *Enterococcus faecium* (particularly vancomycin-resistant strains [VRE]), as well as activity against fastidious Gram-negatives, including *Haemophilus* spp., *Moraxella catarrhalis*, *Neisseria* spp., and *Legionella pneumophila* (Rittenhouse et al. 2006; Sader et al. 2012a,b; Paukner et al. 2013c). Pleuromutilins also display potent activity against mycoplasmas, ureaplasmas, chlamydia (Hannan et al. 1997), and *Brachispira hyodysenteriae* (Karlsson

Table 1. Inhibition of bacterial and eukaryotic in vitro transcription–translation by various pleuromutilin derivatives

| Compound | IC$_{50}$ (CI95) (μM) | | |
	Escherichia coli	Staphylococcus aureus	Eukaryotic (reticulocyte lysate system)
Pleuromutilin	0.76 (0.63–0.92)	1.73 (1.22–2.44)	ND
Tiamulin	0.50 (0.44–0.57)	0.36 (0.32–0.42)	ND
Valnemulin	0.59 (0.54–0.66)	0.38 (0.35-0.41)	ND
Retapamulin	0.69 (0.64–0.76)	0.35 (0.32–0.39)	850 (562–1287)
Lefamulin	0.51 (0.45–0.57)	0.31 (0.29–0.33)	952 (732–1238)
BC-3205	0.62 (0.56–0.68)	0.49 (0.44–0.54)	>100
BC-7013	0.74 (0.65–0.83)	0.64 (0.59–0.69)	>100
Cycloheximide	>100	>100	0.44 (0.29–0.68)
Puromycin	0.39 (0.34–0.46)	0.19 (0.16–0.23)	0.31 (0.27–0.36)

Unpublished data (Nabriva).
ND, Not determined.

et al. 2001). Activity against anaerobic organisms has been seen for retapamulin, lefamulin, and BC-7013, including *Propionibacterium acnes* (Goldstein et al. 2006), *Peptostreptococcus* spp., *Prevotella* spp., *Porphyromonas* spp., *Fusobacterium* spp., and *Clostridium perfringens*, whereas pleuromutilins generally show weak activity against strains from the *Bacteroides fragilis* group (Odou et al. 2007; Paukner et al. 2013a). Pleuromutilin activity against *Clostridium difficile* varies and is dependent on the C14 side chain; retapamulin and lefamulin possess no relevant to weak activity, whereas BC-7013 has potent activity with 78% of isolates inhibited at concentrations of ≤1 μg/mL (Nabriva, unpubl.). The lack of lefamulin activity against *B. fragilis* group and Enterobacteriaceae is anticipated to result in limited disruption to the normal gastrointestinal microbiome and potentially a lower propensity to be associated with *C. difficile* infection (CDI). Further studies are warranted to assess the effect of the pleuromutilins on the gut microbiome and its impact on CDI.

No relevant activity was observed against *Enterococcus faecalis*, Enterobacteriaceae, and nonfermenting Gram-negatives, such as *Acinetobacter baumannii* and *Pseudomonas aeruginosa*, although coupled in vitro transcription–translation assay results showed inhibition of the bacterial translation in these organisms and would suggest also antibacterial in vitro

activity (Nabriva, unpubl.). Research at Nabriva revealed that the intrinsic resistance of Enterobacteriaceae is caused by the efflux of pleuromutilins mediated by the AcrAB-TolC efflux pump. This is supported by the fact that AcrAB-TolC deficient *E. coli* strains were susceptible to pleuromutilins (Paukner et al. 2014b) and that the minimum inhibitory concentration (MIC) values against Enterobacteriaceae were significantly reduced by the addition of efflux pump inhibitors (e.g., PAβN). Furthermore, recent new pleuromutilin derivatives partially overcome efflux and show increased activity against Enterobacteriaceae, including carbapenem-resistant isolates. These so-called "extended spectrum pleuromutilins" (ESPs) are characterized by the modification of the tricyclic pleuromutilin core at C12 (Paukner et al. 2014a,b, 2015a,b,c; Strickmann et al. 2014; Wicha and Ivezic-Schoenfeld 2014; Wicha et al. 2015b). Further investigations are needed to identify the mechanism(s) responsible for decreased susceptibility in nonfermenting organisms and *E. faecalis*.

Given that lefamulin is the first in-class IV and oral pleuromutilin antibiotic to advance into late-stage clinical development, additional information is provided on its activity. The antibacterial spectrum of lefamulin against a recent strain collection is well matched to the profile required for the empiric treatment

Cite this article as *Cold Spring Harb Perspect Med* doi: 10.1101/cshperspect.a027110

of patients with CABP and ABSSSI (Table 2). Notably, lefamulin was equipotent against *S. aureus* strains that were community-acquired or health-care-associated and its activity was not negatively influenced by the presence of Panton-Valentine leukocidin (PVL). Compared with macrolides, lincosamides, fluoroquinoloes, tetracyclines, β-lactams, linezolid, and vancomycin, lefamulin was among the most active compounds in vitro and the lefamulin activity was unaffected by resistance or multidrug resistance to these antibiotic classes (Paukner et al. 2013c). Lefamulin's activity is not adversely affected by presence of serum ($\leq$95% v/v) or lung surfactant ($\leq$1 mg/mL Survanta; 4% v/v). Lefamulin has shown high intracellular concentration in macrophages (Paukner et al. 2013b), achieves excellent penetration into human tissues including epithelial lining fluid of the lung (Zeitlinger et al. 2016), has oral bioavailability, potent in vivo efficacy in skin and pulmonary infection mouse models (Wicha et al. 2010, 2013, 2015a), and possesses a low potential for resistance development. Moreover, lefamulin has shown efficacy in patients with ABSSSI infection caused primarily by MRSA (including PVL-producing CA-MRSA) comparable to that of vancomycin (Prince et al. 2013). To date, lefamulin has been well tolerated in phase 1 and 2 clinical studies involving exposure of more than 400 subjects. Lefamulin also possesses potent activity (MIC_{90} values $\leq$2 μg/mL) against organisms causing sexually transmitted infections (STIs), including resistant *Neisseria gonorrhoeae*, *Mycoplasma genitalium*, *Chlamydia trachomatis*, or *Haemophilus ducreyi*, warranting further evaluation of this drug for treatment of STIs (Paukner et al. 2013a).

Resistance and Cross-Resistance

The unique mode of action of pleuromutilins and the binding to highly conserved ribosomal targets implies a low probability of resistance development and lack of cross-resistance with other antibiotic classes including protein synthesis inhibitors, such as macrolides, ketolides, or fusidic acid (Yan et al. 2006). The binding sites and mode of action of pleuromutilins can be clearly differentiated from those of oxazolidinones, lincosamides, phenicols, and streptogramins; however, pleuromutilins also have partly overlapping interaction sites with these antibacterials (Schlunzen et al. 2004). Consequently, resistance mechanisms exist that can mediate cross-resistance with these antibacterials, albeit with an exceedingly low incidence.

Pleuromutilins have shown a low potential for resistance development in vitro as shown in various studies for tiamulin and valnemulin in *Brachyspira* spp. (Karlsson et al. 2001; Pringle et al. 2004), *Mycoplasma* spp. (Long et al. 2009; Li et al. 2011), *S. aureus* and *E. coli* (Miller et al. 2008), for retapamulin in *S. aureus* and *S. pyogenes* (Kosowska-Shick et al. 2006; Gentry et al. 2007), and for lefamulin in *S. aureus*, *S. pneumoniae*, and *S. pyogenes* (Paukner et al. 2012). Generally, the spontaneous mutation frequencies are low ($\leq$10^{-9}) with no stable resistant mutants selected at four- to eightfold MIC. In multipassage experiments, resistance developed in a slow and step-wise manner with multiple mutations required to cause high-level resistance. Mutations in 23S rRNA, *rplC*, and *rplD* genes encoding the large ribosomal proteins L3 and L4, have been identified as the primary resistance mechanism in vitro. In clinical isolates, two additional resistance mechanisms have been identified: the acquisition of *vga*(A) encoded or related ATP-binding cassette (ABC)-F transporters and the acquisition of *cfr* encoding the Cfr methyltransferase. The common denominator is the alteration of the pleuromutilin target site.

Mutations in the 23S RNA gene (*rrn*) at positions G2032A, C2055A, A2058, A2058G, A2059G, G2061U, G2447A/U, C2499A, A2503U, U2504A/G, and A2572U were primarily observed in laboratory-selected *Brachyspira* spp., *Mycoplasma* spp., and in clinical isolates. Mutations in 23S rRNA have been described to confer resistance only in *Mycoplasma* spp. and *Brachyspira* spp., which only have a single copy of 23S rRNA, whereas staphylococcal and streptococcal species have multiple copies (Pringle et al. 2004; Miller et al. 2008; Long et al. 2009; Li et al. 2010; Hillen et al. 2014). Experiments with single-copy *rrn* knockout strains of *E. coli*

Table 2. Antibacterial activity of lefamulin and comparators

Species	n	\(MIC_{90}\) (µg/mL)									
		LMU	ERY	CLI	DOX	TGC	LZD	VAN	DAP	LEV	PEN
Organisms causing predominantly SSSI and bacteremia											
Staphylococcus aureus	5527	0.12	>4	>2	0.25	0.25	1	1	0.5	>4	–
MSSA	3157	0.12	>4	≤0.25	0.25	0.25	2	1	0.5	1	–
MRSA	2370	0.25	>4	>2	1	0.25	1	1	0.5	>4	–
CoNS	878	0.12	>4	>2	2	0.25	1	2	0.5	>4	–
Enterococcus faecium	536	4	>4	–	>8	0.25	1	>16	2	>4	–
Vancomycin-nonsusceptible	304	0.25	>4	–	>8	0.25	1	>16	2	>4	–
β-Haemolytic Streptococcus spp.	763	0.03	>4	>2	8	0.06	1	0.5	0.25	1	0.06
Viridans group Streptococcus spp.	245	0.5	>4	≤0.25	>8	0.06	1	0.5	0.5	2	0.5

Species	n	LMU	ERY	AZI	DOX	TGC	LEV	SXT	CXM	IMI	PEN
Organisms causing predominantly RTI											
Streptococcus pneumoniae[a]	1473	0.25	>4	>4	8	0.06	1	>4	8	0.5	4
Haemophilus influenzae[b]	360	2	8	2	0.5	0.25	–	>4	2	1	–
Moraxella catarrhalis	253	0.25	0.25	≤0.25	0.25	0.25	–	≤0.5	2	≤0.12	>4
Legionella pneumophila[c]	30	0.5	0.25	0.12	–	–	0.12	–	–	–	–
Mycoplasma pneumoniae	50 (4)[d]	0.006	0.0025–0.005[f]	≤0.0003[f]	0.04–0.04[f]	–	–	–	–	–	–
Chlamydia pneumoniae	50 (2)[e]	0.04	0.04–0.16[f]	0.08–0.16[f]	0.04–0.08[f]	–	–	–	–	–	–

Data adapted from Paukner et al. (2013c) and Sader (2012b).

MRSA, Methicillin-resistant S. aureus; MSSA, methicillin-susceptible S. aureus; CoNS, coagulase-negative Staphylococcus spp.; AZI, azithromycin; CLI, clindamycin; CXM, cefuroxime; DAP, daptomycin; DOX, doxycycline; ERY, erythromycin; IMI, imipenem; LEV, levofloxacin; LMU, lefamulin; LZD, linezolid; PEN, penicillin; SXT, trimethoprim-sulfamethoxazole; TGC, tigecycline; VAN, vancomycin.

[a]61.3% of S. pneumoniae isolates were penicillin-susceptible.
[b]23.6% of H. influenzae isolates were β-lactamase-positive.
[c]MICs determined by agar dilution using charcoal-supplemented BCYEα medium.
[d]LMU was tested against n = 50 isolates, whereas comparators were tested against n = 4 isolates.
[e]LMU was tested against n = 50 isolates, whereas comparators were tested against n = 2 isolates.
[f]Range of MICs for isolates tested.

Cite this article as Cold Spring Harb Perspect Med doi: 10.1101/cshperspect.a027110

illustrated that "the copy number of 23S rRNA is the limiting factor in the selection of 23S rRNA mutants" (Miller et al. 2008). In *Brachyspira*, resistance is often associated with additional mutations in *rplC* (Pringle et al. 2004; Long et al. 2009; Li et al. 2010, 2011; Hidalgo et al. 2011).

Mutations and deletions in the *rplC* and *rplD* genes, although L3 and L4 do not directly interact with the pleuromutilins, can cause conformational changes in the PTC and hinder correct positioning of the pleuromutilins in the pocket formed between the nucleotides G2576 with U2506 and G2505 (Eyal et al. 2015). Mutations in *rplC* and *rplD* have been described for *Staphylococcus* spp. (Pringle et al. 2004; Kosowska-Shick et al. 2006; Gentry et al. 2007; Miller et al. 2008; Paukner et al. 2012) and together with mutations in 23S rRNA for *B. hyodysenteriae* (Hillen et al. 2014). Notably, mutations in *rplC* have also been associated with considerable loss of fitness (Gentry et al. 2007). Pleuromutilin resistance by mutational changes in *rplC* and 23S rRNA develops gradually and in a stepwise manner both in vitro and in vivo, suggesting that multiple mutations are needed to achieve high-level resistance. (Karlsson et al. 2001; Gentry et al. 2007; Miller et al. 2008; Hidalgo et al. 2011; Paukner et al. 2012).

ABC-F transporters encoded by *vga*(A) and its variants *vga*(A)v, *vga*(A)LC, *vga*(B), *vga*(C), *vga*(D), *vga*(E), and *lsa*(E) have been described to confer resistance to pleuromutilins, streptogramin A, and lincosamides in *Staphylococcus* spp., *E. faecium*, and *Erysypelothrix rhusiopathiae*. Isolates were collected almost exclusively from animal species, predominantly swine (Kadlec and Schwarz 2009; Kadlec et al. 2010; Overesch et al. 2011; Schwendener and Perreten 2011; Li et al. 2013, 2014a,b; Zhang et al. 2015). MRSA isolates collected from humans appeared to be related to animal-associated lineages of *S. aureus* such as ST398 (Lozano et al. 2012). Recent studies concluded that ABC-F transporters, which lack a transmembrane domain, likely mediate resistance by the interference of translation at the PTC and by the action as efflux transporter. This is based on the homology of *vga*(A) and variants with the ABC-F

transporter EttA, which is a translation factor binding to the tRNA exit site (E-site) (Lenart et al. 2015).

Last, the rarely encountered methyltransferase Cfr, methylating the nucleotide A2503 of 23S rRNA, can confer resistance. Because of steric hindrance, binding of phenicols, lincosamides, oxazolidinones, pleuromutilins, and streptogramins (PhLOPS antibiotics) is prohibited, which results in the PhLOPS-resistance phenotype. The *cfr* gene was originally identified in coagulase-negative staphylococci from animals and has been detected mostly in livestock-associated staphylococci (Kehrenberg et al. 2005, 2009; Alba et al. 2015; Feltrin et al. 2015; Moon et al. 2015) but, more recently, has also been found in a limited number of staphylococcal isolates from humans including one outbreak of a *cfr*-positive MRSA in a Spanish hospital, which was terminated by reduction of linezolid use and infection-control measures (Sanchez et al. 2010; Shore et al. 2010, 2016). Cfr was also found in nonstaphylococcal species collected, with the exception of *E. faecalis*, exclusively from livestock animals and related farm environments: one out of 1230 *E. coli* isolates collected from pigs, ducks, and chickens in China, in one *Proteus vulgaris* out of 557 nasal swabs of Chinese swine and a porcine *Bacillus* spp., as well as a *Macrococcus caseolyticus* and *Jeotgalicoccus pinnipedialis* isolate (Wang et al. 2011, 2012a,b,c,d). Cfr has also been detected in an *E. faecalis* isolate collected from a Chinese animal as well as in an animal-associated isolate from a patient in Thailand. Cfr has been located on the chromosome and on various plasmids or transferrable elements indicating the ability to spread (Locke et al. 2012; Shen et al. 2013; Li et al. 2015; Shore et al. 2016). In vitro, the *cfr*-carrying plasmid isolated from human *E. faecalis* was only transferrable by conjugation to another *E. faecalis* laboratory strain, whereas it was not transferrable to *S. aureus* or *E. faecium* (Diaz et al. 2012). Recently, the transferability of *cfr*-carrying plasmids from *S. epidermidis* to MRSA by conjugation or transduction was shown, indicating a role of *S. epidermidis* as a potential reservoir for *cfr* spread (Cafini et al. 2016).

Table 3. Cfr-positive isolates collected from human in the course of global surveillance studies

Investigation period	Number of resistant *Staphylococcus aureus* isolates per total screened isolates (%)	Number of resistant CoNS isolates per total screened isolates (%)	Country, surveillance program	References
2015	0/2434 (0%)	1/465 (0.22%)	Europe, ZAAPS	Flamm et al. 2016
2014	0/3560 (0%)	0/956 (0%)	Worldwide, ZAAPS	Mendes et al. 2016
2012	1/4077 (0.02%)	3/905 (0.33%)	Worldwide, ZAAPS	Mendes et al. 2014
2011	0/3884 (0%)	3/928 (0.32%)	Worldwide, ZAAPS	Flamm et al. 2013
2010	1/5527 (0.018%)	2/823 (0.24%)	Worldwide, lefamulin, SENTRY	Paukner et al. 2013c
2010	0/2875 (0%)	2/855 (0.23%)	Worldwide, ZAAPS	Flamm et al. 2012
2002–2009	0/5952 (0%)	2/2132 (0.09%)	Worldwide, ZAAPS	Ross et al. 2011
2007	0/3000 (0%)	0/716	Worldwide, ZAAPS	Jones et al. 2009
1999–2010	1/2215 (0.045%)	ND	Spain	Sierra et al. 2013

Note: Annual Appraisal of Potency and Spectrum (ZAAPS) Program and SENTRY Surveillance Program.
CoNS, Coagulase-negative *Staphylococcus* spp.; ND, not determined.

Most importantly, it should be noted that despite the characterization of isolates resistant to pleuromutilins and the descriptions of mechanisms conferring resistance, the rate of resistance to pleuromutilins remains low. In the SENTRY surveillance program conducted with lefamulin in 2010, the total incidence of pleuromutilin resistance was 0.18% for *S. aureus* and 3.4% for coagulase-negative *Staphylococcus* spp. Among the *S. aureus* isolates, 0.018% harbored the *cfr* gene, 0.11% harbored the *vga*(A) gene, and 0.05% had mutations in *rpl*C. Among coagulase-negative *Staphylococcus* spp., the incidences for *cfr*, *vga*(A), and *rpl*D alterations were 0.11%, 2.5%, and 0.34%, respectively (Paukner et al. 2013c). The low prevalence of *cfr* is consistent with data collected in the course of linezolid-resistance monitoring (Table 3).

lefamulin is the most advanced in clinical development. The incidence of pleuromutilin-resistant bacterial isolates is low despite the use of tiamulin and valnemulin in veterinary medicine for more than 30 years. The availability of topical retapamulin in human medicine since 2007 and selection pressure for *cfr* by the use of linezolid over the past two decades does not appear to have had a major effect on the incidence of pleuromutilin-resistant bacterial isolates among organisms causing infections in humans. Nevertheless, close monitoring of resistance development to pleuromutilins is warranted, along with prudent use of oxazolidinones and veterinary pleuromutilins to maintain low resistance rates and retain the potent activity of this novel antibacterial class against pathogens that have acquired resistance to other established antibiotic classes.

CONCLUDING REMARKS

In summary, pleuromutilins display potent antibacterial activity against a variety of Gram-positive, fastidious Gram-negative and atypical respiratory bacterial pathogens—a profile well suited to treat human infections, including CABP, ABSSSI, and STI. Chemical modifications have led to derivatives with optimized physicochemical properties and improved ADME properties, allowing for intravenous and oral dosing in humans; of these analogs,

ACKNOWLEDGMENTS

We thank the Nabriva team and particularly W. W. Wicha for kind support for compilation of this review.

REFERENCES

Alba P, Feltrin F, Cordaro G, Porrero MC, Kraushaar B, Argudin MA, Nykasenoja S, Monaco M, Stegger M, Aarestrup FM, et al. 2015. Livestock-associated methicillin resistant and methicillin susceptible *Staphylococcus aureus* sequence type (CC)1 in European farmed animals: High genetic relatedness of isolates from Italian cattle herds and humans. *PLoS ONE* **10**: e0137143.

Anchel M. 1952. Chemical studies with pleuromutilin. *J Biol Chem* **199**: 133–139.

Andemichael YW, Chen J, Clawson JS, Dai W, Diederich A, Downing SV, Freyer AJ, Liu P, Oh LM, Patience DB, et al. 2009. Process development for a novel pleuromutilin-derived antibiotic. *Org Process Res Dev* **13**: 729–738.

Arigoni D. 1962. The structure of a new class of terpene. *Gazz Chim Ital* **92**: 884–901.

Arigoni D. 1968. Some studies in the biosynthesis of terpenes and related compounds. *Pure Appl Chem* **17**: 331–348.

Bacqué E, Pautrat F, Zard SZ. 2002. A flexible strategy for the divergent modification of pleuromutilin. *Chem Commun* **20**: 2312–2313.

Bacqué E, Pautrat F, Zard SZ. 2003. A concise synthesis of the tricyclic skeleton of pleuromtuilin and a new approach to cycloheptenes. *Org Lett* **5**: 325–328.

Berner H, Vyplel H. 1987. Pleuromutilin derivatives process for their use and their preparation. U.S. Patent 4,675,330.

Berner H, Schulz G, Schneider H. 1980. Synthesis of an AB-*trans*-annulated derivative of the tricyclic diterpene pleuromutilin via an intramolecular 1,5-hydride shift. *Tetrahedron* **36**: 1807–1811.

Berner H, Schulz G, Schneider H. 1981. Chemistry of pleuromutilins. II: Synthesis of 12-desvinylpleuromutilin. *Tetrahedron* **37**: 915–919.

Berner H, Vyplel H, Schulz G. 1983. Chemistry of pleuromutilins. VII: Base-induced transannular 1,4-hydride shift in 8-substituted pleuromutilin derivatives. *Monatshefte für Chemie* **114**: 501–507.

Berner H, Vyplel H, Schulz G, Stuchlik P. 1984. Chemistry of pleuromutilins. VIII: Functionalization at C-13 by intramolecular nitrene insertion. *Tetrahedron* **40**: 919–923.

Berner H, Vyplel H, Schulz G, Stuchlik P. 1986. Chemistry of pleuromutilins. XI: Inversion of configuration of the vinyl group at carbon 12 by reversible retro-en-cleavage. *Monatshefte für Chemie* **117**: 1073–1080.

Berner H, Vyplel H, Schulz G. 1987. Chemistry of pleuromutilins. XII: A cyclopropyl conjugated system within the tricyclic skeleton of the diterpene pleuromutilin: Formation and synthetic use. *Tetrahedron* **43**: 765–770.

Berry V, Dabbs S, Frydrych CH, Hunt E, Woodnutt G, Sanderson FD. 1999. Pleuromutilin derivatives as antimicrobials. Patent WO199921855.

Biedenbach DJ, Jones RN, Ivezic-Schoenfeld Z, Paukner S, Novak R. 2009. In vitro antibacterial spectrum of BC-7013, a novel pleuromutilin derivative for topical use in humans. *49th Interscience Conference on Antimicrobial Agents and Chemotherapy (ICAAC)*. San Francisco, CA, September 12–15.

Birch AJ, Cameron DW, Holzapfel CW, Richards RW. 1963. The diterpenoid nature of pleuromutilin. *Chem Ind* **5**: 374–375.

Birch AJ, Holzapfel CW, Richards RW. 1966. The structure and some aspects of the biosynthesis of pleuromutilin. *Tetrahedron* **8**: 359–387.

Boucher HW, Talbot GH, Bradley JS, Edwards JE, Gilbert D, Rice LB, Scheld M, Spellberg B, Bartlett J. 2009. Bad bugs, no drugs: No ESKAPE! An update from the Infectious Diseases Society of America. *Clin Infect Dis* **48**: 1–12.

Brooks G, Hunt E. 2000. Mutilin 14-ester derivatives having antibacterial activity. Patent WO2000037074.

Brooks G, Burgess W, Colthurst D, Hinks JD, Hunt E, Pearson MJ, Shea B, Takle AK, Wilson JM, Woodnutt G. 2001. Pleuromutilins. Part 1: The identification of novel mutilin 14-carbamates. *Bioorg Med Chem* **9**: 1221–1231.

Cafini F, Nguyen le TT, Higashide M, Roman F, Prieto J, Morikawa K. 2016. Horizontal gene transmission of the *cfr* gene to MRSA and *Enterococcus*: Role of *Staphylococcus epidermidis* as a reservoir and alternative pathway for the spread of linezolid resistance. *J Antimicrob Chemother* **71**: 587–592.

Davidovich C, Bashan A, Auerbach-Nevo T, Yaggie RD, Gontarek RR, Yonath A. 2007. Induced-fit tightens pleuromutilins binding to ribosomes and remote interactions enable their selectivity. *Proc Natl Acad Sci* **104**: 4291–4296.

Diaz L, Kiratisin P, Mendes RE, Panesso D, Singh KV, Arias CA. 2012. Transferable plasmid-mediated resistance to linezolid due to *cfr* in a human clinical isolate of *Enterococcus faecalis*. *Antimicrob Agents Chemother* **56**: 3917–3922.

Dornhelm P, Hogenauer G. 1978. The effects of tiamulin, a semisynthetic pleuromutilin derivative, on bacterial polypeptide chain initiation. *Eur J Biochem* **91**: 465–473.

Egger H, Reinshagen H. 1978. Pleuromutilin esters. U.S. Patent 4,208,326.

EMA. 2008. EPAR for Altargo (retapamulin). European Medicines Agency, London.

EMA. 2014a. Reflection paper on use of pleuromutilins in food-producing animals in the European Union: Development of resistance and impact on human and animal health. European Medicines Agency, London.

EMA. 2014b. Sales of veterinary antimicrobial agents in 26 EU/EEA countries in 2012. European Medicines Agency, London.

Eyal Z, Matzov D, Krupkin M, Wekselman I, Paukner S, Zimmerman E, Rozenberg H, Bashan A, Yonath A. 2015. Structural insights into species-specific features of the ribosome from the pathogen *Staphylococcus aureus*. *Proc Natl Acad Sci* **112**: E5805–E5814.

FDA. 2007. Retapamulin. U.S. Food and Drug Administration, Silver Spring, MD.

Feltrin F, Alba P, Kraushaar B, Ianzano A, Argudin MA, Di Matteo P, Porrero MC, Aarestrup FM, Butaye P, Franco A, et al. 2016. A livestock-associated, multidrug-resistant, methicillin-resistant *Staphylococcus aureus* clonal complex 97 lineage spreading in dairy cattle and pigs in Italy. *Appl Environ Microbiol* **82**: 816–821.

Flamm RK, Farrell DJ, Mendes RE, Ross JE, Sader HS, Jones RN. 2012. ZAAPS Program results for 2010: An activity and spectrum analysis of linezolid using clinical isolates from 75 medical centres in 24 countries. *J Chemother* **24**: 328–337.

Flamm RK, Mendes RE, Ross JE, Sader HS, Jones RN. 2013. An international activity and spectrum analysis of linezolid: ZAAPS Program results for 2011. *Diagn Microbiol Infect Dis* **76**: 206–213.

Flamm RK, Mendes RE, Streit JM, Sader HS, Hogan PA, Jones RN. 2016. Activity of linezoli when tested against contemporary European bacterial clinical isolates (2015). *26th European Congress of Clinical Microbiology and Infectious Diseases, ESCMID.* Amsterdam, April 9–12.

Gentry DR, Rittenhouse SF, McCloskey L, Holmes DJ. 2007. Stepwise exposure of *Staphylococcus aureus* to pleuromutilins is associated with stepwise acquisition of mutations in *rplC* and minimally affects susceptibility to retapamulin. *Antimicrob Agents Chemother* 51: 2048–2052.

Gibbons EG. 1982. Total synthesis of (+)-pleuromutilin. *J Am Chem Soc* 104: 1767–1769.

Goldstein EJ, Citron DM, Merriam CV, Warren YA, Tyrrell KL, Fernandez HT. 2006. Comparative in vitro activities of retapamulin (SB-275833) against 141 clinical isolates of *Propionibacterium* spp., including 117 *P. acnes* isolates. *Antimicrob Agents Chemother* 50: 379–381.

Hannan PC, Windsor HM, Ripley PH. 1997. In vitro susceptibilities of recent field isolates of *Mycoplasma hyopneumoniae* and *Mycoplasma hyosynoviae* to valnemulin (Econor), tiamulin and enrofloxacin and the in vitro development of resistance to certain antimicrobial agents in *Mycoplasma hyopneumoniae. Res Vet Sci* 63: 157–160.

Hanson RL, Matson JA, Brzozowski DB, LaPorte TL, Springer DM, Patel RN. 2002. Hydroxylation of mutilin by *Streptomyces griseus* and *Cunninghamella echinulata. Org Process Res Dev* 6: 482–487.

Heilmann C, Jensen L, Jensen JS, Lundstrom K, Windsor D, Windsor H, Webster D. 2001. Treatment of resistant mycoplasma infection in immunocompromised patients with a new pleuromutilin antibiotic. *J Infect* 43: 234–238.

Hidalgo A, Carvajal A, Vester B, Pringle M, Naharro G, Rubio P. 2011. Trends towards lower antimicrobial susceptibility and characterization of acquired resistance among clinical isolates of *Brachyspira hyodysenteriae* in Spain. *Antimicrob Agents Chemother* 55: 3330–3337.

Hillen S, Willems H, Herbst W, Rohde J, Reiner G. 2014. Mutations in the 50S ribosomal subunit of *Brachyspira hyodysenteriae* associated with altered minimum inhibitory concentrations of pleuromutilins. *Vet Microbiol* 172: 223–229.

Hogenauer G. 1975. The mode of action of pleuromutilin derivatives. Location and properties of the pleuromutilin binding site on *Escherichia coli* ribosomes. *Eur J Biochem* 52: 93–98.

Hogenauer G, Ruf C. 1981. Ribosomal binding region for the antibiotic tiamulin: Stoichiometry, subunit location, and affinity for various analogs. *Antimicrob Agents Chemother* 19: 260–265.

Hogenauer G, Egger H, Ruf C, Stumper B. 1981. Affinity labeling of *Escherichia coli* ribosomes with a covalently binding derivative of the antibiotic pleuromutilin. *Biochemistry* 20: 546–552.

Hunt E. 2000. Pleuromutlin antibiotics. *Drugs Future* 25: 1163–1168.

Jones RN, Kohno S, Ono Y, Ross JE, Yanagihara K. 2009. ZAAPS International Surveillance Program (2007) for linezolid resistance: Results from 5591 Gram-positive clinical isolates in 23 countries. *Diagn Microbiol Infect Dis* 64: 191–201.

Kadlec K, Schwarz S. 2009. Novel ABC transporter gene, *vga*(C), located on a multiresistance plasmid from a porcine methicillin-resistant *Staphylococcus aureus* ST398 strain. *Antimicrob Agents Chemother* 53: 3589–3591.

Kadlec K, Pomba CF, Couto N, Schwarz S. 2010. Small plasmids carrying *vga*(A) or *vga*(C) genes mediate resistance to lincosamides, pleuromutilins and streptogramin A antibiotics in methicillin-resistant *Staphylococcus aureus* ST398 from swine. *J Antimicrob Chemother* 65: 2692–2693.

Karlsson M, Gunnarsson A, Franklin A. 2001. Susceptibility to pleuromutilins in *Brachyspira* (*Serpulina*) *hyodysenteriae. Anim Health Res Rev* 2: 59–65.

Kavanagh F, Hervery A, Robbins WJ. 1951. Antibiotic substances from basidiomycetes. VIII: *Pleurotus mutilus* (Fr.) Sacc. and *Pleurotus Passeckerianus* Pilat. *Proc Natl Acad Sci* 37: 570–574.

Kehrenberg C, Schwarz S, Jacobsen L, Hansen LH, Vester B. 2005. A new mechanism for chloramphenicol, florfenicol and clindamycin resistance: Methylation of 23S ribosomal RNA at A2503. *Mol Microbiol* 57: 1064–1073.

Kehrenberg C, Cuny C, Strommenger B, Schwarz S, Witte W. 2009. Methicillin-resistant and -susceptible *Staphylococcus aureus* strains of clonal lineages ST398 and ST9 from swine carry the multidrug resistance gene *cfr. Antimicrob Agents Chemother* 53: 779–781.

Kosowska-Shick K, Clark C, Credito K, McGhee P, Dewasse B, Bogdanovich T, Appelbaum PC. 2006. Single- and multistep resistance selection studies on the activity of retapamulin compared to other agents against *Staphylococcus aureus* and *Streptococcus pyogenes. Antimicrob Agents Chemother* 50: 765–769.

Lenart J, Vimberg V, Vesela L, Janata J, Balikova Novotna G. 2015. Detailed mutational analysis of Vga(A) interdomain linker: Implication for antibiotic resistance specificity and mechanism. *Antimicrob Agents Chemother* 59: 1360–1364.

Li BB, Shen JZ, Cao XY, Wang Y, Dai L, Huang SY, Wu CM. 2010. Mutations in 23S rRNA gene associated with decreased susceptibility to tiamulin and valnemulin in *Mycoplasma gallisepticum. FEMS Microbiol Lett* 308: 144–149.

Li BB, Wu CM, Wang Y, Shen JZ. 2011. Single and dual mutations at positions 2058, 2503 and 2504 of 23S rRNA and their relationship to resistance to antibiotics that target the large ribosomal subunit. *J Antimicrob Chemother* 66: 1983–1986.

Li B, Wendlandt S, Yao J, Liu Y, Zhang Q, Shi Z, Wei J, Shao D, Schwarz S, Wang S, et al. 2013. Detection and new genetic environment of the pleuromutilin-lincosamide-streptogramin A resistance gene *lsa*(E) in methicillin-resistant *Staphylococcus aureus* of swine origin. *J Antimicrob Chemother* 68: 1251–1255.

Li J, Li B, Wendlandt S, Schwarz S, Wang Y, Wu C, Ma Z, Shen J. 2014a. Identification of a novel *vga*(E) gene variant that confers resistance to pleuromutilins, lincosamides and streptogramin A antibiotics in *Staphylococci* of porcine origin. *J Antimicrob Chemother* 69: 919–923.

Li XS, Dong WC, Wang XM, Hu GZ, Wang YB, Cai BY, Wu CM, Wang Y, Du XD. 2014b. Presence and genetic environment of pleuromutilin-lincosamide-streptogramin A resistance gene *lsa*(E) in *Enterococci* of human and swine origin. *J Antimicrob Chemother* 69: 1424–1426.

Cite this article as *Cold Spring Harb Perspect Med* doi: 10.1101/cshperspect.a027110

Li D, Wu C, Wang Y, Fan R, Schwarz S, Zhang S. 2015. Identification of multiresistance gene *cfr* in methicillin-resistant *Staphylococcus aureus* from pigs: Plasmid location and integration into a staphylococcal cassette chromosome *mec* complex. *Antimicrob Agents Chemother* **59**: 3641–3644.

Lindsey J. 2006. Ecology of Commanster, www.commanster .eu/commanster.html.

Liu J, Lotesta SD, Sorensen EJ. 2011. A concise synthesis of the molecular framework of pleuromutilin. *Chem Commun* **47**: 1500–1502.

Locke JB, Rahawi S, Lamarre J, Mankin AS, Shaw KJ. 2012. Genetic environment and stability of *cfr* in methicillin-resistant *Staphylococcus aureus* CM05. *Antimicrob Agents Chemother* **56**: 332–340.

Long KS, Hansen LH, Jakobsen L, Vester B. 2006. Interaction of pleuromutilin derivatives with the ribosomal peptidyl transferase center. *Antimicrob Agents Chemother* **50**: 1458–1462.

Long KS, Poehlsgaard J, Hansen LH, Hobbie SN, Bottger EC, Vester B. 2009. Single 23S rRNA mutations at the ribosomal peptidyl transferase centre confer resistance to valnemulin and other antibiotics in *Mycobacterium smegmatis* by perturbation of the drug binding pocket. *Mol Microbiol* **71**: 1218–1227.

Lotesta SD, Liu J, Yates EV, Krieger I, Sacchettini JC, Freundlich JS, Sorensen EJ. 2011. Expanding the pleuromutilin class of antibiotics by de novo chemical synthesis. *Chem Sci* **2**: 1258–1261.

Lozano C, Aspiroz C, Rezusta A, Gomez-Sanz E, Simon C, Gomez P, Ortega C, Revillo MJ, Zarazaga M, Torres C. 2012. Identification of novel *vga*(A)-carrying plasmids and a Tn5406-like transposon in methicillin-resistant *Staphylococcus aureus* and *Staphylococcus epidermidis* of human and animal origin. *Int J Antimicrob Agents* **40**: 306–312.

Mang R, Heilmayer W, Badegruber R, Strickmann D, Novak R, Ferencic M, Bulusu ARCM. 2008. Pleuromutilin derivatives for the treatment of diseases mediated by microbes. Patent WO2008113089.

Mendes RE, Hogan PA, Streit JM, Jones RN, Flamm RK. 2014. Zyvox Annual Appraisal of Potency and Spectrum (ZAAPS) program: Report of linezolid activity over 9 years (2004–12). *J Antimicrob Chemother* **69**: 1582–1588.

Mendes RE, Hogan PA, Jones RN, Sader HS, Flamm RK. 2016. Surveillance for linezolid resistance via the Zyvox Annual Appraisal of Potency and Spectrum (ZAAPS) programme (2014): Evolving resistance mechanisms with stable susceptibility rates. *J Antimicrob Chemother* doi: 10.1093/jac/dkw052.

Miller K, Dunsmore CJ, Fishwick CW, Chopra I. 2008. Linezolid and tiamulin cross-resistance in *Staphylococcus aureus* mediated by point mutations in the peptidyl transferase center. *Antimicrob Agents Chemother* **52**: 1737–1742.

Moon DC, Tamang MD, Nam HM, Jeong JH, Jang GC, Jung SC, Park YH, Lim SK. 2015. Identification of livestock-associated methicillin-resistant *Staphylococcus aureus* isolates in Korea and molecular comparison between isolates from animal carcasses and slaughterhouse workers. *Foodborne Pathog Dis* **12**: 327–334.

Naegeli P. 1961. "Zur Kenntnis des Pleuromutilins" [Notice about pleuromutilin]. PhD thesis, ETH, Zurich.

Novak R. 2011. Are pleuromutilin antibiotics finally fit for human use? *Ann NY Acad Sci* **1241**: 71–81.

Odou MF, Muller C, Calvet L, Dubreuil L. 2007. In vitro activity against anaerobes of retapamulin, a new topical antibiotic for treatment of skin infections. *J Antimicrob Chemother* **59**: 646–651.

Overesch G, Buttner S, Rossano A, Perreten V. 2011. The increase of methicillin-resistant *Staphylococcus aureus* (MRSA) and the presence of an unusual sequence type ST49 in slaughter pigs in Switzerland. *BMC Vet Res* **7**: 30.

Paquette LA, Bullman-Page PC. 1985. A relay approach to (+)-pleuromutilin. II: Preparation of an advanced optically pure intermediate. *Tetrahedron Lett* **26**: 1607–1611.

Paquette LA, Wiedemann PE. 1985. A relay to (+)-pleuromutilin. I: De novo synthesis of a levorotatory tricyclic lactone subunit. *Tetrahedron Lett* **26**: 1603–1606.

Paukner S, Clark C, Ivezic-Schoenfeld Z, Kosowska-Shick K. 2012. Single- and multistep resistance selection with the pleuromutilin antibiotic BC-3781. *52nd Interscience Conference on Antimicrobial Agents and Chemotherapy (ICAAC)*. San Francisco, September 9–12.

Paukner S, Gruss A, Fritsche TR, Ivezic-Schoenfeld Z, Jones RN. 2013a. In vitro activity of the novel pleuromutilin BC-3781 tested against bacterial pathogens causing sexually transmitted diseases (STD). *Abstracts of the Fifty-Third Interscience Conference on Antimicrobial Agents and Chemotherapy.* Denver, September 10–13.

Paukner S, Krause K, Gruss A, Keepers T, Gomez M, Bischinger A, Strickmann DB, Ivezic-Schoenfeld Z. 2013b. Accumulation of the pleuromutilin antibiotic BC-3781 in murine macrophages and effect of lung surfactant on the BC-3781 in vitro activity. *53rd Interscience Conference on Antimicrobial Agents and Chemotherapy (ICAAC).* Denver, CO, September 10–13.

Paukner S, Sader HS, Ivezic-Schoenfeld Z, Jones RN. 2013c. Antimicrobial activity of the pleuromutilin antibiotic BC-3781 against bacterial pathogens isolated in the SENTRY antimicrobial surveillance program in 2010. *Antimicrob Agents Chemother* **57**: 4489–4495.

Paukner S, Kollmann H, Thirring K, Heilmayer W, Ivezic-Schoenfeld Z. 2014a. Antibacterial in vitro activity of novel extended spectrum pleuromutilins against Gram-positive and -negative bacterial pathogens. *24th European Congress of Clinical Microbiology and Infectious Diseases (ECCMID).* Barcelona, Spain, May 10–13.

Paukner S, Strickmann D, Ivezic-Schoenfeld Z. 2014b. Extended spectrum pleuromutilins: Mode-of-action studies. *24th European Congress of Clinical Microbiology and Infectious Diseases (ECCMID).* Barcelona, Spain, May 10–13.

Paukner S, Kollmann H, Riedl R, Ivezic-Schoenfeld Z. 2015a. Kill curves of the novel extended-spectrum pleuromutilin antibiotic BC-9529. *25th European Congress of Clinical Microbiology and Infectious Diseases (ECCMID).* Copenhagen, Denmark, April 25–28.

Paukner S, Wicha WW, Heilmayer W, Thirring K, Riedl R. 2015b. Extended spectrum pleuromutilins: potent translation inhibitors with broad-spectrum antibacterial activity in vitro and in vivo. *ICAAC/ICC 2016.* San Diego, CA, September 17–21.

Paukner S, Wicha WW, Thirring K, Kollmann H, Ivezic-Schoenfeld Z. 2015c. *In vitro* and *in vivo* efficacy of novel extended spectrum pleuromutilins against *S. aureus* and *S. pneumoniae. 25th European Congress of Clinical Microbiology and Infectious Diseases (ECCMID).* Copenhagen, Denmark, April 25–28.

Poulsen SM, Karlsson M, Johansson LB, Vester B. 2001. The pleuromutilin drugs tiamulin and valnemulin bind to the RNA at the peptidyl transferase centre on the ribosome. *Mol Microbiol* **41:** 1091–1099.

Prince WT, Ivezic-Schoenfeld Z, Lell C, Tack KJ, Novak R, Obermayr F, Talbot GH. 2013. Phase II clinical study of BC-3781, a pleuromutilin antibiotic, in treatment of patients with acute bacterial skin and skin structure infections. *Antimicrob Agents Chemother* **57:** 2087–2094.

Pringle M, Poehlsgaard J, Vester B, Long KS. 2004. Mutations in ribosomal protein L3 and 23S ribosomal RNA at the peptidyl transferase centre are associated with reduced susceptibility to tiamulin in *Brachyspira* spp. isolates. *Mol Microbiol* **54:** 1295–1306.

Rittenhouse S, Biswas S, Broskey J, McCloskey L, Moore T, Vasey S, West J, Zalacain M, Zonis R, Payne D. 2006. Selection of retapamulin, a novel pleuromutilin for topical use. *Antimicrob Agents Chemother* **50:** 3882–3885.

Ross JE, Farrell DJ, Mendes RE, Sader HS, Jones RN. 2011. Eight-year (2002–2009) summary of the linezolid (Zyvox Annual Appraisal of Potency and Spectrum; ZAAPS) program in European countries. *J Chemother* **23:** 71–76.

Ruscoe RE, Fazakerley NJ, Huang H, Flitsch S, Procter DJ. 2015. Copper-catalyzed double additions and radical cyclization cascades in the re-engineering of the antibacterial pleuromutilin. *Chem Eur J* **21:** 1–5.

Sader HS, Biedenbach DJ, Paukner S, Ivezic-Schoenfeld Z, Jones RN. 2012a. Antimicrobial activity of the investigational pleuromutilin compound BC-3781 tested against Gram-positive organisms commonly associated with acute bacterial skin and skin structure infections. *Antimicrob Agents Chemother* **56:** 1619–1623.

Sader HS, Paukner S, Ivezic-Schoenfeld Z, Biedenbach DJ, Schmitz FJ, Jones RN. 2012b. Antimicrobial activity of the novel pleuromutilin antibiotic BC-3781 against organisms responsible for community-acquired respiratory tract infections (CARTIs). *J Antimicrob Chemother* **67:** 1170–1175.

Sanchez GM, De la Torre MA, Morales G, Pelaez B, Tolon MJ, Domingo S, Candel FJ, Andrade R, Arribi A, Garcia N, et al. 2010. Clinical outbreak of linezolid-resistant *Staphylococcus aureus* in an intensive care unit. *JAMA* **303:** 2260–2264.

Schlunzen F, Pyetan E, Fucini P, Yonath A, Harms JM. 2004. Inhibition of peptide bond formation by pleuromutilins: The structure of the 50S ribosomal subunit from *Deinococcus radiodurans* in complex with tiamulin. *Mol Microbiol* **54:** 1287–1294.

Schwendener S, Perreten V. 2011. New transposon Tn6133 in MRSA ST398 contains *vga*(E), a novel streptogramin A-, pleuromutilin-, and lincosamide-resistance gene. *Antimicrob Agents Chemother* **55:** 4900–4904.

Shen J, Wang Y, Schwarz S. 2013. Presence and dissemination of the multiresistance gene *cfr* in Gram-positive and Gram-negative bacteria. *J Antimicrob Chemother* **68:** 1697–1706.

Shore AC, Brennan OM, Ehricht R, Monecke S, Schwarz S, Slickers P, Coleman DC. 2010. Identification and characterization of the multidrug resistance gene *cfr* in a Panton-Valentine leukocidin-positive sequence type 8 methicillin-resistant *Staphylococcus aureus* IVa (USA300) isolate. *Antimicrob Agents Chemother* **54:** 4978–4984.

Shore AC, Lazaris A, Kinnevey PM, Brennan OM, Brennan GI, OC B, Fessler AT, Schwarz S, Coleman DC. 2016. First report of *cfr*-encoding plasmids in the pandemic sequence type 22 methicillin-resistant *Staphylococcus aureus* Staphylococcal cassette chromosome *mec* type-IV clone. *Antimicrob Agents Chemother* **22:** 3007–3015.

Sierra JM, Camoez M, Tubau F, Gasch O, Pujol M, Martin R, Dominguez MA. 2013. Low prevalence of Cfr-mediated linezolid resistance among methicillin-resistant *Staphylococcus aureus* in a Spanish hospital: Case report on linezolid resistance acquired during linezolid therapy. *PLoS ONE* **8:** e59215.

Springer DM, Sorenson ME, Huang S, Connolly TP, Bronson JJ, Matson JA, Hanson RL, Brzozowski DB, LaPorte TL, Patel RN. 2003. Synthesis and activity of a C-8 keto pleuromutilin derivative. *Bioorg Med Chem Lett* **13:** 1751–1753.

Springer DM, Goodrich JT, Luh B-Y, Bronson JJ, Gao Q, Huang S, DenBleyker K, Dougherty TJ, Fung-Tomc J. 2008. Antibacterial pleuromutilin derivatives based on alternate core structures: Arigoni and Birch chemistry revisited. *Lett Drug Des Discov* **5:** 327–331.

Strickmann D, Wicha WW, Kollmann H, Ivezic-Schoenfeld Z. 2014. In vitro metabolism and in vivo pharmacokinetics of novel extended spectrum pleuromutilin antibiotics. *24th European Congress of Clinical Microbiology and Infectious Diseases (ECCMID).* Barcelona, Spain, May 10–13.

Takadoi M, Sato T, Fukuda Y. 2007. Synthesis and antibacterial activity of novel C-12 substituted mutilins. *47th Interscience Conference on Antimicrobial Agents and Chemotherapy (ICAAC).* Chicago, IL, September 17–20.

Thirring K, Ascher G, Paukner S, Heilmayer W, Novak R. 2007. Mutilin derivatives and their use as pharmaceutical. Patent WO2007079515.

Wang H, Andemichael YW, Vogt FG. 2009. A scalable synthesis of 2S-hydroxymutilin via modified rubottom oxidation. *J Organ Chem* **74:** 478–481.

Wang Y, Wang Y, Wu CM, Schwarz S, Shen Z, Zhang W, Zhang Q, Shen JZ. 2011. Detection of the staphylococcal multiresistance gene *cfr* in *Proteus vulgaris* of food animal origin. *J Antimicrob Chemother* **66:** 2521–2526.

Wang Y, He T, Schwarz S, Zhou D, Shen Z, Wu C, Wang Y, Ma L, Zhang Q, Shen J. 2012a. Detection of the staphylococcal multiresistance gene *cfr* in *Escherichia coli* of domestic-animal origin. *J Antimicrob Chemother* **67:** 1094–1098.

Wang Y, Schwarz S, Shen Z, Zhang W, Qi J, Liu Y, He T, Shen J, Wu C. 2012b. Co-location of the multiresistance gene *cfr* and the novel streptomycin resistance gene *aadY* on a small plasmid in a porcine *Bacillus* strain. *J Antimicrob Chemother* **67:** 1547–1549.

Wang Y, Wang Y, Schwarz S, Shen Z, Zhou N, Lin J, Wu C, Shen J. 2012c. Detection of the staphylococcal multiresistance gene *cfr* in *Macrococcus caseolyticus* and *Jeot-*

galicoccus pinnipedialis. J Antimicrob Chemother **67:** 1824–1827.

Wang Y, Zhang W, Wang J, Wu C, Shen Z, Fu X, Yan Y, Zhang Q, Schwarz S, Shen J. 2012d. Distribution of the multidrug resistance gene *cfr* in *Staphylococcus species* isolates from swine farms in China. *Antimicrob Agents Chemother* **56:** 1485–1490.

Wicha WW, Ivezic-Schoenfeld Z. 2014. In vivo activity of extended spectrum pleuromutilins in murine sepsis model. *24th European Congress of Clinical Microbiology and Infectious Diseases (ECCMID)*. Barcelona, Spain, May 10–13.

Wicha WW, Ivezic-Schoenfeld Z, Novak R. 2010. Pharmacokinetics, mass balance and tissue distribution of [^{14}C]-BC-3781 in non-pigmented rats. *20th European Congress of Clinical Microbiology and Infectious Diseases (ECCMID)*. Vienna, April 10–13.

Wicha WW, Fischer E, Kappes BC, Ivezic-Schoenfeld Z. 2013. Comparative pharmacodynamics of BC-3781 in murine *Streptococcus pneumoniae*—Thigh and lung infection models. *53rd Interscience Conference on Antimicrobial Agents and Chemotherapy (ICAAC)*. Denver, May 18–21.

Wicha WW, Paukner S, Strickmann D, Bhavnani SM, Ambrose PG. 2015a. Pharmacokinetics-pharmacodynamics of lefamulin in a neutropenic murine lung infection model. *ICAAC/ICC 2015*. San Diego, September 17–21.

Wicha WW, Paukner S, Strickmann D, Thirring K, Kollmann H, Heilmayer W, Ivezic-Schoenfeld Z. 2015b. Efficacy of novel extended spectrum pleuromutilins against *E. coli* in vitro and in vivo. *25th European Congress of Clinical Microbiology and Infectious Diseases (ESCMID)*. Copenhagen, April 25–28.

Yan K, Madden L, Choudhry AE, Voigt CS, Copeland RA, Gontarek RR. 2006. Biochemical characterization of the interactions of the novel pleuromutilin derivative retapamulin with bacterial ribosomes. *Antimicrob Agents Chemother* **50:** 3875–3881.

Zeitlinger M, Schwameis R, Burian A, Burian B, Matzneller P, Muller M, Wicha WW, Strickmann DB, Prince W. 2016. Simultaneous assessment of the pharmacokinetics of a pleuromutilin, lefamulin, in plasma, soft tissues and pulmonary epithelial lining fluid. *J Antimicrob Chemother* **71:** 1022–1026.

Zhang A, Xu C, Wang H, Lei C, Liu B, Guan Z, Yang C, Yang Y, Peng L. 2015. Presence and new genetic environment of pleuromutilin-lincosamide-streptogramin A resistance gene *lsa*(E) in *Erysipelothrix rhusiopathiae* of swine origin. *Vet Microbiol* **177:** 162–167.

Fusidic Acid: A Bacterial Elongation Factor Inhibitor for the Oral Treatment of Acute and Chronic Staphylococcal Infections

Prabhavathi Fernandes

Cempra Incorporated, Chapel Hill, North Carolina 27517

Correspondence: prabha@fernandes-domain.com

Fusidic acid is an oral antistaphylococcal antibiotic that has been used in Europe for more than 40 years to treat skin infections as well as chronic bone and joint infections. It is a steroidal antibiotic and the only marketed member of the fusidane class. Fusidic acid inhibits protein synthesis by binding EF-G-GDP, which results in the inhibition of both peptide translocation and ribosome disassembly. It has a novel structure and novel mode of action and, therefore, there is little cross-resistance with other known antibiotics. Many mutations can occur in the *FusA* gene that codes for EF-G, and some of these mutations can result in high-level resistance (minimum inhibitory concentration [MIC] > 64 mg/L), whereas others result in biologically unfit staphylococci that require compensatory mutations to survive. Low-level resistance (<8 mg/L) is more common and is mediated by *fusB*, *fusC*, and *fusD* genes that code for small proteins that protect EF-G-GDP from binding fusidic acid. The genes for these proteins are spread by plasmids and can be selected mostly by topical antibiotic use. Reports of resistance have led to combination use of fusidic acid with rifampin, which is superseded by the development of a new dosing regimen for fusidic acid that can be used in monotherapy. It consists of a front-loading dose to decrease the potential for resistance development followed by a maintenance dose. This dosing regimen is now being used in clinical trials in the United States for skin and refractory bone and joint infections.

The ribosome is the target for many of the commonly used natural product–derived antibiotics, such as the macrolides, tetracyclines, phenicols, and aminoglycosides, all of which inhibit protein synthesis. However, there are many other potential antibiotic targets in the protein synthesis pathway, such as the tRNA synthetases and the peptide elongation factors, elongation factor EF-Tu and EF-G (Wilson 2014). Fusidic acid, a steroidal antibiotic (Fig. 1) synthesized by the fungus *Fusidium coccineum* (Godtfredsen et al. 1962), is a specific inhibitor of EF-G. EF-G performs an essential function in shifting the nascent polypeptide chain from the A site on the 30S subunit to the P site, a process called peptide translocation, and also interacts with ribosome release factor (RRF) to release the ribosome complex on reaching the stop codon during protein synthesis (Chen et al. 2010b; Guo et al. 2012). Although there are similarities with mammalian elongation factor (EF-2), bacterial EF-G and protein synthesis, in general, are less complex than in mammalian cells, which makes EF-G a selective antibiotic target.

A

B

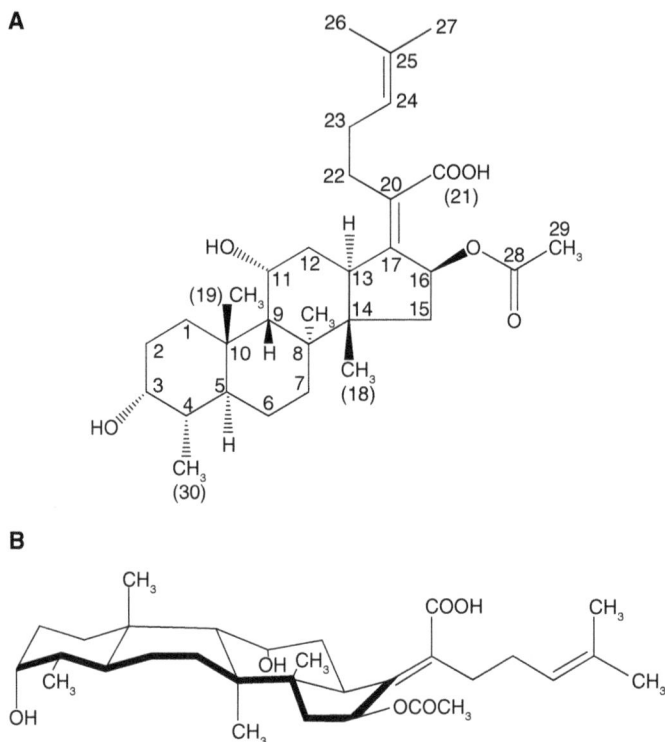

Figure 1. Structural formula of fusidic acid. (*A*) Steroidal conformation. (*B*) Chair-boat-chair conformation.

Although fusidic acid is known to be steroidal, its ring conformation is unlike that of natural steroids in that it is characterized as a chair-boat-chair conformation (Fig. 1A,B). Some of the other steroidal antibiotics include cephalosporin P1, viridian, and helvolic acid (Fig. 2A–C) (von Daehne et al. 1979; Betina 1983), but fusidic acid is the only antibiotic marketed in this class. Unlike most antibiotics, fusidic acid is not synthesized by *Streptomyces* spp. Another unique feature of fusidic acid is that it is not secreted into the fermentation medium after synthesis and, therefore, must be extracted from the mycelium. This makes the isolation and purification more difficult than those antibiotics that are secreted.

SPECTRUM OF ACTIVITY

After the discovery of fusidic acid in the early 1960s, its activity against a panel of bacteria was published (von Daehne et al. 1979). The results showed mostly a narrow spectrum of activity against Gram-positive bacteria, including staphylococci, and thereafter most of the subsequent studies focused on this pathogen. Following the initial discovery in 1962, it was introduced into clinical practice to treat staphylococcal skin infections (Godtfredsen et al. 1962). The in vitro evaluation of the spectrum and potency of fusidic acid against bacterial clinical isolates has been extensive (Verbist 1990) and includes studies conducted in both academic and clinical laboratories in Canada, Europe, and Australia (Collignon and Turnidge 1999; Oliva et al. 2004). Although used successfully in treating skin infections and staphylococcal bone and joint infections, the potential for the emergence of resistance during therapy has led to the use of fusidic acid in combination with rifampin (Collignon and Turnidge 1999; Howden and Grayson 2006).

Fusidic acid when formulated as sodium fusidate is well absorbed after oral administra-

A

B

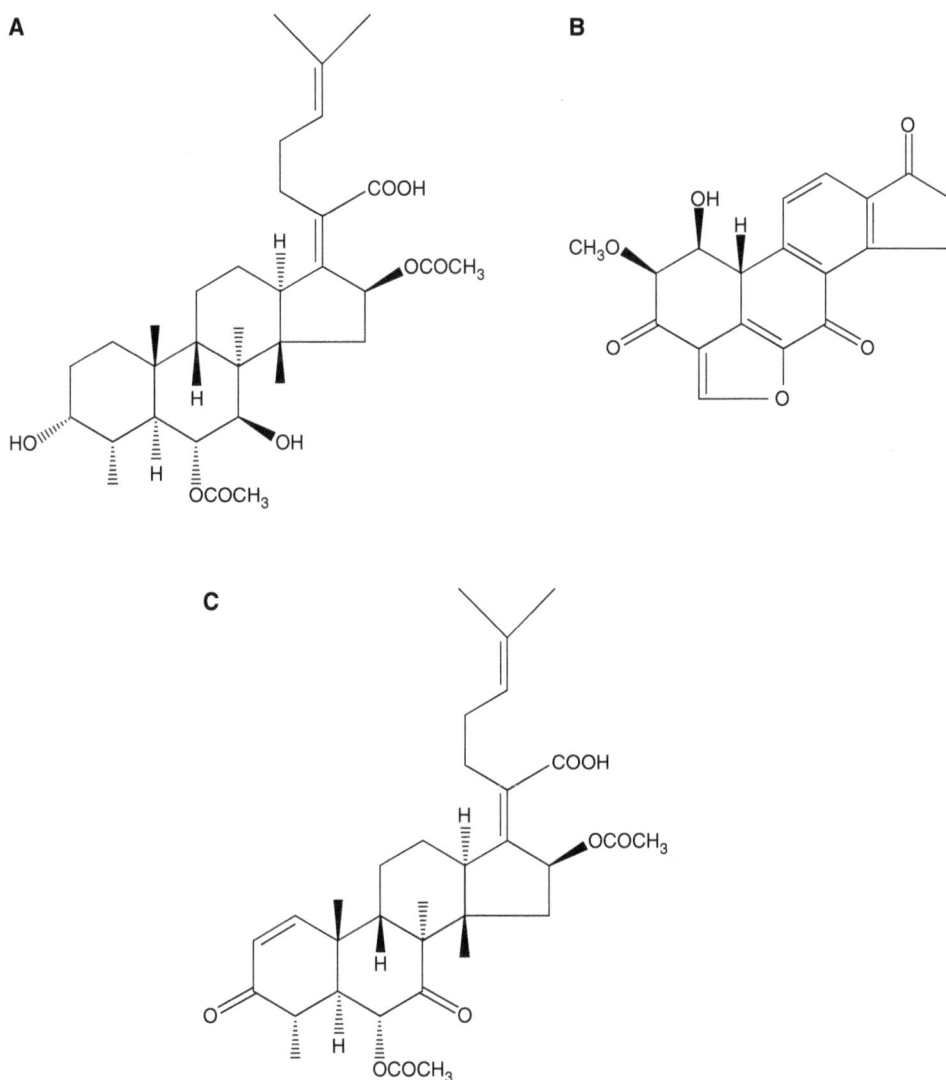

C

Figure 2. Structures of other steroidal antibiotics. (*A*) Cephalosporin P1. (*B*) Viridin. (*C*) Helvolic acid.

tion, with >90% bioavailability (Still et al. 2011). In addition, it has a long plasma half-life of ~10–14 h and is highly (>95%) but reversibly protein bound (Reeves 1987; Turnidge 1999). Attempts have been made to obtain fusidic acid analogs with improved pharmacokinetic/pharmocodynamic properties and a large number of semisynthetic analogs have been synthesized (Godtfredsen and Vangedal 1966). Some modifications were made that retain activity including saturation of the C24–C25 dou-

ble bond, replacement of the 16α-acetoxyl group by other groups, and conversion of the 11-hydroxyl group into the corresponding ketone. The Z configuration at the C17–C20 double bond and a 16β-acetyl group were identified as being essential for antibacterial activity. Modifications to the rings, the C17–C20 double bond, or functional group changes all resulted in reduced activity. Interestingly, none of the analogs are more active than fusidic acid itself. Thus, even 46 yr after discovery, fusidic acid remains the only member of this class of antibiotics to be clinically useful.

As newer, more potent antistaphylococcal agents were developed and marketed, the clinical use of oral tablets of fusidic acid decreased. Furthermore, the need to administer this drug in combination with another antibiotic, such as rifampin, was not favorable owing to the potential for drug–drug interactions. However, fusidic acid as a topical ointment is broadly used in monotherapy and still has widespread use in Europe, Canada, and other parts of the world. Although marketed in Europe and Canada, fusidic acid has never received marketing approval in the United States (Fernandes and Pereira 2011). In 2007, when methicillin-resistant *Staphylococcus aureus* (MRSA) spread widely, there were few safe oral options for the treatment of MRSA. At this time, Cempra began development of fusidic acid for clinical use as an oral, safe alternative for the treatment of MRSA and other staphylococcal infections in the United States.

PREVALENCE OF FUSIDIC ACID RESISTANT GRAM-POSITIVE BACTERIA

Although fusidic acid is known to have a high rate of resistance, a recent study (Jones et al. 2006) showed that, in clinical strains of staphylococci from medical centers in Australia (2002–2003) that included methicillin-susceptible *S. aureus* (MSSA) (100 isolates) and methicillin-resistant *S. aureus* or MRSA (100 isolates), ~88% and 81% were susceptible to fusidic acid, respectively, despite its use for decades in Australia. International surveillance studies conducted at Cempra from 2008 through 2011 indicated that fusidic acid had potent activity against *Staphylococcus* spp., including methicillin-resistant strains. It was noted in these studies that, after four decades of use in Europe, 89.3% of 2700 *S. aureus* strains in 2008, 93.5% of 2166 strains in 2009, 92.8% of 2263 strains in 2010, and 93.4% of 1938 strains in 2011 were susceptible to fusidic acid based on the European Committee on Antimicrobial Susceptibility Testing (EUCAST) susceptibility breakpoint of minimum inhibitory concentra-

tion (MIC) $\leq$ 1 mg/L (Jones et al. 2010a). Thus, the resistance rate was stable over many years and was <10% in countries where the antibiotic has been used for a long time. In 2011, virtually all *S. aureus* strains in the United States were susceptible to fusidic acid; collectively, MSSA and MRSA strains were 99.7% susceptible. For coagulase-negative staphylococci (CoNS) collected globally in 2008–2009, 75%–81% of strains had MIC values $\leq$1 mg/L. In the same years, *Streptococcus pyogenes* and *Streptococcus agalactiae* have MIC_{90} values of 8 mg/L (Collignon and Turnidge 1999; Oliva et al. 2004) and resistant strains have not been reported. In the United States, the MIC_{90} for fusidic acid is 8 mg/L against *S. pyogenes* (Jones et al. 2011).

Although fusidic acid has moderate in vitro activity against *S. pyogenes*, it has been shown to have clinical efficacy in patients (Spelman 1999). These results can be explained by the relatively high and sustained blood levels achieved after oral dosing with fusidic acid. Plasma concentrations of 102 mg/L and mean steady-state trough concentrations of 66 mg/L have been reported following oral dosing of 500 mg twice a day for 6 d (Vaillant et al. 2000; Still et al. 2011). The mean peak skin blister concentration was 79 mg/L and the steady-state mean trough skin blister concentration was 39 mg/L (Vaillant et al. 1992). The sustained blood and tissue levels along with the plasma trough levels are all above the MIC_{90} of fusidic acid for *S. pyogenes*. The distribution of MICs of *S. aureus* is contrasted to the narrow MIC distribution for *S. pyogenes* (or group A *Streptococcus*) in Figure 3. Although *S. pyogenes* is inherently less susceptible to fusidic acid than *S. aureus* (4–8 mg/L versus 0.25 mg/L, respectively), resistant strains are more often found with *S. aureus* than with *S. pyogenes*. Interestingly, fusidic acid is bactericidal for *S. pyogenes* (Okusanya et al. 2011; Tsuji et al. 2011), although bacteriostatic for the more susceptible *S. aureus*. It is not yet known why the inhibition of EF-G in *S. pyogenes* results in cell death. Once the structure of *S. pyogenes* EF-G is elucidated, it could provide an explanation for these differences.

After 2010, when fusidic acid development was initiated in the United States, it was

Cite this article as *Cold Spring Harb Perspect Med* doi: 10.1101/cshperspect.a025437

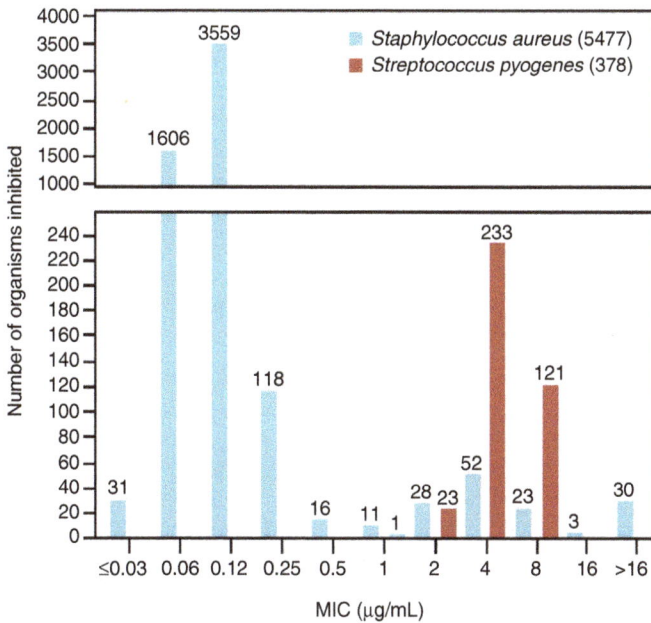

Figure 3. Minimum inhibitory concentration (MIC) distributions of fusidic acid against *S. aureus* (5477 strains) and *S. pyogenes* (378 strains). Data were obtained from an international resistance surveillance study conducted on strains collected in 2011 by Jones Microbiology Institute (North Liberty, Iowa) under contract to Cempra Inc., Chapel Hill, NC. Nonduplicated *S. aureus* and *S. pyogenes* strains were collected from 49 medical centers located in the 26 clinical laboratories in the United States and 23 clinical laboratories in Europe. Susceptibility testing was by Clinical Laboratory Standards Institute (CLSI) approved methodology by broth-dilution methods.

tested broadly against a variety of pathogens using standardized methods (Biedenbach et al. 2010). A summary of the in vitro activity from several studies is shown in Table 1. Fusidic acid is active against staphylococci and has potent activity (MIC$_{90}$ 0.12 mg/L) against MRSA, community-acquired MRSA (CA-MRSA), such as USA300 strains, hospital-acquired MRSA (HA-MRSA), and linezolid-resistant strains (MIC$_{90}$ <0.25 mg/L) (Sahm et al. 2013). CoNS with or without methicillin resistance, including *Staphylococcus epidermidis* and other species, are also susceptible to fusidic acid (MIC$_{90}$ 0.25 mg/L) (Jones et al. 2011). Among other Gram-positive bacteria, fusidic acid has activity against enterococci (MIC$_{90}$ 4 mg/L), *Corynebacterium* spp. (MICs ≤0.12 mg/L), *S. pyogenes* (MIC$_{90}$ 8 mg/L), and viridans streptococci (MICs 1 to >8 mg/L) (Biedenbach et al. 2010; Jones et al. 2011). In addition, antibacterial activity

against *Propionibacterium* spp. has been reported with MIC$_{90}$ of 1 mg/L (Hardy et al. 2014). Fusidic acid has limited activity against Gram-negative bacteria owing to its large size and lipophilicity, which does not favor transport through porins in the outer membrane of these bacteria. However, *Neisseria gonorrhoeae* is susceptible, with an MIC$_{90}$ of 1 mg/L (Jones et al. 2010b). Fusidic acid has also shown activity against *Chlamydia trachomatis* (MIC$_{90}$ 0.5 mg/L). Fusidic acid has moderate activity against some Gram-positive anaerobic bacteria, such as *Clostridium* species (*C. difficile* MIC$_{90}$ 2 mg/L) (Collignon and Turnidge 1999), and has been used to treat *C. difficile* enterocolitis (Wullt and Odenholt 2004). The unique mechanism of action of fusidic acid has limited its cross-resistance with other antibiotics and, therefore, this has not been commonly observed in clinical isolates.

Table 1. In vitro susceptibility of fusidic acid against a wide variety of pathogens

Organism	MIC$_{90}$ (μg/mL)
Staphylococcus aureus (7339)[a]	0.25
Methicillin-resistant S. aureus (MRSA) (3876)[a]	0.25
Methicillin-susceptible S. aureus (MSSA) (3463)[a]	0.25
Coagulase-negative S. aureus (1352)[a]	0.25
Enterococci (2448)[a]	4
Streptococcus pyogenes (131)[a]	8
Propionibacterium acnes (51)[b]	1
Neisseria gonorrhoeae (35)[c]	1
Chlamydia trachomatis (10)[c]	0.5
Clostridium difficile (200)[d]	2

[a]Data from Jones et al. (2011).
[b]Data from Hardy et al. (2014).
[c]Data from Jones et al. (2010b).
[d]Data from Collignon and Turnidge (1999).

pH EFFECTS AND INTRACELLULAR ACTIVITY

Fusidic acid binds protein, primarily albumin and, therefore, the MIC is increased up to 4- to 16-fold in the presence of 10% serum (Biedenbach et al. 2010; Lemaire et al. 2011). However, the influence of protein binding on the in vitro activity is markedly reduced at an acidic pH. When tested at pH 5.5, fusidic acid activity was enhanced relative to standard in vitro testing media (pH 7.2), whereas the activity of other staphylococcal antibiotics, such as linezolid and clindamycin were either not affected or negatively impacted (Lemaire et al. 2011). The activity of fusidic acid changed from 0.25 mg/L at pH 7.2 to 0.0058 mg/L at pH 5.5. The MIC of linezolid remained at 2 mg/L at the acidic and neutral pH, whereas clindamycin had decreased activity of 0.125 mg/L to 4 mg/L at pH 7.2 (Fig. 4). Although fusidic acid has ~97% protein binding, it is active in infected sites in both the skin and bone and has been able to distribute well to the dermis (Vaillant et al. 1992). Fusidic acid also penetrates macrophages and inhibits growth of staphylococci within cells. Interestingly, these studies showed that in macrophages, the intracellular concentration of fusidic acid was high (Lemaire et al. 2011). These

data suggest that fusidic acid is only loosely bound to plasma protein. Thus, the in vitro activity may not accurately predict its activity at the sites of infections.

MECHANISM OF PROTEIN SYNTHESIS INHIBITION

Protein synthesis involves initiation, elongation, translocation, and release, which are catalyzed by four proteins: IF-2, EF-Tu, EF-G, and RRF. Each of these proteins possesses GTPase activity and if any of them are inhibited, protein synthesis will be blocked. Fusidic acid binds to EF-G-GDP that is bound to the ribosome and inhibits protein synthesis by inhibiting translocation of the growing polypeptide as well as the recycling of the ribosomal subunits when the stop codon on the mRNA is reached (Gao et al. 2009). In the protein synthesis cycle, the nascent growing peptide chain is bound to the P site of the ribosome, whereas the tRNA with a new amino acid is brought to the A site (Fig. 5). The first GTPase protein EF-Tu, brings the nascent peptide from the P site to the A site with the hydrolyzed GTP bound to it. The developing peptide is now attached to the new amino acid on the A site. The complex of tRNA and the newly forming peptide must move one codon on the mRNA and also transfer the tRNA-peptide to the P site. This process is catalyzed by EF-G-GTP in a process called translocation. Translocation of the peptide chain to the P site is mediated by EF-G bound to GTP, accompanied by the release of tRNA from the A site. This vacates the A site to allow tRNA to bring a new amino acid to the A site and continue the cycle of peptide synthesis. The translocation step is also accompanied by the hydrolysis of EF-G-GTP to form EF-G-GDP. The translocation step is strictly regulated by EF-G to ensure that there is no slippage that could result in mistranslation as the peptide chain is lengthened. Once the peptide is safely transferred to the P site, EF-G releases GDP and leaves the ribosome. The cycle of peptide chain synthesis and elongation continues until a stop codon on the mRNA is reached. At this time, EF-G interacts with another GTPase protein, RRF, which cata-

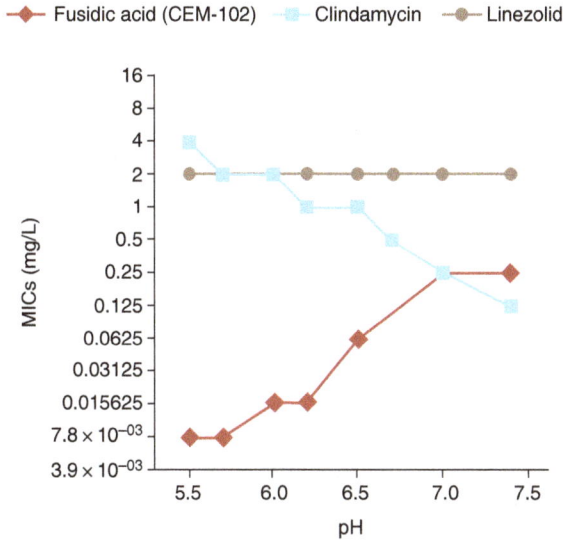

Figure 4. Influence of pH on the minimum inhibitory concentration (MIC) of fusidic acid, clindamycin, and linezolid.

lyzes the dissociation of the ribosomal subunits, thereby releasing the newly formed peptide and mRNA.

Fusidic acid binds to EF-G-GDP in the translocation step, which results in the EF-G's inability to release the bound GDP and in turn

the EF-G-GDP complex is unable to vacate the P site of the ribosome (Fig. 5). Fusidic acid has little affinity for free EF-G or EF-G-GTP and allows multiple rounds of GTP hydrolysis before locking the EF-G-GDP onto the ribosome and blocking completion of translocation and

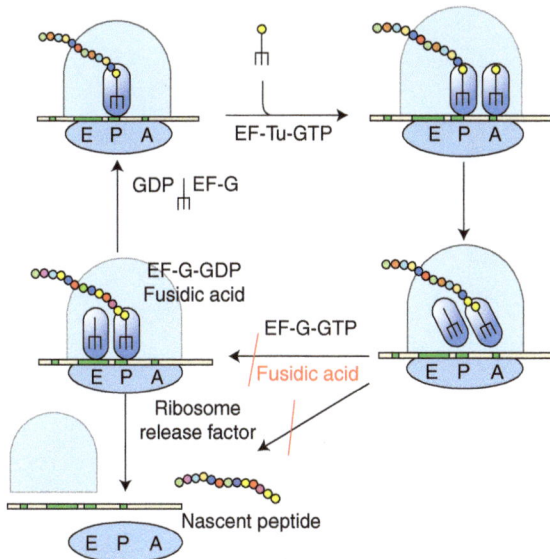

Figure 5. Cartoon demonstrating the two steps at which fusidic acid blocks peptide synthesis by binding to the EF-G-GDP complex.

any further ribosome cycling. The K_{50} value for FA inhibition of the EF-G cycle was estimated at 200 μmol (Borg et al. 2015). The inhibition of ribosome recycling by fusidic acid has been further characterized and shown to be a key target, as 1000 times less fusidic acid is sufficient to inhibit ribosome release versus peptide chain elongation (Savelsbergh et al. 2009). The K_{50} value for fusidic acid action on ribosomal recycling was calculated to be 0.1 μmol. The concentration of fusidic acid required to block ribosomal recycling is closer to the concentration required to inhibit *S. aureus* growth (MIC$_{90}$ 0.12 mg/L *S. aureus*).

Structural studies were first reported using cryoelectron microscopy of EF-G-GDP bound to the ribosome and free EF-G (Agrawal et al. 1998; Frank and Agrawal 2000; Besier et al. 2003). Further structural studies were conducted on crystal forms of *Thermus thermophilus* EF-G-GDP–L-9 ribosomal protein mutant-ribosome complex, which was stabilized using fusidic acid (Gao et al. 2009). Later the crystal structure of apo EF-G (free EF-G) of *S. aureus* (at 1.9 Å) was determined (Chen et al. 2010b; Guo et al. 2012). Although there are some differences between the EF-G of *T. themophilus* and that of *S. aureus*, especially in the orientation

of the structures between the bound and free forms, overall, there is a general agreement that EF-G has five domains (Fig. 6A,B). As with all GTPase proteins there is a G domain (domain I). In EF-G, the G domain has a P site that interacts with the α and β phosphates of GTP and two switch regions that interact with the γ phosphate. The switch regions change conformation from a tense, GTP-bound state to a relaxed GDP-bound state. The GTP-binding site and GTPase activity of EF-G are similar to other proteins that have GTPase activity, such as EF-Tu (Laurberg et al. 2000; Guo et al. 2012). The change in conformation of EF-G as GTP is hydrolyzed to GDP is also reminiscent of the change in structure observed for other GTPase proteins. In addition, the following interactions have been noted between EF-G and the ribosome based on crystallographic work and studies with mutants: domain I with the L-12 ribosomal protein; domain II with ribosomal protein S12; domains II and III with fusidic acid; domain IV with the decoding center; and domain V with 23S RNA as well as ribosomal protein L6. The structure of EF-G bound to the ribosome is different from that of free EF-G, with domain IV undergoing dramatic conformational change in the ribosome-bound form

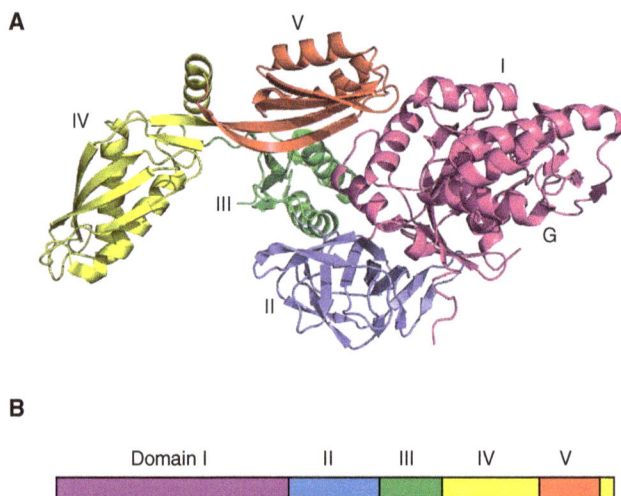

Figure 6. Structure of *S. aureus* EF-G. (*A*) Illustration of the five domains of EF-G (each shown in a different color). (*B*) Topology of EF-G. (From Guo et al. 2012; reproduced, with permission, from Open Biology © 2012 per the terms of reuse by The Royal Society Publishing open access license, rsob.royalsocietypublishing.org/content/2/3/120016.figures-only.)

with domains III and IV rotated relative to domains I and II. It was determined that EF-G-GDP is held between the 30S and 50S subunits. X-ray crystallography of the *T. thermophilus* cocrystals (described at 3.6 Å), shows that domain IV of EF-G is inserted into the 30S decoding center. At that point, it makes contact with both the P site and mRNA, and is believed to aid in locking the translocation complex. Domain III has GTPase activity, is required for GTP hydrolysis, and makes contact with both the 30S and the 50S ribosomal subunits.

The crystal structure has revealed that fusidic acid is lodged between domains II and III, surrounded by switch II near the G domain. Fusidic acid binds with high affinity only to ribosome-bound EF-G-GDP. It appears that fusidic acid cannot fit into the closed conformation structure of the switch 1 loop of EF-G when it is bound to GTP and can only bind after the switch loop opens. Once the switch loop opens, this allows water to enter and subsequently hydrolyze the GTP. Therefore, fusidic acid can only bind after GTP is hydrolyzed. Fusidic acid binds directly to domain II, stabilizing the EF-G-GDP in a conformation that resembles that of EF-G-GTP instead of the normal EF-G-GDP bound form and, therefore, prevents the release of GDP. Based on this data, it has been proposed that the main mechanism of fusidic acid action is to lock the switch 2 loop in a conformation that is similar to its GTP-bound state. This causes the translocation machinery to lock into a specific conformation, thereby preventing the release of GDP from EF-G. The structure also shows that fusidic acid interacts with L11 and L12 regions of the 50S subunit as well as domain V of EF-G.

The crystal structure data are supported by work with EF-G mutants (Hansson et al. 2005; Chen et al. 2010a). Mutant T84A, which is in the fusidic acid–binding region, displays resistance, whereas a G16V opens switch loop II and allows fusidic acid to bind, creating a hypersensitive mutant (Martemyanov et al. 2001). More than 40 EF-G mutants have been mapped and their interactions with fusidic acid have been defined. These mutants have been classified into classes A–D and correlated to the degree of resistance based on drug binding (group A), ribosome–EF-G interactions (group B), EF-G conformation (group C), and EF-G stability (group D) (Chen et al. 2010b). Several mutations could affect more than one feature, such as conformation, and may have an impact on stability and binding.

Fusidic acid inhibits bacterial EF-G specifically and does not interact with mammalian elongation factors and other GTPase proteins. The EF-G of mitochondria was shown to be functional in *Escherichia coli* protein synthesis but the *E. coli* EF-G was not functional in mitochondrial protein synthesis (Chung and Spremulli 1990). Also, mitochondrial EF-G (EF-G$_{mt}$) is completely resistant to fusidic acid when tested with its mitochondrial ribosomes, but is partially active when tested with the bacterial ribosome.

RESISTANCE DEVELOPMENT AND MECHANISMS

Resistance to fusidic acid can occur by mutation of genes (*fusA*) that encode the drug target EF-G as noted above, but can also occur by the acquisition of genes (*fusB*, *fusC*) that encode cytoplasmic proteins that protect the target site of the drug (Castanheira et al. 2010; Pfaller et al. 2010). It can also occur rarely by a mutation in the gene *fusE*, which codes for a ribosomal protein. Mutation frequencies for *S. aureus* following a single exposure to fusidic acid at concentrations 2 to 16 times the MIC range from 10^{-6} to 10^{-8}, respectively (Biedenbach et al. 2010). Mutations have been reported to occur at a much lower rate in both MSSA and MRSA at therapeutic concentrations of fusidic acid, which are significantly higher than 16 times the MIC (O'Neill et al. 2001). Therefore, high plasma concentrations of fusidic acid would be expected to produce lower mutation rates in the clinical setting. Experiments conducted with therapeutically achievable plasma concentrations showed that resistance is rare when *S. aureus* strains are exposed to such concentrations (O'Neill et al. 2001).

Resistance to fusidic acid can be high-level resistance (i.e., >64 mg/L) or low-level

2–32 mg/L (Castanheira et al. 2010; Jones et al. 2011). High-level resistance is attributable to mutations in the gene *fusA*. Mutations within the *fusA* gene have been identified in the laboratory to select for resistance to fusidic acid as well as in naturally occurring clinical strains. The mutations in clinical isolates include V90L, H457Q, L461K, and A655V, and some strains with single mutations like L461K can have high-level resistance (>256 mg/L). However, not all mutations in EF-G result in high-level resistance as the actual location of the mutation in the fusidic acid–binding site of EF-G is pertinent to the specific level of resistance (Castanheira et al. 2010; Chen et al. 2010a). Mutants obtained in vitro generally have low-level resistance and can have P406L, H475Y, and R464C mutations, along with several others. In some clinical strains, more than one mutation has been identified, and some of these mutations could be compensatory through their restoration of biological fitness (Besier et al. 2005). Resistance mediated by *fusA* mutations that are amino acid substitutions like P406L and H457Y in EF-G show reduced growth, and biological fitness can be restored in these strains by second site mutations within EF-G, such as A67T and S416F (Nagaev et al. 2001; Besier et al. 2005; Koripella et al. 2012). The latter mutations do not contribute to resistance but rather stabilize the original mutation, which is believed to play a role in maintaining high-level resistant strains within a population. Resistance mediated by *fusA* can be transferred to a new host by a plasmid carrying a *fusA* mutation (Besier et al. 2003). Compensatory mutations that can allow peptide translocation have also been described and these mutations that restore the function of EF-G in translocation could likely bring back the balance between GTP- and GDP-bound conformations of EF-G (Johanson et al. 1996).

It is of interest that fusidic acid is less potent in vitro against *S. pyogenes* (MIC$_{90}$ 4–8 mg/L compared with 0.125 mg/L for *S. aureus*) but resistance to fusidic acid is not reported. The EF-G of *S. pyogenes* appears to be unable to mutate in either clinical situations or the laboratory setting. There is only one *fusA* gene in both of

these species (Margus et al. 2007) and, therefore, the higher MIC but decreased resistance rate is not related to the multiplicity of *fusA* genes. The GTP-binding site is similar in all bacteria. It is possible that the EF-G of *S. pyogenes* has already evolved such that it renders streptococci less susceptible and further mutations could be nonviable. These mutations could result in biologically unfit strains as has been noted with some fusidic acid–resistant mutants of staphylococci. The fact that fusidic acid is bactericidal for *S. pyogenes*, whereas bacteriostatic for *S. aureus* also indicates additional targets that remain to be characterized in *S. pyogenes*.

Although mutations in *fusA* can result in high-level resistance, *fusB*, *fusC*, and *fusD* give rise to low-level resistance. The genes for *fusB*, *fusC*, and *fusD* code for closely related small proteins that bind to EF-G and block the binding of fusidic acid to EF-G (O'Neill and Chopra 2006). Another mechanism of resistance, which is rare but occurs in *fusE*, is related to frameshift or truncation of the *rplF* gene that codes for ribosomal protein L6 (a contact area for EF-G), resulting in low-level resistance. *fusB* was first identified on a plasmid, pUB101, whereas *fusC* is a chromosomally encoded homolog. The gene for *fusB* can be found on plasmids or on the chromosome of *S. aureus* and other staphylococcal species (O'Neill and Chopra 2006) and is transferred horizontally on a transposon-like element. This gene codes for a small EF-G-binding protein that is ~25 kDa in size. The third member of this group, *fusD*, confers inherent resistance among *S. saprophyticus* strains. The mechanism of a small protein blocking antibiotic-target interaction to give low-level resistance is reminiscent of the *qnr* resistance observed in the DNA gyrase and fluoroquinolone area as well as ribosomal protection proteins Tet(M) and Tet(O) that release tetracyclines from the ribosome (Thaker et al. 2010). In the case of Qnr, the Qnr protein binds to DNA gyrase, altering its structure such that fluoroquinolones can no longer bind to the DNA–DNA gyrase complex (Tran et al. 2005). In the case of Tet(M) and Tet(O), the binding of tetracycline to the ribosome is inhibited. In both cases, the level of resistance is low yet trans-

Cite this article as *Cold Spring Harb Perspect Med* doi: 10.1101/cshperspect.a025437

missible and the protein that confers such resistance binds to the target to block the activity of the antibiotic. Qnr proteins belong to a pentapeptide repeat family of proteins and differ on a structural basis from FusB.

Crystallographic analysis of FusB and FusC binding to EF-G has shown that these proteins have the same architecture and have a two-domain zinc-binding protein with a carboxy-terminal portion, which contains a fold of helices and β sheets that make up the zinc-binding domain (Fig. 7A,B) (Gao et al. 2009; Cox et al. 2012). This zinc-binding domain interacts with high-affinity 1:1 with the carboxy-terminal region of EF-G. The carboxy-terminal domain of FusB has a protein fold that is not found in any other structure in the protein database (Guo et al. 2012). Because fusidic acid binds to the amino-terminal region of EF-G, steric inhibition of binding is unlikely to be the mechanism of inhibition. In addition, FusB was unable to block E. coli EF-G binding to the ribosome and is specific to the binding of the carboxy-terminal domain of EF-G of staphylococci (Guo et al.

2012). These results are explained by using hybrid protein constructs of staphylococcal and E. coli EF-G, where it was shown that the carboxy-terminal domains were different between the two organisms. Data show that FusB-type proteins dissociate the stalled ribosome-EF-G-GDP complex, shifting the balance to favor dissociation, which explains the mechanism of fusidic acid resistance. A greater concentration of fusidic acid is, therefore, needed to result in fusidic acid-EF-G-GDP ribosome complexes to halt further peptide chain elongation and inhibit the growth of staphylococci carrying *fusB*.

FusC and FusD have ≤80% homology at the amino acid level with each other and with FusB (O'Neill et al. 2007). The increase in staphylococcal resistance found in regional outbreaks in Europe has been principally associated with the spread of a single fusidic acid–resistant clone, in which plasmid-borne *fusB* has been recruited to the *S. aureus* chromosome (O'Neill et al. 2007). *S. epidermidis* resistance in clinical isolates is also commonly associated with acquisition of the *fusB* determinant. Use of topical

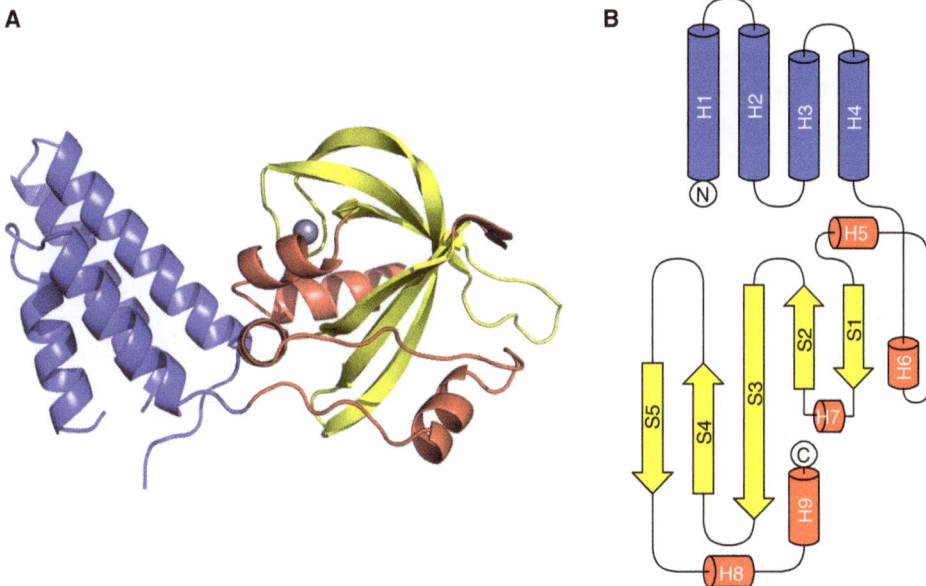

Figure 7. Structure of FusB protein showing the carboxy-terminal region that has a fold of α helices and β sheets making up the zinc-binding domain. (*A*) Cartoon diagram of the FusB structure. (*B*) Domain I is shown in blue and domain II is indicated by yellow and red. (From Guo et al. 2012; reproduced, with permission, from Open Biology © 2012 per the terms of reuse by The Royal Society Publishing open access license, rsob .royalsocietypublishing.org/content/2/3/120016.figures-only.)

fusidic acid is believed to enhance this resistance transfer (Dobie and Gray 2004; Williamson et al. 2014). *fusB* is currently the most common type of resistance in Europe and has been found in other staphylococcal species, such as *S. aureus* and *Staphylococcus lugdunensis*. Of 1649 *S. aureus* strains tested from Denmark in 2003–2005, 291 strains (17.6%) had MICs of >1 mg/L, and 87% carried *fusB* and *fusC* genes (McLaws et al. 2011). Thus, only 13% of the strains in this collection carried *fusA* mutations, which confer high-level resistance. FusB protein is more efficient than FusC and FusD proteins in its ability to protect EF-G from fusidic acid binding. Unlike *fusA*, *fusB*, *fusC*, and *fusD* do not affect the fitness of the host strain and appear to be housekeeping genes. Furthermore, because they have no disadvantage to the host, they are able to survive and disseminate. It has been proposed that *fusB* is the ancestral gene found in *Staphylococcus saprophyticus* that has given rise to *fusC* and *fusD* in other staphylococcal species (O'Neill et al. 2007). At the gene level, the extent of homology is ~60% among these genes and, therefore, they can be missed when tested by hybridization methods (O'Neill et al. 2007). Mutations in *fusA* and *fusB* or *fusC* are not commonly found in the same organism (O'Neill and Chopra 2006). Corroborating this is the report from a collection of MRSA and MSSA in Taiwan. MRSA that harbored *fusA* did not contain *fusB*, and MSSA with *fusB* did not contain *fusA* (Chen et al. 2010a). FusB, FusC, and FusD are thought to be accessory proteins in bacterial protein synthesis whose natural role has yet to be characterized (Cox et al. 2012). By homology, FusB is related to *Listeria monocytogenes* fibronectin-binding protein, but FusB does not bind fibronectin (Guo et al. 2012).

NOVEL CLASS AND NOVEL STRUCTURE: IS THERE CROSS RESISTANCE?

Although cross resistance to other classes of antibiotics is not expected because fusidic acid belongs to a unique class, such resistance has been noted in the aminoglycosides, like kanamycin (Johanson and Hughes 1994; Norstrom

et al. 2007). This cross resistance is noted in small colony variants of *S. aureus* that have mutations in the structural domain V of EF-G and sometimes in domains I or III. There may also be mutations in *rplF* coding for ribosomal protein L6. The *rplF* mutant is also known as *fusE*. In addition, resistance owing to decreased permeability has also been reported (Collignon and Turnidge 1999), but is not likely a major mechanism of resistance.

Fusidic acid is also known for having particularly protective effects on linezolid by extending the resistance development window in *S. aureus* from five transfers to 25 transfers when the two antibiotics are tested in combination (Miller et al. 2008). These investigators showed that this delay in resistant mutant selection to linezolid could be through an effect on recombination as the protection was lost in *recA* mutants.

PHARMACOKINETIC AND PHARMACODYNAMIC APPROACH TO MODIFIED DOSING OF FUSIDIC ACID TO DECREASE RESISTANCE DEVELOPMENT

Because resistance in staphylococcal strains to fusidic acid was believed to occur rapidly, fusidic acid has been used clinically in combination with rifampin for several decades (Howden and Grayson 2006). Surprisingly, it has been noted recently that coadministration of rifampin with fusidic acid results in drug–drug interactions, presumably by the induction of the liver enzyme CYP3A4 by rifampin. These interactions subsequently caused a decrease in the blood levels of fusidic acid to sufficiently low plasma levels, allowing for possible resistance development. Even more importantly, the low fusidic acid plasma levels could result in exposure of staphylococci solely to rifampin (Oldach et al. 2015). It is well known that rifampin must only be used in combination with another antibiotic to protect it from selecting for RNA polymerase mutants. Rifampin is known to induce CYP3A4 and fusidic acid is metabolized by cytochrome P450 enzymes (Mandell 2010). Therefore, it may be deduced that rifampin induces CYP3A4 in the liver, which then metabo-

Cite this article as *Cold Spring Harb Perspect Med* doi: 10.1101/cshperspect.a025437

lizes fusidic acid. As noted in an earlier section, the frequency of resistance decreases with increasing concentrations of the drug. Therefore, modern pharmacokinetic/pharmacodynamic modeling studies were pursued to determine if the dose could be adjusted to decrease resistance development such that it would allow clinical use of fusidic acid in monotherapy.

Fusidic acid has a long history of clinical use in Europe, Canada, and Australia. It has been used for staphylococcal and streptococcal skin infections with shown efficacy. Although oral tablets are used for 10–14 d for skin and soft structure infections, it has also been used safely for months to years for treating chronic bone and joint infections. In these cases, fusidic acid is often combined with rifampin (Drancourt et al. 1997; Zimmerli et al. 2004; Trampuz and Widmer 2006; Aboltins et al. 2007; Ferry et al. 2010; Chiang et al. 2011). In Europe, the current dosing regimen is 500 mg administered three times a day, orally, which delivers high plasma concentration after a few days, but low plasma levels after the first dose. This low exposure during the first 24 h is likely to result in mutant selection as evidenced by the in vitro resistance development studies. As noted above, the frequency of resistance is ~10^{-6} to 10^{-8} at 2 and 16 times the MIC, respectively, but resistance rates were shown to be $\geq 10^{-11}$ when the exposure to fusidic acid is higher, for example, at 15

and 30 mg/L (O'Neill et al. 2001). The pharmacokinetics (PKs) of fusidic acid in rodents is not an accurate predictor of the corresponding PK parameters in humans, as the oral bioavailability of fusidic acid in mice and rats is limited and the amount that is absorbed is rapidly excreted (Degenhardt et al. 2009). Therefore, dose optimization studies cannot be conducted using rodent models, which then led to the development of a hollow fiber infection model (HFIM) using *S. aureus* and *S. pyogenes* for optimization of the dosing regimen (Okusanya et al. 2011; Tsuji et al. 2011).

Monte Carlo simulations using human population PKs (Still et al. 2011; Tsuji et al. 2011), an in vitro pharmacodynamic infection model, and kill kinetics from the HFIM predicted the following front-loaded dosing regimens: fusidic acid dosed at either 1200 or 1500 mg orally every 12 h on day 1 followed by 600 mg every 12 h on subsequent days. This would likely be an effective dosing regimen for acute bacterial skin and skin structure infections (ABSSSIs) and thereby substantially reduce the emergence of resistant bacteria. Figure 8 shows the difference in plasma levels obtained from the European dosing schedule versus the new dose regimen being used in clinical trials in the United States.

A loading and maintenance dose regimen was selected from hollow fiber studies and test-

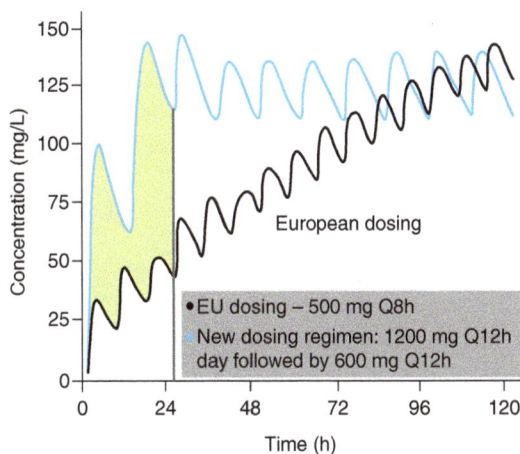

Figure 8. Plasma levels modeled from population pharmacokinetic data showing the high plasma levels achieved in the first 24 h for the new loading dose–maintenance dose regimen versus the European dose regimen.

ed in a phase 2 clinical trial of ABSSSI in the United States (Craft et al. 2011). This regimen precluded the need for using rifampin or other antibiotics to help decrease resistance development. In this study, fusidic acid was tested as a loading dose and maintenance dose for 10–14 d, in which it showed comparable efficacy to linezolid. Thus, a loading dose–maintenance dose strategy has been selected to overcome resistance development in monotherapy in ABSSSI as well as in bone and joint infections (Wolfe 2011).

Fusidic acid has been used in treating chronic bone and joint infections, in which biofilm is thought to play a role in limiting antibiotic penetration. An in vitro model of biofilm was used to test fusidic acid's activity alone and in combination with traditional staphylococcal antibiotics (Siala et al. 2015). *S. aureus* reference strain ATCC25923 and clinical strains isolated from medical devices or from chronic tissue infections were used. Biofilms were grown for 24 h in 96 well plates and then exposed for 48 h to increasing concentrations (0.25–64 mg/L) of fusidic acid (to obtain full concentration-response curves), combined with concentrations corresponding to the combination drug. Combining fusidic acid with daptomycin, vancomycin, or linezolid improved its efficacy with reduction in staphylococcal viability in the biofilm model as shown in Figure 9 and appears to be a useful strategy to increase its antibacterial activity in biofilms.

Fusidic acid is active against *E. coli* EF-G but does not have significant activity against Gram-negative bacteria, as it is thought to be excluded by their cell membranes. The recent spread of multi-drug-resistant Gram-negative pathogens and the inability to treat these infections even with colistin has led to testing fusidic acid in combination with colistin (Phee et al. 2015). Fusidic acid is synergistic with colistin against *Acinetobacter baumannii* in vitro and the combination prevented resistance selection to both agents. Moreover, fusidic acid was active against *A. baumannii* (even when the strain was resistant to colistin), leaving the unanswered question of how this antibiotic enters the Gram-negative cell. There is one report of the successful

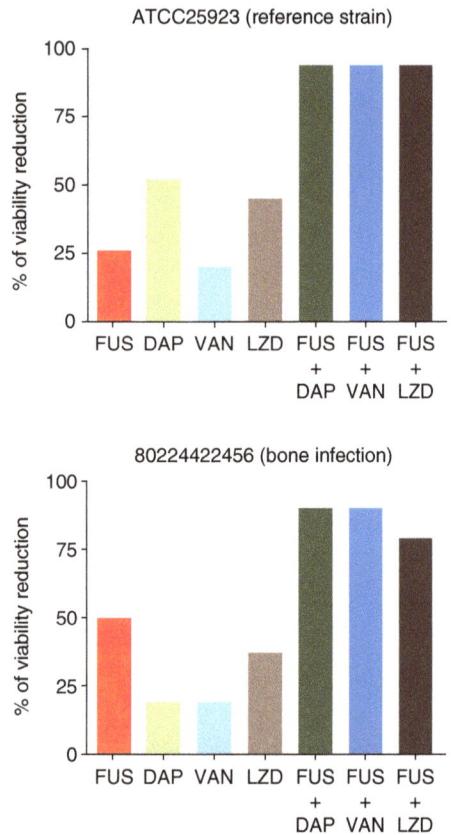

Figure 9. Percent reduction in viability of *S. aureus* within biofilms for fusidic acid, daptomycin, vancomycin, and linezolid or for fusidic acid at free drug C_{max} combined with other antibiotics. Reduction in viability was compared with the untreated control.

treatment of a patient with an infection with multidrug-resistant *A. baumannii* (Phee et al. 2014). This exciting potential for clinical use of fusidic acid will be of interest as the combination use is expanded to Gram-negative multidrug-resistant pathogens.

Fusidic acid formulated as an ointment is used more commonly than orally administered tablets. Topical use of fusidic acid is not recommended to minimize the transfer of resistance genes from commensal skin bacteria. The association between exposure to topical fusidic acid and development of resistance is attributed to the high frequency of spontaneous mutations (Mason and Howard 2004).

CONCLUDING REMARKS

Antibiotics that bind to the ribosome to inhibit protein synthesis have been successfully used for treating bacterial infections. Fusidic acid has an interesting and novel mode of action in that it shows selective inhibition of EF-G in its GDP-bound state, thereby blocking both translocation and ribosome disassembly (two key steps in protein synthesis). High-level resistance is rare and the development of a new dosing regimen could decrease the risk of such resistance and the need to use fusidic acid in combination with other drugs. With the implementation of this new dosing strategy, fusidic acid could become a useful oral antistaphylococcal antibiotic for treating both complicated skin infections and chronic bone and joint infections. As shown by recent data, its use could be further expanded to include Gram-negative bacteria once the permeability hurdle is overcome with antibiotic combinations.

Structural studies on the mechanism of fusidic acid's interaction with EF-G-GDP could be useful in developing new/better fusidanes along with other molecules. In addition, revelation of common features in the structural changes in EF-G as well as EF-Tu, two GTPase proteins that function in sequence in protein synthesis, could aid in the design of inhibitors that block both these proteins. Such an inhibitor would be capable of blocking three proteins—IF-2, EF-G, and RRF—in the same pathway, increasing potency and further decreasing the likelihood of resistance development.

ACKNOWLEDGMENTS

I thank Dr. Ronald Jones, Dr. Paul Ambrose, and their teams for the design of the new dosing regimen. I also thank the staff at Cempra, Dr. Sherwood Gorbach, and Dr. Carl Craft who have helped to develop this drug in the United States. I also acknowledge the help of Dr. David Pereira for drawing the chemical structures and the support of Evan Martens, Biologist at Cempra, who has helped organize the references and has given me so much of his time to carefully edit this manuscript.

REFERENCES

Aboltins CA, Page MA, Buising KL, Jenney AW, Daffy JR, Choong PF, Stanley PA. 2007. Treatment of staphylococcal prosthetic joint infections with debridement, prosthesis retention and oral rifampicin and fusidic acid. *Clin Microbiol Infect* **13**: 586–591.

Agrawal RK, Penczek P, Grassucci RA, Frank J. 1998. Visualization of elongation factor G on the *Escherichia coli* 70S ribosome: The mechanism of translocation. *Proc Natl Acad Sci* **95**: 6134–6138.

Besier S, Ludwig A, Brade V, Wichelhaus TA. 2003. Molecular analysis of fusidic acid resistance in *Staphylococcus aureus*. *Mol Microbiol* **47**: 463–469.

Besier S, Ludwig A, Brade V, Wichelhaus TA. 2005. Compensatory adaptation to the loss of biological fitness associated with acquisition of fusidic acid resistance in *Staphylococcus aureus*. *Antimicrob Agents Chemother* **49**: 1426–1431.

Betina V. 1983. Chemical classification of antibiotics. In *The chemistry and biology of antibiotics* (ed. Nauta WTh, Rekker RF), pp. 154–156. Elsevier, Amsterdam.

Biedenbach DJ, Rhomberg PR, Mendes RE, Jones RN. 2010. Spectrum of activity, mutation rates, synergistic interactions, and the effects of pH and serum proteins for fusidic acid (CEM-102). *Diagn Microbiol Infect Dis* **66**: 301–307.

Borg A, Holm M, Shiroyama I, Hauryliuk V, Pavlov M, Sanyal S, Ehrenberg M. 2015. Fusidic acid targets elongation factor G in several stages of translocation on the bacterial ribosome. *J Biol Chem* **290**: 3440–3454.

Castanheira M, Watters AA, Mendes RE, Farrell DJ, Jones RN. 2010. Occurrence and molecular characterization of fusidic acid resistance mechanisms among *Staphylococcus* spp. from European countries (2008). *J Antimicrob Chemother* **65**: 1353–1358.

Chen HJ, Hung WC, Tseng SP, Tsai JC, Hsueh PR, Teng LJ. 2010a. Fusidic acid resistance determinants in *Staphylococcus aureus* clinical isolates. *Antimicrob Agents Chemother* **54**: 4985–4991.

Chen Y, Koripella RK, Sanyal S, Selmer M. 2010b. *Staphylococcus aureus* elongation factor G-structure and analysis of a target for fusidic acid. *FEBS J* **277**: 3789–3803.

Chiang ER, Su YP, Chen TH, Chiu FY, Chen WM. 2011. Comparison of articulating and static spacers regarding infection with resistant organisms in total knee arthroplasty. *Acta Orthop* **82**: 460–464.

Chung HK, Spremulli LL. 1990. Purification and characterization of elongation factor G from bovine liver mitochondria. *J Biol Chem* **265**: 21000–21004.

Collignon P, Turnidge J. 1999. Fusidic acid in vitro activity. *Int J Antimicrob Agents* **12**: S45–S58.

Cox G, Thompson GS, Jenkins HT, Peske F, Savelsbergh A, Rodnina MV, Wintermeyer W, Homans SW, Edwards TA, O'Neill AJ. 2012. Ribosome clearance by FusB-type proteins mediates resistance to the antibiotic fusidic acid. *Proc Natl Acad Sci* **109**: 2102–2107.

Craft JC, Moriarty SR, Clark K, Scott D, Degenhardt TP, Still JG, Corey GR, Das A, Fernandes P. 2011. A randomized, double-blind phase 2 study comparing the efficacy and safety of an oral fusidic acid loading-dose regimen to oral linezolid for the treatment of acute bacterial skin and skin structure infections. *Clin Infect Dis* **52:** S520–S526.

Degenhardt TP, Still JG, Clark K, Fernandes P. 2009. From mouse to man: The pharmacokinetics of CEM-102 (fusidic acid). In *49th Interscience Conference on Antimicrobial Agents Chemotherapy*, Poster A1-1930. San Francisco, CA.

Dobie D, Gray J. 2004. Fusidic acid resistance in *Staphylococcus aureus*. *Arch Dis Child* **89:** 74–77.

Drancourt M, Argenson J, Roiron R, Groulier R, Raoult D. 1997. Oral treatment of *Staphylococcus* spp. infected orthopaedic implants with fusidic acid or ofloxacin in combination with rifampicin. *J Antimicrob Chemother* **39:** 235–240.

Fernandes P, Pereira D. 2011. Efforts to support the development of fusidic acid in the United States. *Clin Infect Dis* **52:** S542–S546.

Ferry T, Uckay I, Vaudaux P, Francois P, Schrenzel J, Harbarth S, Laurent F, Bernard L, Vandenesch F, Etienne JP, et al. 2010. Risk factors for treatment failure in orthopedic device-related methicillin-resistant *Staphylococcus aureus* infection. *Eur J Clin Microbiol Infect Dis* **29:** 171–180.

Frank J, Agrawal RK. 2000. A ratchet-like inter-subunit reorganization of the ribosome during translocation. *Nature* **406:** 318–322.

Gao YG, Selmer M, Dunham CM, Weixlbaumer A, Kelley AC, Ramakrishnan V. 2009. The structure of the ribosome with elongation factor G trapped in the posttranslational state. *Science* **326:** 694–699.

Godtfredsen WO, Vangedal S. 1966. On the metabolism of fusidic acid in man. *Acta Chem Scand* **20:** 1599–1607.

Godtfredsen W, Roholt K, Tybring L. 1962. Fucidin: A new orally active antibiotic. *Lancet* **1:** 928–931.

Guo X, Peisker K, Backbro K, Chen Y, Koripella RK, Mandava CS, Sanyal S, Selmer M. 2012. Structure and function of FusB: An elongation factor G-binding fusidic acid resistance protein active in ribosomal translocation and recycling. *Open Biol* **2:** 120016.

Hansson S, Singh R, Gudkov AT, Liljas A, Logan DT. 2005. Structural insights into fusidic acid resistance and sensitivity in EF-G. *J Mol Biol* **348:** 939–949.

Hardy DJ, Vicino D, Keedy K, Fernandes P. 2014. Susceptibility of *Propionibacterium acnes* to fusidic acid. In *54th Interscience Conference on Antimicrobial Agents Chemotherapy*, Abstr. D-867.7. Washington, DC.

Howden BP, Grayson ML. 2006. Dumb and dumber—The potential waste of a useful antistaphylococcal agent: Emerging fusidic acid resistance in *Staphylococcus aureus*. *Clin Infect Dis* **42:** 394–400.

Johanson U, Hughes D. 1994. Fusidic acid-resistant mutants define three regions in elongation factor G of *Salmonella typhimurium*. *Gene* **143:** 55–59.

Johanson U, Aevarsson A, Liljas A, Hughes D. 1996. The dynamic structure of EF-G studied by fusidic acid resistance and internal revertants. *J Mol Biol* **258:** 420–432.

Jones RN, Li Q, Kohut B, Biedenbach DJ, Bell J, Turnidge JD. 2006. Contemporary antimicrobial activity of triple antibiotic ointment: A multiphased study of recent clinical isolates in the United States and Australia. *Diagn Microbiol Infect Dis* **54:** 63–71.

Jones RN, Castanheira M, Rhomberg PR, Woosley LN, Pfaller MA. 2010a. Performance of fusidic acid (CEM-102) susceptibility testing reagents: Broth microdilution, disk diffusion, and Etest methods as applied to *Staphylococcus aureus*. *J Clin Microbiol* **48:** 972–976.

Jones RN, Biedenbach DJ, Roblin PM, Kohlhoff SA, Hammerschlag MR. 2010b. Update on fusidic acid (CEM-102) tested against *Neisseria gonorrhoeae* and *Chlamydia trachomatis*. *Antimicrob Agents Chemother* **54:** 4518–4519.

Jones RN, Mendes RE, Sader HS, Castanheira M. 2011. In vitro antimicrobial findings for fusidic acid tested against contemporary (2008–2009) Gram-positive organisms collected in the United States. *Clin Infect Dis* **52:** S477–S486.

Koripella RK, Chen Y, Peisker K, Koh CS, Selmer M, Sanyal S. 2012. Mechanism of elongation factor-G-mediated fusidic acid resistance and fitness compensation in *Staphylococcus aureus*. *J Biol Chem* **287:** 30257–30267.

Laurberg M, Kristensen O, Martemyanov K, Gudkov AT, Nagaev I, Hughes D, Liljas A. 2000. Structure of a mutant EF-G reveals domain III and possibly the fusidic acid binding site. *J Mol Biol* **303:** 593–603.

Lemaire S, Van Bambeke F, Pierard D, Appelbaum PC, Tulkens PM. 2011. Activity of fusidic acid against extracellular and intracellular *Staphylococcus aureus*: Influence of pH and comparison with linezolid and clindamycin. *Clin Infect Dis* **52:** S493–S503.

Mandell LA. 2010. Fusidic acid. In *Principles and practice of infectious diseases* (ed. Mandell GL, et al.), pp. 355–357. Churchill Livingstone, Philadelphia.

Margus T, Remm M, Tenson T. 2007. Phylogenetic distribution of translational GTPases in bacteria. *BMC Genomics* **8:** 15.

Martemyanov KA, Liljas A, Yarunin AS, Gudkov AT. 2001. Mutations in the G-domain of elongation factor G from *Thermus thermophilus* affect both its interaction with GTP and fusidic acid. *J Biol Chem* **276:** 28774–28778.

Mason BW, Howard AJ. 2004. Fusidic acid resistance in community isolates of methicillin susceptible *Staphylococcus aureus* and the use of topical fusidic acid: A retrospective case-control study. *Int J Antimicrob Agents* **23:** 300–303.

McLaws FB, Larsen AR, Skov RL, Chopra I, O'Neill AJ. 2011. Distribution of fusidic acid resistance determinants in methicillin-resistant *Staphylococcus aureus*. *Antimicrob Agents Chemother* **55:** 1173–1176.

Miller K, O'Neill AJ, Wilcox MH, Ingham E, Chopra I. 2008. Delayed development of linezolid resistance in *Staphylococcus aureus* following exposure to low levels of antimicrobial agents. *Antimicrob Agents Chemother* **52:** 1940–1944.

Nagaev I, Bjorkman J, Andersson DI, Hughes D. 2001. Biological cost and compensatory evolution in fusidic acid-resistant *Staphylococcus aureus*. *Mol Microbiol* **40:** 433–439.

Norstrom T, Lannergard J, Hughes D. 2007. Genetic and phenotypic identification of fusidic acid-resistant mutants with the small-colony-variant phenotype in *Staphylococcus aureus*. *Antimicrob Agents Chemother* **51:** 4438–4446.

Cite this article as *Cold Spring Harb Perspect Med* doi: 10.1101/cshperspect.a025437

Okusanya OO, Tsuji BT, Bulitta JB, Forrest A, Bulik CC, Bhavnani SM, Fernandes P, Ambrose PG. 2011. Evaluation of the pharmacokinetics-pharmacodynamics of fusidic acid against *Staphylococcus aureus* and *Streptococcus pyogenes* using in vitro infection models: Implications for dose selection. *Diagn Microbiol Infect Dis* **70:** 101–111.

Oldach D, Stolarski E, Murphy B, Mould D, Das A, Fernandes P. 2015. Rifampin (RIF) significantly reduces plasma concentrations of fusidic acid (FA) when used in combination for treatment of prosthetic joint infection (PJI). In *25th European Congress of Clinical Microbiology and Infectious Diseases*, Poster 2352. Copenhagen, Denmark.

Oliva B, O'Neill AJ, Miller K, Stubbings W, Chopra I. 2004. Anti-staphylococcal activity and mode of action of clofazimine. *J Antimicrob Chemother* **53:** 435–440.

O'Neill AJ, Chopra I. 2006. Molecular basis of fusB-mediated resistance to fusidic acid in *Staphylococcus aureus*. *Mol Microbiol* **59:** 664–676.

O'Neill AJ, Cove JH, Chopra I. 2001. Mutation frequencies for resistance to fusidic acid and rifampicin in *Staphylococcus aureus*. *J Antimicrob Chemother* **47:** 647–650.

O'Neill AJ, Larsen AR, Skov R, Henriksen AS, Chopra I. 2007. Characterization of the epidemic European fusidic acid-resistant impetigo clone of *Staphylococcus aureus*. *J Clin Microbiol* **45:** 1505–1510.

Pfaller MA, Castanheira M, Sader HS, Jones RN. 2010. Evaluation of the activity of fusidic acid tested against contemporary Gram-positive clinical isolates from the USA and Canada. *Int J Antimicrob Agents* **35:** 282–287.

Phee L, BB, Wareham DW. 2014. Successful treatment of multi-drug resistant *Acinetobacter baumannii* ventilator-associated pneumonia with a novel colistin and fusidic acid combination therapy. In *ID Week*, Poster 481. Philadelphia, PA.

Phee LM, Betts JW, Bharathan B, Wareham DW. 2015. Colistin and fusidic acid: A novel potent synergistic combination for the treatment of multi-drug resistant *Acinetobacter baumannii* infections. *Antimicrob Agents Chemother* **59:** 4544–4550.

Reeves DS. 1987. The pharmacokinetics of fusidic acid. *J Antimicrob Chemother* **20:** 467–476.

Sahm DF, Deane J, Pillar CM, Fernandes P. 2013. In vitro activity of CEM-102 (fusidic acid) against prevalent clones and resistant phenotypes of *Staphylococcus aureus*. *Antimicrob Agents Chemother* **57:** 4535–4536.

Savelsbergh A, Rodnina MV, Wintermeyer W. 2009. Distinct functions of elongation factor G in ribosome recycling and translocation. *RNA* **15:** 772–780.

Siala W, R-V H, Fernandes P, Tulkens PM, Van Bambeke F. 2015. Activity of fusidic acid (FUS) alone or in combination with daptomycin (DAP), vancomycin (VAN), or linezolid (LZD) in an in vitro model of *Staphylococcus aureus* biofilm. In *the 25th European Congress on Clinical Microbiology and Infectious Diseases*, Abstr. EV0056. Copenhagen, Denmark.

Spelman D. 1999. Fusidic acid in skin and soft tissue infections. *Int J Antimicrob Agents* **12:** S59–S66.

Still JG, Clark K, Degenhardt TP, Scott D, Fernandes P, Gutierrez MJ. 2011. Pharmacokinetics and safety of single, multiple, and loading doses of fusidic acid in healthy subjects. *Clin Infect Dis* **52:** S504–S512.

Thaker M, Spanogiannopoulos P, Wright GD. 2010. The tetracycline resistome. *Cell Mol Life Sci* **67:** 419–431.

Trampuz A, Widmer AF. 2006. Infections associated with orthopedic implants. *Curr Opin Infect Dis* **19:** 349–356.

Tran JH, Jacoby GA, Hooper DC. 2005. Interaction of the plasmid-encoded quinolone resistance protein Qnr with *Escherichia coli* DNA gyrase. *Antimicrob Agents Chemother* **49:** 118–125.

Tsuji BT, Okusanya OO, Bulitta JB, Forrest A, Bhavnani SM, Fernandez PB, Ambrose PG. 2011. Application of pharmacokinetic-pharmacodynamic modeling and the justification of a novel fusidic acid dosing regimen: Raising Lazarus from the dead. *Clin Infect Dis* **52:** S513–S519.

Turnidge J. 1999. Fusidic acid pharmacology, pharmacokinetics and pharmacodynamics. *Int J Antimicrob Agents* **12:** S23–S34.

Vaillant L, Machet L, Taburet AM, Sorensen H, Lorette G. 1992. Levels of fusidic acid in skin blister fluid and serum after repeated administration of two dosages (250 and 500 mg). *Br J Dermatol* **126:** 591–595.

Vaillant L, Le Guellec C, Jehl F, Barruet R, Sorensen H, Roiron R, Autret-Leca E, Lorette G. 2000. Comparative diffusion of fusidic acid, oxacillin, and pristinamycin in dermal interstitial fluid after repeated oral administration. *Ann Dermatol Venereol* **127:** 33–39.

Verbist L. 1990. The antimicrobial activity of fusidic acid. *J Antimicrob Chemother* **25:** 1–5.

von Daehne W, Godtfredsen WO, Rasmussen PR. 1979. Structure-activity relationships in fusidic acid-type antibiotics. *Adv Appl Microbiol* **25:** 95–146.

Williamson DA, Monecke S, Heffernan H, Ritchie SR, Roberts SA, Upton A, Thomas MG, Fraser JD. 2014. High usage of topical fusidic acid and rapid clonal expansion of fusidic acid-resistant *Staphylococcus aureus*: A cautionary tale. *Clin Infect Dis* **59:** 1451–1454.

Wilson DN. 2014. Ribosome-targeting antibiotics and mechanisms of bacterial resistance. *Nat Rev Microbiol* **12:** 35–48.

Wolfe CR. 2011. Case report: Treatment of chronic osteomyelitis. *Clin Infect Dis* **52:** S538–S541.

Wullt M, Odenholt I. 2004. A double-blind randomized controlled trial of fusidic acid and metronidazole for treatment of an initial episode of *Clostridium difficile*-associated diarrhea. *J Antimicrob Chemother* **54:** 211–216.

Zimmerli W, Trampuz A, Ochsner PE. 2004. Prosthetic-joint infections. *N Engl J Med* **351:** 1645–1654.

Antibacterial Antifolates: From Development through Resistance to the Next Generation

Alexavier Estrada, Dennis L. Wright, and Amy C. Anderson

Department of Pharmaceutical Sciences, University of Connecticut, Storrs, Connecticut 06269

Correspondence: dennis.wright@uconn.edu; amy.anderson@uconn.edu

The folate cycle is one of the key metabolic pathways used by bacteria to synthesize vital building blocks required for proliferation. Therapeutic agents targeting enzymes in this cycle, such as trimethoprim and sulfamethoxazole, are among some of the most important and continually used antibacterials to treat both Gram-positive and Gram-negative pathogens. As with all antibacterial agents, the emergence of resistance threatens the continued clinical use of these life-saving drugs. In this article, we describe and analyze resistance mechanisms that have been clinically observed and review newer generations of preclinical compounds designed to overcome the molecular basis of the resistance.

Inhibitors of the folate biosynthetic pathway have been successful drugs since the 1940s when sulfa powder was first applied topically to soldier's wounds to prevent infection on the battlefield. The folate pathway is essential in the synthesis of one-carbon donors needed for the production of deoxythymidine monophosphate (dTMP), purine nucleotides, and the amino acids methionine and histidine. In bacteria, dihydrofolate is first synthesized from early precursor molecules. Most notably, the bacterial enzyme, dihydropteroate synthase (DHPS) catalyzes the formation of 7,8-dihydropteroate from *para*-aminobenzoic acid and 6-hydroxymethyl-7,8,-dihydropterin pyrophosphate (Fig. 1). One additional biosynthetic transformation in the bacteria results in the formation of dihydrofolate, which is reduced by the essential enzyme dihydrofolate reductase (DHFR), to tetrahydrofolate. DHFR uses the cofactor, nicotinamide adenine dinucleotide phosphate (NADPH), as the stoichiometric reducing agent bound in a site immediately adjacent to the folate-binding site. These early steps in the formation of dihydrofolate are selective for bacteria as mammalian cells obtain folic acid from dietary sources and then sequentially reduce it to tetrahydrofolate using DHFR. Therefore, antibacterial agents targeting steps before DHFR have inherent selectivity, whereas subsequent targets require the design of species-selective agents. The folate cofactors are then part of a critical cycle from which the essential metabolites dTMP, purine nucleotides, histidine, and methionine required for bacterial growth and division are produced.

Blockade of the folate pathway using inhibitors called "antifolates" results in an effective "thymine-less death" for the bacterial cell. The earliest programs to discover antibacterials commenced in the early 1930s and centered on antibacterial activity observed with synthetic dyes. Bayer chemists discovered Prontosil, one of the first systemic antibacterial agents to become widely used (van Miert 1994). Later, it was found that Prontosil is a prodrug and became metabolized to the active agent, sulfanilamide, which was identified as a mimic of the *para*-aminobenzoic acid substrate and a potent inhibitor of DHPS. Many additional inhibitors of DHPS, such as sulfamethoxazole (Fig. 2), followed and were synthesized to improve pharmacokinetic parameters, reduce side effects, and overcome early resistance (Skold 2010). During the 1940s, George Hitchings and Gertrude Elion determined that substituted 2,4-diaminopyrimidines also interfere with folate metabolism. They synthesized trimethoprim, the first inhibitor of DHFR (Fig. 2). Trimethoprim is very selective for some bacterial DHFR enzymes such as *Staphylococcus aureus* and *Escherichia coli* over human DHFR, a highly potent antibacterial and orally bioavailable. One noteworthy feature of trimethoprim (TMP) affinity is its significant cooperativity in binding with the NADPH cofactor. Soon after, Bactrim, a combination of trimethoprim and sulfamethoxazole, became a widely used therapy to treat infections caused by both Gram-positive (skin and soft-tissue infections and pneumonia) and Gram-negative (urinary tract infections) bacteria. To date, tri-

methoprim and sulfamethoxazole have been one of the few truly synergistic antibacterial combinations and the only clinically used antifolates to treat bacterial infections (Darmstadt 1997; Stevens et al. 2005; Gorwitz et al. 2006; Nathwani et al. 2008; Frei et al. 2010).

In this review, we will describe the common mechanisms of resistance to both trimethoprim and sulfamethoxazole, drawing comparisons between resistant proteins when possible to understand the limitations of the enzymes in developing resistance and to pose possibilities to use that knowledge to overcome resistance. New research to develop DHFR and DHPS inhibitors has focused on overcoming current mechanisms of resistance and providing a next generation of antifolates.

RESISTANCE TO TRIMETHOPRIM

During the 1980s, resistance to Bactrim began to occur in both the hospital and community settings (Then et al. 1992; Dale et al. 1997). Two prevalent modes of resistance arose: chromosomal mutations in the *dfrB* and *folP* genes of *S. aureus*, which encode for DHFR and DHPS, respectively, and the horizontal transfer of plasmid-encoded, trimethoprim-resistant *dfr* genes or sulfamethoxazole-resistant *sul* genes in both

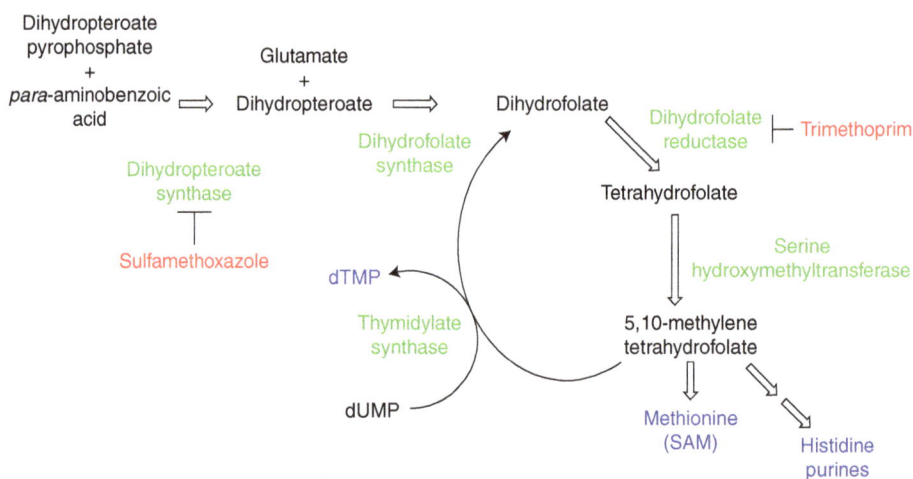

Figure 1. The folate cycle in bacteria. Enzymes are shown in green, key products in blue, and inhibitors in red. Dihydropteroate synthase (DHPS), a critical enzyme in the formation of dihydrofolate, is inhibited by sulfamethoxazole, and dihydrofolate reductase (DHFR) is inhibited by trimethoprim. dTMP, Deoxythymidine monophosphate; dUMP, deoxyuridine monophosphate; SAM, *S*-adenosylmethionine

Cite this article as *Cold Spring Harb Perspect Med* doi: 10.1101/cshperspect.a028324

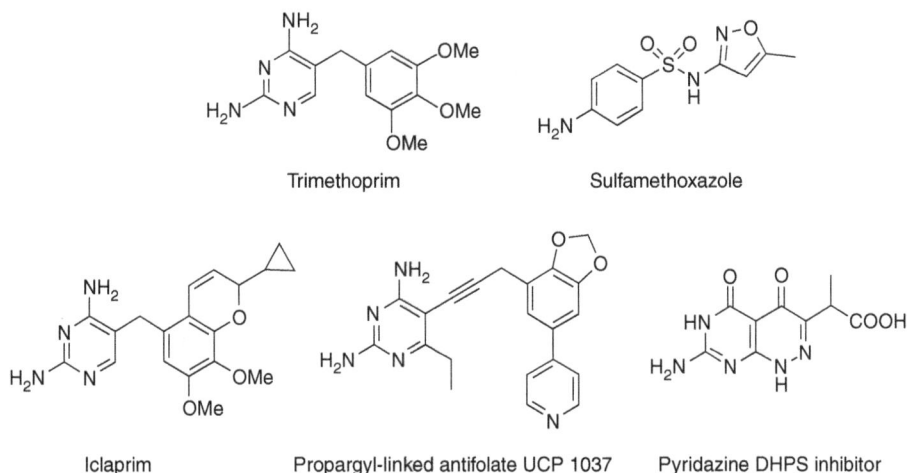

Figure 2. Antifolates. Clinically approved antibacterial antifolates: trimethoprim and sulfamethoxazole. Preclinical and experimental antifolates: dihydrofolate reductase (DHFR) inhibitors iclaprim and propargyl-linked antifolates, and dihydropteroate synthase pyridazine (DHPS) inhibitors.

Gram-positive and Gram-negative bacteria (Amyes 1989; Amyes et al. 1992; Skold 2010). A discussion of the chromosomal mutations and plasmid-encoded resistant proteins follows; comparisons of these resistant proteins (Fig. 3) may shed light on common mechanisms to decrease drug affinity.

In 1997, Dale et al. (1997) reported that chromosomal mutations in the *dfrB* gene encoding DHFR in *S. aureus* were responsible for minimum inhibitory concentration (MIC) changes of ∼256-fold. Of the chromosomal mutations, the active-site mutation, F98Y, posed the greatest problem, conferring >400-fold resistance to trimethoprim (IC$_{50}$ values for TMP against the wild-type (wt) and F98Y enzymes are 0.01 and 4.1 μM, respectively). The F98Y mutation most frequently occurs with an additional compensatory mutation, either H30N or H149R, that bring the fitness of the enzyme to near wild-type levels and increase the degree of resistance to trimethoprim by 800- to 2400-fold over wt. At the time the study was published, 88% of resistant strains possessed MIC changes of ∼256-fold, suggesting the widespread presence of chromosomal mutations in the *dfrB* gene.

In parallel, a transposon-encoded trimethoprim-resistant enzyme encoded by the gene *dfrA* and called S1 DHFR was isolated and shown to confer resistance to strains in Australia and the United States (Amyes 1989). S1 DHFR, which may originate from *Staphylococcus epidermis*, possesses three key residue substitutions, F98Y, V31I, and G43A, that have been shown to be responsible for the majority of TMP resistance (Fig. 3) (Dale et al. 1995). It is noteworthy that this plasmid-encoded DHFR recapitulates the key F98Y mutation observed in the chromosome. A detailed kinetic and structural comparative study of the wt and S1 enzymes revealed that the S1 enzyme independently reduces the interactions of both TMP and NADPH such that the binding synergy between the two is significantly reduced by 750-fold. A crystal structure of S1 with TMP revealed six molecules in the asymmetric unit, only three of which were the ternary structure. The other three S1 proteins bound only TMP, indicating the altered interactions of the S1 enzyme with NADPH (Heaslet et al. 2009).

In 2005, a second exogenous DHFR encoded by *dfrG* and called S3 DHFR, was discovered in a collection of 43 clinical isolates of TMP-resistant strains of methicillin-resistant *Staphylococcus aureus* (MRSA) from Thailand and 244 isolates from Japan (243 of these were TMP-sensitive; one was resistant) (Sekiguchi

```
                                                   28 31
SaWT        -----MTLSILVAHDLQRVIGFENQLPWHLPNDLKHVKKLSTGHTLVMGRKTFESIGKPL    55
DfrA (S1)   -----MTLSIIVAHDKQRVIGYQNQLPWHLPNDLKHIKQLTTGNTLVMARKTFNSIGKPL    55
DfrG (S3)   -----MKVSLIAAMDKNRVIGKENDIPWRIPKDWEYVKNTTKGHPIILGRKNLESIGRAL    55
KpWT        ------MISLIAALAVDRVIGMENAMPWNLPADLAWFKRNTLNKPVVMGRLTWESIGRPL    54
DfrA1       -----MKISLMVAISKNGVIGNGPDIPWSAKGEQLLFKAITYNQWLLVGRKTFESMGA-L    54
DfrA12      MNSESVRIYLVAAMGANRVIGNGPNIPWKIPGEQKIFRRLTEGKVVVMGRKTFESIGKPL    60
DfrA17      -----MKISLISAVSENGVIGSGPDIPWSVKGEQLLFKALTYNQWLLVGRKTFDSMGV-L    54

                                                         92    98
SaWT        PNRRNVVLTSDT-SFNVEGVDVIHSIEDIYQLP----GHVFIFGGQTLFEEMIDKVDDMY   110
DfrA (S1)   PNRRNVVLTNQA-SFHHEGVDVINSLDEIKELS----GHVFIFGGQTLYEAMIDQVDDMY   110
DfrG (S3)   PDRRNIILTRDK-GFTFNGCEIVHSIEDVFELCKNE-EEIFIFGGEQIYNLFFPYVEKMY   113
KpWT        PGRKNIVISSKP-GS-DDRVQWVSSVEEAIAACGDV-EEIMVIGGGRVYEQFLPKAQKLY   111
DfrA1       PNRKYAVVTRSSFTSDNENVLIFPSIKDALTNLKKITDHVIVSGGGEIYKSLIDQVDTLH   114
DfrA12      PNRHTLVISRQA-NYRATGCVVVSTLSHAIALASELGNELYVAGGAEIYTLALPHAHGVF   119
DfrA17      PNRKYAVVSKNGISSSNENVLVFPSIENALKELSKVTDHVYVSGGGQIYNSLIEKADIIH   114

SaWT        ITVIEGK-FRGDTFFPPYTFEDWEVASSVEGKLDEKNTIPHTFLHLIRKK--------   159
DfrA (S1)   ITVIDGK-FQGDTFFPPYTFENWEVESSVEGQLDEKNTIPHTFLHLVRRKGK------   161
DfrG (S3)   ITKIHHE-FEGDTFFPEVNYEEWNEVFAQKGIKNDKNPYNYY-FHVYERKNLLS----   165
KpWT        LTHIDAE-VEGDTHFPDYDPDEWESVFSEFHDADAQNSHSYCFEILERR--------   159
DfrA1       ISTIDIE-PEGDVYFPEI-PSNFRPVFTQDFASNI----NYSYQ-IWQKG-------   157
DfrA12      LSEVHQT-FEGDAFFPMLNETEFELVSTETIQAVI----PYTH-SVYARRNG------   165
DfrA17      LSTVHVE-VEGDIKFPIM-PENFNLVFEQFFMSNI----NYTYQ-IWKKG-------   157
```

Figure 3. Sequence alignment for trimethoprim-sensitive (SaWT and KpWT) and trimethoprim-resistant dihydrofolate reductases (DHFRs). S1 DHFR and S3 DHFR are found in Gram-positive bacteria. DfrA1, DfrA12, and DfrA17 are found in Gram-negative bacteria. Residues colored in red are located in the inhibitor-binding pocket. As *Klebsiella pneumoniae* and *Escherichia coli* sequences are almost identical, only one sequence is shown here.

et al. 2005). TMP-resistant isolates possessing *dfrG* showed greatly elevated MIC values of at least 512 μg/mL. The *dfrG* gene, while originally horizontally transferred via plasmid, incorporated into the genome. Isolation and characterization of the S3 enzyme showed that, although K_M values for DHF and NADPH were similar to the wt enzyme, TMP affinity was weakened by ~20,000-fold (IC$_{50}$ values of the wt and S3 enzyme were 0.013 and 254 μM, respectively). MIC values for TMP using *E. coli* transformants with the new plasmid were significantly increased, validating that the presence of *dfrG* confers TMP resistance. All isolates from Thailand were resistant to TMP and possessed *dfrG*; only one isolate from Japan was resistant to TMP and contained *dfrG* (Sekiguchi et al. 2005). In a second publication, expansion of *dfrG* to Africa and transfer to European travelers from Africa was described (Nurjadi et al. 2014).

As Bactrim is first-line treatment for common infections caused by Gram-negative

bacteria, surveillance of resistance in this population is also actively monitored. In Gram-negative bacteria, especially the Enterobacteriaceae *E. coli* and *Klebsiella pneumoniae*, a different group of trimethoprim-resistant extrachromosomal DHFR proteins are horizontally transferred between bacteria. In Gram-negative bacteria, plasmid-encoded TMP resistance was first reported in 1992 (Amyes et al. 1992) with the description of two types of plasmid-encoded DHFR enzymes. Each type was highly resistant to trimethoprim, conferring MIC values >1 mM, but differed in their protein expression levels, with type 1 enzymes producing ~10-fold greater levels of protein than the chromosomal enzyme. Two studies, one conducted in Sweden (Brolund et al. 2010) and the other in Western Europe and Canada (Blahna et al. 2006), surveyed the prevalence of type 1 or type 2 integron-encoded trimethoprim-resistant *dfr* genes. The Swedish study concurred that the type 1 gene *dfrA1* is most prevalent

Cite this article as *Cold Spring Harb Perspect Med* doi: 10.1101/cshperspect.a028324

and occurred in 34% of resistant isolates. Of the 350 isolates studied in the European/Canadian study, 66% possessed an integron-associated *dfr* allele. The gene *dfrA1* was again the most prevalent (40%), followed by a second gene, *dfrA17* (31%).

Interestingly, although the sequences of the trimethoprim-resistant DHFR proteins diverge, the proteins share several motifs that distinguish the TMP-resistant proteins from the TMP-sensitive (Fig. 3). For example, position 28 is a leucine in *S. aureus* (wt), *E. coli*, and *K. pneumoniae* (wt) but changes to a tryptophan in S3 DHFR and is consistently a glutamine in all of the TMP-resistant Gram-negative Dfr proteins. This switch from a leucine to a bulky tryptophan or polar glutamine is likely to affect positioning of the diaminopyrimidine ring as well as the hydrophobic trimethoxyphenyl ring (Fig. 4A,B). These effects were in fact observed in the structure of DfrA1 bound to an experimental antifolate (Lombardo et al. 2015).

Position 31 is valine in *S. aureus* (wt) but a phenylalanine in *K. pneumoniae* (wt) or any of the Gram-negative DfrA proteins. The valine in *S. aureus* and S3 is balanced by Phe 92, which is on the opposite side of the active site, forming hydrophobic interactions that support the diaminopyrimidine ring and linker (Fig. 4A). Interestingly, this residue is an isoleucine in S1 DHFR, which is known to contribute to TMP resistance (Heaslet et al. 2009). In the Gram-negative enzymes, Phe at position 31 is balanced by less bulky residues on the opposite side of the active site (Fig. 4B): Ile in the wt, Ser in DfrA1, Ala in DfrA12, and Ser in DfrA17. These smaller residues (serine and alanine) in the DfrA1, DfrA12, and DfrA17 are likely to decrease affinity with the antifolates relative to the isoleucine in the wild-type.

Interestingly, position 98 is a phenylalanine in *S. aureus* (wt) and corresponds to a tyrosine in S1 and S3 DHFR that are responsible for TMP-resistance in Gram-positive bacteria (Fig. 4A,B). From the early clinical observations, the chromosomal F98Y mutation is also known to cause significant TMP resistance in *S. aureus* DHFR. More detailed studies have shown that this residue change greatly affects the position and affinity of the cofactor, NADPH (Heaslet et al. 2009) as well as the kinetics and entropy of binding. Recent structural and biochemical studies investigating the wild-type and F98Y enzymes show that the enzyme

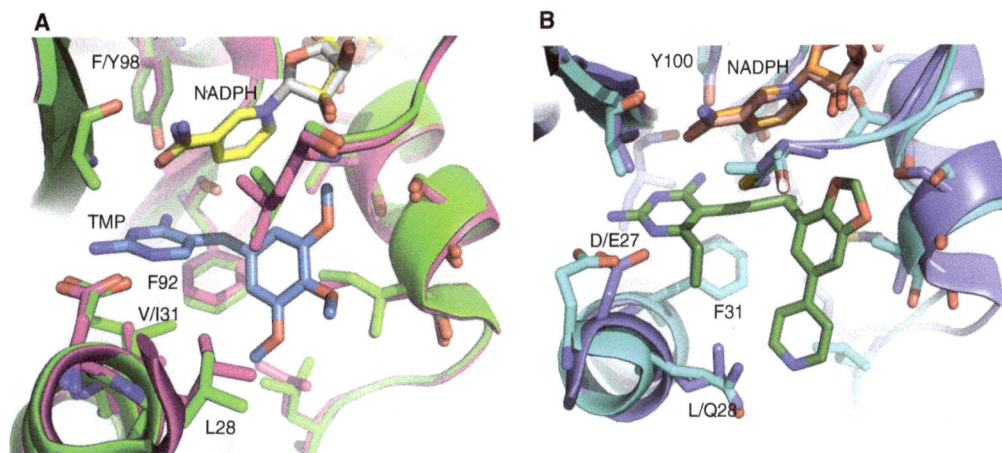

Figure 4. Structural comparisons of wild-type and trimethoprim-resistant DHFR. Structural superposition of (*A*) *Staphylococcus aureus* (wild-type [wt]) (pink) bound to nicotinamide adenine dinucleotide phosphate (NADPH) (yellow) and trimethoprim (TMP) (blue) with S1 dihydrofolate reductase (DHFR) (green) bound to NADPH and TMP (not shown) (Heaslet et al. 2009), and (*B*) *Klebsiella pneumoniae* (wt) (dark blue) bound to NADPH (orange) and a propargyl-linked antifolate (PLA) (not shown) with DfrA1 (cyan) bound to NADPH and a PLA (green) (Lombardo et al. 2015).

displays plasticity in the cofactor binding site, which is enhanced with the F98Y mutation. This plasticity affects a distribution of ternary states that includes an alternate bound configuration of the cofactor (Keshipeddy et al. 2015). In contrast, both the wild-type *E. coli* and *K. pneumoniae* DHFR enzymes that are sensitive to TMP as well as the plasmid-encoded TMP-resistant DfrA1, DfrA12, and DfrA17 all natively possess tyrosine at position 98, suggesting that compensatory and subtle differences exist between the Gram-positive and the Gram-negative DHFR enzymes.

RESISTANCE TO SULFAMETHOXAZOLE

Resistance to sulfamethoxazole also arose early after introduction and began to be monitored. As with trimethoprim, resistance to sulfamethoxazole arises through combinations of chromosomal mutations and plasmid-encoded copies of resistant genes. Several point mutations in the *folP* gene encoding DHPS in *S. aureus*, have been reported in clinical isolates. Seven of the nine resistant strains analyzed showed at least 13 residues that differ from wild-type; the remaining two isolates have two residue differences. Crystal structures of *S. aureus* (Hampele et al. 1997), *Yersinia pestis* (Yun et al. 2012), or *Mycobacterium tuberculosis* DHPS (Baca et al. 2000) show that most of the mutations implicated in resistance are spread across the surface of the protein or in flexible loops. Resistance mutations F28L/I and P64S (*E. coli* numbering) are found in loop 1 and loop 2, respectively, close to the active site and affect the *para*-aminobenzoic acid (PABA) binding site. Yun et al. (2012) further notes that mutations at positions F33 and P69 are likely to affect inhibitor but not substrate binding, which may make these sites more susceptible to resistance-conferring mutations.

In addition to chromosomal gene mutations, sulfamethoxazole resistance can be transferred by integrons carrying the *sul1*, *sul2*, or *sul3* genes (Skold 1976; Blahna et al. 2006) encoding insensitive forms of DHPS. In fact, a collection of 106 clinical isolates of *Stenotrophomonas maltophilia* from India, for which TMP-SMX is first-line therapy, showed that 24 out of 106 isolates were resistant to TMP-SMX. The *sul1* or *sul2* gene was present in 12 or 14 isolates, respectively (Kaur et al. 2015).

Resistance to antibacterial agents can take many forms including efflux, drug-modifying enzymes, changes in cell membrane permeability, and alterations of the drug target. It appears that the majority of resistance to currently used antifolates is heavily dependent on alterations of the drug target and the introduction of mobile elements that carry a resistant copy of the target as opposed to mechanisms that alter drug concentration. This knowledge of the mechanisms of resistance coupled with detailed structural and biochemical analysis underlies the design of next-generation antifolates.

COMPOUNDS IN DEVELOPMENT

Fortunately, there continues to be significant development of new antifolates intended to overcome existing resistance mechanisms. Development has focused on new DHFR inhibitors including iclaprim and the propargyl-linked antifolates as well as new DHPS inhibitors intended to bind and inhibit the pterin pocket of the enzyme, thus avoiding resistance mutations already present.

Iclaprim (Fig. 2), originally developed by Roche and licensed by Arpida for clinical advancement (Hawser et al. 2006), was designed to overcome the resistance conferred by point mutations in the chromosomal gene and to be effective against other TMP-resistant organisms. The compound showed potent activity against MRSA, TMP-resistant MRSA, *Streptococcus pyogenes* (MIC values between 0.03 μg/mL and 0.06 μg/mL), as well as Gram-negative bacteria such as *Enterococcus* and *Haemophilus influenza* (MIC values 0.12 and 0.5 μg/mL, respectively) (Schneider et al. 2003). Furthermore, the propensity of iclaprim to induce the formation of resistance mutations is reported to be very low. Iclaprim advanced through phase I, II, and III clinical trials focused on complicated skin and soft tissue infections. Unfortunately, some cardiotoxicity issues presented during the trials (Sincak and Schmidt 2009) and more importantly, the drug did not meet the regulatory metrics of non-inferiority relative to linezolid in place at the time. Recently, a biotechnology company, Motif Bio (see motifbio.com/iclaprim), received FDA approval to reopen the phase III clinical trials for acute bacterial skin and skin structure infections as well as hospital-acquired bacterial pneumonia.

A series of compounds known as the propargyl-linked antifolates (PLAs; example shown in Fig. 2) is currently under development to inhibit both Gram-positive and Gram-negative bacteria; they are specifically designed to target both trimethoprim-sensitive and trimethoprim-resistant enzymes. Using a structure-based approach founded on the determination of several tens of structures of DHFR from many species, including *S. aureus* (Frey et al. 2009), mutant forms of *S. aureus* (Frey et al. 2009, 2010; Keshipeddy et al. 2015), *E. coli*, *K. pneumoniae* (Lamb et al. 2014), and plasmid-encoded trimethoprim-resistant DHFR such as DfrA1 (Lombardo et al. 2015), new DHFR inhibitors were designed to overcome mutations that cause TMP resistance as well as whole enzymes that are horizontally transferred for resistance. The development of these compounds has led to very potent inhibitors of wild-type and mutant forms of *S. aureus* DHFR as well as the growth of the *S. aureus* (wild-type and mutant) bacteria. Some of the most potent compounds have MIC values of 0.0391 μg/mL against the wild-type and 0.625 μg/mL against *S. aureus* with the F98Y mutation (Keshipeddy et al. 2015). In addition, many of the compounds show activity against Gram-negative bacteria and show potency against the trimethoprim-resistant DfrA1 (Lombardo et al. 2015). Early investigations into the propensity to incur resistance mutations show that the compounds generally have very low resistance frequencies between 1.2×10^{-9} and 8.9×10^{-10} (Frey et al. 2012).

There has also been considerable effort to develop new inhibitors of DHPS. Switching attention from the *para*-aminobenzoic acid site where the sulfa compounds bind, medicinal chemistry work has focused on the pocket that binds the partner substrate, dihydropterin pyrophosphate (DHPP) (Zhao et al. 2012). Using a crystal structure of *Bacillus anthracis* DHPS bound to a lead pterin (Hevener et al. 2012), Lee and coworkers synthesized several pyridazine derivatives intended to optimize different functional groups in the lead (Fig. 2). Using a combination of enzyme assays, ITC, SPR, and crystallography, they show that a demethylated pyridazine core is a pterin mimic for DHPS (Zhao et al. 2012). In parallel work, they show that a compound targeted to the DHPS dimer interface acts as an allosteric inhibitor of the enzyme, opening a new area of design and intervention (Hammoudeh et al. 2014).

CONCLUSION

Despite the fact that trimethoprim and sulfamethoxazole have been in continued use for over 60 years, these classes of antibacterials have not undergone the sustained medicinal chemistry efforts that have led to multiple generations of other important classes such as β-lactams or fluoroquinolones. Although resistance to these agents has appeared, they remain some of the most important first-line therapeutics for common Gram-positive and Gram-negative pathogens. However, increased use coupled with globalization demands the development of new antifolates that will continue to allow this important class to be clinically effective.

ACKNOWLEDGMENTS

The authors acknowledge support from the National Institutes of Health (NIH) (AI104841 and AI111957).

REFERENCES

Amyes S. 1989. The success of plasmid-encoded resistance genes in clinical bacteria. *J Med Microbiol* **28:** 73–83.

Amyes S, Towner K, Young H. 1992. Classification of plasmid-encoded dihydrofolate reductases conferring trimethoprim resistance. *J Med Microbiol* **36:** 1–3.

Baca A, Sirawaraporn R, Turley S, Sirawaraporn W, Ho W. 2000. Crystal structure of *Mycobacterium tuberculosis* 6-hydroxymethyl-7,8-dihydropteroate synthase in complex with pterin monophosphate: New insight into the enzymatic mechanism and sulfa-drug action. *J Mol Biol* **302:** 1193–1212.

Blahna M, Zalewski C, Reuer J, Kahlmeter G, Foxman B, Marrs C. 2006. The role of horizontal gene transfer in the spread of trimethoprim-sulfamethoxazole resistance among uropathogenic *Escherichia coli* in Europe and Canada. *J Antimicrob Chemother* **57:** 666–672.

Brolund A, Sundqvist M, Kahlmeter G, Grape M. 2010. Molecular characterisation of trimethoprim resistance in *Escherichia coli* and *Klebsiella pneumoniae* during a two-year intervention on trimethoprim use. *PLoS ONE* **5:** e9233.

Dale G, Broger C, Hartman P, Langen H, Page M, Then R, Stuber D. 1995. Characterization of the gene for the chromosomal dihydrofolate reductase (DHFR) of *Staphylococcus epidermis* ATCC 14990: The origin of the trimethoprim-resistant S1 DHFR from *Staphylococcus aureus*? *J Bacteriol* **177:** 2965–2970.

Dale G, Broger C, D'Arcy A, Hartman P, DeHoogt R, Jolidon S, Kompis I, Labhardt A, Langen H, Locher H, et al. 1997. A single amino acid substitution in *Staphylococcus aureus* dihydrofolate reductase determines trimethoprim resistance. *J Mol Biol* **266:** 23–30.

Darmstadt G. 1997. Oral antibiotic therapy for uncomplicated bacterial skin infections in children. *Pediatr Infect Dis J* **16:** 227–240.

Frei C, Miller M, Lewis J, Lawson K, Hunter J, Oramasionwu C, Talbert R. 2010. Trimethoprim-sulfamethoxazole or clindamycin for community-associated MRSA (CA-MRSA) skin infections. *J Am Board Fam Med* **23:** 714–719.

Frey K, Liu J, Lombardo M, Bolstad D, Wright D, Anderson A. 2009. Crystal structures of wild-type and mutant methicillin-resistant *Staphylococcus aureus* dihydrofolate reductase reveal an alternative conformation of NADPH that may be linked to trimethoprim resistance. *J Mol Biol* **387:** 1298–1308.

Frey K, Lombardo M, Wright D, Anderson A. 2010. Towards the understanding of resistance mechanisms in clinically isolated trimethoprim-resistant, methicillin-resistant *Staphylococcus aureus* dihydrofolate reductase. *J Struc Biol* **170:** 93–97.

Frey K, Viswanathan K, Wright D, Anderson A. 2012. Prospectively screening novel antibacterial inhibitors of dihydrofolate reductase for mutational resistance. *Antimicrob Agents Chemother* **56:** 3556–3562.

Gorwitz RJ, Jernigan DB, Powers JH, Jernigan JA; Participants in the CDC-Convened Experts' Meeting on Management of MRSA in the Community. 2006. *Strategies for clinical management of MRSA in the community: Summary of an experts' meeting convened by the Centers for Disease Control and Prevention*, www.cdc.gov/mrsa/community/clinicians.

Hammoudeh D, Date M, Yun MK, Zhang W, Boyd V, Follis A, Griffith E, Lee R, Bashford D, White S. 2014. Identification and characterization of an allosteric inhibitory site on dihydropteroate synthase. *ACS Chem Biol* **9:** 1294–1302.

Hampele I, D'Arcy A, Dale G, Kostrewa D, Nielsen J, Oefner C, Page M, Schonfeld HJ, Stuber D, Then R. 1997. Structure and function of the dihydropteroate synthase from *Staphylococcus aureus*. *J Mol Biol* **268:** 21–30.

Hawser S, Luciuro S, Islam K. 2006. Dihydrofolate reductase inhibitors as antibacterial agents. *Biochem Pharmacol* **71:** 941–948.

Heaslet H, Harris M, Fahnoe K, Sarver R, Putz H, Chang J, Subramanyam C, Barreiro G, Miller JR. 2009. Structural comparison of chromosomal and exogenous dihydrofolate reductase from *Staphylococcus aureus* in complex with the potent inhibitor trimethoprim. *Proteins* **76:** 706–717.

Hevener K, Yun MK, Qi J, Kerr I, Baboglu K, Hurdle J, Blakrishna K, Balakrishna K, White S, Lee R. 2012. Structure-based design of novel pyrimidopyridazine derivative as dihydrofolate synthase inhibitors with increased affinity. *J Med Chem* **53:** 166–177.

Kaur P, Gautam V, Tewari R. 2015. Distribution of class 1 integrons, *sul1* and *sul2* genes among clinical isolates of *Stenotrophomonas maltophilia* from a tertiary care hospital in North India. *Microb Drug Resist* **21:** 380–385.

Keshipeddy S, Reeve S, Anderson A, Wright D. 2015. Nonracemic antifolates stereoselectively recruit alternate cofactors and overcome resistance in *S. aureus. J Am Chem Soc* **137:** 8983–8990.

Lamb K, Lombardo M, Alverson J, Priestley N, Wright D, Anderson A. 2014. Crystal structures of *Klebsiella pneumoniae* dihydrofolate reductase bound to propargyl-linked antifolates reveal features for potency and selectivity. *Antimicrob Agents Chem* **58:** 7484–7491.

Lombardo M, G-Dayanandan N, Wright D, Anderson A. 2016. Crystal structures of trimethoprim-resistant DfrA1 rationalize potent inhibition by propargyl-linked antifolates. *ACS Inf Dis* **2:** 149–156.

Nathwani D, Morgan M, Masterton R, Dryden M, Cookson B, French G, Lewis D; British Society for Antimicrobial Chemotherapy Working Party on Community-Onset MRSA Infections and Infections. 2008. Guidelines for UK practice for the diagnosis and management of methicillin-resistant *Staphylococcus aureus* (MRSA) infections presenting in the community. *J Antimicrob Chemother* **61:** 976–994.

Nurjadi D, Olalekan A, Layer F, Shittu A, Alabi A, Ghebremedhin B, Schaumber F, Hofmann-Eifler J, Genderen P, Caumes E, et al. 2014. Emergence of trimethoprim resistance gene *dfrG* in *Staphylococcus aureus* causing human infection and colonization in sub-Saharan Africa and its import to Europe. *J Antimicrob Chemother* **69:** 2361–2368.

Schneider P, Hawser S, Islam K. 2003. Iclaprim, a novel diaminopyrimidine with potent activity on trimethoprim sensitive and resistant bacteria. *Bioorg Med Chem Lett* **13:** 4217–4221.

Sekiguchi J, Tharavichitkul P, Miyoshi-Akiyama T, Chupia V, Fujino T, Araake M, Irie A, Morita K, Kuratsuji T, Kirikae T. 2005. Cloning and characterization of a novel trimethoprim-resistant dihydrofolate reductase from a nosocomial isolate of *Staphylococcus aureus* CM.S2 (IMCJ1454). *Antimicrob Agents Chemother* **49:** 3948–3951.

Sincak C, Schmidt J. 2009. Iclaprim, a novel diaminopyrimidine for the treatment of resistant Gram-positive infections. *Ann Pharmacother* **43:** 1107–1114.

Skold O. 1976. R-factor mediated resistance to sulfonamides by a plasmid-borne, drug-resistant dihydropteroate synthase. *Antimicrob Agents Chemother* **9**: 49–54.

Skold O. 2010. Sulfonamides and trimethoprim. *Exp Rev Anti Infect Ther* **8**: 1–6.

Stevens D, Bisno A, Chambers H, Everett E, Dellinger P, Goldstein E, Gorbach S, Hirschmann J, Kaplan E, Montoya J, et al. 2005. Practice guidelines for the diagnosis and management of skin and soft-tissue infections. *Clin Infect Dis* **41**: 1373–1406.

Then R, Kohl I, Burdeska A. 1992. Frequency and transferability of trimethoprim and sulfonamide resistance in methicillin-resistant *Staphylococcus aureus* and *Staphylococcus epidermis*. *J Chemother* **4**: 67–71.

van Miert A. 1994. The sulfonamide-diaminopyrimidine story. *J Vet Pharmacol* **4**: 309–316.

Yun MK, Wu Y, Li Z, Zhao Y, Waddell MB, Ferreia A, Lee R, Bashford D, White S. 2012. Catalysis and sulfa drug resistance in dihydropteroate synthase. *Science* **335**: 1110–1114.

Zhao Y, Hammoudeh D, Yun M, Qi J, White S, Lee R. 2012. Structure-based design of novel pyrimido[4,5-*c*] pyridazine derivatives as dihydropteroate synthase inhibitors with increased affinity. *ChemMedChem* **7**: 861–870.

Antibacterials Developed to Target a Single Organism: Mechanisms and Frequencies of Reduced Susceptibility to the Novel Anti-*Clostridium difficile* Compounds Fidaxomicin and LFF571

Jennifer A. Leeds

Infectious Disease Area, Novartis Institutes for BioMedical Research, Emeryville, California 94608

Correspondence: jennifer.leeds@novartis.com

Clostridium difficile is the most common cause of antibacterial-associated diarrhea. Clear clinical presentation and rapid diagnostics enable targeted therapy for *C. difficile* infection (CDI) to start quickly. CDI treatment includes metronidazole and vancomycin (VAN). Despite decades of use for CDI, no clinically meaningful resistance to either agent has emerged. Fidaxomicin (FDX), an RNA polymerase inhibitor, is also approved to treat CDI. Mutants with reduced susceptibility to FDX have been selected in vitro by single and multistep methods. Strains with elevated FDX minimum inhibitory concentrations (MICs) were also identified from FDX-treated patients in clinical trials. LFF571 is an exploratory agent that inhibits EF-Tu. In a proof-of-concept study, LFF571 was safe and effective for treating CDI. Spontaneous mutants with reduced susceptibility to LFF571 were selected in vitro in a single step, but not via serial passage. Although there are several agents in development for treatment of CDI, this review summarizes the frequencies and mechanisms of *C. difficile* mutants displaying reduced susceptibility to FDX or LFF71.

Therapy for a suspected infection is often started empirically, based on the most likely syndrome and local epidemiological data well before the causative organism is identified. Because many clinical syndromes can be caused by a number of different pathogens, singly or in combination, the initial empiric therapy of severe infection is selected to cover the most likely organisms; this often involves broad-spectrum coverage to ensure that the condition of the patient does not worsen while the precise etiology is being determined microbiologically. However, in some circumstances, a physician can narrow the list of potential pathogens quickly and choose an initial therapy that is more organism specific. One of those indications is antibiotic-associated diarrhea, in which the leading cause is infection with *C. difficile* and rapid diagnostics can quickly identify whether it is likely to be the pathogen.

Clostridium difficile infection (CDI) can cause mild to severe intestinal disease (Cohen et al. 2010). A recently published survey of hospital inpatients from geographically diverse sites in the United States indicated that *C. difficile* was the most commonly reported nosocomial pathogen, and was the cause of 12.1% of health care–related infections in the institutions studied (Magill et al. 2014). The ability of *C. difficile* to form spores makes it difficult to remove from surfaces and allows it to spread easily and quickly within a health care setting. Over the past decade, the incidence of *C. difficile* infection has increased, and hypervirulent strains, such as B1/NAP1/027, have become more prevalent (McDonald et al. 2005). The epidemiology of CDI is also changing, as the disease is now more commonly seen outside the hospital environment and in some patients with no obvious underlying risk factors, compared with what was observed historically (Hensgens et al. 2014; Evans and Safdar 2015; Lessa et al. 2015).

Because the major risks for acquiring CDI are well defined, those, in combination with the clear clinical presentation and the availability of relatively rapid and accurate diagnostic tools for identifying toxigenic *C. difficile* (reviewed in Bagdasarian et al. 2015), allow for a presumptive diagnosis to be made quite confidently within a few hours following the onset of symptoms with appropriate therapy started quickly, and a definitive diagnosis can be made within a day or two, depending on the testing algorithm applied. The current guidelines for diagnosis and treatment of CDI (Cohen et al. 2010; Debast et al. 2014), depending on disease severity and number or risk of recurrences, include the generic antibiotics vancomycin and metronidazole, the former of which is active against Gram-positive bacteria only, and the latter having activity limited to obligate anaerobes (Baxter Healthcare Corp. Metronidazole Injection, UST RTU Product Insert, Deerfield, IL, 2009 [www.accessdata.fda .gov/drugsatfda_docs/label/2009/018657s029lbl .pdf]; and Baxter Healthcare Corp. Vancomycin Injection, USP Product Insert, Deerfield, IL, 2011 [www.accessdata.fda.gov/drugsatfda_docs/ label/2012/050671s018lbl.pdf]). The antibacterial actions of these agents are owing to inhibition of functions that are essential to bacterial growth and survival, and the molecular target(s) of these antibacterials are synthesized from more than one gene product (reviewed in Healy et al. 2000). This multigene/target feature coincides with the observation that there are no reports of single-step spontaneous mutations in *C. difficile* that confer high-level clinically relevant loss of susceptibility to either agent. Recently, a pan-European survey of *C. difficile* isolates from 2011 to 2012 showed >96% susceptibility to vancomycin and metronidazole, despite many decades of use of these agents to treat *C. difficile* infection (Freeman et al. 2015). Accordingly, there is little compelling clinical data from adequately powered and controlled studies to indicate that treatment failure for CDI is owing to antimicrobial resistance to either of these agents, and there are no breakpoints or routine susceptibility testing used to guide the choice of treatment for CDI (Bauer et al. 2009; Cohen et al. 2010; Debast et al. 2014). Unfortunately, recurrent disease occurs in ~20%–25% of patients whose symptoms resolved following treatment with VAN and MET (reviewed in Debast et al. 2014). As such, current clinical development programs monitor the impact of exploratory interventions on the rate of CDI recurrence following initial clinical cure.

In 2011, the first drug from a new class of antibacterials, fidaxomicin (FDX), was approved by the U.S. Food and Drug Administration (FDA) for the treatment of *C. difficile*–associated diarrhea in adults ≥18 years of age (Optimer Pharmaceuticals, Dificid Package Insert, San Diego, CA, 2011 [www.accessdata.fda .gov/drugsatfda_docs/label/2011/201699s000 lbl.pdf]). Marketing authorization followed shortly thereafter throughout the European Union, Japan, Australia, and Canada. Based on data from two phase III trials, FDX is also recommended, with moderate evidence, by the European Society of Clinical Microbiology and Infectious Diseases (ESCMID) as an alternative to vancomycin when patients are at risk of first recurrence or have had multiple recurrences (Debast et al. 2014). Infectious Disease Society of America (IDSA) has not yet updated its guidance to include FDX (Cohen et al. 2010); however, clinical practice advisory groups have published recommendations for its use (Surawicz et al. 2013). FDX inhibits the growth of *C. difficile*, and other susceptible organisms, via inhibition of the clinically validated target bacterial RNA polymerase (Coronelli et al. 1975). The mechanism of inhibition and the predicted binding site for FDX on RNA polymerase (RNAP) is distinct from that of the rifamycins (reviewed in Srivastava et al. 2011) and there is no cross-resistance between FDX and the clinically used agents from this drug class (Goldstein et al. 2011).

There are multiple exploratory compounds with excellent in vitro potency against *C. difficile*, some of which have entered clinical development and have shown efficacy in patients with CDI (reviewed recently in Tsutsumi et al. 2014 and Ivarsson et al. 2015). Several of the antibacterials in clinical trials for treatment of CDI, including surotomycin and cadezolid, are derivatives of clinically used drugs (daptomycin and fluoroquinolones/oxazolidinones, respectively) and are inhibitors of clinically validated molecular targets (Locher et al. 2014b; Yin et al. 2015). LFF571 is an exploratory anti–*C. difficile* compound that has shown safety and efficacy in a phase II trial in humans for treatment of mild to moderate *C. difficile* infection (Mullane et al. 2015); however, LFF571 is derived from a novel scaffold that is not in clinical use and inhibits a molecular target, elongation factor Tu, that is also clinically unprecedented (LaMarche et al. 2012; Leeds et al. 2012).

LFF571 and FDX both interfere with the growth of *C. difficile* by inhibiting essential pathways that involve large, multimolecular complexes; however, because the molecular mechanisms by which they inhibit their respective targets can be interrupted by single amino acid substitutions within single-gene products, both are subject to reduction in antibacterial potency via selection for spontaneous, single-step mutants.

This review will summarize the mechanisms of action and review the literature reports of the frequencies and mechanisms of reduced susceptibility to FDX and LFF571. These agents exemplify relatively narrowly focused therapeutics in that, although both show antibacterial activity in vitro against a range of Gram-positive species, they have both been studied in humans

exclusively against *C. difficile* because of their remarkable potency against this pathogen (minimum inhibitory concentrations [MICs]$_{90}$ $\leq 0.5\ \mu/mL$) (Citron et al. 2012; Goldstein et al. 2012; Hecht and Gerding 2012; Debast et al. 2013) and because their physicochemical properties result in the ability to achieve high drug concentrations in the target organ (colon) with very little systemic exposure (Sears et al. 2012; Ting et al. 2012; Bhansali et al. 2015).

ANTIMICROBIAL MECHANISM OF ACTION OF FDX

FDX is an 18-membered macrocyclic natural bacterial fermentation product that is identical in structure to lipiarmycin A3, tiacumicin B, and clostimicin B1 (Fig. 1) (Omura et al. 1986; Theriault et al. 1987; Bedeschi et al. 2014). For simplicity, this review will refer to this structure only as "fidaxomicin." Published reports of the isolation of this chemical structure and the description of its mechanism of action dates back to 1975 (Coronelli et al. 1975). Macromolecular synthesis inhibition assays indicated that the mechanism of antimicrobial action of FDX was via rapid inhibition of RNA synthesis, which led to an almost simultaneous loss of protein synthesis (Coronelli et al. 1975; Sergio et al. 1975). However, the drug does not inhibit the translation machinery directly and only inhibits protein synthesis as a result of inhibition of transcription via binding to RNA polymerase (Sergio et al. 1975). FDX inhibits the very early steps in initiation of RNA synthesis, before the formation of a stable open complex, but does not inhibit elongation by RNAP (Sergio et al. 1975; Artsimovitch et al. 2012). Based on mapping of residues conferring loss of susceptibility to FDX, the drug is thought to bind portions of the β- and β'-subunits found in the RNA exit channel of RNAP and part of the switch region, including parts of switch 2 and switch 3 of the RNA polymerase (Srivastava et al. 2011; Artsimovitch et al. 2012).

Although FDX inhibits the function of RNAP from Gram-positive and Gram-negative organisms (Coronelli et al. 1975; Sergio et al. 1975; Sonenshein et al. 1977), it is active in vitro

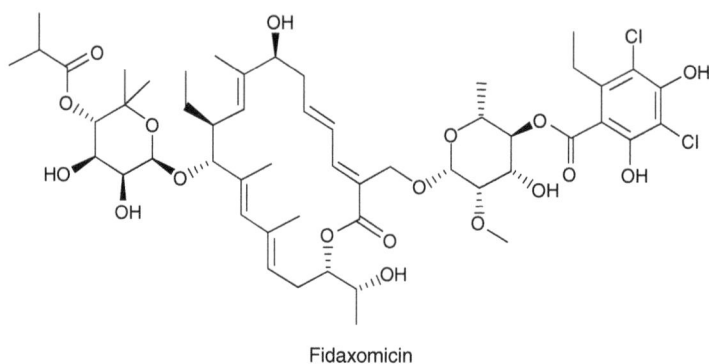

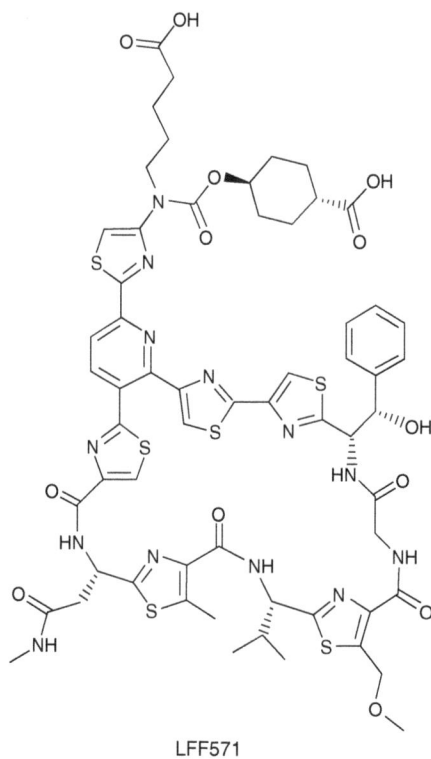

Figure 1. Chemical structures of fidaxomicin (FDX) and LFF571.

against only a limited range of Gram-positive aerobes and anaerobes, including *C. difficile* (Goldstein et al. 2012). FDX is a time-dependent inhibitor of *C. difficile* growth (Babakhani et al. 2011), is more potent in vitro against *C. difficile* than vancomycin or metronidazole (Goldstein et al. 2012), and in vitro FDX inhibits *C. difficile* toxin A/B production (Allen et al. 2013) and the outgrowth of *C. difficile* spores

(Allen et al. 2013). FDX is effective against all ribotypes of *C. difficile* tested in vitro (Goldstein et al. 2012), although the FDX MIC_{90} is higher against B1/NAP1 strains of *C. difficile* than against other ribotypes (Goldstein et al. 2011), and clinical trials showed that FDX was less likely to prevent relapse from NAP1 strains of *C. difficile* (Louie et al. 2011; Cornely et al. 2012).

FREQUENCIES AND MECHANISMS OF REDUCED SUSCEPTIBILITY TO FDX

Point mutations conferring loss of susceptibility of *Bacillus subtilis*, *Mycobacterium tuberculosis*, *Enterococcus faecalis*, and *C. difficile* to FDX were typically selected in the genes encoding the β- and β'-subunits of RNAP (encoded by *rpoB* and *rpoC*, respectively) (Gualtieri et al. 2006; Kurabachew et al. 2008; Gualtieri et al. 2009; Seddon et al. 2011; Srivastava et al. 2011; Seddon and Sears 2012; Babakhani et al. 2014; Leeds et al. 2014). Shortly after the mechanism of action of FDX was first reported, Sonenshein et al. (1977) published the first report of *B. subtilis* mutants selected on superinhibitory concentrations of the compound that helped to validate RNAP as the target of the antibacterial activity. In early studies with *C. difficile*, Swanson et al. (1991) reported attempts to select mutants of a strain of *C. difficile* with reduced susceptibility to FDX; however, they were unable to do so using the reported experimental design

(the frequency of resistance was reported to be $<2.8 \times 10^{-8}$).

Seddon et al. (2011) first reported the in vitro selection of spontaneous single-step *C. difficile* mutants on superinhibitory concentrations FDX in a conference poster (Table 1). Based on the methods reported, mutant frequencies of $>1 \times 10^{-8}$ were observed. The study included two reference strains (ATCC 9689 and ATCC 700057), which were selected on four or eight times the MIC of FDX. Selected mutants had single amino acid changes in the β-subunit (Gly1074Lys, Val1143Phe, or Val1143Asp) or the β'-subunit (Asp237Tyr) of RNAP, which shifted the FDX MIC by 16- to >64-fold above that observed for the parental strains. In a conference poster the following year, Seddon and Sears (2012) reported additional in vitro selection data for *C. difficile* (Table 1). The study included one reference strain (ATCC 9689) and one clinical isolate (ORG916) of *C. difficile*. Mutants were selected on 8–64 times the MIC of FDX at a frequency of

Table 1. Frequencies and mechanisms of spontaneous *C. difficile* mutants selected in a single step on FDX

Study	Strain	Mutant frequency	Selecting concentration	Amino acid change	Fold change in FDX MIC
Swanson et al. 1991	ATCC 9689	$<2.8 \times 10^{-8}$	4–8× MIC	None	None
Seddon et al. 2011	ATCC9689	$>1 \times 10^{-8}$	4–8× MIC	β-subunit G1074K; V1143F; V1143D	32 to $>64 \times$ MIC
	ATCC 700057	$>1 \times 10^{-8}$	4–8× MIC	β'-subunit D237Y	16 × MIC
Seddon and Sears 2012	ATCC 9689	$\leq 4 \times 10^{-10a}$	16–64× MIC	β-subunit V1143G; V1143D; β'-subunit R89G	64–8192× MIC
	ORG 916	$\leq 5 \times 10^{-10a}$	8–32× MIC	β-subunit Q1074K; Q1074H; V1143-G; β'-subunit I10R; R89G	16–64× MIC
Babakhani et al. 2014	ATCC 9689	$\leq 1.28 \times 10^{-8}$	4–8× MIC	β-subunit Q1074K; V1143F; β'-subunit D237Y	16× MIC
	ATCC 700057	$<1.41 \times 10^{-9}$	4–8× MIC	None	
	ORG911	$<3.58 \times 10^{-9}$	4–8× MIC	None	
	ORG916	$<2.71 \times 10^{-9}$	4–8× MIC	None	
Locher et al. 2014a	ATCC 9689	$\leq 1.3 \times 10^{-8}$	2–8× MIC	ND	$\geq 64 \times$ MIC
	NCTC 13366	$\leq 7.2 \times 10^{-7}$	4–8× MIC	ND	16× MIC
	A-1410	1.1×10^{-6}	8× MIC	ND	Not reported

[a]Calculated from the methods/results.

Cite this article as *Cold Spring Harb Perspect Med* doi: 10.1101/cshperspect.a025445

$\leq 4 \times 10^{-10}$, based on the methods reported. The FDX MIC against the selected mutants shifted 16–8192 times the MIC reported for the parental strains. Selected mutants had single amino acid substitutions within the RNAP β-subunit (Gln1074Lys, Gln1074His, Val1143Gly, and Val1143Asp) or β′-subunit (Ile10Arg and Arg89Gly).

In a more recent publication, the same group (Babakhani et al. 2014) described the results of selecting for spontaneous single-step mutants of four strains of *C. difficile* on 4× and 8× MIC of FDX (Table 1). The study included two reference strains (ATCC 700057 and ATCC 9689) and two clinical isolates (ORG911 and ORG916). Rifaximin and VAN were included as comparators for the selections. Although rifaximin selected for spontaneous single-step mutants of *C. difficile* at a frequency of $\leq 1.73 \times 10^{-7}$ on 8× MIC of drug, no mutants were selected on 8× MIC of FDX or vancomycin (VAN) and the resistance frequency was reported to be $< 1.4 \times 10^{-9}$. When assays were conducted with 4× MIC of FDX, mutants of ATCC 9689 were selected at a frequency of 1.28×10^{-8} and the MIC of FDX against these mutants shifted 16-fold. The mutants harbored amino acid substitutions in RNAP subunit β (Gln1074Lys or Val1143Phe) or in subunit β′ of Asp237Tyr. The same or greater magnitude of MIC shift (16- to ≥ 64-fold) was observed for two mutants of *C. difficile* strains ATCC 700057 and ATCC 43255 that were selected via serial passage (Babakhani et al. 2014). The investigators also describe a mutant strain of ATCC 43255 with a ≥ 64-fold shift in FDX MIC over the parental strain, but no mutant frequencies or amino acid substitutions were reported for this strain. None of the mutants selected on FDX showed reduced susceptibility to the clinically used RNAP inhibitor rifampin (RIF), supporting the model that the binding site of FDX does not overlap with that of RIF or other RNAP inhibitors that function through sites other than the switch region of RNAP (Srivastava et al. 2011). The investigators (Babakhani et al. 2014) described in vitro synergy studies with FDX and RIF, emphasizing the nonoverlapping mechanisms of

resistance, which raises the potential to combine the two agents to suppress resistance development to either agent alone; the outcomes of such combination studies have not been reported.

Locher et al. selected for single-step spontaneous mutants of *C. difficile* on 2–8× MIC of FDX, as well as by serial passage (Table 1) (Locher et al. 2014a). Mutants were selected on 8× MIC of FDX at much higher frequencies ($\leq 1.1 \times 10^{-6}$) than those reported by Babakhani et al. (2014). Spontaneous mutants were selected on all four strains of *C. difficile* that were tested including ATCC 9689 and three clinical isolates. Mutants selected in a single step were then subjected to two additional sequential rounds of selection on higher concentrations of FDX. The MICs of FDX against mutants selected in a single step or in multiple steps ranged from 2 μg/mL to >128 μg/mL and accounted for MIC shifts of eight- to >4000-fold over the parental strains. The mechanisms of resistance to FDX were not reported in this publication.

Leeds et al. subjected *C. difficile* strain ATCC 43255 and three clinical isolates of *C. difficile* to 10 passages on a range of concentrations of FDX (Leeds et al. 2014). The genomes of selected *C. difficile* mutants with reduced susceptibility to the selecting agents and their isogenic parental strains were fully sequenced to determine the single-nucleotide polymorphisms (SNPs) that were observed in the mutant background compared with the susceptible parent. When passaged on FDX, a 16-fold decrease in susceptibility was observed for clinical strain NB95013, and a 64-fold decrease in susceptibility was observed for clinical strain NB95026. There was a twofold change in FDX susceptibility for clinical strain NB95031 and a fourfold shift in FDX susceptibility for clinical strain NB95047. The NB95026 mutant showing a 64-fold decrease in susceptibility to FDX harbored a Gln1073Arg substitution in RNAP subunit β. Previous work by Seddon et al. (2011) had identified lysine and histidine substitutions at the same residue (numbered residue 1074 in Seddon et al. 2011 and Seddon and Sears 2012). The NB95013 mutant showing a 15-fold decreased susceptibility to FDX harbored a deletion in CD22120 (*marR* homolog), resulting in a frameshift after amino acid 117 of a homolog of the MarR family of transcriptional regulators. This finding represented the first report of a mechanism outside RNAP that may alter susceptibility to FDX. Although MarR is associated with multidrug resistance, the investigators did not observe reduced susceptibility to antibiotics other than FDX in that study.

Although notable, the high variability observed for frequencies of selecting mutants of *C. difficile* with reduced susceptibility to FDX (Table 1) may be owing to variations in methodology combined with too few replicates in any one study; therefore, it is not clear if the differences are biologically meaningful.

C. difficile WITH REDUCED SUSCEPTIBILITY TO FDX ISOLATED FROM PATIENTS TREATED WITH FDX IN PHASE III CLINICAL TRIALS (STUDIES OPT-80-003 AND -004)

Goldstein et al. reported that the FDX MIC for all baseline isolates from the modified intent-to-treat (mITT) and per protocol populations across both phase III studies was 0.25 μg/mL (Goldstein et al. 2011). Although there were no instances of FDX MIC >0.5 μg/mL among the baseline *C. difficile* isolates from the 26 patients in the per protocol population who were clinical failures following treatment with FDX, a single strain of *C. difficile* with reduced susceptibility to FDX was isolated from a FDX-treated, clinically cured patient who suffered a recurrence. The strain, which harbored a Val1143Gly substitution in the RNAP β-subunit was isolated at the time of recurrence and the FDX MIC was 16 μg/mL (Goldstein et al. 2011).

Eyre et al. reported whole genome sequences of the 28 pairs of isolates available at baseline and recurrence in FDX-treated patients from the ITT population in the pooled phase III studies (~41% of the total intent-to-treat (ITT) population receiving FDX who suffered recurrences of CDI during the study) (Eyre et al. 2014). In this analysis, isolates from two ITT patients who suffered recurrences were reported

to have mutations in the target gene *rpoB*. One of those isolates is the same one reported previously by Goldstein et al. (2011) against which FDX had a MIC of 16 μg/mL. In addition, a second strain was isolated from a patient with recurrence that showed a 10-fold reduction in susceptibility to FDX (FDX MIC = 0.015 μg/mL at baseline and 0.125 μg/mL at follow-up). The strain harbored a Val1143Leu substitution in RNAP β-subunit. Eyre et al. (2014) indicated that because neither mutation was observed in isolates from any other patient, it suggests that the mutants are uncommon. However, sequencing data from only 28 baseline and matched-patient postbaseline isolates from the FDX arm of the combined phase III trials, which included a total of 572 subjects who received FDX, have been reported, so more surveillance and posttreatment isolates would need to be tested to understand the true prevalence of such mutants. Eyre et al. (2014) also reported other nonsynonymous SNPs, including one in a hypothetical protein and one in a putative PTS system, which were identified in strains from patients treated with FDX who experienced a recurrence of disease following clinical cure at end of treatment. The extent to which any of these substitutions, whether in the target of FDX or elsewhere, contribute to recurrence of disease is unknown. Regular postmarketing surveillance data will be required to inform whether the use of FDX to treat *C. difficile* infection results in the selection for additional mutants with reduced susceptibility in patients or in the environment.

ANTIMICROBIAL MECHANISM OF ACTION OF LFF571

LFF571 is a semisynthetic derivative of the natural product GE2770A (Fig. 1) (LaMarche et al. 2012). Scientists at Le Petit Research Center identified GE2270A, a thiopeptide-based secondary metabolite isolated from the fermentation broth of *Planobispora rosea*, in a screen for *Staphylococcus aureus* growth inhibitors that were antagonized by the addition of exogenous elongation factor Tu (EF-Tu) into the assay system (Selva et al. 1991, 1997; Landini et al. 1996);

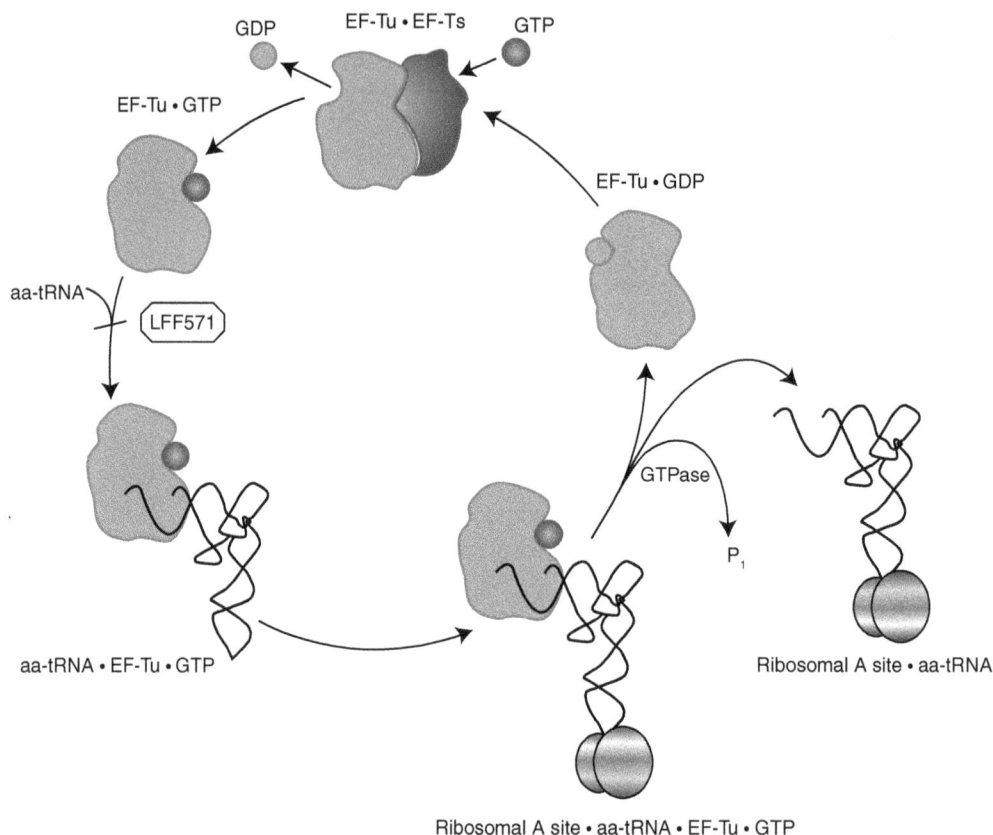

Figure 2. EF-Tu functional cycle with point of LFF571 inhibition indicated. GDP, Guanosine diphosphate; GTP, guanosine triphosphate.

the structure assignment for GE2270A was corrected in subsequent publications (Tavecchia et al. 1994, 1995). As a human therapeutic candidate, LFF571 has a novel mechanism of action, in that, like the natural product precursor GE2270A, it binds to and inhibits the function of bacterial EF-Tu (Landini et al. 1996; Leeds et al. 2012).

EF-Tu is one of three proteins (along with EF-Ts and EF-G) that are required, after formation of the 70S initiation complex of the ribosome, to catalyze the subsequent steps in peptide chain elongation (Fig. 2) (reviewed in Voorhees and Ramakrishnan 2013). EF-Tu is the chaperone that delivers aminoacylated tRNAs to the A site of the ribosome in the form of the ternary complex EF-Tu • guanosine triphosphate (GTP) • aa-tRNA (Fahnestock et al. 1972; Stark et al. 1997). Subsequent GTP hydrolysis causes

the release of EF-Tu in complex with guanosine diphosphate (GDP). EF-Tu is then reactivated by the action of EF-Ts, which catalyzes the GDP/GTP nucleotide exchange and renders EF-Tu competent for carrying out the next round of tRNA delivery (Miller and Weissbach 1970).

Because the binding site overlaps with that of tRNA, LFF571 inhibits protein synthesis by interfering with the ability for EF-Tu to deliver aminoacylated tRNA to the ribosome (Deng et al. 2011; Leeds et al. 2012). There are several other natural products that are known to inhibit the function of EF-Tu by the same or other mechanisms including kirromycin, pulvomycin, thiomuracin, amithiamycin, and efrotomycin (Parmeggiani and Nissen 2006; Morris et al. 2009). Synthetic small molecule inhibitors of EF-Tu have also been disclosed in publications (Deibel et al. 2004; Jayasekera et al. 2005).

Cite this article as *Cold Spring Harb Perspect Med* doi: 10.1101/cshperspect.a025445

To date, no EF-Tu inhibitors have been registered for clinical use in humans, although a different analog of GE2270A with an unreported structure (BI-K-0376 or "VIC-acne") was investigated as a topical agent for the treatment of acne (reviewed in Butler 2005).

LFF571, like FDX, is active in vitro against its molecular target (EF-Tu) from Gram-positive and Gram-negative organisms (Deng et al. 2011); however, the antibacterial spectrum of LFF571 is limited to Gram-positive aerobes and anaerobes (Citron et al. 2012). The K_d of LFF571 binding to *C. difficile* EF-Tu • GDP is ~130 nM (Deng et al. 2011). LFF571 is effective against all ribotypes of *C. difficile* tested (MIC$_{90}$ = 0.25 μg/mL) and is more potent in vitro against *C. difficile* than vancomycin and metronidazole (Citron et al. 2012; Debast et al. 2013). LFF571 was shown to be efficacious and safe in a proof-of-concept study in humans with mild to moderate CDI (Mullane et al. 2015), and to have low oral bioavailability and high fecal concentrations in infected patients (Bhansali et al. 2015). Anticipated by the novel mechanism of antibacterial activity, LFF571 is potent against bacteria with reduced susceptibility or resistance to marketed antibiotics (Citron et al. 2012). Supporting its use as a therapeutic for CDI, LFF571 reduces *C. difficile* toxin A/B levels in culture supernatants even at sub-MIC concentrations (Sachdeva and Leeds 2015).

FREQUENCIES AND MECHANISMS OF REDUCED SUSCEPTIBILITY TO LFF571

LFF571 is a semisynthetic derivative of the natural EF-Tu inhibitor GE2270A. The GE2270A-producing strain of *P. rosea* has a naturally occurring, functional EF-Tu variant (EF-Tu1) that is not susceptible to inhibition by GE2270A (Mohrle et al. 1997). Zuurmond et al. (2000) systematically introduced individual amino acid substitutions that are found in the native *P. rosea* EF-Tu1 into the GE2270A-sensitive EF-Tu protein from *E. coli*. They found that only two individual substitutions could account for the naturally occurring resistance to GE2270A, namely, Gly257Ser or Gly275Ala (*E. coli* num-

bering), both of which are found in GE2270A-binding region of EF-Tu domain II. The locations of the substitutions in the protein are supported by the cocrystal data showing that GE2270A binds to domain II of bacterial EF-Tu, making contact with residues 215–230, 256–264, and 273–277 (Heffron and Jurnak 2000).

EF-Tu is one of the most abundant proteins in the cell, representing up to 10% of the total protein content (Ishihama et al. 2008). There is approximately one EF-Tu molecule for every tRNA and this ratio is constant under different growth conditions (Furano 1975). In many organisms, including *C. difficile*, there are two copies of the genes encoding EF-Tu (*tufA* and *tufB*) (Jaskunas et al. 1975; Sela et al. 1989; Ke et al. 2000; Leeds et al. 2012). One altered copy of EF-Tu is sufficient for reduced susceptibility to GE2270A or LFF571 (Zuurmond et al. 2000; Leeds et al. 2012), unlike the mechanism of resistance to kirromycin, an inhibitor of the GTPase activity of EF-Tu, against which reduced susceptibility is only achieved when both copies of *tuf* are mutated (reviewed in Parmeggiani and Nissen 2006). Nevertheless, despite the high copy number of the target protein and the dominance of resistance, single-step spontaneous mutants of *C. difficile* with reduced susceptibility to LFF571 are selected in vitro at low frequencies ($\leq 1.2 \times 10^{-9}$) (Table 2) (Leeds et al. 2012). Mutant selections were attempted with four strains of *C. difficile* (three clinical strains and one high-toxin expressing-type strain ATCC 43255); reduced susceptibility to LFF571 was observed at the following frequencies: 1.7×10^{-10} (NB95002 selected at 0.5 and 1 μg/mL LFF571), 1.2×10^{-9} and $<6.2 \times 10^{-10}$ (NB95013 at 0.5 and 1 μg/mL, respectively), and 3.0×10^{-11} and $<3.0 \times 10^{-11}$ (NB95026 at 0.5 and 1 μg/mL, respectively). No mutants of NB95031 were selected on LFF571 under the conditions tested ($<4.5 \times 10^{-11}$).

All *C. difficile* mutants selected in vitro on LFF571 showed *tufB* mutation G782A, resulting in amino acid substitution G260E; NB95013-JAL0759 harbored the G782A change in both *tufA* and *tufB*, which are identical in sequence in this pathogen (Leeds et al. 2012). Residue

Table 2. Frequencies and genotypes of spontaneous *C. difficile* mutants selected in a single step on LFF571

Strain	Selecting concentration	Mutant frequency	Mutant selected on LFF571	Amino acid change	Fold change in LFF571 MIC
NB95002	$4\times$ MIC	1.7×10^{-10}	NB95002-JAL0777	*tufB*: G260E	$>256\times$ MIC
NB95013	$1\times$ MIC	1.2×10^{-9}	NB95013-JAL0758	*tufB*: G260E	$>512\times$ MIC
			NB95013-JAL0759	*tufA*: G260E *tufB*: G260E	$>512\times$ MIC
	$2\times$ MIC	$<6.2 \times 10^{-10}$	None	None	
NB95026	$1\times$ MIC	3.0×10^{-11}	NB95026-JAL0792	*tufB*: G260E	$>512\times$ MIC
	$2\times$ MIC	$<3.0 \times 10^{-11}$	None	None	
NB95031	$1\times$ MIC	$<4.5 \times 10^{-11}$	None	None	
	$2\times$ MIC	$<4.5 \times 10^{-11}$	None	None	

G260 in *C. difficile* EF-Tu is synonymous with G257 in *E. coli* EF-Tu, one of the two residues in which amino acid substitutions resulted in GE2270A-resistant EF-Tu. The proposed mechanism of resistance to LFF571 is that, rather than displacing the inhibitor, the G260E gain-of-function substitution restores the interactions between aa-tRNA and EF-Tu • GTP • LFF571. Several lines of evidence support the model that LFF571 and tRNA bind simultaneously to the mutant EF-Tu • GTP complex. First, LFF571 retains weak antibacterial activity against the EF-Tu G260E mutants (MIC 128–256 μg/mL) (Leeds et al. 2012). Second, Zuurmond et al. (2000) reported that the affinity of the natural product inhibitor GE2270A for wild-type *E. coli* EF-Tu • GTP is the same as for the G260S mutant, and in the presence of the GE2270A, the affinity of aa-tRNA for the G260S mutant increases $100\times$ over its affinity for wild-type EF-Tu • GTP/GE2270A. Finally, the affinity of GE2270A for *E. coli* EF-Tu • GDP is dramatically decreased by G260S, so on GTP hydrolysis at the ribosome, the inhibitor more quickly dissociates, leaving the aa-tRNA in the A site. Therefore, the mechanism of resistance to this thiopeptide class, in general, may not be a result of decreased affinity of the inhibitor to EF-Tu • GTP; rather, the mutant restores the ability of the GTP-bound enzyme to bind the substrate aa-tRNA and the mutant confers a more rapid dissociation of the inhibitor from the mutant EF-Tu/ribosomal complex once the aa-tRNA has engaged the A site (Zuurmond et al. 2000). This proposed mechanism of resistance to LFF571 is a model based on the behavior of the natural analog GE2270A and would require structural and affinity studies with LFF571 to be validated.

CONCLUDING REMARKS

FDX and LFF571 are novel antibacterial agents with clinically unprecedented mechanisms of target inhibition that exemplify compounds with a clinical focus directed at a single organism—*C. difficile*. Both compounds select for reduced susceptibility in *C. difficile* in vitro, and mutants with reduced susceptibility to FDX have been reported to be isolated at the time of recurrence from two patients who received FDX as a treatment for CDI. Because *C. difficile* is a spore-forming species, selection for mutants in patients could lead to dissemination of mutant spores, which, over time, could shift the susceptibility of the compounds for *C. difficile* strains in the general population. Routine susceptibility testing is not used to guide the choice of appropriate therapy for CDI. Therefore, it would fall to regular surveillance studies or epidemiological data from outbreaks to inform trends in the loss of susceptibility to such agents. Whether such a shift in susceptibility would have an impact on the ability to treat *C. difficile* with such agents is unknown, especially given that the concentrations of these agents in the colon, at the site of antibacterial action, is typically well above the MIC for the mutants (Louie et al. 2011; Bhansali et al. 2015).

Although not approved for use, other drugs, such as rifaximin and fusidic acid, against which high-frequency single-step resistance can be selected in *C. difficile*, are occasionally used to treat CDI. Attempts to understand the impact of mutants with reduced susceptibility to these agents on therapeutic outcome have been reported (Noren et al. 2006; Mattila et al. 2013). For example, a study comparing fusidic acid with metronidazole for the treatment of CDI showed that although all isolates from patients randomized to fusidic acid were susceptible at baseline, 55% (11/20) of posttreatment isolates were resistant to fusidic acid (median MIC >256 µg/mL) (Noren et al. 2006). Clinical failure and relapse, among those treated with fusidic acid, occurred in 12% and 34% of the patients with fusidic acid–resistant strains posttreatment versus 6% and 17% of the patients with fusidic acid–sensitive strains posttreatment. Persistence of *C. difficile* in stool culture at follow-up predicted failure/relapse regardless of treatment assignment. One or two nonsynonymous substitutions in EF-G (the target of fusidic acid) were responsible for the shift in susceptibility in *C. difficile* (Noren et al. 2007). Fusidic acid fecal concentrations have not been directly reported; however, ∼2% of an oral dose is excreted unchanged (Reeves 1987).

Recently, Adams et al. reported that the serial passage of *C. difficile* in subinhibitory concentrations of the phase III CDI compound surotomycin resulted in mutants with an eight- to16-fold shift in susceptibility to both surotomycin and daptomycin, the clinically used natural product antibiotic from which surotomycin is derived (Adams et al. 2015). In addition, surotomycin also selected for mutants of *Enterococcus faecium* and *E. faecalis* that are cross-resistant to daptomycin (Mascio et al. 2014). Therefore, even with antibiotics for which the mechanism of inhibition involves a complex membrane target, rather than a single-gene product, one can select for loss of susceptibility that is stable and imparts cross-resistance of other pathogens to the drugs used clinically to treat infections with those organisms.

Perhaps with the increasing focus on new agents for CDI, especially in the era of target-based drug discovery, the paradigm for guiding appropriate treatment will shift, and will encompass some form of susceptibility testing to insure adequate therapy. Or perhaps the high drug exposures will obviate the need. The data coming from the many trials in CDI, as well as the surveillance studies for newly registered drugs, will help to inform this decision.

ACKNOWLEDGMENTS

I am grateful to Johanne Blais for helpful suggestions and assistance with the presentation of data.

REFERENCES

Adams HM, Li X, Mascio C, Chesnel L, Palmer KL. 2015. Mutations associated with reduced surotomycin susceptibility in *Clostridium difficile* and *Enterococcus* species. *Antimicrob Agents Chemother* **59:** 4139–4147.

Allen CA, Babakhani F, Sears P, Nguyen L, Sorg JA. 2013. Both fidaxomicin and vancomycin inhibit outgrowth of *Clostridium difficile* spores. *Antimicrob Agents Chemother* **57:** 664–667.

Artsimovitch I, Seddon J, Sears P. 2012. Fidaxomicin is an inhibitor of the initiation of bacterial RNA synthesis. *Clin Infect Diseases* **55:** S127–S131.

Babakhani F, Gomez A, Robert N, Sears P. 2011. Killing kinetics of fidaxomicin and its major metabolite, OP-1118, against *Clostridium difficile*. *J Med Microbiol* **60:** 1213–1217.

Babakhani F, Seddon J, Sears P. 2014. Comparative microbiological studies of transcription inhibitors fidaxomicin and the rifamycins in *Clostridium difficile*. *Antimicrob Agents Chemother* **58:** 2934–2937.

Bagdasarian N, Rao K, Malani PN. 2015. Diagnosis and treatment of *Clostridium difficile* in adults: A systematic review. *JAMA* **313:** 398–408.

Bauer MP, Kuijper EJ, van Dissel JT. 2009. European Society of Clinical Microbiology and Infectious Diseases (ESCMID): Treatment guidance document for *Clostridium difficile* infection (CDI). *Clin Microbiol Infect* **15:** 1067–1079.

Bedeschi A, Fonte P, Fronza G, Fuganti C, Serra S. 2014. The co-identity of lipiarmycin A3 and tiacumicin B. *Nat Prod Commun* **9:** 237–240.

Bhansali SG, Mullane K, Ting LS, Leeds JA, Dabovic K, Praestgaard J, Pertel P. 2015. Pharmacokinetics of LFF571 and vancomycin in patients with moderate *Clostridium difficile* infections. *Antimicrob Agents Chemother* **59:** 1441–1445.

Butler MS. 2005. Natural products to drugs: Natural product derived compounds in clinical trials. *Nat Prod Rep* **22:** 162–195.

Citron DM, Tyrrell KL, Merriam CV, Goldstein EJ. 2012. Comparative in vitro activities of LFF571 against *Clos-*

tridium difficile and 630 other intestinal strains of aerobic and anaerobic bacteria. *Antimicrob Agents Chemother* **56:** 2493–2503.

Cohen SH, Gerding DN, Johnson S, Kelly CP, Loo VG, McDonald LC, Pepin J, Wilcox MH. 2010. Clinical practice guidelines for *Clostridium difficile* infection in adults: 2010 update by the society for healthcare epidemiology of America (SHEA) and the infectious diseases society of America (IDSA). *Infect Control Hosp Epidemiol* **31:** 431–455.

Cornely OA, Crook DW, Esposito R, Poirier A, Somero MS, Weiss K, Sears P, Gorbach S. 2012. Fidaxomicin versus vancomycin for infection with *Clostridium difficile* in Europe, Canada, and the USA: A double-blind, non-inferiority, randomised controlled trial. *Lancet Infect Dis* **12:** 281–289.

Coronelli C, White RJ, Lancini GC, Parenti F. 1975. Lipiarmycin, a new antibiotic from Actinoplanes. II: Isolation, chemical, biological and biochemical characterization. *J Antibiot (Tokyo)* **28:** 253–259.

Debast SB, Bauer MP, Sanders IM, Wilcox MH, Kuijper EJ. 2013. Antimicrobial activity of LFF571 and three treatment agents against *Clostridium difficile* isolates collected for a pan-European survey in 2008: Clinical and therapeutic implications. *J Antimicrob Chemother* **68:** 1305–1311.

Debast SB, Bauer MP, Kuijper EJ. 2014. European Society of Clinical Microbiology and Infectious Diseases: Update of the treatment guidance document for *Clostridium difficile* infection. *Clin Microbiol Infect* **20:** 1–26.

Deibel MR Jr, Bodnar AL, Yem AW, Wolfe CL, Heckaman CL, Bohanon MJ, Mathews WR, Sweeney MT, Zurenko GE, Marotti KR, et al. 2004. Immobilization of a novel antibacterial agent on solid phase and subsequent isolation of EF-Tu. *Bioconjug Chem* **15:** 333–343.

Deng GLL, Palestrant DJ, Whitehead L, Sachdeva M, Dzink-Fox J, LaMarche MJ, Leeds JA. 2011. Investigation of mode of binding of elongation factor Tu inhibitor LFF571. In *51st Interscience Conference on Antimicrobial Agents and Chemotherapy.* Chicago, IL.

Evans CT, Safdar N. 2015. Current trends in the epidemiology and outcomes of *Clostridium difficile* infection. *Clin Infect Dis* **60:** S66–S71.

Eyre DW, Babakhani F, Griffiths D, Seddon J, Del Ojo Elias C, Gorbach SL, Peto TE, Crook DW, Walker AS. 2014. Whole-genome sequencing demonstrates that fidaxomicin is superior to vancomycin for preventing reinfection and relapse of infection with *Clostridium difficile*. *J Infect Dis* **209:** 1446–1451.

Fahnestock S, Weissbach H, Rich A. 1972. Formation of a ternary complex of phenyllactyl-tRNA with transfer factor Tu and GTP. *Biochim Biophys Acta* **269:** 62–66.

Freeman J, Vernon J, Morris K, Nicholson S, Todhunter S, Longshaw C, Wilcox MH. 2015. Pan-European longitudinal surveillance of antibiotic resistance among prevalent *Clostridium difficile* ribotypes. *Clin Microbiol Infect* **21:** 248.e9–248.e16.

Furano AV. 1975. Content of elongation factor Tu in *Escherichia coli*. *Proc Natl Acad Sci* **72:** 4780–4784.

Goldstein EJ, Citron DM, Sears P, Babakhani F, Sambol SP, Gerding DN. 2011. Comparative susceptibilities to fidaxomicin (OPT-80) of isolates collected at baseline, recurrence, and failure from patients in two phase III trials of fidaxomicin against *Clostridium difficile* infection. *Antimicrob Agents Chemother* **55:** 5194–5199.

Goldstein EJ, Babakhani F, Citron DM. 2012. Antimicrobial activities of fidaxomicin. *Clin Infect Dis* **55:** S143–S148.

Gualtieri M, Villain-Guillot P, Latouche J, Leonetti JP, Bastide L. 2006. Mutation in the *Bacillus subtilis* RNA polymerase β'-subunit confers resistance to lipiarmycin. *Antimicrob Agents Chemother* **50:** 401–402.

Gualtieri M, Tupin A, Brodolin K, Leonetti JP. 2009. Frequency and characterisation of spontaneous lipiarmycin-resistant *Enterococcus faecalis* mutants selected in vitro. *Int J Antimicrob Agents* **34:** 605–606.

Healy VL, Lessard IA, Roper DI, Knox JR, Walsh CT. 2000. Vancomycin resistance in enterococci: Reprogramming of the D-ala-D-Ala ligases in bacterial peptidoglycan biosynthesis. *Chem Biol* **7:** R109–R119.

Hecht DOD, Gerding D. 2012. Activity of LFF571 against 103 clinical isolates of *C. difficile*. In *22nd European Congress on Clinical Microbiology and Infectious Diseases*. London, UK.

Heffron SE, Jurnak F. 2000. Structure of an EF-Tu complex with a thiazolyl peptide antibiotic determined at 2.35 A resolution: Atomic basis for GE2270A inhibition of EF-Tu. *Biochemistry* **39:** 37–45.

Hensgens MP, Dekkers OM, Demeulemeester A, Buiting AG, Bloembergen P, van Benthem BH, Le Cessie S, Kuijper EJ. 2014. Diarrhoea in general practice: When should a *Clostridium difficile* infection be considered? Results of a nested case-control study. *Clin Microbiol Infect* **20:** O1067–O1074.

Ishihama Y, Schmidt T, Rappsilber J, Mann M, Hartl FU, Kerner MJ, Frishman D. 2008. Protein abundance profiling of the *Escherichia coli* cytosol. *BMC Genomics* **9:** 102.

Ivarsson ME, Leroux JC, Castagner B. 2015. Investigational new treatments for *Clostridium difficile* infection. *Drug Discov Today* **20:** 602–608.

Jaskunas SR, Lindahl L, Nomura M. 1975. Identification of two copies of the gene for the elongation factor EF-Tu in *E. coli*. *Nature* **257:** 458–462.

Jayasekera MM, Onheiber K, Keith J, Venkatesan H, Santillan A, Stocking EM, Tang L, Miller J, Gomez L, Rhead B, et al. 2005. Identification of novel inhibitors of bacterial translation elongation factors. *Antimicrob Agents Chemother* **49:** 131–136.

Ke D, Boissinot M, Huletsky A, Picard FJ, Frenette J, Ouellette M, Roy PH, Bergeron MG. 2000. Evidence for horizontal gene transfer in evolution of elongation factor Tu in enterococci. *J Bacteriol* **182:** 6913–6920.

Kurabachew M, Lu SH, Krastel P, Schmitt EK, Suresh BL, Goh A, Knox JE, Ma NL, Jiricek J, Beer D, et al. 2008. Lipiarmycin targets RNA polymerase and has good activity against multidrug-resistant strains of *Mycobacterium* tuberculosis. *J Antimicrob Chemother* **62:** 713–719.

LaMarche MJ, Leeds JA, Amaral A, Brewer JT, Bushell SM, Deng G, Dewhurst JM, Ding J, Dzink-Fox J, Gamber G, et al. 2012. Discovery of LFF571: An investigational agent for *Clostridium difficile* infection. *J Med Chem* **55:** 2376–2387.

Landini P, Soffientini A, Monti F, Lociuro S, Marzorati E, Islam K. 1996. Antibiotics MDL 62,879 and kirromycin bind to distinct and independent sites of elongation factor Tu (EF-Tu). *Biochemistry* 35: 15288–15294.

Leeds JA, Sachdeva M, Mullin S, Dzink-Fox J, Lamarche MJ. 2012. Mechanism of action of and mechanism of reduced susceptibility to the novel anti-*Clostridium difficile* compound LFF571. *Antimicrob Agents Chemother* 56: 4463–4465.

Leeds JA, Sachdeva M, Mullin S, Barnes SW, Ruzin A. 2014. In vitro selection, via serial passage, of *Clostridium difficile* mutants with reduced susceptibility to fidaxomicin or vancomycin. *J Antimicrob Chemother* 69: 41–44.

Lessa FC, Mu Y, Bamberg WM, Beldavs ZG, Dumyati GK, Dunn JR, Farley MM, Holzbauer SM, Meek JI, Phipps EC, et al. 2015. Burden of *Clostridium difficile* infection in the United States. *N Engl J Med* 372: 825–834.

Locher HH, Caspers P, Bruyere T, Schroeder S, Pfaff P, Knezevic A, Keck W, Ritz D. 2014a. Investigations of the mode of action and resistance development of cadazolid, a new antibiotic for treatment of *Clostridium difficile* infections. *Antimicrob Agents Chemother* 58: 901–908.

Locher HH, Seiler P, Chen X, Schroeder S, Pfaff P, Enderlin M, Klenk A, Fournier E, Hubschwerlen C, Ritz D, et al. 2014b. In vitro and in vivo antibacterial evaluation of cadazolid, a new antibiotic for treatment of *Clostridium difficile* infections. *Antimicrob Agents Chemother* 58: 892–900.

Louie TJ, Miller MA, Mullane KM, Weiss K, Lentnek A, Golan Y, Gorbach S, Sears P, Shue YK. 2011. Fidaxomicin versus vancomycin for *Clostridium difficile* infection. *N Engl J Med* 364: 422–431.

Magill SS, Edwards JR, Bamberg W, Beldavs ZG, Dumyati G, Kainer MA, Lynfield R, Maloney M, McAllister-Hollod L, Nadle J, et al. 2014. Multistate point-prevalence survey of health care-associated infections. *N Engl J Med* 370: 1198–1208.

Mascio CT, Chesnel L, Thorne G, Silverman JA. 2014. Surotomycin demonstrates low in vitro frequency of resistance and rapid bactericidal activity in *Clostridium difficile, Enterococcus faecalis,* and *Enterococcus faecium. Antimicrob Agents Chemother* 58: 3976–3982.

Mattila E, Arkkila P, Mattila PS, Tarkka E, Tissari P, Anttila VJ. 2013. Rifaximin in the treatment of recurrent *Clostridium difficile* infection. *Aliment Pharmacol Ther* 37: 122–128.

McDonald LC, Killgore GE, Thompson A, Owens RC Jr, Kazakova SV, Sambol SP, Johnson S, Gerding DN. 2005. An epidemic, toxin gene-variant strain of *Clostridium difficile. N Engl J Med* 353: 2433–2441.

Miller DL, Weissbach H. 1970. Interactions between the elongation factors: The displacement of GPD from the TU-GDP complex by factor Ts. *Biochem Biophys Res Commun* 38: 1016–1022.

Mohrle VG, Tieleman LN, Kraal B. 1997. Elongation factor Tu1 of the antibiotic GE2270A producer *Planobispora rosea* has an unexpected resistance profile against EF-Tu targeted antibiotics. *Biochem Biophys Res Commun* 230: 320–326.

Morris RP, Leeds JA, Naegeli HU, Oberer L, Memmert K, Weber E, LaMarche MJ, Parker CN, Burrer N, Esterow S, et al. 2009. Ribosomally synthesized thiopeptide antibiotics targeting elongation factor Tu. *J Am Chem Soc* 131: 5946–5955.

Mullane K, Lee C, Bressler A, Buitrago M, Weiss K, Dabovic K, Praestgaard J, Leeds JA, Blais J, Pertel P. 2015. Multicenter, randomized clinical trial to compare the safety and efficacy of LFF571 and vancomycin for *Clostridium difficile* infections. *Antimicrob Agents Chemother* 59: 1435–1440.

Noren T, Wullt M, Akerlund T, Back E, Odenholt I, Burman LG. 2006. Frequent emergence of resistance in *Clostridium difficile* during treatment of *C. difficile*-associated diarrhea with fusidic acid. *Antimicrob Agents Chemother* 50: 3028–3032.

Noren T, Akerlund T, Wullt M, Burman LG, Unemo M. 2007. Mutations in fusA associated with posttherapy fusidic acid resistance in *Clostridium difficile. Antimicrob Agents Chemother* 51: 1840–1843.

Omura S, Imamura N, Oiwa R, Kuga H, Iwata R, Masuma R, Iwai Y. 1986. Clostomicins, new antibiotics produced by *Micromonospora echinospora* subsp. *armeniaca* subsp. nov. I: Production, isolation, and physico-chemical and biological properties. *J Antibiot (Tokyo)* 39: 1407–1412.

Parmeggiani A, Nissen P. 2006. Elongation factor Tu-targeted antibiotics: Four different structures, two mechanisms of action. *FEBS Lett* 580: 4576–4581.

Reeves DS. 1987. The pharmacokinetics of fusidic acid. *J Antimicrob Chemother* 20: 467–476.

Sachdeva M, Leeds JA. 2015. Subinhibitory concentrations of LFF571 reduce toxin production by *Clostridium difficile. Antimicrob Agents Chemother* 59: 1252–1257.

Sears P, Crook DW, Louie TJ, Miller MA, Weiss K. 2012. Fidaxomicin attains high fecal concentrations with minimal plasma concentrations following oral administration in patients with *Clostridium difficile* infection. *Clin Infect Dis* 55: S116–S120.

Seddon JBF, Sears P. 2012. Mutant prevention concentration of fidaxomicin for *Clostridium difficile.* In *52nd Interscience Conference on Antimicrobial Agents and Chemotherapy.* San Francisco, CA.

Seddon JBF, Gomez A, Artsimovitch I, Sears P. 2011. RNA polymerase target modification in *Clostridium difficile* with reduced susceptibility to fidaxomicin. In *51st Interscience Conference on Antimicrobial Agents and Chemotherapy.* Chicago, IL.

Sela S, Yogev D, Razin S, Bercovier H. 1989. Duplication of the *tuf* gene: A new insight into the phylogeny of eubacteria. *J Bacteriol* 171: 581–584.

Selva E, Beretta G, Montanini N, Saddler GS, Gastaldo L, Ferrari P, Lorenzetti R, Landini P, Ripamonti F, Goldstein BP, et al. 1991. Antibiotic GE2270 a: A novel inhibitor of bacterial protein synthesis. I: Isolation and characterization. *J Antibiot (Tokyo)* 44: 693–701.

Selva E, Montanini N, Stella S, Soffientini A, Gastaldo L, Denaro M. 1997. Targeted screening for elongation factor Tu binding antibiotics. *J Antibiot (Tokyo)* 50: 22–26.

Sergio S, Pirali G, White R, Parenti F. 1975. Lipiarmycin, a new antibiotic from *Actinoplanes.* III: Mechanism of action. *J Antibiot (Tokyo)* 28: 543–549.

Sonenshein AL, Alexander HB, Rothstein DM, Fisher SH. 1977. Lipiarmycin-resistant ribonucleic acid polymerase mutants of *Bacillus subtilis. J Bacteriol* 132: 73–79.

Srivastava A, Talaue M, Liu S, Degen D, Ebright RY, Sineva E, Chakraborty A, Druzhinin SY, Chatterjee S, Mukhopadhyay J, et al. 2011. New target for inhibition of bacterial RNA polymerase: "Switch region." *Curr Opin Microbiol* **14:** 532–543.

Stark H, Rodnina MV, Rinke-Appel J, Brimacombe R, Wintermeyer W, van Heel M. 1997. Visualization of elongation factor Tu on the *Escherichia coli* ribosome. *Nature* **389:** 403–406.

Surawicz CM, Brandt LJ, Binion DG, Ananthakrishnan AN, Curry SR, Gilligan PH, McFarland LV, Mellow M, Zuckerbraun BS. 2013. Guidelines for diagnosis, treatment, and prevention of *Clostridium difficile* infections. *Am J Gastroenterol* **108:** 478–498; quiz 499.

Swanson RN, Hardy DJ, Shipkowitz NL, Hanson CW, Ramer NC, Fernandes PB, Clement JJ. 1991. In vitro and in vivo evaluation of tiacumicins B and C against *Clostridium difficile*. *Antimicrob Agents Chemother* **35:** 1108–1111.

Tavecchia P, Gentili P, Kurz M, Sottani C, Bonfichi R, Lociuro S, Selva E. 1994. Revised structure of the antibiotic GE 2270A. *J Antibiot (Tokyo)* **47:** 1564–1567.

Tavecchia P, Gentili P, Kurz M, Sottani C, Bonfichi R, Selva E, Locurio S, Restelli E, Ciabatti R. 1995. Degradation studies of antibiotic MDL 62,879 (GE2270A) and revision of the structure. *Tetrahedron* **51:** 4867–4890.

Theriault RJ, Karwowski JP, Jackson M, Girolami RL, Sunga GN, Vojtko CM, Coen LJ. 1987. Tiacumicins, a novel complex of 18-membered macrolide antibiotics. I: Tax-

onomy, fermentation and antibacterial activity. *J Antibiot (Tokyo)* **40:** 567–574.

Ting LS, Praestgaard J, Grunenberg N, Yang JC, Leeds JA, Pertel P. 2012. A first-in-human, randomized, double-blind, placebo-controlled, single- and multiple-ascending oral dose study to assess the safety and tolerability of LFF571 in healthy volunteers. *Antimicrob Agents Chemother* **56:** 5946–5951.

Tsutsumi LS, Owusu YB, Hurdle JG, Sun D. 2014. Progress in the discovery of treatments for *C. difficile* infection: A clinical and medicinal chemistry review. *Curr Top Med Chem* **14:** 152–175.

Voorhees RM, Ramakrishnan V. 2013. Structural basis of the translational elongation cycle. *Annu Rev Biochem* **82:** 203–236.

Yin N, Li J, He Y, Herradura PS, Pearson A, Mesleh MF, Mascio CT, Howland K, Steenbergen J, Thorne GM, et al. 2015. Structure-activity relationship studies of a series of semi-synthetic lipopeptides leading to the discovery of surotomycin, a novel cyclic lipopeptide being developed for the treatment of *Clostridium difficile*-associated diarrhea. *J Med Chem* **58:** 5137–5142.

Zuurmond AM, Martien de Graaf J, Olsthoorn-Tieleman LN, van Duyl BY, Morhle VG, Jurnak F, Mesters JR, Hilgenfeld R, Kraal B. 2000. GE2270A-resistant mutations in elongation factor Tu allow productive aminoacyl-tRNA binding to EF-Tu.GTP.GE2270A complexes. *J Mol Biol* **304:** 995–1005.

Index